The Physiology of Vegetable Crops

The Physiology of Vegetable Crops

Edited by

H.C. Wien
Department of Fruit and Vegetable Science
Cornell University, Ithaca, NY, USA

CAB INTERNATIONAL

CAB INTERNATIONAL
Wallingford
Oxon OX10 8DE
UK

Tel: +44 (0)1491 832111
Fax: +44 (0)1491 833508
E-mail: cabi@cabi.org

CAB INTERNATIONAL
198 Madison Avenue
New York, NY 10016-4341
USA

Tel: +1 212 726 6490
Fax: +1 212 686 7993
E-mail: cabi-nao@cabi.org

A catalogue record for this book is available from the British Library, London, UK
A catalogue record for this book is available from the Library of Congress, Washington DC, USA

ISBN 0 85199 146 7

Typeset in 10/12pt Palatino by Columns Design Ltd., Reading
Printed and bound in the UK at the University Press, Cambridge

Contents

Contributors

F. Azanza, *Department of Natural Resources and Environmental Science, University of Illinois, 1201 West Gregory Drive, Urbana, Illinois 61801, USA*

L.R. Benjamin, *Crop and Weed Science Department, Horticulture Research International, Wellesbourne, Warwick CV35 9EF, UK*

J.L. Brewster, *Crop and Weed Science Department, Horticulture Research International, Wellesbourne, Warwick CV35 9EF, UK*

P.Q. Craufurd *The University of Reading, Department of Agriculture, Plant Environment Laboratory, Cutbush Lane, Shinfield, Reading RG2 9AD, Berkshire, UK*

J.H.C. Davis, *PBI Cambridge Ltd, Maris Lane, Trumpington, Cambridge CB2 2LQ,UK*

D.T. Drost, *Department of Plants, Soils, and Biometeorology, Utah State University, Logan, Utah 84322–4820, USA*

R.H. Ellis, *The University of Reading, Department of Agriculture, Plant Environment Laboratory, Cutbush Lane, Shinfield, Reading RG2 9AD, Berkshire, UK*

E.E. Ewing, *Department of Fruit and Vegetable Science, Cornell University, 134A Plant Science Building, Ithaca, New York 14853–5908, USA*

D. Gray, *Crop and Weed Science Department, Horticulture Research International, Wellesbourne, Warwick CV35 9EF, UK*

J.A. Juvik, *Department of Natural Resources and Environmental Science, University of Illinois, 1201 West Gregory Drive, Urbana, Illinois 61801, USA*

J.M. Kinet, *Laboratoire de Cytogénétique, Département de Biologie, Université*

Catholique de Louvain, Place Croix du Sud, 5 bte 13, B-1348 Louvain-la-Neuve, Belgium

H. Krug, *Institute of Vegetable Crops, University of Hannover, Herrenhäuser Strasse 2, D-30419, Hannover, Germany*

A. McGarry, *Crop and Weed Science Department, Horticulture Research International, Wellesbourne, Warwick CV35 9EF, UK*

K.E. McPhee, *USDA/ARS, Grain Legume Genetics and Physiology Research Unit, Washington State University, Pullman, Washington 99163, USA*

F.J. Muehlbauer, *USDA/ARS, Grain Legume Genetics and Physiology Research Unit, Washington State University, Pullman, Washington 99163, USA*

M.M. Peet, *Department of Horticultural Science, Box 7609, North Carolina State University, Raleigh, North Carolina 27695–7609, USA*

E. Pressman, *ARO the Volcani Center, Institute for Field and Garden Crops, PO Box 6, Bet Dagan 50250, Israel*

E.H. Roberts, *The University of Reading, Department of Agriculture, Plant Environment Laboratory, Cutbush Lane, Shinfield, Reading RG2 9AD, Berkshire, UK*

R.J. Summerfield, *The University of Reading, Department of Agriculture, Plant Environment Laboratory, Cutbush Lane, Shinfield, Reading RG2 9AD, Berkshire, UK*

A.G. Taylor, *Department of Horticultural Sciences, New York State Agricultural Experiment Station, Cornell University, Geneva, New York 14456, USA*

T.R. Wheeler, *The University of Reading, Department of Agriculture, Plant Environment Laboratory, Cutbush Lane, Shinfield, Reading RG2 9AD, Berkshire, UK*

H.C. Wien, *Department of Fruit and Vegetable Science, Cornell University, 134A Plant Science Building, Ithaca, New York 14853–5908, USA*

D.W. Wolfe, *Department of Fruit and Vegetable Science, Cornell University, 134A Plant Science Building, Ithaca, New York 14853–5908, USA*

Preface

Plant physiology is defined as the study of the functions and activities of plants. This book on the physiology of vegetable crops focuses more narrowly on those growth and development processes that determine the formation of the harvested product. It emphasizes processes at the organ and whole-plant level, rather than the cellular or molecular scale. Vegetable crop physiology seeks to explain the functions and activities of the major plant parts (leaves, stems, roots, flowers, fruits) in the development of yield, and to detail the influence of environmental factors on major ontogenetic processes such as flowering, and the formation of the harvested products.

To determine which species would be considered as vegetables in this book, I was guided by the definition of a vegetable crop as 'a herbaceous plant harvested for its edible parts, which can be consumed fresh or with little preparation as a human food' (Krug, 1986, p. 12). This includes more than 23 species in the individual crop chapters, in some cases grouped together according to the part of the plant harvested, as in the root vegetables (Chapter 16). The book overlaps in subject matter and crop species to a small extent with Evans' *Crop Physiology, Some Case Histories* (1975), and Goldsworthy and Fisher's *The Physiology of Tropical Field Crops* (1984). A major difference with these earlier books is that here the quality of the yield of vegetable products acceptable to consumers is emphasized as much as the quantity of yield. Because most vegetables are directly consumed, the taste and appearance of the harvested product, and factors determining these attributes, are important considerations in physiological studies.

The present volume begins with a general survey of the growth stages and processes of importance in vegetable production, namely seed germination and seedling growth, transplanting, the induction of flowering, and the influence of environmental factors on growth and yield. Considerations on how growth is coordinated among the different plant parts in vegetables

(Chapter 5) is followed by the detailed physiology of specific vegetable crops in 13 crop chapters. In all chapters, the crops are assumed to be growing in a temperate environment, but enough information is given on the influence of environmental factors that an extrapolation to locations with short photoperiods and higher temperatures will be possible.

The intended audience for this book is the horticultural researcher and farm adviser who is working with vegetable crops and wants an understanding of the factors governing the principal growth processes. It also meant as a reference work for students in horticulture, and the associated fields of vegetable breeding, entomology and plant pathology of vegetable crops. To keep the book to a reasonable length, it was necessary to leave out important areas such as plant nutrition, postharvest physiology and details of vegetable production practices. Fortunately, these areas have been adequately covered in other recent literature.

The production of this book would have been impossible without the dedication and hard work of many individuals. I am particularly grateful for Margaret Schoneman's patient and careful transformation of the rough manuscripts to finished chapters. The assistance of Tim Hardwick, Amanda Horsfall, Emma Critchley and Pippa Smart at CAB International facilitated the publishing process. My most grateful thanks go to the authors of these chapters, who volunteered many hours of their time to contribute to this book. Without them, it would never have been completed. I also appreciate the sacrifice and patience of my wife Betty, who saw vacations indefinitely delayed 'until the book is finished'.

H.C. Wien
Ithaca, New York
September, 1996

REFERENCES

Evans, L.T. (ed.) (1975) *Crop Physiology, Some Case Histories*. Cambridge University Press, Cambridge.

Goldsworthy, P.R. and Fisher, N.M. (eds) (1984) *The Physiology of Tropical Field Crops*. John Wiley, Chichester.

Krug, H. (1986) *Gemüseproduktion. Ein Lehr- und Nachschlagewerk für Studium und Praxis*. Paul Parey, Berlin.

Seed Storage, Germination and Quality

A.G. Taylor

Department of Horticultural Sciences, New York State Agricultural Experiment Station, Cornell University, Geneva, New York 14456, USA

Vegetable crop growth and development begins with seeds. It is therefore appropriate that the first chapter of this book is devoted to aspects of seed physiology related to storage, germination, quality and enhancements. Vegetable crop seeds are diverse in size, morphology and composition. This makes a comprehensive overview of vegetable seeds difficult to achieve, especially since there is little scientific literature on seed physiology of small-seeded crops of minor economic importance. Due to these limitations different approaches will be used to develop a coherent picture of vegetable seeds. Relative differences between crop seeds are shown with regard to composition, storage longevity, seed and seedling morphology, temperature requirements for germination and the use of reserve materials during emergence. More generalized information is presented on factors and events associated with storage, germination and ageing. Finally the process of ageing and other aspects of seed physiology and technology are illustrated using specific crop examples.

The focus of this chapter is on postharvest aspects of seeds; the period of seed development and production stages will not be addressed. Seeds are stored at low water content during storage and they have unique characteristics that permit them to withstand desiccation (Leopold and Vertucci, 1986). Desiccation tolerance is essential for long-term survival and allows for a time interval between seed production and crop production. Most vegetable seeds can imbibe water readily and, given a conducive environment, can germinate and resume active growth. The seed uses its reserve materials following germination and then becomes an active photosynthetic seedling fixing its own carbon and producing energy.

Seed quality is a broad term and encompasses several attributes of seeds including the germination and seedling performance. In this chapter, certain physiological and biochemical processes associated with loss of seed

© CAB INTERNATIONAL 1997. *The Physiology of Vegetable Crops*
(H.C. Wien, ed.)

quality will be described, and symptoms of seed deterioration presented at physiological and whole plant levels. The sowing environment has a direct effect on germination and stand establishment, and under severe stress seedling performance may be seriously impaired. To ensure stand establishment, high quality seeds are needed for transplant production and for direct seeding. Seed performance can be improved by pre-sowing hydration methods to achieve maximum emergence when sown in suboptimal conditions. In addition, seed-coating technologies may be used to improve precision placement of seeds, provide a delivery system and improve adherence of additives to protect seed and seedlings, and upgrade seed quality by facilitating detection of non-viable seeds in a seed lot.

STORAGE

The period of time from harvesting to sowing seeds may vary from a few months to several years. It is during this time that the quality of seeds, defined as the ability to germinate, may decline. In this section, the description of water in seeds during storage and an explanation of methods to determine moisture content are presented. The protein, lipid and starch content of many vegetable seeds is tabulated since seed composition (in particular lipid content) influences the equilibrium moisture content. The effect of relative humidity and temperature will be discussed with respect to the rate of ageing. Finally, differences in longevity among vegetable seeds are illustrated from long-term storage studies.

STATUS OF WATER IN SEEDS

Seeds in storage are said to be in a dry condition. However, dry is a relative term and does not mean that water is absent from seeds. Water is present in seed tissue, and the status of water is related to many aspects of seed physiology, including seed longevity. First, it is necessary to quantify the water in seeds and express the water in meaningful units.

Standardized gravimetric methods to determine the moisture content of seeds have been published by the International Seed Testing Association (ISTA, 1985). Other primary and secondary methods have also been developed and have been reviewed by Grabe (1989). The moisture content may be calculated and presented in different ways. Seed moisture content calculated on a fresh or wet weight (fw) basis is commonly used in seed testing and in commerce, while moisture content presented on a dry weight (dw) basis may be found in physiological or biophysical literature. The following formulas are used to determine the percentage of seed moisture content on a fresh and dry weight basis:

% moisture, fw basis = [weight of water (dry weight of seed +
weight of water)$^{-1}$] × 100

% moisture, dw basis = [weight of water (dry weight of seed)$^{-1}$] × 100.

For comparison, the two methods of expressing the percentage of moisture content can be converted by the following equations:

% moisture, dw basis = 100 × % moisture, fw basis
(100 − % moisture fw basis)$^{-1}$.

% moisture, fw basis = 100 × % moisture, dw basis
(100 + % moisture, dw basis)$^{-1}$.

The seed moisture content may also be expressed as g H_2O g^{-1} fresh or dry weight.

The seed moisture content eventually comes into equilibrium with the relative humidity of the air. The relationship between the equilibrium moisture content and relative humidity reveals a negative sigmoidal-shaped curve known as a moisture isotherm (Iglesias and Chirife, 1982). This relationship is determined by placing seeds in a range of known relative humidities produced in closed containers with the use of saturated salt solutions (Taylor *et al.*, 1992). We have prepared a number of saturated salt solutions and desiccants to achieve a range of humidities from *c.* 0 to 89% and equilibrated snap beans (*Phaseolus vulgaris*) and broccoli (*Brassica oleracea* Botrytis group) seeds. Three regions of water binding or types of water are observed: type 1, < 20% RH; type 2, 25–65% RH; and type 3, > 70% RH (Fig. 1.1). The curves are similar for both kinds of seeds, however, the equilibrium moisture content at a given relative humidity is always greater for snap beans compared to broccoli. Differences between these two species is largely attributed to seed composition, in particular lipid content, as lipids have little affinity for water. Thus seeds with high lipid content have a lower equilibrium moisture content at a given relative humidity than a seed with low lipid content.

In our study, the lipid content was determined for the snap bean and broccoli samples and found to be 1 and 31%, respectively (A.G. Taylor, 1994, Cornell University). Since composition of seeds is important to many aspects of seed physiology, the protein, lipid and starch content of many vegetable seeds was compiled from the literature (Table 1.1). There are two groups of seeds based on their storage reserves: starch and lipid storing. The large-seeded crops such as peas, beans and corn contain high levels of starch with little lipid, while most of the small-seeded crops are lipid storing. The large-seeded cucurbits are exceptions, having nearly half of their weight as lipids. Spinach and beet both contain starch which is stored in the perisperm (see Seed and seedling morphology section below). All seeds contain proteins and values range between 12 and 31% for intact seeds.

Another method used to quantify the water status of seeds that is not affected by seed composition is to measure the water activity (a_w). Water

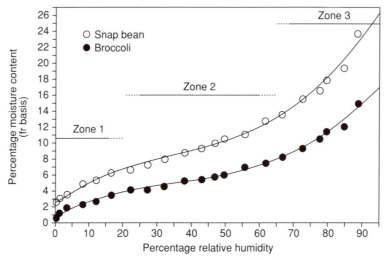

Fig. 1.1. Moisture isotherms for snap bean and broccoli seeds (A.G. Taylor and D.H. Paine, unpublished data).

activity is defined as the ratio of the vapour pressure of water in a seed to the vapour pressure of pure water at the same temperature (Bourne, 1991). Water activity is measured by determining the equilibrium relative humidity of the seed's headspace in a closed container and is expressed as a decimal. For example, a seed (any seed) equilibrated to 50% relative humidity will have a water activity of 0.50. Water activity measurements are commonly used in food science and are currently finding application in seed science.

MOISTURE CONTENT AND TEMPERATURE

Two major environmental factors that influence seed storage are seed moisture content and temperature. Moisture content, as previously shown, is determined by the storage relative humidity and by seed characteristics, largely lipid content. The concentration of water in the seed tissue directly affects the rate of deterioration at a given temperature. The moisture isotherm reveals three types (see Fig. 1.1) in which the water binding to seed tissues differ. Type 1 water is bound tightly and water interacts very strongly with charged groups of proteins (Vertucci, 1993). Type 2 water is less tightly bound and condenses over the hydrophilic sites of macromolecules (Leopold and Vertucci, 1989). Type 3 water is bound with negligible energy and forms bridges over hydrophobic moieties (Vertucci, 1993). The status of water in the seed tissue governs the type of reactions (enzymatic or non-enzymatic) that may occur in storage (Vertucci, 1993).

Table 1.1. The percentage protein, lipid and starch content of intact vegetable seeds (unless otherwise noted). Data from Earle and Jones (1962) and other cited sources.

Crop	Genus species	Protein (%)	Lipid (%)	Starch* (%)
Asparagus	*Asparagus officinalis*	16	15	0
Bean, lima	*Phaseolus lunatus*	24	1	+
Bean, common	*Phaseolus vulgaris*[†]	22	2	42
Bean, runner	*Phaseolus coccineus*	24	2	+
Beet	*Beta vulgaris*	14	5	+
Cabbage	*Brassica oleracea* var. *capitata*	28	38	0
Carrot	*Daucus carota*	28	24	0
Celery	*Apium graveolens*	19	32	0
Leek	*Allium porrum*	27	15	0
Lettuce	*Lactuca sativa*	29	38	0
Pea	*Pisum sativum*[†]	22	1	48
Pepper, tobasco	*Capsicum frutescens*	18	22	0
Radish	*Raphanus sativus*	31	36	0
Spinach	*Spinacia oleracea*	24	7	+
Squash	*Cucurbita maxima*[**]	39	50	0
Squash	*Cucurbita moschata*[**]	39	50	0
Squash	*Cucurbita pepo*[**]	39	49	0
Sweet corn	*Zea mays*[‡]	12	6	52
Watermelon	*Citrullus lanatus*[**]	38	52	0

* A starch test was performed and 0 or + indicate a negative or positive test, respectively.
[†] Data taken from Norton *et al.* (1985).
[‡] Data taken from Bewley and Black (1978).
[**] All species in the *Cucurbitaceae* were analysed without seed coats.

Temperature has a direct influence on longevity, and the rate of deterioration increases as temperature increases at a given relative humidity. Seeds may be stored over a wide range of temperatures depending on the particular needs and available conditions. Seeds for most applications are stored above 0°C; however, for long-term preservation seeds may be stored below 0°C. Seeds with water contents in zones 1 or 2 will not be injured since the water is non-freezable in the seed tissue. Long-term germplasm preservation has taken advantage of this ability to withstand freezing injury as seeds are stored above liquid nitrogen in the vapour phase at −150 to −180°C (Roos, 1989).

The need to keep seeds cool and dry for long-term preservation has been known for centuries. In the 1960s, Harrington's Rules of Thumb were developed as guidelines for storage (cited by Justice and Bass, 1978). The first two rules relate the influence of moisture content and temperature independently on longevity. The life of seed is halved by a 1% increase in seed moisture, and this rule applies when seed moisture content ranges from 5 to 14%. The life of seed is halved by a 5°C increase in storage temperature, and this applies to storage between 0 and 50°C. The most widely quoted rule relates both temperature and relative humidity to storage. 'The sum of the temperature in °F and the percentage of relative humidity should not exceed 100'.

Since the 1960s mathematical equations have been developed to model seed ageing for a particular species maintained at a given condition of temperature and moisture content (fw basis) known as the Ellis and Roberts equations (cited by Priestley, 1986). These equations take into consideration the initial germinability of the seed lot and will predict the germination after a specified period of time in storage. The general equation for modelling the loss of germination in storage is as follows (Ellis *et al.*, 1982):

$$v = K_i - \frac{p}{10^{[K_E - (C_w \times \log m) - (C_H \times t) - (C_q \times t^2)]}}$$

where v represents the probit of the percentage germinability after a storage period of p days; K_i is the probit of the initial germinability for the seed lot; K_E, C_w, C_H and C_q are species-specific constants; m is the seed moisture content (expressed on a fresh weight basis); and t is the storage temperature (°C). The species-specific constants for onion are; $K_E = 6.975$, $C_w = 3.470$, $C_H = 0.040$ and $C_q = 0.000428$ (Ellis and Roberts, 1981). We have used this equation to develop a computer program to plot onion seed ageing curves (D.H. Paine and A.G. Taylor, 1993, Cornell University). For illustration, the influence of temperature, moisture content and initial viability were studied as variables for ageing of onion seeds. Increasing temperature or moisture content independently had a profound effect on ageing (Figs 1.2a and b). The useful range of the program is considered to extend from −20 to 90°C and from 5 to 18% moisture content. The curves generally reveal a sigmoidal shape, especially when the initial germination is high, while the initial plateau phase is not observed with seed lots of low germination (Fig. 1.2c).

SPECIES DIFFERENCES IN STORABILITY

Although the storage environment is important, differences exist between species held under the same conditions. Results from a number of earlier storage studies on different species of seeds including vegetable seeds have been summarized (Justice and Bass, 1978; Priestley, 1986). Attempts have been made to relate seed longevity to other aspects of seeds such as composition. Many seeds with high lipid content are short-lived; however, tomato seed with 25% lipid content is a notable exception (Priestley, 1986). At this time, the physiological basis for differences in longevity between species is not understood.

Long-term seed storage experiments have been initiated in the past, and data from these studies have been compiled and analysed. As shown earlier, a

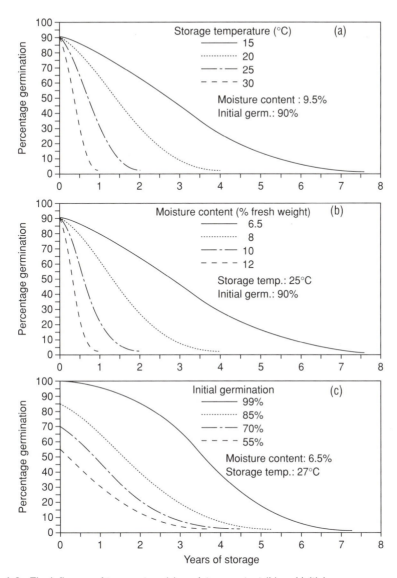

Fig. 1.2. The influence of temperature (a), moisture content (b) and initial germination (c) on onion seed ageing. Ageing curves were developed from the equation and constants of Ellis and Roberts (1981).

sigmoidal relationship is revealed with a loss of germination in time. The sigmoidal curve is then transformed by probit analysis to reveal a linear relationship and from these data the P50 is calculated, defined as the period of time in years for viability to be reduced by half. The first set of P50 values were derived from tests of seeds held in open storage conditions in the temperate zone throughout the world (Priestley *et al.*, 1985). The second set of P50 values were obtained from seeds stored in semi-controlled conditions in the USA, and the initial samples were obtained from the US Department of Agriculture's Horticultural Field Station in Cheyenne, Wyoming (Roos and Davidson, 1992). This collection was moved to the National Seed Storage Laboratory in Fort Collins, Colorado, in 1962 and stored at 5°C and < 40% relative humidity. In about 1977, most of the samples were transferred to sealed moisture-proof bags at −18°C and maintained under those conditions.

The P50 for the study conducted under open storage conditions ranged from 3 to 25 years, and seeds of asparagus, celery, parsley and parsnip were short lived, while tomato and pea were long lived (Table 1.2). The P50 for the second study, in which the latter portion of the study was performed under controlled conditions, ranged from 29 to 130 years. In the second study, onion and pepper were short lived and okra was long

Table 1.2. The half-viability periods (P50) in years for different vegetable seeds.

Seed	Genus species	P50*	P50†
Asparagus	*Asparagus officinalis*	3.92	–
Bean, lima	*Phaseolus lunatus*	13.12	–
Bean, runner	*Phaseolus coccineus*	7.99	–
Bean, snap	*Phaseolus vulgaris*	15.97	46
Beet	*Beta vulgaris*	16.51	43
Cabbage	*Brassica oleracea*	7.15	–
Carrot	*Daucus carota*	6.63	35
Celery	*Apium graveolens*	4.11	–
Corn	*Zea mays*	9.6	65
Cucumber	*Cucumis sativus*	4.92	45
Eggplant	*Solanum melongena*	–	54
Leek	*Allium porrum*	5.30	–
Lettuce	*Lactuca sativa*	6.42	–
Muskmelon	*Cucumis melo*	–	61
Okra	*Abelmoschus esculentus*	–	125
Onion	*Allium cepa*	5.43	29
Parsley	*Petroselinum crispum*	3.41	–
Parsnip	*Pastinaco sativa*	4.04	–
Pea	*Pisum sativum*	15.86	130
Pepper	*Capsicum annuum*	–	27
Radish	*Raphanus sativus*	13.82	–
Spinach	*Spinacia oleracea*	12.76	37
Tomato	*Lycopersicon esculentum*	24.52	124
Watermelon	*Citrullus lanatus*	–	43

* P50 values obtained from open storage conditions (Priestley *et al.*, 1985).
† P50 values obtained from semi-controlled storage conditions (Roos and Davidson, 1992). Original samples from open storage in Cheyenne, Wyoming, were later transferred to the National Seed Storage Laboratory in 1962 and stored at 5°C and < 40% RH. In *c.* 1977, most samples were then stored at −18°C

lived. A significant linear relationship ($r = 0.68^*$) was found between the P50s for the first and second study in which data was available for the same species. The regression equation revealed that the P50 values were approximately four times greater for the second compared to the first study, and, for example, the P50 of onion, the shortest lived seed in the second study, was greater than tomato, the longest lived seed in the first study. In conclusion, even though differences exist in the longevity among species, the major factor influencing storage life is the storage environment.

GERMINATION AND SEEDLING GROWTH

Germination is the transition period between the resting and the growth stages of the plant and is considered to be completed at the time of visible radicle emergence (Bewley and Black, 1994). In this discussion, post-germination events will be included to the point where the seed becomes a functional seedling. Definitions are needed to describe the resting seed condition with respect to environmental conditions favourable to support growth. Seeds in storage with low moisture content are in a state of quiescence, defined as the absence of growth because of environmental conditions that do not favour growth (Copeland and McDonald, 1985). The environmental factors needed to overcome this state of arrested development are water, oxygen and a suitable temperature. Seed dormancy, in contrast, is a physical or physiological condition of a living seed that prevents germination even in the presence of otherwise favourable environmental conditions (Copeland and McDonald, 1985). Seed dormancy mechanisms will not be discussed in this chapter, but will be addressed in the individual crop chapters for vegetable species in which dormancy may be a problem.

The diversity in the seed and seedling morphology found in temperate vegetable crops is described below. The remaining portions of this section will discuss aspects of germination that are relevant to all seeds, with special reference to vegetable seeds where appropriate. For additional information, the reader is referred to books devoted to the subject area of seed physiology (Mayer and Poljakoff-Mayber, 1982; Bewley and Black, 1994).

SEED AND SEEDLING MORPHOLOGY

The botanical term 'seed' refers to the mature ovule from the mother plant which contains the embryonic plant or embryo, the integuments which become the seed coat and additional storage tissue which may also be present such as the endosperm (Copeland and McDonald, 1985). Many so-called seeds are enclosed in remnants of the fruit and are not technically true seeds. We will take a broader interpretation and consider the dispersal units or propagules from the mother plant including dry indehiscent fruits as seeds. The embryo morphology is generally similar within a plant family, however, vegetable seeds include both monocots and dicots, and each class contains different families (Lorenz and Maynard, 1980). Seeds that contain reserve materials in a well-developed endosperm are called 'endospermic'. If they contain most of their reserve materials in the cotyledons (cotyledons are embryonic tissue), they are called 'non-endospermic' (Bewley and Black, 1994). After the completion of germination, the embryo develops the root and shoot system of the seedling. Seedlings can be categorized into two groups depending on the orientation of the cotyledons with regards to the soil or growing media. Seedlings in which the cotyledons are raised above the soil by the expansion of the hypocotyl are termed epigeal. Those seedlings in which the hypocotyl does not elongate appreciably, which results in the cotyledons remaining in the soil, are termed hypogeal (Bewley and Black, 1994). The cotyledons often become photosynthetic in epigeous seedlings, while expansion of the epicotyl or mesocotyl is responsible for shoot development in hypogeous seedlings.

The following discussion will group selected vegetable seeds by their seed and seedling morphology (Table 1.3 and Fig. 1.3). Sweet corn and asparagus have reserves in the endosperm and exhibit hypogeal seedling growth. The sweet corn seed is a caryopsis in which the pericarp is tightly fused to the seed coat (Copeland and McDonald, 1985), and the embryo is oriented in a lateral position as is typical of the grasses (Martin, 1946). The endosperm is non-living in sweet corn, and the scutellum, which is a part of the embryo, is considered analogous to the cotyledon. Many endospermic seeds, such as asparagus, have live endosperm tissue. The endosperm is considered living if there is a positive test with the vital stain 2,3,5-triphenyl tetrazolium chloride (Moore, 1985). The solanaceous and umbelliferous crops and genus *Allium* seeds all contain linear embryos embedded in a living endosperm with epigeal germination. Though the embryo is considered linear, it is commonly found coiled as in the case of tomato and pepper (Martin, 1946). In this group, carrot and celery are examples of schizocarps that have two fused carpels separating at maturity to form one-seeded mericarps (Copeland and McDonald, 1985). Beet and spinach have embryos that surround non-living nutritive tissue, and the nutritive tissue is a well-developed perisperm rather than endosperm (Hayward, 1938; Heydecker and Orphanos, 1968). Beet seed is a fruit and in many cultivars a 'seed ball' is formed by the aggregation of two or three flowers to produce a multigermed propagule (Hayward, 1938). Non-endospermic seeds may contain remnants of endosperm that is adjacent to the testa; however, little reserve material is present. The endosperm may be specialized as in the case of lettuce in which a two-layer envelope surrounds the embryo and serves as a semi-permeable barrier to solute diffusion (Hill and Taylor, 1989). Non-endospermic seeds can exhibit hypogeal germination (found in pea and runner bean) or epigeal germination (represented by seeds from four different families). The bent embryo orientation is found in the large-seeded legumes and brassicas and is termed the 'jack-knife' position of the embryonic axis with the cotyledons. Lettuce and the cucurbits have spatulate

Table 1.3. The embryo orientation, the presence of a well-developed endosperm for storage of reserve materials and cotyledon orientation after germination of different vegetable seeds.

Seed kind	Embryo orientation	Location of reserves	Cotyledon orientation
Sweet corn	Lateral*	Endospermic	Hypogeal
Asparagus	Linear†	Endospermic	Hypogeal
Tomato, pepper, eggplant, onion, carrot, celery	Linear†	Endospermic	Epigeal
Beet and spinach	Peripheral‡	Endospermic	Epigeal
Pea and runner bean	Bent	Non-endospermic**	Hypogeal
Snap bean and *Brassicas*	Bent	Non-endospermic	Epigeal
Lettuce and cucurbits	Spatulate	Non-endospermic	Epigeal

* Embryo partially surrounded by non-living tissue.
† Embryo surrounded by living nutritive tissue.
‡ Embryo surrounds non-living nutritive tissue.
** Non-endospermic seeds have little or no nutritive tissue as endosperm.

embryos, and lettuce is an achene where the ribs on the seed surface are formed by the pericarp (Hayward, 1938).

WATER

Germination begins with water uptake. This essential process can best be explained by classical water relations terminology, and is adapted from Bewley and Black (1994). A more detailed explanation of water relations in seeds may be found in Koller and Hadas (1982). Water potential is an expression of the energy status of water. Water moves in a passive manner from a high to a low potential. The water potential of pure water is 0, and a decrease in water potential (less water available) is denoted by more negative values. The components of the water potential are the algebraic sum of the matric, osmotic and pressure potentials. The matric or suction potential refers to the ability of the matrices in cells to bind water and is a negative value. The osmotic potential refers to the contribution of dissolved solutes to decrease water potential and is also a negative value. The pressure potential is a positive value and occurs when water enters the cells and creates an internal force on the cell walls. Water potential measurements can also be made on seeds in storage, and the water activity, described previously, is related to the water potential in a log-linear relationship (Taylor *et al.*, 1992).

To illustrate the process of water uptake in vegetable seeds, water uptake curves were developed by imbibing cabbage and tomato seeds on moistened blotters maintained at 25°C in the dark (Fig. 1.4). Water uptake in seeds reveals a triphasic pattern with an initial rapid uptake phase, followed by a lag phase and then a second increase in moisture content.

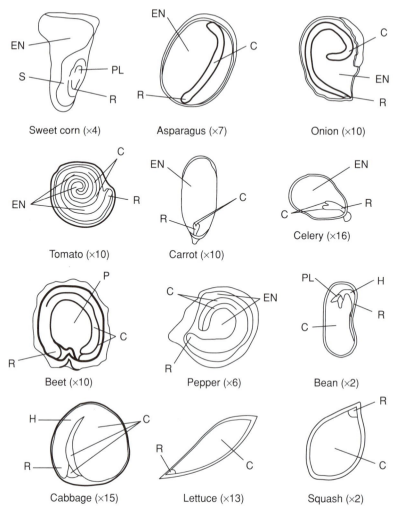

Fig. 1.3. Internal morphology of selected vegetable seeds (courtesy of D.H. Paine). EN, endosperm; S, scutellum; PL, plumule; R, radicle; C, cotyledon; P, perisperm; H, hypocotyl.

Phase I is known as imbibition and is a physical process that occurs in both living and dead seeds (Bewley and Black, 1994). Seeds of most vegetable crops take up water readily, and in our study, seeds were fully imbibed in a period of 4–8 h (Fig. 1.4). The rapid water uptake is attributed to the negative matric potential of the seed which is caused by cell wall components and protein (Leopold, 1983). Swelling occurs during imbibition due to the expansion of hydrophilic compounds such as proteins, cellulose, pectic substances and mucilage (Mayer and Poljakoff-Mayber, 1982). The rate of water uptake is influenced by a number of factors including temperature, initial moisture content, seed composition and morphology. Some vegetable seeds, such as okra, do not imbibe water readily due to impermeable seed coverings (Anderson *et al.*, 1953). A lag in the imbibition time has been shown in snap bean seeds with the semi-hard seed characteristic only when the initial seed moisture content is low (Taylor and Dickson, 1987). During the second phase (lag phase), there is little uptake of water. The matric potential is negligible during this period, and the osmotic and pressure potentials regulate the total water potential. During this phase, enzymes and membranes are functional in the fully hydrated cells as the seed advances to the completion of germination. The duration of phase II is dependent on species and is influenced by environmental conditions (see Water and Temperature stress sections below), and in our example, tomato has a longer lag period than cabbage even though tomato had a greater moisture content after imbibition than cabbage (Fig. 1.4). Phase III of water uptake commences with visible germination (see arrow on Fig. 1.4) in which the seed coat is ruptured by the emerging radicle which forms the root system of the plant. Radicle growth is caused by cell elongation, and is then followed by shoot growth. Greater uptake of water is caused by a further reduction in osmotic potential caused by degradation of reserve materials into osmotically active smaller molecules (discussed below) and elongation of seedling tissue. The increase in the moisture content is much slower in phase III (seedling growth) than the physical process of water uptake in phase I (Fig. 1.4). In phase III, the seed becomes a seedling and also loses its ability to withstand desiccation. Therefore, drying seeds after visible germination will result in death of the radicle, while drying seeds during phase I or II is not injurious to the viability of the seed.

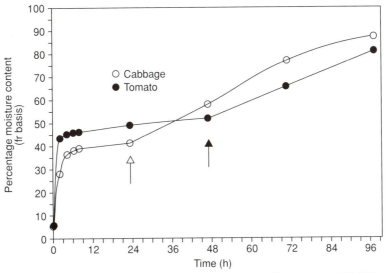

Fig. 1.4. Triphasic water uptake curves for imbibing cabbage and tomato seeds (A.G. Taylor and D.H. Paine, unpublished data). Phase I (imbibition) occurs from 0–6 h, Phase II (lag phase) from *c.* 6 to 24 h in cabbage and 6 to 48 h in tomato, and Phase III (seedling growth) occurs after visible germination as indicated by the arrow for each species.

OXYGEN

Oxygen is necessary for germination and is the substrate required for respiration to produce energy in the form of adenosine triphosphate (ATP). Seeds with low moisture content exhibit negligible gas exchange, and under a static environment respiration is not appreciable until type III water binding in which the water activity is greater than 0.9 (Vertucci and Roos, 1990). Germination is a dynamic process, and gas exchange increases rapidly as seeds imbibe water. The different phases of water uptake (previously described) can be used to describe respiration patterns. In phase I, there is a rapid increase in respiration which is attributed to the activation of existing enzymes. For example, in peas the respiration rate has been shown to increase linearly with the degree of swelling (Kolloffel, 1967). In phase II, a lag phase in oxygen uptake is observed in many large-seeded species and has been attributed to the relative impermeability of the seed coat or seed coverings to gas diffusion since removal of the seed coat reduced this lag (Bewley and Black, 1994). In phase III, visible germination results in piercing of the seed coat, allowing ample oxygen for a second increase in the respiration rate. Also, there is an increase in the number of mitochondria resulting in greater total respiratory activity. Finally, a phase IV has been described only for storage tissue, such as the cotyledons, in which there is a decline in respiration rate attributed to the exhaustion of reserve materials.

MOBILIZATION OF RESERVES

Germination has been previously described in three phases with respect to water uptake and gas exchange. The source of substrates for respiration and/or growth are different before and after radicle emergence. Readily available substrates are needed for the early phases of germination since mobilization of reserve materials to produce smaller molecules does not occur until phase III. Dry seeds have been shown to store sugars that would be a source of soluble carbohydrates needed for respiration during phase I and II. Sucrose is commonly found in dry seeds, and other oligosaccharides such as raffinose and stachyose may also be present (Amuti and Pollard, 1977).

Mobilization of reserves is a post-radicle emergence event and can be studied at the morphological as well as the biochemical levels. Early studies were performed with germinating seeds in time-course experiments by removing samples, dissecting seed and seedling parts and weighing each component. There was a general loss of weight in storage tissue such as the cotyledons or endosperm with an increase in weight of roots and shoots (Mayer and Poljakoff-Mayber, 1982). These studies illustrated trends for utilization of reserves in germinating seed and provided the foundation for more detailed biochemical studies. Unfortunately, our understanding of biochemical events associated with mobilization of reserves in vegetable seeds is fragmentary. Most information is available on agronomic crops and especially those seeds that are also used for human consumption. The following section will briefly outline biochemical events associated with the catabolism of the reserve compounds starch, lipids, proteins and phosphorus. This information was largely obtained from books on seed physiology, and this subject is covered in much greater detail by Bewley and Black (1994).

Starch is the common form of storage carbohydrate and is found in the cotyledons or endosperm of large seeds (see Tables 1.1 and 1.3). Starch occurs in starch grains and is found as a straight chain polymer of glucose called amylose or the branched polymer, amylopectin. Starch is enzymatically degraded by amylase and other enzymes to form monomeric units of glucose. Glucose can be respired or converted to the disaccharide sucrose since sucrose is the form that is transported to the growing regions of the seedling.

Seeds store lipids or triacylglycerols in oil bodies; however, lipids are not commonly found in growing plant tissue. Many small-seeded crops store lipids (see Table 1.1) since lipids are more chemically reduced than starch, and have a higher energy content. The lipid must be converted to a form that can be transported from the storage tissue to the growing points. Lipid degradation occurs by unique biochemical pathways and also utilizes a specialized organelle, the glyoxysome. Glyoxysomes are either present in dry seeds and enlarge during germination or are formed by *de novo* synthesis. The integration of the biochemical pathways with the organelles results in a process known as gluconeogenesis or 'making new sugar'. Briefly, lipids are degraded by lipases to produce three free fatty acids and glycerol. Breakdown products from fatty acids are utilized ultimately to form sucrose that can be transported as described for starch.

Proteins are ubiquitous in seeds (see Table 1.1) and occur as enzymes or storage proteins. The storage proteins are found in protein bodies and degraded by proteinases to form different amino acids. The fate of the released amino acids is complex, and amino acids can be converted to other amino acids or be transported to the growing points. Organic acids can also be formed from amino acids and later respired. Phosphorus is stored in seeds in an organic form called phytic acid or phytin. Phytic acid or myo-inositol hexaphosphate occurs as a salt and can contain potassium, magnesium and calcium as well as the minor elements iron, manganese and copper. Phytase releases phosphate which is used for synthesis of nucleic acids, ATP and phospholipids for membrane synthesis. Macro- and micronutrients are also released and can be used for cell growth and development.

In conclusion, reserve materials in seeds provide a source of carbon in the form of sugars, nitrogen in the form of amino acids, phosphorus in the form of phosphate along with other elements. Seeds are dependent on these stored reserve materials to support initial growth and development of the seedling. After seedling emergence and subsequent growth, the seedling becomes self-supporting by producing its own carbon in photosynthesis and through the uptake of other nutrients via the root system.

SEED QUALITY

Seed quality can encompass many parameters of a seed lot, but this discussion will focus on germination and seedling performance. One method used to estimate seed quality is the standard germination test which is conducted under ideal environmental conditions in the laboratory (Association of Official Seed Analysts, 1993). The criterion for germination used by physiologists is radicle emergence; however, the seed analyst extends this interpretation by classification of seedlings as either normal or abnormal. Abnormal seedlings are those seedlings with an impaired root and/or shoot development or other seedling defects (AOSA, 1992). Only normal seedlings are considered when reporting the actual germination of a lot after sufficient time has been provided in this test to allow for complete germination. The standard germination test is the only test accepted for commercial labelling. The term 'seed vigour' has been used to describe seed quality, and the following definition has been adopted by the AOSA (1983): 'Seed vigor comprises those properties which determine the potential for rapid, uniform emergence and development of normal seedlings under a wide range of field conditions'. Seed vigour differs from germination in that vigour emphasizes the germination rate (rapid and uniform) and the application of the results to forecast field emergence rather than laboratory performance. Seed vigour is generally associated with a higher level of plant performance, and during the early phases of seed deterioration, a reduction in seed vigour occurs prior to a reduction in germination (AOSA, 1983).

The effect of seed ageing can be considered on a population or sample basis as well as on a single seed basis. The decline in the germination of a seed lot maintained under the same environmental conditions reveals a sigmoidal pattern with time (see Fig. 1.2). This curve indicates that all seeds do not die at the same time and that, in general, most seed lots are composed of a mixture of viable and non-viable seeds. The sigmoidal pattern is most evident from lots with an initial high germination (Fig. 1.2c), and these lots are most desirable from a horticultural perspective since these lots can withstand stress or ageing before showing a marked reduction in germination.

On a single seed basis, seeds placed in a conducive environment will either germinate or fail to germinate. This categorical judgement may be inadequate since a number of events can occur to a viable seed before it is rendered non-viable. These events during deterioration also support the concept of a loss of vigour prior to a loss in germination. Different schemes or models have been developed to illustrate changes associated with the loss of viability, and a proposed sequence of changes in seeds during ageing has been described by Delouche and Baskin (1973) and is shown in Fig. 1.5. This sequential model has been challenged by Coolbear (1994) as it appears that one process must be affected prior to the next process and so on. In its defence, the model can be used to illustrate the relative sensitivity of each event to ageing rather than a domino effect. Other merits of this scheme are that both physiological and whole plant responses are presented and the

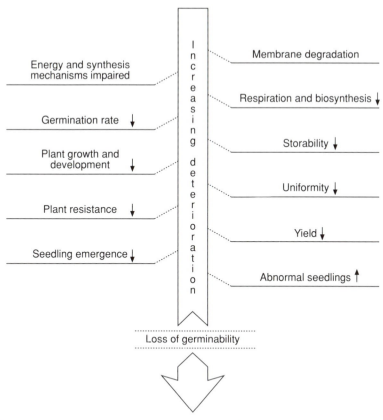

Fig. 1.5. Probable sequence of changes in seed during deterioration (adapted from Delouche and Baskin, 1973).

model does provide a framework to describe a loss of vitality prior to the loss of viability. A review of other events associated with seed ageing and another model of seed ageing based only on biochemical and physiological changes associated with the loss of viability has been described by Priestley (1986).

PHYSIOLOGICAL ASPECTS

Membrane degradation was shown to be the earliest event to occur during seed deterioration (Fig. 1.5). Functional membranes are a prerequisite for cell viability and efficient metabolism. The plasmalemma and tonoplast are essential for compartmentalization of internal constituents, and the mitochondrial membrane is required for respiration and electron transport (Goodwin and Mercer, 1983). As seeds imbibe water, cell membrane components undergo a transition from a gel phase to a liquid-crystalline state

(Crowe *et al.*, 1989). This transition is associated with a period of leakiness caused by the lack of membrane integrity; therefore, for successful germination reorganization of membranes must occur rapidly during phase I. Membrane reorganization is slower or may be prevented as a consequence of ageing and death.

Cell membrane integrity may be tested directly or indirectly as a means to predict seed quality. Certain vital stains have been used in seed testing and provide a direct method to assess cellular integrity (Overaa, 1984). Indigo carmine, a cell membrane permeability stain, is excluded from living cells and is not injurious to subsequent embryo growth. Measuring leakage of compounds during the early stages of germination provides an indirect test of cellular integrity. Leakage tests are commonly performed on intact seeds and compounds such as sugars, amino acids, electrolytes, phenolic compounds and others are measured from the imbibing solution (Priestley, 1986).

Research has been performed on cabbage seeds of different viability exhibiting leakage during the early phases of germination (Hill *et al.*, 1988) (Fig. 1.6). Electrolyte leakage was assessed by measuring the conductivity of the soak solution. Sinapine leakage was estimated by measuring the absorbance of the solution in the ultraviolet (UV) region. Sinapine, the choline ester of a phenolic compound, is a storage material found only in cruciferous seeds (Taylor *et al.*, 1993b). In this study, seeds were removed from the soak water and germinability was determined on a single seed basis. Electrolytes were shown to leak from germinable seeds; however, non-germinable seeds leaked approximately 2.5-fold more than germinable seeds. Sinapine leaked only in trace amounts from seeds of high quality while much greater quantities were detected diffusing from non-germinable seeds. Sinapine leakage was therefore better able to discriminate germinable from non-germinable seeds than the conductivity test. Heat-killed seeds leaked the greatest quantity of both electrolytes and sinapine compared to the other treatments. Non-germinable seeds have dead tissue in critical areas associated with embryo growth but may contain living cells in the storage tissue (Lee and Taylor, 1995). The living cells would not contribute appreciably to the total leakage.

Leakage tests have found application in seed quality programmes, and methods to assess seed vigour by measuring conductivity have been described (AOSA, 1983). This type of biochemical test can be performed in a relatively short period of time compared to a germination test, and commercial instrumentation has been developed to measure the conductivity of 100 single seeds (Wavefront Inc., Ann Arbor, Michigan). Unfortunately, there are serious limitations for the wide-scale use of these methods since seeds of many species possess an inner semi-permeable seed coat layer that restricts leakage of solutes (Beresniewicz *et al.*, 1995b). For example, seeds of onion and leek have been shown to contain a semi-permeable layer composed of cutin, while tomato and pepper have a suberized layer (Beresniewicz *et al.*, 1995a).

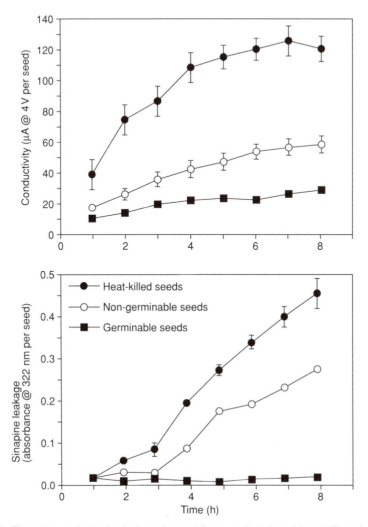

Fig. 1.6. Electrolyte and sinapine leakage from cabbage seeds (adapted from Hill *et al.*, 1988).

Energy synthesis pathways and respiration are vital to a viable seed, and both processes are shown to be associated with early events during deterioration (see Fig. 1.5). All biochemical pathways require enzymes to catalyse reactions within the cell. The relationship of enzyme activity with seed ageing has been studied. Many enzymes remain active after all viability is lost; however, the dehydrogenases are one group of enzymes that were shown to be directly related with a loss of viability (MacLeod, 1952). Dehydrogenase enzymes are found in several steps in respiratory pathways and catalyse oxidation-reduction reactions (Goodwin and Mercer, 1983). As previously discussed, respiration rate increases rapidly during imbibition, and maximal respiration is needed for the completion of germination. Therefore, dehydrogenase activity can best be determined during phase II germination, a period of active metabolism. The most common method to measure dehydrogenase activity is with the vital stain 2,3,5-triphenyl tetrazolium chloride (TTC or TZ). Tetrazolium salts in the oxidized form are colourless and water soluble and can be reduced by dehydrogenase enzymes to the water insoluble red stain, formazan (Moore, 1985). Tetrazolium salts were first used in the 1940s, and the TZ test is still used as a viability test for many species (Copeland and McDonald, 1985). Limitations are that the test is subjective and staining patterns may be difficult to interpret resulting in inaccurate assessment of viability. However, a trained seed analyst can distinguish seed deterioration from other problems such as mechanical injury or hydration damage (Moore, 1972). Other vital stains have been used in seed testing (Overaa, 1984) and provide an alternative to the TZ method.

WHOLE PLANT RESPONSES

A primary interest in horticulture is to determine the consequence of seed ageing on the whole seed and whole plant level. The germination rate is the most sensitive index of seed quality at the whole seed level (see Fig. 1.5); however, the standard germination test does not measure the rate but only the total germination of a lot. Actually, the only step of the germination process which can be accurately measured is the onset of phase III (visible germination). Therefore, the time to radicle emergence provides information on the relative vigour of a seed lot. To illustrate this effect, lettuce seed was aged for a short period by first increasing the moisture content to 20% (fw basis) and then incubating the seeds at 40°C for 24 h (Tomas, 1990). This procedure was adapted from the controlled deterioration test developed as a method to assess seed vigour (Powell et al., 1984). The time to radicle emergence was studied using time-sequence photography using a 35 mm camera to record germination at 2-h intervals (Tomas et al., 1992a). Aged seeds germinated approximately 6 h later than non-aged seeds and produced a greater percentage of necrotic seedlings (Fig. 1.7). Necrotic seedlings are symptomatic of a disorder associated with ageing in lettuce seeds called physiological necrosis (Tomas et al., 1992b) and are classified as abnormal seedlings (AOSA, 1992). These data reveal that a mild ageing treatment decreases the germination rate and increases the incidence of abnormal seedlings with physiological necrosis although the total number of germinable seedlings is not changed.

Field emergence is of major horticultural importance for direct seeded crops since seeds are sown to achieve a desired plant population for optimal harvest efficiency. Seed vigour has been shown to be related to field emergence (Roberts, 1972; Heydecker, 1977) with poor seed quality resulting in

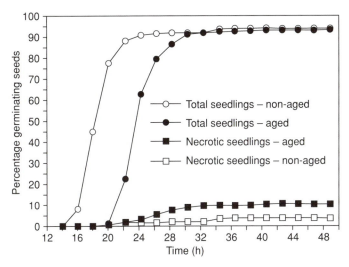

Fig. 1.7. The influence of ageing on germination rate of lettuce seeds including the presence of necrotic seedlings (adapted from Tomas, 1990).

poor stand and reduced yield. Although the effect of plant population on yield has been frequently studied and reviewed (Wiley and Heath, 1969), another more interesting question – Does a reduction in seed vigour affect yield if plant population is not a factor? – has been considered less often. Data from many separate studies have been summarized to address this question, and the following information has been adapted from the review by Tekrony and Egli (1991). Crops were grouped based on the time of harvest of the harvestable product, either as vegetative, early reproductive or full reproductive stages. Sources of seed quality differences were obtained by ageing. In general, crops harvested in the vegetative stage showed the greatest response to ageing with loss of vigour resulting in reduced uniformity of seedling emergence. Fairly consistent beneficial responses from sowing high quality seeds were also measured on crops that were harvested in the early reproductive stage. Crops harvested at full maturity, including most agronomic crops, did not show a positive response to seed vigour, and results from dry beans presented in Table 1.4 were variable. In conclusion, seed vigour is important for yield potential in many vegetable crops since most of these crops are harvested in the vegetative or early reproductive stages.

SOWING ENVIRONMENT

The soil environment must finally be considered as it can greatly influence germination and seedling establishment of vegetable crops. In practice, direct seeding is often performed early in the growing season in soil

Table 1.4. Relationship of seed vigour to yield when crops are harvested vegetatively and at early and full reproductive maturity. Adapted from Tekrony and Egli, 1991.

Crop name	Source of seed quality differences*	Yield	
		Plant component measured	Response to seed vigour[†]
Vegetative harvest			
Lettuce	Long-term storage	Head fresh weight	None
	Long-term storage	Head fresh weight	Positive
Onion	Long-term storage	Bulb fresh weight	Positive
Radish	Long-term storage	Root fresh weight	Positive
Early reproductive harvest			
Pea	Short-term storage	(i) Plant height	(i) Positive
		(ii) Seed dry weight	(ii) None
Tomato	Long-term storage	Fruit fresh weight	Positive
	Artificial ageing	Fruit fresh weight	Positive
	Long-term storage	Fruit fresh weight and no./plant	Positive
Full reproductive harvest			
Field bean	Long-term storage	Seed dry weight	None
Red kidney bean	Artificial ageing	Seed dry weight	Positive

* Artificial ageing: seed exposed to high temperatures for short time interval; short-term storage: seed stored at specified temperature and seed moisture for < 6 months; long-term storage: seed stored at range of temperatures, seed moistures for several years.
[†] Positive indicates positive yield response to seed vigour; none indicates either no yield response or a negative response to vigour.

conditions that are less than optimal for a particular species. Various abiotic stresses including water, temperature and physical impedance and biotic stresses such as soil pathogens, insects and other predators may be present from the time of sowing to seedling establishment. The following discussion will briefly review the role of the three major abiotic stresses and describe the sensitive phase for each.

WATER STRESS

Water is essential for germination; however, water stress may occur as either an excess or deficit in the field for direct seeded crops or in the greenhouse for transplant production. In the case of water excess, oxygen can be limiting for the completion of germination or seedling growth, because phases II and III in germination are sensitive to oxygen deficiencies. Seeds of different species were subjected to anoxia by soaking in water and germination was recorded after 72 h (Crawford, 1977). Of the vegetable seeds tested, lettuce was found to be tolerant to soaking while pea was most sensitive, and it was shown that sensitivity to soaking was associated with a large production of the fermentation product ethanol. Seeds are generally sown initially under favourable soil moisture conditions and then rain or irrigation can create a flooding condition. It has been shown that a period of low oxygen (hypoxia) can condition germinating seedlings so that seedling roots can survive for a longer period of time in a subsequent anaerobic condition compared to those that were not conditioned (Hole *et al.*, 1992).

Seeds are commonly sown at a shallow depth, and the soil may dry to a water potential below that necessary for the completion of germination. Seeds due to their tremendous matric potential can imbibe or at least partially imbibe water even in dry soils and may enter phase II germination (Bewley and Black, 1994). However, due to the dry conditions the seed cannot achieve sufficient moisture content to complete germination. Phase III is associated with cell elongation and later cell division, and these processes have been shown to be most sensitive to water stress in growing tissue (Hsiao, 1973). A higher water potential (more water available) is needed for the initiation of cell elongation than for the maintenance of radicle growth after visible germination (reviewed by Hegarty, 1978). This indicates that a threshold moisture level must be achieved before the seed will complete germination (phase III). Since seeds are desiccation-tolerant in phases I and II, the inability to complete germination under water deficits would help to ensure survival under these stressful environments.

TEMPERATURE STRESS

Temperature regulates all aspects of biology including the germination of seeds. The cardinal temperatures of germination (minimum, optimum and maximum) have been summarized for vegetable crops seeds (Lorenz and Maynard, 1980). Though temperature does affect the time for full imbibition, temperature primarily influences the germination rate by regulating the duration of the lag phase or phase II. Germination can be predicted by incorporating a heat sum in degree days (S) and the minimum temperature for germination (T_{min}) (Bierhuizen and Wagenvoort, 1974). The S and T_{min} are shown for 31 vegetable crop seeds (Table 1.5) and the predictive equations were highly correlated over the temperature range tested. In general, the warm season crops have a higher T_{min} than the cool season crops, while the value for S varies with species.

Temperature stress may occur, and temperatures can be suboptimal or supra-optimal for a particular species. The effect of elevated temperatures on lettuce seed germination will be presented in Chapter 14 and is not discussed here. Seeds of many warm season crops are negatively influenced by low temperature, and this physiological disorder is known as chilling injury (reviewed by Herner, 1990). There are two groups of chilling sensitive seeds:

Table 1.5. Minimum germination temperature (T_{min}) and heat sum (S) in degree days for seedling emergence, and the applicable temperature (T) range for germination of various vegetables. Crops are ranked within groups by heat sum (S) in degree days. Adapted from Bierhuizen and Wagenvoort, 1974.

Group	Crop	Genus species	T_{min} (°C)	S (degree days)	T (°C)
Leaf vegetables and *Brassica* crops	Purslane	*Portulaca oleracea*	11.0	48	15–25
	Cress	*Lepidium sativum*	1.0	64	3–17
	Lettuce	*Lactuca sativa*	3.5	71	6–21
	Witloof, chicory	*Cichorium sativa*	5.3	85	9–25
	Endive	*Cichorium endiva*	2.2	93	3–17
	Savoy cabbage	*B. oleracea var. sabauda*	1.9	95	3–17
	Turnip	*B. campestris var. rapa*	1.4	97	3–17
	Borecole, kale	*B. oleracea var. acephala*	1.2	103	3–17
	Red cabbage	*B. oleracea var. purpurea*	1.3	104	3–17
	White cabbage	*B. oleracea var. capitata*	1.0	106	3–17
	Brussels sprouts	*B. oleracea var. gemmifera*	1.1	108	3–17
	Spinach	*Spinacea oleracea*	0.1	111	3–17
	Cauliflower	*B. oleracea var. botrytis*	1.3	112	3–17
	Corn salad	*Valerianella olitoria*	0.0	161	3–17
	Leek	*Allium porrum*	1.7	222	3–17
	Celery	*Apium graveolens*	4.6	237	9–17
	Parsley	*Petroselinum crispum*	0.0	268	3–17
Fruit vegetables	Tomato	*Lycopersicon esculentum*	8.7	88	13–25
	Eggplant	*Solanum melongena*	12.1	93	15–25
	Gherkin	*Cucumis sativus*	12.1	108	15–25
	Melon	*Cucumis melo*	12.2	108	15–25
	Sweet pepper	*Capsicum annuum*	10.9	182	15–25
Leguminous crops	Garden pea	*Pisum sativum*	3.2	86	3–17
	French sugar pea	*P. sativum var. sacharatum*	1.6	96	3–17
	Bean (French)	*Phaseolus vulgaris*	7.7	130	13–25
	Broad bean	*Vicia faba*	0.4	148	3–17
Root crops	Radish	*Raphanus sativus*	1.2	75	3–17
	Scorzonera	*Scorzonera hispanica*	2.0	90	3–17
	Beet	*Beta vulgaris*	2.1	119	3–17
	Carrot	*Daucus carota*	1.3	170	3–17
	Onion	*Allium cepa*	1.4	219	3–17

(i) those that are injured during phase III germination such as the solanaceous crops and the cucurbits; and (ii) those seeds that are sensitive during phase I such as snap and lima beans. In the second group, damage occurs during hydration of the seed tissue and is referred to as imbibitional chilling injury. Seeds become more susceptible to this type of injury as seed moisture content decreases. Seed coat permeability and integrity are also important factors as seeds that imbibe water rapidly, especially seeds with cracked seed coats are more prone to imbibitional chilling injury (Taylor *et al.*, 1992).

To understand further the role of seed moisture content and temperature in imbibitional chilling injury, two cultivars (chilling sensitive and tolerant) of snap bean seeds were adjusted to a range of moisture contents from 5 to 25% (dw basis) and germinated at 20 or 5°C (Wolk *et al.*, 1989). Those seeds that were germinated at 5°C were transferred to 20°C after 24 h. A critical seed moisture content was determined for each treatment that marked the onset of imbibitional chilling injury and was termed the breakpoint. Only seeds with moisture contents below the breakpoint showed a reduction in germination. At 20°C, the moisture content breakpoints for the chilling sensitive and tolerant cultivars were 15 and 11%, respectively. When seeds were tested at 5°C, the breakpoints were 19 and 16% for the sensitive and tolerant cultivar, respectively. Thus at either temperature, the breakpoint moisture content was always greater for the sensitive compared to the tolerant cultivar. Decreasing the temperature from 20 to 5°C shifted the breakpoint to a higher level for each cultivar; however, the deleterious effect of imbibing seeds at 5°C could be totally overcome by the elevated moisture content. Below the breakpoint for all treatments, there was an average 4.6% decrease in germination for each 1% decrease in moisture content. In conclusion, imbibitional chilling injury is influenced by the interaction of several environmental and seed factors. The initial seed moisture content is the primary factor that determines the incidence of imbibitional chilling for a particular seed lot, while temperature has a moderating effect.

PHYSICAL IMPEDANCE

The soil can act as a physical barrier to seedling emergence and may decrease or even prevent seedling establishment especially under conditions of soil crusting (Goyal *et al.*, 1980). Germinating seedlings must produce sufficient force to overcome this barrier. The sensitive period to this type of stress is late in phase III when these seeds have already completed radicle emergence. There are several factors that influence the emergence ability of a seedling including the speed of germination and morphological characteristics (Inouye *et al.*, 1979). A faster emerging seedling has a better chance of escaping the physical barrier of a hard crust since it may emerge before soil crusting conditions occur. Two primary factors which determine the rate of emergence are soil temperature and species (see Table 1.5). Considering a particular crop, seed quality may play a role because the rate of germination is influenced by vigour. Seedling morphology is also important (Table 1.3), since seeds with hypogeal germination have a smaller crosssectional area to penetrate the soil than those with epigeal germination, which must also pull the cotyledons through the soil.

Seedling emergence forces have been quantified for a number of vegetable seeds and are shown in Table 1.6 (Taylor and Ten Broeck, 1988). A positive relationship ($r = 0.98**$) was determined between seed weight and force for the eight species tested, and among small, medium and large snap bean seeds of the same lot. The pressure (force per unit cross-sectional area of emerging hypocotyl or cotyledon in onion) was positively correlated ($r = 0.85*$) with time to achieve maximum force. Seedlings with the ability to continue to generate forces may have a better chance of establishment than those that produced pressure for a short time. The energy content is related to seed composition as seeds generally store either starch or lipid (see Table 1.1). The energy content of the starch storing snap beans yielded 17 kJ g^{-1} in comparison to the lipid containing seeds that ranged from 22 to 26 kJ g^{-1} of seed. The use of reserves varies by seed type, but does provide some relative information on the efficiency of mobilization. The study on seed size in snap bean revealed that small seeds contained less reserves than large seeds; however, the small seeds were more efficient in utilization of their reserves in seedling emergence forces (Taylor and Ten Broeck, 1988). A subsequent study measured emergence forces from snap bean seeds subjected to imbibitional chilling injury (Taylor *et al.*, 1992). The seeds subjected to chilling produced less force per seedling and required a longer period of time to generate the maximum force indicating that the injury sustained during imbibition reduced subsequent seedling growth potential.

SEED ENHANCEMENTS

The last section of this chapter turns from the theme of seed biology to seed technology. 'Seed enhancements' include several methods that can be used to accelerate the rate of germination and increase the percentage germination or seedling growth. All enhancements are applied prior to sowing and may be

Table 1.6. Seedling emergence force characteristics and energy contents of vegetable crop seeds. Crops are ranked by maximum seedling emergence forces generated. Adapted from Taylor and Ten Broek, 1988.

Crop	Seed wt. (mg)	Maximum force (mN)*	Time to achieve maximum force (h)	Pressure exerted (kPa)†	Energy content (J seed^{-1})‡	Use of reserves (N kJ^{-1})
Snap bean	268.0	3400 ± 360**	21 ± 1	234	4554 ± 12	0.75
Radish	10.0	558 ± 88	19 ± 8	317	231 ± 1	2.40
Cucumber	31.4	241 ± 49	9 ± 3	63	801 ± 1	0.30
Cabbage	4.20	157 ± 24	11 ± 2	241	111 ± 1	1.41
Onion	4.04	83 ± 11	19 ± 4	259	90.4 ± 0.2	0.92
Tomato	2.90	44 ± 5	10 ± 2	96	71.8 ± 0.1	0.61
Carrot	1.00	35 ± 9	5 ± 1	117	23.8 ± 0.1	1.47
Lettuce	1.07	29 ± 6	7 ± 2	89	27.3 ± 0.1	1.06
Beets	–	26 ± 6	4 ± 2	62	–	–

* Maximum force achieved per seedling in millinewtons.
† Pressure exerted in kilopascals.
‡ Energy content per seed in joules.
** Mean ± standard error.

performed on a commercial basis by the seed industry. Three topics will be addressed under this heading: hydration treatments, coating technologies and the integration of these methods to upgrade seed quality. Although coating technologies are not considered as enhancement processes (Kaufman, 1991), they will be discussed in this section since coatings act as delivery systems for materials beneficial to seeds. Other pre-sowing treatments are reviewed by Heydecker and Coolbear (1977) and Khan (1992).

HYDRATION TREATMENTS

The importance of water in seeds has been described in relation to storage, germination and seedling establishment. Two hydration methods to improve germination and seedling establishment especially under stressful conditions, namely moisturization and priming will be presented. Moisturization has been developed for large-seeded legumes, in particular snap beans; however, the technique may be used for other leguminous vegetable seeds. Priming has been performed on many small-seeded vegetable seeds, and this method is being adapted for a wide range of species.

The importance of seed moisture content on the resistance of bean seeds to imbibitional chilling injury has been discussed (see Temperature stress section above). Moisturization improved field emergence of early plantings, especially if the soil was wetted immediately after sowing (Wilson and Trawatha, 1991; Taylor et al., 1992). The percentage of initial moisture content can affect germination of the same bean seed lot in the laboratory, and low moisture content samples had lower germination results than high moisture samples (Pollock and Manalo, 1970). Another benefit of moisturization is increased resistance to mechanical damage as moist seeds are less brittle (Bay, 1994; Bay et al., 1995). However, moisturization to too high a level can be deleterious as the rate of ageing in storage will increase dramatically.

The moisture content of bean seeds may be adjusted prior to packaging, and seeds are then stored in this condition. In practice, seed moisture content is adjusted to the upper level of region 2, which corresponds to a seed moisture content of 12–13% (fw basis) or an a_w from 0.6 to 0.65 (see Fig. 1.1). Seeds may be moisturized by passing humidified air through the seed mass to increase the seed moisture content or by incubation with a moist solid media (Wilson and Trawatha, 1991).

'Seed priming' is a general term that refers to several different techniques used to hydrate seeds under controlled conditions, but preventing the completion of germination (phase III). During priming, seeds are able to imbibe or partially imbibe water and achieve an elevated seed moisture content usually in phase II (lag phase) germination (see Fig. 1.4). Seeds may be kept in this condition for a period of time that may range from less than 1 day to several weeks (Taylor and Harman, 1990). Priming temperatures range from 10 to 35°C, but 15 to 20°C is most commonly used (Bradford, 1986). Since seeds have not completed germination, they remain desiccation-tolerant and can still be dried for long-term storage. All priming techniques rely on the controlled uptake of water to achieve a critical moisture content that will activate metabolic activity in a controlled environment. There is a considerable number of terms used in the literature and in the industry to describe these methods, and attempts will be made to clarify this terminology.

There are three priming techniques that have been employed to advance the germination of seeds. The most studied technique utilizes aqueous solutions as the priming medium. In large-scale priming, a ratio of approximately 10 parts priming solution to 1 part seeds is used (Nienow *et al.*, 1991). Therefore, due to the large reservoir of priming solution, the water uptake by seeds is regulated by the water potential of the solution which varies by species and ranges from -0.5 to -2.0 MPa (Khan *et al.*, 1980/81). Many compounds have been used to achieve a solution of known water potential and include inorganic salts such as $NaCl$, KNO_3, K_3PO_4, KH_2PO_4 and $MgSO_4$, low molecular weight organic compounds such as glycerol and mannitol and large molecular weight polymers such as polyethylene glycol (PEG) (Khan, 1992). The 6000 or 8000 molecular weight PEGs have been widely used to regulate water potential and formulas have been developed to calculate the water potential of a solution of known concentration and temperature (Michel, 1983). Since gas diffusion is limited in solution, aeration is needed during the priming process. A number of terms that have been used to describe this technique include liquid priming, osmotic conditioning and osmoconditioning. The term osmo- or osmotic may be misleading when PEG is used since the water potential of PEG solutions is controlled primarily by matric forces (Steuter *et al.*, 1981).

Two other proprietary priming techniques, solid matrix priming (Eastin *et al.*, 1993) and drum priming (Rouse cited by Gray, 1994), have also gained attention as an alternative to liquid priming. In solid matrix priming (SMP), seeds are mixed with a solid particulate material and water (Taylor *et al.*, 1988). Seeds, due to their negative matric potentials, are able to imbibe water from the solid material. A number of materials have been used in this process including a leonardite shale, diatomaceous silica, exfoliated vermiculite and expanded calcined clay (Khan, 1992). The amount of solid carrier required for a particular species depends on its water holding capacity, and has been reported from 2 to 0.2 times the weight of the seed (Taylor *et al.*, 1988; Khan, 1992). The amount of water needed to achieve an equilibrium water potential conducive for priming is determined on an empirical basis as described for liquid priming. Leonardite shale is a material that regulates water potential by its osmotic potential (Taylor *et al.*, 1988). Diatomaceous silica materials regulate water potential by their matric properties and their use in priming has been termed matriconditioning (Khan, 1992). Drum priming was developed in the UK and involves hydration by misting seeds with water during a 1- or 2-day period in a revolving drum (Gray, 1994). The level of hydration is controlled for each species, and tumbling ensures uniformity of moisture distribution.

The benefits of priming by different techniques have been documented for a number of vegetable seeds and have been reviewed by Heydecker and Coolbear (1977), Bradford (1986), Khan (1992) and Gray (1994). In general, priming hastens the rate of germination and seedling emergence, especially under suboptimal temperatures for germination. In the case of lettuce, priming ameliorates the deleterious effect of high temperatures causing thermoinhibition and thermodormancy. Seeds primed by the seed industry are dried back to their original moisture for storage until utilization. Drying may reduce some of the benefits gained by priming, and optimal drying conditions of temperature, relative humidity and air flow are needed to retain these beneficial effects (Ptasznik and Khan, 1993). Excessive drying of certain lots of lettuce seeds after priming can negate the acquired tolerance to high temperatures (Weges and Karssen, 1990). Seed storage life after priming is an important consideration for utilization of this technology in commercial practice. Research conducted on lettuce and tomato has shown that the enhancement of germination rates by priming also accelerates the ageing rate compared to non-primed seeds (Argerich *et al.*, 1989; Tarquis and Bradford, 1992).

COATING TECHNOLOGIES

Seed coatings have evolved with time and have been used for a number of purposes to improve agricultural productivity and reduce environmental hazards. Early emphasis was placed on the development of coatings for small and irregularly shaped seeds to facilitate precision placement during sowing (Tonkin, 1979). Coatings were later developed to act as a delivery system for a number of materials required at time of sowing (Scott, 1989; Taylor and Harman, 1990). Finally, coating technologies have been further refined to reduce worker exposure to seed treatments during handling (Halmer, 1994; Robani, 1994). Two coating technologies, pelleting and film-coating, will be discussed with respect to general methodology and application to vegetable seeds.

Seed pelleting consists of the application of solid particles that act as a filler with a binder or adhesive to form a more or less spherically shaped dispersal unit. Pelleting is routinely performed by the seed industry on high-value, small-seeded horticultural seeds. In this process, seeds are generally pelleted on a batch basis in a coating pan or tumbling drum. Pellets are sized during and at the end of the process and then dried. The materials and techniques used are proprietary; however, a number of ingredients have been listed in the literature (Halmer, 1988; Scott, 1989; Taylor and Harman, 1990). Onion seeds have been pelleted in a laboratory scale coating pan by the application of a mixture of diatomaceous earth (filler) with finely ground polyvinyl alcohol (binder) (Taylor and Eckenrode, 1993). Seeds were tumbled with repeated applications of the coating formulation followed by intermittent spraying of seeds with water to activate the binder and cause the formation of the pellet around each seed. The pellets improved the plantability of the seeds and also acted as a delivery system for plant protectants needed to control soil-borne insects and diseases.

Film-coating of seeds is a more recent development than pelleting and is derived from techniques originally developed for the pharmaceutical industry (Porter and Bruno, 1991). Film-coating consists of spraying a solution or suspension of film-forming polymer onto a mass of seeds to achieve a uniform deposition of materials. A number of film-forming polymers and pigments have been used (Halmer, 1988; Robani, 1994). Coating pans described for pelleting may also be used; however, in contrast to the wet operation of pelleting, the aqueous film-forming formulation must be dried immediately after spraying to avoid agglomeration. Perforated pans have been used to allow for rapid drying and continuous flow methods have also been developed (Halmer, 1994; Robani, 1994). Fluidized bed technology has also been used for film-coating (Halmer, 1988), however, this technology has limitations for large-scale use. The benefits of film-coating include uniform placement of seed treatment chemicals onto seeds, essentially a dust-free environment, enhanced appearance due to the addition of pigments and finally an excellent medium to deliver a number of plant protectants (Robani, 1994). Film-coating and pelleting are not mutually exclusive, and seeds may be film-coated after pelleting to provide coloration as well as to reduce dust from the pellet.

Hydration and coating technologies may be integrated with the goal of improving the germination of a seed lot. A novel system was developed to exploit sinapine leakage as a method of non-destructively detecting and upgrading *Brassica* seed quality (Taylor *et al.*, 1991). The sequence of events in this process is shown in Fig. 1.8. Seeds are first hydrated in water or primed. During that process, sinapine leakage commences only from non-viable seeds. The hydrated seeds are coated with an adsorbent that retains the sinapine leakage on a single seed basis. The coated seeds are then dried,

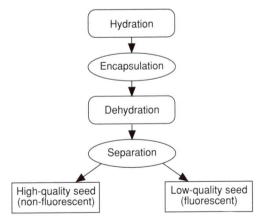

Fig. 1.8. The sequence of events in upgrading *Brassica* seed quality (adapted from Taylor *et al.*, 1991).

exposed to UV light and sorted based on the fluorescence of the coatings. Sowing the non-fluorescent fraction resulted in a greater percentage of normal seedlings than the control (non-sorted) seeds, while the fluorescent fraction contained a high percentage of dead seeds and seeds that produced abnormal seedlings in greenhouse studies (Taylor *et al.*, 1991). An electronic colour sorter equipped with a UV source to illuminate coated seeds was successfully demonstrated to upgrade cabbage, cauliflower and broccoli seed lots (Taylor *et al.*, 1993a).

CONCLUDING REMARKS

An understanding of vegetable seeds is an important first step for the subsequent study of vegetable physiology and culture. Vegetable seeds are a diverse group with respect to botanical classification, morphology and composition, and this diversity impedes rapid progress in our understanding of how vegetable seeds should be handled for optimum performance.

The high value of the harvested product increases the demand for vegetable seeds of high quality and maximum performance. The goal is for each seed sown to develop into a usable transplant or productive plant in the field. To achieve this goal, a seed lot should have complete, uniform and rapid germination and seedling emergence. This goal is seldom achieved due to one or a combination of factors including small seed size, slow germination rate, low inherent seed quality and sensitivity to environmental stresses at the time of sowing. Unlike most agronomic crops, most vegetable seeds are not directly consumed, and have not been selected for large seed size. In the case of one large-seeded vegetable, sweet corn, the seed has been selected for its high sugar content, which also makes it susceptible to seed- and soil-borne pathogens.

To overcome some of the seed quality and slow germination problems, commercial companies provide sized and graded seeds to facilitate sowing. Seed enhancements, such as priming have improved seedling emergence of warm-season crops under cool conditions, and lettuce under high temperatures. Priming also accelerates the rate of germination, especially in cool conditions, when seedlings may fail to emerge due to formation of a soil crust. Combining seed enhancements with coating technologies has been shown to improve germination by detecting poor quality seeds and eliminating them from the seed lot.

Continued effort will be needed by seed scientists to develop the knowledge of physiological mechanisms associated with seed ageing, and the factors limiting seed performance. Some of these basic studies will be performed with vegetable species, and others adapted from agronomic crops. Seed quality criteria may be incorporated into breeding and selection programmes, in the development of improved cultivars. In conclusion, by integrating several approaches, the seed quality of vegetable crops in future should improve as we gain a better understanding of seed physiology.

REFERENCES

Amuti, K.S. and Pollard, C.J. (1977) Soluble carbohydrates of dry and developing seeds. *Phytochemistry* 16, 529–532.

Anderson, W.H., Carolus, R.L. and Watson, D.P. (1953) The germination of okra seed as influenced by treatment with acetone and alcohol. *Proceedings of the American Society for Horticultural Science* 62, 427–432.

Argerich, C.A., Bradford, K.J. and Tarquis, A.M. (1989) The effects of priming and ageing on resistance to deterioration of tomato seeds. *Journal of Experimental Botany* 40, 593–598.

Association of Official Seed Analysts (1983) *Seed Vigor Testing Handbook*. Association of Official Seed Analysts, Contribution No. 32.

Association of Official Seed Analysts (1992) *Seedling Evaluation Handbook*. Association of Official Seed Analysts, Contribution No. 35.

Association of Official Seed Analysts (1993) Rules for testing seeds. *Journal of Seed Technology* 16(3), 1–113.

Bay, A.P.M. (1994) Comparative study of mechanical damage resistance of two isogenic lines and a commercial cultivar of snap bean seeds. PhD thesis, Cornell University.

Bay, A.P.M., Taylor, A.G. and Bourne, M.C. (1995) The influence of water activity on three genotypes of snap beans in relation to mechanical damage. *Seed Science and Technology* 23, 583–593.

Beresniewicz, M.M., Taylor, A.G., Goffinet, M.C. and Koeller, W.D. (1995a) Chemical nature of a semipermeable layer in seed coats of leek, onion (*Liliaceae*), tomato and pepper (*Solanaceae*). *Seed Science and Technology* 23, 135–145.

Beresniewicz, M.M., Taylor, A.G., Goffinet, M.C. and Terhune, B.T. (1995b) Characterization and location of a semipermeable layer in seed coats of leek, onion (*Liliaceae*), tomato and pepper (*Solanaceae*). *Seed Science and Technology* 23, 123–134.

Bewley, J.D. and Black, M. (1978) *Physiology and Biochemistry of Seeds*, Springer-Verlag, Berlin.

Bewley, J.D. and Black, M. (1994) *Seeds: Physiology of Development and Germination*, 2nd edn. Plenum Press, New York.

Bierhuizen, J.F. and Wagenvoort, W.A. (1974) Some aspects of seed germination in vegetables. 1. The determination and application of heat sums and minimum temperature for germination. *Scientia Horticulturae* 2, 213–219.

Bourne, M.C. (1991) Water activity: food texture. In: Hui, Y.H. (ed.) *Encyclopedia of Food Science and Technology*. John Wiley, New York, pp. 2801–2815.

Bradford, K.J. (1986) Manipulation of seed water relations via osmotic priming to improve germination under stress conditions. *HortScience* 21(5), 1105–1112.

Coolbear, P. (1994) Mechanisms of seed deterioration. In: A.S. Basra (ed.) *Seed Quality: Basic Mechanisms and Agricultural Implications*. Food Products Press, New York, pp. 223–277.

Copeland, L.O. and McDonald, M.B. (1985) *Principles of Seed Science and Technology*, 2nd edn. Burgess Publishing, Minneapolis.

Crawford, R.M.M. (1977) Tolerance of anoxia and ethanol metabolism in germinating seeds. *New Phytology* 79, 511–517.

Crowe, J.H., Crowe, L.M., Hoekstra, F.A. and Wistrom, C.A. (1989) Effects of water on the stability of phospholipid bilayers: the problem of imbibition damage in dry organisms. In: Stanwood, P.C. and McDonald, M.B. (eds) *Seed Moisture*. Crop Science Society of America, Madison, pp. 1–14.

Delouche, J.C. and Baskin, C.C. (1973) Accelerated aging techniques for predicting the relative storability of seed lots. *Seed Science and Technology* 1, 427–452.

Earle, F.R. and Jones, Q. (1962) Analyses of seed samples from 113 plant families. *Economic Botany* 16, 221–250.

Eastin, J.A., Vendeland, J.S. and Kubick, K.K. (1993) Solid matrix priming (SMP) and effects on onion germination and emergence. *Proceedings of the National Onion Conference*, 151–156.

Ellis, R.H., Osei-Bonsu, K. and Roberts, E.H. (1982) The influence of genotype, temperature and moisture on seed longevity in chickpea, cowpea, and soya bean. *Annals of Botany* 50, 69–82.

Ellis, R.H. and Roberts, E.H. (1981) The quantification of ageing and survival in orthodox seeds. *Seed Science and Technology* 9, 373–409.

Goodwin, T.W. and Mercer, E.I. (1983) *Introduction to Plant Biochemistry,* 2nd edn. Pergamon Press, Oxford.

Goyal, M.R., Carpenter, T.G. and Nelson, G.L. (1980) Soil crusts versus seedling emergence. *American Society of Agricultural Engineers.* Paper no. 80–1009, pp. 1–33.

Grabe, D.F. (1989) Measurement of seed moisture. In: Stanwood, P.C. and McDonald, M.B. (eds) *Seed Moisture*. Crop Science Society of America, Madison, pp. 69–92.

Gray, D. (1994) Large-scale seed priming techniques and their integration with crop protection. In: Martin, T. (ed.) *Seed Treatment: Progress and Prospects*. British Crop Protection Council, Surrey, UK, pp.353–362.

Halmer, P. (1988) Technical and commercial aspects of seed pelleting and film-coating. In: Martin, T.J. (ed.) *Application to Seeds and Soil*. British Crop Protection Council, Surrey, UK, pp. 191–204.

Halmer, P. (1994) The development of quality seed treatments in commercial practice – objectives and achievements. In: Martin, T.J. (ed.) *Seed Treatment: Progress and Prospects*. British Crop Protection Council, Surrey, UK, pp. 363–374.

Harmond, J.E., Brandenburgh, N.R. and Klein, L.M. (1968) *Mechanical Seed Cleaning and Handling*. United States Department of Agriculture, Washington, DC.

Hayward, H.E. (1938) *The Structure of Economic Plants*. Macmillan, New York.

Hegarty, T.W. (1978) The physiology of seed hydration and dehydration, and the relation between water stress and the control of germination: a review. *Plant, Cell and Environment* 1, 101–119.

Herner, R.C. (1990) The effects of chilling temperatures during seed germination and early seedling growth. In: Wang, C.Y. (ed.) *Chilling Injury of Horticultural Crops*. CRC Press, Boca Raton, pp. 51–69.

Heydecker, W. (1977) Stress and seed germination: an agronomic view. In: Khan, A.A. (ed.) *The Physiology and Biochemistry of Seed Dormancy and Germination,* North-Holland, Amsterdam, pp. 237–282.

Heydecker, W. and Coolbear, P. (1977) Seed treatments for improved performance – survey and attempted prognosis. *Seed Science and Technology* 5, 353–425.

Heydecker, W. and Orphanos, P.I. (1968) The effect of excess moisture on the germination of *Spinacia oleracea* L. *Planta* 83, 237–247.

Hill, H.J. and Taylor, A.G. (1989) Relationship between viability, endosperm integrity and imbibed lettuce seed density and leakage. *HortScience* 24(5), 814–816.

Hill, H.J., Taylor, A.G. and Huang, X.L. (1988) Seed viability determinations in cabbage utilizing sinapine leakage and electrical conductivity measurements. *Journal of Experimental Botany* 39(207), 1439–1448.

Hole, D.J., Cobb, B.G. and Drew, M.C. (1992) Enhancement of anaerobic respiration in root tips of *Zea mays* following low-oxygen (hypoxic) acclimation. *Plant Physiology* 99, 213–218.

Hsiao, T.C. (1973) Plant responses to water stress. *Annual Review of Plant Physiology* 24, 519–570.

Iglesias, H.A. and Chirife, J. (1982) *Handbook of Food Isotherms: Water Sorption Parameters for Food and Food Components*, Academic Press, New York.

Inouye, J., Tankamaru, S. and Hibi, K. (1979) Elongation of seedlings of leguminous crops. *Crop Science* 19, 599–602.

International Seed Testing Association (1985) Determination of moisture content. *Seed Science and Technology* 13(2), 338–495.

Justice, O.L. and Bass, L.N. (1978) *Principles and Practices of Seed Storage*. USDA, Washington, DC.

Kaufman, G. (1991) Seed coating: a tool for stand establishment; a stimulus to seed quality. *HortTechnology* 1(1), 98–102.

Khan, A.A. (1992) Preplant physiological seed conditioning. In: Janick, J. (ed.) *Horticultural Reviews*, Vol. 13. John Wiley, New York, pp. 131–181.

Khan, A.A., Peck, N.H. and Samimy, C. (1980/81) Seed osmoconditioning: physiological and biochemical changes. *Israel Journal of Botany* 29, 133–144.

Koller, D. and Hadas, A. (1982) Water relations in the germination of seeds. In: Lange, D.L., Nobel, P.S., Osmond, C.B. and Ziegler, H. (eds) *Encyclopedia of Plant Physiology*, New series, Vol. 12B. Springer-Verlag, Berlin, pp. 401–431.

Kolloffel, C. (1967) Respiration rate and mitochondrial activity in the cotyledons of *Pisum sativum* L. during germination. *Acta Botanica Neerlandica* 16(3), 111–122.

Lee, P.C. and Taylor, A.G. (1995) Accuracy of sinapine leakage in *Brassica* as a method to detect seed germinability. *Plant Varieties and Seeds* 8, 17–18.

Leopold, A.C. (1983) Volumetric components of seed imbibition. *Plant Physiology* 73, 677–680.

Leopold, A.C. and Vertucci, C.W. (1986) Physical attributes of desiccated seeds. In: Leopold, A.C. (ed.) *Membranes, Metabolism, and Dry Organisms*. Comstock Publishing, Ithaca, New York, pp. 22–34.

Leopold, A.C. and Vertucci, C.W. (1989) Moisture as a regulator of physiological reaction in seeds. In: Stanwood, P.C. and McDonald, M.B. (eds) *Seed Moisture*. Crop Science Society of America, Madison, pp. 51–68.

Lorenz, O.A. and Maynard, D.N. (1980) *Knott's Handbook for Vegetable Growers*, 2nd edn. Wiley-Interscience, New York.

MacLeod, A.M. (1952) Enzyme activity in relation to barley viability. *Transactions of the Botanical Society of Edinbourgh* 36, 18–33.

Martin, A.C. (1946) The comparative internal morphology of seeds. *American Midland Naturalist* 36(3), 513–660.

Mayer, A.M. and Poljakoff-Mayber, A. (1982) *The Germination of Seeds*, 3rd edn. Pergamon Press, Oxford.

Michel, B.E. (1983) Evaluation of the water potentials of solutions of polyethylene glycol 8000 both in the absence and presence of other solutes. *Plant Physiology* 72, 66–70.

Moore, R.P. (1972) Effects of mechanical injuries on viability. In: Roberts, E.H. (ed.) *Viability of Seeds*. Syracuse University Press, Syracuse, New York, pp. 94–113.

Moore, R.P. (1985) *Handbook on Tetrazolium Testing*. International Seed Testing Association, Zurich, Switzerland.

Nienow, A.W., Bujalski, W., Petch, G.M., Gray, D. and Drew, R.L.K. (1991) Bulk priming and drying of leek seeds: the effects of two polymers of polyethylene glycol and fluidised bed drying. *Seed Science and Technology* 19, 107–116.

Norton, G., Bliss, F.A. and Bressani, R. (1985) Biochemical and nutritional attributes of grain legumes. In: Summerfield, R.J. and Roberts, E.H. (eds) *Grain Legume Crops*. Sheridan House, London, pp. 73–114.

Overaa, P. (1984) Distinguishing between dormant and inviable seeds. In: Dickie, J.B., Linington, S. and Williams, J.T. (eds) *Seed Management Techniques for Genebanks*. International Board for Plant Genetic Resources, Rome, pp. 182–196.

Pollock, B.M. and Manalo, J.R. (1970) Simulated mechanical damage to garden beans during germination. *Journal of the American Society for Horticultural Science* 95, 415–417.

Porter, S.C. and Bruno, C.H. (1991) Coating of pharmaceutical solid-dosage forms. In: Lieberman, H.A., Lachman, L. and Schwartz, J.B. (eds) *Pharmaceutical Dosage Forms*. Marcel Dekker, New York, pp. 77–159.

Powell, A.A., Don, R., Haigh, P., Phillips, G., Tonkin, J.H.B. and Wheaton, O.E. (1984) Assessment of the repeatability of the controlled deterioration vigour test both within and between laboratories. *Seed Science and Technology* 12, 421–427.

Priestley, D.A. (1986) *Seed Aging Implications for Seed Storage and Persistence in the Soil*. Comstock Publishing, Ithaca, New York.

Priestley, D.A., Cullinan, V.I. and Wolfe, J. (1985) Differences in seed longevity at the species level. *Plant Cell and Environment* 8(8), 557–562.

Ptasznik, W. and Khan, A.A. (1993) Retaining the benefits of matriconditioning by controlled drying of snap bean seeds. *HortScience* 28(10), 1027–1030.

Robani, H. (1994) Film-coating horticultural seed. *HortTechnology* 4, 104–105.

Roberts, E.H. (1972) Loss of viability and crop yields. In: Roberts, E.H. (ed.) *Viability of Seeds*. Syracuse University Press, Syracuse, pp. 307–320.

Roos, E.E. (1989) Long-term seed storage. In: Janick, J. (ed.) *Plant Breeding Reviews*, Vol. 7. AVI Publishing, Westport, Connecticut, pp. 129–158.

Roos, E.E. and Davidson, D.A. (1992) Record longevities of vegetable seeds in storage. *HortScience* 27(5), 393–396.

Scott, J.M. (1989) Seed coatings and treatments and their effects on plant establishment. *Advances in Agronomy* 42, 43–83.

Steuter, A.A., Mozafar, A. and Goodin, J.R. (1981) Water potential of aqueous polyethylene glycol. *Plant Physiology* 67, 64–67.

Tarquis, A.M. and Bradford, K.J. (1992) Prehydration and priming treatments that advance germination also increase the rate of deterioration of lettuce seeds. *Journal of Experimental Botany* 43, 307–317.

Taylor, A.G., Churchill, D.B., Lee, S.S., Bilsland, D.M. and Cooper, T.M. (1993a) Color sorting of coated *Brassica* seeds by fluorescent sinapine leakage to improve germination. *Journal of the American Society for Horticultural Sciences* 118(4), 551–556.

Taylor, A.G. and Dickson, M.H. (1987) Seed coat permeability in semi-hard snap bean seeds: its influence on imbibitional chilling injury. *Journal of Horticultural Science* 62(2), 183–190.

Taylor, A.G. and Eckenrode, C.J. (1993) Seed coating technologies to apply Trigard for the control of onion maggot and to reduce pesticide application. New York State Integrated Pest Management Publication, no. 117, pp. 73–78.

Taylor, A.G. and Harman, G.E. (1990) Concepts and technologies of selected seed treatments. *Annual Review of Phytopathology* 28, 32–339.

Taylor, A.G., Klein, D.E. and Whitlow, T.H. (1988) SMP: solid matrix priming of seeds. *Scientia Horticulturae* 37(1–2), 1–12.

Taylor, A.G., Min, T.G. and Mallaber, C.A. (1991) Seed coating system to upgrade

Brassicaceae seed quality by exploiting sinapine leakage. *Seed Science and Technology* 19(2), 423–434.

Taylor, A.G., Paine, D.H. and Paine, C.A. (1993b) Sinapine leakage from *Brassica* seeds. *Journal of the American Society for Horticultural Science* 118(4), 546–550.

Taylor, A.G., Prusinski, J., Hill, H.J. and Dickson, M.D. (1992) Influence of seed hydration on seedling performance. *HortTechnology* 2(3), 336–344.

Taylor, A.G. and Ten Broeck, C.W. (1988) Seedling emergence forces of vegetable crops. *HortScience* 23(2), 367–369.

Tekrony, D.M. and Egli, D.B. (1991) Relationship of seed vigor to crop yield a review. *Crop Science* 31(3), 816–822.

Tomas, T.N. (1990) Studies on the relationship of physiological necrosis and seedlot quality in lettuce using rate of radicle emergence, response to force and controlled deterioration and the evolution of acetaldehyde during germination as measures of vigor and viability. PhD thesis, Cornell University.

Tomas, T.N., Taylor, A.G. and Ellerbrock, L.A. (1992a) Time-sequence photography to record germination events. *HortScience* 27(4), 372.

Tomas, T.N., Taylor, A.G., Ellerbrock, L.A. and Chirco, E.M. (1992b) Lettuce seed necrosis. *Seed Science and Technology* 20, 539–546.

Tonkin, J.H.B. (1979) Pelleting and other presowing treatments. *Advances in Research and Technology of Seeds* 4, 84–105.

Vertucci, C.W. (1993) Predicting the optimum storage conditions for seeds using thermodynamic principles. *Journal of Seed Technology* 17, 41–53.

Vertucci, C.W. and Roos, E.E. (1990) Theoretical basis of protocols for seed storage. *Plant Physiology* 94, 1019–1023.

Weges, R. and Karssen, C.M. (1990) The influence of redesiccation on dormancy and potassium ion leakage of primed lettuce seeds. *Israel Journal of Botany* 39(4–6), 327–336.

Wiley, R.W. and Heath, S.B. (1969) The quantitative relationships between plant population and crop yield. *Advances in Agronomy* 21, 281–321.

Wilson, D.O. and Trawatha, S.E. (1991) Enhancement of bean emergence by seed moisturization. *Crop Science* 31, 1648–1651.

Wolk, W.D., Dillon, P.F., Copeland, L.F. and Dilley, D.R. (1989) Dynamics of imbibition in *Phaseolus vulgaris* L. in relation to initial seed moisture content. *Plant Physiology* 89(3), 805–810.

Transplanting   2

H.C. Wien

*Department of Fruit and Vegetable Science, Cornell University, 134A
Plant Science Building, Ithaca, New York 14853–5908, USA*

The term transplanting refers to raising seedlings in specialized containers
or confined field areas and then transferring them to the place where they
will produce the harvested product. It is a practice commonly used with
small-seeded vegetable crops, particularly those which are slow or difficult
to germinate, or require special germination conditions. Transplanting is
also common in areas where the growing season is short, because planting
seedlings rather than seed gives the grower earlier harvests. Transplanting
maximizes use of available water resources, and allows more latitude on
measures used for weed control. It permits more precise control of plant
population and spacing, and is more efficient in use of today's increasingly
expensive seed than direct seeding (Salter, 1985; Klassen, 1993).

In recent years, these advantages have led to the increasing use of transplants to produce a range of vegetable crops. At present, virtually all the
cauliflower and celery grown in the US is raised from transplants (Klassen,
1993; H.C. Wien, personal observations). The situation is similar in the UK,
where much of the leek, brussels sprouts and about 60% of the lettuce are
transplanted (Salter, 1985). Fresh market tomatoes, bell peppers and egg-
plants are almost always transplanted in North America.

Technological advances in transplanting have contributed to the growth
of the industry, by reducing costs and increasing reliability of production.
Plastic seedling trays, with individual cell sizes as small as 3 ml, permit
short plant-raising periods and have reduced costs. Artificial soil-less media
make possible close and consistent control of plant growth rate. They provide an ideal medium for root growth, and supply water and nutrients in
regulated amounts. The automation of many operations, including seeding
of the containers and transplanting to the field, has further reduced costs.
Increasing specialization by some growers in areas where year-round production is possible has led to the establishment of some enterprises that produce

more than 500 million transplants per year (Klassen, 1992). Such companies typically have facilities in Florida or the central coast of California, and supply both local growers and producers in northern areas.

Although these intensive systems are becoming more common, many transplants for long season vegetable crops in Europe and North America are less intensively grown in outdoor or protected seedbeds (Fordham and Biggs, 1985). Similarly, vegetables grown for processing in northern states of the US are frequently started from 'bare-root' transplants produced in outdoor seedbeds in Georgia and shipped (Orzolek, 1986).

Whether for intensive transplant production in containers, or more extensive, economical seedbed culture, it is necessary to understand the physiological processes that occur in the growing and transplanted seedling if plant performance is to be optimized. Transplant production requires an understanding of the growth of seedlings in dense plant communities, the factors that regulate the growth rate of these plants, and the stresses that they are subjected to when moved to the production environment, as will be described below. Decisions such as choice of container cell size or seedbed spacing, nutrient supply to the seedlings, and efforts to harden the plants before transplanting are frequently made on the basis of economics. These decisions also have profound effects on plant performance and yield. By knowing more about how transplants grow and how transplant shock can be minimized, vegetable producers, their advisers and researchers can make more informed choices on the optimum cultural practices to be used for seedlings to be transplanted.

GROWTH IN CONTAINERS

CONTAINER CELL SIZE

When vegetable seedlings are grown in typical transplant containers, growth rate tends to be proportional to the volume of the container cell (Fig. 2.1). The more the space available to the plant, the larger it becomes, and the more quickly it attains particular growth stages. For instance, Saito et al. (1963) have shown that wide spacing of tomato seedlings in the seedbed resulted in first flower clusters occurring at a lower node than on seedlings crowded in the bed. Accordingly, many researchers have found that early yield of tomatoes is enhanced by producing seedlings in large rather than smaller containers (Knavel, 1965; Liptay et al., 1981; Weston and Zandstra, 1986; Marr and Jirak, 1990; Beverly et al., 1992). They also generally found that total yield was not significantly different between plants raised in large and small cells before transplanting.

The response of other fruit-bearing vegetables to container cell size appears to be similar to tomato. Weston (1988) grew bell peppers in cells ranging in size from 6 to 40 ml before transplanting. Early yields were

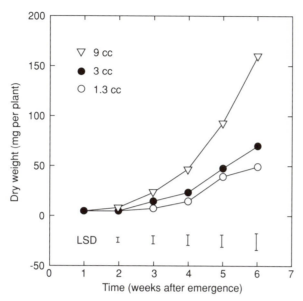

Fig. 2.1. Influence of container cell size on dry matter gain of tomato seedlings (Marr and Jirak, 1990).

proportional to the volume of the individual container cells, but total yields did not differ between treatments.

Thus in situations where growers obtain highest prices for the earliest fruit, the transplants should be grown in the largest cell size that is economically feasible. For production systems in which earliness is not important, such as production for processing, tomato transplants grown in cell volumes as small as 3 ml have given satisfactory yields (Garton, 1992; Bennett and Grassbaugh, 1992).

Transplants that produce the earliest fruit do not always give total yields that are equal or superior to those of less developed transplants. Nicklow and Minges (1962) showed that for the early determinate cultivar Fireball, presence of flowers and fruit at the time of transplanting may result in a restriction of further vegetative growth and reduction in total yield (Fig. 2.2). To avoid such problems, removal of any open flowers and fruit from transplants is often advocated at planting. Thus the advantage of large transplants with regard to early and total yield is greatest if plants are transplanted well before flowering. The amount of space available to broccoli and cauliflower seedlings before transplanting has had variable influence on performance and yield. Dufault and Waters (1985) found that, despite a four-fold difference in seedling size at transplanting, yields and earliness did not differ between treatments. On the other hand, Magnifico *et al.* (1980) found that large seedlings gave higher and earlier yield of broccoli than did smaller transplants. Large seedling size combined with adverse growing conditions

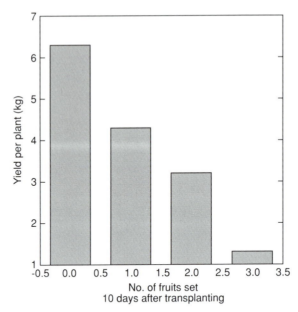

Fig. 2.2. Reduction in tomato yield in the determinate cultivar Fireball due to presence of flowers and fruits on plants at transplanting (Nicklow and Minges, 1962).

in the field at transplanting can result in preferential development of the cauliflower curd and the broccoli head at the expense of further vegetative growth (Skapski and Oyer, 1964; Baggett and Mack, 1970; Whitwell and Crofts, 1972). This process, termed 'buttoning', drastically reduces marketable yield through a reduction in head size (see Chapter 15).

Transplanting larger seedlings of head-forming vegetables such as lettuce and cabbage has generally resulted in production of larger heads and increased yields (Kratky *et al.*, 1982; Csizinsky and Schuster, 1993). This is particularly true when the transplants are provided with high fertility levels and adequate moisture in the field (Csizinsky and Schuster, 1993).

TRANSPLANT AGE

Larger transplant sizes can of course be achieved by growing seedlings for a longer time in the transplant containers. Experiments that compared the effect of transplant age have generally corroborated the findings for container sizes (Weston, 1988; Weston and Zandstra, 1989). Within limits, older plants have generally produced higher early yields. As growth is prolonged in seedling trays, however, the plants become increasingly constrained by both the aerial and the below-ground environment, and growth rate decreases. Thus if plants are kept in seedling containers past the time in which they are making active growth, growth after transplanting and yield may suffer (Casseres, 1947; NeSmith, 1993). This growth decrease is particularly noticeable in the smaller cell sizes (see Fig. 2.1), and may not be alleviated by transplanting into non-limiting growing environments (Marr and Jirak, 1990). Thus each container cell size has an optimum growth duration for a particular crop under recommended environmental conditions, and this duration will be shorter for smaller cell sizes than for larger ones. Another important consideration is the length of time transplants of a particular species can be held in the containers without adverse effects on yield, if the growing conditions do not permit timely transplanting (NeSmith, 1993). Again, this holding period is shorter for smaller than for larger containers.

FACTORS REGULATING GROWTH IN CONTAINERS

The reduction in growth rate of plants growing in close proximity in containers is due to the interaction of several environmental factors. Above-ground, closely spaced individuals have less total light available (Harper, 1977) which is richer in far-red light (Kasperbauer, 1987), than do individuals farther apart. Below ground, the supply of essential mineral nutrients and water may be limiting growth of the plants whether plants are grown separately in individual containers or competing in a common seedbed. Since plants crowded together above ground are generally also limited in below-ground space, there may not be a single factor constraining growth in such situations. Where attempts have been made to separate the soil factors from those acting above ground, it has indeed been shown that plants grown in deeper containers, but with limited space for the tops, show little benefit from the additional medium for root growth (Sayre, 1948). At the other extreme, Weston and Zandstra (1986) found that tomato seedlings given wide spacing above ground but only limited root space in plant container cells produced leaf area in proportion to the below-ground volume available.

The physiological mechanisms that may restrict growth when there is limited space for root growth has recently been more closely examined. In these experiments, widely spaced plants were grown in solution culture, with adequate levels of mineral nutrients and aeration, and individuals with roots restricted to a volume of less than 450 ml were compared to others given 1500 ml or more. When tomatoes were grown with such root restrictions, plant growth was severely curtailed, with proportional decreases in size of most plant parts (Fig. 2.3) (Hameed *et al.*, 1987; Ruff *et al.*, 1987; Peterson *et al.*, 1991a).

The mechanism by which the growth reduction comes about is not yet clear, and must be pieced together from experiments with several species. Work with tomatoes indicates that root-restricted plants have lower root respiration rates (Peterson *et al.*, 1991b), and greater specific leaf weight than plants with ample root space (Hameed *et al.*, 1987). Robbins and Pharr (1988) found that cucumbers with restricted root space transport less starch from their leaves at night than unrestricted controls. The smaller below-ground sink may therefore lead to accumulation of assimilates in the above-

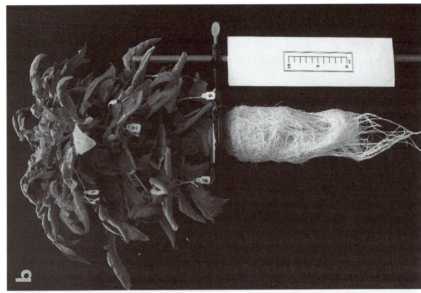

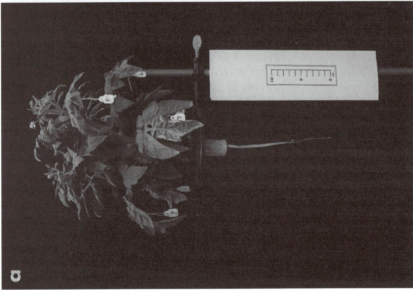

Fig. 2.3. Top and root growth of 'Better Bush' tomatoes as a result of root restriction. Plants grown at wide above-ground spacing, in solution culture, with 25 ml (a) or 1500 ml (b) available for root growth (Peterson *et al.*, 1991a).

ground parts. Both cucumbers and tomatoes were found to have low leaf net assimilation rates when roots were confined (Hameed *et al.*, 1987; Robbins and Pharr, 1988). Research with beans and carrots indicates that restricted root systems may produce lower levels of growth-promoting cytokinins and gibberellins, and more ethylene than roots of unrestricted plants, resulting in inhibition of top growth (Carmi and Heuer, 1981; Thomas, 1993). The effect of root confinement on plant water relations is not clear. Hameed *et al.* (1987) showed increases in root hydraulic resistance and leaf transpiration rates, whereas Peterson *et al.* (1991b) found no change in the same species. In summary, there is probably no single process that controls the reduction in top growth rate when the root system is physically restricted. Perhaps further research will clarify this complex picture. A better understanding of the physiological processes inhibited by root restriction of transplants could aid in the control of transplant growth, both during and after confinement. It might be possible to use the root confinement brought about by small transplant container volume to act as the principal means of regulating transplant size.

CONTROL OF TRANSPLANT SIZE

Vegetable seedlings growing at high population densities under conditions of adequate moisture and mineral nutrients become increasingly tall and etiolated. In greenhouses, the high relative humidity and lack of air movement contribute to the problem. When such spindly, tender plants are transplanted, they are susceptible to mechanical damage in the transplant operation, and to high stress levels in the field. Working out practical methods of keeping plants short and stocky before transplanting has been the subject of much recent research.

Mechanical stress

Plants subjected to motion tend to be short in stature. This concept has been known since trees in windy habitats were observed to be stockier than the same species in nearby windstill environments. While some of this difference is probably due to drought effects, it has been shown that movement *per se*, and other forms of mechanical stimulation, can have remarkable dwarfing effects. Jaffe (1973) observed that stem growth of a number of herbaceous species was significantly reduced by 10 s of gentle daily rubbing for 1 week, and termed the response thigmomorphogenesis. Gentle shaking or flexing of etiolated pea seedlings produced similar results (Mitchell *et al.*, 1975). Others have induced plant movement by brushing with horizontal wooden sticks, stiff paper or cardboard (Biddington, 1986; Latimer, 1991), or blowing on seedlings with fans (Liptay, 1985).

　　The effects of these mechanical perturbations on herbaceous plants vary among species, and even among cultivars, but in general result in a reduction of plant stature (Fig. 2.4). Most often, stem elongation is reduced, petioles are shorter, and chlorophyll content of leaves increases compared to undisturbed plants as a result of mechanical stress (Biddington and Dearman, 1985a). There is frequently also a lower rate of increase in dry weight and leaf area increase, while stems become thicker, as in the case of beans (Jaffe, 1973), or stay thin, as with tomato (Heuchert *et al.*, 1983). In the latter case, however, the mechanical stress resulted in an increase in elastic strength of the stem, and greater ability to withstand mechanical damage. Such shorter stems and petioles and greater elastic strength indicate that mechanical stress would improve the survival and growth potential of transplants in the field when exposed to the force of wind or rain. It is much less clear, however, whether plants hardened by mechanical stress would show improved growth under other stress conditions, such as drought after transplanting. Water use per plant could decrease with smaller leaf area of mechanically stressed plants, as found for brushed seedlings of lettuce, celery and cauliflower by Biddington and Dearman (1985b). Water loss per unit leaf area, however, was actually increased over control plants by the brushing treatment. With tomato, gentle shaking had no effect on transpiration rates, although stomatal resistance increased temporarily immediately after the plants were moved (Mitchell *et al.*, 1977). This temporary stomatal closure, combined with smaller plant size, may explain the improved survival of mechanically stressed bean plants in unwatered pots (Jaffe and Biro, 1979; Suge, 1980).

Few studies have compared field establishment rates of mechanically stressed seedlings with those hardened by other means. Latimer (1990) found that brushed broccoli seedlings had significantly faster growth after transplanting than unhardened seedlings in two out of three field seasons, or than those hardened by mild drought stress or by wind. Yields were not significantly influenced by any of the conditioning treatments. Tomato seedlings

Fig. 2.4. Effect of brushing tomato seedlings 0, 10 or 20 times daily. (Garner and Björkman, 1996).

conditioned by wind were smaller at transplanting, but caught up to the larger controls 3 weeks later (Liptay, 1985).

In some cases, yields of mechanically stressed transplants have been similar to plants grown without conditioning (Wurr *et al.*, 1986; Latimer and Beverly, 1992); in others lower yields of the conditioned plants have been recorded (Pardossi *et al.*, 1988; Latimer *et al.*, 1991). It is not clear to what extent physical damage of the developing flowers might be responsible for the reduced yield.

The physiological effects of mechanical stress on the young plant are not yet well understood, but are currently under intensive investigation. Within seconds of being mechanically stimulated, the Ca^{2+} ion concentration of the cytoplasm of tobacco plants increases sharply (Knight *et al.*, 1992). This appears to trigger an increased level of expression of genes that encode calmodulin and related proteins, and thus increases the concentration of this calcium-transporting protein in the cell (Braam and Davis, 1990; Braam, 1992). How calmodulin and calcium levels influence the growth of plants, and bring about the many changes documented in mechanically stressed plants, is not clear at present.

The longer-term effects of mechanical stimulation on plants have often been linked with hormone activities. Ethylene is commonly emitted by plants that have been shaken or rubbed (Biro and Jaffe, 1984), and the morphological changes that occur on ethylene exposure are at least partly like those seen on mechanically stressed plants (Biro and Jaffe, 1984; Biddington and Dearman, 1986). Both research groups have shown, however, that when ethylene formation or action was inhibited in plants, only some of the morphological changes could be prevented. This implies that other factors are involved in the changes in growth elicited by mechanical stimulation.

In spite of the often-measured decreased total dry matter produced by mechanically stressed seedlings, comparisons of photosynthetic rates per unit leaf area of stressed and normal seedlings of tomato showed no significant differences, and respiration rates were 15% lower for stressed plants (Mitchell *et al.*, 1977). In summary, it is evident that a better understanding of the physiological changes that occur with mechanical stress is needed to predict how plant growth rates can be controlled for more successful transplant handling and field establishment.

Growth-retardant chemicals

Preventing the rapid increase in plant height of vegetable seedlings crowded in transplant containers by the use of growth-retarding chemicals has long seemed an attractive possibility. Researchers envisioned the possibility of applying one foliar spray to seedlings in seedbeds or containers, to inhibit their growth temporarily and increase their resistance to adverse conditions during and after transplanting. The compounds selected for this purpose have generally acted by reducing stem elongation rate, while maintaining the production of leaf area by the plants (Cathey, 1964).

One group of chemicals acted by inhibiting the production of gib- berellins by the plant (Graebe, 1987). Among these, daminozide (*N*-dimethyl amino succinamic acid) when applied to seedling tomato plants, kept intern- odes short and thickened stems. Its effect on reproductive growth varied with the age of reproductive structure. Daminozide treatment of plants with flower buds, flowers and fruit resulted in more severe abscission among buds than flowers, with nearly complete retention of fruits (Read and Fieldhouse, 1970; Veliath and Ferguson, 1973). When applied to vegetative tomato seedlings, reproductive growth was not inhibited (Taha *et al.*, 1980). Nevertheless, treatment of plants near anthesis of the first cluster resulted in a 10-day delay in flowering and ripening dates of early clusters (Pisarczyk and Splittstoesser, 1979). Daminozide was widely tested as a growth-retar- dant for transplants produced in seedbeds in the southern US for shipment to growers in the northern US and Canada (Jaworski *et al.*, 1970). For this application, it compared favourably with mechanical pruning of the plants, and with use of ethephon (see below). Use of the chemical on food crops was halted in 1989, when it was withdrawn from production by its manufacturer over allegations that it might cause cancer.

Other gibberellin synthesis inhibitors have also been found promising for use as transplant growth-retardants. Paclobutrazol was effective at much lower concentrations than daminozide (Aloni and Pashkar, 1987; Latimer, 1992). When applied on pepper seedlings, it inhibited stem growth more than root growth (Aloni and Pashkar, 1987), and Latimer (1992) found that it improved field establishment of tomato transplants compared to untreated, unhardened plants. Uniconazole, another triazole derivative, reduced stem extension and had the unforeseen benefit of reducing the incidence of blos- som-end rot in greenhouse tomatoes (Wang and Gregg, 1990).

The ethylene-releasing compound ethephon [(2-chloroethyl)phosphonic acid] has been widely tested as a growth inhibitor of tomato transplants (Campbell, 1976; Taha *et al.*, 1980; Liptay *et al.*, 1988). When treated just as plants were showing flower buds, the chemical caused cluster abortion, retardation of stem elongation and thickening of stems. The active ingredi- ent ethylene probably acts in a manner similar to its role in mechanical stress effects described above. Root growth of treated seedlings was increased both below ground and along the stem (Campbell, 1976; Taha *et al.*, 1980), so that plants were difficult to untangle in the seedbeds. To avoid the problem, researchers advocated that plants should be left less than 2 weeks in the seedbed or flat after ethephon treatment. Early yields were lower than untreated, pruned controls, but total yields were similar or higher (Taha *et al.*, 1980).

Additional growth-retardant chemicals for controlling transplant height have been, and will continue to be identified. It is doubtful, however, if any of these will be submitted by chemical companies to government agencies for the necessary permits for use in North America. The potential market is much too small for the testing costs involved. Controlling transplant growth will therefore probably continue to rely on non-chemical means.

Water stress

When plants are subjected to mild drought stress, the rate of stem elonga-tion and leaf area expansion decreases, and assimilates accumulate in the leaves. Water stress therefore induces changes in plant growth that are help-ful in preparing the plant for transplanting. Drought-hardening indeed led to improved field establishment with tomatoes (Latimer, 1992), and improved growth in cool greenhouses through a reduction in chilling injury (Pardossi *et al.*, 1988). It may be difficult to induce only a mild drought stress, particularly with transplants growing in containers of root volume of 10 ml or less. With such containers, plants require watering three or more times a day to remain turgid, and on sunny days, can become severely water stressed in a few hours. Alternatively, it may be possible to induce lower levels of stress by increasing the osmotic potential of the growing medium, or of the nutrient solution a few days before transplanting (Rosa, 1921).

Nutrient conditioning

Adequate levels of nitrogen, phosphorus and potassium are necessary for satisfactory growth of plants in transplant containers. Since the nutrient content of the growing medium is often quite low, the composition and fre-quency of application of nutrient solutions determine seedling nutrition. Masson *et al.* (1991) found that growth rate of vegetable seedlings was directly related to the nitrogen concentration of the nutrient solution within the range of 100–400 mg l^{-1} (Fig. 2.5). Others have made similar findings (Dufault, 1985; Weston and Zandstra, 1989; Melton and Dufault, 1991). As nutrient concentrations exceeded these levels, however, growth was decreased, presumably because of salt effects (Kratky and Mishima, 1981). At nitrogen levels that are high but not toxic, growth of tomato seedlings may be so rapid that field survival is adversely affected (Liptay and Nicholls, 1993).

Increasing the level of phosphorus in the transplant has led to less marked increases in plant growth than with nitrogen (Dufault, 1985, 1986; Widders, 1989; Melton and Dufault, 1991), with responses to concentrations of 15–60 mg l^{-1}. Melon and tomato seedlings showed little response to varia-tions in potassium from 10 to 250 mg l^{-1} (Dufault, 1986; Melton and Dufault, 1991).

It should therefore be possible to regulate the growth rate of vegetable seedlings by controlling the concentration of nitrogen and perhaps other nutrients in the growing medium. Two approaches have been advocated. One suggests that transplant growth be kept slow with low nutrient levels, and then be increased just before transplanting (Widders and Garten, 1992). This scheme has the disadvantage of a long transplant growth duration at the lower nutrient level, and possible plant size regulation problems if transplanting is delayed by bad weather. A more common approach is to do the opposite: provide adequate nutrient levels during early growth, and

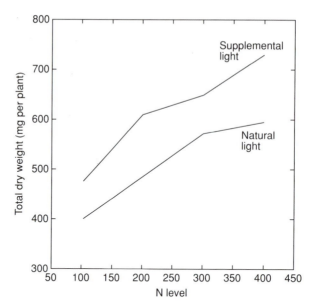

Fig. 2.5. Effect of nitrogen level in the nutrient solution and supplementary lighting on broccoli dry weight at transplanting. Data averages of three sowing dates (Masson *et al.*, 1991).

then reduce supply before transplanting. As long as the transplants are not completely starved of the major nutrients by this procedure, there should be little problem with the resumption of growth after transplanting. Low levels especially of nitrogen can, however, result in significantly lower growth rates after field setting, and reduced yields (Widders, 1989; Widders and Garton, 1992).

Light quality

Seedling plants develop abnormally elongated stems and petioles when illuminated with light having relatively high amounts in the far-red (725 nm) compared to the red (650 nm) part of the spectrum (Zack and Loy, 1980; Decoteau *et al.*, 1988; Decoteau and Friend, 1991). Since light transmitted through and reflected off leaves is high in far-red wavelengths, it is no coincidence that these plants resemble plants grown in dense stands (Kasperbauer, 1987). Thus most plants sense the presence of neighbouring plants through the change in light quality, which allows them to avoid the shading influence through the stem elongation response. Recent research has shown that this elongation can be inhibited by light treatments. Decoteau and Friend (1991) demonstrated that exposing tomato seedlings in relatively large transplant containers to red light at the end of the day can reduce stem extension rate. More work is needed to determine if this treatment will also work in high density small cell container-grown seedlings. Plant species that normally grow in the shade of other plants often do not have a pronounced shade-avoiding reaction (Morgan and Smith, 1979). It may be possible in the future to transfer genes from these shade-tolerating species to our vegetable crops, to reduce the stem elongation when crowded in seedboxes. Such a transfer was recently accomplished by McCormac *et al.* (1991) from a shade-tolerant barley line to tobacco.

Transplant pruning

Instead of controlling plant size through manipulation of the growing conditions or the imposition of stress, direct removal of excess leaf and stem tissue through pruning has been practised for many years (Kraus, 1942). In the early days, leaves were removed from transplants in the belief that this would reduce moisture loss from the plant in the field and result in more rapid establishment. However, in 14 experiments over 3 years, Kraus found that leaf pruning did not increase plant survival because defoliation deprived the plant of an important source of carbohydrates needed to resume root growth.

More recently, mowing of the tops of seedbed-grown seedlings has been used to control plant height, increase uniformity, and reduce plant size prior to shipping (Jaworski and Webb, 1967, 1971; Jaworski *et al.*, 1969). Pruning was an important way of preventing early fruiting on small plants, a significant source of yield loss (Nicklow and Minges, 1962). The treatment significantly delayed flowering and yield if the pruning removed the apex and developing reproductive structures of tomato as well as leaves (Jaworski *et al.*, 1969).

Timing of the pruning or clipping operation relative to the time of transplant harvest may influence yield. Peppers pruned 12 days before shipping made some axillary growth and gave increased yield in comparison to plants topped at 6 days or not pruned (Jaworski and Webb, 1971). The time of pruning of tomato did not significantly influence yields (Jaworski and Webb, 1967). Pruning is advisable only in healthy stands, for it could facilitate the spread of diseases throughout the planting, as has been shown for tomato (McCarter and Jaworski, 1969).

Day–night temperature differences

When growing outdoors, plants are normally exposed to day temperatures (DT) that are warmer than night temperatures (NT). Went (1944, 1945) demonstrated in his pioneering studies in air-conditioned greenhouses that tomato stem extension rate was maximized by having DT higher than NT, and that the reverse markedly reduced stem growth rates. For the cultivar San Jose Canner, for example, stem growth rates were 26.9 mm day^{-1} with DT/NT of 26.5/20°C, and 19.4 mm day^{-1} for the reverse combination (Went, 1944). Forty years passed before this information was put to practical use in the control of plant height in herbaceous flowering plants such as

Lilium, *Fuchsia* and chrysanthemum (Erwin *et al.*, 1992). These researchers discovered that stem growth rates in *Lilium* were directly proportional to the difference in DT and NT, which they termed DIF (DIF = DT − NT). Keeping plants cooler during the day than at night reduced height increases in the range of temperatures between 10 and 30°C.

From a practical standpoint, maintaining low day temperatures in greenhouses would be difficult to accomplish. However, Went (1944) and Erwin *et al.* (1992) found stem elongation to be most rapid during the predawn period and the first hours of light. Erwin *et al.* (1989) confirmed that cooling plants during the first 2 h after dawn was almost as effective as maintaining low temperatures the entire day.

Little work has been reported so far on the use of this temperature manipulation on vegetable transplants. Erwin and Heins (1990) showed that tomato, watermelon, sweet corn and bean had significantly shorter internodes with 17/23°C (DT/NT) than the reverse, but stem elongation of some cucumber and pea cultivars was not inhibited. Furthermore, Went (1944, 1945) showed that tomato flower development and fruit set were inhibited by warm nights and cool days. It is therefore possible that vegetable seedlings subjected to negative DIF treatments will be delayed in flowering once transplanted, compared to plants grown at warm day and cool night conditions.

STORAGE AND TRANSPORT OF TRANSPLANTS

For many years, vegetable growers in northern areas of the USA and in Canada have relied on planting materials produced in southern states such as Georgia and Florida. Seedlings of tomato, pepper, eggplant and the cole crops were traditionally produced in field beds, uprooted, bundled together, packed in boxes and shipped by truck (Moran *et al.*, 1962). By 1984, it was estimated that 1.25 billion tomato plants were grown in Georgia and shipped north annually (Orzolek, 1986). Some of the southern transplant production has intensified in recent years, with the use of transplant containers, artificial soil and plant culture in greenhouses (Klassen, 1992). Similar transplant production facilities have been established in California. While some of these plants are used by local growers, part of the production is shipped to more distant growing areas. Typically, the plants are removed from the transplant containers and packed in cardboard cartons (Leskovar and Cantliffe, 1991). Whether produced by traditional or more modern methods, we must understand the effect of shipping and storage conditions on the capability of the transplant to resume growth and produce adequate yield.

Research on dark storage of transplants has generally concluded that plants can be held up to a week without adverse effects on yield (Risse *et al.*, 1979; Hall, 1985; Brydon and Orzolek, 1992). Deterioration in storage was closely related to storage temperature. Successful storage durations declined above 15°C, with accelerated yellowing and abscission of leaves, leading to higher mortality of the transplants in the field (Risse *et al.*, 1979). For species susceptible to chilling injury, storage at low temperatures also reduced storage life. Tomatoes suffered less yield decline after storage at 13°C than at 4°C (Risse *et al.*, 1979), and sweet potato rooted cuttings survived better after 16°C than 5°C storage temperatures (Hall, 1985). On the other hand, tomato plants survived without yield loss if dark storage periods at 2°C were alternated with daily light periods at 29°C (Dufault and Melton, 1990). Deterioration at high temperatures may be due at least in part to the evolution of ethylene by the stored transplants (Kays *et al.*, 1976). Pepper transplants dark-stored for 48 h at 14–20°C released sufficient ethylene to cause abscission of the expanded leaves. Inclusion of the ethylene-absorbent potassium permanganate in the storage container largely prevented leaf loss (Kays *et al.*, 1976).

The carbohydrate status of transplants after dark storage at various temperatures has received little attention, but may be a key factor in plant survival. Indirect evidence comes from the work of Smith and Zink (1951), who showed that tomato transplants survived 72 h of dark storage at 20°C better if sprayed with 10% sucrose solution before storage than untreated seedlings (Table 2.1). Went and Carter (1948) had previously demonstrated positive growth responses in tomato plants with applied sucrose.

Plant moisture status is another factor which must be considered in successful storage of seedlings. In storage and shipment of bare-rooted plants, roots were wrapped in moist sphagnum moss, or packed in containers lined with polyethylene sheeting (Thomas and Moore, 1947; Moran *et al.*, 1962). Container-grown plants in artificial medium are presumably easier to keep well hydrated than bare-rooted plants even after plant removal from the containers.

During storage and transport, the rate at which packed transplants reach the desired storage temperature depends on the density of packing in the container, and the air circulation in storage (Moran *et al.*, 1962; Risse *et*

Table 2.1. Survival percentage and plant height increase of tomato transplants after 72 h of dark storage and treatment with water or 10% sucrose solution (Smith and Zink, 1951).

Experiment no.	Treatment	Survival (%)*	Height increase (cm)*
1	Sucrose	100	3.86
	None	100	2.61
2	Sucrose	44	–
	None	0	–
3	Sucrose	100	4.13
	None	86	3.03

* Taken 10 days after transplanting

al., 1985; Leskovar and Cantliffe, 1991). Too many seedlings in a tight package prevented rapid plant cooling, and reduced survival of the transplanted tomatoes (Risse *et al.*, 1985).

TRANSPLANTING

The most detailed study of the physiological effects of transplanting on growth and yield of vegetables was conducted by Loomis (1925), and it still provides much of our knowledge of the processes involved, even though transplanting methods have greatly changed since that time.

PLANT CHARACTERISTICS

Loomis recognized that the act of transplanting had an immediate negative effect on the plant's water status. The severity of this effect depended on the degree of hydration of the plant before the transfer, the amount of root damage in the transplant operation, the climatic conditions of the post-transplant environment, and the species of vegetable being transplanted. Regaining a fully hydrated status required not only an ample water supply in the environment, but also conditions that permitted the transplant to make active root growth. Each of these topics will be considered in more detail.

Transplant water relations

The vegetable seedling is dependent on the water content of the medium around its roots, and in the immediate soil environment when planted in a new location. Without supplementary watering, the transplant medium water content is most important, because water transfer from the soil to the transplant medium takes place slowly (Kratky *et al.*, 1980). In the opposite situation, if plants in dry medium are transplanted into wet soil, growth is initially inhibited as the plants slowly regain turgor. In experiments with cauliflower, root growth into the soil was dictated more by transplant medium moisture content than by the degree of dryness of the soil (Kratky *et al.*, 1980).

Even without the presence of plants in the transplant container, the water content of the container soil or artificial medium tends to be dictated by the surrounding soil after transplanting (Nelms and Spomer, 1983). Drainage into and through the surrounding soil resulted in loss of between 30 and 85% of the container soil's water content in a few hours. Similar findings were made by Costello and Paul (1975) working with nursery stock transplants. This emphasizes the need for frequent watering to maintain plant water status after transplanting.

The importance of maintaining plant water relations after transplanting has prompted a number of studies on the use of antitranspirants (Whitlock,

1980; Berkowitz and Rabin, 1988; Nitzsche *et al.*, 1991). In general, these have concluded that materials that coat the leaf surfaces to reduce water loss have not resulted in improved stands or increased recovery. In some cases, phytotoxic effects from surfactants used with the wax-based antitranspirants resulted in leaf abscission (Nitzsche *et al.*, 1991). In addition, materials that reduce water loss could also be reducing leaf carbon dioxide uptake and photosynthesis rate.

Another approach has been to treat transplants with abscisic acid (ABA), the natural growth regulator that causes stomatal closure in drought-stressed plants (Mansfield, 1976). When pepper transplants were dipped in ABA solution before transplanting in dry soil, seedling survival and yield were significantly improved (Berkowitz and Rabin, 1988). The growth regulator treatment also increased leaf water potential and leaf diffusive resistance. It is important to note, however, that prompt watering after transplanting both in this and in Whitlock's study (1980) resulted in greater improvement in stand and yield than the use of antitranspirants.

The transpiration rate of a transplant can also be reduced by removal of leaves. In a detailed series of studies, Kraus (1942) demonstrated that plant survival was not improved by defoliation. Although leaf removal reduced plant water use initially, it greatly impeded the production of new roots and the capacity of the plant to obtain water from the soil (Table 2.2). As will be explained further below, the leaves act both as a source of stored carbohydrates, and of new photosynthate to bring about root growth.

Transplant top/root relations

It should be the objective of the grower to encourage active transpiration and growth of the plants once the excessive dehydration of plants during the transplanting operation has been prevented. A high rate of transpiration is an indication of an actively growing plant with functioning roots. The reestablishment of an active root system depends on the rate at which new roots are formed, and how fast root system growth resumes after transplanting. For vegetable crops such as tomato, the carbohydrate status of the plant

Table 2.2. Increase in root volume after various degrees of leaf pruning of 'Danish Giant' cauliflower transplants (Kraus, 1942).

Treatment	Gain in root volume (ml) after:			Dry weight per plant at 21 days (g)	
	6 days	13 days	21 days	tops	roots
Unpruned	0.51	2.38	2.78	1.11	0.95
Lightly pruned	0.23	0.77	1.25	0.68	0.47
Heavily pruned	0.09	0.74	1.05	0.52	0.35
$LSD_{0.05}$	0.29	0.52	0.58	0.19	0.21

Table 2.3. Reduction of early or of total yield from heavy pruning of foliage at transplanting for five vegetable species (Kraus, 1942).

		Yield reduction from heavy pruning (%)	Carbohydrate loss by pruning (%)*
	Yield		
Head lettuce	Early	69	53
	Total	24	
Cauliflower	Early	59	48
	Total	11	
Celery	Total	13	46
Peppers	Early	13	16
	Total	0	
Onion	Total	1	22

* Percentage loss in tops.

is a primary determinant of the production of new roots (Kraus and Kraybill, 1918; Reid, 1924). Plants which have been grown under high light intensities, with relatively low levels of available nitrogen produce a dense mat of adventitious roots in comparison to plants of low carbohydrate status. Smith and Zink (1951) showed that rooting potential could be enhanced artificially by treating tomato transplants with 10% sucrose solution. The concept was further supported by the defoliation studies of Kraus (1942), where recovery from leaf removal was slowest for species such as lettuce, cauliflower and celery that stored the largest proportion of total carbohydrates in the leaves, stems and petioles that were removed by pruning (Table 2.3).

It is thus apparent that the relations between tops and roots in transplants are important in the understanding of the plant's response to transplanting. Brouwer (1962) summarized previous findings, and set up the following 'rules of thumb', which are helpful in visualizing top/root relations of vegetative herbaceous plants. When plants are growing under optimum conditions, above-ground and below-ground growth takes place in constant proportion. If leaf tissue is removed, assimilates are preferentially translocated to the tops for new leaf growth and root growth declines, presumably due to a shortage of translocated carbohydrates. Conversely, pruning of a part of the root system leads to reduced water and mineral uptake, a reduction in top growth and use of stored carbohydrates for root growth until the equilibrium has been restored. In transplanting, both these situations could arise.

When vegetable seedlings are transplanted, they may wilt temporarily. A hormonal mechanism may aid in plant recovery. Aloni *et al.* (1991) have found that there is a temporary stimulation of root growth, and a decrease in top/root ratio, if pepper plants wilt at transplanting. Translocation of assimilates was inhibited in the water-stressed plants; transport to the shoot was more strongly inhibited than to the root system. Recent work with maize seedlings indicates that this preferential translocation may be mediated by ABA levels (Saab *et al.*, 1990), with a relative increase in root ABA content contributing to a maintenance of root growth.

Species differences

There are marked differences in the survival rate of different vegetable species after transplanting, particularly if roots have been damaged, or weather conditions are unfavourable. In his classic work on transplanting, Loomis (1925) established three classes of transplanting ease. A commonly transplanted group, that usually survives the process well includes cabbage, tomato, lettuce, cauliflower and beets. An intermediate group contains celery, eggplant, onions and peppers. The third category comprises species that are difficult to transplant and includes beans, corn, cucumbers and melons. Primarily it is the below-ground characteristics that differentiate these categories (Loomis, 1925). Ease of root replacement correlated well with transplanting survival (Table 2.4). Suberization and formation of cutin on the endodermal layer inhibited root formation and reduced water uptake by the roots remaining after transplanting. Species that were difficult to transplant also had a greater amount of suberization (Table 2.4).

Table 2.4. Degree of root suberization and percentage of root replacement 8 days after transplanting for three categories of vegetable species (Loomis, 1925).

| Transplanting | Age (weeks) | Root replacement (%)* Plant size | | Root suberization rating[†] |
		Small	Large	
Easy	5	217 ± 15	95 ± 2	0.75
Intermediate	7	59 ± 2	26 ± 4	2.25
Difficult	3	72 ± 11	39 ± 8	2.75

* Root replacement, as a percentage of root weight before transplanting, taken 8 days later.
[†] Rating: 0 = none, 1 = light, 2 = moderate, 3 = heavy.

Ease of transplanting decreases with age, and the differences between species become magnified (Fig. 2.6). At 1 week of age, most species could be transplanted with little check in growth, whereas by 4 weeks the hard-to-transplant species were set back more severely. Performance was again related to the relative ease of root replacement, presumably due to the rate of suberization of the endodermal layer (Loomis, 1925).

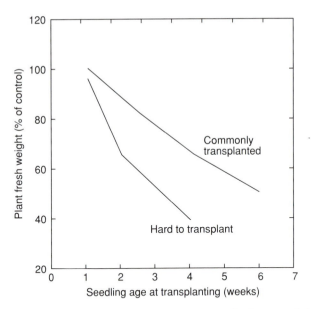

Fig. 2.6. Influence of age at transplanting on subsequent growth of commonly transplanted vegetables (average of cabbage, cauliflower, lettuce and tomato), and crops difficult to transplant (cucumber, muskmelon, corn) (Loomis, 1925).

FIELD CONDITIONS

The success of a transplanting operation depends on the characteristics of the plant being grown, and the climatic conditions to which it will be exposed after transplanting. While some factors such as moisture are often under the grower's control, the temperatures in the field may be difficult to alter. The effect of high temperatures is often manifested in the moisture status of the plants, and can result in water stress as discussed above. In temperate areas, low temperatures at transplanting are more likely, and their effects on seedling vegetables need to be understood.

As most gardeners in temperate regions have observed, there are vast differences between vegetable species in their tolerance of low temperatures (Table 2.5). In general, most vegetable species that originated in tropical or subtropical areas are susceptible to temperatures between 0 and 10°C (chilling injury). Vegetables of temperate origin can generally withstand 0–25°C without injury. The effects of freezing on tender crops are readily apparent in collapse and death of organs or whole plants while the symptoms of chilling injury are not as obvious.

Chilling injury

When chilling-sensitive plants are exposed to temperatures between 0 and 10°C for several hours, the first symptoms to appear are a loss of turgor of

Table 2.5. Classification of vegetable crops according to their tolerance of temperatures just above and below 0°C.

Lowest temperature tolerated*	Species showing injury as growing plants
10°C	Most legume vegetables (except *Vicia faba*), Cucurbits Solanaceous sp.
−5°C	*Vicia faba*, Crucifers, lettuce
−10°C	Garlic, onions, leek, spinach
−20°C	Rhubarb, asparagus (below-ground organs only)

* Temperature tolerated without injury to vegetative parts of plant.

the stem and leaves, followed by death and drying out of the leaf tissue (Rikin *et al.*, 1976). If windy conditions exist in the field during the cool temperature, the flaccid seedling may be twisted around repeatedly until the stem is severed. Such symptoms are particularly prevalent with cucurbit transplants, whose stems lack sufficient fibre to resist the twisting action.

Although chilling injury research has focused primarily on postharvest chilling of fruits and vegetables, our understanding of the effects of chilling on young plants has improved greatly in recent years. Wilting and desiccation from chilling are brought about by a decreased water flow through the root system when sensitive plants are subjected to cool temperatures (Bagnall *et al.*, 1983; Markhart, 1986). In addition, many chilling-sensitive plants lose stomatal control when exposed to drying conditions at temperatures that cause chilling. Somates fail to close as the leaves of these plants lose turgor, thereby worsening the water deficit (Guye and Wilson, 1987).

The second major source of chilling injury comes when plants are chilled under direct sunlight (Wise and Naylor, 1987; Peeler and Naylor, 1988). Under those conditions, the chloroplasts of chilling-sensitive species such as cucumber are injured by photo-oxidation and the generation of superoxide anion radicals (Wise and Naylor, 1987). Two hours of exposure to high light and 5°C were sufficient to reduce cucumber photosynthetic rates to zero, whereas the chilling-resistant pea was much less affected.

The adverse effects of chilling can be counteracted in sensitive species. It has been known for some time that hardening reduces chilling injury (extensively reviewed by Levitt, 1980), but the mechanisms involved are not clear. Hardening occurs when environmental factors reduce active growth and induce the accumulation of carbohydrates and other compounds. A number of conditions can result in increased hardening, including mild water stress, a slight deficiency of mineral nutrients such as nitrogen or phosphorus, excess salts in the nutrient solution, and cool temperatures above the chilling temperature range (Rosa, 1921). Cool temperatures only harden against chilling injury if the treatment induces water stress in the plants (Wilson, 1976). This implies that production of ABA by the plant may be a key process in protecting against chilling injury. Several workers have demonstrated a reduction in chilling injury with added ABA in crops such as soyabean and cucumber (Rikin *et al.*, 1976; Markhart, 1986). Applications of synthetic compounds that may induce plant ABA production, such as mefluidide, have recently been found to alleviate chilling injury symptoms in tomato, pepper and cucumber (Li, 1991). The evidence for ABA's role in reducing chilling injury is not all positive, however. For instance, only a weak correlation was found between root hydraulic conductance and endogenous ABA content in mung beans subjected to cool root temperatures, with no improvement in conductance with ABA additions (Bagnall *et al.* 1983).

Another promising avenue of research exploits the range of genetic variation in chilling sensitivity. Efforts are under way to transfer the superior chilling resistance of *Lycopersicon hirsutum* to tomato (Patterson and Payne, 1983). *Phaseolus coccineus* is also more chilling-resistant than *Phaseolus vulgaris* (Guye and Wilson, 1987). Genetic engineering techniques will probably accelerate the transfer of chilling and freeze tolerance into now susceptible species.

Freezing injury

Our understanding of the physiological effects of temperatures below freezing on crop plants has grown rapidly in recent years. The existence of a number of excellent reviews of this topic allows a very abbreviated treatment here (Levitt, 1980; Li, 1989; Li and Christersson, 1993; Steponkus, 1990). Freezing injury results from both the direct damage of plant cell membranes and the indirect injury brought about by cell dehydration (Steponkus, 1990). In a freeze event to temperatures no lower than $-5°C$, ice crystals form outside the cells and the cells become dehydrated. Cell damage most often occurs during the subsequent thaw, when the cell plasma membrane is unable to reform and ruptures as the cell contents rehydrate. If tissue is frozen to lower temperatures, the dehydration of cell contents becomes more severe, and both cell contents and the plasma membrane are directly damaged by high osmotic concentrations. The acclimation to cold involves the alteration of plasma membrane properties to allow contraction and expansion of cells without damage, through changes in membrane lipid composition.

The scope for preventing the damaging effects of freezing in vegetable seedlings is limited, but several approaches have been tried. In situations where the temperature drops only a few degrees below freezing, research into the avoidance of freezing has held initial promise. Formation of ice is facilitated by the presence of nuclei around which the ice first forms. If nucleators are not present, it may be possible to undercool the plant significantly without ice formation. Lindow and Panopoulos (1988) found that five bacterial species catalyse ice formation at temperatures near $0°C$ in potato fields. With genetic engineering techniques bacterial strains were developed that lacked the specific protein necessary for active nucleation. However, attempts to protect plants from freezing by flooding the environment with these inactive strains were only partially successful.

In another approach, spray programmes to reduce the numbers of nucleating bacteria have been tried, with variable success (Anderson *et al.*, 1984). The antibiotic spectinomycin, but not streptomycin, reduced the temperature at which tomato plants froze. The difference in effectiveness may be related to location of ice-nucleating bacteria in the leaves, with only spectinomycin penetrating to the intercellular spaces. Field-grown tomato plants may have other nucleation sites besides the ice-nucleation-active bacteria (Anderson and Ashworth, 1985). The practical importance of preventing ice-nucleating bacteria from building up on or in transplanted vegetables thus requires further clarification.

As with chilling injury, differences between species in tolerance of freezing temperatures exist, and may be exploited in breeding programmes. In *Solanum*, species differ in the degree to which they can be acclimated to freezing temperatures (Chen *et al.*, 1983). Potato showed little change in frost tolerance with hardening treatments of 2°C, but *Solanum commersonii* could survive progressively lower freezing temperatures, from −3 to −12°C, as hardening was prolonged. This difference was correlated with a temporary increase in plant ABA levels, and the rise of a soluble storage protein in *S. commersonii* but not potato. It may thus be possible to develop more freeze-tolerant potato lines through interspecific crosses.

CONCLUDING REMARKS

Using transplants rather than seeds to produce vegetables requires an understanding of the physiological processes that govern plant growth in containers, and the ways in which the growth inhibition of transplanting can be minimized. Generally, economic considerations dictate that seedlings be grown for as short a time as possible, and in minimal space, before transplanting. Research has indicated that this strategy is also the one most likely to be successful with crops such as cauliflower and broccoli. With others, such as tomato and pepper, early yield will be sacrificed.

Significant advances in transplant production systems have been made in recent years, but our ability to inhibit plant growth temporarily before transplanting has not kept pace with these developments. The objective should be to increase the seedling's capability to withstand stress during transplanting, and to hasten the recovery of active growth in the field. Unfortunately, weather conditions in the field often prohibit timely transplanting. As a result, seedlings remain crowded in small containers for too long. Our understanding of the changes that such plants undergo, both in the containers, and after they have finally been transplanted, is still fragmentary. The process of hardening is probably most important, but there has been little work in this area since the 1920s (Rosa, 1921; Crist, 1928; Babb, 1940). Recent research on the physiological effects of root restriction may be relevant (e.g. Peterson *et al.*, 1991a), but need to be further pursued, and expanded to include study of the changes that occur when restriction is relieved.

Finally, recent advances in the techniques of plant breeding may provide opportunities to develop vegetable cultivars that are better adapted to transplant production systems than are current materials. In particular, lines which exhibit less stem elongation when crowded in seedling containers, and cultivars with resistance to chilling and freezing injury would fulfil a significant need. Other attributes that need modification will probably be identified as we improve our understanding of the physiological processes involved in plant growth in containers and in the transition to field establishment.

ACKNOWLEDGEMENTS

I am grateful to Drs Pamela Ludford and David Wolfe for their helpful suggestions on this chapter. Thanks also to Ms Lauren Garner and Dr Thomas Björkman for supplying Fig. 2.4, and Dr Donald Krizek for making Fig. 2.3 available.

REFERENCES

Aloni, B., Daie, J. and Karni, L. (1991) Water relations, photosynthesis and assimilate partitioning in leaves of pepper (*Capsicum annuum*) transplants: effect of water stress after transplanting. *Journal of Horticultural Science* 66, 75–80.

Aloni, B. and Pashkar, T. (1987) Antagonistic effects of paclobutrazol and gibberellic acid on growth and some biochemical characteristics of pepper (*Capsicum annuum*) transplants. *Scientia Horticulturae* 33, 167–177.

Anderson, J.A. and Ashworth, E.N. (1985) Ice nucleation in tomato plants. *Journal of the American Society of Horticultural Science* 110, 291–296.

Anderson, J.A., Buchanan, D.W. and Stall, R.E. (1984) Reduction of bacterially induced frost damage to tender plants. *Journal of the American Society of Horticultural Science* 109, 401–405.

Babb, M.F. (1940) Residual effect of forcing and hardening of tomato, cabbage and cauliflower plants. *USDA Technical Bulletin* 760, 1–34.

Baggett, J.R. and Mack, H.J. (1970) Premature heading of broccoli cultivars as affected by transplant size. *Journal of the American Society of Horticultural Science* 95, 403–407.

Bagnall, D., Wolfe, J. and King, R.W. (1983) Chill-induced wilting and hydraulic recovery in mung bean plants. *Plant Cell and Environment* 6, 457–464.

Bennett, M.A. and Grassbaugh, E.M. (1992) Influence of transplant characteristics on processing tomato seedling development and yields. *Proceedings of the National Symposium on Stand Establishment in Horticultural Crops* 301–310.

Berkowitz, G.A. and Rabin, J. (1988) Antitranspirant associated abscisic acid effects on the water relations and yield of transplanted bell pepper. *Plant Physiology* 86, 329–331.

Beverly, R.B., Latimer, J.G. and Oetting, R.D. (1992) Effect of root cell size and brushing on transplant growth and field establishment of 'Sunrise' tomato under a line-source irrigation variable. *Proceedings of the National Symposium on Stand Establishment in Horticultural Crops*, 249–258.

Biddington, N.L. (1986) The effects of mechanically induced stress in plants – a review. *Plant Growth Regulation* 4, 103–123.

Biddington, N.L. and Dearman, A.S. (1985a) The effect of mechanically induced stress on the growth of cauliflower, lettuce and celery seedlings. *Annals of Botany* 55, 109–119.

Biddington, N.L. and Dearman, A.S. (1985b) The effect of mechanically induced stress on water loss and drought resistance in lettuce, cauliflower and celery seedlings. *Annals of Botany* 56, 795–802.

Biddington, N.S. and Dearman, A.S. (1986) A comparison of the effects of mechanically induced stress, ethephon and silver thiosulphate on the growth of cauliflower seedlings. *Plant Growth Regulation* 4, 33–41.

Biro, R.L. and Jaffe, M.J. (1984) Thigmomorphogenesis: ethylene evolution and its role in the changes observed in mechanically perturbed bean plants. *Physiologia Plantarum* 62, 289–296.

Braam, J. (1992) Regulated expression of the calmodulin-related TCH genes in cultured *Arabidopsis* cells: induction by calcium and heat shock. *Proceedings of the National Academy of Science* 89, 3213–3216.

Braam, J. and Davis, R.W. (1990) Rain-, wind-, and touch-induced expression of calmodulin and calmodulin-related genes in *Arabidopsis. Cell* 60, 357–364.

Brouwer, R. (1962) Distribution of dry matter in the plant. *Netherlands Journal of Agricultural Science* 10, 361–376.

Brydon, K. and Orzolek, M.D. (1992) Storage of field-grown cabbage transplants. *Proceedings of the National Symposium on Stand Establishment in Horticultural Crops* 259–263.

Campbell, G.M. (1976) Effect of ethephon and SADH on quality of clipped and non-clipped tomato transplants. *Journal of the American Society of Horticultural Science* 101, 648–651.

Carmi, A. and Heuer, B. (1981) The role of roots in control of bean shoot growth. *Annals of Botany* 48, 519–527.

Casseres, E.H. (1947) Effect of date of sowing, spacing, and foliage trimming of plants in flats on yield of tomatoes. *Proceedings of the American Society of Horticultural Science* 50, 285–289.

Cathey, H.M. (1964) Physiology of growth retarding chemicals. *Annual Review of Plant Physiology* 15, 271–302.

Chen, H.H., Li, P.H. and Brenner, M.L. (1983) Involvement of abscisic acid in potato cold acclimation. *Plant Physiology* 71, 362–365.

Costello, L. and Paul, J.L. (1975) Moisture relations in transplanted container plants. *HortScience* 10, 371–372.

Crist, J.W. (1928) Ultimate effect of hardening tomato plants. *Michigan Agricultural Experiment Station Technical Bulletin* 89, 1–22.

Csizinsky, A.A. and Schuster, D.J. (1993) Impact of insecticide schedule, N and K rates, and transplant container size on cabbage yield. *HortScience* 28, 299–302.

Decoteau, D.R. and Friend, H.H. (1991) Phytochrome-regulated growth of young watermelon plants. *Journal of the American Society of Horticultural Science* 116, 512–515.

Decoteau, D.R., Kasperbauer, M.J., Daniels, D.D. and Hunt, P.G. (1988) Plastic mulch color effects on reflected light and tomato plant growth. *Scientia Horticulturae* 34, 169–175.

Dufault, R.J. (1985) Relationship among nitrogen, phosphorus and potassium fertility regimes on celery transplant growth. *HortScience* 20, 1104–1106.

Dufault, R.J. (1986) Influence of nutritional conditioning on muskmelon transplant quality and early yield. *Journal of the American Society of Horticultural Science* 111, 698–703.

Dufault, R.J. and Melton, R.R. (1990) Cyclic cold stresses before transplanting influence tomato seedling growth, but not fruit earliness, fresh-market yield or quality. *Journal of the American Society of Horticultural Science* 115, 559–563.

Dufault, R.J. and Waters, L.J. (1985) Container size influences broccoli and cauliflower transplant growth but not yield. *HortScience* 20, 682–684.

Erwin, J.E. and Heins, R.D. (1990) Don't use B-nine on your vegetable plugs – it's illegal. Try temperature for height control. *Grower Talks* 50(10), 60–61.

Erwin, J.E., Heins, R.D., Berghage, R.D., Kovanda, B.J., Carlson, W.H. and Biernbaum, J.A. (1989) Cool mornings control plant height. *Grower Talks* 52(9), 73–74.

Erwin, J.E., Heins, R.D., Carlson, W. and Newport, S. (1992) Diurnal temperature fluctuations and mechanical manipulation affect plant stem elongation. *Plant Growth Regulator Society of America Quarterly* 20, 1-17.

Fordham, R. and Biggs, A.G. (1985) *Principles of Vegetable Crop Production.* Collins, London.

Garner, L.C. and Björkman, T. (1996) Mechanical conditioning for controlling excessive elongation in tomato transplants: sensitivity to dose, frequency and timing of brushing. *Journal of the American Society of Horticultural Science* 121, 894–900.

Garton, R.W. (1992) Field performance of 406 cell tomato transplants. *Proceedings of the National Symposium of Stand Establishment in Horticultural Crops* 295–299.

Graebe, J. (1987) Gibberellin biosynthesis and control. *Annual Review of Plant Physiology* 38, 419–465.

Guye, M.G. and Wilson, J.M. (1987) The effects of chilling and chill-hardening temperatures on stomatal behaviour in a range of chill-sensitive species and cultivars. *Plant Physiology and Biochemistry* 25, 717–721.

Hall, M.R. (1985) Influence of storage conditions and duration on weight loss in storage, field survival and root yield of sweet potato transplants. *HortScience* 20, 200–203.

Hameed, M.A., Reid, J.B. and Rowe, R.N. (1987) Root confinement and its effects on the water relations, growth and assimilate partitioning of tomato (*Lycopersicon esculentum* Mill.). *Annals of Botany* 59, 685–692.

Harper, J.L. (1977) The limiting resources of the environment. In: *Population Biology of Plants.* Academic Press, London, pp. 305–323.

Heuchert, J.C., Marks, J.S. and Mitchell, C.A. (1983) Strengthening of tomato shoots by gyratory shaking. *Journal of the American Society of Horticultural Science* 108, 801–805.

Jaffe, M.J. (1973) Thigmomorphogenesis: the response of plant growth and development to mechanical stimulation. *Planta* 114, 143–157.

Jaffe, M.J. and Biro, R.L. (1979) Thigmomorphogenesis: the effect of mechanical perturbation on the growth of plants with special reference to anatomical changes, the role of ethylene, and interaction with other environmental stresses. In: Mussell, H. and Staples, R.(eds) *Stress Physiology in Crop Plants.* John Wiley, New York, pp. 25–59.

Jaworski, C.A. and Webb, R.E. (1967) Preliminary tests on the performance of clipped tomato transplants. *Proceedings of the American Society of Horticultural Science* 91, 550–555.

Jaworski, C.A. and Webb, R.E. (1971) Pepper performance after transplant clipping. *HortScience* 6, 480–482.

Jaworski, C.A., Webb, R.E., Garrison, S.A., Bergman, E.L. and Shannon, S. (1970) Growth-retardant-treated tomato transplants. *HortScience* 5, 255–256.

Jaworski, C.A., Webb, R.E., Wilcox, G.E. and Garrison, S.A. (1969) Performance of tomato cultivars after various types of transplant clipping. *Journal of the American Society of Horticultural Science* 94, 614–616.

Kasperbauer, M.J. (1987) Far-red light reflection from green leaves and effects on phytochrome-mediated assimilate partitioning under field conditions. *Plant Physiology* 85, 350–354.

Kays, S.J., Jaworski, C.A. and Price, H.C. (1976) Defoliation of pepper transplants in transit by endogenously evolved ethylene. *Journal of the American Society of Horticultural Science* 101, 449–451.

Klassen, P. (1992) Refining proven transplant technology. *American Vegetable Grower* 40(12), 10–11.

Klassen, P. (1993) Transition to transplants. *American Vegetable Grower* 41(4), 19–21.

Knavel, D.E. (1965) Influence of container, container size, and spacing on growth of transplants and yields in tomato. *Journal of the American Society of Horticultural Science* 86, 582–586.

Knight, M.R., Smith, S.M. and Trewavas, A.J. (1992) Wind-induced plant motion immediately increases cytosolic calcium. *Proceedings of the National Academy of Science* 89, 4967–4971.

Kratky, B.A., Cox, E.F. and McKee, J.M.T. (1980) Effects of block and soil water content on establishment of transplanted cauliflower seedlings. *Journal of Horticultural Science* 55, 229–234.

Kratky, B.A. and Mishima, H.Y. (1981) Lettuce seedling and yield response to preplant and foliar fertilization during transplant production. *Journal of the American Society of Horticultural Science* 106, 3–7.

Kratky, B.A., Wang, J.K. and Kubojiri, K. (1982) Effect of container size, transplant age, and spacing on Chinese cabbage. *Journal of the American Society of Horticultural Science* 107, 345–347.

Kraus, E.J. and Kraybill, H.R. (1918) Vegetation and reproduction with special reference to the tomato. *Oregon Agricultural Experiment Station Bulletin* 149, 1–90.

Kraus, J.E. (1942) Effects of partial defoliation at transplanting time on subsequent growth and yield of lettuce, cauliflower, celery, peppers and onion. *USDA Technical Bulletin* 829, 1–35.

Latimer, J.G. (1990) Drought or mechanical stress affects broccoli transplant growth and establishment but not yield. *HortScience* 25, 1233–1235.

Latimer, J.G. (1991) Mechanical conditioning for control of growth and quality of vegetable transplants. *HortScience* 26, 1456–1461.

Latimer, J.G. (1992) Drought, paclobutrazol, abscisic acid, and gibberellic acid as alternatives to daminozide in tomato transplant production. *Journal of the American Society of Horticultural Science* 117, 243–247.

Latimer, J.G. and Beverly, R.B. (1992) Brushing controls growth of cucurbit transplants without affecting yields. *Proceedings of the National Symposium on Stand Establishment in Horticultural Crops* 231–235.

Latimer, J.G., Johjima, T. and Harada, K. (1991) The effect of mechanical stress on transplant growth and subsequent yield of four cultivars of cucumber. *Scientia Horticulturae* 47, 221–230.

Leskovar, D.I. and Cantliffe, D.J. (1991) Tomato transplant morphology affected by handling and storage. *HortScience* 26, 1377–1379.

Levitt, J. (1980) Responses of plants to environmental stresses. *Chilling, Freezing and High Temperature Stresses*, 2nd edn, Vol. 1. Academic Press, New York.

Li, P.H. (ed.) (1989) *Low Temperature Stress Physiology in Crops*. CRC Press, Boca Raton, Florida.

Li, P.H. (1991) Mefluidide-induced cold hardiness in plants. *Critical Reviews of Plant Science* 9, 497–517.

Li, P.H. and Christersson, L. (eds) (1993) *Advances in Plant Cold Hardiness*. CRC Press, Boca Raton, Florida.

Lindow, S.E. and Panopoulos, N.J. (1988) Field tests of recombinant ice-minus *Pseudomonas syringae* for biological frost control in potato. In: Sussman, M. (ed.) *The Release of Genetically Engineered Microorganisms*. Academic Press, London, pp. 121–138.

Liptay, A. (1985) Reduction in spindliness of tomato transplants grown at high densities. *Canadian Journal of Plant Science* 65, 797–801.

Liptay, A., Jaworski, C.A. and Phatak, S.C. (1981) Effect of tomato transplant stem diameter and ethephon treatment on tomato yield, fruit size and number. *Canadian Journal of Plant Science* 61, 413–415.

Liptay, A. and Nicholls, S. (1993) Nitrogen supply during greenhouse transplant production affects subsequent tomato root growth in the field. *Journal of the American Society of Horticultural Science* 118, 339–342.

Liptay, A., Phatak, S.C. and Jaworski, C.A. (1988) Ethephon treatment of tomato transplants improves frost tolerance. *HortScience* 17, 400–401.

Loomis, W.E. (1925) Studies in the transplanting of vegetable plants. *Cornell Agricultural Experiment Station Memoir* 87, 1–63.

Magnifico, V., Bianco, V. and Fortunato, I.M. (1980) The effect of seedling size at transplanting on the production characteristics of broccoli. *Annali Facolta di Agraria Universita di Bari* 31, 717–731.

Mansfield, T.A. (1976) Chemical control of stomatal movements. *Philosophical Transactions of the Royal Society of London* 273B, 541–550.

Markhart, A.H. III (1986) Chilling injury: a review of possible causes. *HortScience* 21, 1329–1333.

Marr, C.W. and Jirak, M. (1990) Holding tomato transplants in plug trays. *HortScience* 25, 173–176.

Masson, J., Tremblay, N. and Gosselin, A. (1991) Nitrogen fertilization and HPS supplementary lighting influence vegetable transplant production. I. Transplant growth. *Journal of the American Society of Horticultural Science* 116, 594–598.

McCarter, S.M. and Jaworski, C.A. (1969) Field studies on spread of *Pseudomonas solanacearum* and tobacco mosaic virus in tomato plants by clipping. *Plant Disease Reporter* 53, 942–946.

McCormac, A.C., Cherry, J.R., Hershey, H.P., Vierstra, R.D. and Smith, H. (1991) Photoresponses of transgenic tobacco plants expressing an oat phytochrome gene. *Planta* 185, 162–170.

Melton, R.R. and Dufault, R.J. (1991) Nitrogen, phosphorus and potassium fertility regimes affect tomato transplant growth. *HortScience* 26, 141–142.

Mitchell, C.A., Dostal, H.C. and Seipel, T.M. (1977) Dry weight reduction in mechanically dwarfed tomato plants. *Journal of the American Society of Horticultural Science* 102, 605–608.

Mitchell, C.A., Severson, C.J., Wott, J.A. and Hammer, P.A. (1975) Seismomorphogenic regulation of plant growth. *Journal of the American Society of Horticultural Science* 100, 161–165.

Moran, C.H., Hardenburg, R.E., Peel, R.D. and Moore, J.F. (1962) Commercial packaging and truck transport of bare root tomato plants in polyethylene-lined crates. *Proceedings of the American Society of Horticultural Science* 81, 458–466.

Morgan, D.C. and Smith, H. (1979) A systematic relationship between phytochrome-controlled development and species habitat, for plants grown in simulated natural radiation. *Planta* 145, 253–258.

Nelms, L.R. and Spomer, L.A. (1983) Water relations of container soils transplanted into ground beds. *HortScience* 18, 863–866.

NeSmith, D.S. (1993) Transplant age influences summer squash growth and yield. *HortScience* 28, 618–620.

Nicklow, C.W. and Minges, P.A. (1962) Plant growing factors influencing the field performance of the Fireball tomato variety. *Proceedings of the American Society of Horticultural Science* 81, 443–450.

Nitzsche, P., Berkowitz, G.A. and Rabin, J. (1991) Development of a seedling-applied antitranspirant formulation to enhance water status, growth and yield of transplanted bell pepper. *Journal of the American Society of Horticultural Science* 116, 405–411.

Orzolek, M.D. (1986) Use of growth retardants for tomato transplant production. *Applied Agricultural Research* 1, 168–171.

Pardossi, A., Togoni, F. and Lovemore, S.S. (1988) The effect of different hardening treatments on tomato seedling growth, chilling resistance and crop production in cold greenhouse. *Acta Horticulturae* 229, 371–379.

Patterson, B.D. and Payne, L.A. (1983) Screening for chilling resistance in tomato seedlings. *HortScience* 18, 340–341.

Peeler, T.C. and Naylor, A.W. (1988) A comparison of the effects of chilling on leaf gas exchange in pea (*Pisum sativum* L.) and cucumber (*Cucumis sativus* L.). *Plant Physiology* 86, 143–146.

Peterson, T.A., Reinsel, M.O. and Krizek, D.T. (1991a) Tomato (*Lycopersicon esculentum* Mill. cv. Better Bush) plant response to root restriction. I. Alteration of plant morphology. *Journal of Experimental Botany* 42, 1233–1240.

Peterson, T.A., Reinsel, M.O. and Krizek, D.T. (1991b) Tomato (*Lycopersicon esculentum* Mill. cv. Better Bush) plant response to root restriction. II. Root respiration and ethylene generation. *Journal of Experimental Botany* 42, 1241–1249.

Pisarczyk, J.M. and Splittstoesser, W.E. (1979) Controlling tomato transplant height with chlormequat, daminozide and ethephon. *Journal of the American Society of Horticultural Science* 104, 342–344.

Read, P.E. and Fieldhouse, D.J. (1970) Use of growth retardants for increasing tomato yields and adaptation for mechanical harvest. *Journal of the American Society of Horticultural Science* 95, 73–78.

Reid, M.E. (1924) Relation of kind of food reserves to regeneration in tomato plants. *Botanical Gazette* 77, 103–110.

Rikin, A., Blumenfeld, A. and Richmond, A.E. (1976) Chilling resistance as affected by stressing environments and abscisic acid. *Botanical Gazette* 137, 307–312.

Risse, L.A., Kretchman, D.W. and Jaworski, C.A. (1985) Quality and field performance of densely packed tomato transplants during shipment and storage. *HortScience* 20, 438–439.

Risse, L.A., Moffit, T. and Bryan, H.H. (1979) Effect of storage temperature and duration on quality, survival and yield of containerized tomato transplants. *Proceedings of the Florida State Horticultural Society* 92, 198–200.

Robbins, N.S. and Pharr, D.M. (1988) Effect of restricted root growth on carbohydrate metabolism and whole plant growth of *Cucumis sativus* L. *Plant Physiology* 87, 409–413.

Rosa, J.T. Jr (1921) Investigations on the hardening process in vegetable plants. *Missouri Agricultural Experiment Station Research Bulletin* 48, 1–97.

Ruff, M.S., Krizek, D.T., Mirecki, R.M. and Inouye, D.W. (1987) Restricted root zone volume: influence on growth and development of tomato. *Journal of the American Society of Horticultural Science* 112, 763–769.

Saab, I.N., Sharp, R.E., Pritchard, J. and Voetberg, G.S. (1990) Increased endogenous abscisic acid maintains primary root growth and inhibits shoot growth of maize seedlings at low water potentials. *Plant Physiology* 93, 1329–1336.

Saito, T., Konno, Y. and Itoh, H. (1963) Studies on the growth and fruiting in the tomato. IV. Effect of the early environment on the growth and fruiting. 4. Fertility of bed soil, watering and spacing. *Journal of the Japanese Society of Horticultural Science* 32, 186–196.

Salter, P.J. (1985) Crop establishment: recent research and trends in commercial practise. *Scientia Horticulturae* 36, 32–47.

Sayre, C.B. (1948) Early and total yields of tomatoes as affected by time of seeding, topping the plants, and space in the flats. *Proceedings of the American Society of Horticultural Science* 51, 367–370.

Skapski, H. and Oyer, E.B. (1964) The influence of pre-transplanting variables on the growth and development of cauliflower plants. *Proceedings of the American Society of Horticultural Science* 85, 374–385.

Smith, P.G. and Zink, F.W. (1951) Effect of sucrose foliage spray on tomato transplants. *Proceedings of the American Society of Horticultural Science* 58, 168–178.

Steponkus, P.L. (1990) Cold acclimation and freezing injury from a perspective of the plasma membrane. In: Katterman, F. (ed.) *Environmental Injury to Plants.* Academic Press, San Diego, pp. 1–16.

Suge, H. (1980) Dehydration and drought resistance in *Phaseolus vulgaris* as affected by mechanical stress. *Report of the Institute of Agricultural Research of Tohoku University* 31, 1–10.

Taha, A.A., Kretchman, D.W. and Jaworski, C.A. (1980) Effect of daminozide and ethephon on transplant quality, plant growth and development, and yield of processing tomatoes. *Journal of the American Society of Horticultural Science* 105, 705–709.

Thomas, H.R. and Moore, W.D. (1947) Influence of the length and manner of storage of tomato seedlings on stand, early growth and yield. *Proceedings of the American Society of Horticultural Science* 49, 264–266.

Thomas, T.H. (1993) Effects of root restriction and growth regulator treatments on the growth of carrot (*Daucus carota* L.) seedlings. *Plant Growth Regulation* 13, 95–101.

Veliath, J.A. and Ferguson, A.C. (1973) A comparison of ethephon, DCIB, SADH, and DPA for abscission of fruits, flowers, and floral buds in determinate tomatoes. *Journal of the American Society of Horticultural Science* 98, 124–126.

Wang, Y.T. and Gregg, L.L. (1990) Uniconazole controls growth and yield of greenhouse tomato. *Scientia Horticulturae* 43, 55–62.

Went, F.W. (1944) Plant growth under controlled conditions: II. Thermoperiodicity in growth and fruiting of the tomato. *American Journal of Botany* 31, 135–150.

Went, F.W. (1945) Plant growth under controlled conditions. V. The relation between age, light, variety and thermoperiodicity of tomatoes. *American Journal of Botany* 32, 469–479.

Went, F.W. and Carter, M. (1948) Growth response of tomato plants to applied sucrose. *American Journal of Botany* 35, 95–106.

Weston, L.A. (1988) Effect of flat cell size, transplant age, and production site on growth and yield of pepper transplants. *HortScience* 23, 709–711.

Weston, L.A. and Zandstra, B.H. (1986) Effect of root container size and location of production on growth and yield of tomato transplants. *Journal of the American Society of Horticultural Science* 111, 498–501.

Weston, L.A. and Zandstra, B.H. (1989) Transplant age and N and P nutrition effects on growth and yield of tomatoes. *HortScience* 24, 88–90.

Whitlock, A. (1980) The effect of antitranspirants and postplanting watering on the establishment and yield of celery transplants. *Experimental Horticulture* 31, 21–25.

Whitwell, J.D. and Crofts, J. (1972) Studies on the size of cauliflower transplants in relation to field performance with particular reference to date of maturity and length of cutting season. *Experimental Horticulture* 23, 34–42.

Widders, I.E. (1989) Preplant treatments of N and P influence growth and elemental accumulation in tomato seedlings. *Journal of the American Society of Horticultural Science* 114, 416–420.

Widders, I.E. and Garton, R.W. (1992) Effects of preplant nutrient conditioning on elemental accumulation in tomato seedlings. *Scientia Horticulturae* 52, 9–17.

Wilson, J.M. (1976) The mechanism of chill- and drought-hardening of *Phaseolus vulgaris* leaves. *New Phytologist* 76, 257–270.

Wise, R.R. and Naylor, A.W. (1987) Chilling-enhanced photo-oxidation. The peroxidative destruction of lipids during chilling injury to photosynthesis and ultrastructure. *Plant Physiology* 83, 272–277.

Wurr, D.C.E., Fellows, J.R. and Hadley, P. (1986) The influence of supplementary lighting and mechanically induced stress during plant raising on transplant and maturity characteristics of crisp lettuce. *Journal of Horticultural Science* 61, 325–330.

Zack, C.D. and Loy, J.B. (1980) The effect of light quality and photoperiod on vegetative growth of *Cucurbita maxima*. *Journal of the American Society of Horticultural Science* 105, 939–943.

The Induction of Flowering 3

E.H. Roberts, R.J. Summerfield, R.H. Ellis, P.Q. Craufurd and T.R. Wheeler

The University of Reading, Department of Agriculture, Plant Environment Laboratory, Cutbush Lane, Shinfield, Reading RG2 9AD, Berkshire, UK

The first step towards maximizing yield by breeding is to assure that the phenology of the crop is well matched to the target environment. ... Matching phenology with the environment may be achieved either by genetically modifying the crop through the manipulation of photo-period or vernalization sensitive/insensitive genes or by modifying management.

(Richards, 1989)

The task of the scientist is to describe, predict, understand and apply – not necessarily in that order. Prediction and application may precede understanding, and much understanding may be inapplicable to prac-tical problems.

(Thornley, 1987)

It might be said that a plant, like a coastal yachtsman, proceeds on its genetically charted course whilst its dead-reckoning is up-dated from time to time by environmental fixes so that it reaches predetermined waypoints and its final destination under favourable circumstances (i.e. when the tide is on the flood and the pubs are opening!).

(S. Rogers, 1987, Extract from his answer to a question in an MSc paper at Reading)

The title of this chapter duplicates that of a much-cited book edited by L.T. Evans and published in 1969. The 20 case histories reviewed there occupied 442 printed pages and were the synthesis of more than 1000 literature cita-tions. It was clear then, and so it remains, that two environmental factors are of overriding importance in the induction of flowering – photoperiod (daylength) and temperature.

There is now at least some information on the photothermal flowering responses of thousands of species (Rees, 1987), albeit with a strong bias towards photoperiodism (e.g. Withrow, 1959; Vince-Prue, 1975; Vince-Prue *et al.*, 1984; Atherton, 1987). Physiological signals that induce or inhibit flowering have been described (Kinet *et al.*, 1985; Bernier *et al.*, 1993); metabolic mutants in which the response to inductive daylengths is altered are increasingly targeted (Bernier *et al.*, 1993; Chassan, 1994); but progress in understanding the physiological, genetic and molecular bases of induction is slow (Evans, 1993). Intense research efforts have often focused on a single clone or cultivar; experiments are commonly done at a single temperature; and there is still no means of assaying the quantity of stimulus (to flower or not to flower) produced in response to a given treatment.

Given these constraints, and the range and variety of flowering responses to photothermal conditions (Salisbury, 1982), it has proved difficult 'to find patterns or to make generalisations that have some meaning and predictive value' (Rees, 1987). We do not here wish to lose sight of that range of responses and their ecological significance (Evans, 1975). We shall, nevertheless, attempt a synthesis and suggest generalizations. Practical utility demands both simplicity and a proven ability to anticipate biological events (Waggoner, 1974).

THE TIMING OF FLOWERING AND CLIMATIC ADAPTATION

People probably began to domesticate plants about 10,000 years ago and more or less simultaneously in at least three different regions – the Near East, northern China and Meso-America (Hawkes, 1983). Thus, agriculture began, and other crops were later domesticated in regions with pronounced wet and dry seasons, neither of which were particularly long (Harris, 1969). By the time 'seed agriculture' had come to replace what Hawkes describes as 'colonization and gathering' it must have been clear that crops could only be grown satisfactorily during relatively short, wetter seasons.

In discussing the genetic adaptation of field crops to the variable and frequently stressful conditions under which they are grown, Lawn and Imrie (1994) have expressed the view of many plant breeders: 'The key aim is to optimise productivity by matching the ontogeny [sequence of developmental stages] to the weather resources of the environment [e.g. duration of favourable temperature or water supply] and, where unfavourable extremes are unavoidable, to minimize their coincidence with more vulnerable stages. Not surprisingly, therefore, phenology [the influence of environment on ontogeny] is the most important single factor influencing genotypic adaptation.' Richards (1991) and Hamblin (1994) come to similar conclusions.

Flowering is a particularly important event in crop development since it is a phase which is especially vulnerable to environmental stress. Furthermore, it is the timing of this stage of plant development that is very largely responsible for determining when cereal, pulse, oilseed crops and many vegetables will subsequently be ripe for harvest.

It was 'fairly certain' about 50 years ago that a given genotype of any annual crop 'has its own definite optimal requirements of temperature and light [photoperiod] without which it cannot proceed at an economically desirable rate to flower formation, flowering and the production of seeds' (Whyte, 1946). Later, the photothermal control of the timing of flowering came to be recognized as 'especially important in crop adaptation' (e.g. Nuttonson, 1953). Since then, however, successive attempts to correlate the timing of flowering in a wide range of crops with various weather variables tended to confirm the influential viewpoint of Went (1957): 'It is obvious that the problem of flowering … is a very complicated phenomenon, and … I do not believe that anything is gained by considering it simple.' Indeed, because the reliable prediction of flowering in fluctuating field environments has proved so difficult, a common viewpoint has been that 'photothermal effects are strongly curvilinear and interactive' (e.g. Aitken, 1974; Salisbury, 1982; Robertson, 1983; Wallace, 1985). However, we now believe that the responses are amenable to relatively simple analysis to provide predictive models which are valuable in breeding, agronomy and genetic analysis of responses.

THE ECOGEOGRAPHY OF PHOTOTHERMAL RESPONSES

Because the earth's axis is inclined at 23.5° towards the plane of the earth's orbit around the sun, there is a regular annual variation in daylength in any part of the world, although the extent of that variation depends on latitude (Fig. 3.1 and Table 3.1). Given the relatively small seasonal differences in tropical daylengths (Table 3.1) it is not surprising that some of the most sensitive flowering responses to photoperiod are found in tropical species: differences of 10–20 min day^{-1} can be critical in rice (*Oryza sativa*) and cowpea (*Vigna unguiculata*), for example (Roberts, 1991).

Current knowledge indicates that all crops of tropical or subtropical origin are essentially short-day plants (SDPs), whereas the vast majority of crops of Mediterranean or temperate origin are long-day plants (LDPs). The only exception to this generalization of which we are aware are the SDPs sunflower (*Helianthus annuus*) and soyabean (*Glycine max*) which originated between 30 and 35°N in the USA and between 34 and 40°N in China, respectively (Roberts and Summerfield, 1987; Summerfield and Roberts, 1987). The exception of soyabean can be explained by a major perturbation of the intertropical convergence zone (ITCZ) in summer in north-east China which results in a late growing season (Roberts, 1991).

Many species of crops also include genotypes which are insensitive to photoperiod, i.e. day-neutral plants (DNPs), but these are often the product of special selection – either deliberate, or as an inevitable consequence of adaptation to regions outside the latitude of origin. For example, if tropical

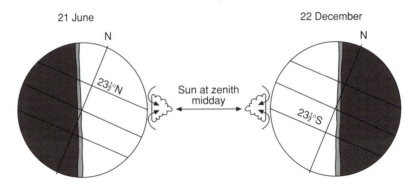

Fig. 3.1. The earth at the summer and winter solstices showing rains tending to centre on the Tropic of Cancer in June and on the Tropic of Capricorn in December. The hatched area between day and night is Civil Twilight, the period between sunset and when the centre of the sun is 6° below the horizon, when irradiance is still sufficient to be perceived by plants and therefore should be considered as part of the natural photoperiod (from Roberts, 1991).

Table 3.1. Variation in daylength (h:min) with latitude.

Latitude (°N or °S) with exemplar locations		Daylength inclusive of Civil Twilight (h : min)*		
		Longest	Shortest	Annual variation
0	(Entebbe, Uganda)	12:50	12:50	0:00
10	(Costa Rica)	13:29	12:19	1:10
20	(Hawaii)	14:11	11:44	2:27
30	(Cairo)	15:01	11:04	3:57
40	(Philadelphia, USA)	16:01	10:22	5:39
50	(South Cornwall, UK)	17:53	9:20	8:33
60	(Shetland Isles)	22:25	7:48	14:37
70	(North Alaska)	24:00	0:00	24:00

* Civil Twilight begins before sunrise and ends after sunset when the true centre of the sun is 6° below the horizon; it corresponds to an illuminance of about 4 lux.

plants typically have short-day responses flowering would be excessively delayed in the long days of temperate summers unless, in the course of their spread from the low latitudes, strains were selected which were much less sensitive or insensitive to photoperiod. Conversely, species of temperate origin with long-day responses could not adapt to the short days of the tropics unless strains with little or no sensitivity were selected. Thus the rice landraces of tropical origin (*Oryza sativa* ssp. *indica*) are typically very sensitive SDPs, while those adapted to the temperate latitudes of Japan, Europe and USA (ssp. *japonica*) are typically much less sensitive or insensitive. Examples from vegetables are not always so clear but tomato

(*Lycopersicon esculentum*), for example, was probably domesticated in Central America (Rick, 1976) and is essentially a short-day species (Atherton and Harris, 1986) but those cultivars in commercial production tend to be relatively insensitive to daylength, probably as a result of intensive breeding in temperate latitudes. The same applies to *Phaseolus vulgaris*. On the other hand the photoperiod sensitivity of lentil (*Lens culinaris*) which is a long-day species originating in West Asia (Fig. 3.2) is less in tropical landraces (Erskine *et al.*, 1990). The same applies for wheat (*Triticum aestivum*) (Curtis, 1988).

Another feature in the control of flowering is that many LDP species include genotypes in which flowering is advanced by cool temperatures – a response described as vernalization. Such a response appears to be extremely rare amongst SDPs , and in fact only one confirmed example is known, namely *Chrysanthemum moriflorum,* which originated in China between 20 and 35°N (Roberts and Summerfield, 1987).

It is plausible that SDPs are typical of the tropics because the growing season is limited by rain (not by temperature or irradiance) and that the rainy season is a product of the ITCZ. This zone tends to coincide with the latitude where the sun is overhead at midday, and the sun's zenith moves from 23.5°N in June to 23.5°S in December (see Fig. 3.1). Crops ideally ripen towards the end of the rainy season after the accumulation of sufficient vegetative growth, and when the conditions are becoming drier and more suitable for the maturation of seeds. In these circumstances only a short-day response could result in appropriate phenology.

In Mediterranean and temperate latitudes it is also an advantage for crops to ripen towards the end of the growing season, and in a warm and not unduly wet season for seed maturation. The growing season in these regions is primarily limited by cool temperatures and, to a lesser extent, by low irradiance during the winter. In Mediterranean climates it is also curtailed by summer drought. Accordingly most crops flower in the spring in the Mediterranean and in the early summer in temperate latitudes – in both cases when the days are lengthening. Under these circumstances only a long-day response can ensure appropriate phenology; and, in the case of autumn-sown annuals or biennials, this long-day response needs to be supplemented by a prior vernalization requirement in order to prevent the plants responding to the relatively long days of autumn.

RATES OF PROGRESS TOWARDS FLOWERING

With annual crops, the time taken from sowing to flower, f, is of special interest and therefore needs to be predicted. But, between measurements of times to flower in various experimental treatments and their reliable prediction in other circumstances, comes the analysis of responses and the synthesis of models. For these two purposes it has proved profitable to analyse photothermal flowering responses in terms of rates, i.e. $1/f$, the rate of

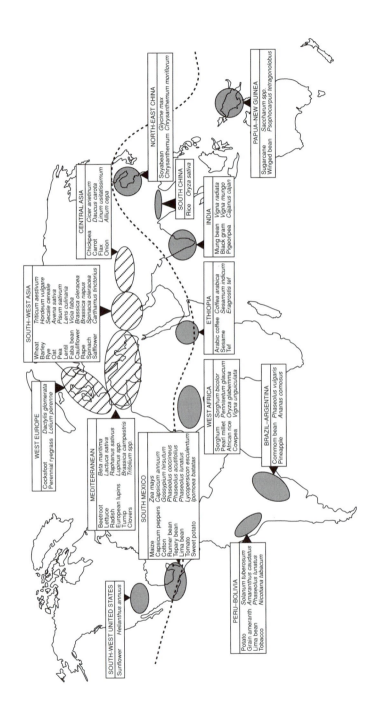

Fig. 3.2. Origins of crop plants. Those with long-day photoperiodic responses originate from hatched areas, and those with short-day responses originate from stippled areas. The broken line shows the position of the inter-tropical convergence zone in July (from Roberts, 1991).

progress towards flowering (see Roberts and Summerfield, 1987; Summerfield *et al.*, 1991a). These rates cannot be measured directly but, as in measuring the rate of progress in space (speed) or the rate of enzyme reactions, they can be calculated by taking the reciprocal of the time taken to reach an endpoint – in this case the appearance of the first flower.

The main advantages of this approach are as follows:

1. Typically, the separate responses to both photoperiod and temperature become linear over wide ranges of conditions if phenology data are transformed to rates.

2. Whereas large, significant interactions occur between temperature and photoperiod responses if data are analysed as times to flower, these interactions often disappear if rates are used.

3. A consequence of (1) and (2) is that simple equations without interaction terms within certain well-defined limits can be developed to predict rates of progress towards flowering and therefore, indirectly, the times taken from sowing to flowering.

4. As there are no interaction terms in such equations it is possible to identify and measure the separate genetic control of photoperiod-sensitivity and of temperature-sensitivity.

5. The recognition that linearized rate equations may be applied to the data permits the application of the concept of thermal time (for photoperiod-insensitive genotypes) or an analogous concept of photothermal time (for photoperiod-sensitive genotypes). The use of these concepts allows prediction of times to flowering in environments which are not constant, e.g. in natural environments where temperature fluctuates and photoperiod changes systematically. It is not always realized that the concepts of thermal time and photothermal time are *only* legitimate if *rates* of development are *linearly* related to temperature and/or photoperiod.

6. As a consequence of (3) and (4), the genotypically controlled values of the parameters which describe sensitivity to photoperiod and temperature can, where appropriate models have been developed, be estimated from responses in only a few, controlled but carefully chosen, environments (in some cases, in theory, no more than four may be essential). As a consequence of (5), it is possible to use fluctuating field conditions, or to integrate the use of controlled and natural environments to estimate these parameters.

7. A consequence of (4) and (6) is that simple and economic techniques may prove feasible for screening large germplasm collections for sensitivity to photoperiod and temperature in a way which allows rational genetic analysis and predictions of time of flowering in a wide range of environments.

The experimental results and arguments which led to these conclusions are described in detail elsewhere (Roberts and Summerfield, 1987; Summerfield *et al.*, 1991b). Only the principal features of the photothermal models are summarized here, and are illustrated using data for various annual crops.

TEMPERATURE RESPONSES

Temperature may affect times from sowing to flowering in three distinct ways: there may be a specific cold-temperature hastening of flowering known as vernalization; otherwise over a wide range of temperatures the rate of progress towards flowering increases with increase in temperature to an optimum temperature at which flowering is most rapid; and at supra-optimal temperatures flowering is progressively delayed as temperatures get warmer (see Roberts and Summerfield, 1987).

In photoperiod-insensitive genotypes (of which there are relatively few in any of the major annual crops) and in photoperiod-sensitive ones in circumstances where photoperiod, P, does not interfere (see subsequent sections), the rate of progress towards flowering is a positive linear function of temperature from a base temperature, T_b, at which the rate is zero, up to an optimum temperature, T_o, at which it is maximal. Between these limits the relation may be described as:

$$1/f = a + bT \tag{1}$$

in which T is mean temperature and f is the time (days) from sowing to the first open flower in the dicotyledons (e.g. food legumes) or to well-defined stages of apical reproductive development in monocotyledons, e.g. double ridges in wheat, tassel initiation in maize (*Zea mays*), panicle initiation in rice or awn emergence in barley (*Hordeum vulgare*). The values of a and b are specific to the genotype. Above T_o the values of the constant will differ from those for the suboptimal range and the relation (and so the sign of the coefficient b) will be negative.

The value of the base temperature, at and below which there is no progress made towards flowering (i.e. the time taken to flower is infinite), is given by:

$$T_b = -a/b \tag{2}$$

This is because when $T = T_b$ then $1/f = 0$ and so, from equation 1, $a + bT_b = 0$. If equation 1 applies it follows that flowering occurs when the thermal time, Θ, appropriate for a given genotype has been accumulated. Thermal time is measured in day-degrees (°Cd) above the base temperature and is calculated on successive days by subtracting the base temperature from the mean daily temperature and adding each value to the subtotal accumulated since the seed was sown:

$$\theta = \sum_{i=1}^{f} (T_i - T_b) \tag{3}$$

where i is the ith day from sowing and T_i is the mean temperature for that day. It follows from equation 3 that in environments where the mean daily temperature (T_i) remains constant, and where the diurnally changing temperature values neither exceed T_o nor fall below T_b, the thermal time accumulated may be expressed as:

$$\theta = f(T - T_b) \tag{4}$$

Accordingly, if the time taken to flower is observed in any two different natural environments (for photoperiod-insensitive genotypes), and the mean daily temperature is recorded or, if this is not possible, estimated from [(daily max. + daily min.)/2], and T is calculated as:

$$\left(\sum_{i=1}^{f} T_i \right) / f$$

then the values of θ and T_b may be solved by simultaneous equations (equation 4). Alternatively, if the values of the constants a and b in equation 1 have been determined, then the thermal time may be calculated from the expression:

$$\theta = 1/b \tag{5}$$

Equation 1 may be visualized as a three-dimensional graph in which the rate of progress towards flowering is shown as a linear function of temperature and in which photoperiod has no effect. This produces what may be described as the thermal response plane which could be pictured as a single-pitch roof, as for example in soyabean cv. Fiskeby (Fig. 3.3a).

PHOTOPERIOD RESPONSES

Most if not all flowering plants, so far as we know, show the fundamental temperature response described above. But it may not always be detected in plants which respond to photoperiod given that the photoperiod-sensitivity genes are expressed in certain photoperiods and cause a delay in flowering in a simple and predictable way.

In environments where the daylengths are longer than the critical photoperiod in SDPs, or shorter than the critical photoperiod in LDPs, flowering is delayed according to the equation:

$$1/f = a' + b'\,T + c'\,P \tag{6}$$

in which P is the mean pre-flowering photoperiod (h day^{-1}) and a', b' and c' are genetically determined coefficients. This equation quantifies what can be described as the photothermal plane since both photoperiod and temperature determine its position. It intersects with the thermal plane at the critical photoperiod, i.e. along a line where the solutions to equations 1 and 6 produce the same value (see cv. Biloxi in Fig. 3.3d). Thus we define the critical photoperiod, P_c, as that photoperiod above which in SDP flowering is delayed and below which in LDP flowering is delayed.

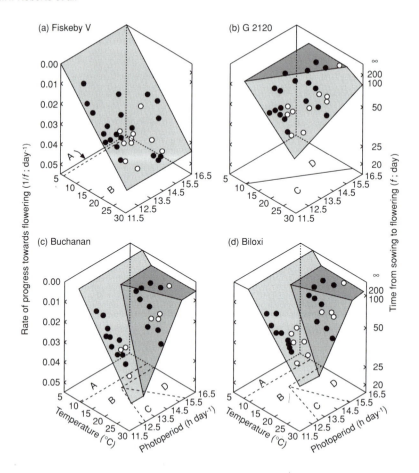

Fig. 3.3. The effects of photoperiod and mean temperature on rate of progress to flowering in four cultivars of soyabean, $1/f$ (left vertical axis); this is transformed to time to first flower, days (f) on the reciprocal scale on the right vertical axis. Cultivar Fiskeby V is typical of photoperiod-insensitive cultivars; cvs Biloxi and Buchanan are typical of strongly photoperiod-sensitive cultivars; whilst G2120 shows intermediate photoperiod sensitivity. The graphs were fitted to data derived from different sowing dates at six contrasting sites in Australia in 1986–88 (●), and from one site in Australia and two in Taiwan in 1989–90 (○). Where points lie above or below the calculated response surface, this is shown by a vertical line from the point to the surface. The intersections marking boundaries between the planes of the response surface are projected vertically to the base to show the environmental domains where: (A) the temperature is below the base temperature and so no progress to flowering is possible; (B) the rate of progress towards flowering is solely dependent on temperature; (C) both photoperiod and temperature independently affect rate of progress towards flowering; and (D) maximum photoperiodic delay occurs and variations in photoperiod and temperature no longer affect the rate. The boundary between domains (A) and (B) is the base temperature (T_b), that between (B) and (C) is the critical photoperiod (P_c), that between (C) and (D) is the ceiling photoperiod (P_{ce}), and that between (B) and (D) marks the temperature below which the photoperiodic response is not expressed (T_p) (from Summerfield *et al.*, 1993).

Since at P_c the right-hand side of equations 1 and 6 give the identical solution in terms of $1/f$ then in these environments:

$$(a + bT) - (a' + b'\,T + c'\,P) = 0$$

Under these conditions $P = P_c$ and so, rearranging,

$$P_c = [a - a' + T\,(b - b')]/c' \tag{7}$$

When $P < P_c$ in SDP the rate of development is limited by the underlying and fundamental temperature response and when $P > P_c$ then the rate of development is limited primarily by photoperiod, although in most cases, i.e. when $b' > 0$, temperature still affects the rate of development.

If the photoperiod is increased sufficiently in SDP it ultimately reaches a value beyond which there is no further delay in flowering. Similarly in LDP if daylength is decreased sufficiently there comes a point below which there is no further delay. In each case we have referred to this limit as the ceiling photoperiod, P_{ce}. Beyond this limit is the plane of maximum delay within which, in soyabean at least (the only species so far for which we have sufficient information), the rate of progress towards flowering is unaffected by either photoperiod or temperature (see cv. Biloxi in Fig. 3.3d). This plane may be simply defined therefore as:

$$1/f = d' \tag{8}$$

where d' is a genetically determined characteristic of photoperiod-sensitive genotypes. In those cases where $d' = 0$ then $f = \infty$, and so the photoperiodic response can then be described as obligate since photoperiods less than P_{ce} in SDP, or greater than P_{ce} in LDP are an obligate requirement if the plant is not to remain permanently vegetative. On the other hand if $d' > 0$ then f has some finite value and consequently, if given sufficient time, the plant will eventually flower in any daylength. Thus when $d' > 0$ the photoperiodic response may be described as quantitative since photoperiod is a determinant of the time taken to flower but cannot prevent flowering. The ceiling photoperiod, P_{ce}, is defined by a line where the solution to equation 6 is identical to that for equation 8. And accordingly through algebraic arguments similar to those used to define P_c it can be shown that:

$$P_{ce} = [d' - (a' + b'\,T)]/c' \tag{9}$$

Clearly if $b' > 0$, which it usually is, then the value of the ceiling photoperiod, P_{ce}, varies with temperature as can be clearly seen on Fig. 3.3. On the other hand the extent to which the critical photoperiod, P_c, is temperature-dependent depends on the extent to which b' differs from b. If they are identical the value of P_c is invariant.

When the geometry of the triple-plane model is examined (see Fig. 3.3) another boundary becomes apparent. This is the minimum temperature necessary for the expression of photoperiod-sensitivity (T_p): it is where the delay demanded by the photoperiod-sensitivity genes (equation 6) is no greater than that called for by the underlying temperature response. Again using similar algebraic arguments to those used to define P_c and P_{ce} it can be defined by the equation:

$$T_p = (d' - a)/b \tag{10}$$

The three intersecting planes described by equations 1, 6 and 8 which meet at the boundaries defined by equations 7, 9 and 10 form the triple-plane rate model of development which has now been shown to be applicable to a very wide range of species, e.g. cowpea (*Vigna unguiculata*) (Hadley *et al.*, 1983; Ellis *et al.*, 1994a; Craufurd *et al.*, 1997); soyabean (Hadley *et al.*, 1984; Summerfield *et al.*, 1993); mung bean (*Vigna radiata*) and related *Vigna* spp. (Imrie and Lawn, 1990; Ellis *et al.*, 1994b); Bambara groundnut (*Vigna subterranea*) (Linneman and Craufurd, 1994); common bean (*Phaseolus vulgaris*); faba bean (*Vicia faba*) (Ellis *et al.*, 1988a, 1990); pea (*Pisum sativum*) (Summerfield *et al.*, 1991); chickpea (*Cicer arietinum*) (Roberts *et al.*, 1985; Ellis *et al.*, 1994c); subterranean clover (*Trifolium subterraneum*) (Evans *et al.*, 1992); barley (Ellis *et al.*, 1988c, 1989); wheat (Loss *et al.*, 1990); rice (Summerfield *et al.*, 1992) and lentil (Summerfield *et al.*, 1985; Erskine *et al.*, 1990, 1994).

In some cases only one of the three planes of the model is required, e.g. only the thermal plane (equation 1) is needed when dealing with photoperiod-insensitive plants at suboptimal temperatures. In other cases in photoperiod-sensitive plants it may be that the range of relevant natural environments fall mostly within the photothermal plane (equation 6), as has been found when applying the model to lentil (Erskine *et al.*, 1990).

In some cases, however, the range of conditions which have to be considered spreads over two or three of the planes, and when dealing with multi-location sowing date trials there is a potential difficulty of deciding which observations near a boundary should be allocated to each plane on either side of it. However, the problem has been solved statistically and operationally by the development of an iterative computer procedure called RoDMod (Watkinson *et al.*, 1994). Examples of applying this to field data for soyabean are shown in Fig. 3.3.

Figure 3.3 demonstrates that the photothermal environment can be divided into different domains according to their influence on development. In two of these planes (the thermal and photothermal) there is ample evidence that the response to temperature is linear; whereas in the other (the plane of maximum delay) there is no response in soyabean. There is also considerable evidence that the rate of progress of a number of developmental processes is a linear function of temperature. The response has been examined in greatest detail for seed germination (Hegarty, 1973; Bierhuizen and Wagenwoort, 1974; Garcia-Huidobro *et al.*, 1982; Covell *et al.*, 1986; Ellis *et al.*, 1986). It is by no means clear why the response of developmental rate to temperature between the base and optimum values should be linear, even though there has been some speculation (Roberts *et al.*, 1988). It is even less clear why the rate of progress towards flowering should be a linear function of photoperiod between the critical and ceiling values. But it is a fortunate

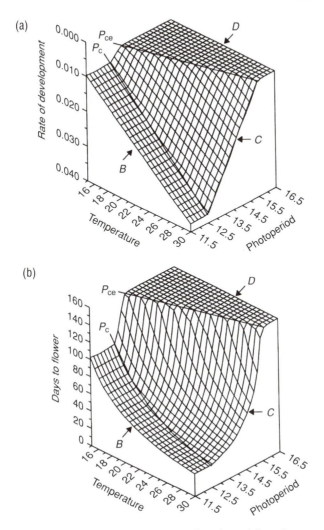

Fig. 3.4. A graphical comparison between (a) the photothermal flowering responses expressed in terms of rate (1/*f*), and (b) in terms of the time taken to flower (*f*). Response surfaces calculated from field data for soyabean cv. UFV 1. Note the complex interaction between temperature and photoperiod within the photothermal plane (C) which is evident in (b) but not in (a). (From Watkinson *et al.*,1994).

circumstance that response to both temperature and photoperiod within any domain are linear and without interaction since this simplifies the mathematics. If the raw data, i.e. times from sowing to first flower, are not transformed to rates then clearly powerful interactions occur – as indicated by the curvature of response surfaces in Fig. 3.4b. The transformation to rates removes the need for an interaction term and also has another statistical advantage: times to flower show non-homogeneous variance – the variance increasing with delay in flowering – a common observation of experimentalists. Transformation to rates removes this problem. Not only does the use of rates legitimize the statistical analysis of responses but it also takes the investigation nearer to the underlying physiology, for it is difficult to avoid the deduction that the time taken to reach an endpoint such as flowering depends on the rate of the processes leading to it. The converse would be illogical.

Although these analytical and theoretical advantages of using a model based on rates are important, the main virtue of the triple-plane rate model is that it involves relatively few coefficients (a, b, a', b', c', d'), all of which, and their derivatives, have clearly defined biological meaning (e.g. c' indicates relative photoperiod sensitivity; $-a/b$ is the base temperature) and, of paramount importance, the six coefficients are not affected by the environment but are genotypic characters which determine phenotypic response to the environment, and which can predict this response quantitatively.

LINKAGE BETWEEN THE MODEL AND THE ENVIRONMENT

The projection of the triple-plane photothermal model identifies environments into four domains of flowering response as shown in Fig. 3.3, namely no flowering (A), the thermal response (B), the photothermal response (C), and the domain of maximum delay to flowering which is insensitive to variation in either temperature or photoperiod (D). It is possible to subdivide further these domains by isochrones calculated from equations 1, 6 and 8 which indicate the time taken to flower in any combination of photoperiod and temperature if a plant were to be maintained continuously in those conditions (Fig. 3.5).

In the real world, however, the photothermal environment changes with the march of the seasons and this can be illustrated by a photothermograph (first devised by Fergusson, 1952) which can be accommodated on the isochrone projection described above. Our version of the photothermograph joins 12 points representing the mean temperature each month and the photoperiod on day 21 of that month (the latter device although not exactly the mean monthly photoperiod enables the photothermograph to include the longest and shortest day in the annual cycle).

Figure 3.5 includes four genotypes of soyabean and five natural environments which in combination can be used to illustrate the adaptation of flowering responses to different regions. First, the upper two diagrams compare the adaptation, or lack of it, *cf.* cvs Biloxi and Hill to three areas of the USA typified by Mobile (Alabama), Nashville (Tennessee) and Chicago (Illinois). In North America soyabean cultivars are classified into various maturity groups (MG) according to the geographical zones to which they are suited. Typically each zone extends over about 2–3° of latitude (about 220–330 km). There are ten of these zones from 00 in the north to VIII in the south (although other MGs adapted to environments beyond this range are

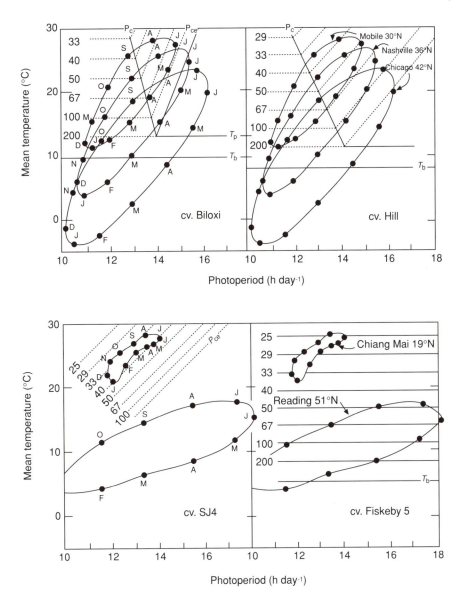

Fig. 3.5. Photothermal flowering response isochrones for four cultivars of soyabean on which photothermographs of various climates have been superimposed. The isochrones are shown as broken lines with figures showing the number of days from sowing to flowering assuming the cultivars were maintained continuously in the conditions indicated. Where there was sufficient information to calculate them, the main boundaries of the photothermal domains identified in Fig. 3.3 (and defined by equations 7, 9 and 10) are shown as solid straight lines, namely base temperature (T_b), the lower temperature limit for photoperiod-sensitivity expression (T_p), the critical photoperiod (P_c) and the ceiling photoperiod (P_{ce}). A query (?) after any symbol indicates it is an approximation. The photothermographs for the locations indicated are solid lines describing diagonal ovoids with letters indicating mid-month conditions, as described in the text (from Roberts *et al.*, 1993).

now in use for more extreme climates). Biloxi is classified as MG VIII and is suitable for climates represented here by the photothermograph for Mobile (30°N), whereas cv. Hill is classified as MG V which is suitable for climates represented here by the photothermograph for Nashville (36°N). For comparison, the photothermograph for Chicago (42°N), a climate suitable for cultivars in MG II, has also been superimposed on the photothermal response isochrones for both cvs Biloxi and Hill in Fig. 3.5.

Notice that if cv. Biloxi were sown in mid-May in Chicago the conditions at that time would be in the domain of the maximum photoperiodic delay – and if the crop were to be maintained in such conditions it would not flower for 200 days. These conditions, however, would only continue until almost mid-August. Thus after the first 3 months the crop would have progressed by about $3 \times 30/200$ days towards flowering, i.e. about 45% of the way. Conditions would then become less inhibitory until early September, but then more inhibitory again; but from mid-August to mid-September the average conditions would suggest about 70 days to flower of which 30 would be experienced, i.e. the crop would have progressed by another $30/70 = 43\%$ of the way towards flowering, and the accumulated progress by mid-September would be 88% of the way towards flowering. Continuing with similar arguments it would be seen that the crop might flower towards the end of September (if not frosted), but now too late in the season to ripen.

In contrast if cv. Biloxi were sown near Mobile in mid-May it will be seen that by mid-July it will have progressed about 70% of the way towards flowering and should flower soon afterwards. On the other hand, cv. Hill, which is adapted to about 36°N in North America, would flower prematurely in Mobile but too late in Chicago. But it does not follow that these cultivars would be well adapted to similar latitudes outside the USA because it is clear that temperature will alter the path of the photothermographs.

The lower two diagrams in Fig. 3.5 show the responses of two very different cultivars – cv. SJ4, which is adapted to an area around Chiang Mai in Thailand, and cv. Fiskeby V which was bred in Sweden for cool temperate conditions. As is typical for short-day species in general, only those in which photoperiod-sensitivity has been eliminated are suitable for the long summer days of the high latitudes; otherwise the crop would never flower. Consequently it is not surprising that, as the horizontal isochrones in Fig. 3.5 show, cv. Fiskeby V is completely photoperiod-insensitive. Following the photothermograph for Reading, UK (51°N) shows that it is possible to grow crops which flower early enough to mature in such climates (although, for other reasons, this and other similar soyabean cultivars do not crop sufficiently well in such conditions to be grown commercially). On the other hand, if cv. Fiskeby V were grown in Chiang Mai (19°N), then it would flower too soon to produce an adequate yield.

Although the information available for cv. SJ4 was not sufficient to complete its photothermal-response diagram (e.g. it is not possible to define the critical photoperiod), nevertheless there is sufficient information to determine the isochrones in the photothermal-sensitive plane. And so it can be seen that conditions in Chiang Mai are always such that flowering would occur after about 30–40 days – a long enough period to accumulate sufficient leaf canopy and photosynthate to produce an adequate yield. Furthermore, it is particularly interesting to see that providing water is available (e.g. by irrigation in the dry season), the cultivar would be well adapted to growing at any time of the year because the crop duration would be relatively stable, irrespective of the time of sowing; this is because the isochrones for cv. SJ4 more or less follow the diagonal trajectory of the photothermograph for Chiang Mai (which is typical of many other regions at this latitude). It is often argued that stability of crop duration depends on insensitivity to photoperiod; but in fact photoperiod-insensitivity (horizontal isochrones as for cv. Fiskeby) would result in considerable variation in time to flower according to whether the crop were sown in the wet or dry season. For maximum stability it can be seen that some sensitivity to photoperiod provides a compensating mechanism to allow for the universal sensitivity to temperature.

THE USE OF THE PHOTOTHERMAL COEFFICIENTS IN GENETIC ANALYSIS

Before moving on to discuss the genetical implications, a limitation to the model should be recognized. As it has been described the model implies that photoperiod sensitivity is expressed throughout the period from sowing to first flower. This is not true. There is an initial pre-inductive phase (sometimes referred to as the juvenile phase) when plants of most species are insensitive to photoperiod. Likewise there is often a period immediately before flowering by which time the onset of flowering has been determined and so far as the timing of the appearance of the first flower is concerned, the plant is once again photoperiod-insensitive. This is the post-inductive phase. The photoperiod-sensitive inductive phase is sandwiched between these two.

The length of these three phases can be accurately estimated by observing the time taken to flower in reciprocal-transfer experiments in which samples of plants are successively transferred from long days to short days and other samples are successively transferred from short days to long days. Graphical methods of analysis (Kiniry et al., 1983; Roberts et al., 1988a), or a combination of such approaches with regression analysis of partial datasets (Wilkerson et al., 1989), have been superseded by a more rigorous and holistical method of analysis (Ellis et al., 1992; Collinson et al., 1992, 1993). It is the duration of the photoperiod-insensitive 'pre-inductive' or 'juvenile' phase which is of greatest concern: it may be relatively short, 8–10 days as in spring barley (Roberts et al., 1988a), or relatively long, and at least as long as 21 days in some cultivars of rice (Collinson et al., 1992).

Although it is possible to investigate the genetics of photoperiodism without a knowledge of the quantitative nature of the photoperiod and temperature responses outlined here, the results could be – and in some cases have been – misleading. For example, if photoperiod sensitivity were to be assessed based on the difference in time to flower in two contrasting daylengths, the value obtained would depend not only on temperature, as Fig. 3.4 or Fig. 3.5 makes evident, but also on whether the two photoperiods selected are within environmental domain B or one observation is in domain A or C. Even more problems of interpretation can arise, as they often did in earlier work, if comparisons are made on the basis of two photoperiodic regimes in which the number of plants which have flowered after an arbitrary time are recorded, rather than measuring the time taken for each sample to flower from which rates can be derived. Furthermore, since temperature as well as photoperiod affects the flowering responses, it is important to compare responses at different temperatures to determine whether or not a gene which influences photoperiod sensitivity also affects the response, or is affected by the response, to temperature.

In the soyabean genome five so-called maturity loci have been identified with two alternative alleles at each locus, namely E_1/e_1, E_2/e_2, E_3/e_3, E_4/e_4 and E_5/e_5. Upadhyay *et al.* (1994a, b) have investigated the effects of all eight possible isolines of three of these loci E_1/e_1, E_2/e_2 and E_3/e_3 in a cv. Clark background (cv. Clark has the following complement of maturity genes: e_1 E_2 E_3 E_4 e_5). It was shown that none of the genes affects temperature sensitivity (b or b') and that the major effect of the dominant alleles is on photoperiod sensitivity, i.e. on c', although additional (pleiotropic?) effects can also be recognized. There were three categories of increasing sensitivity to photoperiod: (i) least sensitivity (but nevertheless still slightly sensitive, probably due to the presence of E_4 in the cv. Clark background) represented by e_1 e_2 e_3, e_1 E_2 e_3 and e_1 e_2 E_3; (ii) intermediate sensitivity represented by E_1 e_2 e_3 and e_1 E_2 E_3; and (iii) most sensitivity represented by E_1 E_2 e_3, E_1 e_2 E_3 and E_1 E_2 E_3. Thus photoperiod sensitivity is dominant. Increase in sensitivity never accelerates flowering but delays it in any photoperiod greater than the critical photoperiod. E_1 has the greatest effect. While neither E_2 nor E_3 have any effect independently, together they show epistasis and produce roughly the same effect as E_1 on its own. Furthermore, either E_2 or E_3 can enhance the effect of E_1 (Fig. 3.6).

With the increase in photoperiod sensitivity induced by these gene combinations, there is also a concomitant decrease in the value of d', i.e. an increase in the time taken to flower in unfavourable photoperiods greater than the ceiling value. Figure 3.6 shows the negative linear correlation between c' and d' and that the maximum time taken to flower (at a mean temperature of 25°C) in these isolines varies from about 50 days in the least sensitive genotypes to about 100 days in the most sensitive one.

All these effects of the *E*-maturity genes which result in different phenotypic responses in different environments are genetically characterized mostly by the correlated values of the c' and d' coefficients although a' is also affected. There is in addition a correlated effect in which, paradoxically, the more sensitive genotypes exhibit longer phases of photoperiod insensitivity; e.g. the pre-inductive phase increased from 3 to 10 days and the post-

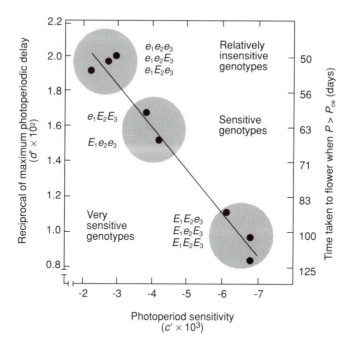

Fig. 3.6. The relation between photoperiod sensitivity (c') and the reciprocal of the maximum time taken to flower (d') among eight maturity isolines of soyabean in the genetic background of cv. Clark. The genotypes fall into three main groups of photoperiod sensitivity (stippled) (data from Upadhyay *et al.*, 1994a).

inductive phase increased from 12 to 18 days with increase in photoperiod sensitivity of the gene combinations (Upadhyay *et al.*, 1994b).

We believe, however, that the most important effects of the genes for practical applications reside in the values of the c' and d' coefficients and, since the values are correlated (see Fig. 3.6), it might well prove adequate when screening germplasm to concentrate on c', thus reducing the number of environments required.

Although the model described here was developed from research in controlled environments, there is now considerable evidence that it can be applied to a very wide range of natural environments in several species. Multi-locational trials augmented by successional sowings and, if considered necessary, supplementary illumination in the field to increase daylength, can be used to estimate the values of the model coefficients: (i) to characterize germplasm collections and so predict flowering behaviour elsewhere; (ii) for interpreting and understanding crop adaptation (Erskine *et al.*, 1990; Roberts *et al.*, 1993; Lawn and Imrie, 1994); and (iii) for genetic analysis of photoperiod sensitivity (Upadhyay *et al.*, 1994a, b). We do not yet know whether the model has any contribution to make to the understanding of the biochemical mechanisms of photoperiod and temperature responses. But at the very least, it should provide the basis for indicating the most appropriate environmental conditions, genotypes and physiological stage of the plants for such investigations.

VERNALIZATION IN ANNUAL CROPS

In the responses discussed earlier we have only touched upon the special effect of cool temperatures known as vernalization which is common in species which are adapted to temperate and Mediterranean climates. This response may be found in winter annuals, biennials and perennials and it is particularly important in biennials which are common in the vegetable cuisine of temperate latitudes.

Although this book is concerned with vegetables we first need to consider temperate cereals since it was in these species that vernalization was first recognized and investigated. In fact it was recognized long before the name vernalization was assigned to the phenomenon: in the middle of the last century it was known that winter wheat could be made to behave like spring wheat by exposing the germinating seeds to cool temperatures (Evans, 1969). Gassner (1918) was the first to make a systematic analysis of the response and found that the spring variety of Petkus rye (*Secale cereale*) had no chilling requirement for flowering, but the winter variety did. His work was extended by a number of Russian scientists, including L.S. Lysenko (the proponent of the inheritance of acquired characteristics) who first coined the term 'Jarovizacija' which means 'to make like spring', the Anglicized equivalent of which is vernalization. In the early 1930s Lysenko was probably the first to realize that the cold treatment needs to be followed by long days if flowering is to occur sufficiently early.

Much of our understanding of vernalization derives from the work, started in 1931, of Gregory and Purvis who also worked on Petkus rye. Some of their main findings may be summarized as follows. The spring variety of Petkus rye is a typical quantitative LDP. Under long days the apex becomes a reproductive structure after about seven leaves have been initiated; but under short days (10 h) only after about 12 leaves have been formed. The winter variety, when germinated under warm temperatures (e.g. 18°C), produces 22 leaves before flowering, irrespective of daylength. But if the germinating seed is held close to 1°C for several weeks it behaves like the spring variety. The effectiveness of the chilling treatment was proportional to the period for which it was applied – the half-maximum effect occuring after about 15–20 days. Vernalization could be reversed (devernalization) if germinating seeds were exposed to 25–40°C for about 4 days. The optimum temperature for vernalization was reported to be between 1 and 7°C, though some response was shown from −4 to 15°C. Where there is sufficiently reliable information, subsequent investigations have usually confirmed that these temperatures are typical for species which respond to vernalization. However, it may well be that earlier work has concluded that the optimum temperature is a little higher than it really is: the possibility of confounding vernalization with subsequent positive responses to temperature according to equations 1 or 6 has seldom been taken into account.

In addition to rye, all the other temperate cereals – wheat, barley, oats (*Avena sativa*) – have two categories of cultivars – winter (i.e. autumn sown) and spring (spring sown). In barley it has been shown that vernalization advances flowering in winter cultivars by advancing ear initiation and that before initiation the plants are insensitive to photoperiod. The photoperiod-insensitive phase which ends at ear initiation is relatively short in spring cultivars (typically about 10 days at 15°C) and is not affected by vernalization. It is about the same length in winter cultivars providing the seeds have been fully vernalized but, if not, this pre-inductive period may be 32 days or more. In some cases short days may partially substitute for cold-temperature vernalization. Once the ears have been initiated barley becomes photoperiod-sensitive and the subsequent rate of progress to flowering is increased with increase in daylength (Roberts *et al.*, 1988a).

In spring cultivars or in winter cultivars which have been vernalized, barley conforms to the triple-plane rate model in its responses to photoperiod and temperature (Roberts *et al.*, 1988a; Ellis *et al.*, 1988c). It is possible to modify the triple-plane model to incorporate vernalization using two additional equations to quantify the vernalization response (Ellis *et al.*, 1989). The first of these defines the time taken to saturate the vernalization response at different temperatures:

$$S = a_s + b_s T_v \qquad (11)$$

where S is the time (days) taken to saturate the vernalization response at a vernalization temperature T_v, a_s is the time (days) required to saturate the vernalization response at 0°C, and b_s is the time required per 1°C increase in vernalization temperature; a_s and b_s are genotypic constants.

The second equation defines the effect of vernalization temperature on the time taken to flower:

$$d_v = a_v - b_v T_v^2 \qquad (12)$$

where d_v is the reduction in time to flower (days) at vernalization temperature T_v and a_v and b_v are genotypic constants. The upper temperature limit for vernalization is given by $\sqrt{(a_v/b_v)}$.

The combined model (i.e. a combination of equations 1, 6, 8, 11 and 12) was tested on barley cv. Gerbel when seeds had been treated at 1, 5 or 9°C for 10, 30 or 60 days in photoperiods of either 8 or 16 h day^{-1} before transferring seedlings to factorial combinations of two photoperiods, 11 or 16 h day^{-1}, and two mean temperatures, 12 or 19°C; i.e. there were a total of 72 treatments. The agreement of the actual times to flowering (awn emergence) compared with those predicted by the model were very satisfactory (r^2 = 0.807) (Ellis *et al.*, 1989).

It is difficult if not impossible to quantify the vernalization response in such cases without detailed controlled environment experiments and adequate methods of analysis because of the way it interacts with the subsequent photothermal flowering responses. In particular it is important to recognize that any temperature above the base temperature and below, say, 15°C could be contributing to both vernalization (in which cooler temperatures are generally more effective) and the universal effect of temperature on rate of development which increase with increases in temperature over the range T_b to T_o. Thus in order to demonstrate a true vernalization response it is not sufficient to show that a preceding cool period decreases the subsequent time required at a warmer temperature for flowering to occur. When proper account has been taken of the entire photothermal experience it has been shown, for example, that some previous reports of vernalization responses in chickpea are unfounded (Summerfield *et al.*, 1989) and this is also the case for lentil (Roberts *et al.*, 1988b). It is also doubtful whether many cultivars of faba bean show true vernalization responses if those responses are adequately analysed (Ellis *et al.*, 1988a, b).

VERNALIZATION IN BIENNIAL CROPS

All the cases of vernalization discussed so far relate to winter annual crops but many temperate crops in which vernalization is of much greater significance are biennials. The biological strategy of such species is to store photosynthate in the first season's growth and use it for flowering and fruit production in the second season – except that when cropped, man intervenes to harvest the storage organs before their contents are dissipated in sexual reproduction. Almost any organ, or part thereof, has been adopted in different species as the preferred repository for photosynthate: e.g. hypocotyl combined with root – carrots (*Daucus carota*), radish (*Raphanus sativus*), parsnip (*Pastinaca sativa*), beetroot (*Beta vulgaris*); stem tuber – potatoes (*Solanum tuberosum*); petioles – chard (*Beta vulgaris* var. *cicla*), celery (*Apium graveolens*); leaves – cabbage (*Brassica oleracea*), lettuce (*Lactuca sativa*), spinach (*Spinacea oleracea*); modified leaves – onions (*Allium cepa*); modified buds – Brussels sprouts (*Brassica oleracea* var. *gemmifera*); modified stems – kohlrabi (*Brassica oleracea* var. *caulo-rapa*); and arrested inflorescences – cauliflower (*Brassica oleracea* var. *botrytis*). (See Fig. 15.1 for variation within the brassicas.)

Temperate vegetables such as these often tend to bolt, i.e. elongate their stems and then flower in response to cool temperatures followed by long days. (In cauliflower and broccoli, because of the nature of the storage organ, flowering does not involve vegetative stem elongation, but there is an elongation of peduncles and flower buds instead.) Such vernalization responses provide an added complication to simple photothermal models of the rate of progress to flowering in these vegetables. For example, floral initiation at the apex of carrot precedes visible internode appearance and subsequent flower appearance. However, the response to temperature of the rate of elongation of the seedstalk is different to that of floral initiation (Hiller and Kelly, 1979). Nevertheless, it appears that this discrepancy may be small for the purpose of predictive models of bolting in carrot (for example, Craigon *et al.*, 1990). Attempts to model visible spathe appearance in onion have considered the effects of environment separately on each of the three phases before inflorescence appearance: that is, floral initiation at the apex, spathe formation, and flower stalk elongation. Floral initiation at the apex is affected by both temperature and photoperiod (Brewster, 1983) and so appears to conform to a photothermal plane relation with a long-day response (equation 6). Spathe appearance is then solely a function of subsequent temperature (Brewster, 1987; see Chapter 18), but the rate of flower stalk elongation is affected by both subsequent photoperiod and temperature (Brewster 1983, 1987). Most studies of vegetable flowering (in those species which bolt) have recorded the appearance of the elongating seedstalk or the appearance of the inflorescence. This duration, then, comprises both developmental changes at the apex and growth processes during seedstalk elongation. Thus, simple responses between the rate of progress to flower appearance and environment may be difficult to quantify. The relative importance of cool temperatures and long days varies amongst species but the ecological principle is clear: the double stimulus ensures that plants do not flower until the benign conditions for seed production occur in the spring following the preceding season of vegetative growth.

However, selection for many modern vegetable cultivars has reduced the expression of the photoperiod-sensitivity genes. From a geographical centre of origin of the western Mediterranean region (Heywood, 1983), the European-type of carrot is expected to show a long-day response of flowering to photoperiod. Nevertheless, some modern cultivars have been reported to be day neutral (El Sayed Sakr and Thompson, 1942; Hiller and Kelly, 1979). Furthermore, it has been suggested that the response of flowering to photoperiod in carrot refutes the hypothesis that all biennials require long days to flower (Hiller and Kelly, 1979). Nevertheless, Atherton *et al.* (1984) clearly demonstrated that shorter photoperiods do delay the time of flowering in some cultivars of carrot, albeit at much shorter photoperiods than normally experienced: fully vernalized plants subsequently grown at 16-h photoperiods flowered within 45 days compared with those grown at 8-h photoperiods which were still vegetative after 140 days. Thus, selection for suitable cultivars for northern-latitude root crops seems to have favoured a much reduced sensitivity of flowering to photoperiod in carrot.

The *Brassica oleracea* complex, from which modern cultivars of cauliflower and broccoli are derived, is of polyphyletic origins in the eastern Mediterranean and western European regions (Gray, 1989) and so a long-day reponse to photoperiod is also expected. Most European winter cultivars are biennial but annual summer varieties have been developed. Any sensitivity of the rate of curd initiation to photoperiod seems to have been bred out of modern cultivars, particularly summer varieties, although the few studies in which the effects of differences in photoperiod on curd formation have been investigated have only recorded the number of flowering plants after a set time interval rather than the rate of progress to flowering (e.g. Sadik, 1967).

In general, the rate of progress towards flowering of most biennial vegetable species is increased if cold temperatures are experienced early in growth, and the term vernalization is used to describe such observations. Early evidence for a vernalization requirement in winter cauliflowers was provided by Sadik (1967) who found that plants kept at 8°C for 6 weeks initiated curds after returning to a glasshouse whereas those plants kept continually in the glasshouse did not initiate curds. The temperature response of vernalization in cauliflower has now been quantified in a manner analogous to the thermal plane for the rate of progress to flowering (equation 1): values of T_o, T_b and T_{ce} (the ceiling temperature above which the rate of vernalization is zero) for an early summer cultivar were 5.5, 21.3 and 24.5°C, respectively (Atherton *et al.*, 1984), and for a summer/autumn cultivar they were 9–9.5, 9 and 21°C, respectively (Wurr *et al.*, 1993). Similar responses were derived for data from many cultivars of onion with a range of T_o of 7–12°C (Brewster, 1987) and leek (*Allium porrum* L.) with T_o of 5°C (Wiebe, 1994). The various vernalization functions which have been applied to quantify low-temperature vernalization have a range of forms, from the inverted triangle more usually fitted to the rate of progress to flowering (leek), to a 'flat-top' function with a range of temperatures over which the rate of vernalization is maximal (onion), to a function with no suboptimal range for vernalization (cauliflower) (Wurr *et al.*, 1993). Nevertheless, all seem to support the use of simple functions between temperature and the rate of vernalization. However, because of the relatively low values of T_o, the fitting of such functions usually requires the effects of supra-optimal temperatures to be accounted for. Analytical techniques to incorporate the effects of supra-optimal temperatures on the rate of vernalization or progress to flowering (i.e. equation 1 in which $T_b < T < T_o$) have been developed (Craigon *et al.*, 1990; Pearson *et al.*, 1993; Wheeler *et al.*, 1995). These approaches calculate the suboptimal equivalent of any supra-optimal temperature and combine the results with the suboptimal range of observations to fit a single linear function.

Early work on a possible vernalization response in carrot identified an optimum temperature of about 5°C for subsequent bolting (El Sayed Sakr and Thompson, 1942; Hiller and Kelly, 1979). Atherton *et al.* (1990) refined this estimate to 6.5 ± 1°C by using linear relations between temperature and the rate of progress to either internode appearance or flower appearance following a vernalization treatment. As the cardinal temperatures of this function were the same following different durations of vernalization, they were able to propose a thermal time-based model of the effects of vernalization on bolting in carrot. Such a model, therefore, can account for the effects of fluctuating temperature regimes in a similar manner to the thermal plane flowering models. Model predictions compared well with the time of bolting of carrot crops grown in a subsequent field trial (Craigon *et al.*, 1990).

In a few of the vegetable cultivars which do show a long-day response to photoperiod, the rate of progress to bolting was also affected by differences in photoperiod during the vernalization treatment. An increase in photoperiod during chilling increased the number of days to visible internodes in carrot (Atherton *et al.*, 1984) and to visible flower formation in leeks (Wiebe, 1994). These observations provide evidence for 'short-day vernalization', and hence are similar to those reported in barley (Roberts *et al.*, 1988b).

The analysis of Pearson *et al.* (1994) provides an alternative explanation for the potential vernalization effects observed in cauliflower. They derived a single temperature response of the rate of curd initiation from transplanting without assuming a vernalization requirement or a juvenile period (although the seedlings may have progressed beyond any potential juvenile period at the time of transplanting) for a summer/autumn cultivar of cauliflower. The estimates of T_o, T_b, and T_{ce} obtained were 14, 2.8 and 25°C, respectively (Pearson *et al.*, 1994). Such comparatively cool estimates of T_o for the entire period from transplanting to curd initiation were confirmed in a different cultivar of cauliflower by Wheeler *et al.* (1995) who reported a value of 15.5°C. Thus, an alternative interpretation of the results of Sadik (1967) is possible: the temperature of the glasshouse in which the control plants did not flower (day temperatures of 24–35°C) may have been supraoptimal for the rate of curd initiation and hence less inductive than that of the putative vernalization treatment – in which case vernalization need not be involved to explain the results. For the reasons discussed earlier in other crops, caution may also be necessary in interpreting the potential vernalization response of biennial vegetables. Only by quantifying the entire photothermal response of the rate of progress to flowering will a true vernalization response be exposed and quantified.

For the typical vernalization response to be effective it is ecologically important that, unlike winter annuals, the biennial crops should not become very sensitive to vernalization in the early part of vegetative growth, otherwise plants would bolt in the first season thus diverting photosynthate to sexual reproduction and denying the development of the economically important vegetative storage organ. So although there is sometimes evidence of some vernalization in the seed – even while on the mother plant, e.g. in some cultivars of chicory and lettuce (Wiebe, 1989) – it is nevertheless true that, in general, biennials do not respond significantly to vernalization until a later stage of development, i.e. after what is sometimes called a juvenile phase (see Chapter 15 for a further discussion on these responses in brassicas).

The few studies on the response of flowering to the environment in carrot provide evidence for a juvenility requirement in this crop. Days to internode appearance (bolting) and flower appearance were progressively reduced as older roots were vernalized (Atherton *et al.*, 1990). From those results, the end of juvenility was defined to occur when between six and eight leaves had been initiated at the apex and this coincided with the accumulation of 756°C days (above a base temperature of 0°C). This contradicts a previous report that imbibed seeds of carrot could be vernalized (Vince-Prue, 1975). Seedlings of cauliflower of less than 6 weeks old did not form a curd, within the time of the experiment, even after chilling (Sadik, 1967). The duration of this juvenile period was associated with the production of a set number of leaves, a criterion which has been refined and commonly adopted more recently (for example, Wiebe, 1972; Hand and Atherton, 1987). Alternatively, others have associated a minimum plant dry mass to coincide with the end of juvenility (for example, Brewster, 1987; Wiebe, 1994).

Some of the flowering responses to temperature and photoperiod outlined above appear complex, especially in vegetables with a biennial habit. Nevertheless, there is sufficient evidence, especially from the research on annual agricultural crops, to suggest we can be optimistic about the resolution in simple terms of the photothermal responses of vegetables – providing research on these topics is encouraged and funded.

ACKNOWLEDGEMENTS

We thank the Overseas Development Administration of the UK and Foreign and Commonwealth Office for financial support, several teams of research collaborators in the warmer regions, and the following technical and engineering colleagues: K.E. Chivers, S.D. Gill, A. Pilgrim and C.J. Hadley.

REFERENCES

Aitken, Y. (1974) *Flowering Time, Climate and Genotype.* Melbourne University Press, Melbourne.

Atherton, J.G. (ed.) (1987) *Manipulation of Flowering.* Butterworths, London.

Atherton, J.G., Basher, E.A. and Brewster, J.L. (1984) The effects of photoperiod on flowering in carrot. *Journal of Horticultural Science* 59, 213–215.

Atherton, J.G., Craigon, J. and Basher, E.A. (1990) Flowering and bolting in carrot. I. Juvenility, cardinal temperatures and thermal times for vernalization. *Journal of Horticultural Science* 65, 423–429.

Atherton, J.G. and Harris, G.P. (1986) Flowering. In: Atherton, J.G. and Rudich, J. (eds) *The Tomato Crop.* Chapman & Hall, London, pp. 167–200.

Bernier, G., Havelange, A., Houssa, C., Petitjean, A. and Lejeune, P. (1993) Physiological signals that induce flowering. *The Plant Cell* 5, 1147–1155.

Bierhuizen, J.F. and Wagenwoort, W.A. (1974) Some aspects of seed germination in vegetables. I. The determination and application of heat sums and minimum temperature for germination. *Scientia Horticulturae* 2, 213–219.

Brewster, J.L. (1983) Effects of photoperiod, nitrogen nutrition and temperature on inflorescence initiation and development in onion (*Allium cepa* L.). *Annals of Botany* 51, 429–440.

Brewster, J.L. (1987) Vernalization in the onion – a quantitative approach. In Atherton, J.G. (ed.) *Manipulation of Flowering.* Butterworths, London. pp. 171–183.

Chassan, R. (1994) A time to flower. *The Plant Cell* 6, 1–3.

Collinson, S.T., Summerfield, R.J., Ellis, R.H. and Roberts, E.H. (1992) Durations of the photoperiod-sensitive and photoperiod-insensitive phases of development to flowering in four cultivars of rice (*Oryza sativa* L.). *Annals of Botany* 70, 339–346.

Collinson, S.T., Summerfield, R.J., Ellis, R.H. and Roberts, E.H. (1993) Durations of the photoperiod-sensitive and photoperiod-insensitive phases of development to flowering in four cultivars of soyabean [*Glycine max* (L.) Merrill]. *Annals of Botany* 71, 389–394.

Covell, S., Ellis, R.H., Roberts, E.H. and Summerfield, R.J. (1986) The influence of temperature on seed germination rate in grain legumes. I. A comparison of chickpea, lentil, soyabean and cowpea at constant temperatures. *Journal of Experimental Botany* 37, 705–715.

Craigon, J., Atherton, J.G. and Basher, E.A. (1990) Flowering and bolting in carrot. II. Prediction in growth room, glasshouses and field environments. *Journal of Horticultural Science* 65, 547–554.

Craufurd, P.Q., Summerfield, R.J., Ellis, R.H. and Roberts, E.H. (1997) Photoperiod, temperature and the growth and development of cowpea (*Vigna unguiculata*). In: *Proceedings of the World Cowpea Research Conference. II.* Accra, Ghana (in press).

Curtis, B.C. (1988) The potential for expanding wheat production in marginal tropical environments. In: Klatt, A.R. (ed.) *Wheat Production Constraints in Tropical Environments.* CIMMYT, Mexico, pp. 3–11.

El Sayed Sakr and Thompson, H.C. (1942) Effect of temperature and photoperiod on seedstalk development in carrots. *Proceedings of the American Society for Horticultural Science* 41, 343–346.

Ellis, R.H., Covell, S., Roberts, E.H. and Summerfield, R.J. (1986) The influence of temperature on seed germination rate in grain legumes. II. Intraspecific variation in chickpea (*Cicer arietinum* L.) at constant temperatures. *Journal of Experimental Botany* 37, 1503–1515.

Ellis, R.H., Summerfield, R.J. and Roberts, E.H. (1988a) Effects of temperature, photoperiod and seed vernalization on flowering in faba bean (*Vicia faba*). *Annals of Botany* 62, 17–27.

Ellis, R.H., Roberts, E.H. and Summerfield, R.J. (1988b) Photothermal time for flowering in faba bean *Vicia faba* and the analysis of potential vernalization responses. *Annals of Botany* 61, 73–82.

Ellis, R.H., Roberts, E.H., Summerfield, R.J. and Cooper, J.P. (1988c) Environmental control of flowering in barley (*Hordeum vulgare*). II. Rate of development as a function of temperature and photoperiod and its modification by low-temperature vernalization. *Annals of Botany* 62, 145–158.

Ellis, R.H., Summerfield, R.J., Roberts, E.H. and Cooper, J.P. (1989) Environmental control of flowering in barley (*Hordeum vulgare*). III. Analysis of potential vernalization responses, and methods of screening germplasm for sensitivity to photoperiod and temperature. *Annals of Botany* 63, 687–704.

Ellis, R.H., Summerfield, R.J. and Roberts, E.H. (1990) Flowering in faba bean: genotypic differences in photoperiod sensitivity, similarities in temperature sensitivity, and implications for screening germplasm. *Annals of Botany* 65, 129–138.

Ellis, R.H., Collinson, S.T., Hudson, D. and Patefield, W.M. (1992) The analysis of reciprocal transfer experiments to estimate the durations of the photoperiod-sensitive and photoperiod-insensitive phases of plant development: an example in soyabean. *Annals of Botany* 70, 87–92.

Ellis, R.H., Lawn, R.J., Summerfield, R.J., Qi, A., Roberts, E.H., Chay, P.M., Brouwer, J.B., Rose, J.L. and Yeates, S.J. (1994a) Towards the reliable prediction of time to flowering in six annual crops. III. Cowpea (*Vigna unguiculata*). *Experimental Agriculture* 30, 17–29.

Ellis, R.H., Lawn, R.J., Summerfield, R.J., Qi, A., Roberts, E.H., Chay, P.M., Brouwer, J.B., Rose, J.L., Yeates, S.L. and Sandover, S. (1994b) Towards the reliable prediction of time to flower in six annual crops. IV. Cultivated and wild mung bean. *Experimental Agriculture* 30, 31–43.

Ellis, R.H., Lawn, R.J., Summerfield, R.J., Qi, A., Roberts, E.H., Chay, P.M., Brouwer, J.B., Rose, J.L., Yeates, S.J. and Sandover, S. (1994c) Towards the reliable prediction of time to flowering in six annual crops. V. Chickpea (*Cicer arietinum*). *Experimental Agriculture* 30, 271–282.

Erskine, W., Ellis, R.H., Summerfield, R.J., Roberts, E.H. and Hussain, A. (1990) Characterization of responses to temperature and photoperiod for time to flowering in a world lentil collection. *Theoretical and Applied Genetics* 80, 193–199.

Erskine, W., Hussain, A., Tahir, M., Bahksh, A., Ellis, R.H., Summerfield, R.J. and Roberts, E.H. (1994) Field evaluation of a model of photothermal flowering responses in a world lentil collection. *Theoretical and Applied Genetics* 88, 423–428.

Evans, L.T. (ed.) (1969) *The Induction of Flowering.* Macmillan, Melbourne, Australia.

Evans, L.T. (1975) *Daylength and the Flowering of Plants.* W.A. Benjamin, Menlo Park, California, USA.

Evans, L.T. (1993) The physiology of flower induction – paradigms lost and paradigms regained. *Australian Journal of Plant Physiology* 20, 655–660.

Evans, P.M., Lawn, R.J. and Watkinson, J.R. (1992) Use of linear models to predict flowering in subterranean clover (*Trifolium subterraneum* L.). *Australian Journal of Agricultural Research* 43, 1547–1558.

Fergusson, J.H.A. (1952) Photothermographs: a tool for climate studies in relation to the ecology of vegetable varieties. *Euphytica* 6, 97–105.

Garcia-Huidobro, J., Monteith, J.L. and Squire G.R. (1982) Time, temperature, and germination of pearl millet (*Pennisetum typhoides* S & H). I. Constant temperature. *Journal of Experimental Botany* 33, 288–296.

Gassner, G. (1918) Beiträge zur physiologischen Charakteristik sommer- und winterannueller Gewächse, ins besondere der Getreidepflanzen. *Zeitschrift für Botanik* 10, 417–430.

Gray, A.R. (1989) Taxonomy and evolution of broccolis and cauliflower. *Baileya* 23, 28–46.

Hadley, P.H., Roberts, E.H., Summerfield, R.J. and Minchin, F.R. (1983) A quantitative model of reproductive development in cowpea (*Vigna unguiculata* (L.) Walp.) in relation to photoperiod and temperature, and implications for screening germplasm. *Annals of Botany* 51, 531–543.

Hadley, P., Roberts, E.H., Summerfield, R.J. and Minchin F.R. (1984) Effects of temperature and photoperiod on flowering in soyabean [*Glycine max* (L.) Merrill.]: a quantitative model. *Annals of Botany* 53, 669–681.

Hamblin, J. (1994) Can resource capture principles assist plant breeders or are they too theoretical? In: Monteith, J.L., Scott, R.K. and Unsworth, M.H. (eds) *Resource Capture by Crops.* Nottingham University Press, Nottingham, pp. 211–232.

Hand, D.J. and Atherton, J.G. (1987) Curd initiation in the cauliflower. I. Juvenility. *Journal of Experimental Botany* 38, 2050–2058.

Harris, D.R. (1969) Agricultural systems, ecosystems and the origins of agriculture.

In: Ucko, P.J. and Dimbleby, G.W. (eds) *The Domestication of Plants and Animals.* Duckworth, London, pp. 3–15.

Hawkes, J.G. (1983) *The Diversity of Crop Plants.* Harvard University Press, London.

Hegarty, T.W. (1973) Temperature relations of germination. In: Heydecker, W. (ed.) *Seed Ecology.* Butterworths, London, pp. 411–432.

Heywood, V.H. (1983) Relationships and evolution in the *Daucus carota* complex. *Israel Journal of Botany* 32, 51–65.

Hiller, L.K. and Kelly, W.C. (1979) The effects of post-vernalization temperature on seedstalk elongation and flowering in carrots. *Journal of the American Society for Horticultural Science* 104, 253–257.

Imrie, B.C. and Lawn, R.J. (1990) Time to flowering of mung bean (*Vigna radiata*) genotypes and their hybrids in response to photoperiod and temperature. *Experimental Agriculture* 26, 307–318.

Kinet, J., Sachs, R.M. and Bernier, G. (1985) *The Physiology of Flowering,* Vol. 3, *The Development of Flowers.* CRC Press, Baton Rouge, Florida, USA.

Kiniry, J.R., Ritchie, J.T., Musser, R.L., Flint, E.P. and Iwig, W.C. (1983) The photoperiod sensitive period in maize. *Agronomy Journal* 75, 687–690.

Lawn, R.J. and Imrie, B.C. (1994) Exploiting physiological understanding in crop improvement. In: *Proceedings of the 10th Australian Plant Breeding Conference,* Gold Coast, Queensland, p. 23.

Lawn, R.J., Summerfield, R.J., Ellis, R.H., Qi, A., Roberts, E.H., Chay, P.M., Brouwer, J.B., Rose, J.L. and Yeates, S.J. (1995) Towards the reliable prediction of flowering in six annual crops. VI. Applications in crop improvement. *Experimental Agriculture* 31, 89–108.

Linneman, A.R. and Craufurd, P.Q. (1994) Effects of temperature and photoperiod on phenological development in three genotypes of Bambara groundnut (*Vigna subterranea*). *Annals of Botany* 74, 675–681.

Loss, S.P., Perry, M.W. and Anderson, W.K. (1990) Flowering times in wheat in South-western Australia: a modelling approach. *Australian Journal of Agricultural Research* 41, 213–23.

Nuttonson, M.Y. (1953) *Phenology and Thermal Environment as a Means for a Physiological Classification of Wheat.* American Institute of Crop Ecology, Washington.

Pearson, S., Hadley, P. and Wheldon, A.E. (1993) A reanalysis of the effects of temperature and irradiance on time to flowering in Chrysanthemum (*Dendranthema grandiflora*). *Journal of Horticultural Science* 67, 1–9.

Pearson, S., Hadley, P. and Wheldon, A.E. (1994) A model of the effects of temperature on the growth and development of cauliflower (*Brassica oleracea L. botrytis*). *Scientia Horticulturae* 59, 91–106.

Rees, A.R. (1987) Environmental and genetic regulation of photoperiodism – a review. In: Atherton, J.G. (ed.) *Manipulation of Flowering.* Butterworths, London, pp. 187–202.

Richards, R.A. (1989) Breeding for drought resistance – physiological approaches. In: Baker, F.W.G. (ed.) *Drought Resistance in Cereals.* CAB International, Wallingford, pp. 65–79.

Richards, R.A. (1991) Crop improvement for temperate Australia: future opportunities. *Field Crops Research* 26, 141–169.

Rick, C.M. (1976) Tomato. In: Simmonds, N.W. (ed.) *Evolution of Crop Plants.* Longman, London, pp. 268–272.

Roberts, E.H. (1988) Temperature and seed germination. In: Long, S.P. and

Woodward, F.E. (eds) *Plants and Temperature.* Symposia of the Society of Experimental Biology, Vol. 42. Company of Biologists Ltd, Cambridge, pp. 109–132.

Roberts, E.H. (1991) How do crops know when to flower? The importance of daylength and temperature. *Biological Sciences Review* 3, 2–7.

Roberts, E.H., Hadley, P. and Summerfield, R.J. (1985) Effects of temperature and photoperiod on flowering in chickpeas (*Cicer arietinum* L.). *Annals of Botany* 55, 881–892.

Roberts, E.H. and Summerfield, R.J. (1987) Measurement and prediction of flowering in annual crops. In: Atherton, J.G. (ed.) *Manipulation of Flowering.* Butterworths, London, pp. 17–50.

Roberts, E.H., Summerfield, R.J., Ellis, R.H. and Stewart, K.A. (1988a) Photothermal time for flowering in lentils (*Lens culinaris*) and the analysis of potential vernalization responses. *Annals of Botany* 61, 29–39.

Roberts, E.H., Summerfield, R.J., Cooper, J.P. and Ellis, R.H. (1988b) Environmental control of flowering in barley (*Hordeum vulgare* L.). II. Rate of development as a function of temperature and photoperiod and its modification by low-temperature vernalization. *Annals of Botany* 62, 127–144.

Roberts, E.H., Summerfield, R.J., Ellis, R.H. and Qi, A. (1993) Adaptation of flowering in crops to climate. *Outlook on Agriculture* 22, 105–110.

Robertson, G.W. (1983) Weather-based mathematical models for estimating development and ripening of crops. *World Meteorological Organization Technical Note,* No. 180. World Meteorological Organization, Geneva.

Sadik S. (1967) Factors involved in curd and flower formation in cauliflower. *American Society for Horticultural Science* 90, 252–259.

Salisbury, F.B. (1982) Photoperiodism. *Horticultural Reviews* 4, 66–105.

Summerfield, R.J. and Roberts, E.H. (1987) Effects of illuminance on flowering in long- and short-day grain legumes: a reappraisal and unifying model. In: Atherton, J.G. (ed.) *Manipulation of Flowering.* Butterworths, London, pp. 203–223.

Summerfield, R.J., Roberts, E.H., Erskine, W. and Ellis, R.H. (1985) Effects of temperature and photoperiod on flowering in lentils (*Lens culinaris* Medic.). *Annals of Botany* 56, 659–671.

Summerfield, R.J., Ellis, R.H. and Roberts, E.H. (1989) Vernalization in chickpea (*Cicer arietinum*): fact or artefact? *Annals of Botany* 64, 599–603.

Summerfield, R.J., Roberts, E.H., Ellis, R.H. and Lawn, R.J. (1991a) Towards the reliable prediction of time to flowering in six annual crops. I. The development of simple models for fluctuating field environments. *Experimental Agriculture* 27, 11–31.

Summerfield, R.J., Ellis, R.H., Roberts, E.H. and Qi, A. (1991b) Measurement, prediction and genetic characterization of flowering in *Vicia faba* and *Pisum sativum. Aspects of Applied Biology* 27, 253–261.

Summerfield, R.J., Collinson, S.T., Ellis, R.H., Roberts, E.H. and Penning de Vries, F.W.T. (1992) Photothermal responses of flowering in rice (*Oryza sativa*). *Annals of Botany* 69, 101–112.

Summerfield, R.J., Lawn, R.J., Qi, A., Ellis, R.H., Roberts, E.H., Chay, M.W., Brouwer, J.B., Rose, J.L., Shanmugasundaram, S., Yeates, S.J. and Sandover, S. (1993) Towards the reliable prediction of time to flowering in six annual crops. II. Soyabean (*Glycine max*). *Experimental Agriculture* 29, 253–289.

Thornley, J.H.M. (1987) In: Atherton, J.G. (ed.) *Manipulation of Flowering.* Cambridge University Press, Cambridge, pp. 67–79.

Upadhyay, A.P., Ellis, R.H., Summerfield, R.J., Roberts, E.H. and Qi, A. (1994a) Characterization of photothermal flowering responses in maturity isolines of soyabean [*Glycine max* (L.) Merrill] cv. Clark. *Annals of Botany* 74, 87–96.

Upadhyay, A.P., Summerfield, R.J., Ellis, R.H., Roberts, E.H. and Qi, A. (1994b) Variations in the durations of the photoperiod-sensitive and photoperiod-insensitive phases of development to flowering among eight maturity isolines of soyabean [*Glycine max* (L.) Merrill]. *Annals of Botany* 74, 97–101.

Vince-Prue, D. (1975) *Photoperiodism in Plants*. McGraw-Hill, London.

Vince-Prue, D., Thomas, B. and Cockshull, K.E. (eds) (1984) *Light and the Flowering Process*. Academic Press, London.

Waggoner, P.E. (1974) Using models of seasonability. In: Lieth, H. (ed.) *Phenology and Seasonability Modelling*. Chapman & Hall, London, pp. 401–405.

Wallace, D.H. (1985) Physiological genetics of plant maturity, adaptation and yield. *Plant Breeding Reviews* 3, 21–166.

Watkinson, A.R., Lawn, R.J., Ellis, R.H., Qi, A. and Summerfield, R.J. (1994) *RoDMod, a Computer Program for Characterising Genotypic Variation in Flowering Responses to Photoperiod and Temperature*. CSIRO, 47pp.

Went, F.W. (1957) *The Experimental Control of Plant Growth*. Ronald Press, New York.

Wheeler, T.R., Ellis, R.H., Hadley, P. and Morison, J.I.L. (1995) Effects of CO_2, temperature and their interaction on the growth, development and yield of cauliflower (*Brassica oleracea* L. *botrytis*). *Scientia Horticulturae* 60, 181–197.

Whyte, R.O. (1946) *Crop Production and Environment*. Faber & Faber, London.

Wiebe, H.-J. (1972) Wirkung von Temperatur und Licht auf Wachstum und Entwicklung von Blumenkohl. I. Dauer der Jugendphase für die Vernalisation. *Gartenbauwissenschaft* 37, 165–178.

Wiebe, H.-J. (1989) Effects of low temperature during seed development on the mother plant on bolting of vegetable crops. *Acta Horticulturae* 253, 25–30.

Wiebe, H.-J. (1994) Effects of temperature and daylength on bolting of leek (*Allium porrum* L.). *Scientia Horticulturae* 59, 177–185.

Wilkerson, C.G., Jones, J.W., Boote, K.J. and Buol, G.S. (1989) Photoperiodically sensitive interval in time to flower of soyabean. *Crop Science* 29, 721–726.

Withrow, R.B. (1959) A kinetic analysis of photoperiodism. In: Withrow, R.B. (ed.) *Photoperiodism*. American Association for the Advancement of Science, Washington, USA, pp. 439–447.

Wurr, D.C.E., Fellows, J.R., Phelps, K. and Reader, R.J. (1993) Vernalization in summer/autumn cauliflower (*Brassica oleracea* var. *botrytis* L.). *Journal of Experimental Botany* 44, 1507–1514.

Environmental Influences on Development, Growth and Yield

H. Krug

Institute of Vegetable Crops, University of Hannover,
Herrenhäuser Strasse 2, D-30419, Hannover, Germany

THE ENVIRONMENT–MAN–PLANT SYSTEM

Environmental factors influence development, growth and biological yield of plants primarily by affecting their physiology, and form the central focus of this chapter. In the context of crop production, however, economic, technical and phytopathological factors also have to be taken into account, expanding the scope of environment to a rather comprehensive system (Fig. 4.1). The ecological environment, in terms of plant production, the habitat, is the result of the combined effects of climate, soil and biotic factors. Each of these subsystems consists of many elements which may affect plants by different properties and intensities. The subsystems in Fig. 4.1, for example, may be subdivided as follows:

1. Climate
- radiation (irradiance, daylength)
- temperature (effects of means and extremes, diurnal and seasonal cycles, fluctuations)
- water (precipitation, evapotranspiration, seasonal distribution and intensity)
- wind (mean velocity, extremes, directions)
- CO_2 concentration
- air pollution

2. Soil
- structure, texture
- chemical components
- nutrients
- soil water (infiltration rate, storage capacity, ground water)
- air content

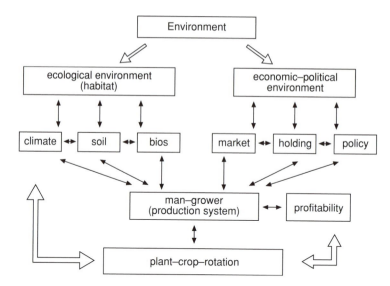

Fig. 4.1. Structure of the environment–man–plant system at high abstraction level with regard to plant production.

3. Bios

- microorganisms (nutrient availability, diseases, symbionts)
- weeds
- pests and beneficial organisms.

These factors and others contribute to the growing conditions of the habitat. Due to limited space, this chapter cannot deal with the comprehensive habitat system, but is restricted to the factors of radiation, temperature and CO_2, which are not as extensively discussed in the literature as nutrient elements and water.

The economic environment involves influences of the market, such as absorption capacity and prices, factors such as investment costs, wages, prices of the production factors, quality and quantity aspects of yield as well as timing. Policy may control competition by restrictions or transport costs, it may support production by financial subsidies or impose restrictions concerning the use of fertilizers, herbicides or pesticides.

Man, concerning crop production is the grower, and is positioned at the centre of the system. He depends on and acts in the ecological, economic and political environments. He decides which species and cultivars to grow and on the rotations. His decisions are based on his experience and information concerning the crop, the habitat and the market, they are based on his own preferences and, at least in the short term, bound to his existing production system. In the medium to long term, profitability has to be the primary criterion.

The subsystem plant, in production mainly grown as a crop, again is a very complex dynamic subsystem. Its productivity is governed by numerous processes, each with special reactions to environmental factors. Moreover, the plants in a crop are subjected to canopy effects with intra- and interspecific competition. The crop, however, is not only the result of the environment and activities of the grower, but feeds back to them.

To understand the dynamics of such complex systems, they must first be analysed at a highly abstract level, focusing on the main and most influential elements which involves a process of selection. The arrangement of these elements in space and time and the main relations between them must then be developed. This results in a diagram or model which should be parameterized for quantification and computerization. Restrictions and simplifications may cause inaccuracies and may result in considerable variability. Therefore the model has to be validated, to test for deviations from reality. Due to our limited understanding it is appropriate to start with analysis and control of integrated structures and processes at high abstraction level. For a more advanced and causal view these models should be extended and completed by more detailed analyses thus reducing the level of abstraction.

PLANT REACTIONS TO ENVIRONMENTAL FACTORS

The primary goal of production and the most important factor of profitability is the yield of the crops and cropping systems. Growers, therefore, want to know the relationship between the magnitude of yield fluctuations and environmental influences of the season, during both short periods and over longer durations. Relationships between yields in the open field and weather conditions or intensities of distinct growth factors were tested statistically. This methodology was later supplemented by experiments in greenhouses or growth chambers with more or less controlled environments to disentangle the input complex, to ensure extremes and to facilitate statistical analysis by simplifications. In the last decades of this century analysis and control of systems have been considerably strengthened by system analysis theories, initiated by Wiener (1965), Bertalanffy (1973) and others.

The plants and crops were analysed in more detail by sophisticated methodology in order to understand the reaction mechanisms and discover causal relationships. Analysis of this complex set of factors contributes to an understanding of the system and may offer new insights and enable more efficient manipulations. However, if, due to our limited understanding, production control is based on one or only a few processes, neglecting other actions and interactions (reductionism), failures will frequently result. To avoid this problem, it is preferable to determine at which abstraction level of the plant system the best understanding and the best basis for planning and management of production can be obtained. This approach is aided by distinguishing between plant development and growth.

PLANT DEVELOPMENT

Basis and definitions

Development can be defined as the process and timing of differentiation. The component events cause qualitative changes in form and function of the plant and therefore in formation of yield.

The general pattern of the development of cultivated monocot and dicot plants is shown in Fig. 4.2. Sowing is the beginning of the process, after which plants emerge, grow and develop with or without transplanting and may be harvested in the vegetative (e.g. lettuce, radish) or reproductive phase. If grown for reproductive organs, the vegetative phase is followed by flower bud differentiation, anthesis, fertilization, concentrated or continuous fruiting and ripening followed by new growth, dormancy or death. These and other phenological features may be called the development stages, representing stages in the life cycle. They are primarily physiological characteristics, supplemented by characteristics of the production system (transplanting, pruning, etc.), which may cause distinct effects on development and growth.

The periods between the development stages are called development phases. Amongst others the predominant phases are as follows: (i) the vegetative and reproductive phases; (ii) the heterotrophic (germination or emergence phase) and autotrophic phases; (iii) the transplant growth phase; (iv) the exponential growth phase; (v) the linear growth phase; (vi) the fruit growth or embryo phase; and (vii) the grain filling phase. These phases may overlap and there is some mixture with growth. Thus the best suited phases should always be chosen according to the data available and the method of analysis.

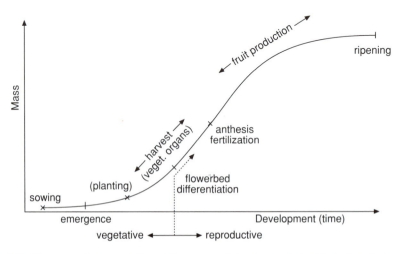

Fig. 4.2. Diagrammatic presentation of the course of development of cultivated angiosperms.

A numerical scheme that provides a more detailed description of development and growth characteristics has been developed called the BBCH scale (Compendium of growth stage identification keys for mono- and dicotyledonous plants) (Fig. 4.3; Feller *et al.*, 1995). The macro-classification defines 10 principal phenological stages together with intermediate and secondary growth stages, providing a more detailed, species-specific subdivision, represented by a two- or three-digit code. Examples for digitizing physiological stages are illustrated in Fig. 4.4. This scale can be used in development-oriented production procedures such as land cultivation, irrigation or estimation of crop mass as a criterion of nutrient uptake and needs for fertilization.

To quantify or parameterize environmental influences on development the following criteria must be considered.

1. The differentiation rate (DiR = $n \times t^{-1}$ where n is the number of organs differentiated and t is the unit of time) in phases with dominating differentiation processes, such as the differentiation of leaves, flowers, pods, seeds and other organs.

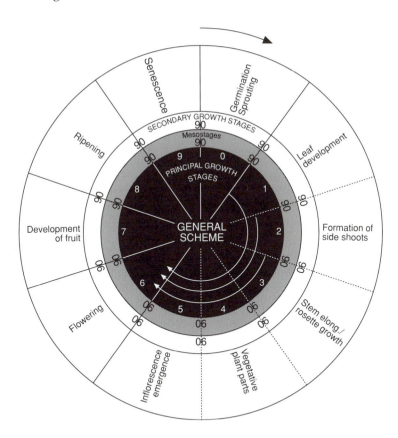

Fig. 4.3. Phenological stages of monocot and dicot plants described in a three-digit decimal code (Bleiholder *et al.*, 1994).

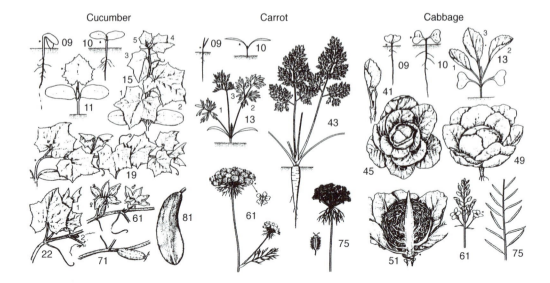

Fig. 4.4. Examples for phenological stages according to the BBCH scale. (a) Cucumber: 09 = emergence, 10 = cotyledons completely unfolded, 11 = first true leaf on main stem fully unfolded, 15 = 5th leaf …, 19 = 9 and more leaves …, 22 = second primary side shoot visible, 61 = first flower open on main stem, 71 = first fruit on main stem has reached typical size and form, 81 = 10% of fruits show typically full ripe colour. (b) Carrots: as in (a) plus 13 = third true leaf unfolded, 43 = 30% of the expected root diameter reached, 61 = beginning of flowering, 10% flowers open, 75 = 50% of fruits have reached typical size. (c) Cabbage: as in (a) and (b) plus 41 = heads begin to form, the two youngest leaves do not unfold, 45 = 50% of the expected head size reached, 49 = typical size, form and firmness of heads, 51 = main shoot inside head begins to sprout (Feller *et al.*, 1995).

2. The minimum number of organs (e.g. leaves) to be differentiated before entering the next development phase. These hurdles are developed during evolution to adapt to special environmental cycles, to ensure that plants reach an adequate size to start reproduction and have a chance to finish seed formation in the rest of the growing season. The period which must be overcome before entering into the reproductive phase is called the juvenile phase.

3. Processes where differentiation is difficult to measure or where a strong connection to growth or elongation exists (e.g. emergence phase, grain filling phase); the time that the plant needs to pass through these phases is a useful criterion. To quantify the efficiency of development, only a relative parameter, the development rate (DeR = t^{-1}) as the reciprocal value of duration is available. The development rate gives the portion of the total development progress accomplished in the unit of time (t). This parameter has the advantage of being linearly related to the temperature demand of many processes (temperature sum concept, see below).

The duration or the development rate can also be used for phases with different processes, with several phases and over the whole life cycle. In these cases, however, they are overall criteria useful for parameterization, but with lower explanatory value. More comprehensive descriptions of developmental processes can be performed using models. Examples are presented in the section on prediction, planning and production analysis below.

Development–growth interactions

Temperature affects the development and growth of plants primarily by controlling the manner and speed of biochemical processes (Q_{10} rule), but it may also affect plants by the induction of stimuli, such as vernalization (see section on control by environmental stimuli below).

Increasing temperature – within species-specific limitations – generally increases the development rate except for stimuli effects. If the growth rate (GR) is accelerated to the same degree as the DeR, earlier and coincident yields will result. Many species or cultivars, however, follow another reaction pattern: the DeR and the GR react differently to temperature (van Dobben, 1962). A good example is the pea plant. H. Krug (Hannover, unpublished observations) cultivated the very early cultivar Hunter and the late cultivar Atlas in growth chambers at 16 and 24°C (18 klx fluorescent light for 12 h plus 150 lx incandescent light for 6 h = 18-h day). Development and growth criteria are shown in Fig. 4.5.

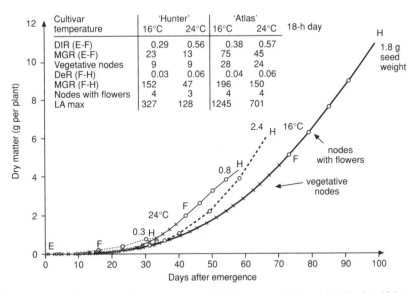

Cultivar temperature	'Hunter'		'Atlas'		18-h day
	16°C	24°C	16°C	24°C	
DIR (E-F)	0.29	0.56	0.38	0.57	
MGR (E-F)	23	13	75	45	
Vegetative nodes	9	9	28	24	
DeR (F-H)	0.03	0.06	0.04	0.06	
MGR (F-H)	152	47	196	150	
Nodes with flowers	4	3	4	4	
LA max	327	128	1245	701	

Fig. 4.5. Interaction of growth and development of an early (Hunter - - -) and a late (Atlas —) pea cultivar, grown at 16°C (thick lines) and 24°C (thin lines). E = emergence, F = first node with flower, H = harvesting, DiR = differentiation rate, MGR = mean growth rate (g day^{-1}), DeR = development rate; LA$_{max.}$ = maximum leaf area (Krug unpublished data).

The graph confirms that an early cultivar as well as higher temperatures result in earlier harvests with lower total dry matter and lower yields. In addition, the graph presents some explanations to the physiological background: the DiR of nodes (leaves) in the period of emergence to the appearance of the first node with flowers (E–F) is markedly increased by the higher temperature, whereas the mean growth rate (MGR) is reduced. The number of vegetative nodes, which have to be differentiated before flower initiation can occur, was not much influenced by temperature, but was much lower with the early cultivar Hunter.

In the period from first flowering node to harvest (F–H), the DeR (DiR is not easy to measure in the grain filling period) was again increased and the MGR reduced by the higher temperature. The MGR was lower with the early cultivar compared to the late one. The number of nodes with flowers and with that the number of inflorescences were similar with both temperatures and with both cultivars. Consequently the main components influenced by the temperature–genotype interaction were the quotient DiR/DeR and GR. The accelerated DiR/DeRs combined with reduced GRs result in lower leaf area, a shorter period of photosynthesis and consequently lower biomass and lower yields. The harvest index, another component of yield, was higher with Hunter (16°C = 0.39; 24°C = 0.30) than with Atlas (0.17 and 0.19, respectively) and compensated to some extent.

Cultivars with a strong increase of DiR and decrease of GR with increasing temperature are sensitive to later spring sowings in temperate climates (van Dobben, 1963) and less adapted to hot climates.

An inverse relationship between DiR and GR at low and high temperature is demonstrated for beans by van Dobben (1962). With this species, increasing temperature promotes GR more than DiR, resulting in taller plants in 25°C than in 16°C (Fig. 4.6).

Control by environmental stimuli

During their evolution plants developed mechanisms to adapt to the prevailing environments for survival, reproduction and dispersal of the species. Besides the development of tolerances, reactions to environmental stimuli were manifested. The main stimuli are light cycles (photoperiod), temperature cycles (e.g. vernalization) or in arid climates changes in water potential. These reactions are very important properties of species and cultivars to control timing and yield. They can be manipulated by breeding as well as by timing the production with respect to the season or by controlling the environment, e.g. in greenhouses. The control of plant development by photoperiodism and vernalization is described in Chapter 3 (see also Krug and Wiebe in Krug, 1991a).

Activity cycle – dormancy

Lang *et al.* (1987) define dormancy as a 'temporary suspension of visible growth of any structure containing a meristem' and designate these

(a)

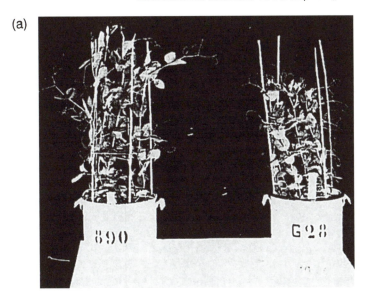

(b)

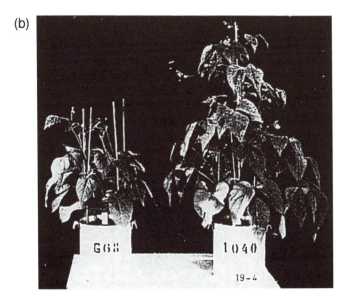

16°C   25°C

Fig. 4.6. Growth × development interactions with (a) peas and (b) bush beans in 16 and 25°C (van Dobben, 1962).

phenomena as endodormancy and ecodormancy. A third category of dormancy, called paradormancy, is regulated by physiological factors outside the affected structure, as for example apical dominance (see Chapter 5).

Endodormancy may be induced endogenously without any synchronizing external stimulus. This occurs with plants, when transferred from seasonally limited growing conditions to which they have adapted to unlimited ones. In habitats with unfavourable seasons for growth, the stimulus indicates the stress periods and triggers reactions which enable the plant to survive. These reactions may involve the transfer of metabolites into storage organs, such as tubers, bulbs, rhizomes, roots or stems. The leaves, as very sensitive organs, are abscised. The storage organs are frequently covered by cell layers that retard water losses and frost resistance is increased by the build up of metabolites and low metabolic activity.

The most important stimuli to induce endodormancy are photoperiod and temperature; in dry areas of the tropics or subtropics it may be water potential. With regard to photoperiod, endodormancy is induced by a daylength forecasting the stress period. With species originating from the temperate zone, this is a short day preceding the cold winter period. This

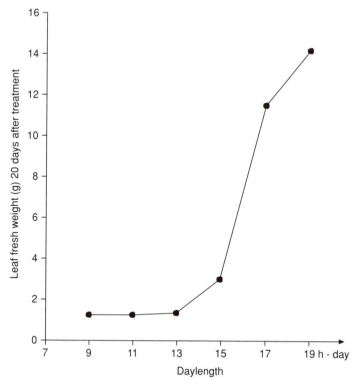

Fig. 4.7. Induction of endodormancy in chives by daylength as reflected by leaf growth rate over 20 days (temperature about 14°C, induction period 6 weeks; Krug and Fölster, 1976).

reaction is often coupled to a corresponding temperature range, generally 5–15°C.

Examples of vegetable species that exhibit endodormancy are the perennials chives and rhubarb. As shown in Fig. 4.7 for chives, the critical daylength is about 13 h with an increasing effect up to more than 18 h. The optimal temperature for chives and rhubarb is about 14°C with decreasing effects to lower and higher temperatures (Fig. 4.8). Chives stay active in low (2°C) and high (22°C) temperature as well as in long day. At least 50 lx are needed to lengthen the day with incandescent light. The same is true for rhubarb, but for day lengthening at least 500 lx incandescent light should be applied. The essential duration of the stimulus varies between 3 and 4 weeks (for rhubarb) to 6 and 8 weeks (for chives).

With onions, however, originating in the hot and dry summer climate of Middle Asia, long days signal the stress period. Accordingly after completion of the juvenile phase, bulb formation and subsequently endodormancy are induced and enhanced by long days coupled with high temperature (see Chapter 18).

With the perennial, day-neutral asparagus plant, 'ripening' of the ferns in autumn is induced by temperatures below approximately 15°C with an optimum of about 8–12°C. It is enhanced with preceding high temperatures (25–30°C). At 2°C and 18°C the ferns did not show any sign of 'ripening' (Krug, 1996). After fern cutting the crowns of 'ripened' and of active plants

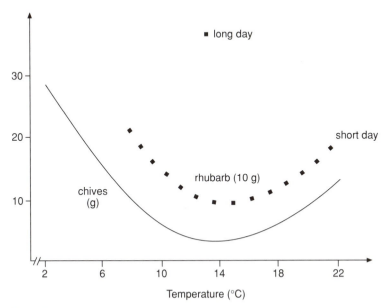

Fig. 4.8. Induction of endodormancy in chives and rhubarb in short day as a function of temperature. Chives mean of 6 and 8 weeks induction period, leaf fresh weight in grams per cluster 20 days after treatment in 20–22°C (Krug and Fölster, 1976); rhubarb mean of 4–8 weeks induction period, cumulative leaf weight in dag (decagram) per plant (Krug, 1991b).

were able to sprout at an incubation temperature of 20°C all year round. At 15°C incubation temperature, however, sprouting ability was very low from June until September. This indicates that crown activity/dormancy is more or less independent of fern 'ripening' and is induced by a decrease in temperature (H. Krug, 1997, Hannover, in preparation).

The duration of endodormancy is also a consequence of evolution. It differs between species and cultivars corresponding to the origin and breeding, respectively. For many species it is overcome by temperatures in the range of 0–15°C for 11–42 days. Chives plants grown in the open field in Germany enter endodormancy in September and are released from November to December (Fig. 4.9). That means a dormant period of 6–8 weeks.

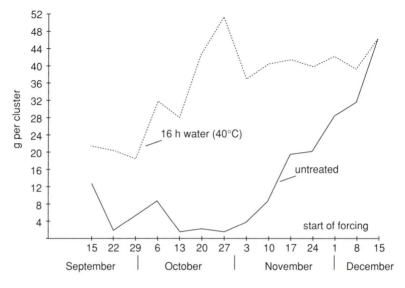

Fig. 4.9. Leaf weight of chives in 20°C as a function of the date of digging in the field without and with warm water treatment to break endodormancy, indicating the duration of endodormancy in a moderate climate in Hannover (Fölster, 1967).

With rhubarb the release of endodormancy is a function of low temperature (Loughton 1961, 1965). The cold sum needed for regrowth can be estimated by the sum of the daily minima at the rhizome (10 cm depth at 9 o'clock) as degree Celsius from a base temperature of 10°C down to –2°C (e.g. 3°C has an effective value of 7°, –2° one of 12°C). The early cv. Timperly Early for example requires a cold sum of 110–150°C and the medium cv. Holsteiner Blut of 235°C. Accordingly regrowth in 20°C after successive field diggings is earlier with Timperly Early (Fig. 4.10).

With onion bulbs endodormancy is overcome earliest in medium temperatures with an optimum of 9–13°C, in a broader range of 5–20°C. Storing the bulbs in 15°C and planting in humid peat at 22°C, the period to 50% of

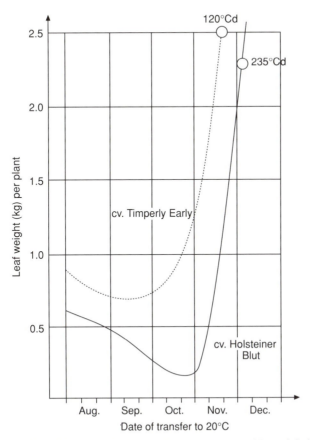

Fig. 4.10. Cumulative leaf fresh weight of an early and a medium cultivar of rhubarb indicating the duration of endodormancy. Means of 1987 and 1988 in Hannover. ○ = cold sums of 120 and 235°C, respectively (Krug, 1991b).

the bulbs with regrowth after planting varies among cultivars from about 10 days (i.e. no endodormancy) to about 90 days (Fig. 4.11) (see also Chapter 18).

Endodormancy can be broken by several means. Most important are (i) extreme temperatures, such as cold treatment (rhubarb, asparagus, less effective with chives) and heat treatment (chives); (ii) soaking in water, which may be connected with abscisic acid release or root growth with cytokinin production (chives, onion); (iii) wounding by cutting or shaking (rhubarb, onion, chives, chicory, potato); and (iv) growth regulators, especially phytohormones, such as gibberellins, cytokinins and ethylene.

In moderate and cool climates endodormancy is often followed by ecodormancy, imposed by environments unfavourable to growth (Fig. 4.12). Endodormancy is induced in September (onions in July), is released in November/December and is replaced by ecodormancy until temperature reaches the critical value (3–5°C) in spring.

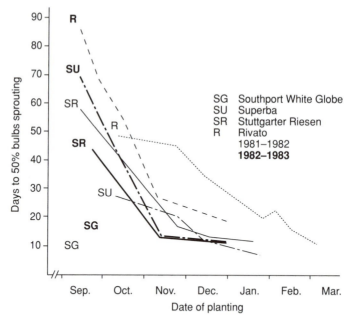

Fig. 4.11. Duration of endodormancy of onion cultivars in 2 years, after dry storage at 15°C successively planted into humid peat at 22°C (Krug, 1983).

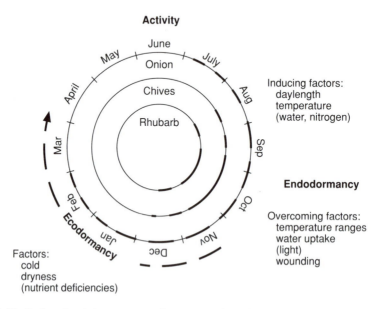

Fig. 4.12. Cycles of endodormancy, ecodormancy and activity of vegetable species in a moderate climate (Krug, 1991a).

Endodormancy is an important attribute of cultivars. Strong endodormancy assists long-term storage and may replace the need for refrigeration. It may prevent frost damage during the winter or in spring by late regrowth and it has to be considered in the timing of cultural practices such as forcing.

PLANT GROWTH

At high abstraction level growth can be defined as an irreversible increase of dry matter or volume as a function of genotype and the environmental complex. Since quantitative and especially integrated knowledge of this system is limited, the general aspects of the influences of the growth factors considered are at first discussed separately.

Radiation

Irradiances[1] at the tops of canopies or at ground level depend on many factors. The extraterrestrial global irradiance (300–2600 nm) can be calculated from the solar constant (mean value 1.39 kWm^{-2}) times sine of sun height (Fig. 4.13). This graph demonstrates the primary radiation conditions at different latitudes and seasons of the northern hemisphere.

Passing the atmosphere, radiation is scattered by molecules, scattered and absorbed by aerosols and absorbed by water vapour, ozone and oxygen. In the summer season on sunny days at the equator, about 78% of the extraterrestrial radiation reaches the surface of the earth, with decreasing values towards the poles (50°N 70–76%; Schulze, 1970).

The reduction of irradiance by clouds is regionally very different and cannot be generalized. During the growing season of the Netherlands (Spitters, 1986) the transmission on a day without sunshine averages at 20%. The mean atmospheric transmission amounts to about 42%. The isopyres (lines of equal radiation sums per day) of the month with highest irradiance in the northern hemisphere and lowest irradiance in the southern hemisphere are presented in Fig. 4.14. Highest irradiances are measured in 30–40°N in the cloudless deserts and at the pole. At the equator and up to about 30° n.l. on cloudless days at noon time (average 1000–1400 hours) mean radiant flux densities of global radiation amount to about 1 kWm^{-2} (105 klx) with peaks of about 1.112 kWm^{-2} (117 klx).

The energy usage of crops by photosynthesis is restricted to the incident radiant flux density (Wm^{-2}) in the range of chlorophyll absorption (400–700 nm). According to the process of absorption the most adequate dimension is photon flux density (mol m^{-2} s^{-1}). Both values, however, are

---

[1] Irradiances are cited as W m^{-2}; 1 W m^{-2} PAR = 4.6 µmol m^{-2} s^{-1} PAR for natural light and 5 µmol m^{-2} s^{-1} for high pressure sodium light; 1 W m^{-2} PAR natural light = 0.25 klx, high pressure sodium light 0.36 klx (McCree, 1972).

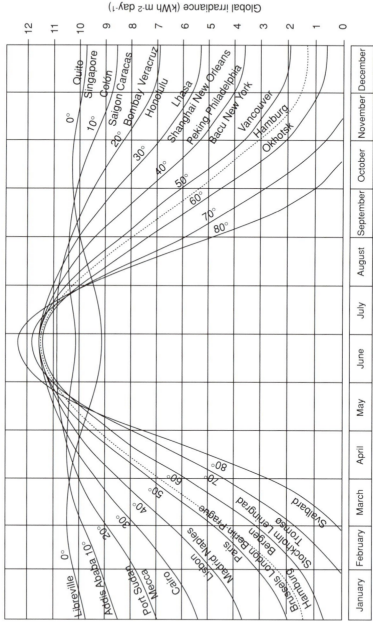

Fig. 4.13. Extraterrestrial irradiances (300–2600 nm) of the northern hemisphere, calculated by the solar constant (mean 1.39 kWm^{-2}) multiplied by inclination and daylength (Schulze, 1970).

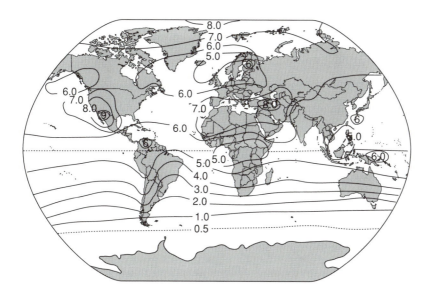

Fig. 4.14. Daily sums of measured global irradiance ($kWhm^{-2} day^{-1}$) at the earth's surface in June (Schulze, 1970).

close together and each of them is called photosynthetic active radiation (PAR, about 400–700 nm). A general view of spectral distribution of global radiation, PAR as well as the sensitivities of the human eye and radiation sensors are presented in Fig. 4.15. The amount of PAR of global radiation reaching the earth's surface is a function of scattering and absorption of radiation in the atmosphere and may vary in northern Germany between 0.38 to 0.69 (Beinhauer, 1977). Mean values are between 0.45 and 0.55 (for a review of the literature see Brückner, 1990). In general 50% of global radiation is considered as PAR.

The absorption of PAR by the canopy is difficult to quantify. In general it depends on the following:

1. The diurnal irradiances entering the top of the canopy, the diurnal partition of direct and diffuse radiation and the course of the angle of incident direct radiation.
2. The absorbing surface (leaf area index, LAI) and its architecture, described by the extinction coefficients of the two radiation components.

A more detailed treatment of this topic is found in Spitters (1986), Spitters *et al*. (1986) and for greenhouses in Gijzen (1992).

The instantaneous assimilation rate of leaf area unit is a function of the light absorbed, the light use efficiency of the plant, depending on the photosynthetic pathway, the levels of other growth factors (e.g. temperature, CO_2 concentration), water potential and the use of photosynthetic capacity by source-sink relationships. Most crops grown in temperate climates are C_3 plants. In warmer climates and mostly with drier sites C_4 plants have

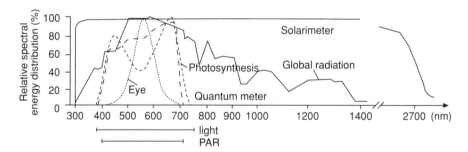

Fig. 4.15. Relative spectral energy distribution of natural radiation, curve of photosynthetic efficiency (Geutler and Krochmann, 1978), spectral sensitivity of the human eye (corresponding to lux meters), quantum meter and solarimeter (Krug, 1991a).

developed with (i) higher net photosynthesis (60–100 mg CO_2 dm^{-2} h^{-1} compared to ≤ 30 mg CO_2 dm^{-2} h^{-1} of C_3 plants); (ii) higher light saturation values of horizontal leaves (400–600 : ≤ 200 Wm^{-2}); (iii) lower CO_2 compensation point (< 10 : 30–60 ml l^{-1} CO_2); (iv) lack of photorespiration (0: about 30% of gross photosynthesis); and (v) a high temperature optimum (30–45 : 10–25°C) and other differences. Among others maize, sugarcane, sorghum, *Amaranthaceae, Portulacaceae,* some *Chenopodiaceae,* some *Euphorbiaceae* and some *Atriplex* species are C_4 type plants (Mohr and Schopfer, 1992).

To quantify the assimilation rate of crops Spitters (1986) in his 'sucros 87 model' divided the canopy into leaf layers. For shaded leaves he used the function:

$$P_{sh} = P_{max.} \cdot [1 - \exp(-\varepsilon \cdot I_{sha} \cdot P_{max}^{-1})] \tag{1}$$

where P_{sh} is the assimilation rate of shaded leaf area (g CO_2 m^{-2} leaf h^{-1}); $P_{max.}$ is the asymptote or assimilation rate at light saturation (about 4.0 g CO_2 m^{-2} h^{-1} for ruderal C_3 species); ε is the initial slope of light use efficiency (g CO_2 J^{-1} absorbed, about $12.5 \cdot 10^{-6}$ g CO_2 J^{-1} at 20°C); and I_{sha} is the absorbed light energy per unit leaf area (J m^{-2} leaf s^{-1}).

For directly sunlit leaf area the leaf angle has to be considered (Spitters, 1986). To calculate the assimilation rate of a canopy with shaded and unshaded leaves the percentage of sunlit leaf area must be known and the assimilation rates of the shaded and sunlit leaves of all layers can then be summed up and integrated over time.

According to Spitters *et al.* (1989), using the sucros 87 model, a typical crop that fully covers the ground, in a moderate climate (3.9 $kWhm^{-2}$ day^{-1} global radiation) produces about 210 kg dry matter per hectare per day. That means an energy input of about 185 Wh global radiation per gram of dry matter.

The calculation procedure of assimilation rates is a rather sophisticated and laborious approach, but it still includes many simplifications. Nevertheless the results are considered to be quite realistic for crops with high LAI and especially growing in field conditions. The consequences of further simplifications desirable for easier handling and computing are discussed by Spitters (1986).

In glasshouses the climatic environments differ from those in the fields in the following ways.

1. Lower outside irradiances in the winter season and losses by reflection and absorption of the glass/film covering and construction elements, amounting to about 30–50%.
2. The correlations between climatic factors can be at least partially broken and their intensities can be improved by techniques such as heating, CO_2 enrichment and/or artificial light.

Consequently in this steeper part of the reaction curve variations in irradiance cause stronger reactions in the crops and interactions with other growth factors, such as temperature or CO_2, and are more important than in open fields. Moreover in glasshouses frequently young plants are grown with expanding leaf areas which are difficult to parameterize by mechanistic models. Furthermore, for optimization procedures the number of parameters has to be restricted. To simulate growth these challenges are often better met with descriptive models by calculating response surfaces (see Fig. 4.24).

The effect of artificial illumination, a very expensive input factor, can be integrated into the response surfaces and thereby 'optimized' according to ecological and/or economic criteria. In experiments with kohlrabi (Brückner, 1990), lettuce transplants (Krug and Liebig, 1994) and ornamentals (Ludolph, 1995) it was demonstrated that the effect of high pressure sodium vapour light supplementing natural light is generally equal to natural light (measured as PAR). This was not true in long-term experiments during the dark season, due to more substantial morphogenetic effects of light quality. In this situation, Brückner achieved a better approximation, when the artificial irradiance was multiplied by a factor of 1.4, which means it was more effective than PAR of natural light (see equation 14 below).

The possibility of integrating various applications of artificial light by mean irradiances was confirmed by Ludolph (1995), who demonstrated that effects on ornamental plants exposed to natural light and a mixture of natural and high pressure sodium vapour light during different periods of the diurnal cycle and with variations in intensities and durations could be approximated by total light quantity.

With both the mechanistic and the descriptive models, however, some reactions are difficult to parameterize: generally photosynthesis models do not consider adaptation to changing irradiances. Lorenz (1980) raised cucumber plants in growth chambers with five irradiances and measured CO_2 uptake at leaf stage 5 after changing the irradiances (Fig. 4.16). At irradiances during the raising period up to 63 Wm^{-2} global radiation, the maximal photosynthetic rate occurred in the irradiance to which the plants were adapted. Higher irradiances during the raising period reduced subsequent photosynthesis considerably and the highest irradiance for transplant growth (128 Wm^{-2}) resulted in lower photosynthesis than raising in

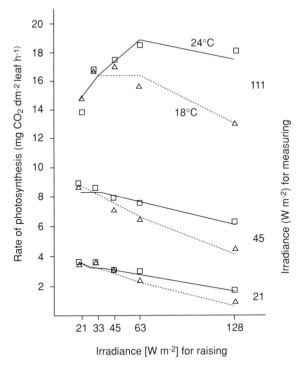

Fig. 4.16. Effect and after-effect of irradiances (global radiation Sylvania Cool White VHO 215 W + 48% incandescent light) on photosynthesis of cucumber plants in leaf stage 5 (Lorenz, 1980).

63 Wm^{-2}. These after-effects may cause strong deviations compared to static irradiances and must be considered, especially when using artificial light for raising transplants.

Fresh weight growth and hence tissue expansion reacts to short-term fluctuating irradiances in ways not connected to photosynthesis. As shown by Liebig (1989), photosynthesis followed changing irradiances immediately, but low irradiances of up to 10 days duration did not affect tuber fresh weight or leaf area of kohlrabi, since the plants used accumulated dry matter for fresh weight growth. Only when dry matter percentage dropped below a critical level was growth reduced. It is evident that plants – at least those with storage organs – possess the property to buffer fluctuating irradiances. This has implications on the adaptation of temperature control to actual irradiances in greenhouses. Since photosynthesis has lower temperature optima than leaf growth and the temperature requirement is already fulfilled by increasing temperature in the greenhouse due to higher irradiances, it is not necessary to control temperature according to short-term fluctuations of irradiance. Long-term adjustment of temperature control to irradiance levels, however, is adequate and growth rates can be correlated to mean irradiances for corresponding periods.

For production planning and control in addition to the adaptation factor the effect of the annual cycle of the climatic factors has to be considered. After-effects of foregoing intensities and seasonal oscillations are not incorporated in mechanistic models. In regression models for some species (radish *R. sativus*: Krug and Liebig, 1979b; lettuce: Mann, 1987), the response surfaces for autumn-grown crops differed from those of spring-grown crops. The attempt to overcome this disagreement by including a term for the season was not always satisfactory. However, other species (radish *R. niger*: Krug and Liebig, 1984a) did not show these deviations.

Independently of these effects the seasonal oscillations of the growth factors cause special 'seasonal effects'. As shown in Fig. 4.17, every change in the intensities of a growth factor, such as higher or lower temperature, irradiance, CO_2 or others, causes a substantial effect in the autumn (a_1), when the crop grows in the less favourable environment. Growth promotion, e.g. by heating, to compensate for later plantings to reach the same date of harvest, however, results in small effects (b_1). The reverse is true in spring with plants growing in more favourable conditions with small effects in promoting growth with equal starting points (a_2), and substantial effects in delaying planting but reaching the same time of harvest (b_2). This means that growth factors always have substantial effects if by their promotion the growth period in the unfavourable season – here low irradiances – is reduced. Promotion during the unfavourable periods, however, are reduced if favourable environments come later, during which one day produces the same progress as several days

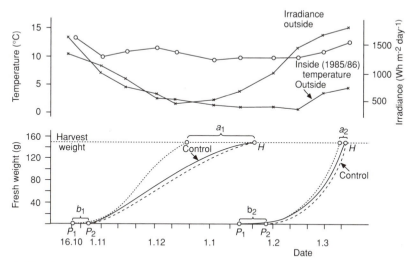

Fig. 4.17. Effect of growth promoting actions depending on the season (data with lettuce, grown at temperature set point 12/8°C; radiation and outside temperature long term means in Hannover. (a) Growth promotion (here CO_2 enrichment) with equal dates of planting (P_1) results in substantial effects in autumn, such as much earlier harvesting (a_1), but in rather small effects in spring (a_2); (b) growth promotion with equal days of harvesting (H) results in small effects in autumn, such as later planting $(P_2$, dotted line), for the same date of harvesting (b_1), but more substantial effects in spring (b_2).

in the preceding low light period. The effect of a growth factor acting in fluctuating environments will thus vary with the degree to which the factor
reduces growth duration in unfavourable conditions.

Radiation affects many other plant processes, such as morphogenesis or
control of development which will not be discussed here.

Temperature

Temperature is not a growth factor supplying energy or constituents but
primarily controls the rates of chemical reactions (Q_{10} rule). Temperature
controls plant development including morphogenesis and plant quality.
These processes make temperature a major growth factor worldwide, determining climatic zones, controlling plant distribution, growth cycles and
growth rates resulting in yields.

Climatic zones are primarily classified by temperature. The tropics are
defined as the area within which the sun will be directly overhead twice a
year (area within the solsticial points), equal to about 23° latitude with a
mean annual temperature of about 25°C and a low seasonal and diurnal
amplitude (Table 4.1). The subtropics expand to about 40°, characterized in
the northern hemisphere by lower mean annual temperatures, at 40°N 14°C,

Table 4.1. Temperature at the earth surface (Blüthgen, 1966 from Caesar, 1986, supplemented).

Latitude		January	July	Mean of year	Amplitude
North pole		−41.0	−1.0	−22.7	40.0
80°N		−32.2	2.0	−17.2	34.2
70	high	−26.3	7.3	−10.7	33.6
60 - - - - - - - -		−16.1	14.1	−1.1	30.2
50	medium	−7.1	18.1	5.8	25.2
40 - - - - - - - -		5.0	24.0	14.1	19.0
30	subtropic	14.5	27.3	20.4	12.8
20 - - - - - - - -		21.8	28.0	25.3	6.2
10		25.8	26.9	26.7	1.4
Equator tropic*		26.4	25.6	26.2	1.0
10		26.3	23.9	25.3	2.4
20 - - - - - - - -		25.4	20.0	22.9	5.4
30	subtropic	21.9	14.7	18.4	7.2
40 - - - - - - - -		15.6	9.0	11.9	6.6
50	medium	8.1	3.4	5.8	4.7
60 - - - - - - - -		2.1	−9.1	−3.4	11.2
70	high	−3.5	−23.0	−13.6	19.5
80°S		−10.8	−39.5	−27.0	28.7
South pole		−13.5	−48.0	−33.1	34.5

* 23° n.l. to 23° s.l.

and larger seasonal (13–19°C) and diurnal amplitudes. In the medium latitudes the mean annual temperature drops to 5–8°C at 50°N with still higher amplitudes. The extremes of the monthly temperature means of all latitudes are 28°C and –48°C.

These general values, predominantly caused by extraterrestrial irradiances, are varied by the distribution of land and water, ocean currents, cloudiness and altitude (1000 m greater altitude reduces temperature by 6°C). Therefore mean values as well as extremes of territories may deviate considerably from the latitude means. Examples are West and Northern Europe, warmed up by the Gulf Stream, the coastal region of California, cooled by the California Current or cool climates and frost at high altitudes in the tropics. Even more deviations occur with the extremes in deserts.

The regional climatic conditions can be derived from maps, if available agroclimatical maps and/or phenological maps integrate the influences of all habitat factors including topography and soils. Moreover, temperature sums may be calculated with data from meteorological stations, indicating the course of temperature with respect to threshold values. Such values, combined with enquiries and practical experience, allow maps to be drawn showing regions with growing conditions suitable for species, as demonstrated by Williams and Joseph (1976) for banana, sugarcane and others.

Within these territories again large differences in the properties of habitats may occur. This is especially pronounced in limiting conditions, such as the low temperatures prevailing in the high latitudes, where the exposure of the terrain, wind protection, maritime sites or even small differences in altitude may cause marked differences in growing conditions. This is also true for opposite extremes in the tropics, where water may be a limiting factor or small differences in temperature may limit production areas of temperature-sensitive or cold-requiring crops.

Plants can only survive and grow within their temperature limits and if during the seasonal cycle sufficient time is supplied and growth is sufficiently efficient. However, the species-specific temperature limits, the minima and maxima, which define the ranges of survival are difficult to specify. Factors affecting this include (i) the tremendous plasticity caused by adaptation (e.g. hardening); (ii) the duration of cold stress; (iii) the developmental stage; (iv) the level of activity or dormancy; and (v) cultivar differences.

Impressive examples of adaptation among vegetables have been published by Kohn and Levitt (1965) and Rietze (1988). Kohn and Levitt showed that the minimum temperature (for 50% survival) of cabbage was decreased by hardening from about –3°C to about –20°C. Rietze demonstrated that with the warm-season crop cucumber the temperature minimum depended on the duration of the cold stress period per day, the number of days with cold stress, irradiances and the position of the cold period during the diurnal cycle. General minima of survival of plants range from –60°C for arctic trees, –20°C for cereals, 0°C for frost-sensitive crops to some degrees above 0°C for some tropical species.

The minima for growth are generally defined to 5°C for cool-season crops, to 10°C for warm-season crops and to 15°C up to 17.5°C (e.g. banana) for tropical crops. The optima for growth, which show rather small deviations among the species range around 25°C. A rough idea of temperature demands of species for growth are given by the cardinal values and derived ranges for 'effective growth' (growth range) and 'optimum growth' (optimum range) shown in Table 4.2.

A more detailed understanding and more exact quantification of the reactions of species and cultivars to temperature is given by growth models (see below). However, growth analysis at medium abstraction level already helps in the understanding of the reactions of the growth components to growth factor intensities. This is shown in Fig. 4.18. For kohlrabi grown with high and low temperature regimes in greenhouses in winter, until about 60 days crop productivity per unit leaf area (net assimilation rate, $NAR = dW \cdot dt^{-1} \cdot LA^{-1}$) shows a rather small temperature reaction, but leaf area (LA) is distinctly promoted by the higher temperature. This is due to a more efficient use of dry matter for leaf expansion by developing thinner leaves, characterized by the higher specific LA ($SLA = cm^2 \cdot g^{-1}$ leaf). This overcompensates the effect of lower leaf weight ratio (LWR = leaf weight $\cdot$ total plant weight^{-1}), resulting from pronounced dry matter partitioning of stems and/or tubers. The early dropping of NAR and LWR in the higher temperature result in an early decline of the absolute growth rate ($AGR = dW \cdot dt^{-1}$) and consequently of total dry matter accumulation (W), (see section on growth models below).

In the low temperature regime, however, slow initial leaf expansion and hence plant growth is followed by a long period of high growth rates, resulting in later but much higher yields. This reaction pattern shows the many influences of temperature and demonstrates the danger of reductionism, as for example to control temperature in greenhouses according to the temperature demands of photosynthesis only (hill climbing method; Lake, 1966). For more details and parameterization see below.

Special processes which should be considered more and which have so far not been sufficiently investigated are as follows.

1. Responses to diurnal temperature amplitudes, called thermoperiodism (Went, 1954). Some species, such as cucumbers or radish (*R. sativus*) do not show specific reactions to day and night temperatures, except to higher night than day temperatures. They react to the mean 24-h temperature and allow for energy-saving temperature strategies in protected cultivation. With other species, such as lettuce, growth rates are promoted by higher day than night temperature while the reverse, higher night than day temperature, causes yellowing and slows down growth (H.J. Wiebe, 1997, Hannover, unpublished observations). Pytlinski and Krug (1993) demonstrated with *Pelargonium* that reactions to diurnal rhythms can be incorporated into a regression model with an additional term. If vernalization-sensitive crops are grown diurnal temperature amplitudes should not be so large as to trigger vernalization effects (see section on prediction, planning and production analysis below).

Table 4.2. Temperature demands and sensitivities of vegetable species (Krug, 1991a).

	Frost sensitivity*
Hot growth range 18–35°C; optimum range 25–27°C	
Okra (*Abelmoschus esculentus*)	+
Roselle (*Hibiscus sabdariffa*)	+
Watermelon (*Citrullus lanatus* var. *vulgaris*)	+
Melon (*Cucumis melo*)	+
Capsicum species	+
Sweet potato (*Ipomea batatas*)	+
Warm growth range (10)12–35°C; optimum 20–25°C	
Cucumber (*Cucumis sativus*)	+
Eggplant (*Solanum melongena*)	+
Sweet pepper (*Capsicum annuum)*	+
Pumpkin, squash (*Cucurbita* species)	+
New Zealand spinach (*Tetragonia tetragonioides*)	
Maize (*Zea mays*)	+
Tomato (*Lycopersicon esculentum*)	+
Phaseolus species	+
Portulaca (*Portulaca oleracea* spp. *sativa*)	
Cool–hot growth range (5)7–30°C; optimum 20–25°C	
Colocasia (*Colocasia esculenta*)	
Globe artichoke (*Cynara scolymus*)	
Onion, shallot (*Allium cepa*)	–
Leek (*Allium porrum*)	–
Garlic (*Allium sativum*)	–
Chicory (*Cichorium intybus* var. *foliosum*)	–
Pak-Choi (*Brassica chinensis*)	–
Scorzonera (*Scorzonera hispanica*)	–
Chives (*Allium schoenoprasum*)	–
Cool–warm growth range (5)7–25°C; optimum 18–25°C	
Pea (*Pisum sativum*)	–
Broad bean (*Vicia faba*)	–
Cauliflower (*Brassica oleracea* convar. *botrytis* var. *botrytis*)	–
Broccoli (*Brassica oleracea* convar. *botrytis* var. *italica*)	–
Cabbage (*Brassica oleracea* convar. *capitata* var. *capitata*)	–
Kohlrabi (*Brassica oleracea* convar. *caulorapa* var. *gongylodes*)	–
Kale (*Brassica oleracea* convar. *acephala* var. *sabellica*)	–
Brussels sprouts (*Brassica oleracea* convar. *fruticosa* var. *gemmifera*)	–
Turnip (*Brassica rapa* ssp. *rapa*)	–
Rutabaga (*Brassica napus* ssp. *rapifera*)	–
Chinese cabbage (*Brassica rapa* ssp. *pekinensis*)	–
Parsley (*Petroselinum crispum*)	–
Fennel (*Foeniculum vulgare*)	(+)
Dill (*Anethum graveolens*)	(+)
Radish (*Raphanus sativus* var. *sativus*)	(+)
Radish (*Raphanus sativus* var. *niger*)	(+)

Continued overleaf

Table 4.2. *Continued*

	Frost sensitivity*
Red beet (*Beta vulgaris* convar. *vulgaris*)	(−)
Swiss chard (*Beta vulgaris* convar. *cicla*)	(−)
Spinach (*Spinacia oleracea*)	−
Lettuce (*Lactuca sativa* var. *capitata*)	(+)
Endive (*Cichorium endivia*)	−
Carrot (*Daucus carota*)	−
Celery, celeriac (*Apium graveolens*)	(+)
Parsnip (*Pastinaca sativa*)	−
Potato (*Solanum tuberosum*)	+
Lambs lettuce (*Valerianella locusta*)	−
Rhubarb (*Rheum rhaponticum*)	−
Asparagus (*Asparagus officinalis*)	(+)
Horse radish (*Armoracia rusticana*)	−
Garden cress (*Lepidium sativum*)	(+)

* +, Sensitive to weak frost; −, relatively insensitive; ( ) uncertain.

2. Responses to fluctuations of growth factor intensities. Plants react quickly in the processes of photosynthesis, respiration and transpiration. They react relatively quickly in morphogenesis to changing temperature and/or air humidity. Abrupt changes in temperature may cause overshoot reactions in growth (Evans, 1963; Fink, 1992), but this does not change significantly the reaction to the 24-h means. Short-term oscillations (in the range of hours) of temperature and water vapour deficit also did not cause specific oscillation effects with kohlrabi (Fink, 1992).

Seasonal adaptation is an important attribute of geographical distribution, even in the subtropics. In temperate and cold climates biennial and perennial species have to be cold tolerant or frost resistant. Some of these species have developed endodormancy periods (rhubarb, chives, asparagus, onion) to strengthen stress resistance for survival. Some have developed systems to control their development by vernalization or photoperiodism and even tropical 'cool' periods may stimulate flowering (citrus, coffee, rice, *Cicer arietinum*). In the tropics and subtropics rainfall or water availability are other major factors controlling seasonal growth. In general, cool-season crops – with a few exceptions such as *Spinacea oleracea* – are geographically widespread throughout the world, but in hot climates are preferentially grown in high altitudes. Warm-season species, however, are restricted to zones with favourable climates and to protected cultivation.

CO_2 concentration

Carbon (C) is a main component of plant biomass, making up 40–45% of total dry matter. It is absorbed as carbon dioxide (CO_2) from the atmosphere by photosynthesis, with only a small part originating from CO_2 solutions in

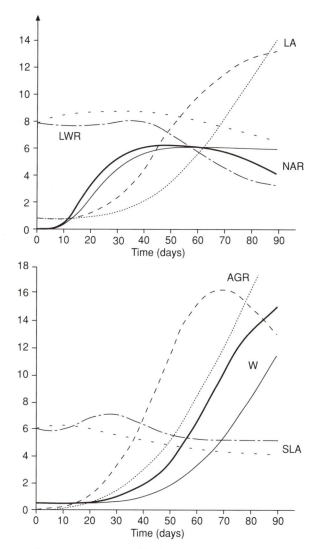

Fig. 4.18. Course of growth parameters of kohlrabi grown in greenhouses with set points of 18/14° (thick lines) and 10/6°C (thin lines), respectively, showing the synergistic and compensating reactions of main growth components to temperature, resulting in early and low and late and high yields, respectively (Krug, 1991a, changed; AGR, g m^{-2} soil area day^{-1}; LA, dm^2; LWR, 0.05 (g g^{-1}); NAR, mg dm^{-2} day; SLA, dm^2 g^{-1}; W, g; dimensions on dryweight basis).

the soil. In 'normal' conditions CO_2 is an odourless, colourless, non-toxic gas, with a molecular weight of 44.01 g mol^{-1}, which is about 1.5 times heavier than air. The CO_2 concentration can be measured as a ratio of CO_2 to air (after Lawlor, 1993, from Nederhoff, 1994):

• in moles (1 μmol mol^{-1} = 1 ppm)

- in volume (1 ml l^{-1} = 1 vpm)
- in weight (mg kg^{-1})
- partial pressure (Pa).

Conversion (20°C, 1013 hPa): 1 ppm = 1 vpm = 1 ml l^{-1} = 1.53 mg kg^{-1} = 1.83 mg m^{-3} = 41.6 µmol m^{-3} = 0.101 Pa.

CO_2 content in the atmosphere is balanced by the output of volcanoes, emissions by industries, households and traffic together with the decomposition of biomass on the one hand, and by the uptake by green plants on the other; it is buffered by the oceans. Due to plant activities it dropped from high original concentrations to about 270 ml l^{-1} in the middle of the nineteenth century. With industrialization it has increased to about 350 ml l^{-1} today and may increase in the future.

The daily volume of CO_2 absorbed by fully expanded crops in favourable conditions corresponds to the CO_2 content of an air column of about 60 m height. This causes CO_2 concentration gradients, with values of less than 250 ml l^{-1} inside field canopies depending on assimilation rate, crop architecture and wind velocity. There is only a limited possibility of increasing the concentration in open fields by manuring or other cultivation practices. In greenhouses with more or less tight covers the CO_2 concentration may be depleted by crops close to the compensation point. Conversely, the protection allows the CO_2 concentration to be increased by artificial supply.

The main effect of CO_2 concentration on plants is on photosynthesis, as demonstrated by Gaastra (1962; Fig. 4.19). With cucumber leaves at a suitable temperature energy supply from light is the limiting factor and photosynthesis increases linearly with irradiance up to 10 Wm^{-2} (PAR). With higher irradiances photosynthesis is limited by CO_2 supply and with high CO_2 concentration also by temperature. Therefore in a steady-state condition, photosynthesis (P) may be described by the CO_2 diffusion process:

$$P = CO_{2ex.} - CO_{2Chl} \cdot (r_a + r_m + r_s)^{-1} \qquad (2)$$

(photosynthesis = cm^3 CO_2 cm^{-2} leaf s^{-1}) where $CO_{2\ ex.}$ is the external CO_2 concentration cm^3 CO_2 cm^{-3}; CO_{2Chl} is the CO_2 concentration near the chloroplasts; r_a is the resistance of the boundary layer; r_m is the mesophyll resistance; and r_s is the stomata resistance; all resistances are expressed in s cm^{-1}.

This function means that photosynthesis can be increased by (i) higher CO_2 concentration in the air; (ii) by a high photosynthetic rate lowering the CO_2 concentration near the chloroplasts; and (iii) by low diffusion resistances, such as small r_a brought about by ventilation or the structure of the leaf surface, small r_m of the genotype and/or small r_s by open stomata.

Besides affecting photosynthesis CO_2 concentration influences morphogenesis. A long-term high CO_2 concentration has effects similar to high irradiances causing shorter internodes and smaller and thicker leaves. It promotes side shoot growth, retards senescence of leaves and may influence sex expression (e.g. female flowers in cucumbers). In general, leaf area ratio

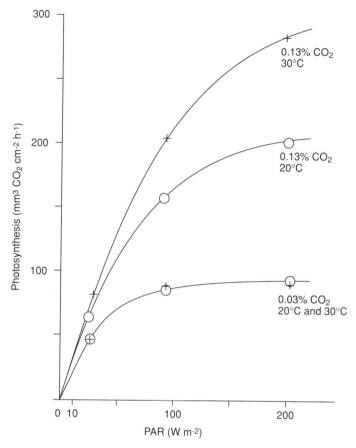

Fig. 4.19. Photosynthesis (P) of a cucumber leaf in relation to irradiance and temperature at a limiting (0.03%) and a saturating (0.13%) CO_2 concentration and two temperatures. Radiation by incandescent lamp 500 W (Gaastra, 1962).

[LAR = leaf area $\cdot$ (total dry weight)$^{-1}$] is reduced, which counteracts increased photosynthetic rates per unit leaf area and reduces long-term CO_2 effects. Therefore CO_2 effects on crops with expanding leaf area can be described only by the parameters of long-term experiments considering leaf growth.

Influences of CO_2 concentration on crop growth can be parameterized by mechanistic or by regression models. Mechanistic models are used by Challa and Schapendonk (1986), Acock (1991), Gijzen (1992), Nederhoff (1994, with literature survey). Since photosynthesis is the dominant process these models reveal realistic results and can be used in practice for CO_2 control. Their validity with young crops with expanding leaves must still be substantiated.

The effect of CO_2 concentration can be introduced into regression models (see equation 13 below) by a multiplicative connected term (Mann, 1987) or by an additive term (Alscher, 1993). According to Alscher, the regression model was more reliable in lettuce for long-term prediction and control but the shortest period is 24 h. On-line control during the 24-h period was performed by the sucros 87 model and its daily results were adapted to those of the regression model. The advantage of short-term optimization, however, was small. The results of Nederhoff (1994) confirmed that plants react primarily to CO_2 concentration means rather than to oscillations and fluctuations.

The effect of CO_2 concentration on the growth of lettuce as a function of irradiance and temperature, calculated by the regression response curve (see equation 13 below) is shown in Fig. 4.20. The MGR of plants follow higher CO_2 concentrations as a saturation function, starting with the compensation point. The compensation point may vary between 30 and 60 ml l[-1] (single leaves) and 150 ml l[-1] in canopies, depending on irradiance, LAI, temperature, CO_2 concentration and other factors. The promotion of MGR by higher CO_2 concentration is higher the more favourable the other growth factors are.

The ecological optimum of CO_2 concentration – here defined as reaching about 98% of the maximum growth rate – therefore ranges roughly between 600 and 1000 ml l[-1]. CO_2 concentrations above this range have small effects on MGR, resulting in saturation curves up to 5000 ml l[-1] for some species (e.g. spinach) and up to 20,000 ml l[-1] in others (Holländer and Krug, 1992). In concentrations over 5000 ml l[-1], which may be reached or overcome when combustion gases are vented into greenhouses, slight growth reductions and physiological disorders may occur. The sensitivity to high CO_2 concentrations depends on the duration of exposure and on species and cultivars (Table 4.3). With a constant 1% CO_2 concentration

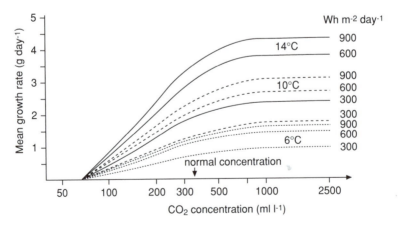

Fig. 4.20. Effect of CO_2 concentration on mean growth rate of lettuce (0–100 g fw, October planting; Mann, 1987).

Table 4.3. Critical duration (days) of CO_2 enrichment for vegetable species using CO_2 concentrations of 1–3% continuously. Criteria: reduction of growth rates, in brackets the duration of treatments in days (Holländer and Krug, 1991).

Species	CO_2 concentration (%)		
	1	2	3
Spinach	> 36 (36)	>15 (15)	3 (10)
Radish (*R. sativus*)	21 (27)	8 (15)	4–5 (10)
Lettuce	> 42 (42)	6–7 (28)	
Kohlrabi	16 (42)	18 (28)	
Lambs lettuce	16 (36)	Not tested	2–3 (10)
Radish (*R. niger*)	20 (36)	4–5 (15)	2 (10)

growth reduction occurred after 16–21 days with kohlrabi, lambs lettuce, radish (var. *niger* and var. *sativus*). Lettuce and especially spinach (*Spinacea oleracea*) and sweet pepper proved less sensitive. The limits of tolerance are generally between 5000 and 10,000 ml l^{-1}. Warm-season vegetables proved more sensitive than cold-season vegetables.

Another effect of high CO_2 concentration on cucumbers was the opening of stomata at concentrations above 1000 ml l^{-1}. This reversible reaction results in an increased sensitivity of plants to all factors that reduce water potential, such as drought, high water vapour pressure deficit of the air or impeded water uptake by fertilization or root damage (Holländer and Krug, 1992). This effect may be the cause of damages with rather low CO_2 concentrations, claimed for some ornamental species (for literature review see Pfeufer, 1990).

PARAMETERIZATION OF GROWTH/DEVELOPMENT AND ENVIRONMENTAL INFLUENCES

Growth functions and growth rates

The general pattern of growth starts with the dry weight of embryo and endosperm (respiration losses during germination are neglected). With autotrophic growth, weight increases exponentially and after the turning point it declines following a saturation function. For fully grown plants this results in approximately sigmoid curves (Fig. 4.21). The curve may be approximately linear for some species and environmental conditions during a more or less long period around the turning point.

The exponential range of the growth curve, that occurs during the growth of young plants, can be described by a physiologically based simple exponential function:

$$W(t) = W_o \cdot e^{bt} \tag{3}$$

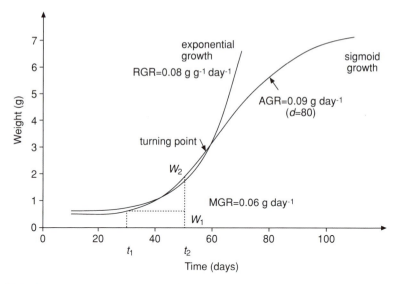

Fig. 4.21. Diagram presenting exponential growth (equation 3), sigmoid growth (equation 5) and parameters of growth efficiency.

where W is weight; W_o is weight at $t = 1$ (first measurement); b is the coefficient of the slope and relative growth rate; and t is time.

By conversion to logarithms the exponential curve is linearized and easier to handle:

$$\ln W(t) = \ln W_o + bt \qquad (4)$$

Attempts to elaborate physiologically based functions for sigmoid growth curves have not been successful (Lioret, 1974). Therefore adaptive functions have been elaborated, which are reviewed by Richards (1969), Bazlen (1985) and others. A classification of major growth functions is presented in Table 4.4.

Polynomial equations are very flexible functions, but should include no more than three or four parameters. Higher order polynomials may display unrealistic indentations and are difficult to interpret. A useful equation is the logistic function developed by Feldmann (1979):

$$W(t) = W_o + A \cdot [\, 1 + e^{b-c \, \ln(t)}]^{-1} \qquad (5)$$

where W is weight; W_o is weight at $t = 1$; A is the maximum value of W; b and c are coefficients; and t is time.

An example is presented in Fig. 4.21 with the sigmoid curve. This function is very flexible, but it cannot follow breaks, which may occur with strong changes in growth factor intensities or by switching into another growth phase. Moreover, growth functions in general equalize not only random variability, but may also smooth out realistic fluctuations. For fresh weight increase, however, fluctuations of irradiances are buffered in a wide

Table 4.4. Growth functions (Bazlen, 1985).

Unlimited growth	
Polynomials	$W = a_0 + a_1 t + a_2 t^2 + \ldots$
Maltus	$W = b \cdot \exp(c \cdot t)$
Time power	$W = b \cdot t^c$
Limited growth	
Without turning point	
Monomolecular	$W = A \cdot (1 - b \cdot \exp(-c \cdot t))$
Mitscherlich	$W = A \cdot (1 - \exp(-c \cdot (t \pm b)))$
Fixed turning point	
Logistic (symmetric)	$W = A \cdot (1 + b \cdot \exp(-c \cdot t))^{-1}$
Gompertz (asymmetric)	$W = A \cdot \exp(-b \cdot \exp(-c \cdot t))$
Variable turning point, asymmetric course	
Feldmann	$W = A \cdot (1 + \exp(b - c \cdot \ln t))^{-1}$
Feldmann, extended	$W = A \cdot (1 + \exp(b - c \cdot \ln t))^{-1} + W_0$
Richards	$W = A \cdot (1 \pm \exp(b - c \cdot t))^{-1/d}$
Including degressive range	
Schneider–von Boguslawski	$W = A \cdot 10^{-z \| \log t \cdot m^{-1/n}}$

A, maximal value of *W*; *m*, period until *A*; *a*, *b*, *c*, *d*, *n* and *z* coefficients.

range (Liebig, 1989) and temperature oscillations affect plant growth according to the respective mean temperature (Fink, 1992).

The Richards function is even more flexible, but needs a second exponent with a fourth parameter (Richards, 1969; Table 4.4). The coefficient *d* determines the form of the growth function:

- $d = +1$ (logistic function)
- $d = -1$ (monomolecular or saturation function)
- $\lim d$ (approximates the Gompertz function).

Criteria of growth efficiency may be parameters of growth functions or the following growth rates:

1. The MGR = $(W_2 - W_1) \cdot (t_2 - t_1)^{-1}$ allows the calculation of growth rates over periods from data of successive measurements (Fig. 4.21). Due to variation of each point measured it is more reliable to use data calculated from the respective growth curve. The range may vary from days to the whole growth period. Interpolation and interpretation of the values must be treated with caution, since these values are linearized.

2. The AGR ($dW \cdot dt^{-1}$) describes the slope in $\Delta t \to 0$ at the respective time and is the first deviation of the growth curve (equation 5) for *t* (Fig. 4.21):

$$\text{AGR}' = [A e^b c\, t^{-c}]^{-1} \cdot [1 + e^b\, t^{-c}]^{-2} \qquad (6)$$

3. The crop growth rate (CGR) expresses AGR or MGR on a soil area basis.

4. The relative growth rate (RGR) relates the growth rate to the weight already reached and indicates the efficiency per total biomass:

$$RGR = (dW \cdot dt^{-1}) \cdot W^{-1} \tag{7}$$

The mean RGR is calculated by:

$$RGR = (\ln W_2 - \ln W_1) \cdot (t_2 - t_1)^{-1} \tag{8}$$

During exponential growth RGR is constant. Beyond the exponential phase RGR decreases due to an increasing portion of unproductive tissue, mutual shading and other factors. Wheeler *et al.* (1993) relate the decline (in a first linear approximation) to the accumulated developmental thermal time:

$$R_{pi} = R_o [1 - (\sum_{t=0}^{t=i} (T - T_b)) \theta_f^{-1}] \tag{9}$$

where R_o is the maximum potential RGR; R_{pi} is the RGR at day i; T is the temperature in °C; T_b is the base temperature for the rate of development; and θ_f is the developmental thermal time at which R_p is zero (°Cd).

When evaluating the effect of environmental influences on growth rates it must be kept in mind that in non-linear ranges only corresponding growth phases should be compared. Detailed reviews of the quantitative analysis of growth are given by Richards (1969), Visser (1963) and others.

Growth models

A precursor of growth models is a descriptive one factor linear approach (OFLA), developed as the temperature sum concept (TSC, also-called heat unit system) by Rèaumur (1735) and improved by de Candolle (1855) by introducing a base temperature. It has been used by numerous authors, especially since the 1920s.

The OFLA has the precondition that the respective factor is dominating and the growth/development rates are linearly related to the respective factor quantity. Domination occurs when

1. The respective factor is far outside and the other factors are inside the optimum range of the response surface.
2. When metabolism draws its supplies from storage organs and is mainly controlled by temperature (e.g. germination, forcing).
3. When environmental stimuli, such as photoperiod, vernalization, induction of dormancy or generally single factors govern the development.

Although biological reactions in general show non-linear responses, linearity may exist for growth/development rates for limited ranges of intensities or may be achieved by appropriate approaches (see below).

The basic function of the TSC is (see Thornley, 1987):

$$S = \sum \Delta t \, (T - T_{min.}) \tag{10}$$

where S is the temperature sum (°Cd); Δt is the time interval in days; T is the mean temperature of Δt; and $T_{min.}$ is the base temperature, equal to T when growth rate or DeR equals 0.

If Δt equals 1 day, it follows that

$$S = \Sigma\ (T{-}T_{min.})\ (°Cd) \tag{11}$$

If detailed meteorological data are lacking, actual temperatures are mostly calculated by $(T_{max.} + T_{min.}) \cdot 2^{-1}$. The ontogenetically relevant, theoretical (extrapolated) $T_{min.}$ of the species has to be subtracted after which the daily values are summed up. The TSC may be extended by introducing correction factors for further inputs, such as water potential or photoperiod. These functions may be applied for growth/development phases or total growth cycles. In some models temperature sums are used to parameterize the course of development, such as for cucumbers – potential photosynthetic rate, potential length growth rate, potential growth rate of fruits, senescence of leaves (Augustin, 1984); fruit development (Marcelis, 1994); ontogenetic decline of RGR (Wheeler *et al.*, 1993) – and for development in general (Spitters *et al.*, 1989). Other examples for application of temperature sums are presented below and in Chapter 3. A critical review of the temperature sum concept is published by Wang (1960).

A more comprehensive analysis and control of growth than using the TSC is possible with growth models, which can be constructed at different abstraction levels. At high abstraction levels the descriptive approach is appropriate. As shown in Fig. 4.22 in the first and most simple approach the plant or crop is considered as a black box, acting as a transfer system and the relations between the intensities of one input factor – the others are considered as constant and close to the optimum – is calculated by a saturation function according to Mitscherlich (1919):

$$W = A \cdot [1 - \exp\ (-\ c\ (x \pm x_{min.}))] \tag{12}$$

where W is the fresh or dry weight; A is the maximum value of W; c is the coefficient of effectiveness of the input factor; x is the intensity of the input factor; and $x_{min.}$ is the value of x for $W = 0$.

This saturation function shown in Fig. 4.23 demonstrates that yield

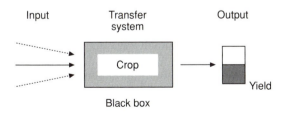

Fig. 4.22. Schematic presentation of the black box approach to parameterize input–output relationships.

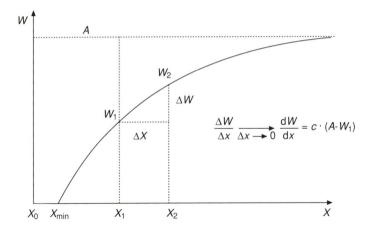

Fig. 4.23. Diagrammatic derivation of the Mitscherlich function (transformed function 12).

increase $(W_2 - W_1)$ per unit input factor $(x_2 - x_1)$ is proportional to the distance of W_1 to the theoretical maximum yield (A). The slope of this relationship is given by the coefficient c, which is characteristic for each input factor. The advantage of this approach, frequently used in agriculture, is that all environmental influences of the site are considered.

Following Mitscherlich, the effect of more than one input factor on plant growth can be parameterized by connecting respective terms for each growth factor by multiplication:

$$MGR = A \cdot [1 - \exp(-c_1 (x_1 - x_{1min.}))] \cdot [1 - \exp(-c_2 \cdot (x_2 - x_{2\,min.}))] \cdot \ldots$$
$$[1 - \exp(-c_n (x_n - x_{nmin}))] \tag{13}$$

where A is the maximum value of MGR; x is the growth factor; x_{min} is the factor intensity when MGR $= 0$; and c_{1-n} are parameters describing factor effectiveness (Krug and Liebig, 1984b, 1988, 1994).

To consider the effect of artificial light, the irradiance term in equation 13 has to be written as

$$[1 - \exp(-c_x (1.4 S_{art.} + S_{nat.} - S_{min.}))] \tag{14}$$

where $S_{art.}$ is artificial light, $S_{nat.}$ is natural irradiance and $S_{min.}$ is the compensation point (Brückner, 1990).

This regression function can be computed for the output factor yield, as preferred by Mitscherlich for agricultural crops, where harvest time is determined by ripening of the crop and the growth period is of minor interest. Many species of vegetable crops are harvested at defined sizes. Yield is composed of a number of pieces of good quality, but with the intensive usage of area and time – caused by high investment and running costs – the length of the growing period is of primary importance. This can be calculated directly or via growth rates.

An example of response surfaces based on growth rates and calculated

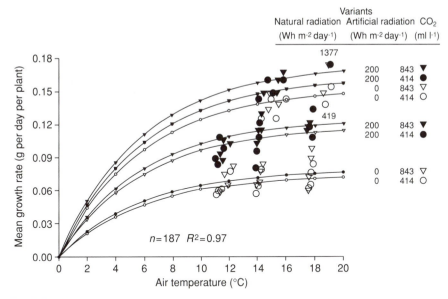

Fig. 4.24. Selection of response surfaces of mean growth rates of lettuce transplants (emergence to 4 g fw), representing the effects of the intensities of four growth factors (soil temperature = 15°C, $T_{min.}$ = 0°C; Krug and Liebig, 1994).

by the multiplicative regression function is given in Fig. 4.24 for lettuce transplants. These curves are more stable than those calculated by multiple regression analysis, especially in the range of low factor intensities. This is achieved by fixing the intercept points on the x-axis (compensation points) which can be determined in separate experiments, by iteration or determined empirically.

The saturation response is realistic for some growth factors, such as radiation and CO_2 concentration in a reasonable range of intensities. It is not for others, such as temperature, for which superoptimal intensities lead to decreasing response. For these the range of testing and interpretation has to be restricted or the decreasing effect may be incorporated with the additional term

$$\cdot \, [1 - \exp{(+ c \, (x - x_{min.}))}] \tag{15}$$

An important precondition of regression models is the independence of the intensities of the input factors which can be at least partially achieved by cultivating the crops in controlled atmospheres. If this is secured, the function results in realistic response surfaces. To check reliability, growth duration until the crops reach marketable size can be compared with actual measurements or reliable growth curves and with simulations. Such comparisons for greenhouse lettuce transplants demonstrated that about 40% of all simulated curves hit the target of 4 g on the same day (Fig. 4.25). If a tolerance of ± 0.2 g fresh weight is accepted, 80% of all curves hit within ± 2

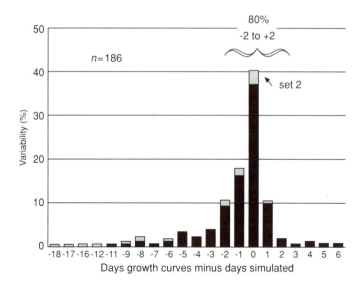

Fig. 4.25. Variability of growth periods estimated by the regression function 13, demonstrated by the frequency distribution of differences between dates of final weight of lettuce transplants (4 g fw) derived from growth curves and simulated (Krug and Liebig, 1995).

days. But some considerable deviations occurred due to CO_2 concentration, insufficiencies during germination in set 2, insufficiencies of the model and due to general variation.

For long-term reactions as used for production planning, MGR was calculated for total growth periods (50% emergence or planting to harvest – the emergence period was always calculated separately – see below). Examples are radish, lettuce and kohlrabi (Table 4.5). If growth passes through different periods such as the vegetative and generative phase, MGR should be calculated separately. To simulate dynamic growth, response surfaces of the RGR are preferred, especially for the exponential phase (Fig. 4.26).

The growing period (t) for final weight (W_f) results from

$$t = W_f \cdot MGR^{-1} \tag{16}$$

To simulate dynamic growth the following functions are used:

exponential growth: $\ln W_f = \ln W_o + RGR_1 + RGR_2 + \dots RGR_n$ (17)
linear growth: $W_f = W_o + AGR_1 + AGR_2 + \dots AGR_n$ (18)

With descending or sigmoid growth it is preferable to split the total duration into shorter periods. The subdivision can be based on formal phases defined by weight, or based on physiological phases. With species with less pronounced development phases, such as species harvested in the vegetative phase, formal phases yielded better results, except with extreme environmental conditions (Bazlen, 1985). The disadvantage of physiologically defined phases is the supplementary need to estimate the phase limits,

Table 4.5. Models for prediction and planning of growth and development (examples with vegetable species).

One growth factor
Linear(ized) relationships
Emergence period
    pea   Ottosson (1958)
    lambs lettuce   Lederle *et al.* (1984)
    many species   Bierhuizen and Wagenvoort (1974);
                                             Krug (1991a); Krug *et al.* (1985)

Regrowth
    asparagus – yield prediction   Liebig and Wiebe (1982)
    onion – growth rate / period   Krug (1983)
    rhubarb – cold requirement   Krug (1991b)
Vernalization
    Chinese cabbage   Nakamura (1977)

Non-linear relationships
    cauliflower   Wiebe (1975, 1979); Wurr *et al.* (1990)
    broccoli   Marshall and Thompson (1987)
    kohlrabi   Wiebe *et al.* (1992); Wiebe (1994)
    Chinese cabbage   Mori *et al.* (1980)
    onion   Brewster (1987)
Diverse phases – development
    pea   Ottosson (1958)
    onion   Brewster (1987)
    cauliflower   Wiebe (1975, 1979)
    broccoli   Marshall and Thompson (1987)
    kohlrabi   Wiebe *et al.* (1992); Wiebe (1994)
Production areas   Seljanow and Nuttonson (1948)
                                               (cited in Walter, 1960)

Diverse growth factors
Linear(ized) relationships
    emergence onion   Finch-Savage and Phelps (1993)
    sprout length growth – asparagus   Kailuweit and Krug (1995)

Non-linear relationships
radish
    R. sativus   Krug and Liebig (1979b)
    R. niger   Krug and Liebig (1984a)
    lettuce transplants   Krug and Liebig (1988, 1994)
    lettuce crops   Scaife *et al.* (1987); Mann (1987);
                                             Seginer *et al.* (1991); Liebig and Alscher
                                             (1993); Alscher (1993); Wheeler *et al.*
                                             (1993)
    kohlrabi transplants   Krug and Liebig (1988)
    kohlrabi crops   Matitschka (1996); Brückner (1990)
    cucumbers   Liebig (1984); Schacht and Schenk (1990);
                                             Schacht (1991); Augustin (1984);
                                             Marcelis (1994); Schapendonk and
                                             Brouwer (1984); Heuvelink and
                                             Marcelis (1989)

Continued overleaf

Table 4.5. *Continued*

tomato	Wolf *et al.* (1986); Jones *et al.* (1991); Bertin and Heuvelink (1994); Seginer *et al.* (1986); Heuvelink and Bertin (1994); Heuvelink and Marcelis (1989); Heuvelink (1996); De Koning (1994); Gary (1988)
cauliflower	Scaife *et al.* (1987); Grevsen and Olesen (1994)
leek	Scaife *et al.* (1987)
celery	Scaife *et al.* (1987)
beans	Doust *et al.* (1983) Hoogenboom *et al.* (1990)
cabbage	Isenberg *et al.* (1975)
various species	Krug *et al.* (1988)

which may cause additional variation. For species with marked physiological growth stages, physiologically based phases are preferred.

An example of growth simulation using radish roots is shown in Fig. 4.26. It shows the fluctuation of irradiance and temperature in greenhouses, the corresponding fluctuation of RGR and cumulative growth simulated using formal and physiologically based phases, respectively.

The descriptive approaches to model plant growth have been improved in recent years and now work satisfactorily (Thornley, 1987), but they are always simplifications and therefore the results are always probabilities. The main constraints are as follows.

1. Insufficient databases, especially insufficient independence of variation of the input factors.
2. Varying environmental rest complex (outside the system considered, e.g. different climates, greenhouses, production techniques).
3. Limitations of statistical methods to describe high-degree interactions, the injury range and feedback relationships.
4. Shortest period of simulation is 24 h.
5. The variability between cultivars and effects of special raising techniques, such as grafting.

In general descriptive models are preferred for modelling over longer periods. Reviews and additional remarks and approaches are presented by Erikson (1976), Hammer and Langhans (1978; response surfaces based on polynomials), Wechsung and Orth (1991) and Krug and Liebig (1994).

To overcome insufficiencies and to obtain a better understanding of causal relationships the black box can be enlightened and the relationships between the growth factors and growth components can be parameterized, first at a rather high abstraction level. Growth of plants can be dissected into the growth components (Watson, 1952; Radford, 1967; Hunt, 1982):

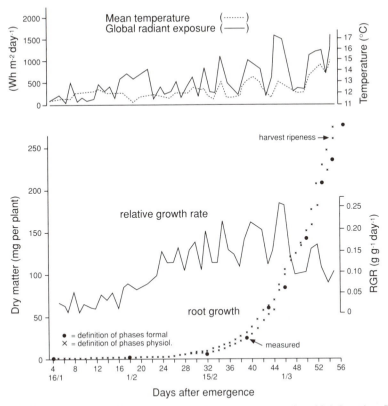

Fig. 4.26. Simulation of growth of radish roots (*R. sativus*) emerged on 13.1, based on RGR as a function of radiation, temperature and ontogenetic stage (data from Bazlen, 1985).

$$RGR = dW \cdot dt^{-1} \cdot LA^{-1} \, (NAR) \cdot LW \cdot W^{-1} \, (LWR) \cdot LA \cdot LW^{-1} \, (SLA) \qquad (19)$$

where W is the weight of the total plant; t is time; LA is leaf area; and LW is leaf weight.

RGR is the product of NAR, LWR and SLA. NAR refers to growth rates, not as RGR of the total, but of the 'productive' tissue (Gregory, 1917):

$$NAR = dW \cdot dt^{-1} \cdot W_{prod.}^{-1} \qquad (20)$$

Productive tissue can be defined as leaf area, leaf weight or parameters such as leaf nitrogen, leaf chlorophyll and so on. According to Gregory (1926) the difference quotient is calculated by

$$NAR = (W_2 - W_1) \cdot (t_2 - t_1)^{-1} \cdot (\ln LA_2 - \ln LA_1) \cdot (LA_2 - LA_1)^{-1} \qquad (21)$$

Like RGR, NAR is a function of genotype, ontogenetic stage and environmental influences. Mean values in the vegetation period of moderate climates vary between 2 and 12 g m^{-2} day^{-1}, with average values of 6–8 g m^{-2} day^{-1} dry matter.

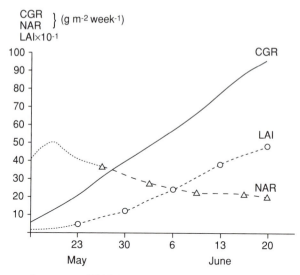

Fig. 4.27. Ontogenetic courses of NAR (net assimilation rate of dry matter), LAI (leaf area index) and the resulting CGR (crop growth rate in dry matter) of white clover, data from Nösberger, 1964: thin lines are extrapolated).

LWR presents the portion of productive tissue to total tissue. SLA designates the leaf area per unit leaf weight, a measure of the efficiency of leaf biomass in relation to its area.

NAR multiplied by LAI [LAI = leaf area per unit soil area ($m^2\ m^{-2}$)] results in the CGR. As shown in Fig. 4.27 in field conditions of a temperate climate, CGR almost parallels LAI. The same was proved for cucumbers in greenhouse production by Liebig (1983). Therefore the light absorbing surface is a very important factor of growth efficiency, which can be more easily controlled by agricultural techniques (e.g. irrigation, fertilization, spacing) than NAR. Mean values of LAI of fully grown crops are presented in Table 4.6. They depend on genotype and environment including crop architecture, which determine leaf size and form, leaf angle and hence the extinction coefficient of the leaf canopy. Due to mutual shading LAI and NAR are often negatively related.

The significance of dissecting the black box 'plant' into growth components to interpret environmental influences is also shown in Fig. 4.28. The RGR of cucumber plants at leaf stage 5 is increased by irradiance and temperature. Increasing irradiance promotes NAR, but decreases SLA and has a slight decreasing effect on LWR. Increasing temperature, however, favours SLA and has no significant effect on NAR at low irradiance level. Therefore high temperature is the most important factor for leaf expansion on young cucumber plants. For fully developed and healthy leaf canopies, however, fruit production is mainly controlled by NAR via irradiance.

At medium abstraction level another retrospective method to analyse

Table 4.6. Mean values of leaf area index of full grown crops (Krug, 1991a).

Field	
Cereals, maize	8–9
Beta species	6–9
Bush beans	5–6
Potatoes	4–6
Cabbage	5
Kale	4
Onions	4
Greenhouse	
Radish	4
Cucumber	3

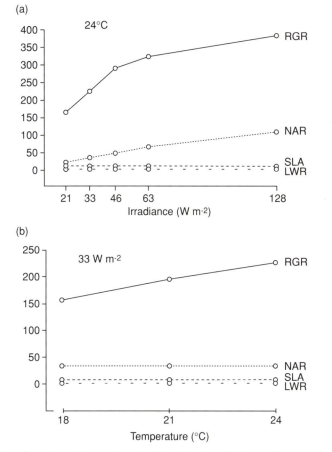

Fig. 4.28. Growth analysis of cucumber plants in leaf-stage 5, (a) growth components as a function of irradiance (Sylvania Cool White VHO 215 W + 6% incandescent light at 24°C; (b) as a function of temperature with 33 W m^{-2} irradiance. [RGR (a) = mgg^{-1} day^{-1}; (b) = cgg^{-1} day^{-1}; SLA = dm^2 g^{-1}; LWR = gg^{-1}; NAR = mg CO$_2$ dm^{-2} day^{-1}; data from Lorenz, 1980].

genotype–environment interactions is the 'morphological yield analysis' developed by Heuser (1927/28). Heuser dissected the yield of cereals into morphological components, as exemplified for peas by Krug (1991a):

(yield dt ha^{-1}) = (shoots ha^{-1}) · (pods shoot^{-1}) · (seeds pod^{-1}) · (mean single seed weight)

$$63 \; dt \; \mathrm{ha}^{-1} = 7 \cdot 10^5 \cdot 5 \cdot 6 \cdot 0.3 \cdot 10^{-5} \, dt$$

A high yield may be based on (i) high crop density (sowing rate, tillering); (ii) high number of pods per shoot; (iii) high number of seeds per pod; and/or (iv) high weight of single seeds. The yield of other species can be dissected similarly. These parameters can be related to environmental factor intensities, to production techniques or to genotypes (species, cultivars).

A more mechanistic approach with another splitting of the black box into growth components regarding plant development and dry matter partitioning is worked out by Stützel (1995) with broad beans (Fags: *Vicia faba* growth stimulator). Daily dry matter production (AGR) is the product of the intensity of a main growth factor in open fields, radiation (PAR), which is multiplied by the proportion of light intercepted (Q) and by light use efficiency (LUE) in grams of dry matter per megajoule of intercepted PAR (Monteith, 1977):

$$AGR = PAR \cdot Q \cdot LUE \tag{22}$$

In closed crops the proportion of PAR intercepted depends among others on canopy architecture and optical properties of the leaves. It can be described by an exponential function of LAI and light extinction coefficient (K):

$$Q = 1 - \exp\left(-K \cdot LAI\right) \tag{23}$$

Light use efficiency comprises a variety of crop and environmental factors and ranges normally between 2 and 3 grams of dry matter per megajoule of PAR intercepted. Decreasing light extinction coefficients characterize a more even light distribution over the canopy and thus increase light use efficiency. It can be calculated by:

$$LUE = (-a + b \cdot K)^{-1} \tag{24}$$

where a and b are coefficients to be determined analytically for the respective cultivars (for broad beans $a = 0.078$; $b = 0.733$). Total dry matter including dropped leaves is the sum of the daily increments of AGR.

Development is expressed by a two-digit scale. From emergence (stage 1) to the onset of flowering (stage 2) the daily progress towards flowering (RFL) is described by decimals (1.1, 1.2 and so on). RFL is related to a multiplicative effect of mean temperature, in relation to an optimum temperature for flowering, and daylength. Leaves are produced until either the maximum node number is reached (input data) or until dry matter partitioning to leaves ceases. The number of leaves is a quadratic function of the temperature sum. After the onset of flowering the growth stage is incremented by 0.1 for each leaf produced. Tillering can be incorporated as well as pod number and seeds bearing pods, dry matter partitioning between shoots

and roots, between vegetative and generative organs, dry matter translocation and dry matter losses.

This model describes growth and development in much more detail
than the model discussed earlier and presented a satisfactory description of
reality, if biotic and abiotic stress is excluded. Weak points are generalization of the parameters and consideration of senescence.

At low abstraction level growth is an open, dynamic, multifactorial and
multi-hierarchical system that is being actively parameterized primarily by
physiologists. These models are called mechanistic or explanatory models,
but it has to be kept in mind that they are still descriptive and when describing whole plants they include subsystems at high abstraction level (e.g. substrate partitioning, morphogenesis). The first prototype ELCROS was
presented by de Wit and Brouwer (1968). More recent approaches have been
reviewed by Loomis *et al.* (1979), Penning de Vries (1983), Thornley and
Johnson (1990), Charles-Edwards *et al.* (1986) and Pakes and Maller (1990).

An example of the schematic structure of a mechanistic growth model is
presented in Fig. 4.29. A central element is a carbohydrate reserve pool. It is
fed by photosynthesis as a function of light and CO_2 and parameterized by
the assimilation rate. The pool is depleted by respiration, delivering CO_2.
This pool supplies photosynthates to shoots and roots, controlled by the
related growth rates. These organs can be described in more detail by criteria of form and size, such as leaf area, specific leaf weight and others. Dotted
lines indicate information flow, especially feedback relations. This model
presents a good summary of the main processes and pathways.
Photosynthesis and growth respiration are realistically parameterized. The
models, however, are still in a preliminary phase with regard to maintenance respiration (Spitters *et al.*, 1989).

Limitations of the mechanistic models are as follows.

1. Insufficient knowledge of some basic physiological processes, for example of distribution patterns of metabolites between different organs (for a
review see Marcelis, 1994) as well as their morphogenetic and feedback
reactions as a function of internal and external conditions, which are especially important for simulating growth in young plants. Therefore these
processes are mainly handled in a descriptive manner.
2. The large number of species-specific parameters, which have to be
worked out experimentally or derived from the literature, which are difficult to handle and hinder optimization.
3. In spite of the numerous parameters not all processes can be included and
these remain as an unconsidered rest complex.
4. The high degree of interaction inside complex and complicated biological
systems, which are difficult to parameterize.

The mechanistic models are very sophisticated and reveal more theoretical possibilities than descriptive models. They are valuable tools for teach-
ing, to uncover gaps in knowledge and to simulate short-term energy flow
and growth processes dominated by photosynthesis. They are especially

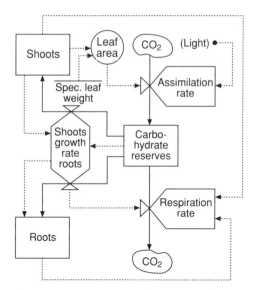

Fig. 4.29. Structure of a mechanistic growth model (de Wit and Penning de Vries, 1983).

important in stimulating integrated thinking and hence to system theoretical approaches. The initial idea, however, of developing a model of general validity for all or at least many species has not been realized.

To overcome some of these problems, the complicated mechanistic models can be simplified in the following ways.

1. By replacing modules such as photosynthesis, by regression functions calculated on the basis of data from the mechanistic models (Matitschka, 1996)
2. By constructing summary models which are restricted to main processes. Such a model at a somewhat higher abstraction level is 'sucros 87' (Spitters *et al.*, 1989). This model has been adapted to kohlrabi (Matitschka, 1996), lettuce (Alscher, 1993), onions (de Visser, 1994) and other crops. The model 'sucam' was developed for greenhouse crops by Gijzen (1992) and is based on sucros 87.

In summary, modelling plant growth and development is still in its early stage and its many gaps and insufficiencies are obvious. However, in the last few decades good progress has been made. There is no doubt that environmental influences and integrated plant reactions can only be understood using system theoretical approaches and this knowledge can be better applied for production management. This directs the grower to a more holistic view of agriculture production and environment protection.

PREDICTION, PLANNING, PRODUCTION ANALYSIS

Quantification and especially parameterization of environment–man–plant relationships enable the prediction of growth and development and hence more precise planning and management of production techniques including rotations. Moreover, retrospective analysis of the causes of yield variability due to weather conditions and/or cultivation practices offer the possibility of increased efficiency by the following methods.

1. Prediction of time of harvest and/or magnitude of yield.
2. Timing of production techniques, such as sowing, planting, fertilizing, irrigation, covering and uncovering with protective covers.
3. Optimization of production techniques, such as row spacing and density (Gray and Finch-Savage, 1994). In greenhouses additional control of light, temperature, CO_2 and other factors (Krug and Liebig, 1979a,b, 1994).
4. Planning and optimization of production techniques for an efficient use of production capacities, such as fields, greenhouses, machines, labour; balanced or controlled market supply; for minimal usage of resources, minimal costs or for highest financial return.
5. Analysis and prediction of physiological disorders, as for example internal browning in lettuce (Wissemeier, 1996).
6. Analysis of the consequences of injuries caused by climatic factors or by pathogens.
7. Assessing the impact of potential climatic changes (Wheeler *et al.*, 1993).

The method of growth prediction, going beyond blue print knowledge, is based on the calculation of growth periods or growth and development rates as a function of one or more dominating growth factors. Most procedures are based on descriptive approaches. In special cases, as for example the after-effects of leaf damage, mechanistic models may be preferred. The crux is less the parameterization of the factor–crop relationships; greater uncertainties are involved in the prediction of the future weather conditions, which are in principle unknown. Estimates can be based on long-term means (normals) or on maximal frequencies of special weather courses. The accuracy of estimation also depends on the length of the period considered and on the season (Krug and Liebig, 1994, 1995).

ONE DOMINATING GROWTH FACTOR

Examples of modelling approaches using vegetables are presented in Table 4.5, arranged according to the number of dominating growth factors involved, the mathematical relationships and the processes. Parameterization and hence prediction is easier, if only one dominating factor controls growth or development (see section on growth models above), especially if there is a linear or linearized relationship.

The emergence period (sowing to emergence) can be calculated rather reliably, if temperature is the ruling factor and no restrictions are acting, such as water shortage, salinity, lack of light stimulus or endodormancy. In this respect greenhouse crops are less problematic than field crops, sowings in maritime climates easier to parameterize than those in continental climates and spring sowings easier than summer sowings.

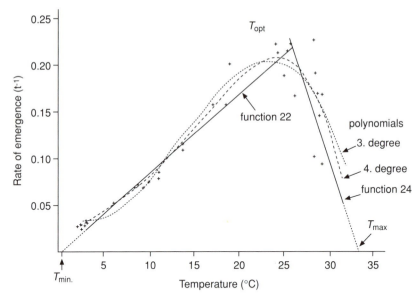

Fig. 4.30. Emergence rate of corn salad as a function of temperature, calculated with functions 22 and 24 compared to polynomials of third and fourth degree (Lederle *et al.*, 1984).

The germination period as a function of temperature can be parameter-ized directly by polynomials (4–5°; Lederle *et al.*, 1984) or by germination rates (t^{-1}; Feddes, 1971; Lederle *et al.*, 1984; Fig. 4.30). The germination rates have the advantage of being approximately linear with temperature in the ranges of $T_{\text{min.}} - T_{\text{opt.}}$ and $T_{\text{opt.}} - T_{\text{max.}}$, respectively. Linearization makes it easier to determine the cardinal points experimentally, affords a lower num-ber of parameters and these are physiologically explainable. Respective functions, based on equation 11, are:

$$T = T_{\text{min.}} + S \cdot t^{-1} \tag{25}$$

$$t = S \cdot (T - T_{\text{min.}})^{-1} \tag{26}$$

$$T = T_{\text{max.}} - S \cdot t^{-1} \tag{27}$$

where T is the mean temperature in °C; $T_{\text{min.}}$ is the minimum theoretical temperature for germination in °C; S is the temperature sum in °Cd for, e.g. 50% emergence or slope of the line; t is the time in days; and $T_{\text{max.}}$ is the maximum theoretical temperature of germination in °C.

These parameters have been determined for many vegetable species and cultivars, allowing for presentation of respective parameters and response surfaces (Bierhuizen and Wagenvoort, 1974; Krug 1991a; Fig. 4.31).

In practice the germination period can be estimated as a function of species and seed or soil temperature. In greenhouses, time to emergence can be calculated by the set point of the heating unit, if the relations between

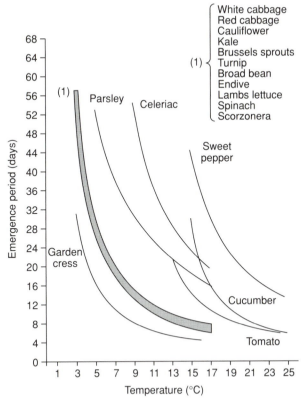

Fig. 4.31. Examples for duration of emergence of vegetable species as a function of temperature of the seeds (calculated with function 26; Krug, 1991a).

outside climate, set point air temperature and soil temperature are modelled. The estimates enable timing and the consideration of physiological as well as economic aspects and in greenhouses additional energy consumption or costs. These calculations demonstrate among other things, that during the cold season with high set points for heating, the energy consumption is lower if plants are cultivated on tables, where the trays are approximately at air temperature. If low set points are used cold-season species germinate faster if sown in the ground, which is warmer due to the heat storage in deeper soil layers (Krug *et al.*, 1985).

In field conditions Ottoson (1958) used the temperature sum concept successfully to estimate the developmental progress of pea seedlings (Fig. 4.32) in order to determine the date of the next sowing to ensure the daily bulk of harvest in summer. In fields the temperature sum concept, however, will only work if a threshold water potential is secured. Gummerson (1986) in his experiments with *Beta vulgaris* used the term hydro-thermal time. A similar approach has been used and extended by Bradford (1990) for lettuce and by Dahal and Bradford (1990) for tomato. In onions a linear relationship

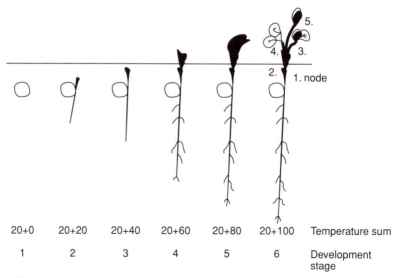

Fig. 4.32. Development of pea seedlings in relation to the temperature sum (Ottosson, 1958).

was shown between the time to 50% emergence and soil water potential by Hegarty (1976). The sensitivity to water potential, however, may differ among the stages of the germination process (Ross and Hegarty, 1979). Onions for example are especially sensitive during the initiation of radicle growth (Finch-Savage and Phelps, 1993).

Regrowth is also mainly controlled by temperature, if water potential is favourable and there is no endogenous dormancy. Examples for linearized relationships to temperature are the regrowth of onion leaves in the range of 7 to about 22°C. This reaction was not significantly influenced by luminous exposure or daylength (Fig. 4.33, not linearized; Krug, 1983).

Length growth of asparagus spears is linearly related to temperature in a wide range (7–24°C), but to estimate growth rates an additive term of actual spear length (Blumenfield *et al.*, 1960) and for white asparagus a third term for soil resistance (Kailuweit and Krug, 1995) have to be added:

$$MGR = a + bT + cH + dR \tag{28}$$

where T is temperature; H is the length of the respective spear; R is soil resistance; and a, b, c and d are coefficients.

Liebig and Wiebe (1982) worked out a model based on the temperature sum concept to predict asparagus yields. The mean daily temperatures (means of 7, 14 and 21 hours) of the previous day with the base temperature of 0°C was linearly related to the daily spear yields. Using a forecast function for 3 days in advance and forecasting temperature data the predicted yields fit the actual yield quite well (Fig. 4.34). In the 6 years of investigation the temperature sums during the harvest season were correlated to the annual yields ($r^2 = 0.61$). The yield per temperature sum, however, declined

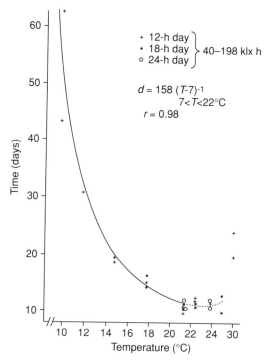

Fig. 4.33. Influence of temperature on the time required to develop 6 g leaf weight (equation 26, non-linear presentation) of the onion cultivar Stuttgarter Riesen (Krug, 1983).

during the harvest period, due to depletion of reserves, or drought, and the yields per final temperature sum varied considerably among the 6 years, caused among other factors by the environmental conditions during the regrowth period in the previous year. These disturbances clearly demonstrate the limits of the temperature sum concept.

In some plants a cold requirement has to be fulfilled before regrowth can commence. With rhubarb the cold requirement is linearly related to temperatures between 10 and –2°C (Loughton, 1965) and varies with cultivars between 110 and 150 °Cd (cv. Timperly Early) and 300–310°Cd (cv. Sutton Seedless). These temperature sums are a criterion of earliness in forcing in the autumn and enable one to predict the start of regrowth as a function of extreme or mean temperature regimes (Fig. 4.35).

Vernalization, the cold requirement for flower bud initiation and/or bolting, is another process dominating the development of sensitive species. Often, however, it is modified by daylength. A linear relationship between temperature and vernalization has been found by Nakamura (1977) for Chinese cabbage. To calculate flower initiation he used the function:

$$t = 87°C \cdot (13°C - T_{0-13°C})^{-1} \tag{29}$$

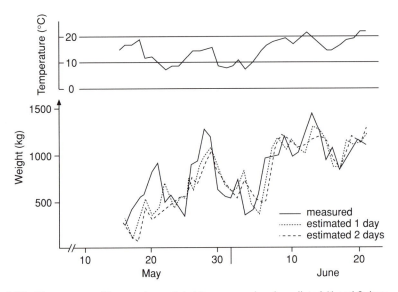

Fig. 4.34. Time course of temperature related to measured and predicted (1 and 2 days before) yield of asparagus (Liebig and Wiebe, 1982).

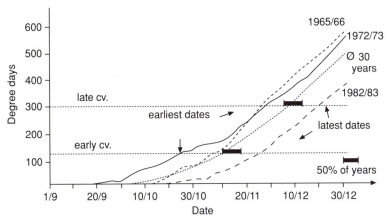

Fig. 4.35. Mean and extreme values of cold units of 30 years (degree days from 10 to −2°C soil temperature, 10 cm depth at 0700 hours in Hannover), indicating the earliest and latest starts of regrowth of two rhubarb cultivars (Krug, 1991b).

where t is the days to flower initiation; and 0–13°C are the limits of vernalizing temperatures.

This approach allows for recommendations concerning temperature control during the raising period in greenhouses or timing of sowing. Nakamura states, however, that sensitivity of the plants increases with their ontogenetic age.

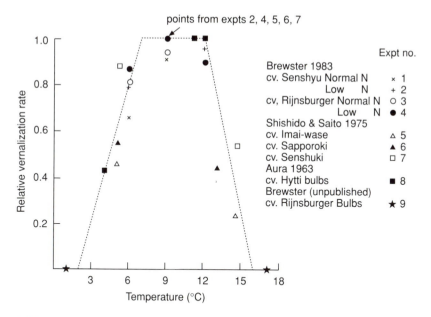

Fig. 4.36. Relative rate of vernalization versus temperature of Japanese and European onion cultivars (Brewster, 1987).

More comprehensive models of vernalization deal with non-linearity, which, however, can be linearized by

- using development rates (see also Chapter 3)
- dividing the temperature scale into more or less linear sections as shown by Brewster (1987) with onions (Fig. 4.36)
- weighting the non-linear temperature effect per unit temperature graphically (Wiebe *et al.*, 1992; Wiebe, 1994)
- weighting by functions. Thornley (1987) proposed the use of a polynomial function:

$$h = \Sigma[a \cdot (T - T_{min.}) + b \cdot (T - T_{min.})^2 + \dots c \cdot (T - T_{min.})^n] \tag{30}$$

where h is the increment of vernalization with time; T is temperature; $T_{min.}$ is the base temperature; and a, b and c are coefficients. Examples of these approaches are discussed below.

More comprehensive vernalization models have to consider devernalization as demonstrated by Wiebe *et al.* (1992). This is true especially for continental climates and greenhouse production with high day temperatures caused by high irradiances. The model worked out for kohlrabi is based on the response curve presented in Fig. 4.37. The number of hours needed for selected temperatures to induce bolting were determined in growth chamber experiments. For 5°C, for example, 6 weeks (1008 h) were required for 100% bolting, resulting in an effect of +0.1% per hour. The same

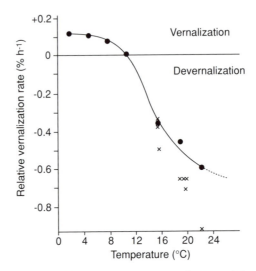

Fig. 4.37. Schematic diagram of the hourly temperature effect on relative vernalization rate of kohlrabi. Line and (•) = marketable yield; (×) = 70% marketable yield (Wiebe *et al.*, 1992).

procedure for temperatures above 11°C resulted in negative effects, compensating preceding cold effects. For 22°C, an effect of –0.6% per hour was derived. Therefore 24 h at 22°C per week cancelled the effect of preceding 6 days of 5°C cold. This model can be used in greenhouses for temperature control to prevent bolting and in combination with a growth model to minimize heating costs.

For longer cycles of development composed of more than one, more or less uniform phases, each phase has to be modelled separately:

Ottosson (1958) divided the development of peas into three phases:
• sowing to the development of the fifth node
• development of additional non-flowering nodes
• first flower until harvest (tenderometer reading 110).

Ottosson states that development until the fifth node required 100–120°Cd (see Fig. 4.32). Thereafter 40°Cd are needed for each node during the vegetative phase. From the first fully developed node with a flower to harvest 310°Cd must be reached. These requirements were found to be equal for all cultivars. These differ only in the number of vegetative nodes, which are a function of genotype and environment – here predominantly vernalization and photoperiod. If the number of non-flowering nodes in a particular site is known, the development of a cultivar can be predicted for an assumed temperature course. This approach may be used for timing, e.g. to predict the time of harvest or to select suitable cultivars for a specific site and to judge sites for their usefulness for pea production (Table 4.7). In the Hannover area the temperature sums of pea cultivars vary between 710 to more than 955°Cd (Zorn, 1962).

Table 4.7. Demarcation of production zones using the temperature sum concept (Seljaninow from Walter, 1960).

Zone	Temperature sums of zones (°Cd)	Crops to grow
I	1000–1400	Root crops, beet, early potato
II	1400–2200	Cereals, potato, flax, fodder crops
III	2200–3500	Maize, sunflower, sugarbeet, winter wheat, soyabean, grape, (melon, rice)
IV	3500–4000	Annual subtropical crops: cotton, tobacco, castor bean, peanut, luffa
V	> 4000	Perennial subtropical crops: fig, laurel, tea, citrus and others

To predict bolting of onions depending on sowing dates, Brewster (1987) simulated the development of onions from sowing to the inflorescence stage 4 (the ridged spate completely covers the apical dome – further development has not yet been worked out). He modelled three separate growth phases:

- sowing to the critical size for vernalization (cv. Senshyu, 0.45 g dry weight)
- vernalization period
- inflorescence development to stage 4.

The growth/development was calculated for each temperature range (for details see Brewster, 1987) and a correction term for photoperiod added. Though other important factors as nitrogen status have been omitted, the model can account for a considerable fraction of the scatter in percentage bolting but by no means all of it.

The development and growth of cauliflower has been simulated by Salter (1960) using the temperature sum concept. Although this approach resulted in satisfactory relationships in the maritime climate in the UK, it failed in continental climates.

Wiebe (1975) therefore developed a simulation model, which takes account of the partially counteracting actions of temperature on growth and development. He divided the growing period of cauliflower into phases with temperature-dependent growth/development rates to determine growth/development stages (Fig. 4.38). In phase 2 (the first phase is not recognized) the leaf development rate from the tenth to the 16th leaf is an approximately linear function of temperature and the relative progress per day related to the duration of this phase is calculated. After reaching 100% the model switches to the vernalization phase. Here, too, the relative progress in vernalization per day is estimated using weighted temperature effects read from interpolated curves elaborated in growth chamber experiments. With 100% in phase 3, characterized by 0.6 mm diameter of the growing point, the model switches to the phase of curd growth (0.6–150 mm diameter) with a linear temperature reaction to the log of curd diameter up

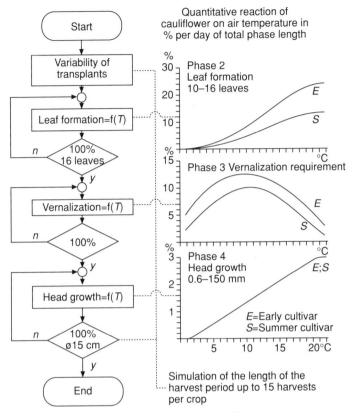

Fig. 4.38. Simulation model of the growing period of cauliflower, using parameters of growth chamber experiments (Wiebe, 1975).

to 22°C. The validation in field conditions necessitated some corrections, as for example different reactions to constant temperatures in growth chambers compared to fluctuating temperatures in fields.

A handicap of the cauliflower model of Wiebe is the requirement of cultivar-specific parameters, especially for vernalization, which have to be derived from growth chamber experiments. To overcome this expense, Wurr *et al.* (1990) avoided phase 2 and 3 of the Wiebe model. They proposed to measure curd size at about 10 mm diameter and to calculate curd diameter using a quadratic relationship between the $\log_e$ of curd diameter and accumulated day-degrees > 0°C from curd initiation. These data were used to predict a specific curd size on the basis of long-term average weather data. This model is simpler and more reliable than the Wiebe model for predicting the time of harvest of a growing crop. It cannot be used, however, to time plantings, since it omits the very sensitive phase of vernalization, which may cause large variations in harvest time and disturbances in the market supply.

A rather old application of the temperature sum concept is the designation of production areas as shown in Table 4.7 and elaborated by Nuttonson (1948). This is a rough approach, which should be replaced by polyfactorial regression or mechanistic models.

DIVERSE GROWTH FACTORS

A useful descriptive approach to model the actions and interactions of more than one growth factor is based on response surfaces (equation 13; see Fig. 4.24), worked out in greenhouse experiments with more or less controlled environments to break – as far as technically possible – correlations among the input factors. The procedure to establish the planning model is illustrated in Fig. 4.39. The estimation of the growth periods (50% emergence until final size) as a function of the input growth factors and the season was calculated by equation 16.

The MGR was calculated by iteration as a function of the intensities of

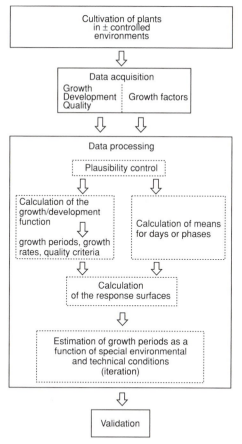

Fig. 4.39. Diagram for modelling long term plant growth (planning model) using regression models (Krug and Liebig, 1994).

the climatic factors during the growth period not yet determined and to be estimated (Krug and Liebig, 1979a, 1994). Validation was performed using greenhouse crops in holdings. Some deviations occurred due to the use of different cultivars, due to the influence of grafting, such as cucumbers on cucurbits or intensities of the climatic factors outside the range tested. In general, however, the models proved reliable (Rippen and Krug, 1994). The available models for vegetable species (see Table 4.6) comprise the input factors air temperature and irradiance, some additional CO_2 concentration, soil temperature and artificial light.

To demonstrate some results of this polyfactorial approach for greenhouse production, the raising period of lettuce transplants as a function of the date of the final weight (4 g), temperature set point and the long-term mean irradiances in Hannover are presented in Fig. 4.40. The two upper curves (1) represent the raising periods for two set points of air temperature without artificial light and without CO_2 enrichment. They reflect the effects of the seasonal outside conditions, especially irradiance and – to a lesser extent – temperature. Moreover, the effect of set point temperature is evident. The four lower curves (2) demonstrate the shortening of the raising period, that means days saved by using higher temperature with artificial light and CO_2 application (a compared to b), by artificial light with CO_2 application (b : c), or by CO_2 application without artificial light (c : d). These are selected examples out of numerous computable combinations between the five input factors and the effects of the season.

For production planning another presentation of these data is prefer-

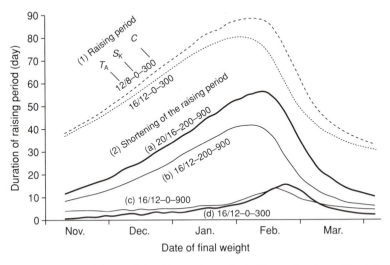

Fig. 4.40. Duration of raising periods of lettuce transplants (1) as a function of the date of final weight (4 g) and temperature set point (T_A) as well as shortening of the raising periods (2) by higher temperature set points, artificial light (SK) and/or CO_2 (C) application (examples, based on equation 13 and Fig. 4.24; for shortening comparison between thick and thin lines only; Krug and Liebig, 1994).

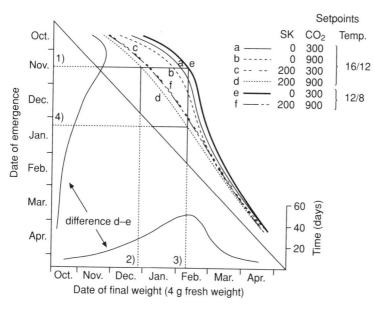

Fig. 4.41. Raising period of lettuce transplants (emergence – 4 g fw) with high and low growth factor intensities in a diagonal presentation, allowing for bilateral readings. At the coordinates differences, transferred from the diagonal graph (SK = artificial light; dimensions see Fig. 4.24, Krug and Liebig, 1994).

able, which allows production periods to be considered as a function of the intensities of the growth factors from the start *and* from the end of the respective period (Fig. 4.41; for details see Weber and Liebig, 1981; Krug and Liebig, 1994). Examples are: (1) Emergence 15.11 – final weight with temperature set point 16/12, artificial light 200, C 900 on 25.12; (2) with temperature set point 12/8, artificial light O, C 300 on 10.2 (3). To reach the final weight on 10.2, the plants have to emerge with temperature set point 16/12, artificial light 200, C 300 on 5.1 (4) and with temperature set point 12/8, artificial light 0, C 300 on 15.11 (1).

Based on growth curves, final transplant weight can be included as another factor. As shown in Fig. 4.42, the time needed for bigger transplants is rather short, due to exponential growth. This is true especially in favourable growing conditions. This also means that timing for a defined transplant weight must be performed very precisely. More substantial difficulties arise to parameterize plant quality. First approaches are presented by Pytlinski and Krug (1993) and Krug and Liebig (1994), but much work is still needed.

These data allow for timing according to the dates of emergence – if a model for the emergence period is included for the dates of sowing – or to the dates of a defined transplant weight. The same holds true for other growth phases, such as planting to harvest (radish *R. sativus* and *R. niger*,

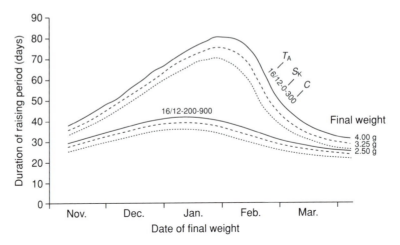

Fig. 4.42. Duration of the raising period of lettuce transplants as a function of final weight, season and climate control (examples, based on equation 13 and Fig. 4.24, T_A = temperature set point, SK = artificial light, C = CO_2 application; Krug and Liebig, 1994).

lettuce, kohlrabi), or planting to start of harvest (e.g. cucumbers) and may be applied to any other phases.

When the calculations for many growing seasons are combined, it is possible to estimate the risk, due to deviating weather conditions. As shown in Fig. 4.43 for lettuce transplants the variability caused by weather is much larger with the low set point of temperature compared to high temperature regime (14/10°C) and the deviations depend on the season. This is illustrated by drawing the differences between the raising periods of the extreme years and 50% of the years with smaller deviations as a function of the date of planting (finish) and the date of emergence (start), respectively. These differences have been calculated for a high (14/10°C) and a low (2/2°C) temperature set point. As already shown for the effects of growth factor intensities, the biggest variations occur referred to the date of emergence in autumn and referred to final transplant size in March–April.

Knowing the growth periods the costs of factors such as heating, CO_2, etc., can be derived using models to estimate the input quantities which then have to be multiplied by the prices. These costs can be graphically demonstrated for the reference points of planning, that means the dates of the start or the dates of finishing (Fig. 4.44). This graph is easier to read than the diagonal graph shown in Figs 4.41 and 4.43, but these latter ones are more informative. These costs are important data for profitable strategies in timing and growth control concerning the application of growth-promoting facilities and of respective intensities.

The planning model is primarily validated for greenhouse production. In principle, however, it can also be adapted to field conditions. E. Bazlen and J. Schlaghecken (Neustadt, 1995, unpublished observations) collected numerous data of lettuce plantings [transplant weight, planting date, date

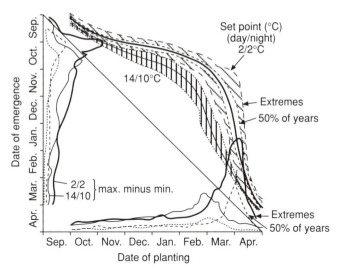

Fig. 4.43. Variation of the growth period of lettuce transplants of 3 g fw as a function of the weather conditions of 27 separate years at Hannover and the set points (diagonal presentation see Fig. 4.41; extremes = years with the longest and shortest raising periods, respectively; 50% of years = 50% of years with larger deviations are omitted; Krug and Liebig, 1988).

of harvest (400–500 g fw)] and estimated the growth periods by (i) using a lettuce model developed by Liebig (Liebig and Alscher, 1993), which is based on equation 13 and regards the three input factors radiation, temperature, transplant weight; and (ii) using long-term empirical data of that production area. Cultivar reactions are taken into account by relating to a standard cultivar.

An example of deviations between the estimated values and the values noted by the growers is presented in Fig. 4.45. Both approaches come to similar results with overestimations of about 5 days (14% of the growing period) from April until August. The predominant overestimation by both approaches indicates a special environmental influence of the year of investigation, which is not covered by the long-term normals and not taken into account by the model. This needs further investigation. Considering the variations caused by determining and noting the time of harvest (varying ± 50 g), the restriction to three input factors only and the measurements of their intensities, these deviations are acceptable and show promise in the improvement of timing in field production. The empirical data, however, will be less reliable in other sites.

A more comprehensive approach but involving greater expenses for simulating growth and development of field plants may be the Stützel Model (see section on growth models above). Applications with vegetables are in progress but not yet finally worked out.

A growth model for transplants of lettuce, cauliflower, leek and celery,

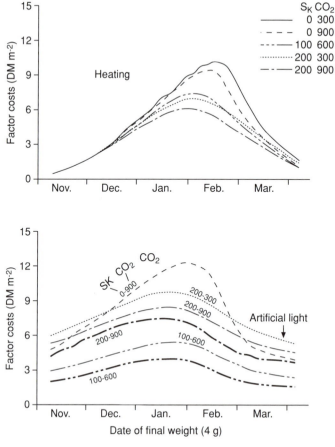

Fig. 4.44. Costs of heating (0.50 DM l⁻¹ oil), of artificial light (SK, 0.152 DM per kWh m⁻² day⁻¹) and of CO₂ (0.80 DM kg⁻¹) for lettuce transplant cultivation as a function of the growth factor intensities and the dates of final weight (4 g). Based on Fig. 4.24, temperature set point 12/8°C; Krug and Liebig, 1994.

considering initial dry weight, temperature sums modified by irradiances and plant densities has been developed by Scaife *et al.* (1987).

Mechanistic growth models have been developed for some agricultural crops (e.g. corn, wheat, potato, soyabean, alfalfa), but are not considered here.

Parameterization of environmental–man–plant relationships can also help to analyse the variances of yield, if the relevant data are available. Van Rijssel (1979a,b) collected a comprehensive set of data for the geographically concentrated horticulture production in the Netherlands and evaluated it by factor analysis. In Germany data from holdings of five different sites were elaborated and used to calculate response surfaces (equation 13) for cucumber, lettuce, radish (*R. sativus*) and other species on the basis of

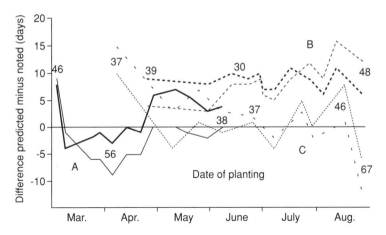

Fig. 4.45. Differences of harvest dates predicted by the Liebig model (Liebig and Alscher, 1993) and by empirical data (long-term local enquiries) to the dates noted by the growers in 1994. A, B, C = holdings, numbers = growing periods in days (Bazlen and Schlaghecken, unpublished data).

temperature and radiation responses (Rippen and Krug, 1994). Influences of watering, fertilizing, CO_2 enrichment and other production techniques were considered, but not yet incorporated into the models. The parameters derived fitted satisfactorily to those derived in the experiments to establish the model. They allowed yield variances to be correlated to the climatic conditions of the sites or to production techniques. These analyses can be useful tools for planning and advice concerning competitive position of sites or holdings and to improve cultivation techniques.

CONTROL OF GROWTH AND DEVELOPMENT

OFF- AND ON-LINE CONTROL

During the production process, deviations from the schedule may arise due to unexpected weather conditions, because of deficiencies of the models or measurements, due to changes of availability of greenhouses or laboratories or due to changes in the options. All these facts may require corrections of the growth rates via control. The potentialities of growth and development control in greenhouses depending on the climate can be classified into the following categories.

1. Assigned long-term fixed set points of the intensities of the environmental factors, which ensure that minima and maxima will not be exceeded (e.g. air temperature, soil temperature, water vapour deficit, CO_2 concentration, irradiances). These set points can be determined empirically or on the basis of planning models. Variations of intensities between minima and maxima by natural inputs may be wanted (e.g. temperature increase in greenhouses by radiation). Such fixed set points do not always mean one constant value, but include different set points for diurnal amplitudes or adaptations to ontogenetic requirements, hardening and others. Pathogen control using threshold values can also be incorporated in this category.

An example of the accuracy of this basically static procedure in timing transplant production in greenhouses is presented in Fig. 4.46. Using a temperature set point of 10/6 (day/night), the raising periods of lettuce transplants were simulated as a function of radiation and outside temperature of 36 individual years in Hannover and as a function of the long-term means of these weather conditions. The differences (single years minus means) show substantial uncertainties with autumn, especially with October sets. However, only a few of the sets emerging from December to February are about 5 days earlier (shorter raising period) than the target and this can be easily controlled by reducing the temperature set points. Longer raising periods than estimated by the long-term normals are due to extreme unfavourable weather conditions, which happen more frequently than favourable ones. Longer raising periods occur in 30–40% of the years with differences of about 5 days and 5–12% of the years with about 10 days. If these deviations can be buffered by final size or tolerated, the use of fixed set points is a most simple procedure.

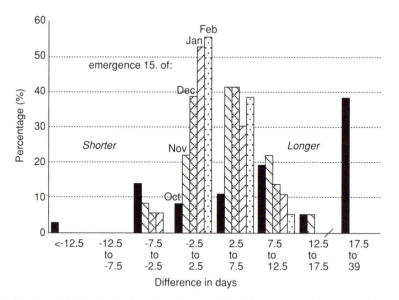

Fig. 4.46. Uncertainties of planning the raising periods of lettuce transplants caused by weather conditions and season, shown by a frequency distribution of the raising periods (4 g fw) simulated for radiation and temperature of 36 single years (1956/7–1991/2) in Hannover minus the raising period simulated for long term means. (Based on equation 13 and Fig. 4.24 and 4.40, temperature set point 10/6°C, Krug and Liebig, 1995.)

2. The procedure described in (1) can be extended by modifying the set points according to other criteria in order to save resources or to reduce the costs. Recommendations are as follows:

- adapting the set point of heating or CO_2 enrichment to wind velocity, compensating lower set points during high wind velocity and therewith heavy leakage periods by increased set points during calm periods;
- to save energy in greenhouse heating by controlling night temperature according to the preceding day temperature or – for longer periods – to assigned temperature sums;
- control of temperature by weekly temperature sums to avoid vernalization: Wiebe *et al.* (1992) demonstrated that kohlrabi can be cultivated in low, vernalizing temperatures, if once a week high temperature compensates the preceding low temperature by devernalization. If this warming is performed by high irradiances, energy is saved. Otherwise, it has to be compensated by heating.

3. A procedure with a higher chance of accuracy is on-line control. This demands measurements of growth/development criteria for feedback, in order to compensate for deviations from the schedule or to react to new targets. Models for on-line control of greenhouse climate aiming at potential growth rates have been developed on the basis of mechanistic approaches (see section on growth models above). These can react to very short periods (e.g. minutes), may gain maximum adaptation and may reduce risks in for example phytomedical aspects. For crops with favourable LAI and continuous fruiting, good results have been reported by Challa and Schapendonk (1986), Jones *et al.* (1991) and Nederhoff (1994).

   With young crops with expanding leaf areas, such as lettuce, Alscher (1993) obtained more reliable simulations and hence control using the regression model (based on equation 13). This, however, is restricted to mean values of 24-h periods. To combine the potentials of both approaches, Alscher simulated daily growth of lettuce plants using the regression model and during the 24-h periods the sucros 87 model. After 24 h the growth rates of the sucros model were adapted to those of the regression model. However, the improvement of on-line control by the sucros model was small. Its significance will depend on the buffer capacity of the species and climatic conditions, but it may be improved by more advanced models or parameters.

   During vegetative growth, especially in periods of expanding leaf area, aiming at potential growth may interfere with timing. If timing is a primary option, it is rather expensive to keep plant growth in the planned schedule by on-line control, since progress by high natural inputs, as sunny or warm periods, is not used and intensities during unfavourable periods have to be compensated by technical inputs causing costs. Therefore Krug and Liebig (1995) cultivated at suboptimal climates and used fixed set points as described in items (1) and (2) which were corrected when it became evident that the target would not be reached by natural compensation. An example was worked out with simulations based on 36-year weather records in

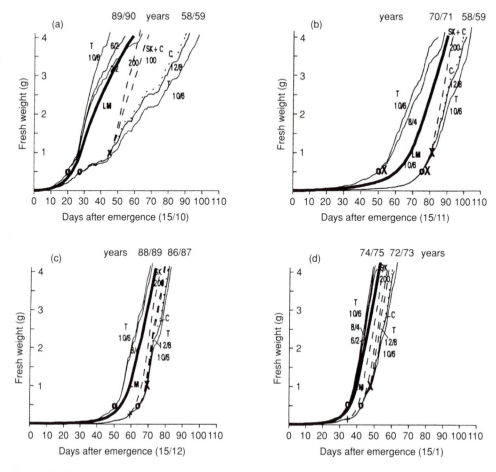

Fig. 4.47. Simulated growth curves of lettuce transplants in the most extreme favourable and unfavourable years and seasons out of 36 years, respectively, in Hannover compared to long-term means (LM), demonstrating the needs and potentials of growth control in different seasons. Temperature set point $T = 10/6°C$. At 0.5 g weight (**o**) control by decreased temperature ($T = 6/2°C$ or $2/2°C$) or increased temperature ($T = 12/8°C$), additionally $C = 1000$ ml l^{-1} CO_2 or at 1 g (**x**) and 0.25 g (**+**) weight, respectively, artificial light SK = 100 and 200 Whm^{-2} day^{-1}, respectively (Krug and Liebig, 1995).

Hannover. Using a temperature set point of 10/6°C, with neither CO_2 enrichment nor artificial light the growth curves of lettuce transplants were calculated for the long-term means of radiation and temperature and additionally for all years separately. Those with the shortest and longest growth periods, respectively, were selected for 5 dates of emergence from October to February (only four presented). These graphs show the extreme deviations in the period tested. From Fig. 4.47 it is obvious that with regard to the date of emergence

- the deviations decrease from autumn to spring (the causes are explained

in the section on radiation above; see Fig. 4.17)
- the extreme deviations, as already shown in Fig. 4.46, are mainly retarda-tions, caused by low irradiances, less by accelerations caused by favourable weather.

To compensate for too rapid growth, growth rates were decreased by lowering temperature set points for heating. Its effect can be intensified by reducing the set point for ventilation. Growth acceleration was achieved by increasing temperature set points to 12/8°C, considered as the highest value for good transplant quality. If this was insufficient, additional growth increase was obtained by enrichment of CO_2 (1000 ml l^{-1}) and by use of arti-ficial light (100 and 200 $Whm^{-2} day^{-1}$, respectively). The latter should only be used sparingly because of cost considerations. The starting point of these actions was estimated by iteration. Figure 4.47 demonstrates that under the conditions tested even in the most extreme years corrections need not be started before plants have attained a weight of 0.5 g and shows, at which weight, depending on the season, the expensive artificial light has to be applied. This procedure uses self compensation, provides a sufficient accu-racy and permits inclusion of quality factors (not yet worked out) and eco-nomic aspects.

Other questions discussed in greenhouse climate control are as follows.

1. Afforded time-dependent constancy of growth factor intensities. There is good evidence, that in conventional ranges there is no disadvantage of short- to medium-term fluctuations or oscillations (Fink, 1992). In contrast, natural variation may be used to reduce financial inputs. Disadvantages, however, may occur by extreme low intensities, such as temperature or even frost damage (Rietze, 1988), by dew promoting fungus attack or by wilting or even breakdown of tissues with high water vapour deficit, especially caused by ventilation after dim weather periods.
2. Required adaptation to uncontrolled intensities, such as irradiance. Seemann (1956) recommended that temperature set points be adapted on-line to irradiances. Lorenz (1980) showed, however, that this is not neces-sary due to the low temperature optimum of photosynthesis in combination with self warming of greenhouses and to the compensating temperature reaction of growth. Therefore adaptation in 24-h periods, as already used in growth models, is adequate.

Energy-saving strategies are beyond the scope of this chapter (see Krug *et al.*, 1988).

FEEDBACK INFORMATION

On-line control is based on the knowledge of the stage already reached com-pared to that scheduled. Criteria of the actual state may concern growth (biomass) and/or correlated quantities of development (e.g. number of leaves/nodes), respectively. For short-term information these have to be

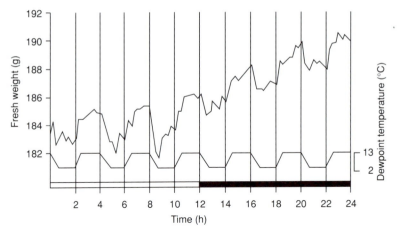

Fig. 4.48. Dynamics of fresh weight of a kohlrabi plant depending on oscillations of water vapour deficit (dew point 2↔13°C, air temperature 14°C (Fink, 1992).

measured automatically. For information in periods of days or longer they can be measured by hand. This should preferentially be performed without damaging the crop. These criteria are as follows.

1. *Biomass.* Gravimetric methods: in greenhouses fresh weight can be measured by weighing plants in trays, positioned on a balance or, with flexible shoots (tomato, cucumber), hanging in a balance. This procedure is easy to handle and is already being used in practice. Accuracy depends on equalizing water control in the substrate and the plant, which varies according to water potential and saturation deficit of the air, respectively (Fig. 4.48). Another precondition is that the growth of the test plants corresponds to that of the crop.

2. *Radiometric methods.* Absorption of ionizing rays by biomass (fresh weight). Gamma-rays, e.g. Americium, are used for scanning plants or crops (Kühn, 1978). For thin layers, e.g. leaves, β-rays may be preferred.

3. Water mass can be measured by absorption of *microwaves* (26.5 GHz) to calculate dry weight by subtracting water from biomass. This procedure, however, can only be used for thin organs such as leaves and has not been worked out satisfactorily (Schulze and Kühn, 1980; Feinhals and Kühn, 1986).

4. Dry weight can be calculated by measuring *photosynthesis* in greenhouses and estimating leakage by N_2O losses. This method, however, is expensive.

5. Crop *silhouettes* may be correlated to biomass and can be measured by video cameras and digital image processing. The pictures and data present the area covered by the crop (pattern area) in contrast to the ground (for details see Evers *et al.*, 1987) or covered by individual plants in pots in front of a vertical background (Hendriks and Scharpf, 1984). Accuracy depends on the contrast between the silhouette and background and hence illuminance (Tantau and Hack, 1990) and on the degree of covering. For lettuce,

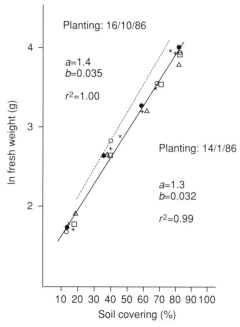

Fig. 4.49. Correlation between degree of soil covering (pattern area) and fresh weight of lettuce with two planting dates and five treatments each (Krug unpublished data).

planted in soil, satisfactory values could only be determined up to about 80% ground cover (Fig. 4.49). The measurements can be improved by 2–3 dimensional silhouettes (2–3 cameras in an angle of 90°) and/or by measurements made at night.

6. *Height* may be a useful criterion if the correlation with the rates of growth/development is satisfactory.

7. *Diameters* of plant organs can be used as feedback information. Examples are diameters of stems (Schoch *et al.*, 1989) or tubers (Fink, 1992). Fink demonstrated that growth of kohlrabi tuber diameter can be very precisely measured by linear voltage displacement transducers to determine progress in about 2-h intervals (Fig. 4.50).

8. *Number of leaves* may be used to determine development stages.

9. *Scales* of growth/development, as presented above (e.g. BBCH scale) may be applied. This criterion is specially useful in open fields.

The procedures described follow the pattern to calculate or measure progress and to keep the plants on schedule by controlling set points of greenhouse climate. Hashimoto and Morimoto (1984) proposed the control of greenhouse climate according to the intensities of physiological processes, which are regarded as set points. This procedure is called the 'speaking plant approach'. Useful applications are stress avoidance, such as extreme temperatures or water vapour deficits by measuring leaf temperature, stomata

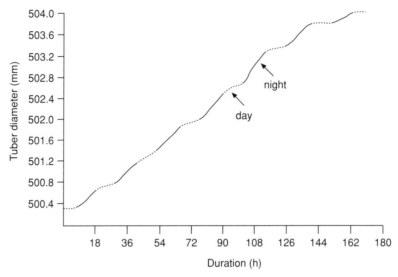

Fig. 4.50. Time course of diameter increase of kohlrabi tubers, measured by a linear voltage displacement transducer (data from Fink, 1992).

opening or leaf water potential. These are situations when one factor and one process are dominating. For optimization of, for example, temperature control in greenhouses the speaking plant approach is not successful, since the complex physiology of growth, governed by numerous processes, cannot be controlled by the one process measured (reductionism). An example is temperature control by the photosynthetic rate ('hill climbing method' by Lake, 1966), which – at least in crops with expanding leaves – neglects temperature demand for leaf growth, which is considerable higher than that of photosynthesis.

In general it can be stated that due to insufficiencies of the growth/development models, feedback information is very important and needs further improvement, which will reduce the demands for accuracy of the models.

DECISION-SUPPORTING APPROACHES

As shown in Fig. 4.1, crop production requires the management of a very complex and complicated system. To ensure high levels of efficiency in harmony with the environment many tough decisions have to be made. Up to the middle of the nineteenth century this was done by rules of thumb, experience and intuition. Therefore plant cultivation was regarded as an art. With increasing knowledge and technical inputs, with higher costs and greater competition more efficiency is required to reach profitability.

A lot of basic knowledge to support decisions is outlined in the literature. But with increasing complexity and amount of information, growers will be incapable of incorporating that flood of information without scientific and technical assistance. Decision-supporting approaches have been developed, therefore, for single operations, such as the application of growth factor intensities, which may be based on the monetary output minus direct costs. Some examples are outlined above. Scientists subsequently modelled subsystems of varying complexity (Challa and van Straten, 1991; Alscher, 1993; Rippen *et al.*, 1995) and some worked on comprehensive systems using mathematical approaches (Reinisch *et al.*, 1984; Lentz, 1987, 1993). Lentz concluded, however, that optimization of comprehensive production systems is not feasible at present.

Complex decisions, however, can be facilitated and improved by a combination of model knowledge and non-parameterized expert knowledge, as realized in the 'knowledge-based decision-supporting systems' (DSS), also called 'expert systems'. These DSSs are based on comprehensive and structured databanks and on model parameters, for evaluation they are based on procedural knowledge and on decision algorithms. The decision parameters elaborated are offered in a dialogue with the user to consider his personal experience, intuition and preferences. This ensures incorporation of the particularities of the respective holding of the grower and encourages his participation.

Decision-supporting systems are successful when broad scale and structured knowledge is available or problems are limited to simpler decision trees. Examples are diagnosis of pests and diseases (e.g. tomatoes by Hoshi, 1988; tomato, potato, melons and others by Le Renard, 1988); herbicide and pesticide application (e.g. Pestemer, 1994; PEMOSYS by Pestemer and Günther, 1995); diagnosis of nutrient deficiency (e.g. cucumbers by Göhler, 1990); and fertilization (e.g. N-Expert by Fink and Scharpf, 1993). For broadscale application DSSs are at present in a preliminary phase and scarcely used in practice. Therefore much is yet to be done in the fields of development and implementation. This will require considerable ability and readiness of scientists, advisers and growers to engage in system theoretical thinking and EDV application.

ACKNOWLEDGEMENT

I thank Ms Anni Romey for support in computing the figures and friendly and helpful assistance. I want to thank the editor for his numerous corrections of my English and his valuable suggestions for writing this chapter.

REFERENCES

Acock, B. (1991) Modeling canopy photosynthetic response to carbon dioxide, light interception, temperature, and leaf traits. In: Boote, K.J. and Loomis, R.S. (eds) *Modeling Crop Photosynthesis from Biochemistry to Canopy.* CSSA Special

Publication no. 19. Crop Science Society of America, Madison, USA, pp. 41–55.

Alscher, G. (1993) Optimierung der CO_2 und Temperaturregelung bei Gewächshaus-kulturen mit Hilfe von Modellansätzen unterschiedlicher Abstraktionsebenen. PhD thesis, University of Hannover.

Augustin, P. (1984) Ein Modell der Ertragsbildung der Gewächshausgurke (*Cucumis sativus* L.). *Archiv für Gartenbau* 32, 1–18.

Bazlen, E. (1985) Modelluntersuchungen zu altersabhängigen Klimareaktionen als Basis für die Simulation des Wachstums am Beispiel von Radies (*Raphanus sativus* var. *sativus*). PhD thesis, University of Hannover.

Beinhauer, R. (1977) Photosynthetisch aktive Strahlung in Gewächshäusern. *Gartenbauwissenschaft* 42, 103–113.

Bertalanffy, L. von (1973) *General System Theory, Foundations, Development, Applications.* Penguin, Harmondsworth.

Bertin, N. and Heuvelink, E. (1994) Dry matter production in a tomato crop: compar-ison of two simulation models. *Journal of Horticulture Science* 68, 995–1011.

Bierhuizen, J.F. and Wagenvoort, W.A. (1974) Some aspects of seed germination in vegetables. I. The determination and application of heat sums and minimum temperature for germination. *Scientia Horticulturae* 2, 213–219.

Bleiholder, H., Buhv, L. and Fellev, C. (1994) *Compendium of Growth Stage Identification Keys for Mono- and Dicotyledonous Plants.* Ciba-Geigy, Basel.

Bluthgen, J. (1966) *Allgemeine Klimageographic.* Walter de Gruyter, Berlin.

Blummenfield, D., Meinken, K.W. and LeComte, S.B.(1960) A field study of aspara-gus growth. *Proceedings of the American Society of Horticultural Science* 77, 386–392.

Bradford, K.J. (1990) A water relations analysis of seed germination rates. *Plant Physiology* 94, 840–849.

Brewster, J.L. (1987) Vernalization in the onion – a quantitative approach. In: Atherton, J.G. (ed) *Manipulation of Flowering.* Butterworths, London.

Brückner, I. (1990) Erfassung und Bewertung der Wirkung einer Zusatzbelichtung mit Hilfe eines Wachstumsmodells am Beispiel von Kohlrabi. PhD thesis, University of Hannover.

Caesar, K. (1986) *Einführung in den tropischen und subtropischen Pflanzenbau.* DLG-Verlag, Frankfurt.

Challa, H. and Schapendonk, A.H.C.M. (1986) Dynamic optimisation of the CO_2 con-centration in relation to climate control in greenhouses. In: Enoch, H.Z. and Kimball, B.A. (eds) *Carbon Dioxide Enrichment of Greenhouse Crops. 1: Status and CO_2 Sources.* CRC Press, Boca Raton, Florida, pp. 147–160.

Challa, H. and van Straten, G. (1991) Reflections about optimal climate control in greenhouse cultivation. In: Hashimoto, Y. and Day, W. (eds) *International Federation of Automatic Control Mathematical and Control Applications in Agriculture and Horticulture.* Matsuyama, Japan.

Charles-Edwards, D.A., Doley, D. and Rimmington, G.M. (1986) *Modelling Plant Growth and Development.* Academic Press, Sydney.

Dahal, P. and Bradford, K.J. (1990) Effects of priming and endosperm integrity on seed germination rates of tomato genotypes. II. Germination at reduced water potential. *Journal of Experimental Botany* 41, 1441–1453.

de Candolle, A. (1855) *Géographie Botanique Raisonnée.* Paris, Masson.

De Koning, A.N.M. (1994) Development and dry matter distribution in tomato: a quantitative approach. PhD thesis, University of Wageningen.

De Visser, C.L.M. (1994) ALCEPAS: an onion growth model based on Sucrun 87. I.

Development of the model. *Journal of Horticultural Science* 69, 501–518.

de Wit, C.T. and Brouwer, R. (1968) Über ein dynamisches Modell des vegetativen Wachstums von Pflanzenbeständen. *Angewandte Botanik* 42, 1–12.

de Wit, C.T. and de Vries, F.W.T. (1983) Crop growth models without hormones. *Netherlands Journal of Agricultural Science* 31, 313–323.

Doust, J.D., Doust, L.L. and Eaton, G.W. (1983) Sequential yield component analysis and models of growth in bush bean (*Phaseolus vulgaris* L.). *American Journal of Botany* 70, 1063–1070.

Erikson, R.O. (1976) Modeling of plant growth. *Annual Review of Plant Physiology* 27, 407–434.

Evans, L.T. (1963) *Environmental Control of Plant Growth.* Academic Press, New York.

Evers, H., Pätzold, G., Seibold, H.W., Liebig, H.P. and Ernst, D. (1987) Die Biomassebestimmung an Pflanzen mit Hilfe der Computerbildanalyse. *Gartenbauwissenschaft* 53, 119–124.

Feddes, R.A. (1971) Effects of water and heat on seed emergence and crop production. Water, heat and crop growth. *Mededelingen Landbouw Hogeschool Wageningen* 71, 1–184.

Feinhals, J. and Kühn, W. (1986) Zerstörungsfreie Bestimmung von Biomasse und Wassergehalt in Pflanzen mit radiometrischen Methoden. *Angewandte Botanik* 60, 337–364.

Feldmann, U. (1979) *Wachstumskinetik. Mathematische Modelle und Methoden zur Analyse Altersabhängiger, Populations-Kinetischer Prozesse, Medizinische Informatik und Statistik 11.* Springer-Verlag, Berlin.

Feller, C., Bleiholder, H., Buhr, L., Boom, T., van den Hack, H., Hess, M., Klose, R., Meier, U., Stauss, R. and Weber, E. (1995) Phänologische Entwicklungsstadien von Gemüsepflanzen I. Zwiebel-, Wurzel-, Knollen- und Blattgemüse. *Nachrichtenblatt des Deutschen Pflanzenschutzdienstes* 47, 193–206, 217–232.

Finch-Savage, W.E. and Phelps, K. (1993) Onion (*Allium cepa* L.) seedling emergence patterns can be explained by the influence of soil temperature and water potential on seed germination. *Journal of Experimental Botany* 44, 407–414.

Fink, M. (1992) Wirkungen kurzfristiger Temperaturschwingungen auf das Planzenwachstum am Beispiel von Kohlrabi (*Brassica oleracea* convar. *acephala* var. *gongylodes* L.). PhD thesis, University of Hannover.

Fink, M. and Scharpf, C. (1993) N-Expert – decision support system for vegetable fertilization in the field. *Acta Horticulturae* 339, 67–74.

Fölster, E. (1967) Die Schnittlauchtreiberei im Herbst. *Gartenbauwissenschaft* 6, 504–511.

Gaastra, P. (1962) Photosynthesis of leaves and field crops. *Netherlands Journal of Agricultural Science* 10, 311–324.

Gary, C. (1988) A simple carbon balance model simulating the short term responses of young vegetative tomato plants to light, CO_2 and temperature. *Acta Horticulturae* 229, 245–250.

Geutler, G. and Krochmann, J. (1978) Die Messung der für die Photosynthese wirksamen Bestrahlungsstärke. *Gartenbauwissenschaft* 43, 271–275.

Gijzen, H. (1992) Simulation of photosynthesis and dry matter production of greenhouse crops. *Simulation Report 28,* Centre Agrobiological Research, Wageningen Agricultural University.

Göhler, A. (1990) Expertensystem Gurke – Düngeplanung und Diagnose von Ernährungsstörungen. PhD thesis, University of Hannover.

Gray, D. and Finch-Savage, W.E. (1994) Timing of vegetable production – the role of crop establishment and forecasting techniques. *Acta Horticulturae* 371, 29–36.

Gregory, F.G. (1917) Physiological conditions in cucumber houses. *Cheshunt (England) Experiment and Research Station, 3rd Annual Report.* pp. 19–28.

Gregory, F.G. (1926) The effect of climatic conditions on the growth of barley. *Annals of Botany* 40, 1–26.

Grevsen, K. and Olesen, J.E. (1994) Modelling cauliflower development from transplanting to curd initiation. *Journal of Horticultural Science* 69, 755–766.

Gummerson, R.J. (1986) The effect of constant temperatures and osmotic potentials on the germination of sugar beet. *Journal of Experimental Botany* 37, 729–734.

Hammer, P.A. and Langhans, R.W. (1978) Modelling of plant growth in horticulture. *HortScience* 13, 456–458.

Hashimoto, Y. and Morimoto, T. (1984) *System Identification of Plant Response for Optimal Cultivation in Greenhouses*, Vol. 9. World Congress: International Federation of Automatic Control, Budapest.

Hegarty, T.W. (1976) Effects of fertilizer on the seedling emergence of vegetable crops. *Journal of Science, Food and Agriculture* 27, 962–968.

Hendriks, L. and Scharpf, H.C. (1984) Ermittlung des Pflanzenwachstums mit Hilfe der Computerbildanalyse. Zierpflanzenversuche 1984, Lehr- und Versuchsanstalt für Gartenbau, Landwirtschaftskammer Hannover.

Heuser, W. (1927/28) Die Ertragsanalyse von Getreidezüchtungen. *Pflanzenbau* 4, 353–357.

Heuvelink, E. (1996) Tomato growth and yield: quantitative analysis and synthesis. PhD thesis, Wageningen.

Heuvelink, E. and Bertin, N. (1994) Dry-matter partitioning in a tomato crop: comparison of two simulation models. *Journal of Horticultural Science* 69, 885–903.

Heuvelink, E. and Marcelis, L.F.M. (1989) Dry matter distribution in tomato and cucumber. *Acta Horticulturae* 260, 149–157.

Holländer, B. and Krug, H. (1991) Wirkungen hoher CO_2-Konzentrationen auf Gemüsearten. I. Symptome, Schadbereiche und Artenreaktionen. *Gartenbauwissenschaft* 56, 193–205.

Holländer, B. and Krug, H. (1992) Wirkungen hoher CO_2-Konzentrationen auf Gemüsearten. Wachstum, CO_2-Gaswechsel und Stomatawiderstand. *Gartenbauwissenschaft* 57, 32–43.

Hoogenboom, G., Jones, J.W. and Boote, K.J. (1990) *Modeling Growth, Development and Yield of Legumes: Current Status of the Soygro, Pnutgro and Beangrow Models.* ASAE Paper no. 90–7060. Paper presented at the ASEA International Summer Meeting, Ohio, USA.

Hoshi, T. (1988) Krankheits- und Schädlingsdiagnose bei Tomaten. In: Deutsche Landwirtschafts-Gesellschaft, Frankfurt (ed.) *Wissensbasierte Systeme in der Landwirtschaft, DLG-Computerkongress.*

Hunt, R. (1982) *Plant Growth Curves. The Functional Approach to Plant Growth Analysis.* Edward Arnold, London.

Isenberg, F.M.R., Pendergress, A., Carroll, J.E., Howell, L. and Oyer, E.B. (1975) The use of weight, density, heat units, and solar radiation to predict the maturity of cabbage for storage. *Journal American Society of Horticultural Science* 100, 313–316.

Jones, J.W., Dayan, E., Allen, L.H., Keulen, H. van and Challa, H. (1991) A dynamic tomato growth and yield model (Tomgro). *Transactions of the American Society of Agricultural Engineers* 34, 663–672.

Kailuweit, H.D. and Krug, H. (1995) Wärme fördert das Längenwachstum, höherer

Bodenwiderstand das Dickenwachstum von Spargel. *Taspo Gartenbaumagazin* 3, 45–46.

Kohn, H. and Levitt, J. (1965) Frost hardiness studies on cabbage grown under controlled conditons. *Plant Physiology* 40, 476–480.

Krug, H. (1983) Ausweitung des Gemüseangebotes im Winter. Das Treiben von Lauchzwiebeln. *Gemüse* 19, 80–82.

Krug, H. (1991a) *Gemüseproduktion*, 2nd edn. Verlag Paul Pavey, Berlin.

Krug, H. (1991b) Aktivitätswechsel von Rhabarber (*Rheum rhaponticumrhabarbarum*) und seine Bedeutung für Anbau und Treiberei. *Gartenbauwissenschaft* 56, 93–98.

Krug, H. (1996) Seasonal growth and development of asparagus (*Asparagus officinalis* L.). I. Temperature experiments in controlled environments. *Gartenbauwissenschaft* 61, 18–25.

Krug, H. and Fölster, E. (1976) Influence of environment on growth and development of chives (*Allium schoenoprasum* L.). I. Induction of the rest period. *Scientia Horticulturae* 4, 211–220.

Krug, H., Lederle, E. and Liebig, H.P. (1985) Modelle zur kultur- und kostengünstigen Temperaturführung während der Auflaufphase. *Gartenbauwissenschaft* 50, 54–59.

Krug, H. and Liebig, H.P. (1979a) Analyse, Kontrolle und Programmierung der Pflanzenproduktion in Gewächshäusern mit Hilfe beschreibender Modelle: I. Das Produktionsmodell. *Gartenbauwissenschaft* 44, 145–154.

Krug, H. and Liebig, H.P. (1979b) Analyse, Kontrolle und Programmierung der Pflanzenproduktion in Gewächshäusern mit Hilfe beschreibender Modelle: II. Produktion von Radies (*Raphanus sativus* var. *sativus*). *Gartenbauwissenschaft* 44, 202–213.

Krug, H. and Liebig, H.P. (1984a) Analyse, Kontrolle und Programmierung der Pflanzenproduktion in Gewächshäusern mit Hilfe beschreibender Modelle. III. Produktion von Rettich (*Raphanus sativus* L. var. *niger*). *Gartenbauwissenschaft* 49, 7–18.

Krug, H. and Liebig, H.P. (1984b) Pflanzenbauforschung mit Hilfe bio-mathematischer Modelle. *Hannover Uni* (2), 15–23.

Krug, H. and Liebig, H.P. (1988) Planungsmodelle für die Jungpflanzenanzucht. *Gartenbauwissenschaft* 53, 206–210.

Krug, H. and Liebig, H.P. (1994) Model for planning and control of transplant production in climate controlled greenhouses. I. Production planning. *Gartenbauwissenschaft* 59, 108–115.

Krug, H. and Liebig, H.P. (1995) Models for planning and control of transplant production in climate controlled greenhouses. II. Production control. *Gartenbauwissenschaft* 60, 22–28.

Krug, H. and Wiebe, H.J. (1991) Entwicklung. In: Krug, H. (ed.) *Gemüseproduktion*, 2nd edn. Verlag Paul Pavey, Berlin.

Krug, H., Wiebe, H.J., Fölster, F., Liebig, H.P., Zabeltitz, Chr.v. and Tantau, H.J. (1988) *Wirtschaftlicher Energieeinsatz im Gemüsebau*. Blickpunkt-Heft, Zentralverband Gartenbau, Bonn.

Kühn, W. (1978) New radiometric and radioanalytical methods in agricultural research. *Atomic Energy Review* 16, 619–655.

Lake, J.V. (1966) Measurement and control of the rate of carbon dioxide assimilation by glasshouse crops. *Nature* 209, 97–98.

Lang, G.A., Early, J.D., Martin, G.C. and Darnell, R.L. (1987) Endo-, para-, and ecodormancy: physiological terminology and classification for dormancy research. *HortScience* 22, 371–377.

Lawlor, D.W. (1993) *Photosynthesis: Molecular, Physiological and Environmental Processes*, 2nd edn. Longman Group, Harlow, UK.

Lederle, E., Krug, H., Liebig, H.P. and Weber, W. (1984) Funktionen zur Quantifizierung der temperaturabhängigen Auflaufdauer am Beispiel von Feldsalat (*Valerianella locusta* L.). *Gartenbauwissenschaft* 49, 262–266.

Lentz, W. (1987) Bestimmung von Temperaturstrategien für Gewächshauskulturen mittels bio-ökonomischer Modelle und numerischer Suchverfahren. PhD thesis, University of Hannover.

Lentz, W. (1993) Neuere Entwicklungen in der Theorie dynamischer Systeme und ihre Bedeutung für die Agrarökonomie. *Volkswirtschaftliche Schriften*, no. 432. Duncker & Humblot, Berlin.

Le Renard, J. (1988) SEPV-Diagnose im Pflanzenschutz für 17 Kulturarten. Deutsche Landwirtschafts-Gesellschaft, Frankfurt (ed.) *Wissensbasierte Systeme in der Landwirtschaft*, Internationaler DLG-Computerkongress.

Liebig, H.P. (1983) Einflüsse endogener und exogener Faktoren auf die Ertragsbildung von Salatgurken (*Cucumis sativus* L.) unter besonderer Berücksichtigung von Ertragsrhythmik, Bestandesdichte und Schnittmassnahmen. PhD thesis, University of Hannover.

Liebig, H.P. (1984) Model of cucumber growth and yield. II. Prediction of yield. *Acta Horticulturae* 156, 139–154.

Liebig, H.P. (1989) Die Quantifizierung der pflanzlichen Stoffproduktion unter fluktuierenden Klimabedingungen. Habilitations thesis, University of Hannover.

Liebig, H.P. and Alscher, G. (1993) Combination of growth models for optimised CO_2- and temperature-control of lettuce. *Acta Horticulturae* 323, 155–162.

Liebig, H.P. and Wiebe, H.J. (1982) Kurzfristige Ertragsprognosen von Bleichspargel. *Gartenbauwissenschaft* 47, 91–96.

Lioret, C. (1974) L'analyse des courbes de croissance. *Physiologie Vegetale* 12, 413–434.

Loomis, R.S., Rabbinge, R. and Ng, E. (1979) Explanatory models in crop physiology. *Annual Review of Plant Physiology* 30, 339–367.

Lorenz, H.P. (1980) Modelluntersuchungen zur Klimareaktion von Wachstumskomponenten am Beispiel junger Salatgurken (*Cucumis sativus* L.) – Ein Beitrag zur Temperaturführung in Gewächshäusern. PhD thesis, University of Hannover.

Loughton, A. (1961) The effect of low temperature before forcing on the behaviour of rhubarb. *Experimental Horticulture* 4, 13–19.

Loughton, A. (1965) Effects of environment on bud growth of rhubarb, with particular reference to low temperatures before forcing. *Journal of Horticultural Science* 40, 325–339.

Ludolph, D. (1995) Untersuchungen über die Wirkung der Lichtmenge auf das Wachstum und Blühen verschiedener Zierpflanzen und deren Anwendung als Steuerungsfaktor. PhD thesis, University of Hannover.

Mann, W. (1987) Wachstumsmodelle zur Wirkung und Bewertung einer CO_2-Anreicherung bei Gewächshauskulturen. *Gartenbauwissenschaft* 52, 241–249.

Marcelis, L.F.M. (1994) Fruit growth and dry matter partitioning in cucumber. PhD thesis, University of Wageningen.

Marshall, B. and Thompson, R. (1987) Applications of a model to predict the time to maturity of calabrese *Brassica oleracea*. *Annals of Botany* 60, 521–529.

Matitschka, G. (1995) Vereinfachte Beschreibung der Bruttophotosynthese des physiologisch-dynamischen Simulationsmodells SUCROS und Weiterentwicklung für spezielle Anwendungen im Gemüsebau (Produktionssysteme Kohlrabi und

Kopfsalat). PhD thesis, University of Stuttgart-Hohenheim, Verlag UE Grauer, Stuttgart.

McCree, K.J. (1972) The action spectrum of photosynthetically active radiation against leaf photosynthesis data. *Agricultural and Forest Meteorology* 10, 443–453.

Mitscherlich, E.A. (1919) Das Gesetz des Planzenwachstums. *Landwitschaftliches Jahrbuch* 53,167–182.

Mohr, H. and Schopfer, P. (1992) *Pflanzenphysiologie.* Springer-Verlag, Berlin.

Monteith, J.L. (1977) Climate and the efficiency of crop production in Britain. *Philosophical Transactions of the Royal Society B* 281, 277–294.

Mori, K., Eguchi, H. and Matsui, T. (1980) Mathematical model of flowerstalk development in Chinese cabbage in relation to high temperature effect combined with low temperature effect. *Environmental Control in Biology* 18, 1–9.

Nakamura, E. (1977) Der Anbau von Chinakohl in Japan. *Gemüse* 13, 194–198.

Nederhoff, E.M. (1994) Effects of CO_2 concentration on photosynthesis, transpiration and production of greenhouse fruit vegetables. PhD thesis, University of Wageningen.

Nösberger, J. (1964) Der Ertragsaufbau von Pflanzenbeständen. *Schweizerische Landwirtschaftliche Forschung* H1, 7–12.

Nuttonson, M.Y. (1948) Some preliminary observations of phenological data as a tool in the study of photoperiodic and thermal requirements of various plant materials. In: Murneek, A.E. and White, R.O. (eds) *Vernalization and Photoperiodism – a Symposium.* Chronica Botanica, Waltham, Massachusetts.

Ottosson, L. (1958) Growth and maturity of peas for canning and freezing. *Växtaodling 9.*

Pakes, A.G. and Maller, R.A. (1990) *Mathematical Ecology of Plant Species Competition: a Class of Deterministic Models for Binary Mixtures of Plant Genotype.* Cambridge University Press, Cambridge.

Penning de Vries, F.W.T. (1983) Modeling of growth and production. In: Lange, O.L., Nobel, P.S., Osmond, C.B. and Ziegler, H. (eds) *Encyclopedia of Plant Physiology, New Series, Vol. 12D, Physiological Plant Ecology IV.* Springer-Verlag, Berlin, pp. 117–150.

Pestemer, W. (1994) Einbindung von phytotoxischen und ökologisch-chemischen Daten zur Wirkung und zum Verhalten von Herbiziden in Expertensystemen für Beratung und monitoring. *Nachrichtenblatt des deutschen Pflanzenschutzdienstes (Braunschweig)* 46, 115–121.

Pestemer, W. and Günther, P. (1995) Use of a Pesticide Monitoring System (PEMO-SIS) for the risk assessement of pesticide leaching potential. *BCPC Monograph No.62: Pesticide Movement to Water.* British Crop Protection Council, Farnham, pp. 351–356.

Pfeufer, B. (1990) Wirkungen hoher CO_2-Konzentrationen auf Gemüsearten: Bedeutung äusserer Faktoren, Schadesymptome und Ursachen. PhD thesis, University of Hannover.

Pressman, E., Arthur, A., Schaffer, D.C. and Eliezer, Z. (1989) The effect of low temperature and drought on the carbohydrate content of asparagus. *Journal of Plant Physiology* 134, 209–213.

Pytlinski, J. and Krug, H. (1993) Modell zur Kulturplanung im Zierpflanzenbau am Beispiel von stecklingsvermehrten Pelargonium – zonale Hybriden. *Gartenbauwissenschaft* 58, 74–81.

Radford, P.J. (1967) Growth analysis formulae – their use and abuse. *Crop Science* 7, 171–175.

Rèaumur, R.A.F. de (1735) *Observation du thermomètre, faites à Paris pendant l'année 1735, comparées avec celles qui ont été faites sous la ligne, à l Isle de France, a Alger et en quelques-unes de nos isles de l' Amérique.* Mémoire d'Academie des Sciences, Paris.

Reinisch, K., Thümmler, C. and Hopfgarten, S. (1984) Hierarchical on-line control algorithm for repetitive optimization with predicted environment and its application to water management problems. *Systems Analysis Modelling Simulation* 1, 263–280.

Richards, F.J. (1969) The quantitative analysis of growth. In: Steward, C. (ed.) *Plant Physiology.* Academic Press, New York, pp. 3–60.

Rietze, E. (1988) Kurzfristige Kälteeinwirkungen bei Gurken (*Cucumis sativus* L.). PhD thesis, University of Hannover.

Rippen, H. and Krug, H. (1994) Analyse biologischer und monetärer Leistungen im Unterglasgemüsebau in Abhängigkeit von Kulturführung und Standort. I Analyse der Wachstumsleistungen. *Gartenbauwissenschaft* 59, 158–166.

Rippen, H., Krug, H. and Storck, H. (1994) Analyse monetärer und biologischer Leistungen im Unterglasgemüsebau in Abhängigkeit von Kulturführung und Standort. II. Ökonomische Analyse. *Gartenbauwissenschaft* 59, 241–248.

Ross, H.A. and Hegarthy, T. A. (1979) Sensitivity of seed germination and seedling radicle growth to moisture stress in some vegetable crop species. *Annals of Botany* 43, 241–243.

Salter, P.J. (1960) The growth and development of early summer cauliflower in relation to environmental factors. *Journal of Horticultural Science* 35, 21–33.

Scaife, A., Cox, E.F. and Morris, G.E.L. (1987) The relationship between shoot weight, plant density and timing during the propagation of four vegetable species. *Annals of Botany* 59, 325–334.

Schacht, H. (1991) Steuerung der Düngung von Gewächshausgurken (*Cucumis saticus* L.) in geschlossenen Nährlösungssystemen mit Hilfe eines Simulationsmodells. PhD thesis, University of Hannover.

Schacht, H. and Schenk, M.K. (1990) Control of nitrogen supply of cucumber (*Cucumis sativus* L.) grown in soilless culture. In: van Beusichem, M.L. (ed.) *Plant Nutrition – Physiology and Applications.* Kluwer Academic Publishers, Dordrecht, pp. 753–758.

Schaffer, G. (1978) Rhubarb – the new way of growing. *Horticulture Industry* 12, 9–10.

Schapendonk, A.H.C.M. and Brouwer, P. (1984) Fruit growth of cucumber in relation to assimilate supply and sink activity. *Scientia Horticulturae* 23, 21–33.

Schoch, P.G., Hold, J.C.L., Dauple, P., Conus, G. and Fabre, M.J. (1989) Microvariation de chambre de tige pour le pilotage de irrigation. *Agronomie* 9, 137–142.

Schulze, E. and Kühn, W. (1980) Kontinuierliche und zerstörungsfreie Messung des Wassergehaltes in Pflanzen durch Absorption von Mikrowellen. *Angewandte Botanik* 58, 465–473.

Schulze, R.W. (1970) *Strahlenklima der Erde.* Dietrich Steinkopff Verlag, Darmstadt.

Seemann, J. (1956) Wärmeregelung in geheizten Gewächshäusern. *Gartenbauwissenschaft* 3, 102–128.

Seginer, I., Angel, A., Gal, S. and Kantz, D. (1986) Optimal CO_2 enrichment strategy for greenhouses: a simulation study. *Journal of Agricultural Engineering Research* 34, 285–304.

Seginer, I., Shina, G., Albright, L.D. and Marsh, L.S. (1991) Optimal temperature set points for greenhouse lettuce. *Journal of Agricultural Engineering Research* 49, 209–226.

Spitters, C.J.T. (1986) Separating the diffuse and direct component of global radiation and its implications for modeling canopy photosynthesis. II. Calculation of canopy photosynthesis. *Agricultural and Forest Meteorology* 38, 231–242.

Spitters, C.J.T., van Keulen, H. and van Kraalingen, D.W.G. (1989) A simple and universal crop growth simulator: sucros 87. In: Rabbinge, R., Ward, S.A. and van Laar, H.H. (eds) *Simulation and System Management in Crop Protection. Simulation Monographs.* Pudoc Wageningen, 147–181.

Spitters, C.J.T., Toussaint, H.A.J.M. and Goudriaan, J. (1986) Separating the diffuse and direct component of global radiation and its implications for modeling canopy photosynthesis. I. Components of incoming radiation. *Agricultural and Forest Meteorology* 38, 217–229.

Stützel, H. (1995) A simple model for simulation of growth and development in faba beans (*Vicia faba* L.). 1. Model description. *European Journal of Agronomy* 4, 175–185.

Tantau, H.J. and Hack, G. (1990) Online-Regelung mit dem Klimacomputer. Arbeitsbericht zum Projekt C 20 im Sonderforschungsbereich 110. Institut für Technik in Gartenbau und Landwirtschaft, Universität Hannover.

Thornley, J.H.M. (1987) Modelling flower initiation. In: Atherton, J.G. (ed.) *Manipulation of Flowering.* Butterworths, London.

Thornley, J.H.M. and Johnson, I.R. (1990) *Plant and Crop Modelling.* Clarendon Press, Oxford.

van Dobben, W.H. (1962) Influence of temperature and light conditions on dry-matter distribution, development rate and yield in arable crops. *Netherlands Journal of Agricultural Science* 10, 377–389.

van Dobben, W.H. (1963) The physiological background of peas, to sowing time. *Jaarbook Instituut voor Biologisch en Scheikundig Onderzoek van Landbouwgewassen.* Wageningen, pp. 41–49.

van Rijssel, E. (1979a) Opbrengstbepalende faktoren bij de teelt van kasrosen in het winterhalfjaar. *Landbouw-Economisch Instituut, The Hague, Publikatie no. 4.84,'s-Gravenhage.*

van Rijssel, E. (1979b) Oorzaken van verschillen in opbrengsten van kasrosen. *Landbouw-Economisch Instituut, The Hague, Publikatie no. 4.97, 's-Gravenhage.*

Visser, W.C. (1963) Formulae for the ecological reaction of crop yields In: *The Water Relations of Plants.* Blackwell Scientific Publishers, Oxford.

Walter, H. (1960) *Einführung in die Phytologie.* III. *Grundlagen der Pflanzenverbreitung,* 1.*Teil Standortslehre.* Verlag Eugen Ulmer, Stuttgart, p. 68.

Wang, J.Y. (1960) A critique of the heat unit approach to plant response studies. *Ecology* 41, 785–790.

Watson, D.J. (1952) The physiological basis of variation in yield. *Advances in Agronomy* 4, 101–145.

Weber, W.E. and Liebig, H.P. (1981) Anpassung einer Ausgleichsfunktion an beobachtete Werte. *EDV in Medizin und Biologie* 12, 88–92.

Wechsung, F. and Orth, W.D. (1991) Heuristische Konstruktion mehrfaktorieller Ertragsmodelle – ein Beitrag zur Erfassung von Wechselwirkungen im landwirtschaftlichen Erzeugungsprozess. In: Buddle, H.J., Geidel, H. and Schiefer, G. (eds) *Agrarinformatik 21,* Eugen Ulmer Verlag, Stuttgart.

Went, F.W. (1954) *The Experimental Control of Plant Growth.* Ronald Press, New York.

Wheeler, T.R., Hadley, P., Morison, J.I.L. and Ellis, R.H. (1993) Effects of temperature on the growth of lettuce (*Lactuca sativa* L.) and the implications for assessing the impacts of potential climate change. *European Journal of Agronomy* 2, 305–311.

Wiebe, H.J. (1974) Zur Bedeutung des Temperaturverlaufes und der Lichtintensität auf den Vernalisationseffekt bei Blumenkohl. *Gartenbauwissenschaft* 39, 1–7.

Wiebe, H.J. (1975) Zur Übertragung von Ergebnissen aus Klimakammern auf Freilandbedingungen mit Hilfe eines Simulations – modells bei Blumenkohl. *Gartenbauwissenschaft* 40, 70–74.

Wiebe, H.J. (1979) Short term forecasting of the market supply of vegetables, especially cauliflower. *Acta Horticulturae* 97, 399–409.

Wiebe, H.J. (1994) Flower formation and timing field production of vegetables. *Acta Horticulturae* 371, 337–343.

Wiebe, H.J., Habegger, R. and Liebig, H.P. (1992) Quantification of vernalization and devernalization effects for kohlrabi (*Brassica oleracea convar. acephala* var. *gongylodes* L.). *Scientia Horticulturae* 50, 11–20.

Wiener, N. (1965) *Cybernetics*. MIT Press, Cambridge, MA.

Williams, C.N. and Joseph, K.T. (1976) *Climate, Soil and Crop Production in the Humid Tropics*. Oxford University Press, Oxford.

Wissemeier, A.H. (1996) Calcium-Mangel bei Kopfsalat, Poinsettie und cowpea: Einfluss von Genotyp und Umwelt. Habilitations thesis, Universität Hannover, Verlag UE Grauer, Stuttgart.

Wolf, S., Rudich, J., Marani, A. and Rekah, Y. (1986) Predicting harvest date of processing tomatoes by a simulation model. *Journal of American Society of Horticultural Science* 111, 11–16.

Wurr, D.C.E., Fellows, J.R., Sutherland, R.A. and Elphinstone, E.D. (1990) A model of cauliflower curd growth to predict when curds reach a specific size. *Journal of Horticultural Science* 65, 555–564.

Zorn, C. (1962) Zur Frage der Anwendung von Wärmesummen zur Erntezeitbestimmung der Erbsen. *Industrielle Obst- und Gemüseverwertung* 52, 124–128.

Correlative Growth in Vegetables   5

H.C. Wien

Department of Fruit and Vegetable Science, Cornell University,
134A Plant Science Building, Ithaca, New York 14853–5908, USA

Growth of herbaceous annual vegetable crops is characterized by the size increase of the plant's organs until they are harvested or senesce. The chief components of plants are carbohydrates, manufactured by the process of photosynthesis. The organ where these carbon compounds are manufactured, the leaves, can thus be considered the principal source organs of the plant. The distribution of assimilates from the mature leaves to developing plant organs is governed by source–sink relations described by the process termed 'correlative growth'. The other main source organs are the roots, which supply other important components of plants, the mineral elements and water, as well as some hormones (Itai and Birnbaum, 1991).

The relations among the principal source and sink organs in vegetable crops have often been described. Unfortunately, these descriptions are, to quote Farrar (1988), 'unpleasantly similar to a mere catalogue of experimental findings. To catalogue is usually a sign of a lack of real understanding'.

Given the complexity of interactions that occur among leaves, roots, stems and fruits, which act as sources in one instance, and sinks in another, it is difficult to study such an interacting system without introducing many artefacts. For instance, although partial leaf removal gives insight into the effect of leaves as sinks, increases in the photosynthetic rate of the remaining leaf area may lead to erroneous assumptions of how many leaves are needed for optimum growth and yield (Wareing *et al.*, 1968). Feedback effects of sink removal may result in stomatal closure, accumulation of assimilates in leaves, and reduced productivity (Neales and Incoll, 1968; Plaut *et al.*, 1987). More subtle interventions are needed that do not cause the large disruptions inherent in organ removal. In recent years, it has become possible to alter the rates of enzyme-regulated reactions in plants, and this promises to allow manipulation of particular growth systems with minimal disruptions (Stitt and Sonnewald, 1995).

© CAB INTERNATIONAL 1997. *The Physiology of Vegetable Crops*
(H.C. Wien, ed.)

Besides describing the interactions of source and sink organs in herbaceous vegetables, this chapter also discusses the relations among competing sinks on the plants. In the first section, the interactions among vegetative shoot apices and the other above-ground growth points are covered. Next, the relations between vegetative and reproductive structures before and after anthesis of the first flowers are detailed. Lastly, the interactions between above- and below-ground organs are described.

The emphasis in the chapter is on vegetable species that produce flowers and fruits. For information on the growth correlations among plant organs in root, bulb or tuber-bearing vegetable crops, the reader should refer to the chapters on those crops.

APICAL DOMINANCE

Gardeners and scientists have for centuries recognized the influence of the vegetative apex on growth of lateral buds and branches on the same shoot. In apically dominant species and cultivars, axillary buds are prevented from growing unless the apex is removed. In both woody and herbaceous species, it is common practice to prune the plant tip to encourage branch growth. On the other hand, the desire to restrict plant growth to one stem, as for instance in glasshouse tomatoes, has led to research on ways to enhance apical dominance (Tucker, 1976). Manual removal of tomato side shoots is a major expense in glasshouse tomato production every year.

The degree of apical dominance can be influenced by environmental factors such as nutrient levels and light (Hillman, 1984). Branch growth tends to be fostered by adequate levels of nutrients, particularly nitrogen, whereas nitrogen-deficient plants may have much reduced branch growth even when apical dominance is reduced by other means. Plants grown in low light intensities also tend to be more apically dominant than those produced in high light environments (Hillman, 1984). Exposure of plants to light rich in far-red light markedly reduced branch growth of tomatoes (Tucker, 1976), but had no effect on branch growth of bush beans (White and Mansfield, 1978).

Genetic differences in degree of apical dominance are also quite marked. In tomato, lines have been developed that have much reduced branching ('Blind'; Brenner et al., 1987), and nearly branchless habit ('Lateral Suppressor'; Tucker, 1981). In peas, similar differences among genotypes have been identified (Stafstrom, 1993). Use of these genotypes has helped to elucidate the control of branch growth, and identify the hormonal factors involved.

Although the environmental factors mentioned above influence the degree of apical dominance, they probably affect correlative growth by altering the level and activity of growth hormones. The hormones thought to be most directly involved are auxins, cytokinins and abscisic acid, with possible additional influence of gibberellins and ethylene in some species (Tamas, 1995).

The central role of indole-3-acetic acid (IAA) in apical dominance has been known for many years (Hillman, 1984). Evidence for its involvement comes from several sources. When apical dominance is eliminated by apex removal, it can be re-established by applying auxin in a lanolin paste on the cut stem. Secondly, axillary buds can be released from apical dominance by placing an inhibitor of polar auxin transport between them and the dominant apex. Thirdly, degree of dominance of shoots is related to capacity for polar auxin transport. In pea plants with one dominant and one suppressed shoot, only the dominant shoot showed measurable capacity for polar auxin transport (Morris and Johnson, 1990).

In spite of the general agreement that auxin plays a central role in apical dominance, the mechanism of action of this hormone in suppression of branch growth is not entirely clear. It is known that auxin diffusing from the shoot apex does not act directly on the inhibited buds, for there is little evidence of auxin movement from the apex to the inhibited bud or subordinate shoot (Morris, 1977). Levels of endogenous auxin in inhibited buds were low, and rose when the apical dominance was removed (Hillman et al., 1977; Gocal et al., 1991). It is therefore thought that auxin exerts control in an indirect manner, either by diverting nutrients to the actively growing apex, and therefore depriving the inhibited buds of nutrients, or by stimulating the production of inhibitory compounds in the buds (Hillman, 1984). A third theory ascribes lack of bud growth to a lack of vascular connections to the inhibited axil. This theory is based on the finding by Aloni (1995) that a polar flow of auxin from a developing organ is necessary for the production of vascular tissue from that organ.

Of the mechanisms envisioned for auxin regulation of bud growth, the generation of an inhibitor has been most generally substantiated. Abscisic acid (ABA) is thought to be the most likely compound to inhibit bud growth. Measurements of ABA levels in tomato lines differing in branching tendency have shown parallel changes in ABA levels in the stem (Tucker, 1981) (Table 5.1). ABA levels in 'Lateral Suppressor' plants were significantly higher than in the normally branching 'Craigella'. If plants were exposed to an end-of-day far-red light treatment which suppressed branching, ABA levels were further increased. Release of *Phaseolus* buds from apical dominance by apex removal resulted in a significant drop in bud ABA concentration (Knox and Wareing, 1984; Gocal et al., 1991). On the other hand, replacing the plant apex with an auxin–lanolin paste produced an increase in bud and stem ABA levels and inhibited branch growth (Knox and Wareing, 1984).

There has been little evidence to link a lack of vascular development to inhibition of axillary buds. Tucker (1976) found little correlation between degree of vascular development and activity of axillary buds in tomato, and Hillman (1984) summarized similar findings by others.

Other hormones may be involved in the regulation of apical dominance. Cytokinins have long been thought to play a role, because application of cytokinins to inhibited buds allowed them to start growing. It is not known

Table 5.1. Concentration of ABA-like substances (µg kg^{-1} dry weight) in tomato stem tissue as influenced by top–root grafting and end-of-day far-red (FR) light treatment (Tucker, 1981). By permission of Academic Press Ltd, London.

Tomato line (top–root)	− FR	+ FR
Intact 'Craigella'	0.45 ± 0.03	10.51 ± 0.5
Intact 'Lateral Suppressor'	8.55 ± 0.4	17.66 ± 0.5
'Lateral Suppressor'/'Craigella'	17.20 ± 0.4	28.44 ± 0.6
'Craigella'/'Lateral Suppressor'	0.51 ± 0.02	9.55 ± 0.5

if the cytokinins are translocated from sites of synthesis in the roots, or are being synthesized in the buds themselves (Tamas, 1995). There is some evidence for production by roots. Tucker (1981) found higher levels of isopentenyl adenosine, an endogenous cytokinin, in roots of freely branching 'Craigella' tomato, than in branchless 'Lateral Suppressor'. Use of 'Craigella' rootstocks with 'Lateral Suppressor' scions failed to stimulate bud growth, however, indicating that root-produced cytokinins may have a secondary role in apical dominance regulation. Recent findings by Bangerth and co-workers may explain how cytokinins are regulated. They discovered that auxin produced in apical regions of vegetative pea and bean plants suppressed root-produced cytokinin production (Bangerth, 1994; Li *et al.*, 1995). This would imply that the auxin produced by the 'Lateral Suppressor' apex prevented the cytokinin production of the root system, and contributed to the inhibition of side shoots. More work is needed to clarify this complex picture.

Evidence for involvement of other growth substances such as gibberellins and ethylene in apical dominance is more conflicting, and shows differences among species (Tamas, 1995).

In summary, our understanding of the hormonal regulation of apical control of axillary bud inhibition is making progress. Recent work on molecular aspects (e.g. Stafstrom, 1993) should advance our knowledge. Much remains to be done, including studies of the processes that govern branch development after the initial release from inhibition.

INTERACTIONS AMONG ABOVE-GROUND PARTS

INTERACTIONS BETWEEN VEGETATIVE AND REPRODUCTIVE PARTS

The interaction between vegetative and reproductive parts is often termed source–sink relations, for the leaves, which make up a large part of the vegetative component of the plant, also serve as the principal source of assimilates for the reproductive tissues. Once rapid growth of the reproductive structures has begun, the developing fruits are generally the major sinks for the products of photosynthesis (Ho, 1992). Flower primordia, and the developing flower buds tend to have lower priority for assimilates than vegetative tissues in many vegetable crops, although there are exceptions. In the paragraphs below, examples of these interactions will be given, and the mechanisms that appear to be operating in each case.

Interactions before anthesis

The primordia of the first-formed flowers are being developed at a time when the plant is forming additional leaves, stem tissue and below-ground structures. If a vegetable crop such as tomato is faced with a limitation in assimilate supply at that time, vegetative primordia have precedence over the reproductive sinks (Ho, 1992). With indeterminate tomatoes growing under low light conditions, such lack of reproductive structure dominance is expressed in a failure of the first cluster to develop to anthesis, and with formation of additional leaves on the main stem (Kinet, 1977a,b; and see Chapter 6). Kinet (1977b) was able to overcome the detrimental effects on cluster development by removing young leaves that were developing at the same time as the cluster. Other research indicated that the competition between developing leaves and clusters was mediated by growth substances (Kinet *et al.*, 1978; Abdul and Harris, 1978). Application of GA_{4+7} (mixture of gibberellins 4 and 7) and benzyl adenine to the developing cluster when the cluster was first macroscopically visible allowed the flower buds to reach anthesis (Kinet *et al.*, 1978). Growth analysis revealed that growth of hormone-treated clusters was occurring at the expense of the plant's apical meristem.

A similar situation occurs in pepper (*Capsicum annuum*), in which assimilate shortages brought about by low light conditions or high temperatures result in flower bud abscission before anthesis (Wien *et al.*, 1993a; and see Chapter 7). Again, vegetative tissues, such as the developing leaves (Aloni *et al.*, 1991), and the expanded leaves (Turner and Wien, 1994) appear to have higher priority for the assimilates than the flower buds. The abscission is mediated by ethylene, and can be reduced by application of ethylene action inhibitors such as silver thiosulphate (Wien and Zhang, 1991). To date, stimulation of reproductive structure growth by direct application of growth-promoting compounds has not been successful.

The balance between growth of vegetative and reproductive tissue can also be controlled by photoperiod. In some lines of peas, such as the photoperiod-sensitive line G-2 for example, long days bring about the cessation of apical vegetative growth, and foster pod and seed development, leading to eventual plant senescence. Under short days this same line supports both vegetative growth and production of flowers and pods. Davies and co-workers demonstrated that the rate at which flowers are developed on the plant lags behind leaf development in short days (Kelly and Davies, 1986, 1988; Sklensky and Davies, 1993). The continued leaf development may be mediated by high gibberellin levels in young leaves, and a relatively lower level of auxin (Sklensky and Davies, 1993). Under long days, lower gibberellin levels in developing leaves, and higher auxin levels in the floral parts may favour assimilate translocation to reproductive rather than vegetative tissues.

Once rapid growth of fruits has begun on the plant, assimilates tend to move preferentially to these reproductive structures (Wardlaw, 1990). This results in the cessation or reduction in the rate of new leaf and stem formation in such crops as cucumber, tomato and eggplant (Claussen, 1976; Hurd et al., 1979; Marcelis, 1993a), and reduced root growth. Removal of fruits before they reach the active growth stage allows a continuation of leaf, stem and root formation on the plant, with little effect on the overall dry matter productivity. This appears to be the situation in some species, such as eggplant (Claussen, 1976). In others, however, removal of the strongest sink for assimilates may lead to reduced photosynthetic activity of leaves (Neales and Incoll, 1968).

The inhibition of leaf photosynthesis rate after sink removal may have several causes. It has been linked in some species with stomatal closure, resulting from a build-up of ABA in the leaf blade (in pepper; Kriedemann et al. 1976; in soyabean; Setter et al., 1980a,b). In others, accumulation of starch in the plastids may distort the membrane structure of the chloroplast enough to lower gas exchange rates (Goldschmidt and Huber, 1992; Stitt et al., 1995). Thirdly, the formation of sugar phosphates after sink removal may in some cases lead to a deficiency of inorganic phosphorus in the leaf (Plaut et al., 1987). Assimilate accumulation in the leaf after sink removal has led in several species to reduced activity of the ribulose bisphosphate carboxylase, the photosynthetic enzyme (in cucumber; Peet et al., 1986; in tomato; Yelle et al., 1989; and see Stitt et al., 1995).

It is important to note that sink removal will not invariably lead to adverse effects on photosynthesis. Most vegetable crops have alternate sinks such as branches, younger fruits, etc., that can become principal sinks after fruit removal. Heuvelink and Buiskool (1995) found, for instance, that dry matter production of glasshouse-grown tomato was not adversely affected until the plants had been reduced to one cluster. Less drastic pruning resulted in increased number and size of the remaining fruits. Similarly, Marcelis (1991) found that photosynthesis rate of glasshouse cucumber was not reduced by partial fruit removal treatments, but required that all the fruits be picked.

There is evidence that the correlative effects of rapidly enlarging fruits on vegetative tissues may be mediated in some cases by hormones. Tamas and co-workers showed that pod growth of determinate bean cultivars inhibited axillary branch growth during the reproductive period (Tamas et al., 1979a). When pods were removed, branch growth resumed, and was accompanied by a decrease in branch ABA content. The branch growth inhibition could be reimposed by substituting lanolin paste containing auxin in place of the growing seeds in the pods (Tamas et al., 1981). The findings imply that the reproductive structures may function like the apical meristem of a vegetative plant, inhibiting the growth of branches through auxin diffusing from the pods. The inhibition may also involve the action of ABA on the axillary branches, although the origin of the ABA is not clear.

Interactions among reproductive parts

On vegetable crops that bear fruits, the developing reproductive structures are the dominant sinks for assimilates after flowering. The first-formed fruits take precedence over those developed later. This process of dominance takes similar form in different species, and has been described by many investigators.

In pod-bearing legumes, pods are located at the nodes of the earliest-opening flowers, generally the lower nodes of the main stem and the basal branches (Fig. 5.1) (De Moura and Foster, 1986a; H.C. Wien, Ibadan, Nigeria). Flowers further up the main stem and at later-formed nodes on the branches open successively later, and may set pods, but their development

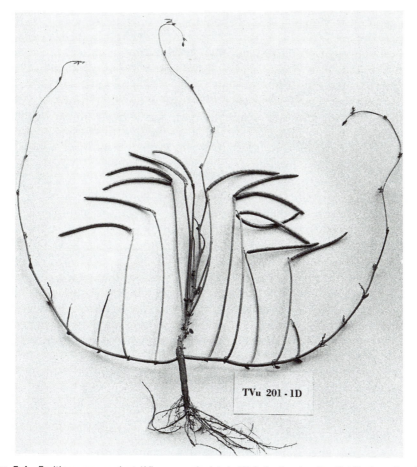

Fig. 5.1. Fruiting cowpea plant (*Vigna unguiculata* L. Walp.), showing the distribution of pods on the main stem and lower branches (Wien, Ibadan, Nigeria, unpublished data).

Table 5.2. Effect of fruit number and temperature on fruit growth rate of 'Corona' cucumber (Marcelis, 1993) with kind permission from Elsevier Science.

| Nodes with fruit (%) | Fresh weight increase rate, g day^{-1} | |
	18°C	25°C
17	27.2	48.1
33	23.5	38.1
100	16.8	24.6

is adversely affected by the presence of the older pods at lower nodes. In cowpea (*Vigna unguiculata* L. Walp.), the younger pods typically abscise, or are inhibited from setting (Ojehomon, 1970). A similar sequence occurs in the individual racemes, on which the oldest flowers set fruit, and later-formed flowers and flower buds are shed (Ojehomon, 1968). If, due to insect damage or environmental stress factors the first flowers are not set, then the younger flowers at higher positions on main stem and branches form the yield on the plant. Typically, the first flowers to open after the stress is removed will then set pods (De Moura and Foster, 1986b; Ojehomon, 1970).

A similar pattern of intermittent fruit production on the plant occurs also in cucumber (De Lint and Heij, 1980) and in tomato (Hurd *et al.*, 1979). Analysis of node-by-node production of seedless cucumber fruits in glasshouse-grown plants indicated that about five fruits are formed in quick succession (De Lint and Heij, 1980). Fruits on the next nodes stay small and wither. The inhibition of later-formed fruits is also expressed in slower development rates of these fruits, and their delay in reaching marketable size. Marcelis (1993b) accomplished similar decreases in fruit growth rate by varying the ratio of fruit to leaves on the plant (Table 5.2). The results imply that the competing demands for assimilate by several simultaneously growing fruits is greater than the capacity of the plant to supply via photosynthesis.

In tomato, the cyclic production of fruits can be dampened somewhat by restricting the number of fruit on the earliest clusters (Hurd *et al.*, 1979). Fruit pruning results in the production of fewer, larger fruits. The degree to which the plants can compensate for reduced fruit numbers by increased fruit size is cultivar-dependent, and may be insufficient in drastic fruit pruning treatments (Favaro and Pilatti, 1987).

The position of individual fruits within a cluster, and the timing of their relative development play an important role in determining final fruit size. Fruits formed from flowers at the proximal position on the cluster tended to have a higher cell number at anthesis of the flower, and to reach anthesis several days before distal flowers (Bangerth and Ho, 1984). If natural pollination was prevented, and the fruits were set using synthetic auxin applied to the ovary, the size of distal and proximal fruits was equalized by triggering fruit set of both at the same time. In that situation, proximal fruits had fewer, and distal fruits more cells than the corresponding controls (Bohner and Bangerth, 1988).

The competition among bell pepper fruits on the plant also is manifested through a reduction in cell numbers in the later-formed fruits. Ali and Kelly (1992) found that removal of the first-formed fruit resulted in an increase in the size of later-formed fruit that was correlated with larger numbers of pericarp cells.

Dominance of the first-formed fruit may be exercised in several ways (Bangerth and Ho, 1984). The earlier fruit may constitute a stronger sink for assimilates, due to a higher pressure gradient between sink and source (Bangerth and Ho, 1984). This gradient may be in part mediated by the action of growth hormones such as auxins and cytokinins active in the growing fruit. However, levels of extractible auxins in the fruit have not correlated well with relative fruit growth (Ho *et al.*, 1982; Bohner and Bangerth, 1988). On the other hand, the rate at which auxin diffused out of the fruit was more closely related to fruit growth, with distal fruits having a lower rate than proximal fruits (Gruber and Bangerth, 1990). This lends support to the theory of 'primigenic dominance', proposed by Bangerth (1989) as a mechanism that regulates the flow of assimilates among competing sinks. According to this theory, the first-formed fruit's greater auxin efflux would inhibit the auxin flow out of later-formed fruits at junction points. Assimilates would flow preferentially to the sink with the greater auxin flow. Much remains to be explained about the nature and location of these junctions, and the possible involvement of other growth hormones in primigenic dominance.

Another possible way in which dominance could be maintained by tomato fruits may be through the production of a growth inhibitor such as ABA. ABA content of competing tomato fruits has, however, not shown any relationship with fruit growth inhibition (Ho *et al.*, 1982; Bohner and Bangerth, 1988).

The importance of seeds in the dominance of competing reproductive structures has been apparent particularly from the work with beans (*Phaseolus vulgaris* L.). Tamas *et al.* (1986) found that seed removal from actively growing pods allowed growth of younger pods to continue, instead of being inhibited. If seeds were replaced by IAA or a synthetic auxin in lanolin, the growth-inhibiting effect of the older pods was restored. The larger fruits were found to export relatively more IAA than smaller ones, in parallel to their greater growth-inhibiting effect. In beans, this inhibition may be mediated by ABA, for Tamas *et al.* (1979b)) found in earlier work that ABA content was higher in the inhibited pods. In this respect, the seed-mediated inhibition has elements in common with apical dominance of vegetative plants (see section on apical dominance above).

Seed-mediated dominance relationships have also been found in squash (*Cucurbita pepo* L.) by Stephenson *et al.* (1988). Fruits containing more seeds inhibited the growth of later-formed fruits more than fruits with fewer seeds. The latter were also more likely to abort before reaching anthesis. It is not known if these effects are mediated by growth hormones. In tomato, the content of auxin and of ABA of seeds is several-fold higher than in the rest of the fruit (Bohner and Bangerth, 1988).

In some species, the functioning of a reproductive sink may be influenced by light, perhaps mediated by growth regulators. If individual fruits in glasshouse-grown cucumber were darkened, for instance, growth rate decreased, and the likelihood of fruit abortion increased (Schapendonk and Brouwer, 1984). More work is needed to elucidate the mechanism of this form of sink regulation.

TOP–ROOT INTERACTIONS

VEGETATIVE STAGE

The growth of annual herbaceous plants after emergence and before flowering consists of the balanced development of above- and below-ground structures. The aerial parts consist of the leaves which form the main photosynthetic organs, borne by the stems and petioles. The plant is anchored by the roots, which also take up water and nutrients essential for growth. The roots depend on assimilates translocated from the leaves to supply the energy and carbon skeletons for growth and uptake of water and nutrient elements. Plant growth thus requires that both the above- and below-ground parts are developed in a balanced fashion, and that mechanisms exist to maintain that balance in spite of alterations in either or both environments.

Studies of the partitioning of assimilates to tops and roots in the early vegetative stage have commonly found that allocation to shoots and roots is nearly constant (Brouwer, 1962a; Aung, 1974) (Fig. 5.2). If the balance is disturbed by leaf or root removal, growth of the affected part is accelerated, with the result that the balance is restored. Thus, defoliation leads to preferential growth of young leaves, rather than roots (Brouwer, 1962a; Shishido *et al.*, 1993). Similarly, root pruning results in a cessation of top growth, and a stimulation of new root formation. These temporary alterations in top : root ratio also occur when plants are subjected to shortages of essential inputs necessary for growth, as will be described below. The mechanisms governing the relative growth of the above-ground parts and the roots may differ depending on the nature of the disturbance, and are in most cases still poorly understood.

Brouwer (1962a) postulated a 'rule of thumb' that helps to predict how the plant will react to perturbations of top or root environments. He stated that in situations of limited supplies of resources, the part of the plant nearest the source will be relatively stimulated, and growth of the part farthest from the source will be relatively inhibited. Thus for instance, when a plant is transferred to low light conditions, growth of the tops (closest to the light source) is relatively less decreased than root growth, and top : root ratio

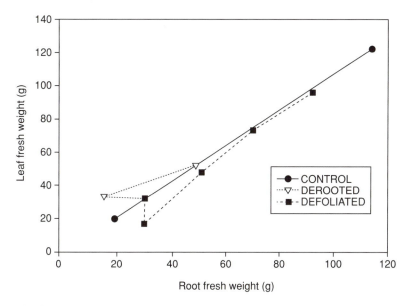

Fig. 5.2. The relationship between leaf and root fresh weight of bean seedlings grown in nutrient solution, as influenced by partial root or leaf removal (Brouwer, 1962a).

increases (Shishido *et al.*, 1993; Minchin *et al.*, 1994). In lower radiant flux densities, leaves become thinner (increased leaf weight ratio), and thus maximize the capacity for light interception per unit plant biomass. Under conditions of limiting light, the relatively low amounts of assimilates produced may be used preferentially by the relatively stronger sinks, the apical meristem and developing leaves and stem tissue (Shishido *et al.*, 1993).

There is evidence that some of these preferential growth processes are linked to differential enzyme activities, perhaps stimulated by the activity of endogenous plant growth regulators. For instance, the transfer of bean seedlings to complete darkness resulted in an increase in internode elongation, and preferential translocation of assimilates to the expanding stem region (Morris and Arthur, 1985a,b). At the same time, relatively less carbohydrate was translocated to the hypocotyl and root. Specific activity of the enzyme acid invertase, which regulates conversion of the translocated sucrose to hexose sugars, was increased in the stem and decreased in hypocotyl and root tissues. Similar effects on these processes could be obtained by applying gibberellic acid as a root drench to the plants (Morris and Arthur, 1985b). The results imply that endogenous gibberellins may regulate light-stimulated stem elongation in the bean plant.

There is evidence from some plants that light quality can influence partitioning of assimilates. In soyabean, Kasperbauer (1987) demonstrated that end of the day exposure to far-red light resulted in enhanced stem and reduced root growth. It is possible that the stem elongation was mediated by gibberellins (Morris and Arthur, 1985b). Similar examples of light quality effects on partitioning have been elucidated for storage root-bearing crops such as radishes and beets (Hole and Dearman, 1993). The example of light quality effects on top : root ratio fits less readily into the resouce limitation scheme postulated by Brouwer.

A common below-ground factor that alters top : root ratios is moisture supply. As soil water level is reduced from that optimum for plant growth, root elongation is less curtailed than shoot growth, and in some cases even slightly stimulated (Brouwer, 1962a; Creelman et al., 1990). In recent years, it has become evident that the relative growth of roots compared to shoots in dry conditions may be mediated by ABA in some plants. Exposure of soyabean and corn seedlings to low moisture potentials resulted in a large increase in ABA content, especially of the root tips (Creelman et al., 1990; Saab et al., 1990). At these ABA levels, root growth was stimulated, and shoot growth inhibited. The rise of ABA levels in plants under drought stress is generally thought to cause stomatal closure in the leaves (Tardieu and Davies, 1993). The hormone thus acts as a feedback control, reducing water loss from leaves during the dry period, while at the same time allowing root growth to proceed. In seedlings, the assimilates needed for this root growth may be coming from remobilization from the growth-inhibited stems and leaves (Sharp, 1990).

Deficiencies of some mineral nutrients also produce a decrease in the shoot : root ratio (Brouwer, 1962b) (Fig 5.3). This phenomenon has been most closely studied with regard to nitrogen deficiency. Given the complex changes that occur as a plant becomes deficient in nitrogen, it has been difficult to determine which of these adjustments are primarily responsible for the top : root ratio change. In the early stages of nitrogen deficiency, carbohydrates accumulate in leaves (Rufty et al., 1988) and in the root system (Chapin et al., 1988b). It is possible that the decline in nitrate uptake associated with transfer of plants to a nitrogen-free medium reduces the need for assimilates otherwise used for nitrate reduction (Rufty et al., 1988). The negative effects of nitrogen deficiency on photosynthetic rate were not apparent until later, so the plants temporarily had a surplus of carbohydrates, available perhaps to continue root growth. Induced nitrogen deficiency led to increases in leaf ABA levels within 2 days in tomato (Daie et al., 1979; Chapin et al., 1988b), and may be responsible for the decrease in leaf expansion common with nitrogen deficiency. Root ABA levels declined with nitrogen stress and are thus not likely to be involved in the root growth stimulation (Chapin et al., 1988a). Increased levels of ABA in the leaves may also be responsible for the reduced stomatal conductance prevalent in nitrogen deficient plants (Chapin et al., 1988b). Further work is needed to determine if the changes measured when plants are suddenly deprived of nitrogen also occur if plants undergo more gradual decrease in nitrogen status.

In phosphorus-deficient seedlings, shoot growth is more inhibited than root growth, resulting in a decreased shoot : root ratio (Fredeen et al., 1989; Cakmak et al., 1994a) (Fig. 5.4). In soyabeans, the reduction in shoot growth

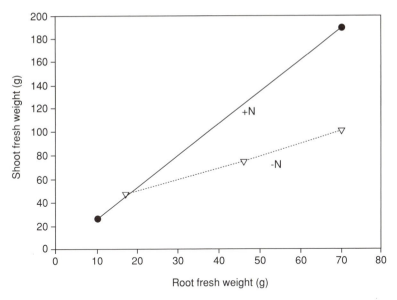

Fig. 5.3. The relationship between shoot and root fresh weight of maize plants grown in nutrient solution containing nitrogen, or transferred to a medium lacking nitrogen (Brouwer, 1962b).

was due primarily to a drastic decline in leaf expansion. Photosynthesis was less inhibited, and starch accumulated in the leaves and fibrous roots (Fredeen *et al.*, 1989). It may be that the reduced leaf growth made assimilates available to the roots, allowing their continued development. In agreement with these results, Cakmak *et al.* (1994b) found that sucrose export from primary leaves of P-deficient beans was not inhibited in comparison to seedlings grown on adequate P levels.

The response of bean seedlings to potassium and magnesium deficiencies contrast sharply with the reactions to –N and –P explained above. Potassium deficiency, and more drastically, magnesium deficiency inhibited assimilate export out of the leaves, and curtailed the growth of the root system (Cakmak *et al.*, 1994a,b) (Fig. 5.4). In general, therefore, the reaction of the plant to deficiencies of particular nutrients will depend on the specific effects that those nutrients have on top growth, the rate of photosynthesis, the export of assimilates, the import of carbohydrates by the root system, and the growth processes of the roots.

REPRODUCTIVE STAGE

Growth of vegetable crops when producing a reproductive structure is characterized by the dominance of that structure over the vegetative part of the plant with regard to assimilates. Typically in fruit-bearing crops such as

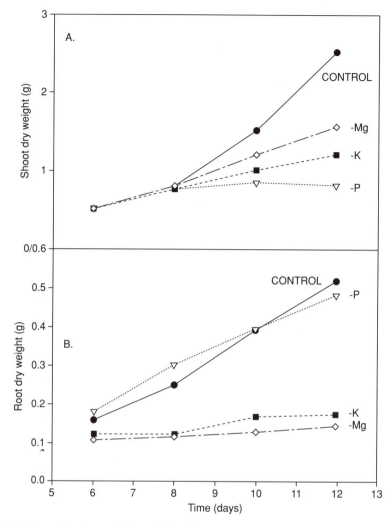

Fig. 5.4. Change in shoot and root weight of bean plants grown in full nutrients, or in solutions lacking phosphorus, potassium or magnesium (Cakmak *et al.*, 1994a). By permission of Oxford University Press.

tomatoes, cucumbers and eggplant, growth of the root system is reduced in rate after flowering (Cooper, 1955; Van der Post, 1968; De Stitger, 1969). On the other hand, if fruits are removed, or when their growth is decreased at ripening, root growth rate increases again (Claussen, 1976) (Fig. 5.5). Many factors can influence the extent to which root growth is inhibited. In simplified plant systems, in which leaf area has been reduced through drastic pruning, root growth can be brought to a halt by growth of one fruit, as was shown by De Stitger (1969) with cucumber. Determinate cultivars of crops with limited leaf area show a similar response, particularly if the developing

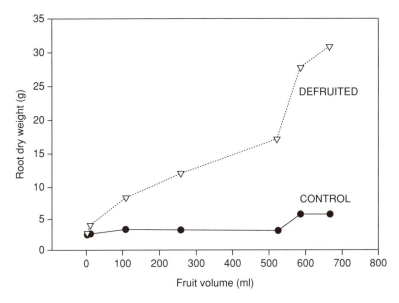

Fig. 5.5. Effect of presence of fruits on root growth in eggplant (Claussen, 1976).

fruit induces the senescence of the leaves, as occurs in grain legume crops such as cowpea or *Phaseolus* beans.

In crops in which the leaf canopy maintains photosynthetic activity over the reproductive period, root systems continue growing in proportion to the above-ground vegetative portion of the crop as successive flushes of fruits develop on the plants. For instance, Hurd *et al.* (1979) found in a glasshouse tomato crop that top : root ratio of the vegetative parts of the plant was maintained at around 4.5 for most of the fruiting period. The ratio was also not significantly altered if flower clusters were only allowed to set three fruits each. Similar results were obtained by Richards (1981) when tomato plant size was varied by growing the plants in containers of different root volumes. Pepper plants on which no, one or three fruits were allowed to develop also showed a constant relation of vegetative above- and below-ground growth (Nielsen and Veierskov, 1988). In all these cases, it is assumed that the demands of the developing fruits for assimilates and nitrogen compounds can be met by current photosynthesis, and does not require the dismantling of leaf proteins. Restricting fruit numbers by pruning, keeping fruit growth rates low by low temperatures, and by optimizing growth conditions with regard to nutrients, abscence of disease and insect pests all serve to maintain photosynthesis and lessen the demand of the dominant reproductive sink.

In other instances, however, presence of fruits had a variable effect on root weight, and on the ratio of root to vegetative shoot weight. For instance, when fruit load was varied on a glasshouse cucumber crop grown at 18°C, proportion of root growth decreased with increasing fruit numbers

(Marcelis, 1994). Paradoxically, the same variation in fruit load resulted in a constant, but lower ratio of roots to vegetative shoot weight when the experiment was repeated at 25°C. In another experiment, glasshouse peppers responded to increased CO_2 in the atmosphere by increasing root growth only if the plants were defruited, but not if four or eight fruits were allowed to remain on the plants (Daunicht and Lenz, 1978). It is difficult to see how partitioning of assimilates to the roots in these cases could be determined by a priority system, in which roots were supplied only if there were surplus assimilates available.

Part of our difficulty in understanding the relation between shoots and roots in plants may be that dry weight changes in these organs do not correspond well with the changes in function of shoots and roots. Hurd *et al.* (1979) noted in glasshouse tomatoes grown in nutrient film culture in which the roots could be viewed, that roots turned brown and stopped growing about 30 days after anthesis of the first flowers. This change was not apparent from the dry matter data. Similarly, Richards (1981) noted a close correlation between the leaf area growth of tomato plants and the number, rather than the weight of roots on the plants. He inferred that hormone production by the roots would be more closely related to the number of growing roots, with functioning root tips, where cytokinins, gibberellins and ABA are manufactured (Itai and Birnbaum, 1991), than to the mass of the root system. Van Nordwijk and De Willigen (1987) have similarly suggested that the functional equilibrium between roots and shoots should depend more on the relationship between root area and leaf area, than the ratios of their dry weights. Until convenient methods are found to measure root area, it will be difficult to test this interesting idea.

The mechanisms which govern the partitioning of assimilates between vegetative and reproductive organs, and among roots and other vegetative parts of the plant after flowering are largely unknown. The interactions of vegetative and reproductive sinks above ground are discussed above. The factors governing sink strength of root systems of reproductive plants are even less well understood. This is because the contributions of the above- and below-ground components to the whole plant can only be assessed by severing top from root. Some insight can be gained by the use of top–root grafts of contrasting plants, but such treatments do not help to identify the cause of the observed changes in top or root growth.

Transferring the tops of plants from their own root systems to those of others has been a common way of overcoming the adverse effects of soil pathogens or cold root zone temperatures (Lee, 1994; and see Chapters 2 and 9). Research by Tachibana (1982, 1987, 1988) indicated that the root system of fig-leaf gourd (*Cucurbita ficifolia*) maintained water and phosphorus uptake in cold soils, and showed continued cytokinin production, even when grafted to shoots of cucumber, that were not capable of such cold soil activity on their own roots. Detached roots of fig-leaf gourd maintained higher rates of respiration in low temperatures than cucumber, and may allow active mineral uptake in cool soils (Tachibana, 1989). Thus the proper-

Table 5.3. Influence of rootstock genotype on first flower date, and on total cluster fresh weight and shoot fresh weight after 3 months of growth of nine tomato genotypes (Zijlstra and den Nijs, 1987) with kind permission from Kluwer Academic Publishers.

Rootstock genotype	Flowering date*	Deviation from mean	
		Cluster fresh weight plant^{-1} (g)	Shoot fresh weight plant^{-1} (g)
Alcobacca	− 3	+ 20	+ 7
Delikatess	− 2	+ 26	+ 1
KNVF	+ 6	− 40	+ 51
IVT line 2	+ 3	− 25	− 13

* Negative = earlier than mean.

ties of the roots *per se* contributed to plant performance independent of the shoot. The improved vegetative growth was sustained into the reproductive period with higher fruit yields in cold soils (den Nijs, 1980), and also under more favourable conditions (Zijlstra *et al.*, 1994).

Top–root grafts among cultivars within a vegetable species can also result in improved performance in cold soils, and modification of flowering and reproductive growth. For instance, Zijlstra and den Nijs (1987) found that the days to first flower could be significantly advanced or delayed by choice of rootstock (Table 5.3). Earlier flowering generally resulted in higher cluster weights after 3 months of growth, but was not correlated to top fresh weight. Work is needed to relate these results to physiological attributes of the rootstocks, such as cytokinin production and root growth capacity. The results point out, however, that a tomato line selected for improved cold weather performance, such as IVT line 2, does not necessarily possess the root characteristics for growth in cold soils.

TEMPERATURE EFFECTS ON TOP-ROOT RELATIONS

The response of plant parts to temperature involves the direct effect of temperature on the specific part, and the influence that this intervention has on the relation among different organs. To understand the effects of temperature on partitioning of assimilates among plant parts, it is necessary to determine the effect on assimilate acquisition, and on the partitioning of the carbohydrate among competing sinks (Farrar, 1988). Each of these processes is governed by the action of key enzymes which have temperature optima, many of which have not yet been characterized. It is beyond the scope of this book to delve into detail on this complex subject. In the succeeding paragraphs, examples of the response of tops and roots of vegetable crops to temperature imposed uniformly on the whole plant will first be presented. We will then consider the effect of differential top–root temperatures on top–root relations.

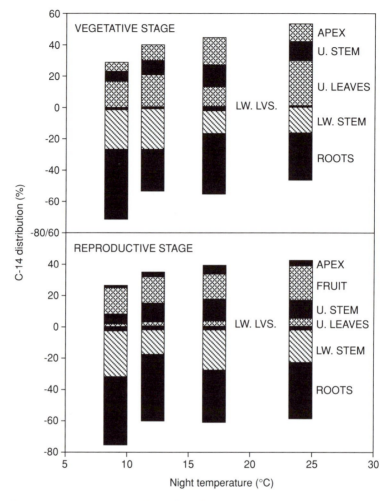

Fig. 5.6. Distribution pattern of carbon-14 labelled assimilates in tomato plants as influenced by the night temperature. Plants were dosed at leaf stage 3 in the vegetative, and leaf stage 7 in the reproductive stage (Hori and Shishido, 1977).

In general, it has been found that as temperature of the whole plant increases, the top : root ratio also goes up (Farrar, 1988) (Fig. 5.6). For instance, tomato plants kept in night temperatures of 24°C translocated a smaller proportion of assimilates to the lower stem and roots than if the plants were kept at 9°C (Hori and Shishido, 1977). The trend was similar in the vegetative and the early reproductive stages, and respiration rates of the different plant parts reflected the translocation patterns (Shishido *et al.*, 1989). Cucumbers in the vegetative stage also showed increased assimilate distribution to tops in higher temperatures (Kanameha and Hori, 1980). In fruit-bearing cucumber plants, shoot : root ratio (including fruits) increased

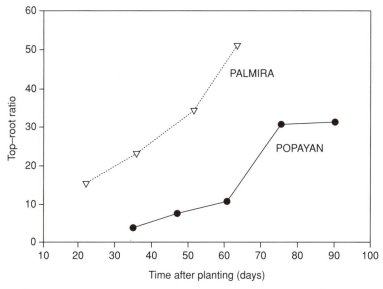

Fig. 5.7. Top–root ratios of bean plants grown under field conditions near Popayan (altitude 1760 m and 18°C mean temperature), and Palmira, Colombia (altitude 1000 m and 24°C mean temperature) (Waters *et al.*, 1983). By permission of Cambridge University Press.

from 13.5 to 31.7 as glasshouse temperature was increased from 18 to 25°C (Marcelis, 1994).

Beans grown at two locations in Colombia differing in elevation, and thus air and soil temperature showed increased nitrogen fixation, and lower top : root ratio at the cooler location (Waters *et al.*, 1983) (Fig. 5.7). Although many aspects other than temperature could have contributed to the difference in performance at the two locations, previous work by Graham (1979) also found increased nodule carbohydrate content in temperature regimes below 24/15°C (day/night).

When shoot temperatures are altered separately from the root zone temperatures, the response depends on the temperature range to which each is subjected. For cucumbers, the optimum temperatures for shoot growth tend to be higher than for root growth, and temperature variation above-ground has more marked effects on productivity than below-ground variations (Kleinendorst and Veen, 1983). Considerable experimentation with glasshouse crops grown at relatively low air temperatures has indicated that increasing root medium temperatures can enhance plant growth and yield (Drews *et al.*, 1980; Gosselin and Trudel, 1983), and leads to a reduction in the shoot : root ratio. A stimulation of root growth was also one of the first signs of growth enhancement when field-grown tomatoes were transplanted into soil covered with a clear plastic mulch (Wien *et al.*, 1993b). The mulch resulted in a 6°C increase in root zone temperatures during the first week after transplanting. Increased root growth was followed by enhanced branch growth and higher fruit yields.

Our understanding of the mechanisms that determine the growth response of shoots and roots to different temperatures is still poor. Since a given temperature could be affecting the physical properties of cells, the enzymes that catalyse chemical reactions controlling growth, the processes that govern assimilation by the leaves, and translocation to the sink organs, be they above- or below-ground, it is almost impossible to avoid complex interactions (Farrar, 1988). With the new methods of enhancing the activity of enzyme systems in transgenic plants, it may be possible in future to get a better understanding of how the partitioning of assimilates is controlled among vegetative and reproductive structures, including roots (Galthier *et al.* 1993; Stitt and Sonnewald, 1995). Galthier *et al.* (1993) showed recently, for instance, that tomato plants with elevated activity of sucrose phosphate synthase, the enzyme key to sucrose synthesis in leaves, had increased top : root ratios compared to normal plants. Manipulation of enzyme systems influential in assimilate transport and accumulation, in shoots and roots separately, will be helpful in improving our understanding of control mechanisms in top : root relations.

CONCLUDING REMARKS

The correlations governing growth of the major organs of fruit-bearing vegetable crops have been characterized frequently, but the mechanisms that determine which part of the plant takes precedence when resources are limiting are still little understood. From the descriptions in the literature, we can predict for instance that developing leaves, stems and roots will receive assimilates in preference to developing reproductive structures in the pre-flowering stage of growth. Once fruits have set, the latter become the dominant sinks, and the root system is lowest in priority in obtaining assimilates. A general sequence that determines sink activity might be initiated by an increase in growth hormone activity in a particular sink. This could lead to stimulation of enzymes catalysing the accumulation or use of assimilates in that sink, resulting in increased translocation to the organ. Under conditions of limited assimilate supply, the activity of a subordinate sink would be inhibited, but the mechanism of that inhibition is even less well understood.

As has been stated repeatedly in this chapter, there are still many gaps in our knowledge on how growth correlations operate. The disruptive nature of experimental manipulation has limited the amount of information that can be obtained from conventional experiments which remove leaves, fruits, etc. Expectations are high that the new genetic interventions, that allow the experimenter to vary the levels or activities of growth hormones and of specific enzymes, will increase our understanding of how plants regulate the relative growth of their principal parts. The frequent use of the tomato in genetic studies, and its relative ease of transformation, should therefore be advantageous for the study of growth correlations in vegetable crops.

REFERENCES

Abdul, K.S. and Harris, G.P. (1978) Control of flower number in the first inflorescence of tomato (*Lycopersicon esculentum* Mill.): the role of gibberellins. *Annals of Botany* 42, 1361–1367.

Ali, A.M. and Kelly, W.C. (1992) The effects of interfruit competition on the size of sweet pepper (*Capsicum annuum*) fruits. *Scientia Horticulturae* 52, 69–76.

Aloni, R. (1995) The induction of vascular tissues by auxin and cytokinin. In: Davies, P.J. (ed.) *Plant Hormones*, 2nd edn. Kluwer, Dordrecht, Netherlands, pp. 531–546.

Aloni, B., Pashkar, T. and Karni, L. (1991) Partitioning of [14C] sucrose and acid invertase activity in reproductive organs of pepper plants in relation to their abscission under heat stress. *Annals of Botany* 67, 371–377.

Aung, L.H. (1974) Root–shoot relationships. In: Carson, E.W. (ed.) *The Plant and its Environment*. University of Virginia Press, Charlottesville, VA, pp. 29–61.

Bangerth, F. (1989) Dominance among fruits/sinks and the search for a correlative signal. *Physiologia Plantarum* 76, 608–614.

Bangerth, F. (1994) Response of cytokinin concentration in the xylem exudate of bean (*Phaseolus vulgaris* L.) plants to decapitation and auxin treatment and relationship to apical dominance. *Planta* 194, 439–442.

Bangerth, F. and Ho, L.C. (1984) Fruit position and fruit set sequence in a truss as factors determining final size of tomato fruits. *Annals of Botany* 53, 315–319.

Bohner, J. and Bangerth, F. (1988) Effect of fruit set sequence and defoliation on cell number, cell size, and hormone levels of tomato fruits (*Lycopersicon esculentum* Mill.) within a truss. *Plant Growth Regulation* 7, 141–155.

Brenner, M.L., Wolley, D.J., Sjut, V. and Salerno, D. (1987) Analysis of apical dominance in relation to IAA transport. *HortScience* 22, 833–835.

Brouwer, R. (1962a) Distribution of dry matter in the plant. *Netherlands Journal of Agricultural Science* 10, 361–376.

Brouwer, R. (1962b) Nutritive influences on the distribution of dry matter in the plant. *Netherlands Journal of Agricultural Science* 10, 399–408.

Cakmak, I., Hengeler, C. and Marschner, H. (1994a) Partitioning of shoot and root dry matter and carbohydrates in bean plants suffering from phosphorus, potassium and magnesium deficiency. *Journal of Experimental Botany* 45, 1245–1250.

Cakmak, I., Hengeler, C. and Marschner, H. (1994b) Changes in phloem export of sucrose in leaves in response to phosphorus, potassium and magnesium deficiency in bean plants. *Journal of Experimental Botany* 45, 1251–1257.

Chapin III, F.S., Clarkson, D.T., Lenton, J.R. and Walter, C.H.S. (1988a) Effect of nitrogen stress and abscisic acid on nitrate absorption and transport in barley and tomato. *Planta* 173, 340–351.

Chapin III, F.S., Walter, C.H.S. and Clarkson, D.T. (1988b) Growth response of barley and tomato to nitrogen stress and its control by abscisic acid, water relations and photosynthesis. *Planta* 173, 352–366.

Claussen, W. (1976) Einfluss der Frucht auf die Trockensubstanzverteilung in der Aubergine (*Solanum melongena* L.). *Gartenbauwissenschaft* 41, 236–239.

Cooper, A.J. (1955) Further observations on the growth of the root and the shoot of the tomato plant. *Proceedings of the International Horticultural Congress* 14, 589–595.

Creelman, R.A., Mason, H.S., Benson, R.J., Boyer, J.S. and Mullet, J.E. (1990) Water deficit and abscisic acid cause differential inhibition of shoot versus root growth in soybean seedlings: analysis of growth, sugar accumulation and gene expression. *Plant Physiology* 92, 205–214.

Daie, J., Seeley, S.D. and Campbell, W.F. (1979) Nitrogen deficiency influence on abscisic acid in tomato. *HortScience* 14, 261–262.

Daunicht, H.J. and Lenz, F. (1978) Das Verhalten von Paprikapflanzen mit unterschiedlichem Fruchtbehang bei Behandlung mit 3 CO_2 Konzentrationen. *Gartenbauwissenschaft* 38, 533–546.

de Lint, P.J.A.L. and Heij, G. (1980) Glasshouse cucumber, effects of planting date and night temperature on flowering and fruit development. *Acta Horticulturae* 118, 123–134.

de Moura, R.L. and Foster, K.W. (1986a) Effect of cultivar and flower removal treatments on the temporal distribution of reproductive structures in bean. *Crop Science* 26, 362–367.

de Moura, R.L. and Foster, K.W. (1986b) Spatial distribution of seed yield within plants of bean. *Crop Science* 26, 337–341.

De Stigter, H.C.M. (1969) Growth relations between individual fruits, and between fruits and roots in cucumber. *Netherlands Journal of Agricultural Science* 17, 209–214.

den Nijs, A.P.M. (1980) The effect of grafting on growth and early production of cucumber at low temperature. *Acta Horticulturae* 118, 57–63.

Drews, M.A., Heissner, A. and Augustin, P. (1980) Die Ertragsbildung der Gewachshausgurke beim Fruhanbau in Abhangigkeit von der Temperatur und Bestrahlungsstarke. *Archiv fur Gartenbau* 28, 17–30.

Farrar, J.F. (1988) Temperature and the partitioning and translocation of carbon. In: Long, S.P. and Woodward, F.I. (eds) *Plants and Temperature*. Company of Biologists, Cambridge, UK, pp. 203–235.

Favaro, J.C. and Pilatti, R.A. (1987) Influence of the number of flowers upon the growth of the first-cluster fruits in tomato. *Scientia Horticulturae* 33, 49–55.

Fredeen, A.L., Rao, I.M. and Terry, N. (1989) Influence of phosphorus nutrition on growth and carbon partitioning in *Glycine max*. *Plant Physiology* 89, 225–230.

Galthier, N., Foyer, C.H., Huber, J., Voelker, T.A. and Huber, S.C. (1993) Effects of elevated sucrose-phosphate synthase activity on photosynthesis, assimilate partitioning and growth in tomato (*Lycopersicon esculentum* var. UC28B). *Plant Physiology* 101, 535–543.

Gocal, G.F.W., Pharis, R.P., Yeung, E.C. and Pearce, D. (1991) Changes after decapitation in concentrations of indole-3-acetic acid and abscisic acid in the larger axillary bud of *Phaseolus vulgaris* L. cv. Tender Green. *Plant Physiology* 95, 344–350.

Goldschmidt, E.E. and Huber, S.C. (1992) Regulation of photosynthesis by end-product accumulation in leaves of plants storing starch, sucrose and hexose sugars. *Plant Physiology* 99, 1443–1448.

Gosselin, A. and Trudel, M.J. (1983) Interactions between air and root temperatures on greenhouse tomato: I. Growth, development and yield. *Journal of the American Society of Horticultural Science* 108, 901–905.

Graham, P.H. (1979) Influence of temperature on growth and nitrogen fixation in cultivars of *Phaseolus vulgaris* L., inoculated with *Rhizobium*. *Journal of Agricultural Science* 93, 365–370.

Gruber, J. and Bangerth, F. (1990) Diffusible IAA and dominance phenomena in fruits of apple and tomato. *Physiologia Plantarum* 79, 354–358.

Heuvelink, E. and Buiskool, R.P.M. (1995) Influence of sink–source interaction on dry matter production in tomato. *Annals of Botany* 75, 381–389.

Hillman, J.R. (1984) Apical dominance. In: Wilkins, M.B. (ed.) *Advanced Plant Physiology*. Pitman, London, pp. 127–148.

Hillman, J.R., Math, V.B. and Medlow, G.C. (1977) Apical dominance and the levels of indole acetic acid in *Phaseolus* lateral buds. *Planta* 134, 191–193.

Ho, L.C. (1992) Fruit growth and sink strength. In: Marshall,C. and Grace, J. (eds) *Fruit and Seed Production. Aspects of Development, Environmental Physiology and Ecology*. Cambridge University Press, Cambridge, UK, pp. 101–124.

Ho, L.C., Sjut, V. and Hoad, G.V. (1982) The effect of assimilate supply on fruit growth and hormone levels in tomato plants. *Plant Growth Regulation* 1, 155–171.

Hole, C.C. and Dearman, J. (1993) The effect of photon flux density on distribution of assimilate between shoot and storage root of carrot, red beet and radish. *Scientia Horticulturae* 55, 213–225.

Hori, Y. and Shishido, Y. (1977) Studies on translocation and distribution of photosynthetic assimilates in tomato plants. *Tohoku Journal of Agricultural Research* 28, 26–40.

Hurd, R.G., Gay, A.P. and Mountifield, A.C. (1979) The effect of partial flower removal on the relation between root, shoot and fruit growth in the indeterminate tomato. *Annals of Applied Biology* 93, 77–89.

Itai, C. and Birnbaum, H. (1991) Synthesis of plant growth regulators by roots. In: Waisel,Y., Eshel, A. and Kafkafi, U. (eds) *Plant Roots the Hidden Half*. Marcel Dekker, New York, pp. 163–178.

Kanameha, K. and Hori, Y. (1980) Time course of export of 14-C assimilates and their distribution patterns as affected by feeding time and night temperature in cucumber plants. *Tohoku Journal of Agricultural Research* 30, 142–152.

Kasperbauer, M.J. (1987) Far-red light reflection from green leaves and effects on phytochrome-mediated assimilate partitioning under field conditions. *Plant Physiology* 85, 350–354.

Kelly, M.O. and Davies, P.J. (1986) Genetic and photoperiodic control of the relative rates of reproductive and vegetative development in peas. *Annals of Botany* 58, 13–21.

Kelly, M.O. and Davies, P.J. (1988) Photoperiodic and genetic control of carbon partitioning in peas and its relationship to apical senescence. *Plant Physiology* 86, 978–982.

Kinet, J.M. (1977a) Effect of light condition on the development of the inflorescence in the tomato. *Scientia Horticulturae* 6, 15–26.

Kinet, J.M. (1977b) Effect of defoliation and growth substances on the development of the inflorescence of the tomato. *Scientia Horticulturae* 6, 27–35.

Kinet, J.M., Hurdebise, D., Parmentier, A. and Stainier, R. (1978) Promotion of inflorescence development by growth substance treatment to tomato plants grown in insufficient light conditions. *Journal of the American Society of Horticultural Science* 103, 724–729.

Kleinendorst, A. and Veen, B.W. (1983) Responses of young cucumber plants to root and shoot temperatures. *Netherlands Journal of Agricultural Science* 31, 47–61.

Knox, J.P. and Wareing, P.F. (1984) Apical dominance in *Phaseolus vulgaris* L.: The possible role of abscisic acid and indole-3-acetic acid. *Journal of Experimental Botany* 35, 239–244.

Kriedemann, P.E., Loveys, B.R., Possingham, J.V. and Satoh, M. (1976) Sink effects on stomatal physiology and photosynthesis. In: Wardlaw, I.F. and Passioura, J.B. (eds) *Transport and Transfer Processes in Plants*. Academic Press, New York, pp. 401–414.

Lee, J.M. (1994) Cultivation of grafted vegetables. I. Current status, grafting methods, and benefits. *HortScience* 29, 235–239.

Li, C.J., Guevara, E., Herrera, J. and Bangerth, F. (1995) Effect of apex excision and replacement by 1-naphthylacetic acid on cytokinin concentration and apical dominance in pea plants. *Physiologia Plantarum* 94, 465–469.

Marcelis, L.F.M. (1991) Effect of sink demand on photosynthesis in cucumber. *Journal of Experimental Botany* 42, 1387–1392.

Marcelis, L.F.M. (1993a) Leaf formation in cucumber (*Cucumis sativus* L.) as influenced by fruit load, light and temperature. *Gartenbauwissenschaft* 58, 124–129.

Marcelis, L.F.M. (1993b) Fruit growth and biomass allocation to the fruits in cucumber. I. Effect of fruit load and temperature. *Scientia Horticulturae* 54, 107–121.

Marcelis, L.F.M. (1994) Effect of fruit growth, temperature and irradiance on biomass allocation to the vegetative parts of cucumber. *Netherlands Journal of Agricultural Science* 42, 115–123.

Minchin, P.E.H., Thorpe, M.R. and Farrar, J.F. (1994) Short-term control of root-shoot partitioning. *Journal of Experimental Botany* 45, 615–622.

Morris, D.A. (1977) Transport of exogenous auxin in two-branched pea seedlings (*Pisum sativum* L.). *Planta* 36, 91–96.

Morris, D.A. and Arthur, E.D. (1985a) Invertase activity, carbohydrate metabolism and cell expansion in the stem of *Phaseolus vulgaris* L. *Journal of Experimental Botany* 36, 623–633.

Morris, D.A. and Arthur, E.D. (1985b) Effects of gibberellic acid on patterns of carbohydrate distribution and acid invertase activity in *Phaseolus vulgaris. Physiologia Plantarum* 65, 257–262.

Morris, D.A. and Johnson, C.F. (1990) The role of auxin efflux carriers in the reversible loss of polar auxin transport in the pea (*Pisum sativum* L.) stem. *Planta* 181, 117–124.

Neales, T.F. and Incoll, L.D. (1968) The control of leaf photosynthesis rate by the level of assimilate in the leaf: a review of the hypothesis. *Botanical Review* 34, 107–125.

Nielsen, T.H. and Veierskov, B. (1988) Distribution of dry matter in sweet pepper plants (*Capsicum annuum* L.) during the juvenile and generative growth phases. *Scientia Horticulturae* 35, 179–187.

Ojehomon, O.O. (1968) Flowering, fruit production and abscission in cowpea, *Vigna unguiculata* L. Walp. *West African Science Association Journal* 13, 227–234.

Ojehomon, O.O. (1970) Effect of continuous removal of open flowers on the seed yield of two varieties of cowpea, *Vigna unguiculata* L. Walp. *Journal of Agricultural Science* 74, 375–381.

Peet, M.M., Huber, S.C. and Patterson, D.T. (1986) Acclimation to high CO_2 in monoecious cucumbers. II. Carbon exchange rates, enzyme activities and starch and nutrient concentrations. *Plant Physiology* 80, 63–67.

Plaut, Z., Mayoral, M.L. and Reinhold, L. (1987) Effect of altered sink:source ratio on photosynthetic metabolism of source leaves. *Plant Physiology* 85, 786–791.

Richards, D. (1981) Root-shoot interactions in fruiting tomato plants. In: Brouwer, R., Gasparikova, O., Kolek, J. and Loughman, B.C. (eds) *Structure and Function of Plant Roots*. Nijhoff-Junk, The Hague, Netherlands, pp. 373–380.

Rufty, T.W. Jr, Huber, S.C. and Volk, R.J. (1988) Alterations in leaf carbohydrate metabolism in response to nitrogen stress. *Plant Physiology* 88, 725–730.

Saab, I.N., Sharp, R.E., Pritchard, J. and Voetberg, G.S. (1990) Increased endogenous abscisic acid maintains primary root growth and inhibits shoot growth of maize seedlings at low water potentials. *Plant Physiology* 93, 1329–1336.

Schapendonk, A.H.C.M. and Brouwer, P. (1984) Fruit growth of cucumber in relation to assimilate supply and sink activity. *Scientia Horticulturae* 23, 21–33.

Setter, T.L., Brun, W.A. and Brenner, M.L. (1980a) Effect of obstructed translocation on leaf ABA and associated stomatal closure and photosynthesis decline. *Plant Physiology* 65, 1111–1115.

Setter, T.L., Brun, W.A. and Brenner, M.L. (1980b) Stomatal closure and photosynthetic inhibition in soybean leaves induced by petiole girdling and pod removal. *Plant Physiology* 65, 884–887.

Sharp, R.E. (1990) Comparative sensitivity of root and shoot growth and physiology to low water potentials. In: Davies, W.J. and Jeffcoat, B. (eds) *Importance of Root to Shoot Communicating in the Responses to Environmental Stress.* British Society of Plant Growth Regulation, Bristol, pp. 29–44.

Shishido, Y., Seyama, N., Imada, S. and Hori, Y. (1989) Carbon budget in tomato plants as affected by night temperature evaluated by steady state feeding with 14-CO_2. *Annals of Botany* 63, 357–367.

Shishido, Y., Kumakura, H. and Hori, Y. (1993) Changes in source–sink relations by defoliation and darkening of source and sink leaf in tomato (*Lycopersicon esculentum* Mill.). *Journal of the Japanese Society of Horticultural Science* 62, 95–102.

Sklensky, D.E. and Davies, P.J. (1993) Whole plant senescence: reproduction and nutrient partitioning. *Horticultural Reviews* 15, 335–366.

Stafstrom, J.P. (1993) Apical bud development in pea: apical dominance, growth cycles, hormonal regulation and plant architecture. In: Amasino, R.M. (ed.) *Cellular Communication in Plants.* Plenum Press, New York, pp. 75–86.

Stephenson, A.G., Devlin, B. and Horton, J.B. (1988) The effects of seed number and prior fruit dominance on the pattern of fruit production in *Cucurbita pepo* (zucchini squash). *Annals of Botany* 62, 653–661.

Stitt, M., Krapp, A., Klein, D., Roper-Schwarz, U. and Paul, M. (1995) Do carbohydrates regulate photosynthesis and allocation by altering gene expression? In: Madore, M.A. and Lucas, W.J. (eds) *Carbon Partitioning and Source–Sink Interactions in Plants.* American Society of Plant Physiology, Rockville, MD, pp. 68–77.

Stitt, M. and Sonnewald, U. (1995) Regulation of metabolism in transgenic plants. *Annual Review of Plant Physiology and Plant Molecular Biology* 46, 341–368.

Tachibana, S. (1982) Comparison of effects of root temperature on the growth and mineral nutrition of cucumber cultivars and figleaf gourd. *Journal of the Japanese Society of Horticultural Science* 51, 299–308.

Tachibana, S. (1987) Effect of root temperature on the rate of water and nutrient absorption in cucumber cultivars and fig leaf gourd. *Journal of the Japanese Society of Horticultural Science* 55, 461–467.

Tachibana, S. (1988) Cytokinin concentrations in roots and root xylem exudate of cucumber and figleaf gourd as affected by root temperature. *Journal of the Japanese Society of Horticultural Science* 56, 417–425.

Tachibana, S. (1989) Respiratory response of detached roots to lower temperatures in cucumber and figleaf gourd grown at 20°C root temperature. *Journal of the Japanese Society of Horticultural Science* 58, 333–337.

Tamas, I.A., Ozbun, J.L., Wallace, D.H., Powell, L.E. and Engels, C.J. (1979a) Effect of fruits on dormancy and abscisic acid concentration in the axillary buds of *Phaseolus vulgaris* L. *Plant Physiology* 64, 615–619.

Tamas, I.A., Wallace, D.H., Ludford, P.M. and Ozbun, J.L. (1979b) Effects of older fruits on abortion and abscisic acid concentration of younger fruits in *Phaseolus vulgaris* L. *Plant Physiology* 64, 620–622.

Tamas, I.A., Engels, C.J., Kaplan, S.L., Ozbun, J.L. and Wallace, D.H. (1981) Role of indoleacetic acid and abscisic acid in the correlative control by fruits of axillary bud development and leaf senescence. *Plant Physiology* 68, 476–481.

Tamas, I.A., Koch, J.L., Mazur, B.K. and Davies, P.J. (1986) Auxin effects on the correlative interaction among fruits in *Phaseolus vulgaris* L. In: Cooke, A.R. (ed.) *Plant Growth Regulator Society of America Proceedings*. Plant Growth Regulator Society of America, Lake Alfred, pp. 208–215.

Tamas, I.A. (1995) Hormonal regulation of apical dominance. In: Davies, P.J. (ed.) *Plant Hormones*, 2nd edn. Kluwer, Dordrecht, Netherlands, pp. 572–597.

Tardieu, F. and Davies, W.J. (1993) Root–shoot communication and whole-plant regulation of water flux. In: Smith, J.A.C. and Griffiths, H. (eds) *Water Deficits: Plant Responses from Cell to Community*. Bios Scientific, Oxford, pp. 147–162.

Tucker, D.J. (1976) Effects of far-red light on the hormonal control of side shoot growth in the tomato. *Annals of Botany* 40, 1033–1042.

Tucker, D.J. (1981) Axillary bud formation in two isogenic lines of tomato showing different degrees of apical dominance. *Annals of Botany* 48, 837–843.

Turner, A.D. and Wien, H.C. (1994) Dry matter assimilation and partitioning in pepper cultivars differing in susceptibility to stress-induced bud and flower abscission. *Annals of Botany* 73, 617–622.

Van der Post, C.J. (1968) Simultaneous observations on root and top growth. *Acta Horticulturae* 7, 138–143.

Van Nordwijk, M. and De Willigen, P. (1987) Agricultural concepts of roots: from morphogenetic to functional equilibrium between root and shoot growth. *Netherlands Journal of Agricultural Science* 35, 487–496.

Wardlaw, I.F. (1990) The control of carbon partitioning in plants. *New Phytologist* 116, 341–381.

Wareing, P.F., Khalifa, M.M. and Treharne, K.J. (1968) Rate-limiting processes in photosynthesis at saturating light intensities. *Nature* 220, 453–457.

Waters Jr, L., Graham, P.H., Breen, P.J., Mack, H.J. and Rosas, J.C. (1983) The effect of plant population density on carbohydrate partitioning and nitrogen fixation of two bean (*Phaseolus vulgaris* L.) cultivars in two tropical locations. *Journal of Agricultural Science* 100, 153–158.

White, J.C. and Mansfield, T.A. (1978) Correlative inhibition of lateral bud growth in *Phaseolus vulgaris* L. Influence of the environment. *Annals of Botany* 42, 191–196.

Wien, H.C., Aloni, B., Riov, J., Goren, R., Huberman, M. and Ho, J.C. (1993a) Physiology of heat stress-induced abscission in pepper. In: Kuo, G.C. (ed.) *Adaptation of Food Crops to Temperature and Water Stress*. Asian Vegetable Research and Development Center, Shanhua, Taiwan, pp. 188–198.

Wien, H.C., Minotti, P.L. and Grubinger, V.P. (1993b) Polyethylene mulch stimulates early root growth and nutrient uptake of transplanted tomatoes. *Journal of the American Society of Horticultural Science* 118, 207–211.

Wien, H.C. and Zhang, Y. (1991) Prevention of flower abscission in bell pepper. *Journal of the American Society of Horticultural Science* 116, 516–519.

Yelle, S., Beeson, R.C., Trudel, M.J. and Gosselin, A. (1989) Acclimation of two tomato species to high atmospheric CO_2. II. Ribulose-1,5-bisphosphate carboxylase/oxygenase and phosphoenolpyruvate carboxylase. *Plant Physiology* 90, 1473–1477.

Zijlstra, S., Groot, S.P.C. and Jansen, J. (1994) Genotypic variation of rootstocks for growth and production of cucumber, possibilities for improving the root system by plant breeding. *Scientia Horticulturae* 56, 185–186.

Zijlstra, S. and den Nijs, A.P.M. (1987) Effects of root systems of tomato genotypes on growth and earliness, studied in grafting experiments at low temperature. *Euphytica* 36, 693–700.

Tomato 6

J.M. Kinet[1] and M.M. Peet[2]

[1]*Laboratoire de Cytogénétique, Département de Biologie, Université Catholique de Louvain, Place Croix du Sud, 5 bte 13, B-1348 Louvain-la-Neuve, Belgium;* [2]*Department of Horticultural Science, Box 7609, North Carolina State University, Raleigh, North Carolina 27695–7609, USA*

… a fruit that is almost universally treated as a vegetable and a perennial plant that is almost universally cultivated as an annual

(Rick, 1978)

No horticultural crop has received more attention and detailed study than that most popular and pampered vegetable – the tomato. It is a model crop for many experimental studies and the knowledge and information gathered from these studies have contributed to the dramatic improvements in production that have occurred during this century

(Tigchelaar and Foley, 1991)

The tomato (*Lycopersicon esculentum* Mill.) is a member of the *Solanaceae*, originating in the western coastal plain of South America, extending from Ecuador to Chile. In pre-Columbian times, the tomato was apparently not known to South American Indians since there is no name for it in their languages, no tradition and no archaeological remains in the Andean region (Rick, 1978; Harlan, 1992). Domestication took place in Mexico where truly wild tomatoes are unknown but weed tomatoes are common in the south of the country (Harlan, 1992). Tomato was first introduced in Europe in the middle of the sixteenth century (Rick, 1978; Kalloo, 1991). An early introduction was probably yellow, since it was named 'pomodoro' (golden apple) in Italy. Because it belongs to the nightshade family, tomato was sometimes considered poisonous, slowing down acceptance and it is only recently that it became a major food crop. Tomato is now one of the most popular vegetables and one of the most important fruit crops. In 1993, world tomato production was 70 million metric tons, overtaking bananas, pome fruits, oranges and grapes (FAO, 1994).

Two main tomato types are currently grown. (i) The determinate or 'bushy' tomato is mainly used for processed food and is the most important outdoor commercial type in USA. It has one time-limited flowering period followed by a period of fruit development. (ii) The indeterminate or 'vine' tomato, largely used for production of fresh fruits in greenhouses and home gardens, produces inflorescences and flowers continuously throughout the plant's life. As a result, total yield of indeterminate cultivars is usually not affected by flower initiation.

In addition to its economic importance, the tomato is an ideal research material for physiological, cellular, biochemical and molecular genetic investigations. It is easy to cultivate, has a short life cycle and is amenable to varied horticultural manipulations, including grafting or cutting. Various types of explants can be cultured *in vitro* and plant regeneration is feasible, allowing the development of transformation procedures (Hille *et al.*, 1989).

A large number of genes have been described and assigned to specific locations among the 12 chromosomes. Numerous unigenic mutants are available. In the last decades, these mutants have begun to play an important role in physiological studies (Sawhney, 1992; Koornneef *et al.*, 1993; Gray *et al.*, 1994) that will further increase with a better knowledge of their biochemical defect(s).

SEED GERMINATION

Tomato germination is strongly influenced by temperature. Optimal temperatures range between 18 and 24°C (Wittwer and Aung, 1969) and the minimum is about 8–11°C (Picken *et al.*, 1986). There is, however, a marked variation in cultivar response to low temperature (Kemp, 1968), as well as to temperatures above the optimum, particularly at and above 35°C (Berry, 1969). Delayed germination and reduced uniformity of emergence in suboptimal temperature conditions may affect the success of processing tomato crops because most are direct seeded.

Although tomatoes will germinate in the dark, phytochrome has been implicated in the germination process (Mancinelli *et al.*, 1966). Far red (FR) light inhibits germination and the effects of FR can be reversed by red (R). The presence of P_{fr} in the seed is thus a prerequisite for germination; this P_{fr} could arise from intermediates accumulated during dehydration of the seed or as a result of hydration. Seeds of the *aurea* (*au*) mutant, a specific phytochrome I mutant (Koornneef *et al.*, 1992), have reduced germination capacity that is enhanced by R (Georghiou and Kendrick, 1991) also indicating that phytochrome is implicated in the control of the germination process in tomato. The inhibitory effect of light is dependent on other environmental parameters, especially temperature, and on the cultivar (Picken *et al.*, 1986).

Exogenous applications of growth regulators, namely gibberellins (GAs) and auxin, stimulate germination (Picken *et al.*, 1986). A role for GAs is further supported by the observation that the dwarf mutant *gib-1* (denoted as *ga-1* in early papers), in which GA biosynthesis is inhibited (Bensen and Zeevaart, 1990), is unable to germinate without application of exogenous GA (Koornneef *et al.*, 1990). GA treatment results in endosperm weakening due to the degradation of the mannan-rich cell walls by GA-induced endo-ß-mannanase, prior to radicle protrusion (Groot and Karssen, 1987; Groot *et al.*, 1988). GA is also required to reinitiate cell-cycle activity during imbibition, as was demonstrated in experiments by Liu *et al.* (1994) who suggested that nuclear replication and endosperm weakening could be parallel events without any causal relationship. In contrast, exogenous applications of abscissic acid (ABA), at low concentrations, inhibit germination without affecting cell-cycle activation (Liu *et al.*, 1994). The primary effect of ABA could be the inhibition of the expression of enzymes responsible for cell wall weakening (Liptay and Schopfer, 1983).

In the cultivated tomato, the seeds are considered non-dormant. However, studies on the rate of germination, the influence of the time interval between harvest and sowing, vivipary in overripe fruits, etc., using the ABA-deficient mutants *sitiens* (*sit*) and *flacca* (*flc*), suggest that the wild type is slightly dormant in comparison to the mutants (Weyers, 1985; Groot and Karssen, 1992). These experiments also indicate that ABA plays a role in tomato seed dormancy.

VEGETATIVE GROWTH

In a strict sense, the vegetative phase is usually short since the floral transition occurs, for most cultivars, when the third leaf is expanding. This is within 3 weeks of cotyledon expansion (Hurd and Cooper, 1970). Usually only six to 11 leaves are produced below the first inflorescence (Picken *et al.*, 1985); they are alternate with a 2/5 phyllotaxy. If too few leaves are produced before floral initiation, assimilate supply may be insufficient to support flower and fruit development. In determinate tomatoes, only two or three inflorescences, separated by one or more leaves, are usually produced on the main stem while, in indeterminate types, inflorescences are continuously initiated at leaf intervals varying with the cultivar and the environmental conditions, three being the most frequent. The indeterminate growth habit is dependent on the activity of the self-pruning (*Sp*) gene; the recessive *sp* allele conferring the determinate phenotype (Rick and Butler, 1956). The expression of *sp* is modified by other genes and even suppressed by the recessive gene jointless (*j*) (Emery and Munger, 1970), which causes the development of leafy inflorescences and prevents the formation of the flower pedicel abscission layer (Rick and Butler, 1956).

In indeterminate tomatoes, vegetative growth and reproductive development are thus proceeding concomitantly during the greatest part of the plant's life. A strong competition between developing leaves and apical meristems influences and conditions both the earliness of harvest and the total yield. High assimilate availability under high light conditions stimulates both meristem activity and leaf growth (Hussey, 1963a), but when plants are source-limited (under high temperatures or low light) young leaf growth is favoured at the expense of apical development. This effect is counteracted by the continuous removal of young leaves (Hussey, 1963b; Kinet 1977b).

Vegetative growth in tomato requires fluctuating environments. Continuous light has a number of injurious effects, inducing leaf chlorosis, hypertrophy of palisade cells, alteration of plastid ultrastructure, starch grains disappearance, etc. (Descomps and Deroche, 1973). If temperature is fluctuating with a sufficiently wide daily change, the detrimental influence of continuous light is suppressed (Hillman, 1956).

When identical culture conditions are maintained throughout, the plastochronic rhythm, i.e. the time interval at which leaves are initiated, is constant (Calvert, 1959; Kinet, 1977a; Picken et al., 1986) despite the fact that the plant is progressing through varied developmental stages. However, the rate of leaf initiation increases with daily irradiance (Fig. 6.1) and temperature.

The effect of daily irradiance upon stem elongation is complex (Picken et al. 1986). In suboptimal light conditions, reducing the daily light integral increases plant height (Hurd and Thornley, 1974), but when daily irradiances are very low, stem elongation is slowed down by a further decrease of light intensity (Kinet, 1977a). The inconsistent effects of daylength upon stem elongation (Picken et al., 1986) probably result from interactions with irradiance, affecting the daily light integral. The growth of the root system is stimulated relative to the growth of the shoot in higher irradiance; as a result, the shoot : root dry weight ratio is decreased (Verkerk, 1955; Kristoffersen, 1963). As the temperature increases, the rate of stem elongation (Calvert, 1964) and the shoot : root dry weight ratio (Verkerk, 1955; Kristoffersen, 1963) are also increased. It appears that conditions that induce higher shoot : root ratios are more detrimental to reproductive development.

The effect of plant growth regulator applications upon the duration of the vegetative phase is unclear (Atherton and Harris, 1986). The various methodologies used to treat the plants along with the different application timings may account for the inconsistencies.

FLOWERING

THE REPRODUCTIVE STRUCTURE

The tomato inflorescence is a cyme initiated by the apical meristem and consisting of a main axis bearing lateral flowers without bracts. Active growth of the adjacent bud in the axil of the last-formed leaf displaces the inflorescence from its terminal position and carries up the subtending leaf until it occupies a position above the inflorescence which is then forced to develop

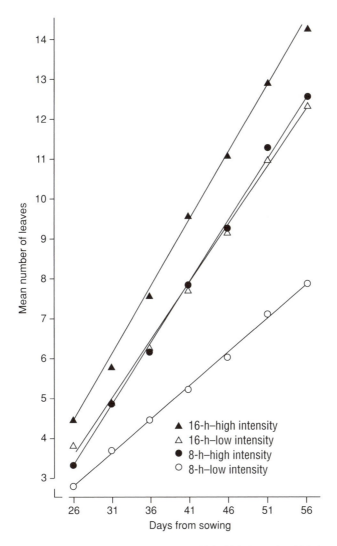

Fig. 6.1. Effect of light integral upon leaf production. 16-h–high intensity: 16-h long days and a daily light integral of 1.04 MJ m^{-2} day^{-1}; 16-h–low intensity: 16-h long days and a daily light integral of 0.52 MJ m^{-2} day^{-1}; 8-h–high intensity: 8 h short days and a daily light integral of 0.52 MJ m^{-2} day^{-1}; 8-h–low intensity: 8-h short days and a daily light integral of 0.26 MJ m^{-2} day^{-1} (Kinet, 1977a).

laterally (Fig. 6.2). This process is repeated for each inflorescence (Calvert, 1965; Sawhney and Greyson, 1972). Sometimes, depending on cultivars, environmental conditions – such as low temperatures – or position on the plant, branched inflorescences having two or more main axes are produced. The number of flowers of these compound inflorescences is increased relative to the single truss (Lewis, 1953). Several mutations affect inflorescence

Fig. 6.2. Diagrammatic representation of leaf and inflorescence position on the tomato stem illustrating the sympodial nature of the growth process. A, B and C are leaves initiated before flower initiation. D is the first initiated leaf on the axillary growth (Calvert, 1965).

structure and flower number in the truss. The compound inflorescence (*s*) mutant produces highly branched inflorescences (Rick and Butler, 1956) that may bear up to several hundred flowers. The anantha (*an*) mutation, which prevents the transition between inflorescence and floral meristems, results in the production of proliferating inflorescential structures without flowers (Rick and Butler, 1956). Single flowers are initiated by uniflora (*uf*) (Fehleisen, 1967) while in terminal flower (*tmf*) the main shoot produces a single flower and axillary branches bear inflorescences (Lukyanenko *et al.*, 1973).

The pendent flowers are hermaphroditic; the anthers form a hollow cone which encloses the pistil. Pollen grains are released introrsely along longitudinal anther slits and fall onto the stigmatic surface (Picken *et al.*, 1985; Kaul, 1991). Tomato flowers are thus essentially self-pollinated. Flower morphogenesis and development are dependent on numerous genes which either assign an identity to organ primordia in the flower whorls such as green pistillate (*Gpi*) and stamenless (*Sl*) that govern determination of petals and stamens (Rasmussen and Green, 1993; Marc *et al.*, 1994) or control further development as revealed by the 35 *ms* (male sterile) mutations that affect microsporogenesis and cause male sterility (Kaul, 1991).

Numerous fundamental studies concern the first inflorescence only, limiting the duration of the investigations. Flowering at the second, and presumably later trusses, appears to be controlled in the same way (Kinet, 1977a) although the influences of the environmental factors are less, probably as a result of the increased photosynthetic leaf area formed before initiation of the upper inflorescences.

TIME TO FLOWERING

Tomato is an autonomous plant, i.e. a plant that does not require particular environmental conditions to initiate reproductive structures. It flowers with time, provided the environmental conditions allow growth. This does not mean that flowering is unaffected by the environment, however. Studies in growth rooms established that when the total daily accumulation of photosynthetic active radiations (PAR) is kept constant, flowering is advanced in short days (Binchy and Morgan, 1970; Kinet, 1977a). Nevertheless, tomato is usually considered as the model day-neutral plant.

The number of days to first anthesis is determined by the number of leaves preceding the first inflorescence and the rate of leaf initiation. Thus, when the rhythm of leaf production is the same, earlier flowering and earlier yield are associated with a lower leaf number below the first inflorescence (Dieleman and Heuvelink, 1992).

The number of leaves produced before floral transition is under genetic control. Comparing two cultivars, Honma *et al.* (1963) found that it is determined by a single gene. Philouze (1978) reported that *j* mutant plants consistently produce one or two more leaves before flowering than the wild type while in combination with *f* (fasciated) and *lf* (leafy), the bifurcate (*bi*) mutation delays first inflorescence until after 28–75 leaves (Mertens and Burdick, 1954). Leaf number below the first inflorescence is also strongly affected by environmental conditions, such as light intensity, temperature, and their interaction, that influence the plastochron rhythm (Calvert, 1959; Kinet, 1977a) (see Fig. 6.1). Increasing light intensity reduces the number of leaves below the inflorescence and stimulates the rate of leaf initiation, resulting in earlier flowering (Calvert, 1959, Kinet, 1977a). By growing single-cluster plants throughout the season, with or without additional lighting, Janes and McAvoy (1991) clearly established that the onset of red fruit production is correlated closely, but negatively with the amount of light the crop received from germination to 5 days before predicted anthesis (Fig. 6.3), i.e. during the vegetative growth phase and the early reproductive development. Assimilate availability is implicated in this accelerated reproductive development since altering

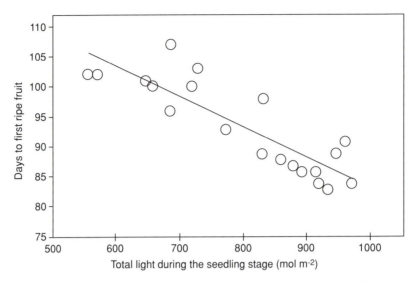

Fig. 6.3. Relationship between the total light received during the seedling stage (emergence to 5 days before anthesis) and the onset of red fruit production in a single-cluster greenhouse tomato crop (Janes and McAvoy, 1991).

starch/sucrose partitioning by increasing the capacity for sucrose synthesis in transgenic tomatoes resulted in a reduced time to 50% blossoming and in an earlier fruit maturation (Micallef *et al.*, 1995). Decreasing temperature also reduces the number of leaves preceding floral initiation but slows down the leaf initiation rate (Calvert, 1959). The effect of other environmental factors, such as CO_2 concentration of the air, nutrition, water availability, etc., appears to be less important (Atherton and Harris, 1986; Dieleman and Heuvelink, 1992).

The inconsistent influence of plant growth regulators is probably due to the various experimental conditions used, especially the site and time of applications. Apparently, the most usual effect of GAs is to increase the number of leaves initiated before floral transition. Since the rate of leaf initiation is concomitantly enhanced, the effect upon flowering time varies (J.M. Kinet, University of Liège, Belgium, 1978, unpublished results).

DEVELOPMENT OF THE REPRODUCTIVE STRUCTURE TO ANTHESIS

Control by environmental factors

The failure of flowers to produce fruits may occur in both determinate and indeterminate cultivars, limiting yield. The available evidence clearly indicates that there is no unique factor responsible for this failure but that they all interact in a complex manner. However, the daily light integral appears to have a central role in the control of reproductive development in tomato

(Fig. 6.4). When plants are grown from sowing under so-called favourable light conditions (16–20-h long days with a photon flux density of 180 µmol m^{-2} s^{-1} at the top of the canopy), floral transition occurs during the fourth week, and the first inflorescence appears macroscopically around the 45th day. Starting about 2 weeks later, i.e. 2 months after sowing, anthesis of most flowers of this inflorescence occurs in sequence; only the one or two last-initiated floral buds invariably abort (Kinet, 1989). When insufficient light conditions, consisting of 8-h short days with a photon flux density of 90 µmol m^{-2} s^{-1}, are given constantly from sowing, floral initiation as well as macroscopic appearance of the inflorescence are later than under high light, indicating that the rate of growth of the truss is slowed down. Furthermore, development never proceeds to anthesis and failure of the whole truss invariably occurs in all individuals. In the first flower of the aborting inflorescence, all floral organs are present, but development of the sporogenous tissue is halted at the pollen mother-cell stage in the anthers while in the ovules, development never goes beyond the early differentiation of the archesporial cell.

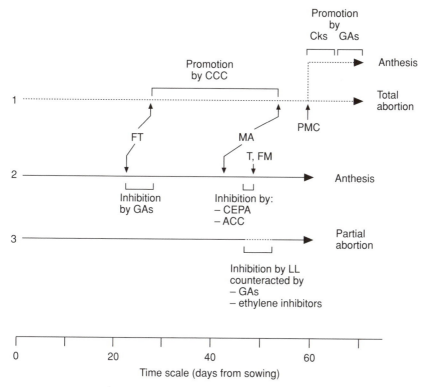

Fig. 6.4. Diagram summarizing the temporal interactions between light conditions and plant growth regulators for the control of inflorescence development. Broken lines, unfavourable light conditions (LL, 0.26 MJ m^{-2} day^{-1}); solid lines, favourable light conditions (1.04 MJ m^{-2} day^{-1}); FT, floral transition; MA, macroscopic appearance of the inflorescence; PMC, pollen mother cell stage; T,FM, tetrad-free microspore stage (Kinet, 1989).

However, sufficiently high light conditions are usually not required from sowing to prevent abortion but they do appear essential for a limited time interval, extending between 5–6 and 10–12 days after the macroscopic appearance of the inflorescence, during meiosis (Howlett, 1936; Kinet and Sachs, 1984). Unfavourable light conditions at that time only cause partial abortion, i.e. failure of the whole inflorescence in some individuals and, in plants which succeed in developing flowers to anthesis, abortion of individual flower buds in unusual positions, i.e. flower buds other than the one or two last ones. This latter situation is most usually encountered in grower conditions.

One of the most obvious effects of temperature is upon the rate of development of the tomato inflorescence: the higher the temperature, the earlier the opening of the flowers (Calvert, 1964; Hurd and Cooper, 1970), provided that premature failure does not occur since abortion is triggered by increasing temperature, particularly when irradiance is limiting (Calvert, 1969). High temperature is particularly detrimental 9–5 days before anthesis during sporogenesis. In contrast, flower buds are rather tolerant to high temperature 1–3 days before anthesis (Sugiyama et al., 1966). Low temperatures during inflorescence initiation increase inflorescence branching (Hurd and Cooper, 1970; Kinet, 1989) and number of floral organs, especially petals, stamens and ovary locules (Sawhney, 1983). Pollen production is impaired at temperatures at or below 10°C after microsporogenesis (Maisonneuve and Philouze, 1982).

Interactions between light and temperature are illustrated by the observation that the detrimental effect of a short stay under insufficient light during sporogenesis is totally counteracted by slightly decreasing the temperature, e.g. from 20 to 16°C in our experiments (Kinet, 1982). Moderate temperatures probably favour assimilate distribution to the reproductive structures at the expense of apical shoot and young leaves. Strong competition between the apical shoot tissues and the inflorescence in tomato plants grown under unfavourable light conditions has been revealed by defoliation experiments. Successive removal of young leaves which develop at the same time as the inflorescence promotes reproductive development (de Zeeuw, 1954; Kinet, 1977b).

High photosynthetic activity in the source leaves appears to be the major contributing factor to high-light induced promotion of flower development. There is no direct photomorphogenetic effect of light on the reproductive structure of tomato since darkening inflorescences from the time of their macroscopic appearance does not prevent anthesis which occurs without delay (Kinet and Sachs, 1984). Under unfavourable light conditions, the plant is strongly source-limited: CO_2 fixation is markedly reduced as compared to favourable light conditions, but the proportion of assimilates exported out of the leaf is the same, reducing the supply to the rest of the plant (Kinet and Sachs, 1984). The importance of photosynthetic assimilates in allowing inflorescence to develop to anthesis is also revealed by experiments manipulating CO_2 concentration of the air (Calvert and Slack, 1975) or plant density (Morris and Newell, 1987). Increasing the CO_2 level in the atmosphere, in low light, also decreases the abortion of the reproductive structures but is less effective in this respect than a temperature drop (Cooper and Hurd, 1968; Calvert and Slack, 1975; Kinet, 1982). High CO_2 either enhances assimilate production in the leaves and/or counteracts ethylene synthesis in reduced light.

Increases in nitrogen supply stimulate reproductive development under high irradiance conditions. In contrast, when light is limiting, excessive N fertilization inhibits flower development, fruit setting and formation. This observation constituted the basis of the C/N theory for the control of flowering (Kraus and Kraybill, 1918). Studying the interactive effects of N levels (5, 10, 15 and 20 meq l^{-1}) and photon flux density (125 and 250 μmol m^{-2} s^{-1}), Larouche et al. (1989) found that, under the lower light conditions, maximum yields were obtained at 5 meq l^{-1} while under the most favourable light regime, higher yields were obtained at 10 and 15 meq l^{-1}. Nitrogen fertilization has thus to be adjusted to the prevailing irradiance regime. Nitrogen in excess of that indicated by the light conditions results in excessive vegetative growth and probably impairs reproductive development by decreasing sink strength of flowers and inflorescences relative to vegetative tissues.

Water availability also affects flower formation and, later, fruit enlargement. The average number of flowers per truss decreases with decreasing water supply (Wudiri and Henderson, 1985). Severe stress, resulting from applying water to meet 25% of the crop evapotranspirative demand reduces fruit set by 40% to more than 90%, depending on the cultivar. Full water supply also affects the reproductive processes, delaying flower initiation and reducing the rates of flowering and fruit setting as well as the number of flowers and fruits. The modulation of the effect of water availability upon floral development by light conditions is also suggested by the finding that, under winter conditions, flower abortion is reduced by water stress (Klapwijk and de Lint, 1974).

Control by growth regulators

Abortion in constant unfavourable light conditions is prevented by applying N^6-benzylaminopurine (BA) and GAs to the inflorescence (see Fig. 6.4) (Kinet, 1977b). These compounds have a sequential effect. The action of the cytokinin (CK) is exerted first and that of the GA only subsequently to the CK action (Kinet et al., 1978; Kinet and Léonard, 1983), indicating that early inflorescence development is dependent on CKs. This conclusion is further supported by the finding that trusses which abort early have markedly reduced CK levels in comparison to levels in inflorescences that develop normally in favourable light conditions (Kinet and Léonard, 1983). Cytokinins also increase the number of flowers initiated in the inflorescence (Kinet et al. 1985a; Atherton and Harris, 1986). In response to a BA + GA treatment, several changes occur in inflorescences targeted for abortion in unfavourable light conditions. Cellular activity in the ovules of the first flower of the inflorescence resumes early (Kinet *et al.*, 1985b), preceding a stimulation of acid invertase activity (Kinet, 1989). This result suggests that the increased import of ^{14}C-assimilates recorded one day after the BA + GA treatment (Léonard *et al.*, 1983) could be indirect.

Gibberellins do not seem to be a limiting factor during the early development of the reproductive structure and the available evidence suggests, in contrast, that high GA activity is inhibitory at that time: their level is elevated in inflorescences undergoing early abortion (Kinet and Léonard, 1983). Moreover, applying the growth retardant 2-chloroethyltrimethylammonium chloride (CCC), from floral transition to macroscopic appearance of the truss, reduces production of diffusible GAs by plant shoot tips while also reducing abortion induced by high temperature and low light treatments (Abdul *et al.*, 1978).

The action of GAs is time-dependent in tomato (Kinet, 1989, 1993) since after macroscopic appearance of the inflorescence, they reduce the proportion of flowers that naturally abort in constant favourable light conditions (Kinet *et al.*, 1985a). They also counteract the detrimental effect of low photon flux densities given during meiosis (5–6 to 10–12 days after macroscopic appearance of the truss). The implication of GAs in the control of the late development of the reproductive structure in tomato is further demonstrated in studies using mutants. In the *gib-1* and *gib-2* tomato mutant, flower buds are initiated but do not develop to anthesis (Nester and Zeevaart, 1988; Jacobsen and Olszewski, 1991). Development is arrested before meiosis and resumes after application of GAs. In the *sl-2/sl-2* mutant, which is structurally and functionally male sterile, flower buds require GA_3 for growth *in vitro*, contain a lower level of GA-like substances than the normal, and young buds treated with GA_3 produce normal stamens with viable pollen (Sawhney *et al.*, 1989). A similar effect of GA_3 application to the short anther parthenocarpic *sha-pat* mutant has been reported (Mapelli *et al.*, 1979). Using the *sl-2/sl-2* mutant, Rastogi and Sawhney (1990a,b) also showed that normal stamen and pollen development is regulated by the level of polyamines.

Ethylene is also involved in the control of the late abortion of the tomato inflorescence, before anthesis. Spraying trusses with 2-chloroethylphosphonic acid (CEPA, the active ingredient of ethephon) or with 1-aminocyclopropane-1-carboxylic acid (ACC, the immediate precursor of ethylene) causes the failure of the truss in high light conditions. In contrast, the inhibition resulting from a short stay in low light is overcome by aminoethoxyvinyl glycine (AVG) or silverthiosulphate (STS) applications (Kinet and El Alaoui Hachimi, 1988), two compounds which inhibit ethylene synthesis or action in the plant.

Reproductive development is thus dependent on relative amounts of a number of endogenous regulators that may act either simultaneously or in sequence (Kinet *et al.*, 1985a; Kinet, 1993). This is also true for specific steps of flower development such as stamen and pollen differentiation (Sawhney and Shukla, 1994).

GROWTH OF THE REPRODUCTIVE STRUCTURE TO MARKETABLE SIZE

Yield in tomato is dependent on both the number and weight of individual fruits, i.e. upon fruit set and development.

FRUIT SET

Fruit set is defined as the proportion of flowers that produce a fruit of a minimal size in the population of flowers which appear to reach anthesis normally (Picken, 1984). This critical stage is affected by environmental factors and plant growth regulators. For fruit set to be high a sequence of processes including pollination, germination of pollen grains, pollen tube growth, fertilization and fruit initiation must take place successfully.

Although pollination is facilitated by flower structure, truss movement either by wind, by cultural activities or by artificial means such as the use of the electric 'bee' is usually required in glasshouses during winter (Picken, 1984). In winter, pollen tends to be sticky and to aggregate. High relative humidity of the air, low light intensity and temperature seem to be involved in the control of this phenomenon (Picken, 1984). Stigma exsertion beyond the anther cone may also account for pollination failure. The length of the style is genetically determined (Rick and Dempsey, 1969) and increased by low light, high temperature, high nitrogen availability and GA treatments (Howlett, 1939; Rudich *et al.*, 1977; see also section on flower and fruit abscission below).

At 18–25°C, pollen grains retain viability from 2–5 days after anthesis. Flowers open in the morning and stigmata are receptive from 16 to 18 h before up to 6 days after anthesis (Kaul, 1991).

Extreme temperatures, above 37.5°C or below 5°C, limit germination of pollen grains and inhibit tube growth which increases in growth rate up to 35°C (Dempsey, 1970). By pollinating low-temperature grown plants with pollen from high-temperature grown plants and vice versa, Levy *et al.* (1978) were able to show that exposing either parent to high temperatures reduces fruit set. Stigma receptivity is impaired by high temperature (Charles and Harris, 1972) and a 4-h period at 40°C between 24 and 96 h after pollination also causes the degradation of the endosperm and damages the pro-embryo (Iwahori, 1966). The influence, if any, of irradiance and relative humidity of the air upon these processes is small (Picken, 1984).

Fertilized ovaries may cease to swell rapidly because of low irradiance, high temperature or interactions between these factors. Cockshull *et al.* (1992) reported that fruit number per truss is positively correlated with solar radiation intercept (most noticeably when less than 1.5 MJ m^{-2} day^{-1} was received) around the time of first anthesis of a truss. McAvoy and Janes (1989) showed that the critical stage is from initial anthesis to early fruit set. Low light during the 2 weeks that follow the anthesis of the first flower of

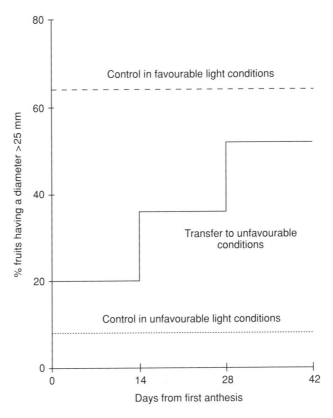

Fig. 6.5. Effect of a 14 day period under unfavourable light conditions (0.26 MJ m^{-2} day^{-1}), given at different times after first anthesis of the first inflorescence, on fruit set. Controls under favourable light conditions: 1.04 MJ m^{-2} day^{-1}.

the inflorescence prevents the growth of most fruits (Fig. 6.5). They are retained on the reproductive structure, but their diameter one month after anthesis is still less than 25 mm. Unfavourable light conditions during the third and fourth weeks after flower opening are still inhibitory. Reducing the temperature by 4°C or increasing CO_2 concentration of the air during the low photon flux density treatment slightly stimulated fruit growth. Continuous leaf excision and plant decapitation also stimulated fruit growth suggesting that the inhibition under low light resulted from a competition for available assimilates between reproductive development and vegetative growth (J.M. Kinet, University of Liège, Belgium, 1978–1980, unpublished results).

Competition with other fruits, either in the same inflorescence where distal fruits are inhibited by the proximal (Bangerth and Ho, 1984), or from different trusses (Hurd *et al.*, 1979), also plays a critical role, especially when light is limiting.

The presence of seeds is not essential for fruit swelling since pollination with pollen from *Lycopersicon peruvianum* causes fruit growth to a size similar to self-pollinated controls, but without the production of seeds (Verkerk, 1957). When present, however, seeds, which are sources of auxin, probably promote fruit set. Endogenous auxin content of the seeds peaks 7–10 days after anthesis (Iwahori, 1967; Mapelli *et al.* 1978) and exogenous application of auxin to flowers stimulates fruit set (Wittwer and Bukovac, 1962). GAs appear also to be implicated in the control of fruit initiation in tomato. When applied to the inflorescence at anthesis, they promote fruit set under low light and their content is high in ovaries of parthenocarpic cultivars (Mapelli *et al.* 1978). In the *gib-1* mutant, small parthenocarpic fruits may occur (Groot *et al.*, 1987) suggesting that GAs are not absolutely required to initiate fruit development in tomato. However, a spray with gibberellin $(GA)_{4+7}$ is needed for flowers to open so exogenously applied GAs still present in the flower at anthesis may trigger fruit set.

FRUIT DEVELOPMENT

Just before anthesis of the first flower, sink activity of the tomato inflorescence is minimal (Kinet, 1989). At anthesis, the growth of the ovary ceases but resumes after fertilization, concomitantly with a large increase in assimilate import in both the ovary (Archbold *et al.*, 1982) and the inflorescence (Kinet, 1989).

Cumulative growth of the fruit is expressed in the form of a sigmoidal curve (Fig. 6.6) with an initial 2-week period during which absolute growth is slow, followed by 3–5 weeks of rapid growth up to the mature green stage and finally a period of slow growth for 2 further weeks (Monselise *et al.*, 1978). Cell division is limited to the early slow growth phase during which cell elongation starts. The relative growth rate of the fruit reaches a peak at the end of the first week and then declines during the period of rapid absolute growth which results from cell elongation only. At the mature green stage, the fruit has almost reached its final weight (Monselise *et al.*, 1978). However, assimilate import in the fruit is still detected at incipient coloration but becomes negligible when the ripe stage is reached (McCollum and Skok, 1960).

The ultimate size of the fruit is correlated to a number of parameters including (i) the number of carpels in the ovary; (ii) the number of seeds; (iii) the position of the fruit; (iv) the sequence of set in a truss; and (v) the environmental conditions prevailing during the growth phase.

The gynoecium of cultivated tomatoes has two to several carpels. Locule number in tomato is under the control of the gene *Lc* (Fryxell, 1954). Ovary structure is also affected by environmental and hormonal factors. Fruits produced at low temperature (18/15°C, day/night) contain a greater number of locules than fruits developing at high temperature (28/23°C) (Sawhney, 1983) and GA_3 increases the number of locules when applied to

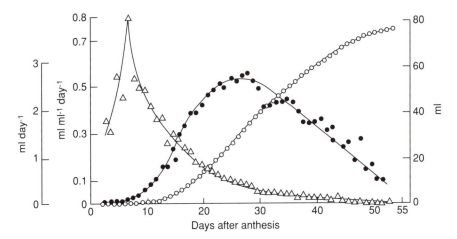

Fig. 6.6. Cumulative growth (●, ml), growth rate (○, ml day^{-1}), and relative growth rate, RGR (△, ml ml^{-1} day^{-1}) of tomato fruits (Monselise *et al.*, 1978).

the plant early during floral transition before the initiation of floral organs (Sawhney and Dabbs, 1978).

Sawhney and Dabbs (1978) established that increasing locule number does not necessarily result in an increase in seed number and that both parameters may independently affect final fruit size. That number of seeds is not entirely responsible for the ultimate size of the fruit is clearly demonstrated by the finding of Verkerk (1957) that fruit growth is possible without seeds. However, when seeds are present, their number is generally positively related to fruit weight (Sawhney and Dabbs, 1978) which is thus indirectly related to pollen production and pollination. Seeds are sources of growth regulators and studies with the GA-deficient *gib-1* mutant revealed that, although fruits may develop in the absence of endogenous GAs, their final weight is increased by the GAs produced by the seeds (Groot *et al.*, 1987). The combination of a mutant mother plant with wild-type embryos, obtained by pollinating the mutant with wild-type pollen, resulted indeed in larger fruits than the combination leading to GA-deficient seeds. Koshioka *et al.* (1994) postulated that GA$_1$ levels may correlate with fruit size but this relationship is obscured by other factors such as fruit position in the truss (Bohner *et al.*, 1988).

Position and set sequence in the truss are critical factors determining final size of tomato fruits. Usually, fruits developing at the proximal position are larger than distal fruits. Final size can be manipulated however by altering set sequence since distal fruits induced first in the truss develop to a larger size than proximal fruits. When all fruits in the same truss are simultaneously induced, their final sizes are similar (Bangerth, 1981). Competition for assimilates between fruits within the truss probably accounts, at least partly, for this phenomenon.

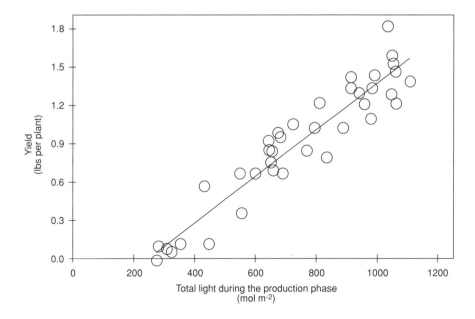

Fig. 6.7. Relationship between yield and the total light received during the 60-day production cycle of a single-cluster glasshouse tomato crop (Janes and McAvoy, 1991).

Yield is positively related to the quantity of solar radiation received by the crop in long season crops (Cockshull *et al.*, 1992) and in single-truss tomatoes (Fig. 6.7) (McAvoy *et al.*, 1989; Janes and McAvoy, 1991). Shading reduced fruit size and so reduced the proportion of fruits in the larger size grades (Cockshull *et al.*, 1992). The greatest increase in average fruit weight occurred when the light was applied from initial fruit set to the mature-green stage (McAvoy and Janes, 1989), that is during the period of rapid absolute growth of the fruit. Light may directly stimulate tomato fruit growth and sugar accumulation (Guan and Janes, 1991). CO_2 enrichment also increases individual fruit weight and total yield (Tripp *et al.*, 1991). The amount of assimilates available thus appears to determine final fruit size (Ho and Hewitt, 1986). Water availability also affects fruit size (Salter, 1958).

Fruit size and yield thus appear to be dependent on assimilate distribution within the fruiting plant which is controlled by the activity of both sources and sinks, and by vascularization. When assimilate availability is lower than total demand, competition between sinks becomes the determinant factor for the control of assimilate distribution. Competition exists between vegetative and reproductive structures, among inflorescences and among fruits on the same truss (Ho and Hewitt, 1986).

The duration of fruit development is unaffected by shade but is influenced by the position of the truss on the plant and of the fruit in the truss in glasshouse environments (Cockshull *et al.* 1992). Under early summer field conditions, exposure of fruit to sunlight may hasten ripening, since rising temperature shortens the growth period of the fruit (Hurd and Graves, 1984). Degree of foliage cover can be manipulated by choice of cultivar, and nitrogen fertility level (Wien and Minotti, 1988) although there is a risk of sun scald damage in high light environments (see section on fruit ripening disorders below).

FRUIT RIPENING

The tomato … within the last 20 years has become the model system for fundamental studies on fruit ripening.

(Grierson and Schuch, 1993)

THE RIPENING CHANGES

During the final period of slow growth of the fruit, colour, flavour, aroma, texture and composition change dramatically. This process is described as fruit ripening and consists of highly coordinated synthetic and degradative reactions.

Degradation of cell walls starts by the action of several enzymes, the most important being polygalacturonase and results in a soft, juicy texture. The fruit colour change starts 2–3 days after the mature green stage and progressively develops from yellow to orange and red. It results from the transformation of chloroplasts to chromoplasts with the accumulation of a number of pigments, most noticeably the red pigment lycopene and ß-carotene (Grierson and Kader, 1986). About 10 days after the start of colour change, an abscission layer forms between the calyx and the fruit cutting assimilate transport to the fruit (McCollum and Skok, 1960).

The dry matter content of the ripe tomato fruit is about 5–7.5% of the total weight. A high proportion of this dry matter is sugar and organic acid (Davies and Hobson, 1981) both of which contribute to the taste of the fruit. Sugars, mainly glucose and fructose, represent 2–4% of fresh weight while sucrose accounts for only 0.1–0.2%. Starch accumulates during the first month of growth and declines thereafter (Ho *et al.*, 1982/83). The organic acid content, mainly citric and malic acids, also increases during fruit development. The ratio of citric acid to malic acid is low during early growth but increases during ripening (Davies and Hobson, 1981). Potassium, nitrogen and phosphorus are the most important mineral elements present in the tomato fruit and account for more than 90% of the total mineral content (Davies and Hobson, 1981).

The tomato fruit has been classified as 'climacteric'. In a climacteric fruit, respiration passes through a minimum before starting to increase at the onset of ripening, reaches a maximum, referred to as the climacteric peak, and then declines (Biale and Young, 1981). This respiratory peak is preceded by a rise in ethylene production.

Ripening mutants have been extensively investigated to identify ripening genes and elucidate the molecular basis of ripening. A number of ripening-related mRNAs have been cloned and the functions of several ripening-related genes have been determined. These genes are involved in cell wall degradation (e.g. polygalacturonase and pectinesterase), ethylene synthesis (such as ACC synthase and ACC oxidase), and carotenoid synthesis (including phytoene synthase) (see review by Gray *et al.*, 1994). Transgenic tomatoes in which these genes have been inhibited, using antisense technology, or overexpressed, exhibited new and stably inherited traits such as improved textural qualities, slower ripening and reduced over-ripening, or enhanced colour (Grierson and Fray, 1994).

THE CONTROL OF RIPENING

Numerous reactions constituting ripening occur at the same time and the detection of the causal factor is thus difficult. The rise in endogenous ethylene production appears, however, to trigger ripening. Exogenous application of ethylene (or ethylene analogues) stimulates ripening of mature green tomatoes (McGlasson *et al.*, 1978) while inhibitors of ethylene synthesis or action delay or prevent the onset of fruit ripening (Hobson *et al.*, 1984).

Ethylene regulates the expression of several ripening-related tomato genes (Gray *et al.*, 1992). Inhibiting 1-aminocyclopropane-1-carboxylate synthase, a key enzyme in ethylene biosynthesis, with antisense RNA resulted in a 99.5% inhibition of ethylene production (Fig. 6.8). The antisense fruits never turn red and soft or develop an aroma for a 90–150-day period (Oeller *et al.*, 1991). This inhibition can be overcome with ethylene or ethylene analogues. Several mutations are known that have pleiotropic effects upon ripening but do not influence growth and development of the rest of the plant (Grierson *et al.*, 1987). They usually affect fruit colour, flavour, softening or firmness and delay ripening. Among these are ripening inhibitor (*rin*), non-ripening (*nor*) and never ripe (*Nr*). The ripening phenotype of these mutants is not significantly affected by exogenous ethylene (Grierson *et al.*, 1987) suggesting that they are either insensitive to ethylene, as recently reported for *Nr* (Lanahan *et al.*, 1994), or defective in an early step regulating the ripening process.

There is evidence that factors in addition to ethylene influence ripening (Grierson and Kader, 1986) and interactions between growth regulators seem to be involved. High CK levels may contribute to delays in the ripening process in the *rin* mutant (Davey and van Staden, 1978) and in fruits of plants transformed with a chimaeric gene construct which causes a ten- to 100-fold elevation of CK levels in fruit tissues (Martineau *et al.*, 1994).

Ripening is not much influenced by light and can occur in the dark, but sugar content is correlated with irradiance during the growth period

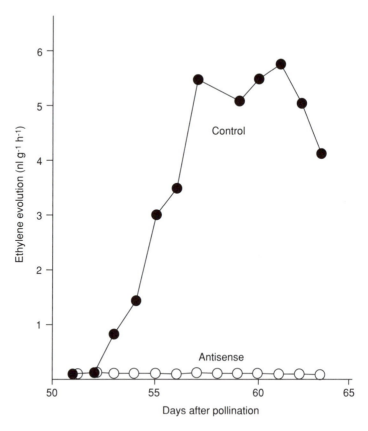

Fig. 6.8. Inhibition of ethylene evolution in detached tomato fruits by antisense ACC synthase RNA (Theologis *et al.*, 1992).

(Winsor and Adams, 1976). Reduced temperature slows down the rate of ripening and reduces lycopene synthesis which is inhibited at temperatures above 30°C. Potassium nutrition strongly influences acid concentration.

These studies of ripening control allowed the manipulation of ripening for commercial purposes. Promoting ripening is feasible by treating harvested mature green fruits with ethylene or ethephon. Retarding ripening is possible by subjecting fruits to reduced temperature in controlled atmosphere (CA; e.g. N_2 at 90%, CO_2 at 5% and O_2 at 5%) to limit respiration and interfere with ethylene synthesis. The antisense technology will also allow the construction of cultivars in which the production of ethylene is inhibited and thus having improved storage life. We are now on the way to developing cultivars with improved fruit quality through biotechnology (Grierson and Schuch, 1993). The use of an antisense technology was the basis of the 'Flavr Savr' tomato, with an antisense polygalacturonase gene conferring resistance to softening to ripe fruit for extended periods of time. Commercialization of this tomato began in the USA, on 21 May 1994, the entrance of a genetically engineered whole food in the public marketplace (Kramer and Redenbaugh, 1994).

PHYSIOLOGICAL DISORDERS

The disorders discussed below have both genetic and environmental components, and in many cases the exact cause of the disorder is not well understood or a complex of factors is involved. Physiological disorders covered in this section have a characteristic set of symptoms whose origin cannot be attributed solely to a biological agent or to a single environmental or cultural factor. The following discussion does not include descriptions of nutrient deficiencies, air pollution damage, herbicide injury or chilling injury. Discussions of flower and fruit abscission in this section partially overlap the detailed developmental descriptions above, but the emphasis here is on practical aspects of diagnosis, control and potential amelioration through breeding.

For many physiological disorders, little in-depth research has been done and the causes are poorly understood both in terms of why cultivars differ in susceptibility and why certain environments or cultural practices predispose plants to the disorder. In the case of blotchy ripening (BR), for example, some early researchers perceived the causal agent to be cultural and environmental factors, and others implicated pathogens. More recent research has not resolved this issue. In the case of another common disorder, recent extensive research by Ho and co-workers has done much to explain the complex interplay of anatomical, physiological, genetic and cultural factors. This disorder serves as a valuable model system to better understand maladies whose genesis may well be equally complex, but are at present more obscure.

BLOSSOM-END ROT (BER)

Blossom-end rot first appears in green tomatoes as areas of white or brown locular tissue. Symptoms next appear in the fruit placenta in the case of internal BER or the blossom-end pericarp in the case of external BER (Adams and Ho, 1992). Externally, a small, water-soaked spot appears at or near the blossom scar. As the spot enlarges, the affected tissue dries out and turns light to dark brown, gradually developing into a well-defined, sunken, leathery spot.

Causes
It has long been known that BER develops when calcium and/or water levels in the root zone are low. Only recently, however, has the complex interplay of anatomical, genetic and environmental factors that determine whether or not a given fruit develops the disorder been explained by Ho

and co-workers. According to Adams and Ho (1993): 'The basic cause of BER is a lack of co-ordination between the transport of assimilates by the phloem and of calcium by the xylem during rapid cell enlargement in the distal placenta tissue, i.e. an interaction between the rates of fruit growth and of calcium acquisition at the distal end of the fruit. Whilst changes in the environment have a marked influence on the incidence of BER, genetic susceptibility is also a major cause of the disorder.'

At the anatomical level, the immediate cause of fruit tissue breakdown is a deficiency of calcium in the distal (blossom end) locular tissue. There are several reasons for these low calcium concentrations. Deposition into the calcium pectate and calcium phosphate fractions in the distal pulp tissue is low under all conditions (Minamide and Ho, 1993). The number of vascular bundles decreases from the proximal (stem) end to the distal end of the fruit (Belda and Ho, 1993) and during the 2 weeks after anthesis, rapid expansion of the fruit means that the density of bundles falls dramatically. When calcium import by the tomato fruit is further reduced by external factors, the calcium requirements of cell walls and cell membranes may not be met. Leakage of cell contents, as a result of a loss of semipermeability of the cell membrane or weakened cell walls may be the direct cause of BER symptoms.

Examples of external factors affecting BER are water availability and relative humidity (e.g. Pill and Lambeth, 1980; Gutteridge and Bradfield, 1983; Banuelos et al., 1985; Tan and Dhanvantari, 1985). Since calcium is transported only in the water-conducting tissues (xylem), when water uptake is reduced, calcium uptake is reduced proportionately. Humidity is important because fruit and leaves compete for water. Low day-time humidity, especially when accompanied by high temperature and high light, increases transpiration, so proportionately more calcium goes to the leaves than the fruit. Conversely, high humidity, low temperature and low light decrease transpiration, increasing the calcium content of the fruit (Adams and Ho, 1993). The increase in calcium uptake at high night-time relative humidities has been attributed to high root pressure (Gutteridge and Bradfield, 1983; Banuelos et al., 1985), so high night-time humidity is particularly effective in preventing BER. Adams and Ho (1993) point out, however, that more calcium is absorbed during the day than at night and the increase in calcium in the fruit due to high humidity at night is relatively small compared to humidity effects on calcium uptake during the day.

High temperatures and high light can also increase BER by increasing fruit growth rates. Ho et al. (1993) reported a positive relationship during the period of rapid fruit growth between BER incidence and the product of the average daily solar radiation integral and temperature. Added heating was found to increase BER incidence to a much greater extent than added sunlight, presumably because high temperatures increased the rate of fruit expansion more than extra light. A rapid rate of fruit enlargement (Ho et al., 1993) would increase the demand for calcium in plasmalemma synthesis because of the higher rate of cellular enlargement.

The incidence of BER also increases in saline conditions. Salinity decreases total calcium uptake and fruit calcium content by restricting water uptake (Adams and Ho, 1993). Xylem development inside the fruit was also restricted by salinity (Belda and Ho, 1993), decreasing the fruit's ability to transport calcium to the distal end. Raising salinity by adding major nutrients such as Mg and K rather than NaCl, increased BER even more (Adams and Ho, 1993), presumably because these elements compete with calcium. Competition for uptake sites may also explain why providing nitrogen in the ammonium, as opposed to the nitrate form, increases BER (Pill and Lambeth, 1980; DeKock et al., 1982a).

Genetic factors that influence the susceptibility of individual cultivars include large fruit size, a rapid rate of cell enlargement and perhaps the inability of the vascular system to rapidly transport calcium to the distal end of the fruit (Ho et al., 1993; Brown and Ho, 1993).

Control

BER is now relatively well understood at the physiological level, but at the practical level, causes and control measures are not always apparent. For example, large fluctuations in soil water deficits increase BER (Pill and Lambeth, 1980; Shaykewich et al., 1971) so mulching should decrease BER by reducing soil moisture losses. Elmer and Ferrandino (1991) however, reported that black plastic mulch increased early season BER and did not significantly affect total seasonal BER. They felt this was because plastic mulch reduced penetration of early season rainwater. As another example, low humidity in the day is known to increase BER and high night-time humidity to decrease it, but constant high (95%) relative humidity in growth chambers reduced calcium concentration and increased BER relative to constant moderate (55%) relative humidity (Banuelos et al., 1985). The authors felt that maintaining constantly high relative humidity prevented the build-up of night-time root pressure, and the associated higher levels of calcium uptake.

The following general guidelines should be helpful in controlling BER: root-zone calcium must be adequate and concentrations of competing cations must not be excessive. The water supply must be conducive to uptake, i.e. not too saline, flooded or otherwise restricted. Water must go to the fruit, as opposed to the leaves. This may be brought about by preventing excessive transpiration. As with all physiological disorders, cultivars differ in susceptibility. Two or three precautionary foliar sprays with calcium chloride or calcium nitrate (4 g l^{-1}) beginning about a month after setting are sometimes recommended (e.g. Konsler and Gardner, 1990). Finally, maintaining a constant, relatively slow fruit growth rate is also advisable. Thinning tomatoes to one or two fruits per truss initially increased the size of the fruit but subsequent trusses were severely affected by BER (DeKock et al., 1982b).

FLOWER AND FRUIT ABSCISSION

Under stress, most commonly from high temperature, buds may abscise prior to anthesis. Early indications are yellowing of the swollen 'joint' area, followed by yellowing from the joint to the flower. Although the flower drops off within a few days of this colour change, the 'stub' remains. If the bud has not developed through anthesis, however, the remaining tissue may become indistinct. Flower development sometimes continues after exposure to high temperatures but is abnormal. Early indications of heat-related flower injuries are: splitting of the antheridial cone (Levy *et al.*, 1978), stigma browning (Dane *et al.*, 1991); and stigma exsertion from the antheridial cone (Levy *et al.*, 1978; Dane *et al.*, 1991). Usually abnormal flowers fall off (Levy *et al.*, 1978). Even after fertilization, flowers sometimes fail to develop because of competition with other fruits or stress, but remain on the plant for some time (Peet, unpublished data). Once the fruit has actually started into its growth phase, however, it usually continues to grow.

Causes

Any environmental or other condition that disrupts the normal course of pollen, ovule or zygote development will predispose the flower or young fruit to abscission. Requirements for normal development are discussed above. Typical causes of poor fruit set in the field or greenhouse are too high or too low temperature or humidity; low light and high winds. For example, day temperatures over 32°C and night temperatures over 21°C reduce fruit set (Moore and Thomas, 1952) as do temperatures below 10°C (Charles and Harris, 1972). As few as 3 h at 40°C on 2 successive days may be sufficient (Picken *et al.*, 1985). This reproductive failure can be associated with one or more of a number of factors including: (i) reduced pollen quantity or quality; (ii) ovule degeneration; (iii) flower malformation; (iv) deficiency of carbohydrates; and (v) imbalances in growth-regulating substances (Aung, 1979).

Effect of adverse temperatures on reproductive structures

There have been many studies on the effects of adverse temperatures on reproductive development at both low and high temperatures. At 6°C stigma configuration, ovule fertility and early embryo development was unaffected (Fernandez-Munoz and Cuartero, 1991) but pollen production, shed, viability and tube growth were all reduced (e.g. Charles and Harris, 1972; Picken, 1984; Kinet *et al.*, 1985a; Fernandez-Munoz *et al.*, 1995). The critical period appears to be 4–6 and 12–14 days before anthesis (Mutton *et al.*, 1987) for low temperature and 9 days before anthesis for high temperature effects. Both these periods of sensitivity probably correspond with sporogenesis, discrepancies in timing reflect the different effects of low and high temperature on developmental rate.

High temperatures disrupt meiosis in both the macro- and microspore mother cells (Iwahori, 1965), even if only applied for 3 h (Sugiyama *et al.*,

1966). Ovules subjected to a temperature of 40°C 18 h after pollination aborted, perhaps due to inhibition of pollen tube growth and endosperm degeneration (Iwahori, 1966).

Other changes in style and stigma from high temperatures also reduce the chances for successful pollination. Charles and Harris (1972) reported reduced stigma receptivity after exposure to high temperature. Levy *et al.* (1978) found that in the field under high temperature days (maximum 36–39°C), there were cultivar differences in susceptibility, but within each cultivar, the amount of abscission was strongly correlated with the amount of style exsertion. No fruit set was ever observed when the style protruded more than 1 mm out of the antheridial cone. Flower abscission also occurred in flowers with normal styles, however. In controlled environment experiments at 33/23°C, bud drop, reduced viable pollen and style exsertion occurred in both susceptible and resistant cultivars, but responses were much more pronounced in the susceptible cultivar. The antheridial cone split, followed by flower abscission, in the susceptible, but not the resistant cultivar. By pollinating low-temperature grown plants with pollen from high-temperature grown plants and vice versa, Levy *et al.* (1978) were able to show that exposing either parent to high temperatures reduced fruit set, but overall, microspores were more affected than macrospores.

To breed more heat-tolerant tomatoes it would be useful to single out the factor most limiting fruit set. Although this has been attempted in numerous studies, no one factor has emerged as most critical (El-Ahmadi and Stevens, 1979; Iwahori, 1965; Charles and Harris, 1972; Dane *et al.*, 1991; Hanna and Hernandez, 1982). Kuo *et al.* (1979) concluded that at high temperatures, flower formation, pollen grain and ovule formation, style elongation, pollen germination, fertilization and seed formation are all adversely affected.

Effects of carbohydrate supply on fruit set

Stephenson (1981) suggests that the rate of abortion of flowers or fruit represents the plant's assessment of its ability to support subsequent fruit development. If conditions are favourable, more fruit will be retained, and if unfavourable, less. The tomato inflorescence is lost if irradiance is low 5–6 and 10–12 days after the truss appears (Kinet and El Alaoui Hachimi, 1988; Kinet, 1989). Hewitt and Curtis (1948) suggested that carbohydrate depletion by high respiration rates promoted abortion of flower buds before anthesis, possibly accounting for observations that low light and high temperatures together are more deleterious to fruit set than either factor alone (Calvert, 1965; Atherton and Harris, 1986).

High temperatures may also reduce source strength of susceptible plants. A tomato cultivar sensitive to heat stress had lower net photosynthetic rates at high temperatures than more resistant cultivars (Bar-Tsur *et al.*, 1985). Dinar *et al.* (1983) demonstrated that export of assimilated carbon from a tomato leaf was reduced under high temperature regimes. Heat sensitivity may also be related to sink strength of the buds relative to competing growing points. The content of reducing sugars in flower buds decreased in response to heat stress and rates of carbon import were low (Dinar and Rudich, 1985a) due to effects on carbohydrate metabolism in the buds (Dinar and Rudich, 1985b). Based on an explant system (fruit with a pedicel and piece of peduncle attached), Starck and Witek-Czuprynska (1993) concluded that cultivar resistance to heat stress is related to maintaining invertase activity and being able to translocate calcium to the fruit at high temperatures.

Hormonal imbalances

Whatever reproductive component most limits fruit set at high temperature, a secondary, if not primary, effect is reduced or altered production of growth-regulating compounds (Kuo and Tsai, 1984; Kuo et al., 1989). Changes in hormone ratios decreased the potential for fruit-setting at high temperatures, but could be overcome by application of exogenous auxin and GA_3 (Kuo et al., 1989). The resulting fruits were parthenocarpic, however, as has been found in other studies in which tomato yield was increased in stress conditions by hormone application (e.g. Watanabe et al., 1989). It is not clear, however, whether altered hormone levels are the cause or result of reduced seed set. Developing seeds are the source of auxins and GAs that promote fruit development (Varga and Bruinsma, 1976). High temperatures prevent normal pollen and ovule development so altered hormone levels are not unexpected.

Control

Using resistant cultivars and optimizing environments at fruit set are the best prevention. The best synthetic growth regulators for promoting fruit development in tomatoes are 4-chlorophenoxyacetic acid and 2-naphthoxyacetic acid (Nickell, 1982). Synthetic auxins are used in unheated greenhouse production to improve fruit set and fruit size at temperatures below 10°C in Spain (Cuartero et al., 1987) and in other areas where conditions for natural pollination are poor (George et al., 1984). However fruit defects and seedless fruits are a problem when synthetic auxins are used (George et al., 1984; Watanabe et al. 1989). Parthenocarpic cultivars are also used when tomatoes are grown at extreme temperatures (George et al., 1984).

PUFFINESS/HOLLOWNESS/BOXINESS

Puffy fruits lack some or all of the 'gel' normally surrounding the seed, leaving a gap between the placental tissue and the outer wall of the locule (Winsor, 1968). Externally, the fruit are angular rather than round. Certain cultivars, such as 'Yellow Stuffer' are naturally hollow.

Causes

The disorder is often associated with low light (Winsor, 1968). It is especially common during autumn, winter and early spring in mild winter areas such as Texas and Israel (Kedar and Palevitch, 1968). Fruit set with the aid of growth regulators are also more likely to be puffy so the condition has been associated with low seed number and poor fruit set by some investigators (Hobson *et al.*, 1977; Abad and Guardiola, 1986) particularly when temperatures are low (Rylski, 1979). However, Kedar and Palevitch (1970) did not find this trait to be necessarily associated with poor pollination, and Kedar and Palevitch (1968) reported no correlation between the average weight of seed or fruit and the frequency or severity of hollowness.

Control
Nutritional and temperature regimes appear to affect the percentage of puffy fruit. High nitrogen increases while high potassium decreases puffiness (Winsor, 1966; Palevitch and Kedar, 1968). Phosphorus is variously reported to increase (Winsor, 1966) or not affect (Palevitch and Kedar, 1968) the proportion of puffy fruit. Plants grown at 22°C had fewer puffy fruit than those grown at 18°C (Davies and Winsor, 1969). Nawata *et al.* (1985) suggest that spraying trusses with CCC at flowering when seed set was expected to be low would increase CK activity in the developing fruit.

FRUIT CRACKING

Cracks of varying size and depth occur in circles around the stem scar (concentric cracking) or radiate from the stem scar (radial cracking). In fruit with 'russetting', minute, hair-line cracks, invisible to the naked eye, cover much of the fruit surface giving the fruit a rough feel. When examined closely the surface appears crazed and fruit are said to have poor skin 'finish'.

Causes
Peet (1992) summarized the causes of radial fruit cracking as follows: cracking occurs when there is a rapid net influx of solutes and water into the fruit at the same time ripening or other factors reduce the strength and elasticity of the tomato skin. Increases in fruit temperature raise gas and hydrostatic pressures of the pulp on the skin, resulting in immediate visible cracking in ripe fruit. In green fruit, minute cracks are created which later expand to become visible. High light conditions, especially on unshaded fruit have also been associated with fruit cracking. High light intensity raises fruit temperatures, fruit soluble solids and fruit growth rates, all of which factors are associated with increased cracking. Russetting increases under conditions conducive to other forms of cracking but is particularly associated with high humidity (Hayman, 1987). It is not clear why cracking sometimes takes one form, and sometimes another.

Kamimura *et al.* (1972) proposed the following explanation for cracking of tomatoes after rain, based on their studies of the effects of soil moisture levels on fruit skin. They showed that high soil moisture lowered the tensile strength of tomato fruit skin. Because of this low tensile strength, the fruit developed many minute cracks when it enlarged rapidly. These minute cracks later developed into visible cracks. Under low soil moisture, they found that the tensile strength of the skin was greater. As a result, fruits grew more slowly and had fewer minute cracks. Changes in soil moisture during fruit growth also affected skin strength. Skin strength increased if soil moisture content decreased. Conversely, skin strength decreased if soil moisture content increased. In fact, changes from low to high soil moisture lowered skin strength compared to continued growth under any moisture regime. Such changes typically occur when drought is relieved by irrigation or rain. Cracking is particularly likely with continued wet weather or overhead irrigation because water enters the fruit through these minute cracks.

Characteristics of radial crack-susceptible cultivars are: (i) large fruit size; (ii) low skin tensile strength and/or low skin extensibility at the turning to pink stage of ripeness; (iii) thin skin; (iv) thin pericarp; (v) shallow cutin penetration; (vi) few fruits per plant; and (vii) fruit not shaded by foliage (Peet, 1992). Although it has been difficult to breed for cracking resistance *per se* (Stevens and Rick, 1986), commercial cultivars bred for firm fruit and tough skin in order to decrease handling and shipping losses in field tomato production in North America are often quite resistant to fruit cracking. This is probably because these qualities are components of resistance to cracking. In USA, cracking problems are most often seen in the large, multilocular (beefsteak-type) cultivars grown by home gardeners, greenhouse tomato growers and other types of growers who supply markets where shipping ability is not a prerequisite.

Control
Cultural practices which result in uniform and relatively slow fruit growth such as constant, relatively low soil moisture, offer some protection against fruit cracking (Peet, 1992). In the greenhouse (Peet and Willits, 1995), reducing watering has been shown to decrease the incidence of radial cracking, and there are also a few reports in field crops of reduced cracking at lower levels of soil moisture. In field crops, however cracking is usually attributed to fluctuations in the water supply. The classic occurrence is when a long period of drought is followed by heavy rain. Cultural practices that reduce diurnal fruit temperature changes such as maintaining vegetative cover to shade the fruit may reduce cracking. Greenhouse growers minimize day and night temperature differences and increase temperatures gradually from night-time to day-time levels. For both field and greenhouse tomato growers, harvesting before the pink stage of ripeness and selection of crack-resistant cultivars probably offer the best protection against cracking.

To avoid russetting in greenhouse crops, the following practices are suggested: (i) selecting a resistant cultivar; (ii) avoiding big fluctuations between day and night temperatures and relative humidities; (iii) avoiding large changes in electrical conductivity (EC) of the nutrient solution; (iv) maintaining a minimum of EC 3.0 in rockwool slabs; and (v) little or no deleafing (Hayman, 1987).

FRUIT RIPENING DISORDERS

Blotchy ripening complex

Of all the physiological disorders of tomato, the ripening disorders (BR, greywall and internal browning) are the least understood. There is disagreement over whether ripening disorders are physiological, biotic or genetic in origin and whether symptoms represent distinct disorders or manifestations of the same disorder. A related disorder, usually referred to as 'irregular ripening', is caused when the sweetpotato (now called silver leaf) whitefly feeds on the plant. This disorder is becoming increasingly important in the southern US, but is not included in the following discussion because the cause is biotic rather than physiological or genetic.

Blotchy ripening
This disorder is characterized by green to greenish-yellow to waxy-white areas near the calyx of otherwise normal red tomato fruit. Affected areas on the fruit surface do not soften as the fruit ripens. The disorder is not apparent in immature fruit (Seaton and Gray, 1936) and appears more often on the lower trusses (Cooper, 1960). Affected areas of the fruit are not sharply separated from normal areas, but rather the colours blend over a distance of 2 mm or so. The discoloration is usually confined to the outer walls, but in extreme cases, radial walls can also be affected (Grierson and Kader, 1986). Internally, the pericarp and placenta tissue have a whitish discoloration. In severe cases, brown lignified vascular strands are found in the outer pericarp. Both the whole fruit and blotchy areas differ from normal in composition of many organic constituents including total solids and organic acids (Grierson and Kader, 1986; Hobson *et al.*, 1977).

Causes. Although the exact cause of BR is not known (Grierson and Kader, 1986), the incidence is highest on soils with low potassium and nitrogen content (Adams *et al.*, 1978). In extreme cases, fruit symptoms are accompanied by foliar symptoms of potassium deficiency (Roorda van Eysinga and Smilde, 1981). Picha and Hall (1981) measured BR symptoms on four tomato cultivars at five levels of potassium fertility. Cultivars differed in susceptibility to BR, but adding potassium reduced BR incidence in all cultivars. The level of pericarp K, Ca, Mg and P in the different cultivars was not associated with their susceptibility to BR.

Boyle (1994) considers BR to be a symptom of tobacco mosaic virus (TMV) infection but most authorities consider BR to be a physiological disorder with symptoms similar to those of TMV (Hobson *et al.*, 1977). There is general agreement that fruit ripen abnormally if the plant is infected with the tomato strain of TMV after fruit are set. Some TMV-resistant cultivars develop BR, however (H.D. Rabinowitch, personal communication, Israel, 1995), and the incidence of ripening disorders varies with certain environmental factors such as temperature and potassium availability, suggesting a physiological component.

Control. Since this disorder is poorly understood, it is difficult to recommend control practices with any confidence. Hobson *et al.* (1977) advise the following as control practices: (i) maintenance of extractable soil potassium at or above 1000 mg l⁻¹; (ii) avoiding large-fruited cultivars; and (iii) keeping air temperatures below 29°C. Using TMV-resistant lines and avoiding infection might be particularly recommended in the case of problems that do not respond to changes in cultural practices.

Greywall

The outer locular wall turns brown or greyish-brown and the area may become slightly depressed and roughened. Internally, severe browning usually appears in the outer pericarp, especially in regions associated with the vascular bundles. Both the cause of the disorder and the names applied to it in the literature are confusing. The term 'vascular browning' has been used to describe this disorder, although a similar term, 'vascular discoloration' can be used to describe TMV damage (Hobson *et al.*, 1977).

Causes. The disorder has been attributed to a number of environmental and pathogenic factors, but is generally considered to be bacterial in origin (Hobson *et al.*, 1977). Stall and Hall (1969) isolated species of *Bacillus*, *Erwinia* and *Aerobacter* from greywall tissue and suggested the discoloration represents a hypersensitive tissue reaction. Injecting *Erwinia herbicola* into fruit produces the symptoms (Hall and Stall, 1967), but bacteria cannot always be isolated from naturally occurring greywall tissue (Stall and Hall, 1969). Boyle (1994) argues that BR, greywall and internal browning are all the result of TMV infection, either alone or complexed with bacterial infections. Timing of virus infection and entry of secondary pathogens determine the subsequent course of symptom development.

Seasonal and nutritional factors also affect greywall incidence. The amount of browning is increased by exposure to low temperature (Stall and Hall, 1969). MacNab *et al.* (1983) consider greywall to be promoted by low light intensity (as within the canopy), low temperatures, excessive soil moisture, excessive soil compaction, high nitrogen levels and low potassium levels. On the other hand, Stevens and Rick (1986) state that high temperature and high light can contribute to the formation of greywall. Picha and Hall (1981) evaluated four tomato cultivars for susceptibility to greywall at five different levels of potassium fertility both in the fall and in the spring. Greywall susceptibility was evaluated both under natural conditions and by injecting *Erwinia herbicola* into the pericarp of immature green fruit. In the spring crop, all four cultivars developed more greywall if no potassium was added to the soil. Also, high levels of added K reduced natural incidence of greywall in sand culture. There were differences between cultivars and between the fall and spring crops in greywall susceptibility, but, as with BR, pericarp levels of K, Ca, Mg or P could not be associated with differences in susceptibility between seasons or between cultivars.

Control. Maintaining adequate K, choosing resistant cultivars and protecting against TMV are probably the best control practices until such time as the disorder is better understood. Boyle (1994) suggests using TMV-resistant lines, or cross-protecting by inoculating young plants with selected mild strains of TMV.

Vascular discoloration or internal browning

Hobson *et al.* (1977) and MacNab *et al.* (1983) apply this term to fruit on TMV-infected plants. The symptoms developing on a plant after TMV infection depend on a number of factors, including age of the plant when infected, environmental stress and level of tissue sensitivity (Taylor *et al.*, 1969). Early infection of young plants give the fruit a mottled appearance (Broadbent and Winsor, 1964) while an attack once the trusses are well formed gives the fruit the appearance of 'bronzing' or 'vascular browning'. These symptoms typically appear 10–25 days after exposure (Jenkins *et al.* 1965). This has been described as a 'shock' reaction of the plant to infection at a susceptible stage of development (Boyle and Wharton, 1957). Plants tend to recover from an early infection but with later attacks, most, although not all fruit on successive trusses may be damaged (Hobson *et al.*, 1977).

Cause. The cause is generally agreed to be TMV infection, but the higher the growing temperature, the lower the incidence (Hobson *et al.* 1977).

Control. The main control is to prevent infection by TMV. This is not always easy, however, unless resistant cultivars are available. Heat and hydrochloric acid can be used to disinfect the seed surface and get rid of internal infections except those originating in the endosperm (Broadbent, 1976; Hobson *et al.* 1977). In some cases protection from severe attacks has been achieved by inoculation of seedling or young plants by an attenuated strain of TMV (Taylor *et al.*, 1969).

Summary. All that can be said to generalize about ripening disorders is: (i) cultivars differ in susceptibility; (ii) the incidence increases when potassium is low and decreases when potassium is raised; (iii) affected areas of the fruit eventually show signs of tissue necrosis, usually involving the vascular system; and (iv) affected areas are usually lower in soluble solids and titratable acidity.

Sunscald, sunburn or sunscorch

Green fruits exposed to direct sunlight ripen unevenly so that yellow patches appear on the side of ripe fruit. Symptoms are most likely to appear at the mature green to breaker stages of development (Retig and Kedar,

1967). The texture of affected areas is leathery and firmer than the surrounding tissues. Yellow areas sometimes have a mottled appearance, and the surface is depressed (Hobson *et al.*, 1977). Depending on the temperature and degree of injury, the area can become white and sunken. The tissue underneath the injured area is whitish and the cells partially collapsed, reducing the normal thickness of the locular walls (Hobson *et al.*, 1977).

Causes

Sunscald is caused by fruit pericarp temperatures exceeding 40°C (Rabinowitch et al., 1974). In bright sunlight, surface temperatures may be more than 10°C higher than the air temperature (Venter, 1970). The increase in temperature is greatest in large, red fruit (Venter, 1970). The degree of injury to the fruit depends on irradiance, spectral quality, temperature and treatment duration (Adegoroye and Jolliffe, 1983). If temperatures are over 30°C, but under 40°C, the area stays yellow (Grierson and Kader, 1986) because temperatures above 30°C prevent lycopene formation, while production of carotene continues up to 40°C (Tomes, 1963).

   Rabinowitch and Sklan (1980) correlated changes in superoxide dismutase levels in the fruit with sensitivity to sunscald. They suggested that the heightened sensitivity to sunscald of fruits which are suddenly exposed to high light and high temperature compared to fruits which develop under high temperature and light was associated with inadequate levels of superoxide dismutase. Conversely, fruit conditioned by high temperature treatment to resist sunscald damage had higher levels of this enzyme (Rabinowitch *et al.*, 1982).

Control

Exposure of fruit pericarp to temperatures of 45°C over a 6-h period, followed by a rest period at a lower temperature conditions fruit to high temperatures and offers some protection against sunscald (Kedar *et al.*, 1975). The protective effect is only temporary, however. The best protection against sunscald is to utilize cultivars with enough foliage to cover the fruit and to provide enough water and pest protection to maintain the foliage. Crops planted at higher densities will also be less susceptible.

GOLD FLECK, FRUIT POX

Gold specks or flecks are often observed around the calyx and shoulders of mature fruit. In green fruit, the specks are white and less abundant. These specks decrease the attractiveness of the fruit and significantly shorten the shelf life (Janse, 1988). Cells with the characteristic gold appearance were identified by Den Outer and van Veenendaal (1988) as containing a granular mass of tiny calcium salt crystals, probably calcium oxalate.

Causes

These specks are considered to be symptoms of excess calcium in the fruit. De Kreij *et al.* (1992) found that under conditions of high air humidity and high Ca : K ratios, more calcium was transported into the fruit and the incidence of gold speck increased. Increasing the P level also increased calcium uptake rate and increased speckling.

Control

The disorder can be reduced by avoiding susceptible cultivars (Ilker *et al.*, 1977). Sonneveld and Voogt (1990) found that raising the electrical conductivity of the nutrient solution reduced gold speck incidence, as did increasing the K : Ca ratio and increasing Mg. Presumably in all three cases, the mechanism was prevention of excess Ca uptake.

GENETIC DISORDERS

Rogues (off-types), silvering (white head or chimaera), leaf distortion and triploids ('Jack' plants) are examples of genetic disorders sometimes seen in tomato populations. Grimbly (1986) describes these disorders in detail. Although these disorders are generally confined to particular lines, environmental factors and cultural practices also affect their incidence (Grimbly, 1986).

Greenback, also-called persistent green shoulder or yellow shoulder, is considered a ripening disorder by Hobson *et al.* (1977), but primarily affects fruit from genotypes lacking the 'uniform ripening' gene (Picha, 1987). Like the ripening disorders, however, composition of affected tissue differs (Hobson *et al.*, 1977; Picha, 1987) and environmental factors have been correlated with disorder incidence including: (i) high relative humidity (Lipton, 1970); (ii) high light (Venter, 1970); and (iii) high temperatures (Venter, 1966). As is also true of the ripening disorders, yellow shoulder is most severe in potassium-deficient plants (Picha and Hall, 1981). The disorder is less at high levels of phosphorus and potassium (Winsor, 1970), but it is difficult to completely control the disorder by increasing these elements (Venter, 1966). Excess nitrogen applied to these cultivars further increases pigment concentration (Venter, 1970). The best control is undoubtedly to choose cultivars with the uniform-ripening gene. In excessive sunlight, even some of these cultivars will develop a proportion of white tissue in the locule walls, although the fruit may be normal externally (Hobson *et al.*, 1977).

OEDEMA

This disorder is often mistaken for bacterial or fungal diseases. Blister-like swellings on the leaf are undifferentiated and callus-like in appearance. The granulated appearance is caused by the splitting of the epidermis, presumably under pressure from within (Grimbly, 1986). Exposed, the turgid parenchyma cells erupt. As the ruptured cells dry out, necrotic areas develop and leaves become twisted and distorted.

Cause
Oedemas are caused by the water provided to the leaves exceeding that used in transpiration for a period of several days. Sagi and Rylski (1978) showed that under high humidity and excess water, the symptoms increased as light intensity decreased, presumably as a result of reduced plant transpiration.

Control
In greenhouses or growth chambers, decreasing watering and promoting transpiration by such measures as increased ventilation, higher temperatures and higher light should be effective. In the field, the disorder only appears when there are long periods of excess water and low transpiration. There is little that can be done in the field, assuming that irrigation has already been stopped, except for trying different cultivars. Sagi and Rylski (1978) found much greater susceptibility to the disorder in an Israeli field cultivar than in a European greenhouse cultivar.

STEM DISORDERS

Pithy stem appears to be associated with water stress (Grimbly, 1986) or with flooding (H.G. Rabinowitch, personal communication, 1995, Israel). Four days after stopping irrigation, Aloni and Pressman (1981) noted that the parenchyma cells of the pith started to die and the cell walls began to degenerate. As the symptoms progressed, large air spaces formed in the pith. Foliar levels of ABA also increased and induced symptoms in non-stressed plants.

Crease-stem sometimes occurs under conditions promoting succulent growth (high nitrogen and high water availability) as a pronounced groove or crease which develops on opposite sides of the stem. Eventually the crease may deepen so it extends completely through the stem (Grimbly, 1986). Although this crease represents a weak point on the stem which is susceptible to breakage, if the plant is otherwise well supported, as in a typical greenhouse training system, transport does not seem to be impeded, and the plant should recover. Plants with this disorder often set fruit poorly, however, which may also be a symptom of excessive vegetative growth (Grimbly, 1986).

BLOSSOM-END SCARRING (ROUGH FRUIT OR CAT-FACING)

Fruit with a long scar at the blossom end are often described as 'cat-faced'. Misshapen, usually large fruits, are described as 'rough'. Both disorders develop under protracted low temperatures. In the field, air temperatures of 17/10°C for a week are sufficient to induce abnormal flower development

(Saito and Ito, 1971). Low temperature treatments during the sensitive period increase the number of locules in the fruit (Wien and Turner, 1994). Foliar sprays of GA_3 also increase locule number and induce the disorder (Wien and Zhang, 1991). The time of greatest sensitivity for an individual flower is well before anthesis. Barten *et al.* (1992) reported greatest sensitivity 19–26 days before anthesis for fruit grown at relatively high temperatures after cold treatment (32/18 or 29/18°C). When fruit were grown at lower temperatures after cold treatment, flowers were most sensitive earlier (26–35 days before anthesis) (Wien and Turner, 1994). In both cases, the target stage is most probably the floral transition (see also section on time of flowering above), the different timings resulting from the different temperatures prevailing after this transition.

ENVIRONMENTAL AND CULTURAL FACTORS AFFECTING PRODUCTIVITY

Tomato crops are most likely to be limited in productivity by light, temperature, nutrition and water supply. The relative importance of these factors will depend on the cropping situation. In heated winter greenhouse crops, as grown in northern Europe and in the northern USA, light is most likely to be limiting. In unheated winter greenhouse crops, as grown in southern Europe and the Middle East, both light and temperature may be limiting. In the field, temperatures (too high or too low); light (short or cloudy days, crowding in the canopy); water; nutrition or lack of protection from weeds and other pests may all be limiting. Overall, Stevens (1986) considers the difficulty of water management to be the most important deterrent to high yields of high-quality tomatoes.

LIGHT

Effects of light on tomato growth are illustrated in Figs 6.1, 6.3 and 6.7 and are also discussed in greater detail in this chapter in the section above on vegetative growth. Light quality and photoperiod are not as important to tomato growth as the daily irradiation integral (Picken *et al.*, 1986). Using only fluorescent light, however, decreases net assimilation rate by 20% relative to the same irradiance with incandescent lights added as a source of far-red (Hurd, 1974). Kronenberg and van de Hulst (1984) analysed the factors accounting for variation in the yields of early tomatoes in Dutch glasshouses. For the period 1970–80, 62–74% of the fluctuation in crop yield could be attributed to differences in average hours of sunlight received by the crop. The single month where yields were most highly correlated with yield was April (69%). For the period March–June the correlation coefficient was 86%.

To overcome light limitation, supplemental lights are sometimes installed in commercial greenhouses. However, this is rarely justifiable in economic terms, except where seedlings are closely packed together, so relatively few lights need be installed (Picken *et al.,* 1986). If supplemental light is used, it should be used to extend the day rather than to raise the light intensity during daylight hours (Hurd, 1973). In terms of flowering and when the daily irradiance integral is kept at a same level, tomato is usually described as a quantitative short-day plant, but vegetative growth is promoted by long days (Picken *et al.,* 1986). The reason that tomatoes grow better under long days is not completely understood, but probably results from a greater daily light integral under long days and possibly involves higher chlorophyll content per unit leaf area, lower respiration losses because of shorter nights, and more efficient use of light when spread out over a longer period (Hurd, 1973). Went (1944) reported no fruiting under continuous light, an effect that is probably related to the fact that tomato development is impaired in non-fluctuating environments (see section on vegetative growth above). Generally, it is better to maximize the natural lighting by careful attention to the greenhouse cover, careful design and optimum winter orientation of the greenhouse and the crops within the greenhouse (Picken *et al.,* 1986).

TEMPERATURE

The optimal temperature for net assimilation rates in tomatoes is between 25 and 30°C (Khavari-Nejad, 1980). It does not seem to matter greatly in young plants if day and night temperatures differ. Relative growth rates of young plants responded relatively independently to the day and night temperatures with an optimum for both at around 25°C (Picken *et al.,* 1986). The effects of changing night temperature on the relative growth rate are about half those of changing day temperature (Picken *et al.,* 1986). If day temperatures are above optimal, a lower than optimal night temperature can compensate. If the day temperature is below optimal, within the range of 5°C, a higher night temperature can compensate (Calvert, 1964).

The benefits of a high night temperature continue for about 6 weeks, but diminish with age (Calvert, 1964). With older plants, there do seem to be benefits of having lower night than day temperatures, and day and night temperature effects are not independent or compensatory (Went, 1944; Picken *et al.,* 1986). There are several possible reasons for this (Picken *et al.,* 1986). In young plants, leaf areas are expanding rapidly and higher night temperatures allow faster growth at night and faster expansion of the leaf area. Also respiratory losses are unaffected by temperature in young plants, but increase with night temperature in older plants. Finally temperature affects stem elongation, which also affects dry weight.

CO_2

As with all C^3 crops, current atmospheric concentrations of CO_2 are limiting to photosynthesis in tomatoes. In phytotrons or when greenhouses are closed and light is high, CO_2 may be depleted below ambient levels, reducing photosynthesis and growth. However, the effects of raising CO_2 concentrations above ambient are more complex. Tomatoes are not particularly responsive to high CO_2, compared to other C^3 plants (Poorter, 1993).

Hurd and Thornley (1974) showed that as CO_2 increases from ambient to 0.2%, the relative growth rate (rate of increase in dry weight per unit dry weight) rises more slowly than the net assimilation rate (rate of dry matter production per unit total leaf area). The relative growth rate in enriched plants is soon similar to that of unenriched plants. CO_2 concentrations exceeding 0.1% are not recommended as 0.5% CO_2 can depress growth relative to 0.1% (Hicklenton and Jolliffe, 1980).

Commonly, in greenhouses in northern Europe, as long as the houses are not venting, CO_2 is added to bring ambient concentrations up to 0.1%. The difficulty with CO_2 enrichment, especially in warm-winter areas, is that plants are more responsive in absolute terms to elevated CO_2 when light is high (Hurd and Thornley, 1974) and temperatures (Mortensen, 1987) are high. Unfortunately, when light and temperatures are high, and plants are potentially most responsive to high CO_2, greenhouse temperatures often rise so much that the greenhouse must be vented, precluding further enrichment. This often means that only a few hours of enrichment are possible each day in the spring or in areas with mild winters (Peet and Willits, 1987).

An additional problem is that, unlike some crops, tomatoes require a fairly long period of enrichment each day (8–10 h) in order to show significant yield increases (Willits and Peet, 1989). Acclimation of photosynthesis has been noted by Besford (1993) and Yelle et al. (1989, 1990). Behboudian and Lai (1994) found tomatoes CO_2 enriched at 22°C did not grow any more rapidly than non-enriched plants in most of the enrichment treatments. At 25°C, there was a growth response overall, but only if enrichment continued for at least 8 h. Four hours of enrichment actually decreased growth significantly compared to the ambient CO_2 treatment. Exposing plants to high CO_2 at 22°C for half the week was also ineffective in terms of increasing growth. They suggested that lower translocation rates in CO_2 enriched plant meant that photosynthates accumulated in leaves rather than contributing to leaf expansion. CO_2 enriched leaves had higher rates of dark respiration, possibly because of the greater presence of photosynthates. At 25°C, higher translocation rates may have improved the plant's ability to use leaf reserves in growth.

Thus CO_2 enrichment or at least CO_2 maintenance is most important in cold climates because enrichment can be continued for long enough to be useful (8–10 h daily) and because greenhouses are increasingly airtight for energy conservation purposes, under these conditions CO_2 can be depleted, slowing growth.

OTHER FACTORS

As previously noted, water management is extremely important in tomatoes for both total and marketable yield. In addition, water quality is important. In many areas of the world, salinity reduces production. Although tomatoes are more salt-tolerant than many other crops, fruit weight is generally reduced by salinity. Nevertheless, saline water up to 7 decisiemens (dS) m^{-1} is used for irrigation in Israel, both because of the scarcity of fresh water for agriculture and because of the effect of salinity in improving soluble solids and other quality components in tomatoes.

In addition to factors which limit vegetative growth, tomatoes can also be said to be limited by their own fruitfulness. As discussed in the sections on vegetative growth and development of the reproductive structure to anthesis high temperatures, high nitrogen and low light, especially in combination, encourage vegetative growth more than reproductive growth. In these conditions the plant may grow well, but be late-yielding or low yielding because vegetative growth is favoured over reproductive growth.

Poor pollination leading to flower abscission, as discussed in the section on physiological disorders, is a frequent contributor to yield and quality problems in tomatoes.

Market demand for large, unblemished fruit also means that fruit suffering from any of a number of defects of the type described under Physiological disorders or damaged by insects will be unsaleable. Even 'cosmetic' defects will lower prices of fresh market tomatoes.

SUMMARY

Tomatoes require relatively high light and high temperatures for rapid growth and good production. In cropping environments, however, productivity in terms of yield is most likely to be limited by the fruitfulness of the plant and the ability of the grower to protect the fruit from damage. As Jones (1980) comments: 'The key to yield success is to obtain a good fruit set on each cluster and to ripen the fruit as quickly as possible … The loss of one or two fruit per cluster or a missing cluster will significantly reduce yield.' Thus, genetic traits, environmental conditions and cultural practices predisposing plants to unfruitfulness, physiological disorders and fruit defects are also significant limitations to productivity.

Comparing production of tomatoes from varying environments is highly subjective. Many people who have worked with tomatoes in many environments feel that given reasonable cultural conditions, tomatoes will produce at about the same rate. For example ' … per plant fruit yields do not vary much for best fruit yields obtained under these wide ranges of growing conditions. Per plant yields will be from 8–12 pounds of fruit during the normal growing season … These yields can be increased only by keeping the plant alive and productive during a longer period of time. One can only conclude that the genetic potential of the tomato plant is being achieved and that any dramatic yield increase per plant must come from some factor or groups of factors which will break the genetic yield potential' (Jones, 1980).

Not surprisingly, highest yields are recorded in situations where the crop stays in production for long periods of time. For example yields from greenhouses in the UK are recorded at 190 tons acre^{-1} (Aikman, 1988). These yields represent the long-season tomato crop however, with the harvest beginning in April and continuing through the end of October. For the field crop, highest yields are seen in arid climates with good soils and ample, properly managed and good quality irrigation water (Stevens, 1986). In the San Joaquin Valley of California and in Israel, yields of 155 tons ha^{-1} have been recorded using drip irrigation (Stevens, 1986). These yields are particularly impressive given the short harvest period of the field crop, and are comparable to those recorded even by good greenhouse tomato growers for the long-season crop (Aikman, 1988).

CONCLUDING REMARKS

Studies on tomato have increased our understanding of many developmental processes. It has long been and is still a model for investigating interorgan correlations, source–sink relationships, and assimilate partitioning. It is also an important model system for geneticists and biotechnologists and for the study of fruit ripening. The numerous available mutants are increasingly exploited to investigate photomorphogenesis, the developmental roles and mechanism of action of plant growth regulators. There is no doubt that tomato will still serve in the future as a 'reference plant'.

REFERENCES

Abad, M. and Guardiola, J.L. (1986) Fruit-set and development in the tomato (*Lycopersicon esculentum* Mill.) grown under protected conditions during the cool season in the south-eastern coast region of Spain. The response to exogenous growth regulators. *Acta Horticulturae* 191, 123–132.

Abdul, K.S., Canham, A.E. and Harris, G.P. (1978) Effects of CCC on the formation and abortion of flowers in the first inflorescence of tomato (*Lycopersicon esculentum* Mill.). *Annals of Botany* 42, 617–625.

Adams, P. and Ho, L.C. (1992) The susceptibility of modern tomato cultivars to blossom-end rot in relation to salinity. *Journal of Horticultural Science* 67, 827–839.

Adams, P. and Ho, L.C. (1993) Effects of environment on the uptake and distribution of calcium in tomato and on the incidence of blossom-end rot. *Plant and Soil* 154, 127–132.

Adams, P., Davies, J.N. and Winsor, G.W. (1978) Effects of nitrogen, potassium and magnesium on the quality and chemical composition of tomatoes grown in peat. *Journal of Horticultural Science* 53, 115–122.

Adegoroye, A.S. and Jolliffe, P.A. (1983) Initiation and control of sunscald injury of tomato fruit. *Journal of the American Society for Horticultural Science* 108, 23–28.

Aikman, D. (1988) How productive is the home-grown glasshouse tomato? *Grower* 109, Jan. 14, 8.

Aloni, B. and Pressman, E. (1981) Stem pithiness in tomato plants: the effect of water stress and the role of abscisic acid. *Physiologia Plantarum* 51, 39–44.

Archbold, D.D., Dennis, F.G. Jr, and Flore, J.A. (1982) Accumulation of ^{14}C-labelled material from foliar-applied ^{14}C sucrose by tomato ovaries during fruit set and initial development. *Journal of the American Society for Horticultural Science* 107, 19–23.

Atherton, J.G. and Harris, G.P. (1986) Flowering. In: Atherton, J. G. and Rudich J. (eds) *The Tomato Crop. A Scientific Basis for Improvement*. Chapman & Hall, London, pp. 167–200.

Aung, L.H. (1979) Temperature regulation of growth and development of tomato during ontogeny. In: Cowell, R. (ed.) *Proceedings of the 1st International Symposium on Tropical Tomato, Shanhua, Taiwan, Republic of China, Oct. 23–27 1978.* Asian Vegetable Research and Development Center Publication no. 78–59, Shanhua, Taiwan, Republic of China, pp. 79–93.

Bangerth, F. (1981) Some effects of endogenous and exogenous hormones and growth regulators on growth and development of tomato fruits. In: Jeffcoat B. (ed.) *Aspects and Prospects of Plant Growth Regulators*. British Plant Growth Regulator Group, Monograph 6, Wantage, pp. 141–150.

Bangerth, F. and Ho, L.C. (1984) Fruit position and fruit set sequence in a truss as factors determining final size of tomato fruits. *Annals of Botany* 53, 315–319.

Banuelos, G.S., Offermann, G.P. and Seim, E.C. (1985) High relative humidity promotes blossom-end rot on growing tomato fruit. *HortScience* 20, 894–895.

Bar-Tsur, A., Rudich, J. and Bravdo, B. (1985) High temperature effects on CO_2 gas exchange in heat-tolerant and sensitive tomatoes. *Journal of the American Society for Horticultural Science* 110, 582–586.

Barten, J.H.M., Scott, J.W., Kedar, N. and Elkind, Y. (1992) Low temperatures induce rough blossom-end scarring of tomato fruit during early flower development. *Journal of the American Society for Horticultural Science* 117, 298–303.

Behboudian, M.H. and Lai, R. (1994) Carbon dioxide enrichment in 'Virosa' tomato plant: responses to enrichment duration and to temperature. *HortScience* 29, 1456–1459.

Belda, R.M. and Ho, L.C. (1993) Salinity effects on the network of vascular bundles during tomato fruit development. *Journal of Horticultural Science* 68, 557–564.

Bensen R.J. and Zeevaart, J.A.D. (1990) Comparison of *ent*-kaurene synthetase A and B activities in cell-free extracts from young tomato fruits of wild-type and *gib-1*, *gib-2*, and *gib-3* tomato plants. *Journal of Plant Growth Regulation* 9, 237–242.

Berry, S.Z. (1969) Germinating response of the tomato at high temperature. *HortScience* 4, 218–219.

Besford, R.T. (1993) Photosynthetic acclimation in tomato plants grown in high CO_2. *Vegetatio* 104/105, 441–448.

Biale, J.B. and Young, R.E. (1981) Respiration and ripening in fruits – retrospect and prospect. In: Friend, J. and Rhodes, M.J.C. (eds) *Recent Advances in the Biochemistry of Fruits and Vegetables*. Academic Press, London, pp. 1–39.

Binchy, A. and Morgan, J.V. (1970) Influence of light intensity and photoperiod on inflorescence initiation in tomatoes. *Irish Journal of Agricultural Research* 9, 261–269.

Bohner, J., Hedden, P., Bora-Haber, E. and Bangerth, F. (1988) Identification and quantification of gibberellins in fruits of *Lycopersicon esculentum*, and their relationship to fruit size in *L. esculentum* and *L. pimpinellifolium*. *Physiologia Plantarum* 73, 348–353.

Boyle, J.S. (1994) Abnormal ripening of tomato fruit. *Plant Disease* 78, 936–944.

Boyle, J.S. and Wharton, D.C. (1957) The experimental reproduction of tomato internal browning by inoculation with strains of tobacco mosaic virus. *Phytopathology* 47, 199–207.

Broadbent, L. (1976) Epidemiology and control of tomato mosaic virus. *Annual Review of Phytopathology* 14, 75–96.

Broadbent, L. and Winsor, G.W. (1964) The epidemiology of tomato mosaic. V. The effect on TMV-infected plants of nutrient foliar sprays and of steaming the soil. *Annals of Applied Biology* 54, 23–30.

Brown, M.M. and Ho, L.C. (1993) Factors affecting calcium transport and basipetal IAA movement in tomato fruit in relation to blossom-end rot. *Journal of Experimental Botany* 44, 1111–1117.

Calvert, A. (1959) Effect of the early environment on the development of flowering in tomato. II. Light and temperature interactions. *Journal of Horticultural Science* 34, 154–162.

Calvert, A. (1964) The effects of air temperature on growth of young tomato plants in natural light conditions. *Journal of Horticultural Science* 39, 194–211.

Calvert, A. (1965) Flower initiation and development in the tomato. *National Agricultural Advisory Service Quarterly Review* 70, 79–88.

Calvert, A. (1969) Studies on the post-initiation development of flower buds of tomato (*Lycopersicon esculentum*). *Journal of Horticultural Science* 44, 117–126.

Calvert, A. and Slack, G. (1975) Effects of carbon dioxide enrichment on growth, development and yield of glasshouse tomatoes. I. Responses to controlled concentrations. *Journal of Horticultural Science* 50, 61–71.

Charles, W.B. and Harris, R.E. (1972) Tomato fruit-set at high and low temperatures. *Canadian Journal of Plant Science* 52, 497–506.

Cockshull, K.E., Graves, C.J. and Cave, C.R.J. (1992) The influence of shading on yield of glasshouse tomatoes. *Journal of Horticultural Science* 67, 11–24.

Cooper, A.J. (1960) The effects of the size, position and maturation period of inflorescences and fruits on abnormal pigmentation in the tomato variety Potentate. *Annals of Applied Biology* 48, 230–235.

Cooper, A.J. and Hurd, R.G (1968) The influence of cultural factors on arrested development of the first inflorescence of glasshouse tomatoes. *Journal of Horticultural Science* 43, 243–248.

Cuartero, J., Costa, J. and Nuez, F. (1987) Problems of determining parthenocarpy in tomato plants. *Scientia Horticulturae* 32, 9–15.

Dane, F., Hunter, A.G. and Chambliss, O.L. (1991) Fruit set, pollen fertility and combining ability of selected tomato genotypes under high-temperature field conditions. *Journal of the American Society for Horticultural Science* 116, 906–910.

Davey, J.E. and Van Staden, J. (1978) Endogenous cytokinins in the fruits of ripening and non-ripening tomatoes. *Plant Science Letters* 11, 359–364.

Davies, J.N. and Hobson, G.E. (1981) The constituents of tomato fruit – the influence of environment, nutrition, and genotype. *CRC Critical Reviews in Food Science and Nutrition* 15, 205–280.

Davies, J.N. and Winsor, G.W. (1969) Tomato fruit quality. The composition of 'hollow' or 'boxy' tomato fruit. *Annual Report of the Glasshouse Crops Research*

Institute for 1968. Littlehampton, UK, pp. 66–67.

DeKock, P.C., Hall, A., Boggie, R. and Inkson, R.H.E. (1982a) The effect of water stress and form of nitrogen on the incidence of blossom-end rot in tomatoes. *Journal of the Science of Food and Agriculture* 33, 509–515.

DeKock, P.C., Inkson, R.H.E. and Hall, A. (1982b) Blossom-end rot of tomato as influenced by truss size. *Journal of Plant Nutrition* 5, 57–62.

De Kreij, C., Janse, J., Van Goor, B.J. and Van Doesburg, J.D.J. (1992) The incidence of calcium oxalate crystals in fruit walls of tomato (*Lycopersicon esculentum* Mill.) as affected by humidity, phosphate and calcium supply. *Journal of Horticultural Science* 67, 45–50.

Dempsey, W.H. (1970) Effects of temperature on pollen germination and tube growth. *Report of the Tomato Genetics Cooperative* 20, 15–16.

Den Outer, R.W. and Van Veenendaal, W.L.H. (1988) Gold speckles and crystals in tomato fruits (*Lycopersicon esculentum* Mill.) *Journal of Horticultural Science* 63, 645–649.

Descomps, S. and Deroche, M.E. (1973) Action de l'éclairement continu sur l'appareil photosynthétique de la tomate. *Physiologie végétale* 11, 615–631.

de Zeeuw, D. (1954) De invloed van het blad op de bloei. *Mededelingen van de Landbouwhogeschool te Wageningen* 54(1), 1–44.

Dieleman, J.A. and Heuvelink, E. (1992) Factors affecting the number of leaves preceding the first inflorescence in the tomato. *Journal of Horticultural Science* 67, 1–10.

Dinar, M. and Rudich, J. (1985a) Effect of heat stress on assimilate partitioning in tomato. *Annals of Botany* 56, 239–248.

Dinar, M. and Rudich, J. (1985b) Effect of heat stress on assimilate metabolism in tomato flower buds. *Annals of Botany* 56, 249–257.

Dinar, M., Rudich, J. and Zamski, E. (1983) Effect of heat stress on carbon transport from tomato leaves. *Annals of Botany* 51, 97–103.

El-Ahmadi, A.B. and. Stevens, M.A. (1979) Reproductive responses of heat-tolerant tomatoes to high temperatures. *Journal of the American Society for Horticultural Science* 104, 686–691.

Elmer, W.H. and Ferrandino, F.J. (1991) Early and late-season blossom-end rot of tomato following mulching. *HortScience* 26, 1154–1155.

Emery, G.C. and Munger, H.M. (1970) Alteration of growth and flowering in tomatoes by the jointless genotype. *Journal of Heredity* 61, 51–53.

FAO (1994) *FAO Production Yearbook*, Vol. 47, 1993. Food and Agriculture Organization of the United Nations, Rome.

Fehleisen, S. (1967) Uniflora and conjunctiflora: two new mutants in tomato. *Report of the Tomato Genetics Cooperative* 17, 26–28.

Fernandez-Munoz, R. and Cuartero, J. (1991) Effects of temperature and irradiance on stigma exsertion, ovule viability and embryo development in tomato. *Journal of Horticultural Science* 66, 395–401.

Fernandez-Munoz, R., Gonzalez-Fernandez, J.J. and Cuartero, J. (1995) Variability of pollen tolerance to low temperatures in tomato and related wild species. *Journal of Horticultural Science* 70, 41–49.

Fryxell, P.A. (1954) Genetics of locule number. *Report of the Tomato Genetics Cooperative* 4, 10–11.

George, W.L. Jr, Scott, J.W. and Splittstoesser, W.E. (1984) Parthenocarpy in tomato. *Horticultural Reviews* 6, 65–84.

Georghiou, K. and Kendrick, R.E. (1991) The germination characteristics of phytochrome-deficient *aurea* mutant tomato seeds. *Physiologia Plantarum* 82, 127–133.

Gray, J.E., Picton, S., Shabbeer, J., Schuch, W. and Grierson, D. (1992) Molecular biology of fruit ripening and its manipulation with antisense genes. *Plant Molecular Biology* 19, 69–87.

Gray, J., Picton, S., Giovannoni, J.J. and Grierson, D. (1994) The use of transgenic and naturally occurring mutants to understand and manipulate tomato fruit ripening. *Plant, Cell and Environment* 17, 557–571.

Grierson, D. and Fray, R. (1994) Control of ripening in transgenic tomatoes. *Euphytica* 79, 251–263.

Grierson, D. and Kader, A.A. (1986) Fruit ripening and quality. In: Atherton, J.G. and Rudich, J. (eds) *The Tomato Crop. A Scientific Basis for Improvement*. Chapman & Hall, London, pp. 241–280.

Grierson, D. and Schuch, W. (1993) Control of ripening. *Philosophical Transactions of the Royal Society, London* B 342, 241–250.

Grierson, D., Purton, M.E., Knapp, J.E. and Bathgate, B. (1987) Tomato ripening mutants. In: Thomas, H. and Grierson, D. (eds) *Developmental Mutants in Higher Plants*. Society for Experimental Biology Seminar Series, Vol. 32. Cambridge University Press, Cambridge, pp. 73–94.

Grimbly, P. (1986) Disorders. In: Atherton, J.G. and Rudich, J. (eds) *The Tomato Crop. A Scientific Basis for Improvement*. Chapman & Hall, London, pp. 369–389.

Groot, S.P.C. and Karssen, C.M. (1987) Gibberellins regulate seed germination in tomato by endosperm weakening: a study with gibberellin-deficient mutants. *Planta* 171, 525–531.

Groot, S.P.C. and Karssen, C.M. (1992) Dormancy and germination of abscisic-acid-deficient tomato seeds. Studies with the *sitiens* mutant. *Plant Physiology* 99, 952–958.

Groot, S.P.C., Bruinsma J. and Karssen, C.M. (1987) The role of endogenous gibberellin in seed and fruit development of tomato: studies with a gibberellin-deficient mutant. *Physiologia Plantarum* 71, 184–190.

Groot, S.P.C., Kieliszewska-Rokicka, B., Vermeer E. and Karssen, C.M. (1988) Gibberellin-induced hydrolysis of endosperm cell walls in gibberellin-deficient tomato seeds prior to radicle protrusion. *Planta* 174, 500–504.

Guan, H.P. and Janes, H.W. (1991) Light regulation of sink metabolism in tomato fruit. II. Carbohydrate metabolizing enzymes. *Plant Physiology* 96, 922–927.

Gutteridge, C.G and Bradfield, E.G. (1983) Root pressure stops blossom-end rot. *Grower* 100, 25–26.

Hall, C.B. and Stall, R.E. (1967) Graywall-like symptoms produced in tomato fruits by bacteria. *Proceedings of the American Society for Horticultural Science* 91, 573–578.

Hanna, H.Y. and Hernandez, T.P. (1982) Response of six tomato genotypes under summer and spring weather conditions in Louisiana. *HortScience* 17, 758–759.

Harlan, J.R. (1992) *Crops and Man*, 2nd edn. American Society of Agronomy, Crop Science Society of America, Madison, WI.

Hayman, G. (1987) The hair-like cracking of last season. *Grower* 107, pp. 3–5.

Hewitt, S.P. and Curtis, O.F. (1948) The effect of temperature on loss of dry matter and carbohydrate from leaves by respiration and translocation. *American Journal of Botany* 35, 746–755.

Hicklenton, P.R. and Jolliffe, P.A. (1980) Alterations in the physiology of CO_2 exchange in tomato plants grown in CO_2-enriched atmospheres. *Canadian Journal of Botany* 58, 2181–2189.

Hille, J., Koornneef, M., Ramanna, M.S. and Zabel P. (1989) Tomato: a crop species amenable to improvement by cellular and molecular methods. *Euphytica* 42, 1–23.

Hillman, W.S. (1956) Injury of tomato plants by continuous light and unfavorable photoperiodic cycles. *American Journal of Botany* 43, 89–96.

Ho, L.C. and Hewitt, J.D. (1986) Fruit development. In: Atherton, J.G. and Rudich, J. (eds) *The Tomato Crop. A Scientific Basis for Improvement.* Chapman and Hall, London, pp. 201–239.

Ho, L.C., Belda, R., Brown, M., Andrews, J. and Adams, P. (1993) Uptake and transport of calcium and the possible causes of blossom-end rot in tomato. *Journal of Experimental Botany* 44, 509–518.

Ho, L.C., Sjut, V. and Hoad, G.V. (1982/83) The effect of assimilate supply on fruit growth and hormone levels in tomato plants. *Plant Growth Regulation* 1, 155–171.

Hobson, G.E., Davies, J.N. and Winsor, G.W. (1977) Ripening Disorders of Tomato Fruit. *Growers' Bulletin No. 4. Glasshouse Crops Research Institute.* Littlehampton, England.

Hobson, G.E., Nichols, R., Davies, J.N. and Atkey, P.T. (1984) The inhibition of tomato fruit ripening by silver. *Journal of Plant Physiology* 116, 21–29.

Honma, S., Wittwer, S.H. and Phatak, S.C. (1963) Flowering and earliness in the tomato. Inheritance of associated characteristics. *Journal of Heredity* 54, 212–218.

Howlett, F.S. (1936) The effect of carbohydrate and of nitrogen deficiency upon microsporogenesis and the development of the male gametophyte in the tomato, *Lycopersicon esculentum* Mill. *Annals of Botany* 50, 767–803.

Howlett, F.S. (1939) The modification of flower structure by environment in varieties of *Lycopersicon esculentum. Journal of Agricultural Research* 58, 79–117.

Hurd, R.G. (1973) Long-day effects on growth and flower initiation of tomato plants in low light. *Annals of Applied Biology* 73, 221–228.

Hurd, R.G. (1974) The effect of an incandescent supplement on the growth of tomato plants in low light. *Annals of Botany* 38, 613–623.

Hurd, R.G. and Cooper, A.J. (1970) The effect of early low temperature treatment on the yield of single-inflorescence tomatoes. *Journal of Horticultural Science* 45,19–27.

Hurd, R.G. and Graves, C.J. (1984) The influence of different temperature patterns having the same integral on the earliness and yield of tomatoes. *Acta Horticulturae* 148, 547–554.

Hurd, R.G. and Thornley, J.H.M. (1974) An analysis of the growth of young tomato plants in water culture at different light integrals and CO_2 concentrations. I. Physiological aspects. *Annals of Botany* 38, 375–388.

Hurd, R.G., Gay, A.P. and Mountifield, A.C. (1979) The effect of partial flower removal on the relation between root, shoot and fruit growth in the indeterminate tomato. *Annals of Applied Biology* 93, 77–89.

Hussey, G. (1963a) Growth and development in the young tomato. I. The effect of temperature and light intensity on growth of the shoot apex and leaf primordia. *Journal of Experimental Botany* 14, 316–325.

Hussey, G. (1963b) Growth and development in the young tomato. II. The effect of defoliation on the development of the shoot apex. *Journal of Experimental Botany* 14, 326–333.

Ilker, R., Kader, A.A. and Morris, L.L. (1977) Anatomical changes associated with the development of gold fleck and fruit pox symptoms on tomato fruit. *Phytopathology* 67, 1227–1231.

Iwahori, S. (1965) High temperature injuries in tomato. IV. Development of normal flower buds and morphological abnormalities of flower buds treated with high temperature. *Journal of the Japanese Society for Horticultural Science* 34, 33–41.

Iwahori, S. (1966) High temperature injuries in tomato. V. Fertilization and development of embryo with special reference to the abnormalities caused by high temperature. *Journal of the Japanese Society for Horticultural Science* 35, 379–386.

Iwahori, S. (1967) Auxin of tomato fruit at different stages of its development with a special reference to high temperature injuries. *Plant and Cell Physiology* 8, 15–22.

Jacobsen, S.E. and Olszewski, N.E. (1991) Characterization of the arrest in anther development associated with gibberellin deficiency of the *gib-1* mutant of tomato. *Plant Physiology* 97, 409–414.

Janes, H.W. and McAvoy, R.J. (1991) Environmental control of a single-cluster greenhouse tomato crop. *HortTechnology* 1, 110–114.

Janse, J. (1988) Goudspikkels bij tomaat: een oplosbaar probleem. *Groenten en Fruit* 43, 30–31.

Jenkins, J.E.E., Wiggell, D. and Fletcher, J.T. (1965) Tomato fruit bronzing. *Annals of Applied Biology* 55, 71–81.

Jones, J.B. (1980) Greenhouse vegetable yields: how high can you go? *American Vegetable Grower* 28, 24–26.

Kalloo, G. (1991) Introduction. In: Kalloo, G. (ed.) *Genetic Improvement of Tomato.* Monographs on Theoretical and Applied Genetics, Vol. 14, Springer-Verlag, Berlin, pp. 1–9.

Kamimura, S., Yoshikawa, H. and Ito, K. (1972) Studies on fruit cracking in tomatoes. *Bulletin Horticultural Research Station, Ministry of Agriculture and Forestry Series C (Morioka) no. 7.* Morioka, Japan.

Kaul, M.L.H. (1991) Reproductive biology in tomato. In: Kalloo, G. (ed.) *Genetic Improvement of Tomato.* Monographs on Theoretical and Applied Genetics, Vol. 14. Springer-Verlag, Berlin, pp. 39–50.

Kedar, N. and Palevitch, D. (1968) Seed number, specific gravity and external appearance of hollow tomato fruits. *Journal of Horticultural Science* 43, 401–407.

Kedar, N. and Palevitch, D. (1970) Structural changes in hollow tomato fruits. *Israel Journal of Agricultural Research* 20, 87–90.

Kedar, N., Rabinowitch, H.D. and Budowski, P. (1975) Conditioning of tomato fruit against sunscald. *Scientia Horticulturae* 3, 83–87.

Kemp, G.A. (1968) Low-temperature growth responses of the tomato. *Canadian Journal of Plant Science* 48, 281–286.

Khavari-Nejad, R.A. (1980) Growth of tomato plants in different oxygen concentrations. *Photosynthetica* 14, 326–336.

Kinet, J.M. (1977a) Effect of light conditions on the development of the inflorescence in tomato. *Scientia Horticulturae* 6, 15–26.

Kinet, J.M. (1977b) Effect of defoliation and growth substances on the development of the inflorescence in tomato. *Scientia Horticulturae* 6, 27–35.

Kinet, J.M. (1982) Un abaissement de la température et une élévation de la teneur en CO_2 de l'atmosphère réduisent l'avortement des inflorescences de tomates cultivées en conditions d'éclairement hivernal. *Revue de l'Agriculture* 35, 1767–1772.

Kinet, J.M. (1989) Environmental and chemical controls of flower development. In: Lord, E. and Bernier, G. (eds) *Plant Reproduction: from Floral Induction to Pollination.* American Society of Plant Physiologists Symposium Series, Vol. I, Rockville, MD, pp. 95–105.

Kinet, J.M. (1993) Environmental, chemical, and genetic control of flowering. *Horticultural Reviews* 15, 279–334.

Kinet, J.M. and El Alaoui Hachimi, H. (1988) Effect of ethephon, 1-aminocyclo-propane-1-carboxylic acid, and ethylene inhibitors on flower and inflorescence development in tomato. *Journal of Plant Physiology* 133, 550–554.

Kinet, J.M. and Léonard, M. (1983) The role of cytokinins and gibberellins in controlling inflorescence development in tomato. *Acta Horticulturae* 134, 117–124.

Kinet, J.M. and Sachs, R.M. (1984) Light and flower development. In: Vince-Prue, D., Thomas, B. and Cockshull, K.E. (eds) *Light and the Flowering Process*. Academic Press, London, pp. 211–225.

Kinet, J.M., Hurdebise, D., Parmentier, A. and Stainier, R. (1978) Promotion of inflorescence development by growth substance treatments to tomato plants grown in insufficient light conditions. *Journal of the American Society for Horticultural Science* 103, 724–729.

Kinet, J.M., Sachs, R.M. and Bernier G. (1985a) *The Physiology of Flowering*. Vol. 3. *The Development of Flowers*. CRC Press, Boca Raton, FL.

Kinet, J.M., Zune, V., Linotte, C., Jacqmard, A. and Bernier, G. (1985b) Resumption of cellular activity induced by cytokinin and gibberellin treatments in tomato flowers targeted for abortion in unfavorable light conditions. *Physiologia Plantarum* 64, 67–73.

Klapwijk, D. and de Lint, P.J.A.L. (1974) Fresh weight and flowering of tomato plants as influenced by container types and watering conditions. *Acta Horticulturae* 39, 237–247.

Konsler, T.R. and Gardner, R.G. (1990) Commercial production of staked tomatoes in North Carolina. *North Carolina Agricultural Extension Service Publication no. AG-60*. Raleigh, NC.

Koornneef, M., Bosma, T.D.G., Hanhart, C.J., van der Veen, J.H. and Zeevaart, J.A.D. (1990) The isolation and characterization of gibberellin-deficient mutants in tomato. *Theoretical and Applied Genetics* 80, 852–857.

Koornneef, M., van Tuinen, A., Kerckhoffs, L.H.J., Peters, J.L. and Kendrick, R.E. (1992) Photomorphogenetic mutants of higher plants. In: Karssen, C.M., van Loon, L.C. and Vreugdenhil, D. (eds) *Progress in Plant Growth Regulation*. Kluwer Academic, The Netherlands, pp. 54–64.

Koornneef, M., Karssen, C.M., Kendrick, R.E. and Zeevaart, J.A.D. (1993) Plant hormone and photomorphogenic mutants of tomato. In: Yoder, J.I. (ed.) *Molecular Biology of Tomato. Fundamental Advances and Crop Improvement*. Technomic Publishing, Lancaster, PA, pp. 119–128.

Koshioka, M., Nishijima, T., Yamazaki, H., Liu, Y., Nonaka, M. and Mander, L.N. (1994) Analysis of gibberellins in growing fruits of *Lycopersicon esculentum* after pollination or treatment with 4-chlorophenoxyacetic acid. *Journal of Horticultural Science* 69, 171–179.

Kramer, M.G. and Redenbaugh K. (1994) Commercialization of a tomato with an antisense polygalacturonase gene: the Flavr Savr™ tomato story. *Euphytica* 79, 293–297.

Kraus, E.J. and Kraybill, H.R. (1918) Vegetation and reproduction with special reference to the tomato. *Oregon Agricultural Experiment Station Bulletin* 149, 1–90.

Kristoffersen, T. (1963) Interactions of photoperiod and temperature in growth and development of young tomato plants (*Lycopersicon esculentum* Mill.). *Physiologia Plantarum* 16 (Suppl.), 1–98.

Kronenberg, H.G. and van de Hulst, H.C. (1984) Stability and instability of production of early glasshouse tomatoes in the Netherlands. *Scientia Horticulturae* 23, 129–136.

Kuo, C.G. and Tsai, C.T. (1984) Alteration by high temperature of auxin and gibberellin concentrations in the floral buds, flowers, and young fruit of tomato. *HortScience* 19, 870–872.

Kuo, C.G., Chen, B.W., Chou, M.H., Tsai, C.L. and Tsay, T.S. (1979) Tomato fruit-set at high temperatures. In: Cowell, R. (ed.) *Proceedings of the 1st International Symposium on Tropical Tomato, Shanhua, Taiwan, Republic of China, Oct. 23–27 1978.* Asian Vegetable Research and Development Center Publication no. 78–59, Shanhua, Taiwan, Republic of China, pp. 94–108.

Kuo, C.G., Chen, H.M., Shen, B.J. and Chen, H.C. (1989) Relationship between hormonal levels in pistils and tomato fruit-set in hot and cool seasons. In: Green, S.K., Griggs, T.D. and McLean, B.T. (eds) *Tomato and Pepper Production in the Tropics: Proceedings of the International Symposium on Integrated Management Practices at Tainan, Taiwan, 21–26 March 1988.* Asian Vegetable Research Development Center Publication no. 89–317. Shanhua, Tainan, pp. 138–149.

Lanahan, M.B., Yen, H-C., Giovannoni, J.J. and Klee, H.J. (1994) The Never ripe mutation blocks ethylene perception in tomato. *Plant Cell* 6, 521–530.

Larouche, R., Gosselin, A. and Vézina, L.P. (1989) Nitrogen concentration and photosynthetic photon flux in greenhouse tomato production: I. Growth and development. *Journal of the American Society for Horticultural Science* 114, 458–461.

Léonard, M., Kinet, J.M., Bodson, M. and Bernier, G. (1983) Enhanced inflorescence development in tomato by growth substance treatments in relation to ^{14}C-assimilate distribution. *Physiologia Plantarum* 57, 85–89.

Levy, A., Rabinowitch, H.D. and Kedar, N. (1978) Morphological and physiological characters affecting flower drop and fruit set of tomatoes at high temperatures. *Euphytica* 27, 211–218.

Lewis, D. (1953) Some factors affecting flower production in the tomato. *Journal of Horticultural Science* 28, 207–220.

Liptay, A. and Schopfer, P. (1983) Effect of water stress, seed coat restraint and abscisic acid upon different germination capabilities of two tomato lines at low temperature. *Plant Physiology* 73, 935–938.

Lipton, W.J. (1970) Effects of high humidity and solar radiation on temperature and color of tomato fruits. *Journal of the American Society for Horticultural Science* 95, 680–684.

Liu, Y., Bergervoet, J.H.W., De Vos C.H.R., Hilhorst, H.W.M., Kraak, H.L., Karssen, C.M. and Bino, R.J. (1994) Nuclear replication activities during imbibition of abscisic acid- and gibberellin-deficient tomato (*Lycopersicon esculentum* Mill.) seeds. *Planta* 194, 368–373.

Lukyanenko, A.N., Ochova, E.P. and Egeyan, M. (1973) A mutant with a single flower terminating the main stem. *Report of the Tomato Genetics Cooperative* 23, 24.

MacNab, A.A., Sherf, A.F. and Springer, J.K. (1983) *Identifying Diseases of Vegetables.* Pennsylvania State University, University Park, Pennsylvania.

Maisonneuve, B. and Philouze, J. (1982) Action des basses températures nocturnes sur une collection variétale de tomate (*Lycopersicon esculentum* Mill.) I. Etude de la production de fruits et de la valeur fécondante du pollen. *Agronomie* 2, 443–451.

Mancinelli, A.L., Borthwick, H.A. and Hendricks, S.B. (1966) Phytochrome action in tomato-seed germination. *Botanical Gazette* 127, 1–5.

Mapelli, S., Frova, C., Torti, G. and Soressi, G.P. (1978) Relationship between set, development and activities of growth regulators in tomato fruits. *Plant and Cell Physiology* 19, 1281–1288.

Mapelli, S., Torti, G., Badino, M. and Soressi, G.P. (1979) Effects of GA$_3$ on flowering and fruit-set in a mutant of tomato. *HortScience* 14, 736–737.

Marc, D., Orban, I. and Kinet, J.M. (1994) Phenotypic characterization of the recessive mutation stamenless (*sl*) in tomato (*Lycopersicon esculentum*). *Biologia Plantarum* 36 (Suppl.), S102 .

Martineau, B., Houck, C.M., Sheehy, R.E. and Hiatt, W.R. (1994) Fruit-specific expression of the *A. tumefaciens* isopentenyl transferase gene in tomato: effects on fruit ripening and defense-related gene expression in leaves. *Plant Journal* 5, 11–19.

McAvoy, R.J. and Janes, H.W. (1989) Tomato plant photosynthetic activity as related to canopy age and tomato development. *Journal of the American Society for Horticultural Science* 114, 478–482.

McAvoy, R.J., Janes, H.W., Godfriaux, B.L., Secks, M., Duchai, D. and Wittman, W.K. (1989) The effect of total available photosynthetic photon flux on single truss tomato growth and production. *Journal of Horticultural Science* 64, 331–338.

McCollum, J.P. and Skok, J. (1960) Radiocarbon studies on the translocation of organic constituents into ripening tomato fruits. *Proceedings of the American Society for Horticultural Science* 75, 611–616.

McGlasson, W.B., Wade, N.L. and Adato, I. (1978) Phytohormones and ripening. In: Letham, D.S., Goodwin, P.B. and Higgins, T.J.V. (eds) *Phytohormones and Related Compounds: a Comprehensive Treatise*, Vol. II. Elsevier/North Holland Biomedical Press, Amsterdam, pp. 447–493.

Mertens, T.R. and Burdick, A.B. (1954) The morphology, anatomy and genetics of a stem fasciation in *Lycopersicon esculentum*. *American Journal of Botany* 41, 726–732.

Micallef, B.J., Haskins, K.A., Vanderveer, P.J., Roh, K.S., Shewmaker, C.K. and Sharkey, T.D. (1995) Altered photosynthesis, flowering, and fruiting in transgenic tomato plants that have an increased capacity for sucrose synthesis. *Planta* 196, 327–334.

Minamide, R.T. and Ho, L.C. (1993) Deposition of calcium compounds in tomato fruit in relation to calcium transport. *Journal of Horticultural Science* 68, 755–762.

Monselise, S.P., Varga, A. and Bruinsma, J. (1978) Growth analysis of the tomato fruit, *Lycopersicon esculentum* Mill. *Annals of Botany* 42, 1245–1247.

Moore, E.L. and Thomas, W.O. (1952) Some effects of shading and para-chlorophenoxy acetic acid on fruitfulness of tomatoes. *Proceedings of the American Society for Horticultural Science* 60, 289–294.

Morris, D.A. and Newell, A.J. (1987) The regulation of assimilate partition and inflorescence development in the tomato. In: Atherton, J.G. (ed.) *Manipulation of Flowering*. Butterworths, London, pp. 379–391.

Mortensen, L.M. (1987) Review: CO$_2$ enrichment in greenhouses. Crop responses. *Scientia Horticulturae* 33, 1–25.

Mutton, L., Patterson, B.D. and Nguyen, V.O. (1987) Two stages of pollen development are particularly sensitive to low temperatures. *Report of the Tomato Genetics Cooperative* 37, 56–57.

Nawata, E., Inden, H. and Asahira, T. (1985) Effects of CCC on the occurrence of tomato puffy fruits and the endogenous cytokinin activities. *Scientia Horticulturae* 26, 119–127.

Nester, J.E. and Zeevaart, J.A.D. (1988) Flower development in normal tomato and a gibberellin-deficient (*ga-2*) mutant. *American Journal of Botany* 75, 45–55.

Nickell, L.G. (1982) *Plant Growth Regulators. Agricultural Uses.* Springer-Verlag, New York.

Oeller, P.W., Wong, L.M., Taylor, L.P., Pike, D.A. and Theologis, A. (1991) Reversible inhibition of tomato fruit senescence by antisense RNA. *Science* 254, 437–439.

Palevitch, D. and Kedar, N. (1968) Effect of fertilizer treatments and manure on hollowness of winter tomatoes. *Israel Journal of Agricultural Research* 18, 113–116.

Peet, M.M. (1992) Fruit cracking in tomato. *HortTechnology* 2, 216–223.

Peet, M.M. and Willits, D.H. (1987) Greenhouse CO_2 enrichment alternatives: effects of increasing concentration or duration of enrichment on cucumber yields. *Journal of the American Society for Horticultural Science* 112, 236–241.

Peet, M.M. and Willits, D.H. (1995) Role of excess water in tomato fruit cracking. *HortScience* 30, 65–68.

Philouze, J. (1978) Comparaison des effets des gènes *j* et *j-2* conditionnant le caractère 'jointless' chez la tomate et relations d' épistasie entre *j* et *j-2* dans les lignées de même type variétal. *Annales de l'Amélioration des Plantes* 28, 431–445.

Picha, D.H. (1987) Physiological factors associated with yellow shoulder expression in tomato fruit. *Journal of the American Society for Horticultural Science* 112, 798–801.

Picha, D.H. and Hall, C.B. (1981) Influences of potassium, cultivar and season on tomato graywall and blotchy ripening. *Journal of the American Society for Horticultural Science* 106, 704–708.

Picken, A.J.F. (1984) A review of pollination and fruit set in the tomato (*Lycopersicon esculentum* Mill.). *Journal of Horticultural Science* 59, 1–13.

Picken, A.J.F., Hurd, R.G. and Vince-Prue, D. (1985) *Lycopersicon esculentum*. In: Halevy, A.H. (ed.) *CRC Handbook of Flowering*, Vol. 3. CRC Press, Boca Raton, FL, pp. 330–346.

Picken, A.J.F., Stewart, K. and Klapwijk, D. (1986) Germination and vegetative development. In: Atherton, J.G. and Rudich, J. (eds) *The Tomato Crop. A Scientific Basis for Improvement.* Chapman and Hall, London, pp. 111–166.

Pill, W.G. and Lambeth, V.N. (1980) Effects of soil water regime and nitrogen form on blossom-end rot, yield, water relations, and elemental composition of tomato. *Journal of the American Society for Horticultural Science* 105, 730–734.

Poorter, H. (1993) Interspecific variation in the growth response of plants to an elevated ambient CO_2 concentration. *Vegetatio* 104/105, 77–97.

Rabinowitch, H.D. and Sklan, D. (1980) Superoxide dismutase: a possible protective agent against sunscald in tomatoes (*Lycopersicon esculentum* Mill.) *Planta* 148, 162–167.

Rabinowitch, H.D., Kedar, N. and Budowski, P. (1974) Induction of sunscald damage in tomatoes under natural and controlled conditions. *Scientia Horticulturae* 2, 265–272.

Rabinowitch, H.D., Sklan, D. and Budowski, P. (1982) Photo-oxidative damage in the ripening tomato fruit: protective role of superoxide dismutase. *Physiologia Plantarum* 54, 369–374.

Rasmussen, N. and Green, P.B. (1993) Organogenesis in flowers of the homeotic *green pistillate* mutant of tomato (*Lycopersicon esculentum*). *American Journal of Botany* 80, 805–813.

Rastogi, R. and Sawhney, V.K. (1990a) Polyamines and flower development in the male sterile stamenless-2 mutant of tomato (*Lycopersicon esculentum*). I. Level of polyamines and their biosynthesis in normal and mutant flowers. *Plant Physiology* 93, 439–445.

Rastogi, R. and Sawhney, V.K. (1990b) Polyamines and flower development in the male sterile stamenless-2 mutant of tomato (*Lycopersicon esculentum*). II. Effects of polyamines and their biosynthetic inhibitors on the development of normal and mutant floral buds cultured *in vitro*. *Plant Physiology* 93, 446–452.

Retig, N. and Kedar, N. (1967) The effect of stage of maturity on heat absorption and sunscald of detached tomato fruits. *Israel Journal of Agricultural Research* 17, 77–83.

Rick, C.M. (1978) The tomato. *Scientific American* 239(8), 67–76.

Rick, C.M. and Butler, L. (1956) Cytogenetics of the tomato. *Advances in Genetics* 8, 267–382.

Rick, C.M. and Dempsey, W.H. (1969) Position of the stigma in relation to fruit setting of the tomato. *Botanical Gazette* 130, 180–186.

Roorda van Eysinga, J.P.N.L. and Smilde, K.W. (1981) *Nutritional Disorders in Glasshouse Tomatoes, Cucumbers and Lettuce*. Centre for Agricultural Publishing and Documentation, Wageningen.

Rudich, J., Zamski, E. and Regev, Y. (1977) Genotypic variation for sensitivity to high temperature in the tomato: pollination and fruit set. *Botanical Gazette* 138, 448–452.

Rylski, I. (1979) Fruit set and development of seeded and seedless tomato fruits under diverse regimes of temperature and pollination. *Journal of the American Society for Horticultural Science* 104, 835–838.

Sagi, A. and Rylski, I. (1978) Differences in susceptibility to oedema in two tomato cultivars growing under various light intensities. *Phytoparasitica* 6, 151–153.

Saito, T. and Ito, H. (1971) Studies on the growth and fruiting in the tomato. XI. Effect of temperature on the development of flower, especially that of the ovary and its locule. *Journal of the Japanese Society for Horticultural Science* 40, 128–138.

Salter, P.J. (1958) The effects of different water-regimes on the growth of plants under glass. IV. Vegetative growth and fruit development in the tomato. *Journal of Horticultural Science* 33, 1–12.

Sawhney, V.K. (1983) The role of temperature and its relationship with gibberellic acid in the development of floral organs of tomato (*Lycopersicon esculentum*). *Canadian Journal of Botany* 61, 1258–1265.

Sawhney, V.K. (1992) Floral mutants in tomato: development, physiology, and evolutionary implications. *Canadian Journal of Botany* 70, 701–707.

Sawhney, V.K. and Dabbs, D.H. (1978) Gibberellic acid induced multilocular fruits in tomato and the role of locule number and seed number in fruit size. *Canadian Journal of Botany* 56, 2831–2835.

Sawhney, V.K. and Greyson, R.I. (1972) On the initiation of the inflorescence and floral organs in tomato (*Lycopersicon esculentum*). *Canadian Journal of Botany* 50, 1493–1495.

Sawhney, V.K. and Shukla, A. (1994) Male sterility in flowering plants: are plant growth substances involved? *American Journal of Botany* 81, 1640–1647.

Sawhney, V.K., Bhadula, S.K., Polowick, P.L. and Rastogi, R. (1989) Regulation and development of male sterility in tomato and rapeseed. In: Lord, E. and Bernier, G. (eds) *Plant Reproduction: From Floral Induction to Pollination*. American Society of Plant Physiologists Symposium Series, Vol. I, pp. 114–120.

Seaton, H.L. and Gray, G.F. (1936) Histological study of tissues from greenhouse tomatoes affected by blotchy ripening. *Journal of Agricultural Research* 52, 217–24.

Shaykewich, C.F., Yamaguchi, M. and Campbell, J.D. (1971) Nutrition and blossom-end rot of tomatoes as influenced by soil water regime. *Canadian Journal of Plant Science* 51, 505–511.

Sonneveld, C. and Voogt, W. (1990) Response of tomatoes (*Lycopersicon esculentum*) to an unequal distribution of nutrients in the root environment. *Plant and Soil* 124, 251–256.

Stall, R.E. and Hall, C.B. (1969) Association of bacteria with graywall of tomato. *Phytopathology* 59, 1650–1653.

Starck, Z. and Witek-Czuprynska, B. (1993) Diverse response of tomato fruit explants to high temperature. *Acta Societatis Botanicorum Poloniae* 62, 165–169.

Stephenson, A.G. (1981) Flower and fruit abortion: proximate causes and ultimate functions. *Annual Review of Ecology and Systematics* 12, 253–279.

Stevens, M.A. (1986) The future of the field crop. In: Atherton, J.G. and Rudich, J. (eds) *The Tomato Crop. A Scientific Basis for Improvement*. Chapman & Hall, London, pp. 559–580.

Stevens, M.A. and Rick, C.M. (1986) Genetics and breeding. In: Atherton, J.G. and Rudich, J. (eds) *The Tomato Crop. A Scientific Basis for Improvement*. Chapman & Hall, London, pp. 35–109.

Sugiyama, T., Iwahori, S. and Takahashi, K. (1966) Effect of high temperature on fruit setting of tomato under cover. *Acta Horticulturae* 4, 63–69.

Tan, C.S. and Dhanvantari, B.N. (1985) Effect of irrigation and plant population on yield, fruit speck and blossom-end rot of processing tomatoes. *Canadian Journal of Plant Science* 65, 1011–1018.

Taylor, G.A., Lewis, G.D. and Rubatzky, V.E. (1969) The influence of time of tobacco mosaic virus inoculation and stage of fruit maturity upon the incidence of tomato internal browning. *Phytopathology* 59, 732–736.

Theologis, A., Zarembinski, T.I., Oeller, P.W., Liang, X. and Abel, S. (1992) Modification of fruit ripening by suppressing gene expression. *Plant Physiology* 100, 549–551.

Tigchelaar, E.C. and Foley, V.L. (1991) Horticultural technology: a case study. *HortTechnology* 1, 7–16.

Tomes, M.L. (1963) Temperature inhibition of carotene synthesis in tomato. *Botanical Gazette* 124, 180–185.

Tripp, K.E., Peet, M.M., Pharr, D.M., Willits, D.H. and Nelson, P.V. (1991) CO_2-enhanced yield and foliar deformation among tomato genotypes in elevated CO_2 environments. *Plant Physiology* 96, 713–719.

Varga, A. and Bruinsma, J. (1976) Role of seeds and auxins in tomato fruit growth. *Zeitschrift für Pflanzenphysiologie* 80, 95–104.

Venter, F. (1966) Investigation on green-back of tomatoes. *Acta Horticulturae* 4, 99–101.

Venter, F. (1970) Beobachtungen über die Temperatur und den Chlorophyllgehalt in Tomatenfrüchten und das Auftreten von 'Grünkragen'. *Angewandte Botanik* 44, 263–270

Verkerk, K. (1955) Temperature, light and the tomato. *Mededelingen van de Landbouwhogeschool te Wageningen* 55, 175–224.

Verkerk, K. (1957) The pollination of tomatoes. *Netherlands Journal of Agricultural Science* 5, 37–54.

Watanabe, A., Beck, J., Rosebrock, H., Huang, J., Busse, U., Luib, M. and Schott, P. (1989) Biological activities of BAS 112W and BAS 113W on fruit-setting and fruit development in tomatoes. In: Green, S.K., Griggs, T.D. and McLean, B.T. (eds) *Tomato and Pepper Production in the Tropics: Proceedings of the International*

Symposium on Integrated Management Practices at Tainan, Taiwan, 21–26 March 1988. Asian Vegetable Research Development Center Publication no. 89–317. Shanhua, Tainan, pp. 174–183.

Went, F.W. (1944) Plant growth under controlled conditions. II. Thermoperiodicity in growth and fruiting of the tomato. *American Journal of Botany* 31, 135–150.

Weyers, J.D.B. (1985) Germination and root gravitropism of *flacca*, the tomato mutant deficient in abscisic acid. *Journal of Plant Physiology* 121, 475–480.

Wien, H.C. and Minotti, P.L. (1988) Response of fresh-market tomatoes to nitrogen fertilizer and plastic mulch in a short growing season. *Journal of the American Society for Horticultural Science* 113, 61–65.

Wien, H.C. and Turner, A.D. (1994) Severity of tomato blossom-end scarring is determined by plant age at induction. *Journal of the American Society for Horticultural Science* 119, 32–35.

Wien, H.C. and Zhang, Y. (1991) Gibberellic acid foliar sprays show promise as screening tool for tomato fruit catfacing. *HortScience* 26, 583–585.

Willits, D.H. and Peet, M.M. (1989) Predicting yield responses to different greenhouse CO_2 enrichment schemes: cucumbers and tomatoes. *Agricultural and Forest Meteorology* 44, 275–293.

Winsor, G.W. (1966) A note on the rapid assessment of 'boxiness' in studies of fruit quality. *Annual Report of the Glasshouse Crops Research Institute for 1965*. Littlehampton, UK, pp. 124–127.

Winsor, G.W. (1968) Potassium and the quality of glasshouse crops. Potassium and the Quality of Agricultural Products. *Proceedings of the 8th Congress of the International Potash Institute, Brussels, 1966*, International Potash Institute, Berne, Switzerland, pp. 303–312.

Winsor, G.W. (1970) A long-term factorial study of the nutrition of greenhouse tomatoes. Fertilization of Protected Crops. *Proceedings of the 6th Colloquium of the International Potash Institute, Florence 1968*, International Potash Institute, Berne, Switzerland, pp. 269–281.

Winsor, G.W. and Adams, P. (1976) Changes in the composition and quality of tomato fruit throughout the season. *Annual Report of the Glasshouse Crops Research Institute for 1975*. Littlehampton, UK, pp. 134–142.

Wittwer, S.H. and Aung, L.H. (1969) *Lycopersicon esculentum* Mill. In: Evans, L.T. (ed.) *The Induction of Flowering. Some Case Histories*. Macmillan, Melbourne, pp. 409–423.

Wittwer, S.H. and Bukovac, M.J. (1962) Exogenous plant growth substances affecting floral initiation and fruit set. *Proceedings Plant Science Symposium Camden 1962*. Campbell Soup Co., Camden, New Jersey, pp. 65–83.

Wudiri, B.B. and Henderson, D.W. (1985) Effects of water stress on flowering and fruit set in processing – tomatoes. *Scientia Horticulturae* 27, 189–198.

Yelle, S., Beeson, R.C. Jr, Trudel, M.J. and Gosselin, A. (1989) Acclimation of two tomato species to high atmospheric CO_2. I. Sugar and starch concentrations. *Plant Physiology* 90, 1465–1472.

Yelle, S., Beeson, R.C. Jr, Trudel, M.J. and Gosselin, A. (1990) Duration of CO_2 enrichment influences growth, yield and gas exchange of two tomato species. *Journal of the American Society for Horticultural Science* 115, 52–57.

Peppers

7

H.C. Wien

*Department of Fruit and Vegetable Science, Cornell University, 134A Plant
Science Building, Ithaca, New York 14853–5908, USA*

The cultivated *Capsicum* peppers are herbaceous, frost-sensitive plants that
in temperate areas are annual in growth duration, but in tropical areas may
continue to grow and produce yield over several years. They are the source
of capsaicin, the most commonly used spice in the world (Andrews, 1984).
Other cultivars are cultivated for production of the green or more mature
fruit, used in salads and as a cooked or raw vegetable. Still others are the
source of red food colouring, made from the dried, ground powdered fruit.

Most of the peppers cultivated in temperate and tropical areas belong to
the botanical species *Capsicum annuum*, thought to originate in Mexico and
Central America (Andrews, 1984). Fruit remnants dating back to 7000 BC
have been found in caves in this area, and evidence for cultivation of this
species has been traced to the period between 5200 and 3400 BC
(Govindarajan, 1985). The explorer Columbus imported peppers into southern Europe, from where they spread to the Middle East, Africa and Asia,
and were introduced back into more northerly parts of North America
(Andrews, 1984).

Four other species are also considered part of the cultivated *Capsicums*,
but play a much smaller role in agriculture and commerce (Smith *et al.*,
1987). They include *C. frutescens*, comprising small, pungent-fruited peppers
that serve as source of tabasco sauce. It, and *C. chinense*, are thought to originate in the Amazon basin of South America. *Capsicum baccatum* also traces
its origin to central South America, and has been selected in Brazil for a
range of fruit sizes and shapes (Nagai, 1989). *Capsicum pubescens* is the only
cultivated pepper species that originated in cooler zones, and is thought to
have come from the highlands of Bolivia (Andrews, 1984).

All cultivated species of *Capsicum* have 2n = 24 chromosomes
(Greenleaf, 1986). Crossability among the species is limited, however, so that
breeders have only been able to make limited use of differences in disease

resistance among species (Greenleaf, 1986). Within *C. annuum*, a tremendous range in size, shape and mature colour of fruits has been selected that now forms the basis for the types used in commerce throughout the world (Andrews, 1984; Greenleaf, 1986; Somos, 1984).

World production of nearly 10 million metric tons of fresh peppers on 1.1 million hectares ranks peppers in the middle range of vegetables in terms of popularity (FAO, 1992). Forty-six per cent of production is centred in Asia, with China the principal producing country. The countries of southern Europe are the second-most important producing region, with 24% of world production. Although Africa and North and Central America are less important in fresh pepper production, total area devoted to peppers is probably underestimated, since cultivation of the crop for dry chilli powder is important in these regions. In Mexico, for instance, about 31% of pepper production area is devoted to the dry product (Laborde and Pozo, 1982). Mexico, as probable country of origin of the crop, also has the most diverse array of cultivars, and wild types that are harvested.

In spite of the rich genetic diversity that exists within the crop, our knowledge of the physiology is limited, and restricted to a few cultivars, primarily of the large-fruited bell peppers within *C. annuum*. This trend has been reinforced by the growing importance of large-fruited peppers in European glasshouse production, and the need to understand the crop's physiology to improve cultivation practices. Nevertheless, production in the field of smaller-fruited types, loosely classified under the common name 'paprika', has also generated important physiological information (Somos, 1984).

SEED GERMINATION

The germination and emergence of pepper is slow at room temperature, and further delayed by cooler conditions. At 25°C, pepper required 3.5 days for radicle emergence, while at 15°C, 9 days were required (Watkins and Cantliffe, 1983). Emergence from 1.2 cm soil depth took 8–9 days at temperatures from 25–35°C (Lorenz and Maynard, 1980), and was prevented altogether at temperatures lower than 15°C. Removal of the testa made little difference in germination rate, but removing the endosperm through which the radicle had to penetrate halved the germination time at 25°C and reduced it from 9 to 3.4 days at 15°C (Watkins and Cantliffe, 1983). It thus appears that the endosperm constitutes the principal barrier to radicle emergence. The growth promoter gibberellic acid may be involved in endosperm penetration during germination. Watkins and Cantliffe (1983) found that soaking the germinating seeds in GA_{4+7} (a mixture of gibberellins 4 and 7) increased the speed of radicle penetration. The activity of endomannanase, a cell wall degrading enzyme, also increased with gibberellin (GA) treatment and during radicle penetration (Watkins *et al.*, 1985). It is possible that GA stimulated enzyme activity that enhanced endosperm breakdown in the area near the radicle tip.

Germination of pepper seeds may be preceded by a period of dormancy. In a survey of several cultivars of two species of *Capsicum*, Randle and Honma (1981) found that emergence took from 20 to 50 days from sowing at room temperature. Emergence times of the slow lines could be shortened to 20 days by storing the freshly harvested seeds at 24°C for 2–3 weeks. The slow emergence could also be eliminated by delaying the extraction of seeds from the fruit for 10 days after the fruit was fully ripe. While different *C. annuum* lines varied in emergence from 20–50 days, *C. frutescens* and *C. chacoense* lines tested required 51–61 days to emerge, respectively. These findings were confirmed by Edwards and Sundstrom (1987), who showed that 3 weeks seed storage of tabasco pepper (*C. frutescens*) increased seed germination from 40–70%. Although details of the dormancy mechanism have not been worked out, it is possible that a lack of GA may prevent prompt germination and emergence. Ilyas (1993) demonstrated that pepper seed that normally would not show dormancy could be prevented from germinating after soaking in the GA synthesis inhibitor tetcyclacis. If this treatment was followed by exposure to GA_{4+7}, germination was restored.

The slow germination and emergence rates of pepper have led to considerable effort in devising seed treatments that speed up these processes. Priming of pepper seed by a variety of techniques (see Chapter 1) has generally demonstrated a considerable reduction in time of germination, but only a small gain in rate of emergence from the soil (Rivas *et al.*, 1984; Bradford *et al.*, 1990; Ilyas, 1993). Sundstrom and Edwards (1989) showed that although the radicle emerged from the pepper seedcoat much earlier after priming, there was little stimulation of hypocotyl elongation rate in tabasco pepper. Addition of GA to primed seed had little added effect on emergence rate, however, indicating that the factor limiting hypocotyl elongation was not a lack of that growth promoting substance (Ilyas, 1993). Further work is needed to translate the benefits from priming in hastened germination to similar stimulation of emergence rate.

VEGETATIVE GROWTH

In addition to a rather slow seed germination and seedling emergence rate, pepper also has a relatively slower seedling growth rate than some other vegetable crops. Comparative growth analysis of tomato, cucumber and pepper indicated that pepper had a 25% lower relative growth rate than the other two species. The slower growth rate of pepper was not due to a lower productivity per unit leaf area (net assimilation rate), but to a reduced production of leaf area (Bruggink and Heuvelink, 1987). Pepper seedlings had significantly thicker leaves (higher specific leaf weight) than the other two species.

It is possible to reduce leaf thickness and increase the proportion of leaf area to total plant mass (leaf area ratio) by reducing incident light (Nilwik, 1981). These changes occur at the expense of the plant growth rate, however, and may thus be counterproductive. Nevertheless, use of light shade

(25–50%) during seedling growth has been advocated to increase yield of pepper in a tropical environment by maximizing leaf area production (Schoch, 1972).

The rate of plant growth is also strongly influenced by the air temperature, which affects both the rate of dry matter production, and the partitioning of that dry matter into leaf tissue. Pepper growth in the vegetative stage has been found to be greatest at 25–27°C day and 18–20°C night temperature (Dorland and Went, 1947; Bakker and van Uffelen, 1988). Total plant mass and leaf area were optimal at 20–22°C mean temperature, and declined outside this range (Bakker and van Uffelen, 1988) (Fig. 7.1). Day temperatures lower than night temperature, and a low night temperature of 12°C were also detrimental to vegetative growth. Lower growth temperature also reduces future productivity by increasing specific leaf weight, and decreasing the ratio of leaf area to total plant dry weight (leaf area ratio) (Nilwick, 1980a).

Root growth by pepper seedlings has been studied in artificial and several field conditions. If allowed to grow undisturbed, pepper plants will produce a prominent tap root during early seedling growth (Stoffella et al., 1988). In the field, direct-seeded peppers grown in a deep soil develop several prominent roots that may reach a depth of 3 m (Weaver and Bruner, 1927). If the plants are transplanted, root growth is shallower and more branched, with 80% of the active root system found in the upper 75 cm of soil (Hammes and Bartz, 1963). Root growth is proportional to shoot growth in the vegetative period in pepper (Nielsen and Veierskov, 1988), as has been found for many herbaceous plants (Brouwer, 1962).

Root growth and distribution are significantly influenced by soil management, including cultivation, and the extent and distribution of irrigation water. Unlike on tomato, shoot cuttings are difficult to root in pepper (I. Rylski, Bet Dagan, Israel 1995 personal communication).

It is likely that root distribution will also be significantly affected by soil structure and density, as well as by the use of polyethylene mulch, a common practice in pepper production. Experiments with tomato have shown that lateral root growth is greatly stimulated by use of plastic mulch (Wien et al., 1993a). No similar studies have been made on this aspect in pepper.

INDUCTION OF FLOWERING

The production of flower primordia in C. annuum appears to be little influenced by daylength, occurring in the same time on plants grown under photoperiods 7–15 h long (Auchter and Hartley, 1924; Cochran, 1942). At a photoperiod of 24 h, flower initiation was delayed by 5–9 days, so the species could be termed a quantitative short-day plant. In these experiments, node number to the first flowering node was not reported, so it is not clear if flowering delay was due to the displacement of flower primordia to a later node, or to a decrease in growth rate (see Chapter 3).

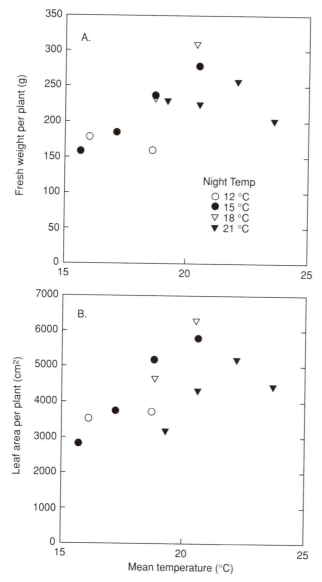

Fig. 7.1. The influence of mean and dark period air temperature on (a) vegetative plant fresh weight and (b) leaf area per plant for pepper plants grown in ambient light conditions in a glasshouse in winter (Bakker and van Uffelen, 1988).

Under most growing conditions, many pepper cultivars produce a terminal flower after forming eight to ten leaves on the main stem (Deli and Tiessen, 1969; Rylski, 1972a). Typically, two or three branches arise at the apical meristem, which again terminate in a flower after producing one node. In the field, the same pattern is then repeated for about five nodes,

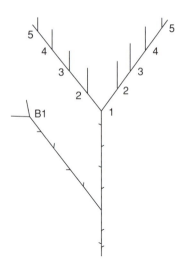

Fig. 7.2. Diagrammatic representation of branching and flowering sequence in pepper. Plants will typically have three or four basal branches; only one is shown (Gaye *et al.*, 1992).

depending on the length of growing season (Somos, 1984) (Fig. 7.2). Under glasshouse conditions, where plants are typically restricted to two stems by pruning, many more nodes may form. At the end of an 11-month growing season, the stem may exceed 3 m in length (M.H. Esmeijer, Naaldwijk, the Netherlands, 1995, personal communication). There is also considerable genetic variation in branching patterns of *C. annum* (Somos, 1984).

The number of nodes formed before flowers are initiated appears to be little influenced by environmental factors. Main stem leaf number showed little variation with changes in photoperiod (Rylski, 1972a), except when short days were combined with warm nights, in which case flower initiation was delayed by one node.

Subjecting pepper seedlings to 10°C night temperatures before flowers have been initiated increased main stem leaf number by one or two leaves (Rylski, 1972a). This contrasts with tomato, in which exposure to 10°C at leaf stage 2 reduces the node to first flower significantly (Wittwer and Teubner, 1956; Calvert, 1957). Covering or removing the plant's cotyledons just after emergence had a similar delaying effect (Table 7.1). These results may indicate that conditions which reduce the assimilatory capacity of the pepper plant can delay the formation of flowers at the plant apex. The assumption that 10°C night temperatures lower plant carbohydrate levels needs to be confirmed.

The stability of node to first flower noted above extends to the influence of mineral nutrient status and plant growth regulators. Even under severe deficiency of the major nutrient elements, node to first flower was not altered (Eguchi *et al.*, 1958), although flowering dates were significantly delayed. Similarly, treatment at the seedling stage with a range of growth

Table 7.1. Influence of cotyledons of 'California Wonder' peppers on leaf size and flowering (Rylski and Halevy, 1972).

Treatment	No. of leaves below 1st flower	Leaf size index*	Days to anthesis[†]
Control	8.1	434	62
Cotyledons removed	9.3	250	73
Cotyledons covered	9.3	276	73
Std. error	0.3	21	1

* Length $\times$ width (cm^2) of all leaves at anthesis.
[†] Days from sowing.

regulating compounds before flower initiation caused no change in main stem leaf number, even though, in the case of GA treatment, flower primordia were aborted (Rylski, 1972b). Only the ethylene-generating compound ethephon brought about a delay of flower initiation of three nodes. The relevance of this finding to the response of pepper to normal environmental factors is not known at present.

The results of experiments to date on flower induction of *C. annuum* thus indicate that the species is day-neutral in the photoperiods and temperatures it would normally encounter in the field. More work is needed to determine if this is also true of other cultivated *Capsicum* species, especially those hitherto largely confined to tropical environments.

FRUIT SET

Flowers of pepper are generally considered to be self-pollinated, although unlike tomato, the anthers and stigma often do not touch each other (McGregor, 1976). On many cultivars, flowers are held horizontally or pendent, so that pollen can fall onto the stigmatic surface. Presumably, in the field insects help to transfer pollen and increase fruit set (Erwin, 1932; Odland and Porter, 1941). Tanksley (1984) found an average of 41% outcrossing among tester lines interplanted in chilli pepper fields. Plant movement by wind probably also contributes to pollination. Under the windstill conditions of glasshouses, the introduction of bees during flowering increased the seed set and fruit size of the fruits produced (de Ruijter *et al.*, 1991; Kristjansson and Rasmussen, 1991; Shipp *et al.*, 1994).

On the day of anthesis, the pepper flower begins to open by dawn, and most new flowers are open by 0800 hours (Erwin, 1932). Anther dehiscence commonly lags behind the flower opening by 1 or 2 h, but in some cultivars has been reported to be delayed by 4 h (Kato, 1989). The significance of such a delay may be questioned, since the stigmata of pepper flowers remained receptive for 3 days at 28/18°C day/night temperatures, and the pollen retained viability for 3 days after anthesis in these studies (Kato, 1989). A similarly long stigma receptivity period was found by Cochran and Dempsey (1966) for pimento pepper.

The evaluation of pollen viability has been hampered by the lack of a reliable *in vitro* germination medium (Song *et al.*, 1976; Kato, 1989). Recently, it was found that a liquid medium containing sucrose, boric acid and calcium chloride improved pollen germination and pollen tube growth over media previously used (Mercado *et al.*, 1994).

There is a considerable lag in seed set after pollen has been placed on the stigma. In detailed anatomical studies, Cochran (1938) found that fertilization occurred 42 h after pollination for plants grown at 27/21°C. Kato (1989) concluded that 36 h were needed for the fertilization process.

Peppers have the capacity to set fruit parthenocarpically, especially under low temperature conditions (12–15°C night temperatures) (Rylski and Spigelman, 1982; Polowick and Sawhney, 1985). Failure of seed set is at least partly due to the formation of abnormal and non-viable pollen, but the mechanism by which the plant is able to retain the fruit in spite of a lack of seed set is not known. The temperature optima for fruit set, and the topic of flower abscission, an important physiological disorder in pepper, is covered in the section on physiological disorders below.

Male sterility is controlled by both nucleic (Shifriss and Rylski, 1972) and cytoplasmic genes (Shifriss and Frankel, 1971). In both types of male sterility, anthers stayed small and shrunken, and were blue–violet in colour, with little or no viable pollen.

FRUIT GROWTH AND MATURATION

The process of fruit growth begins with the formation of the ovary during the early stages of flower differentiation (Cochran, 1938; Munting, 1974). In the period before anthesis of the flower, the basic structure of the ovary is determined, including the number of carpels to be found in the mature fruit. Cell division predominates during this stage, followed by cell enlargement after flowering (Kano *et al.*, 1957; Munting, 1974). Some cell division activity is, however, maintained into later stages of fruit growth in long-fruited pepper types, especially at the basal part of the fruit (Kano *et al.*, 1957; Munting, 1974). Pepper fruit formation differs from that of tomato and squash in that the shape of the ovary at anthesis gives less indication of final fruit shape (Sinnott and Kaiser, 1934; Houghtaling, 1935). From a globular ovary at flowering may form an ovoid or an elongated pepper fruit (Fig. 7.3). Changes in cell shape and the plane and amount of cell division thus influence profoundly final fruit shape.

The temperature at which the plant is growing during this pre-anthesis period can also influence fruit shape. Subjecting pepper seedlings to 35°C from the time the third leaf was longer than 1 cm resulted in a significant increase in fruit locule number (Ali and Kelly, 1993), without increasing fruit size. If pepper plants are grown at low night temperatures (8–10°C) before

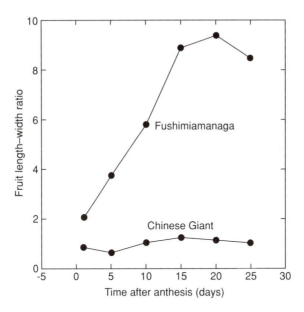

Fig. 7.3. Fruit length:width ratio of developing fruits of 'Fushimiamanaga' and 'Chinese Giant' peppers, an elongated and round-fruited cultivar, respectively (Kano *et al.*, 1957).

flowering, the ovary tends to be larger and broader than that of plants grown at higher (18–20°C) temperatures (Rylski, 1973; Polowick and Sawhney, 1985). Fruits developed from these ovaries also have a decreased length : width ratio, and the style persists and forms a point on the blossom end of the fruit (Rylski, 1973). The increased ovary size of such cool temperature grown plants does not result in bigger fruits at maturity, however, even if normal pollen is used to ensure seed set. In fact, since normal pollen development is impaired at low temperatures, small, seedless fruit are often the result of growing pepper plants in cool conditions (Polowick and Sawhney, 1985).

Conditions after anthesis also play a significant role in pepper fruit development. An important factor is the degree to which seed set has been successful. Rylski (1973) found a direct linear relation between the number of seeds per fruit and final fruit size (Fig. 7.4). Conditions which negatively influence overall plant growth can also reduce final fruit size (see next section). As fruit number per plant increases, the size of individual fruits tends to be smaller. Conversely, restricting fruit set allows the plant to develop the retained fruit to a larger size (Rylski and Spigelman, 1986). Unfortunately, the selection of pepper genotypes with large fruits has probably resulted in cultivars that are very susceptible to flower and flower bud abscission (Wien *et al.*, 1993b).

The changes in carbohydrate levels in pepper fruit during development have been studied by several workers. During the initial growth phase after

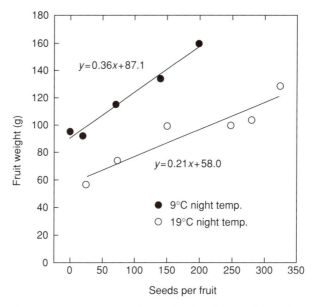

Fig. 7.4. The relation of seed number per fruit and fruit weight for 'California Wonder' pepper grown in a glasshouse at 9 or 19°C night temperature in 1966 (Rylski, 1973).

anthesis, rapid fruit growth coincided with accumulation of glucose and fructose, and lower levels of sucrose and starch (Nielsen *et al.*, 1991; Hubbard and Pharr, 1992) (Fig. 7.5). As growth rate of the fruit slowed, sucrose and starch accumulated. During fruit maturation, there was a further steep rise in reducing sugar content, and reduction in starch and sucrose levels. Fruit growth rate, and the level of hexose sugars was closely related to the activity of acid invertase in the fruit (Nielsen *et al.*, 1991). This indicates that fruit growth may be regulated by the rate at which the imported sucrose is converted to hexose sugars in the fruit. At later stages of fruit development, cleavage of sucrose by the sucrose synthase enzyme plays an important role (Nielsen *et al.*, 1991; Hubbard and Pharr, 1992).

Peppers are important for the production of spice and of red food colouring as well as the fresh vegetable, directly consumed. The pungent compounds in hot peppers are primarily the flavourless, fat-soluble capsaicinoids, composed chiefly of capsaicin and dihydrocapsaicin (Govindarajan, 1985). These compounds are found in the cross wall and placental region of the fruit, and not in the seed or pericarp tissues (Huffman *et al.*, 1978). In pungent pepper types, the epidermal cells of these tissues synthesize the capsaicinoids in the endoplasmatic reticulum, and secrete the materials into a subcuticular cavity (Zamski *et al.*, 1987). At maturity, pungent fruit may be distinguished from the sweet types by the presence of blistered cells on the placenta, whereas the placenta of non-pungent fruits appears smooth. Pungency begins to build up in pepper fruits about 10 days after anthesis,

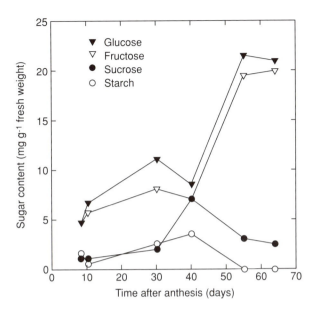

Fig. 7.5. Change in soluble carbohydrate and starch content of sweet pepper fruit (cv. Trophy) with development (Nielsen *et al.*, 1991).

but generally does not reach high values until the fruit is close to maturity (Ohta, 1962; Govindarajan, 1985). There is little definitive information on the influence of environmental factors on the level of capsaicinoids in pepper fruit, except that warm, dry growing conditions favor pungency (Cotter, 1980).

The red pigment in mature pepper fruit that is the source of food colouring is composed of a group of related carotenoids, principally capsanthin, capsorubin and cryptoxanthin (Govindarajan, 1985). These compounds mask the presence of the yellow pigments β-carotene and violaxanthin, which become prominent in yellow-fruited bell peppers at maturity. The colour compounds are found in the outer pericarp layer of the fruit, and their formation is favoured by moderate growing temperatures (Cotter, 1980).

The formation of red colour is one of a number of changes of the maturing fruit of pepper. Pepper is generally not considered to be a climacteric fruit, because it is thought to lack the typical increase in carbon dioxide and ethylene production as it ripens (Saltveit, 1977). More recent research with hot peppers indicates, however, that at least some pepper cultivars do produce low levels of ethylene as the fruit are turning colour (Gross *et al.*, 1986). Many types of peppers can be induced to turn colour more rapidly by treatment of the plants with ethephon once colour development has started (Cantliffe and Goodwin, 1975; Worku *et al.*, 1975), indicating that the ripening processes are not insensitive to ethylene.

FACTORS DETERMINING PRODUCTIVITY

REPRODUCTIVE GROWTH AND PARTITIONING

As with many plants producing reproductive structures, the most actively growing organ on the pepper plant after flowering is the fruit (Hall, 1977; Beese *et al.* 1982) (Fig. 7.6). The characteristic that sets pepper apart from many other fruiting vegetables is that leaf photosynthetic activity is maintained even into late phases of fruit growth (Hall and Brady, 1977). In the field, if environmental conditions permit, additional flushes of vegetative and reproductive growth can occur as the first fruits are harvested.

The most obvious sign of assimilate competition among different organs on the pepper plant is the abscission of flowers and small fruits during the most active fruit growth period, resulting in a cycling of flowering and fruit set (Hall, 1977; Clapham and Marsh, 1987). Preventing the growth of reproductive structures by continuous flower removal eliminates the cycling, and leads to a faster growth rate of vegetative plant parts (Hall, 1977; Clapham and Marsh, 1987). In some cultivars, continuous flower removal resulted in accelerated senescence of leaves, a decrease in leaf photosynthetic rate, and a buildup of carbohydrates in stems and leaves (Hall, 1977; Hall and Milthorpe, 1978). The lowered gas exchange rates were correlated with increases in both stomatal and intracellular resistances to CO_2 diffusion as a result of the deflowering treatment (Hall and Brady, 1977). It is possible that the stomatal closure may be due to an increase in leaf

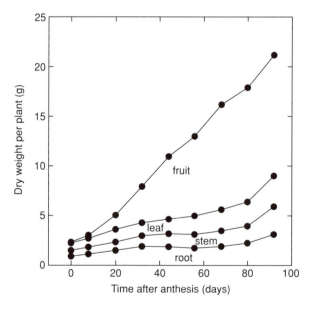

Fig. 7.6. Dry matter partitioning of bell pepper (cv. Market Giant) after flowering for pot-grown plants restricted to one fruit (Hall, 1977).

abscisic acid (ABA) levels upon flower removal, but more work is needed to confirm this mechanism (Kriedemann *et al.*, 1976).

ENVIRONMENTAL CONDITIONS FOR YIELD PRODUCTION

Temperature

The productivity of bell pepper is constrained by the adverse effects of high temperatures on fruit set, and the detrimental influence of low temperatures on fruit shape (Rylski and Spigelman, 1982). For high yields of good quality fruit, Bakker and van Uffelen (1988) found that mean air temperatures of 21–23°C were optimal during vegetative growth, followed by 21°C during the fruit growth period. They also recommended that the amplitude between day and night temperatures be maintained at 7–9°C, for the low light conditions of the North European winter under which the experiment was conducted. As mean temperatures varied from the optimum, they found that the day/night temperature range needed to be increased to maintain high yields. Blondon (1978), working under artificial lights, and Dorland and Went (1947), experimenting in presumably higher light conditions of southern California, reached very similar conclusions. This would imply that these optimum temperatures were valid for a range of light conditions, a conclusion that needs to be checked experimentally.

In current glasshouse production practice, the first-formed flowers are removed to allow the plant to develop an adequate vegetative frame before the start of reproductive growth (Bakker and van Uffelen, 1988). In such systems, it may be more efficient to elevate the temperature during the vegetative period to hasten growth and reduce labour needs for flower removal (Demers *et al.*, 1991).

In glasshouse production systems it is also possible to modulate root temperatures if that is advantageous for growth and yield. Studies have shown that for plants grown at an air temperature of 23/19°C (day/night), fruit weight was maximal at 24 and 30°C root zone temperature (Gosselin and Trudel, 1986). At 30°C root zone temperature, fruit set and early yield were decreased, but the plants made up for the delay with higher later yields. The physiological mechanism of the delayed fruit set with high root temperature was not explained.

Our knowledge of temperature optima for growth of other *Capsicum* species, or the smaller-fruited types within *C. annuum* is very limited. One would expect those types that evolved or were selected in higher temperature environments, such as *C. frutescens*, *C. chinense* and *C. baccatum* to also have a higher optimum temperature, but this requires experimental verification.

Irradiance

The influence of light on the productivity of pepper differs depending on whether one is considering a glasshouse-grown crop during the low light

Table 7.2. Effect of supplementary light treatments on the early and total marketable yield of glasshouse peppers grown in Quebec during winter (Demers *et al.*, 1991).

Lighting treatment	Fruit yield (kg m^{-2})		Fruits m^{-2}	Fruit size (g)
	Early	Total marketable		
Control	0.18 a	0.39 a	4.0 a	117 a
0.9 mJ m^{-2} day^{-1}	0.75 b	1.52 b	14.5 b	127 b
1.4 mJ m^{-2} day^{-1}	1.24 c	2.45 c	22.6 c	128 c
1.8 mJ m^{-2} day^{-1}	1.55 d	2.99 d	27.6 d	128 c

Means in a column followed by different letters are statistically different at the 5% level.

conditions of the temperate winter, or a field of peppers growing in full sun outdoors. Demers *et al.* (1991) demonstrated that supplementing the reduced irradiance of Quebec's winter can significantly increase yield and fruit size (Table 7.2). The high cost of energy makes supplemental lighting prohibitively expensive in many countries, so growers avoid the adverse effects of low light conditions on fruit set and productivity by using that dark time of the year for vegetative growth (Bakker, 1989a).

In the field, we can be faced with the other extreme. Rylski and Spigelman (1986) reported that irradiance levels averaged 28 MJ m^{-2} day^{-1} in the Negev Desert of Israel during the summer growing season, compared to 0.7–1.54 MJ m^{-2} day^{-1} typical of glasshouses growing peppers in Holland in winter (Bakker and van Uffelen, 1988). Under those high light conditions, total fruit yields were reduced 19% compared to plants lightly shaded from transplanting, while marketable yields were decreased by 50% (Table 7.3). The shading treatments sharply reduced sunscald injury of the fruit (see below), and increased fruit size, perhaps as a result of an increase in seed number per fruit. The decrease in overall productivity of the plants grown in full sun may be due to lower leaf area measured for the main fruit-bearing nodes, resulting from a slight drought stress. The foregoing results imply that yield production of pepper will increase with increasing irradiance, only as long as temperatures remain in the optimal range, and plants

Table 7.3. Effect of plant shading level on yield and fruit quality of bell peppers grown in a high light desert climate in Israel. Reprinted from Rylski and Spigelman (1986), by kind permission from Elsevier Science.

Shading (%)	Fruit no. plant^{-1}	Fruit size (g)	Market yield (kg m^{-2})	Sunscald fruit (%)
0	8.1 a	86 c	2.0 c	36 a
12	8.5 a	97 b	3.1 b	20 b
26	7.7 a	108 a	3.7 a	8 c
47	5.4 b	111 a	2.9 b	2 d

Means in a column followed by different letters are statistically different at the 5% level.

have access to sufficient water. Under summer desert conditions, both temperature and water could have been limiting.

Carbon dioxide

A less expensive means of increasing the assimilation rate of glasshouse-grown peppers is to augment the atmospheric CO_2 content. As with many crops, peppers respond to this practice, by increasing the proportion of fruit set (Nederhoff and van Uffelen, 1988) and increasing early production (Daunicht and Lenz, 1973). With regard to fruit yield, raising CO_2 levels by as little as 200 ppm was sufficient to increase the number of fruit harvested by 60% (Nederhoff and van Uffelen, 1988). Carbon dioxide enrichment has become a standard part of glasshouse pepper production in Holland (van Berkel, 1986).

Water

Under field conditions, insufficient moisture supply can adversely affect growth and yield of pepper. Even small deficits, maintained for the entire season, can have large effects (Beese et al., 1982). Early drought stress can restrict vegetative growth and development of leaf area, which then affects the amount of the total plant production that ends up as yield. In an irrigation experiment with chilli peppers in New Mexico, for instance, the application of 20% less water than that given control plants reduced leaf area by about 8%, while total dry matter and fruit yields were reduced by about 10%. Measurements of the moisture status of the plants during the season showed little difference in leaf water potential and stomatal resistance between moisture treatments, indicating that monitoring these characteristics would help little in predicting when more irrigation was needed (Horton et al., 1982).

Shorter, more severe water stress had more drastic effects on plant growth and assimilation rate (Alvino et al., 1990; Katerji et al., 1993). If there was sufficient time for recovery, the plants could resume the production of leaf area and yield, without detrimental effects on the latter (Alvino et al., 1990). Moisture stress imposed at some critical growth stages may have long-term effects from which recovery of full yield may not be possible. Pellitero et al. (1993) found in a field study that imposing stress during fruit set had detrimental effects similar to stressing the plants all season in a field experiment. Katerji et al. (1993) confirmed that pepper plants are particularly sensitive to drought during the fruit set stage, by imposing water stress of equal severity (−1.6 MPa predawn leaf water potential) on pepper plants at several stages of plant development.

The sensitivity of pepper to moisture stress, and the unpredictability of rainfall in many production areas has made the use of irrigation a standard production aid. In irrigation studies, O'Sullivan (1979) found that supplemental watering was needed for bell peppers grown in Southern Ontario in

2 years out of 5. Both O'Sullivan (1979) and Hegde (1987) recommended that water be given to field-grown peppers when soil moisture levels had been depleted to about the 50% available level. In many production areas, irrigation is the sole source of crop water (California's Salinas Valley, Sinaloa area of Mexico).

Variation of relative humidity in the glasshouse production environment can influence pepper fruit size, but appears to have no effect on yield (Bakker, 1989a). High relative humidity, especially at night, led to a 13% increase in fruit size (Bakker, 1989a). Baer and Smeets (1978) showed that the large fruits developed under 95% relative humidity contained significantly more seeds than fruits on plants grown at lower humidities.

ONTOGENY OF YIELD

In trying to understand how yield is developed in pepper plants, it is necessary to consider the field and glasshouse production systems separately, for they are quite different in the early manipulation of the plants, and in the duration of growth.

Field-grown plants

Peppers growing in the field generally are not deflorated or pruned and thus tend to set the first flowers, if environmental conditions permit. Yield on field-grown plants of bell pepper is therefore comprised primarily of fruits on the main stem branches arising beyond the first inflorescence, and the first flowering nodes of basal branches (Gaye et al., 1992) (Fig. 7.7). When fruits on those nodes are growing actively, fruit set at later-formed nodes is inhibited, so that a group of fruits of similar age develop and make up the main yield of the plant. If the season is long enough, the next flush of flowers may set when the first fruits are maturing, forming additional yield. If fruit set during the first flush of bloom is reduced due to adverse climatic factors, the cyclic pattern of yield production is not as pronounced.

Little information is known about the partitioning of fruit yield between main stem nodes and basal branches. Cooksey et al. (1994) found that transplanted chilli pepper plants developed more branches than direct-seeded plants. Stoffela and Bryan (1988) varied in-row spacing from 13 to 50 cm, and left one to three plants per stand with direct-seeded peppers, but these treatments did not alter branch number per plant. Neither of these groups determined the yield of main stem and basal branch nodes separately. Cooksey et al. (1994) did note, however, that the harvestable fruits on transplants were spread over a larger vertical plane than on direct-seeded plants, and judged them to be more difficult to harvest mechanically.

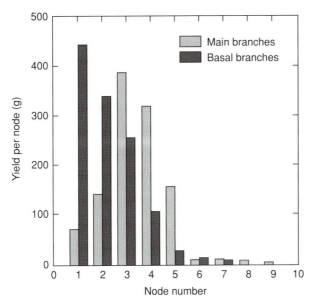

Fig. 7.7. Yield distribution on individual nodes of main and basal branches of bell pepper (cv. Ace). Yield per node of main branches may include fruits on more than one branch (see Fig. 7.2) (Gaye *et al.*, 1992).

Glasshouse pepper

In commercial glasshouse production, pepper plants are typically transplanted into the production area, and pruned to two stems, with all basal and all additional upper branches being removed as they arise (Bakker and van Uffelen, 1988). In addition, fruit set is prevented on the first ten flower nodes to allow an adequate vegetative frame to be built up. The fruiting stems are supported vertically. In comparison to the field-grown plants, glasshouse peppers start producing fruits later, but the production season can extend over a much longer period. Whereas bell peppers in field plantings would typically be harvested over a 2–3-month period, the glasshouse production season can extend for 8 months. Total marketable yields can therefore be considerably higher in the latter case. Even in a shorter harvest period of 4 months, Bakker and van Uffelen (1988) reported marketable yields of 10 kg m^{-2}, compared to typical yields from field experiments of 2–6 kg m^{-2} (Hochmuth *et al.*, 1987; Locascio and Fiskell, 1976; Stoffella and Bryan, 1988).

Chilli or paprika peppers grown for chilli powder used as food colouring were estimated to be grown on 15% of the total pepper hectarage in the USA in 1976 (Andrews, 1984). This crop is generally direct-seeded and harvested when most of the fruits are dry, late in the season. Yields of dry fruits range from 0.1 to 0.7 kg m^{-2} (Beese *et al.*, 1982; Cooksey *et al.*, 1994). This crop often receives a lower intensity of management with regard to inputs of irrigation, fertility, than the bell pepper crop.

Seed yield

When open-pollinated plants are harvested for seed, seed yields tend to be proportional to fruit yield (Osman and George, 1984). In a nitrogen rate experiment with 'Anaheim Chile', Payero *et al.* (1990) found, however, that seed yields were maximal at the lowest N rate used (170 kg ha^{-1}), while fruit yields were not changed as nitrogen rates increased to 310 kg ha^{-1}. The occurrence of cool temperatures during fruit set could also reduce the number of seeds per fruit, and thus lower seed yield (Rylski, 1973).

PHYSIOLOGICAL DISORDERS

FLOWER BUD AND FLOWER ABSCISSION

Loss of flower buds and flowers is a problem occurring primarily in the production of large-fruited bell peppers, which are grown in temperate and subtropical environments. The principal causal factors are high temperature, low light levels, drought stress, the presence of rapidly growing fruit on the plant and biotic agents like certain virus diseases, and insect pests (Wien *et al.*, 1989a). Depending on when the stress occurs, the disorder can result in the delay of anthesis, the prolongation of the flowering period without fruit set, or the early termination of fruit setting. In regions with short growing periods, such as upstate New York, the disorder can result in total yield loss by delaying fruit development into the cool period when fruit grow too slowly to reach market maturity (H.C. Wien, personal observation, Elba, N.Y., August, 1988).

Under conditions of severe stress, the plant abscises open flowers as well as buds of a range of sizes. The reproductive structure loss can be so total, that growers describe the plants as having 'gone vegetative'. In that situation, several weeks may be needed for open flowers to again be developed. More detail on the abiotic causes, the mechanism of the abscission process, and control methods are given below.

Causal factors

The most common cause of flower and flower bud abscission in pepper is high air temperature. Cochran (1936) found that fruit set was reduced at 27/21°C, and that no flowers set if plants were grown at 38/32°C in a glasshouse (Fig. 7.8). High night temperatures are more detrimental to fruit set than high day temperatures (Rylski and Spigelman, 1982; Aloni *et al.*, 1991). In the field, however, even 32/15°C (day/night) sufficed for complete flower and bud abscission of many cultivars (Wien, 1990). When high

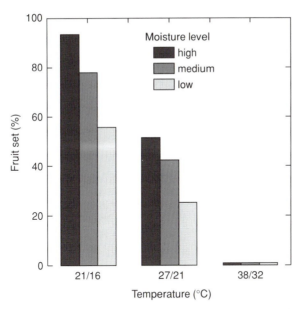

Fig. 7.8. Influence of air temperature and soil moisture on percentage of fruit set of 'World Beater' pepper grown in pots in glasshouse compartments. Fruits were removed after setting. Data are averages of 2 years' experiments, 1932–34 (Cochran, 1936).

temperature is combined with moisture stress, abscission is further increased, although moisture stress at moderate temperatures is generally not sufficient for complete reproductive structure loss (Bereny, 1970; Cochran, 1936) (Fig. 7.8). As stated in the section on fruit set above, low temperatures generally lead to improved fruit set and the production of small, seedless fruit (Rylski and Spigelman, 1982; Polowick and Sawhney, 1985). The production of bell pepper under low light conditions frequently found in glasshouses in the winter often results in poor fruit set (Bakker, 1989b,c). Experimentally, reducing incident light during high light periods with shade cloth can significantly increase reproductive structure abscission (Park and Jeong, 1976; Wien, 1990). For instance, with application of 80% shade for 10 days in the field, abscission increased from 23% in unshaded plants to 60% (Wien, 1990).

Excess nitrogen fertilizer is frequently blamed by growers for poor fruit set under field conditions, yet evidence in the literature for this is difficult to find. Fruit set is generally not directly measured in nitrogen fertilizer trials with peppers, so one must make the risky assumption that yield patterns are due to differences in flower and fruit retention. Even if one accepts that risk, one finds that the most common response to increasing nitrogen rates in field experiments is an increase in yield with higher N rates up to a plateau, and perhaps a slight decline at highest rates (Locascio and Fiskell, 1976; Hochmuth *et al.*, 1987; Crespo-Ruiz *et al.*, 1988; Hartz *et al.*, 1993) (Fig.

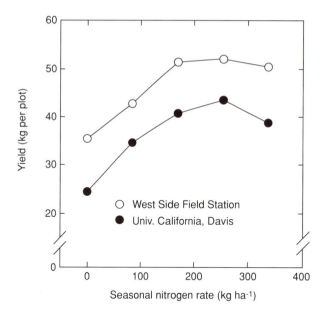

Fig. 7.9. Yield response of bell pepper to nitrogen applied through the trickle irrigation system in two locations in California (Hartz *et al.*, 1993).

7.9). The yield decline at high nitrogen levels might be caused by salt injury of overall plant growth, rather than a direct effect on fruit set (Locascio and Fiskell, 1976). It is possible that the lush, leafy growth that is associated with the effects of excess nitrogen fertilization is the result rather than the cause of poor fruit set. Once flowers and flower buds have abscised due to high temperatures or the other causes listed above, the plants would resume stem and leaf growth, and thus would appear to have been fertilized too heavily.

Nitrogen management under low light conditions of winter glasshouses is likely to be more difficult than in the high light conditions of the field. Although direct experimental evidence for the mechanism of fruit set reduction is still lacking, Knavel (1977) did show a reduced yield of glasshouse-grown peppers as soil nitrogen content was raised from 155 to 265 kg ha^{-1}.

A number of biotic causal factors for pepper reproductive structure abscission have also been identified, including virus diseases, fungal pathogens and insect pests such as leafhoppers (Wien *et al.*, 1989a).

Presence of rapidly growing fruit can reduce the fruit set of later-formed flowers and also lead to flower bud abscission. Flowering and fruit set frequently resumes once the fruits reach mature size, resulting in cycles of fruit setting and abscission (Wien *et al.* 1989a).

Mechanism of abscission

The physiological mechanism of pepper reproductive structure abscission resembles the way in which leaves are thought to abscise (Beyer and Morgan, 1971; Beyer, 1975). While the leaf, or in this case, the flower or flower bud is actively growing, auxin is translocated down the petiole or pedicel, and prevents the formation of an abscission layer at its base. Under stress conditions, ethylene is generated, which both reduces the polar auxin transport down the pedicel, and causes the formation of the abscission layer. There is good evidence that this model is operating in the case of heat stress abscission of pepper flowers (Wien *et al.*, 1993b). Pepper plants exposed to high temperatures showed inhibition of pedicel auxin translocation rates. Abscission of disbudded pepper flower pedicels could be prevented by application of synthetic auxin to the cut stump (Fig. 7.10) whereas controls abscised within 48 h (Wien *et al.*, 1989b). Ethylene and its precursor, aminocyclopropane carboxylic acid (ACC), have been shown to increase in pepper reproductive tissue prior to abscission, when stressed by high temperature or low light (Wien *et al.*, 1989b, 1993b; Aloni *et al.*, 1994). Ethylene and ethylene-generating chemicals such as ethephon are effective abscission-causing agents in pepper (Beaudry and Kays, 1988; Tripp and Wien, 1989).

The way in which high temperature and low light conditions initiate the hormonal changes described above is not entirely clear, but most likely

Fig. 7.10. Prevention of pepper flower pedicel abscission by infusion of synthetic auxin solution into the pedicel by means of an absorbent wick dipped into the hormone solution (Wien *et al.*, 1989b).

involves a reduction of assimilate levels in the reproductive structures (Wien *et al.*, 1993b). Aloni *et al.* (1991) found that heat stress reduced the translocation of assimilates to the reproductive structures, and the conversion of the translocated sucrose to reducing sugars in the flower buds. Acid invertase activity in the buds, but not the young leaves, was inhibited by heat stress. In buds of low light-stressed plants there was a similar lack of reducing sugars, but there was no evidence that invertase activity had been adversely affected (Wien *et al.*, 1989b; Turner and Wien, 1994b). In that case it appeared that mature leaves retained assimilates instead of translocating them to the reproductive structures (Turner and Wien, 1994a,b). It is not clear at present how the low level of assimilates in flowers and flower buds leads to a reduction of auxin and elevation in ethylene levels in these structures.

Control methods

The principal method by which reproductive structure abscission in pepper can be alleviated is through the use of less stress-susceptible cultivars (Wien *et al.*, 1989). There is considerable genetic diversity with regard to reproductive structure abscission even among the bell peppers (Tripp and Wien, 1989; Wien *et al.*, 1993b) (Fig. 7.11). Unfortunately, the cultivars most susceptible to abscission within the bell pepper type also tend to be those with desirable large fruit type. It should be possible, however, to select lines

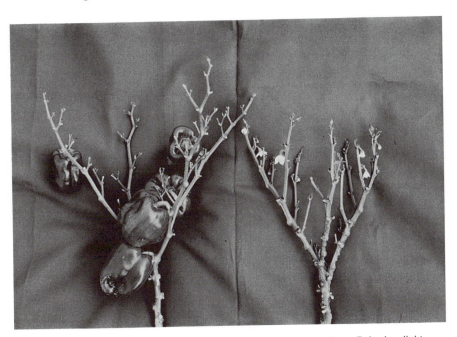

Fig. 7.11. Fruit set of 'Ace' (left) and 'Camelot' (right) bell peppers after a 7-day low light stress period applied at anthesis of the first flowers in the field. Picture taken 45 days after end of shade period (C. Wien, unpublished data).

which resist abscission of reproductive structures past the fruit set stage but then develop only a limited number of fruits to maturity. There are recent indications that some cultivars developed for glasshouse culture may have these attributes (Aloni *et al.*, 1994).

Selection for stress resistance is most directly done by exposing plants to the stress conditions, be they high temperature or low light. If such environments are difficult to create on a large enough scale to permit the screening of breeding materials, it may be possible to use techniques that substitute essential elements of the stress environment, or that elicit the same response as exposure to the stress. For instance, cultivar screening for differences in sensitivity to ethylene use ethephon applications to explants or whole plants, and have shown good correlation with cultivar differences in stress sensitivity (Tripp and Wien, 1989; Aloni *et al.*, 1994). It may also be possible to select lines less inhibited in polar auxin transport by flower pedicels when subjected to stress (Wien *et al.*, 1993b). Cultivar differences in assimilate partitioning to reproductive structures rather than mature leaves have also been demonstrated (Turner and Wien, 1994a).

Other ways in which abscission can be reduced include the moderation of temperatures in the field through frequent sprinkler irrigation, and the mitigation of low light stress in glasshouses by elevation of the CO_2 content (Nederhoff and van Uffelen, 1988). Application of synthetic auxins has not been effective in reducing low light-induced abscission, but application of the ethylene action inhibitor silver thiosulphate has reduced abscission (Wien and Zhang, 1991). The presence of the heavy metal silver limits use of this compound to non-food purposes.

SUNSCALD

When pepper fruit are exposed to high intensity sunlight after they have reached the mature green stage of growth, they are susceptible to tissue damage and bleaching called sunscald. The injury is caused by a combination of heat and light (Rabinowitch *et al.*, 1986). If tissue temperature rises to 50°C, only a 10-min exposure to intense sunlight is sufficient to cause damage (Barber and Sharpe, 1971). The threshold temperature in pepper is 38–40°C (Rabinowitch *et al.*, 1986), but injury at those temperatures requires at least 12 h exposure. The disorder is prevalent in high light environments, and can be a serious detriment to production of bell peppers. For instance, Rylski and Spigelman (1986) reported that unshaded pepper plants grown during the summer in the Negev Desert showed a 36% incidence of sunscald (see Table 7.3).

Damage is caused by a combination of direct injury of the tissue due to the heating effect, and the generation of superoxide anion radicals through the action of light on chlorophyll at high temperatures (Rabinowitch and Sklan, 1981). Heating of the fruit in the dark causes the pericarp of the fruit to turn flaccid and brown, but not become bleached.

Pepper fruits are most susceptible to this disorder at the mature green stage, and when turning colour from green to red. The immature green fruit are less subject to the disorder, and the fully ripe red fruit are not susceptible (Rabinowitch and Sklan, 1981). Chlorophyll must be present in the pericarp for sunscald to occur. Presence of the enzyme superoxide dismutase in the chloroplasts of the fruit can lessen or prevent injury by catalyzing the formation of hydrogen peroxide and oxygen from the superoxide radicals. The increased susceptibility of the mature green fruit to sunscald was correlated with a lower superoxide dismutase activity at this stage (Rabinowitch *et al.*, 1982).

Scientists have also found that pepper fruit could be conditioned to tolerate sun exposure without injury by giving the fruit a heat treatment in the dark. Rabinowitch *et al.* (1986) found that heating peppers for 6 h at 40°C allowed them to tolerate sun exposure with much reduced injury. The potentiating effect of the heat treatment lasts from about 15–36 h after the fruit have been removed from the high temperatures. The mechanism of the protective effect is not entirely clear, but in tomato, Rabinowitch *et al.* (1982) showed that heat treatment significantly increased superoxide dismutase activity.

Pepper breeders have long recognized the need to select plants that are less subject to sunscald, and pay particular attention to the production of adequate leaf area, especially for large-fruited types. These cultivars unfortunately tend to have greater susceptibility to flower and flower bud abscission under stress (Turner and Wien, 1994a).

Another approach is to develop cultivars with yellow instead of green immature fruit. These fruit do not become as hot when exposed to the sun, and their high content of carotenoids helps protect the pericarp from the injurious effects of photo-oxidation (Barber and Sharpe, 1971; Rabinowitch *et al.*, 1982).

Cultural practice management measures that can be used to reduce sunscald damage are to erect shade canopies over the fields that reduce light by 26–36% (Rylski and Spigelman, 1986). In addition, since fruit often become exposed to damaging direct sun after heavy winds have lodged the plants in storms, providing support may reduce the problem.

BLOSSOM-END ROT

Blossom-end rot of pepper appears as sunken dark circular lesions on mature green and riper fruit. The tissue breakdown may occur not only on the stylar end of the fruit, as the name implies, but also on the side. It occurs most frequently on large-fruited cultivars, but has been noted on the smaller-fruited pimentos as well (Hamilton and Ogle, 1962). The disorder appears to be quite similar in cause, occurrence and control methods to blossom-end rot in tomato. Since it has been studied in much more detail in that crop, the reader is referred to Chapter 6 for more information on the physiology of the disorder.

Nutritional studies with pepper have confirmed that blossom-end rot is caused by a localized deficiency of calcium in the fruit, brought about most directly by a lack of calcium in the root zone (Miller, 1961). Given the immobile nature of calcium in the plant, and the difficulty of translocating this element from one part of the plant to the other, blossom-end rot can frequently occur even when there are adequate supplies of calcium in the soil (Bangerth, 1979). Since calcium moves in the xylem, it is translocated preferentially to organs with highest transpiration rates. Thus even in normal situations, fruit calcium levels in pepper are very low, typically in concentrations of 0.2–0.3% (Marti and Mills, 1991b). There is also differential distribution of calcium in the fruit, with values of 0.2% at the stem end, and a range of 0.04–0.07% being found on the flanks and at the blossom end in one study (Marti and Mills, 1991b). This coincided with the symptom location described above. Of the total amount of calcium accumulated by the plant at maturity, only about 6% is found in the fruit (Miller *et al.*, 1979).

Several approaches have been tried to remedy the blossom-end rot problem in pepper. The most important step is to ensure that calcium levels in the soil are adequate, and the levels of other nutrients that might compete with this element for uptake are not excessively high. For instance, it was demonstrated with sand culture that high magnesium levels in the nutrient solution could induce blossom-end rot (Hamilton and Ogle, 1962). Similarly, high nitrogen levels, particularly if supplied by a large proportion of ammonium nitrogen, also increased incidence of the disorder (Miller, 1961; O'Sullivan, 1979; Marti and Mills, 1991a).

Some investigators have attempted to increase fruit calcium content by increasing transpirational flow to the tops or to the fruit, or by reducing transpiration of the foliage by application of antitranspirants. The former method did increase fruit calcium levels, but is difficult to do on a practical scale (Mix and Marschner, 1976). Foliar antitranspirants raised fruit calcium levels and decreased blossom-end rot in only some experiments, but consistently decreased yields (Schon, 1993). As with tomato, providing the plants with even water supply and avoiding moisture stress can greatly minimize the incidence of blossom-end rot (Pill and Lambeth, 1980).

The detailed physiology of this disorder in peppers is largely unexplored, but given the recent studies by Ho and co-workers on blossom-end rot in tomato (e.g. Ho *et al.*, 1993), it should be possible to make rapid progress in our understanding of this malady. Work is needed to determine if root pressure plays any role in calcium distribution in the pepper plant, and if lack of vascular development in the fruit could be inhibiting calcium movement into the distal end, as appears to be the case in tomato (Ho *et al.*, 1993). A better understanding of the physiological mechanism of blossom-end rot could be helpful to devise selection criteria that would allow breeders to develop cultivars with reduced susceptibility to the disorder.

OTHER FRUIT DISORDERS

Pepper fruits may show a number of other disorders, the most common of which are abnormalities in fruit shape or fruit colour, or the formation of cracks in the fruit surface.

Abnormal fruit shape

Fruit shape can be influenced by temperatures during ovary formation (see section on fruit growth and maturation), or by the absence of seeds. The latter condition most commonly occurs if seed set has been inhibited by low temperatures (12–15°C night) (Rylski and Spigelman, 1982; Polowick and Sawhney, 1985). Such seedless fruits are frequently smaller, with thinner than normal pericarps, and of flat, irregular shape. They are often retained on the plant until fruit maturity, and although higher in assimilates, are seldom of marketable shape.

Colour spotting

Several fruit discolorations have been described on bell pepper fruits, of largely unknown cause. Dark, circular spots of 2–7 mm diameter occurring on mature green fruits have been documented on peppers grown in Texas (Villalon, 1975), Queensland, Australia (Hibberd, 1981) and Florida (Ozaki and Subramanya, 1985). Villalon failed to isolate any pathogens from the lesions, which extended into the pericarp tissue. The disorder varied considerably among cultivars, with 'Yolo Wonder' being found susceptible in both the Queensland and Florida studies, and 'Early Calwonder' and 'Sheba', 'Early Bountiful' and 'Superset' showing few symptoms in the Florida and Queensland studies, respectively (Hibberd, 1981; Ozaki and Subramanya, 1985). Incidence of black spot was more prevalent on soils high in calcium in Florida, and the calcium content of the affected areas of the fruit was reportedly higher in the Australian study.

It is not clear whether the black spot disorder is a less severe form of the 'colour spot' disorder described by Aloni *et al.* (1994). This discoloration consists of large, irregular yellow spots, occurring during warm conditions on plants grown in high nitrogen fertility conditions and low light. Examination of the affected tissue revealed high concentrations of calcium oxalate crystals. There were considerable cultivar differences, with 'Maor' being more susceptible than 'Lady Bell'.

More work is needed to elucidate the influence of soil calcium level on occurrence of both these disorders. So far, their occurrence has been too sporadic to allow detailed field investigations. It should be possible, however, to vary the Ca concentration of pepper plants grown in nutrient solutions, and check for fruit symptoms. Miller (1961) conducted such experiments, but did not retain the plants until fruits had reached the susceptible mature green stage.

Fruit cracking

The appearance of cracks on the surface of bell pepper fruits is associated with production of peppers at high night relative humidity and low night temperature (Rylski *et al.*, 1994). Cultivar differences in susceptibility were partly related to pericarp thickness. Somos (1984) found that the disorder was aggravated by uneven watering. In the Jalapeno pepper, the shallow surface cracks are a characteristic of the normal mature fruit (Andrews, 1984).

CONCLUDING REMARKS

This brief look at the physiology of the pepper plant reveals it to be a rather demanding crop. During the long germination and emergence period, a well-watered soil at moderate temperatures is needed to continue plant development. The slow leaf area development rate makes the plant susceptible to competition by other plants. Again, temperatures below 20°C slow down development even further, and reduce leaf area ratio. Although the induction of flowers seems to be relatively unaffected by environmental factors, many cultivars among particularly the bell peppers are very sensitive to the loss of reproductive structures once these have been initiated. On the positive side, once the fruits are growing actively, the plants very efficiently partition assimilates to both the fruits and to maintenance of the vegetative organs. The plants typically retain much of the leaf area in the late reproductive period, and if weather conditions permit, will continue to produce additional flushes of reproductive organs.

In comparison to many other vegetable crops, our depth of knowledge of the physiological processes in pepper is very limited, particularly for the non-bell types. The increasing popularity of the crop, and the fact that it can be easily grown in pots in glasshouses and artificial environments should contribute to a rapid increase in research activity with this fascinating plant.

ACKNOWLEDGEMENTS

The helpful suggestions of Dr Irene Rylski and Ms M.H. Esmeijer are gratefully acknowledged.

REFERENCES

Ali, A.M. and Kelly, W.C. (1993) Effect of pre-anthesis temperature on the size and shape of sweet pepper (*Capsicum annuum* L.) fruit. *Scientia Horticulturae* 54, 97–105.
Aloni, B., Karni, L., Rylski, I. and Zaidman, Z. (1994) The effect of nitrogen fertilization and shading on the incidence of 'colour spots' in sweet pepper (*Capsicum annuum*) fruit. *Journal of Horticultural Science* 69, 767–773.

Aloni, B., Karni, L., Zaidman, Z., Riov, Y., Huberman, M. and Goren, R. (1994) The susceptibility of pepper (*Capsicum annuum*) to heat induced flower abscission: possible involvement of ethylene. *Journal of Horticultural Science* 69, 923–928.

Aloni, B., Pashkar, T. and Karni, L. (1991) Partitioning of [14C]sucrose and acid invertase activity in reproductive organs of pepper plants in relation to their abscission under heat stress. *Annals of Botany* 67, 371–377.

Alvino, A., Centritto, M., Di Lorenzi, F. and Tedeschi, P. (1990) Gas exchange of pepper grown in lysimeter under different irrigation regimes. *Acta Horticulturae* 278, 137–147.

Andrews, J. (1984) *Peppers, the Domesticated Capsicums.* University of Texas Press, Austin.

Auchter, E.C. and Hartley, C.P. (1924) Effect of various lengths of day on development and chemical composition of some horticultural crops. *Proceedings of the American Society of Horticultural Science* 21, 199–214.

Baer, J. and Smeets, L. (1978) Effect of relative humidity on fruit set and seed set in pepper (*Capsicum annuum* L.). *Netherlands Journal of Agricultural Science* 26, 59–63.

Bakker, J.C. (1989a) The effects of day and night humidity on growth and fruit production of sweet pepper (*Capsicum annuum* L.). *Journal of Horticultural Science* 64, 41–46.

Bakker, J.C. (1989b) The effects of temperature on flowering, fruit set and fruit development of glasshouse sweet pepper (*Capsicum annuum* L.). *Journal of Horticultural Science* 64, 313–320.

Bakker, J.C. (1989c) The effects of air humidity on flowering, fruit set, seed set and fruit growth of glasshouse sweet pepper (*Capsicum annuum* L.). *Scientia Horticulturae* 40, 1–8.

Bakker, J.C. and van Uffelen, J.A.M. (1988) The effects of diurnal temperature regimes on growth and yield of sweet pepper. *Netherlands Journal of Agricultural Science* 36, 201–208.

Bangerth, F. (1979) Calcium-related disorders of plants. *Annual Review of Phytopathology* 17, 97–122.

Barber, H.N. and Sharpe, P.J.H. (1971) Genetics and physiology of sunscald of fruits. *Agricultural Meteorology* 8, 175–191.

Beaudry, R.M. and Kays, S.M. (1988) Effect of ethylene source on abscission of pepper plant organs. *HortScience* 23, 742–744.

Beese, F., Horton, R. and Wieringa, P.J. (1982) Growth and yield response of chile pepper to trickle irrigation. *Agronomy Journal* 74, 556–561.

Bereny, M. (1970) Effect of irrigation on flowering and fruit set of red pepper. *Acta Agronomia Academia Scientifica Hungarica* 19, 398–401.

Beyer, E.M. Jr (1975) Abscission: the initial effect of ethylene is in the leaf blade. *Plant Physiology* 55, 322–327.

Beyer, E.M. Jr and Morgan, P.W. (1971) Abscission: the role of ethylene modification of auxin transport. *Plant Physiology* 48, 208–212.

Blondon, F. (1978) Determination des conditions optimales de production et des limites admissibles d'un refroidissement nocturne chez deux varietees extremes de poivron cultivees en conditions artificielles. *Comptes Rendus de L'Academie Agricole de France* 64, 119–208.

Bradford, K.J., Steiner, J.J. and Trawatha, S.E. (1990) Seed priming influence on germination and emergence of pepper seed lots. *Crop Science* 30, 718–721.

Brouwer, R. (1962) Distribution of dry matter in the plant. *Netherlands Journal for Agricultural Science* 10, 361–376.

Bruggink, G.T. and Heuvelink, E. (1987) Influence of light on the growth of young tomato, cucumber and sweet pepper plants in the glasshouse: effects on relative growth rate, net assimilation rate and leaf area ratio. *Scientia Horticulturae* 31, 161–174.

Calvert, A. (1957) Effect of the early environment on development of flowering in the tomato. I. Temperature. *Journal of Horticultural Science* 32, 9–17.

Cantliffe, D.J. and Goodwin, P. (1975) Red color enhancement of pepper fruits by multiple applications of ethephon. *Journal of the American Society of Horticultural Science* 100, 157–161.

Clapham, W.M. and Marsh, H.V. (1987) Relationship of vegetative growth and pepper yield. *Canadian Journal of Plant Science* 67, 521–530.

Cochran, H.L. (1936) Some factors influencing growth and fruit-setting in the pepper (*Capsicum frutescens* L.). *Cornell Agricultural Experiment Station Memoir* 190, 1–39.

Cochran, H.L. (1938) A morphological study of flower and seed development in pepper. *Journal of Agricultural Research* 56, 395–419.

Cochran, H.L. (1942) Influence of photoperiod on the time of flower primordia differentiation in the Perfection pimento (*Capsicum frutescens* L.). *Proceedings of the American Society of Horticultural Science* 40, 493–497.

Cochran, H.L. and Dempsey, A.H. (1966) Stigma structure and period of receptivity in pimentos. *Proceedings of the American Society of Horticultural Science* 88, 454–457.

Cooksey, J.R., Kahn, B.A. and Motes, J.A. (1994) Plant morphology and yield of paprika pepper in response to method of stand establishment. *HortScience* 29, 1282–1284.

Cotter, D.J. (1980) Review of studies on chile. *New Mexico Agricultural Experiment Station Bulletin* 673, 1–29.

Crespo-Ruiz, M., Goyal, M.R., Chao de Baez, C. and Rivera, L.E. (1988) Nutrient uptake and growth characteristics of nitrogen fertigated sweet peppers under drip irrigation and plastic mulch. *Journal of Agriculture of the University of Puerto Rico* 72, 575–584.

Daunicht, H.J. and Lenz, F. (1973) Das Verhalten von Paprikapflanzen mit unterschiedlichem Fruchtbehang bei Behandlung mit drei CO_2-Konzentrationen. *Gartenbauwissenschaft* 38, 533–546.

de Ruijter, A., van den Eijnde, J. and van der Steen, J. (1991) Pollination of sweet pepper (*Capsicum annuum* L.) in glasshouses by honeybees. *Acta Horticulturae* 288, 270–274.

Deli, J. and Tiessen, H. (1969) Interaction of temperature and light intensity on flowering of *Capsicum frutescens* var. *grossum* 'California Wonder'. *Journal of the American Society of Horticultural Science* 94, 349–351.

Demers, D.A., Charbonneau, J. and Gosselin, A. (1991) Effets de l'eclairage d'appoint sur la croissance et la productivite du poivron. *Canadian Journal of Plant Science* 71, 587–594.

Dorland, R.F. and Went, F.W. (1947) Plant growth under controlled conditions. VIII. Growth and fruiting of the chili pepper (*Capsicum annuum* L.). *American Journal of Botany* 34, 393–401.

Edwards, R.L. and Sundstrom, F.J. (1987) Afterripening and harvesting effects of Tabasco pepper seed germination performance. *HortScience* 22, 473–475.

Eguchi, T., Matsumura, T. and Ashizawa, M. (1958) The effect of nutrition on flower formation in vegetable crops. *Proceedings of the American Society of Horticultural Science* 72, 343–352.

Erwin, A.T. (1932) The peppers. *Iowa Agricultural Experiment Station Bulletin* 293, 120–152.

FAO (1992) *FAO Yearbook. Production 1991.* FAO yearbook, Vol. 45. Food and Agriculture Organization of the United Nations, Rome, Italy.

Gaye, M.M., Eaton, G.W. and Joliffe, P.A. (1992) Rowcovers and plant architecture influence development and spatial distribution of bell pepper fruit. *HortScience* 27, 397–399.

Gosselin, A. and Trudel, M.J. (1986) Root-zone temperature effects on pepper. *Journal of the American Society of Horticultural Science* 111, 220–224.

Govindarajan, V.S. (1985) Capsicum – production, technology, chemistry, and quality. Part I: History, botany, cultivation and primary processing. *CRC Critical Reviews in Food Science and Nutrition* 22, 109–176.

Greenleaf, W.H. (1986) Pepper breeding. In: Bassett, M.J. (ed.) *Breeding Vegetable Crops.* Avi Publishing, Westport, Connecticut, pp. 67–134.

Gross, K.C., Watada, A.E., Kang, M.S., Kim, S.D., Kim, K.S. and Lee, S.W. (1986) Biochemical changes associated with the ripening of hot pepper fruit. *Physiologia Plantarum* 66, 31–36.

Hall, A.J. (1977) Assimilate source–sink relationship in *Capsicum annuum* L. I. The dynamics of growth in fruiting and deflorated plants. *Australian Journal of Plant Physiology* 4, 623–636.

Hall, A.J. and Brady, C.J. (1977) Assimilate source–sink relationship in *Capsicum annuum* L. II. Effects of fruiting and defloration on the photosynthetic capacity and senescence of the leaves. *Australian Journal of Plant Physiology* 4, 771–783.

Hall, A.J. and Milthorpe, F.L. (1978) Assimilate source–sink relationship in *Capsicum annuum* L. III. The effects of fruit excision on photosynthesis and leaf and stem carbohydrates. *Australian Journal of Plant Physiology* 5, 1–13.

Hamilton, L.C. and Ogle, W.L. (1962) The influence of nutrition on blossom-end rot of pimiento peppers. *Proceedings of the American Society of Horticultural Science* 80, 457–461.

Hammes, J.K. and Bartz, J.F. (1963) Root distribution and development of vegetable crops as measured by radioactive phosphorus injection technique. *Agronomy Journal* 55, 329–333.

Hartz, T.K., LeStrange, M. and May, D.M. (1993) Nitrogen requirements of drip-irrigated peppers. *HortScience* 28, 1097–1099.

Hegde, D.M. (1987) Effect of soil moisture and N fertilization on growth, yield, N uptake, and water use of bell pepper (*Capsicum annuum* L.). *Gartenbauwissenschaft* 52, 180–185.

Hibberd, A.M. (1981) Symptoms of and variety reaction to green pitting, a non-pathological disorder of red bell peppers in Queensland. *Queensland Journal of Agricultural and Animal Sciences* 38, 47–53.

Ho, L.C., Belda, R., Brown, M., Andrews, J. and Adams, P. (1993) Uptake and transport of calcium and the possible cause of blossom-end rot in tomato. *Journal of Experimental Botany* 44, 509–518.

Hochmuth, G.J., Shuler, K.D., Mitchell, R.L. and Gilreath, P.R. (1987) Nitrogen crop nutrient requirement demonstrations for mulched pepper in Florida. *Proceedings of the Florida State Horticultural Society* 100, 205–209.

Horton, R., Beese, F. and Wierenga, P.J. (1982) Physiological response of chile pepper to trickle irrigation. *Agronomy Journal* 74, 551–555.

Houghtaling, H.B. (1935) A developmental analysis of size and shape in tomato fruits. *Bulletin of the Torrey Botanical Club* 62, 243–251.

Hubbard, N. and Pharr, D.M. (1992) Developmental changes in carbohydrate concentration and activities of sucrose metabolizing enzymes in fruits of two *Capsicum annuum* L. genotypes. *Plant Science* 86, 33–39.

Huffman, V.L., Schadle, F.R., Villalon, B. and Burns, E.E. (1978) Volatile components and pungency in fresh and processed Jalapeno peppers. *Journal of Food Science* 43, 1809–1811.

Ilyas, S. (1993) Invigoration of pepper (*Capsicum annuum* L.) seed by matriconditioning and its relationship with storability, dormancy, aging, stress tolerance and ethylene biosynthesis. Unpublished PhD thesis, Cornell University, Ithaca.

Kano, K., Fugimura, T., Hirose, T. and Tsukamoto, Y. (1957) Studies on thickening growth of garden fruits. I. On the cushaw, eggplant and pepper. *Memoir of the Research Institute of Food Science of Kyoto University* 12, 45–90.

Katerji, N., Mastrorilli, M. and Hamdy, A. (1993) Effects of water stress at different growth stages on pepper yield. *Acta Horticulturae* 335, 165–171.

Kato, K. (1989) Flowering and fertility of forced green peppers at lower temperatures. *Journal of the Japanese Society of Horticultural Science* 58, 113–121.

Knavel, D.E. (1977) The influences of nitrogen on pepper transplant growth and yielding potential of plants grown with different levels of soil nitrogen. *Journal of the American Society of Horticultural Science* 102, 533–535.

Kriedemann, P.E., Loveys, B.R., Possingham, J.V. and Satoh, M. (1976) Sink effects on stomatal physiology and photosynthesis. In: Wardlaw, I.F. and Passioura, J.B. (eds) *Transport and Transfer Processes in Plants*. Academic Press, New York, pp. 401–414.

Kristjansson, K. and Rasmussen, K. (1991) Pollination of sweet pepper (*Capsicum annuum* L.) with the solitary bee *Osmia cornifrons* (Radoszkowski). *Acta Horticulturae* 288, 173–179.

Laborde Cancino, J.A. and Pozo Campodonico, O. (1982) *Presente y pasado del chile en Mexico*. Secretaria de Agricultura y Recursos Hidraulicos, Mexico.

Leskovar, D.I. and Cantliffe, D.J. (1993) Comparison of plant establishment method, transplant, or direct seeding on growth and yield of bell pepper. *Journal of the American Society of Horticultural Science* 118, 17–22.

Locascio, S.J. and Fiskell, J.G.A. (1976) Pepper production as influenced by mulch, fertilizer placement and nitrogen rate. *Soil and Crop Science Society of Florida Proceedings* 36, 113–117.

Lorenz, O.A. and Maynard, D.N. (1980) *Knott's Handbook for Vegetable Growers*, 2nd edn. Wiley, New York.

Marti, H.R. and Mills, H.A. (1991a) Nutrient uptake and yield of sweet pepper as affected by stage of development and N form. *Journal of Plant Nutrition* 14, 1165–1175.

Marti, H.R. and Mills, H.A. (1991b) Calcium uptake and concentration in bell pepper plants as influenced by nitrogen form and stages of development. *Journal of Plant Nutrition* 14, 1177–1185.

McGregor, S.E. (1976) Pepper, green. In: *Insect Pollination of Cultivated Crop Plants*. USDA, ARS, Washington, pp. 292–295.

Mercado, J.A., Fernandez-Munoz, R. and Quesado, M.A. (1994) *In vitro* germination of pepper pollen in liquid medium. *Scientia Horticulturae* 57, 273–281.

Miller, C.H. (1961) Some effects of different levels of five nutrient elements on bell peppers. *Proceedings of the American Society of Horticultural Science* 77, 440–448.

Miller, C.H., McCollum, R.E. and Claimon, S. (1979) Relationship between growth of bell peppers (*Capsicum annuum* L.) and nutrient accumulation during ontogeny in field environments. *Journal of the American Society of Horticultural Science* 104, 852–857.

Mix, G.P. and Marschner, H. (1976) Einfluss exogener und endogener Faktoren auf Calciumgehalt von Paprika und Bohnenfruchten. *Zeitschrift fur Pflanzenernahrung und Bodenkunde* 139, 551–563.

Munting, A.J. (1974) Development of flower and fruit of *Capsicum annuum* L. *Acta Botanica Neerlandica* 23, 415–432.

Nagai, H. (1989) Tomato and pepper production in Brazil. In: Green, S.K., Griggs, T.D. and McLean, B.T. (eds) *Tomato and Pepper Production in the Tropics*. Asian Vegetable Research and Development Center, Shanhua, Tainan, pp. 396–405.

Nederhoff, E.M. and van Uffelen, J.A.M. (1988) Effect of continuous and intermittent carbon dioxide enrichment on fruit set and yield of sweet pepper (*Capsicum annuum* L.). *Netherlands Journal of Agricultural Science* 36, 209–217.

Nielsen, T.H., Skjaerbaek, H.C. and Karlsen, P. (1991) Carbohydrate metabolism during fruit development in sweet pepper (*Capsicum annuum* L.) plants. *Physiologia Plantarum* 82, 311–319.

Nielsen, T.H. and Veierskov, B. (1988) Distribution of dry matter in sweet pepper plants (*Capsicum annuum* L.) during the juvenile and generative growth phases. *Scientia Horticulturae* 35, 179–187.

Nilwik, H.J.M. (1980a) Photosynthesis of whole sweet pepper plants. 2. Response to CO_2 concentration, irradiance and temperature as influenced by cultivation conditions. *Photosynthetica* 14, 382–391.

Nilwik, H.J.M. (1980b) Photosynthesis of whole sweet pepper plants. 1. Response to irradiance and temperature as influenced by cultivation conditions. *Photosynthetica* 14, 373–381.

Nilwik, H.J.M. (1981) Growth analysis of sweet pepper (*Capsicum annuum* L.) 2. Interacting effects of irradiance, temperature, and plant age in controlled conditions. *Annals of Botany* 48, 137–145.

O'Sullivan, J. (1979) Response of peppers to irrigation and nitrogen. *Canadian Journal of Plant Science* 59, 1085–1091.

Odland, M.L. and Porter, A.M. (1941) A study of natural crossing in peppers (*Capsicum frutescens* L.). *Proceedings of the American Society of Horticultural Science* 38, 585–588.

Ohta, Y. (1962) Physiological and genetical studies on the pungency of *Capsicum*. IV. Secretory organs, receptacles and distribution of capsaicin in the *Capsicum* fruit. *Japanese Journal of Breeding* 12, 179–183.

Osman, O.A. and George, R.A.T. (1984) The effect of mineral nutrition and fruit position on seed yield and quality in sweet pepper (*Capsicum annuum* L.). *Acta Horticulturae* 143, 133–141.

Ozaki, H. and Subramanya, R. (1985). Susceptibility of pepper cultivars to black spot fruit disorder. *Proceedings of the Florida State Horticultural Society* 98, 256–258.

Park, S.K. and Jeong, H.J. (1976) The effect of shading on blossom-dropping and fruit-dropping of the hot pepper plant (*Capsicum annuum* L.). *Research Report of the Office of Rural Development (Korea)* 18, 1–8.

Payero, J.O., Bhangoo, M.S. and Steiner, J.J. (1990) Nitrogen fertilization management practices to enhance seed production by 'Anaheim Chili' peppers. *Journal of the American Society of Horticultural Science* 115, 245–251.

Pellitero, M., Pardo, A., Simon, A., Suso, M.L. and Cerrolaza, A. (1993) Effect of irrigation regimes on yield and fruit composition of processing pepper (*Capsicum annuum* L.). *Acta Horticulturae* 335, 257–263.

Pill, W.G. and Lambeth, V.N. (1980) Effect of soil water regime and nitrogen form on blossom end rot, yield, water relations, and elemental composition of tomato. *Journal of the American Society of Horticultural Science* 105, 730–734.

Polowick, P.L. and Sawhney, V.K. (1985) Temperature effects on male fertility and flower and fruit development in *Capsicum annuum* L. *Scientia Horticulturae* 25, 117–127.

Rabinowitch, H.D., Ben-David, B. and Friedmann, M. (1986) Light is essential for sunscald induction in cucumber and pepper fruits, whereas heat conditioning provides protection. *Scientia Horticulturae* 29, 21–29.

Rabinowitch, H.D. and Sklan, D. (1981) Superoxide dismutase activity in ripening cucumber and pepper fruit. *Physiologia Plantarum* 52, 380–384.

Rabinowitch, H.D., Sklan, D. and Budowski, P. (1982) Photo-oxidative damage in the ripening tomato fruit : protective role of superoxide dismutase. *Physiologia Plantarum* 54, 369–374.

Randle, W.M. and Honma, S. (1981) Dormancy in peppers. *Scientia Horticulturae* 14, 19–25.

Rivas, M., Sundstrom, F.J. and Edwards, R.L. (1984) Germination and crop development of hot pepper after seed priming. *HortScience* 19, 279–281.

Rylski, I. (1972a) Regulation of flowering in sweet pepper (*Capsicum annuum* L.) by external application of several plant growth regulators. *Israel Journal of Agricultural Research* 22, 31–40.

Rylski, I. (1972b) Effect of the early environment on flowering in pepper (*Capsicum annuum* L.). *Journal of the American Society of Horticultural Science* 97, 648–651.

Rylski, I. (1973) Effect of night temperature on shape and size of sweet pepper (*Capsicum annuum* L.). *Journal of the American Society of Horticultural Science* 98, 149–152.

Rylski, I., Aloni, B., Karni, L. and Zaidman, Z. (1994). Flowering, fruit set, fruit development and fruit quality under different environmental conditions in tomato and pepper crops. *Acta Horticulturae* 366, 45–55.

Rylski, I. and Halevy, A.H. (1972) The effect of damage to cotyledons or primary leaves on the development of pepper (*Capsicum annuum* L.). *HortScience* 7, 69–70.

Rylski, I. and Spigelman, M. (1982) Effects of different diurnal temperature combinations on fruit set of sweet peppers. *Scientia Horticulturae* 17, 101–106.

Rylski, I. and Spigelman, M. (1986) Effect of shading on plant development, yield and fruit quality of sweet pepper grown under conditions of high temperature and radiation. *Scientia Horticulturae* 29, 31–35.

Saltveit, M.E. (1977) Carbon dioxide, ethylene, and color development in ripening mature green bell peppers. *Journal of the American Society of Horticultural Science* 102, 523–525.

Schoch, P.G. (1972) Effects of shading on structural characteristics of the leaf and yield of fruit in *Capsicum* L. *Journal of the American Society of Horticultural Science* 97, 461–464.

Schon, M.K. (1993) Effects of foliar antitranspirant or calcium nitrate applications on yield and blossom-end rot occurrence in glasshouse-grown peppers. *Journal of Plant Nutrition* 16, 1137–1149.

Shifriss, C. and Rylski, I. (1972) A male sterile (*ms-2*) gene in 'California Wonder' pepper (*Capsicum annuum* L.). *HortScience* 7, 36.

Shifriss, C. and Frankel, R. (1971) New sources of cytoplasmic male sterility in cultivated peppers. *Journal of Heredity* 62, 254–256.

Shipp, J.L., Whitfield, G.H. and Papadopoulos, A.P. (1994) Effectiveness of the bumblebee, *Bombus impatiens* Cr. (Hymenoptera; Apidae) as a pollinator of glasshouse sweet pepper. *Scientia Horticulturae* 57, 29–39.

Sinnott, E.W. and Kaiser, S. (1934) Two types of genetic control over the development of shape. *Bulletin of the Torrey Botanical Club* 61, 1–7.

Smith, P.G., Villalon, B. and Villa, P.L. (1987) Horticultural classification of peppers grown in the United States. *HortScience* 22, 11–13.

Somos, A. (1984) *The Paprika*. Akadediai Kiado, Budapest.

Song, K.W., Park, S.K. and Kim, C.K. (1976) Studies on the flower abscission of hot pepper. *Research Report of the Office of Rural Development (Korea)* 18, 9–32.

Stoffella, P.J. and Bryan, H.H. (1988) Plant population influences growth and yields of bell pepper. *Journal of the American Society of Horticultural Science* 113, 835–839.

Stoffella, P.J., Di Paola, M.L., Pardossi, A. and Tognoni, F. (1988) Root morphology and development of bell peppers. *HortScience* 23, 1074–1077.

Sundstrom, F.J. and Edwards, R.L. (1989) Pepper seed respiration, germination and seedling development following priming. *HortScience* 24, 343–345.

Tanksley, S.D. (1984) High rates of cross-pollination in chile pepper. *HortScience* 19, 580–582.

Tripp, K.E. and Wien, H.C. (1989) Screening with ethephon for abscission resistance of flower buds in bell pepper. *HortScience* 24, 655–657.

Turner, A.D. and Wien, H.C. (1994a) Dry matter assimilation and partitioning in pepper cultivars differing in susceptibility to stress-induced bud and flower abscission. *Annals of Botany* 73, 617–622.

Turner, A.D. and Wien, H.C. (1994b) Photosynthesis, dark respiration and bud sugar concentrations in pepper cultivars differing in susceptibility to stress-induced bud abscission. *Annals of Botany* 73, 623–628.

van Berkel, N. (1986) CO_2 enrichment in the Netherlands. In: Enoch, H.Z. and Kimball, B.A. (eds) *Carbon Dioxide Enrichment of Glasshouse Crops*. Vol. I. *Status and CO_2 Sources*. CRC Press, Boca Raton, pp. 17–33.

Villalon, B. (1975) Black spot – a nonparasitic disease of bell pepper fruit in the lower Rio Grande Valley of Texas. *Plant Disease Reporter* 59, 926–927.

Watkins, J.T. and Cantliffe, D.J. (1983) Mechanical resistance of the seedcoat and endosperm during germination of *Capsicum annuum* at low temperature. *Plant Physiology* 72, 146–150.

Watkins, J.T., Cantliffe, D.J., Huber, D.J. and Nell, T.A. (1985) Gibberellic acid stimulated degradation of endosperm in pepper. *Journal of the American Society of Horticultural Science* 110, 61–65.

Weaver, J.E. and Bruner, W.E. (1927) Pepper. In: *Root Development of Vegetable Crops*. McGraw Hill, New York, pp. 268–273.

Wien, H.C. (1990) Screening pepper cultivars for resistance to flower abscission: a comparison of techniques. *HortScience* 25, 1634–1636.

Wien, H.C., Minotti, P.L. and Grubinger, V.P. (1993a) Polyethylene mulch stimulates early root growth and nutrient uptake of transplanted tomatoes. *Journal of the American Society of Horticultural Science* 118, 207–211.

Wien, H.C., Aloni, B., Riov, J., Goren, R., Huberman, M. and Ho, J.C. (1993b) Physiology of heat stress-induced abscission in pepper. In: Kuo, G.C. (ed.) *Adaptation of Food Crops to Temperature and Water Stress*. Asian Vegetable Research and Development Center, Shanhua, Taiwan, pp. 188–198.

Wien, H.C., Tripp, K.E., Hernandez-Armenta, R. and Turner, A.D. (1989a) Abscission of reproductive structures in pepper: causes, mechanisms and control. In:

Green, S.K. (ed.) *Tomato and Pepper Production in the Tropics*. Asian Vegetable Research and Development Center, Shanhua, Taiwan, pp. 150–165.

Wien, H.C., Turner, A.D. and Yang, S.F. (1989b) Hormonal basis for low light intensity-induced flower bud abscission of pepper. *Journal of the American Society of Horticultural Science* 114, 981–985.

Wien, H.C. and Zhang, Y. (1991) Prevention of flower abscission in bell pepper. *Journal of the American Society of Horticultural Science* 116, 516–519.

Wittwer, S.H. and Teubner, F.G. (1956) Cold exposure of tomato seedlings and flower formation. *Proceedings of the American Society of Horticultural Science* 66, 369–376.

Worku, A., Herner, R.C. and Carolus, R.L. (1975) Effect of stage of ripening and ethephon treatment on color content of paprika pepper. *Scientia Horticulturae* 3, 239–245.

Zamski, E., Shoham, O., Palevitch, D. and Levy, A. (1987) Ultrastructure of capsaicinoid-secreting cells in pungent and nonpungent red pepper (*Capsicum annuum* L.) cultivars. *Botanical Gazette* 148, 1–6.

Potato 8

E.E. Ewing

Department of Fruit and Vegetable Science, Cornell University, 134A Plant Science Building, Ithaca, New York 14853–5908, USA

HISTORY

More than 6000 years before the Spanish came to South America, the so-called 'Irish' potato (*Solanum tuberosum* L.) was under cultivation in the highlands of the Andes (Hawkes, 1992). The potato was a central part of the culture of the Incas and of the civilizations that preceded them. Potatoes could be grown successfully at altitudes up to 3700 m above sea level.

A secondary centre of origin for the potato was farther to the south, in Chile. Here the potato had been selected for ability to tuberize under longer photoperiods than those found in its primary home near the equator. The Chilean form, best represented by potatoes grown in the island of Chiloe (latitude 43°S) is usually designated subspecies *tuberosum* to distinguish it from *andigena*, the subspecies grown in Peru.

Although the *tuberosum* subspecies is far better adapted to the long days of European summers, evidence indicates that it was the *andigena* potato which was introduced to continental Spain in about 1570 (Hawkes and Francisco-Ortega, 1992), having arrived in the Canary Isles some ten years earlier (Hawkes and Francisco-Ortega, 1993). From Spain the potato gradually spread northward throughout Europe. It is likely that the first potatoes grown were poorly adapted to the long summer days of northern Europe and that they would have matured only after daylength had shortened, late in the autumn. Types were gradually selected that were able to tuberize under longer days and that were therefore earlier in maturity. Once such adapted clones had been selected, the climate in northern Europe was found to be almost ideal for potato production, and the crop gained popularity at an extraordinary rate in many countries.

By the end of the eighteenth century the economy of the peasants in Ireland was as fully dependent upon the potato as had been true for the

© CAB INTERNATIONAL 1997. *The Physiology of Vegetable Crops*
(H.C. Wien, ed.)

pre-Spanish peoples of the Andes. The Irish ate enormous quantities of potatoes, often more than 3 kg per person per day. So total was the dependence of the poor farmers upon the potato that when the late blight disease suddenly appeared in 1845, with devastating crop losses, the effects on the population were calamitous. During the next several years perhaps a million people died of starvation or of diseases associated with extreme malnutrition. Another million Irish emigrated to the USA, Canada and various other parts of the world.

After the late blight epidemics, attempts were made to introduce late blight resistance through new germplasm collected in South America. It is probable that the germplasm for the second introduction of potato came from the isle of Chiloe, because the European varieties that had been gradually selected from *andigena* were replaced with subspecies *tuberosum*. Analyses of chloroplast DNA show that almost all cultivars in Europe and North America now belong to the latter subspecies (Hosaka and Hanneman, 1988).

NUTRITIONAL VALUE AND IMPORTANCE

One might wonder how the Irish – even before the late blight organism – were able to survive on potatoes as such a major component of their diet. The nutritional value of the potato is greatly underestimated. Most people realize that potatoes are a cheap source of starch. What is not appreciated is the other nutrients that come with the starch. Although the protein content of the potato is not high, if a person were to eat enough potatoes to meet the daily requirement for calories, the amount of protein required would also be satisfied. Compared to other types of plant protein, the quality of potato protein is relatively good. The vitamin C content is also high; three medium potatoes (90 g each) will supply the adult requirement for vitamin C under average conditions. Potatoes also contain significant amounts of the B vitamins and various minerals. A medium-sized potato contains about the same number of calories as an apple or an orange – it is not 'fattening' except for the gravy, butter, or cooking oil that may be added to it.

In world tonnage the potato ranks after wheat, rice and maize as the fourth most important crop for human consumption. Few if any crops surpass the potato in terms of potential yields of calories or of protein per hectare if calculated per day of growing season, and it ranks among the top half dozen or so food crops in total world production of both protein and energy. The potato is considered an agronomic crop in Europe and South America, and a horticultural crop in Asia and Africa. In the USA it is treated as an agronomic crop in some states and as a horticultural crop in others. If we call it a vegetable, then the potato is the most important vegetable in the world.

Russia leads the world in potato production, and the countries of the former Soviet Union together produce about a quarter of the world total.

Almost all European countries have substantial potato production, but the potato is also an important crop in many parts of Asia, Africa and South America. Most westerners are amazed to learn that the People's Republic of China vies with Poland for second place in potato production. The USA ranks in fifth place with about 7% of world production, and India is a close sixth!

BOTANY

The potato is a member of the Solanaceae family. Thus it is not surprising that its fruit resembles a small green tomato. Seeds can be extracted from the fruits and handled much like tomato seeds. There is usually a period of seed dormancy for several months or more, and treatment with gibberellic acid will improve germination during this time. The seedling will send out stolons from axillary buds just above the soil level (starting from the cotyledonary nodes), the stolon tips will grow under the soil, and tubers will form on the buried stolons. The tubers can then be used for clonal propagation, as described below. In this country the main value of the seeds is to plant breeders who use them to develop new cultivars. As will be discussed in the section on establishment of the leaf canopy it is also possible to grow crops from seeds.

Contrary to popular belief the potato is not a storage root, but rather a specialized underground stem. A potato tuber is a shortened, swollen, starchy stem. The tuber bears minute scale leaves, each with a bud in its axil. These buds form the 'eyes' of the tuber, and the scale leaves form the 'eyebrows'. The sprouts that develop from the eyes come from buds which are in the axils of the leaves. The eyes are arranged in a spiral around the tuber just as leaves on the normal potato plant are arranged in a spiral. Like an ordinary stem, the potato tuber shows apical dominance. The first sprout to develop is from an eye in the apical end of the tuber. If the apical sprout grows, sprouts from the other eyes are suppressed; if it is damaged, the apical dominance is broken and most of the other eyes on the tuber will sprout. The potato tuber develops chlorophyll in the light and has the internal anatomy of an ordinary stem. The epidermis sloughs off early in the development of the tuber and is replaced by a corky periderm. This is the 'skin' of the potato. Under the periderm is the cortex, the vascular ring, and the pith.

If a whole tuber or a piece of tuber containing one or more eyes is planted, the buds sprout and a plant develops above the ground. Well before plant emergence the developing sprout grows adventitious roots, which constitute the root system. Also developing from the underground portion of the stem are stolons (or rhizomes), which may bear new tubers at their tips (Fig. 8.1).

The main stem of the potato plant terminates in a flower cluster. As will be described later, flower bud abortion may occur at a very early stage of development; but in any case apical growth of the main stem ceases with

Fig. 8.1. Roots, stolons and tubers of a 'Désirée' plant, grown in the desert sands of the Negev region, Israel. Note that stolons formed at underground nodes of the main stem, and that tubers formed at stolon nodes.

formation of the flower buds. The cessation of growth of the main shoot axis may not be obvious because sympodial growth of one or more axillary branches just below the apex permits further shoot extension above the flower cluster. After developing up to six or more leaves, the new axillary branch(es) will terminate in a flower cluster in the same manner (Fig. 8.2); but new sympodial growth may again occur (Almekinders and Struik, 1994). In this manner the main axis may be extended by three or more levels of branching.

Axillary branches are not restricted to sympodial growth. Branching may occur at any node, but branching is most common at the base of the plant (Fig. 8.2). Some branches arise from underground nodes on the main stem. Without disturbing the soil it is difficult to distinguish these from stems that have arisen from separate eyes of the seed tuber. Other axillary branches arise from nodes just above the soil level. Cessation of the growth of the main axis associated with flower bud formation encourages basal axillary branching. The extent of axillary branching, both sympodial and basal, is of crucial importance in determining yield potential, and we will be examining the factors that control it. (A good summary of the foliar development of the potato is presented by Vos, 1995a.)

Note that propagation from tubers is vegetative, not sexual. All plants obtained from the offspring of a single tuber are genetically identical, unless chance mutations have occurred. This means that all tubers of a given cultivar should be highly uniform unless they have become infected with a disease organism.

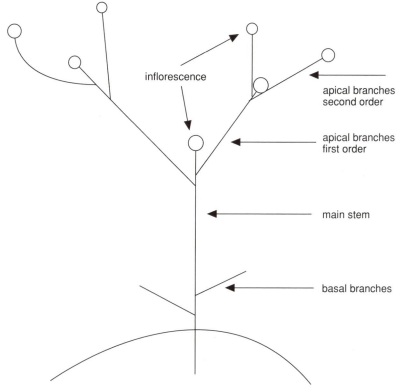

Fig. 8.2. Diagram of a potato shoot, showing positions on the main stem of apical and basal axillary branches. The main stem and all axillary branches eventually terminate in inflorescences, but new branches may form just below the inflorescence to continue the sympodial growth (after Vos, 1995b).

CULTURE

The potato is considered a cool season vegetable, although it possesses only moderate frost tolerance. Best yields are typically obtained in climates where the average growing season temperature is about 15–20°C. The root system on the potato plant is not extensive and ample soil water is necessary whether from rain or supplemental irrigation.

Potato plants require a well-drained soil so that the roots have adequate oxygen. Unless irrigation facilities are available, the soil should not be too droughty. The most attractive tuber shape and skin appearance are achieved with light, sandy soils, or with muck soils. The ideal pH for potatoes grown on mineral soils is between about 4.8 and 5.4. Potatoes grow well at a higher soil pH, but scab can be a problem. If it is not practical to maintain a low pH, scab resistant cultivars must be grown. The rate of N fertilization is a key consideration in managing fertility, because excessive applications delay maturity and reduce the partitioning of dry matter to the tubers, not to mention possible adverse effects on processing quality and on the environment.

In most northern areas potato planting is started as soon as the soil has dried out enough for plowing. Most mineral soils are in prime condition for planting if the cover crop is plowed under in the spring, provided that the soil moisture content was correct for plowing. A light tillage implement dragged behind the plow ordinarily gives all the soil preparation needed prior to planting. Potato planting is usually started in early spring, a week or two before the average date of the last killing frost. Depending upon the weather and the acreage to be planted, planting may continue for three or four weeks. Late plantings limit yield because they limit the total biomass that can be accumulated.

In Europe and many other parts of the world, seed tubers are planted whole. This is partly because for the fresh market Europeans prefer smaller tubers than are commonly eaten in the USA. Cultural practices, including choice of cultivars, are therefore geared to the production of small tubers; and there is an adequate supply of small tubers to be used for seed. Consumers in the USA are accustomed to buying potatoes up to 0.5 kg in weight, and the cultivars grown for fresh market and processing give only a low percentage of tubers in the small size range used for whole seed. When a seed tuber is planted, there is a tendency for only one or two eyes at the apical end to sprout. (This is because of apical dominance.) For this reason a tuber weighing 50 g may give nearly as much production as is obtained from a seed tuber weighing three or four times as much. Because planting whole large tubers would be wasteful and because there are not enough small tubers to meet the demand for seed, most potatoes in the USA are grown from cut seed. Seed pieces should average 40–55 g. Cut seed potatoes will develop wound periderm over the cut surfaces if kept in an environment of high relative humidity, warm temperature, and adequate oxygen supply.

The quantity of seed planted per hectare will have a considerable effect on average tuber size. Cultivars that are inclined to develop large tubers should therefore be spaced more closely. Cut seed tends to give somewhat fewer sprouted eyes per hectare than whole seed and should be spaced a little closer, increasing the quantity of seed per hectare by perhaps 10%. Although the quantity of seed planted per hectare has a considerable effect on yield and tuber size, within reasonable limits the planting arrangement does not seem to have much influence (Sieczka et al., 1986). In the USA rows are most commonly spaced about 86 cm apart; more narrow spacings, such as 71 cm between rows, are common in Europe. If rows are too close together it becomes difficult to operate equipment and to ridge up enough soil for adequate cover of the developing tubers. The distance between seed tubers within the row may range from about 18 cm to as much as 40 cm.

Seed tubers that are planted too deep will be slow to emerge and may be more subject to attack by various diseases. Very shallow planting of seed tubers may result in inadequate soil moisture around the seed piece and in production of tubers so close to the soil surface that greening caused by exposure to light is more of a problem. Planting should be deeper on lighter soils than on heavy ones and deeper late in the season than early. Many growers like to plant seed tubers relatively deep but then cover them with only a shallow layer of soil. More soil covering will then be added as the plant develops. A good rule of thumb is never to have more than 10 cm of soil above the tip of the developing sprout.

Soils should be ridged up along the potato row to provide extra cover for the developing tubers. This tends to reduce the number of tubers that stick out of the soil and are exposed to light. Even diffuse light filtering down through the cracks in the soil will cause tubers to turn green and to develop a bitter flavour. Tubers that turn green in the field are called 'sun-burned' and are unfit for consumption. Secondary benefits of ridging up the soil are that it facilitates harvest and provides weed control.

Before the advent of modern insecticides and fungicides after World War II, potato 'vines' (called 'haulms' in Europe) were dead by the end of summer due to the attack of leafhoppers, aphids, and other pests. Pesticides have made possible much higher yields, but their use sometimes means that potato foliage is still green and vigorously growing when the farmer may desire to start harvesting. Potato tubers dug when vines are still green will have thin periderms that give little protection against skinning and bruising. Use of an appropriate chemical vine killer at least ten days before harvest will cause earlier thickening of the periderm and reduce injuries during harvest. Other advantages of the vine killer may be a reduction in tuber infection when late blight is present and the prevention of oversize tubers. In the case of seed growers, vine killing may reduce the chance of virus spread from the build-up of aphid populations late in the season. A disadvantage of vine killing is that yields are generally reduced. Furthermore, vine killing is difficult and relatively ineffective if the foliage is very immature at time of application. Therefore, vine killing should be considered as a supplement to natural maturation, not as a substitute for it.

TUBER INDUCTION
METHODS OF STUDYING TUBERIZATION

Because tubers form underground, and because exposure to light inhibits their formation, non-destructive observations of the process are difficult. One approach has been to devise special techniques that involve separation of the root zone from the stolon and tuber zone (Struik and Van Voorst, 1986). An alternative is to take cuttings from stems, which under the right conditions are capable of tuberizing (Gregory, 1956; Chapman, 1958). Such cuttings consist of one or more axillary buds that are buried under the soil, and one or more leaves that are exposed to light. Cuttings are placed in a mist bench under continuous light and can be examined in about 12 days for tuber development. The extent to which plants were induced to tuberize at the time cuttings were taken is reflected in the tuberization response of the cuttings (Ewing, 1985). Cuttings taken from strongly induced plants form sessile tubers at the buried node; the buried buds of cuttings taken from plants that were not induced to tuberize remain dormant (if the apex of the cutting is still present) or develop as upright shoots (if the apex of the cutting was removed). Intermediate between these extreme responses, the buried buds of cuttings develop stolons or – with somewhat stronger induction – stolons terminated by tubers. The range of responses is shown in Fig. 8.3.

A third approach to the study of tuberization is to grow plants *in vitro*. Although this technique has yielded useful understanding, extrapolation of results to full-sized plants should be done with caution. *In vitro* plants are ordinarily grown on a sucrose medium, and shoots and roots normally are simultaneously exposed to either light or darkness. Whether for these or other reasons, tuberization responses on *in vitro* plants do not seem to be reliable predictors of whole plant responses.

MORPHOLOGICAL CHANGES

The usual site of tuber formation is at the tip of an underground stolon, either the tip of the main stolon axis or the tip of a branch arising from an axillary bud. Potato stolons are diagravitropic stems with long internodes and scale leaves. Stolons develop as branches from underground nodes (Peterson *et al.*, 1985). Plants that have reached the stage where they are

Fig. 8.3. Apical cuttings after 14 days in a mist chamber, illustrating the range of responses produced by varying levels of induction to tuberize. Cuttings, taken from sibling plants in a population segregating for ability to tuberize under long days, show progressively stronger tuberization responses at the buried basal bud: from left to right, no growth, orthotropic growth, diagravitropic stolon growth, stolon terminated by a tuber, and sessile tuber (Van den Berg *et al.*, 1996).

capable of initiating tubers are said to be induced to tuberize. Induction is brought about by a hypothetical 'tuberization stimulus' that is produced in the leaves and is translocated to the site of tuber initiation (Gregory, 1956). The tuberization stimulus is not necessarily a single compound; more likely it consists of a particular balance between two or more compounds.

Tuber induction leads to tuber initiation, defined to have occurred if the swollen portion of the stolon tip is at least twice the diameter of the stolon (Ewing and Struik, 1992). The tiny swellings are called tuber initials. Many tuber initials fail to develop into tubers, but are resorbed; and their contents are reallocated to other parts of the plant. Such resorption may occur even to tubers that have attained > 2 cm diameter. Tubers that are not resorbed are said to have been 'set'. Even after tuber set, the fate of the tubers is not predictable. Some remain small, while others on the same plant continue to grow. Continued tuber development and maturation generally are favoured by environmental factors that promote tuber induction. Conditions such as high soil temperature can cause even large tubers to revert to stolon growth.

The reversion of tuber growth to stolon growth exemplifies the plasticity displayed by the potato plant (Steward et al., 1981; Clowes and MacDonald, 1987). There are many other examples. Cutting off shoots at soil level causes stolons to turn upward and form new shoots instead of tubers (Sachs, 1893); and girdling of the stem at soil level or other interference with translocation from leaves to stolons can cause tubers to form at above-ground buds (Vöchting, 1887). Even flower buds have been known to tuberize (Werner, 1954); and new tubers can form directly from the buds of the mother tuber (Wellensiek, 1929).

ANATOMICAL AND BIOCHEMICAL CHANGES

Anatomical changes in the stolon during tuber initiation have been thoroughly reviewed by Cutter (1992). It is well known that increases in cell division, cell enlargement and starch deposition all occur before visible swelling of the stolon tip (Plaisted, 1957). It appears that the increase in cell enlargement precedes that in cell division (Booth, 1963; Koda and Okazawa, 1983b; Cutter, 1992).

As the stolon tips begin to develop into tubers, endogenous activity of gibberellin (GA)-like compounds in the stolon tips decreases (Koda and Okazawa, 1983b). Another change associated with the earliest stages of tuberization is an increase in the concentration of 'patatin', a glycoprotein that is one of the major storage proteins present in potato tubers (Paiva et al., 1983). Reviews on other biochemical changes associated with tuberization have been written by Park (1990), Prat et al. (1990), Sanchez-Serrano et al. (1990) and Willmitzer et al. (1990).

ENVIRONMENTAL FACTORS AFFECTING TUBERIZATION

Short days – more accurately, long nights – favour induction to tuberize (Garner and Allard, 1923). Interruption of the long dark period with red light interferes with induction, an effect that is reversible with far-red light (Batutis and Ewing, 1982). This is typical of phytochrome responses. Indeed, it appears that phytochrome B must be present to prevent tuberization of *andigena* under long photoperiods, inasmuch as insertion of an antisense gene for phytochrome B permits such tuberization to occur (Jackson et al., 1996).

Extension of a short photoperiod with dim incandescent light is much more inhibitory to tuberization than is extension with high intensity irradiance from a mixture of fluorescent and incandescent lamps (Wheeler and Tibbitts, 1986; Lorenzen and Ewing, 1990). Even low levels of irradiance from cool white fluorescent lamps are less inhibitory to tuberization than if incandescent lamps are used for the photoperiod extension (Wheeler and Tibbitts, 1986). The exaggerated effects of dim light extension with incandescent lamps should be borne in mind when attempting to translate to field conditions results from controlled environment experiments; but it remains true that even under field conditions, short days favour induction to tuberize.

Another important environmental factor is temperature; cool temperatures promote induction to tuberize (Werner, 1934; Gregory, 1956). It is commonly believed that night temperatures have more influence than day temperatures, but interpretation of experimental evidence is complicated by the fact that diurnal variation in temperature favours tuberization (Steward *et al.*, 1981). Both air and soil temperatures are important: cool air temperatures favour induction of leaves to tuberize, as reflected in cuttings (Gregory, 1956; Reynolds and Ewing, 1989a); and high soil temperatures block the expression of the tuberization stimulus at underground nodes (Reynolds and Ewing, 1989a). There is an interaction between photoperiod and temperature, such that the higher the temperature, the shorter the photoperiod required for tuberization of a given genotype (Snyder and Ewing, 1989).

Other environmental factors that affect induction to tuberize include irradiance and nitrogen nutrition. Low levels of irradiance reduce induction (Bodlaender, 1963), as does increasing the rate of nitrogen fertilization (Werner, 1934; Krauss and Marschner, 1976).

INTERNAL FACTORS

Genetic effects

Under any given set of environmental conditions there are enormous genetic differences in induction to tuberize. Most wild *Solanum* species have a short critical photoperiod for tuberization; they will become induced only if the photoperiod is less than about 12 h. The same is true of most accessions of *S. tuberosum* subsp. *andigena*, which are thus adapted to the short days and cool temperatures of their Andean home. In contrast, *S. tuberosum* subsp. *tuberosum* will become induced even under very long photoperiods, especially if the other environmental factors (temperature, irradiance, and nitrogen level) are favourable for induction. This is particularly true of early maturing cultivars, which have even longer critical photoperiods than do late maturing ones.

Plant breeders have demonstrated that genes for ability to tuberize under longer days are present within the subspecies *andigena* germplasm, even though the trait is usually suppressed. After six to seven cycles of recurrent selection within *andigena* for ability to tuberize under long days, genotypes (designated 'neo-tuberosum') were identified that could not be distinguished from subsp. *tuberosum* in this respect (Simmonds, 1966; Cubillos and Plaisted, 1976). It is theorized that when *andigena* was taken to Europe, a similar selection for earliness among potato seedlings gradually led to the development of cultivars that were adapted to the long summer days. It is likely that subsp. *tuberosum* did not replace the population derived from *andigena* until some two centuries later, after late blight epidemics had led to a search for new germplasm.

Genetic mapping of backcross populations between subsp. *tuberosum* and a wild species has revealed the presence of at least eleven genes for the ability to tuberize under long photoperiods (Van den Berg *et al.*, 1996). At some loci, alleles contributed by the wild species, which has a short critical photoperiod, were more favourable for tuberization than alleles from *tuberosum*. Thus it is understandable that recurrent selection within germplasm that generally required short days for tuberization could eventually lead to a combination of genes that would give tuberization under long days.

Role of the seed tuber

Plants are sometimes grown in the absence of the seed tuber, or with its early removal. Examples of the former are growing plants from true potato seed, from *in vitro* plantlets, or from rooted cuttings. Early removal of the seed tuber takes place when decay organisms attack and destroy the seed tuber soon after plant emergence.

Under short photoperiods, plants grown from rooted cuttings were similar to plants grown from seed tubers (Madec and Perennec, 1962). Both types of plants tuberized strongly, as expected, while the physiological condition of the seed tuber had little effect. Under long photoperiods, tuberization on plants from cuttings was delayed or absent, depending upon the critical photoperiod of the cultivar. By contrast, plants grown from tubers were able to tuberize even under long days, but the condition of the seed tuber was influential. These results led to the hypothesis that there is a 'double induction' to tuberize (Madec and Perennec, 1959, 1962). According to this hypothesis, both the mother tuber and the leaves may contribute to the tuberization stimulus, depending upon the photoperiod and other environmental conditions to which the leaves are exposed, and depending upon the physiological condition of the mother tuber.

Differences in the physiological condition of seed tubers related to their contribution to induction to tuberize are characterized by what is called their physiological age. The physiological age of a tuber is its stage of development, as modified progressively by increasing chronological age, and as affected by environmental and genetic factors. Newly formed tubers go through a dormant stage (or rest period), during which they will not sprout even if placed at warm temperatures. If transferred to warm temperatures by the end of dormancy, sprouting will begin at the apical end of the tuber and there will be few sprouts per eye (Krijthe, 1962). Such tubers are physiologically young. If tubers are stored at cold temperatures, below 4°C, for a long time after dormancy is broken, the sprouts that develop when they are finally moved to warm temperatures (10–15°C) will be noticeably different. Apical dominance will be less pronounced, so that many eyes will have sprouts, there will be more multiple sprouts at the eyes, and the sprouts will be more highly branched (Krijthe, 1962). Such tubers will be physiologically older.

If the storage at cold temperature goes on long enough, when the tubers are finally moved to warm temperatures new tubers begin to form on the

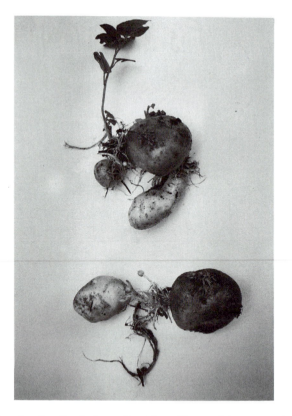

Fig. 8.4. Tubers forming directly from other tubers. The new tubers are variously called 'sprout tubers' or 'little potatoes'. Their development is a sign that the seed tubers are very old physiologically. Leafy shoots that develop from such tubers (above) are too weak to be of value.

sprouts, or in more extreme cases new sessile tubers form at the eyes ('sprout tubers' or 'little potatoes'; Fig. 8.4). Seed tubers that form sprout tubers are extremely old physiologically. The time between the end of tuber dormancy and the beginning of tuber formation on sprouts is termed the 'incubation period' of the tubers (Claver, 1975; Madec, 1978).

Cultivars vary greatly in length of both the dormant period and the incubation period, and this has obvious effects on physiological age. The principal environmental factor affecting physiological age of tubers is the storage temperature. Warmer storage is associated with more rapid ageing. Warm temperatures during the growing season of the crop used for seed may also cause a small increase in physiological age (Iritani, 1968; Claver, 1973). Exposure of seed tubers to diffuse light during sprouting ('green-sprouting') causes a drastic shortening of sprout growth (Scholte, 1989), which reduces the chance that they will break off, decreases water loss by transpiration, and tends to delay ageing.

The general view is that the greater the physiological age of the seed tuber, the greater its potential contribution to induction to tuberize. As already indicated, extremely old tubers tuberize even before sprouts emerge from the soil and before there can be an effect from the leaves. Young tubers may contribute substances (perhaps GAs?) that counteract tuberization stimulus supplied from the leaves, so that plants without seed tubers form new tubers more readily than do plants grown from seed tubers. (This can lead to problems when propagules other than seed tubers are employed, as is discussed below.) Thus induction to tuberize results from the interplay between both positive and negative factors that may be supplied by the leaves alone, by the mother tuber alone, or by both together.

EFFECTS OF INDUCTION ON OVERALL PLANT DEVELOPMENT

The growth and development of the entire plant change with induction to tuberize. Because these effects have been studied most thoroughly with respect to photoperiod, the effects of photoperiod will be described first and then compared to the effects of temperature and other variables.

Photoperiod extensions

The long-term effects of short photoperiods on the growth of the potato plant are well documented. Most comparisons have been made between short photoperiods and long photoperiods obtained by extending the photoperiod with dim incandescent light. Compared to long photoperiods obtained by extending with incandescent lamps, short photoperiods favour tuber growth relative to the growth of all other parts of the plant (Rasumov, 1931; Driver and Hawkes, 1943; Krug, 1960; Madec and Perennec, 1962; Hammes and Nel, 1975; Steward et al., 1981). Not only is stolon growth replaced by tuber growth (Rasumov, 1931), but short-day plants have larger leaflets (Edmundson, 1941; Bodlaender, 1963); shorter stems (Schick,

1931; Edmundson, 1941); a higher dry weight ratio of leaves to stems (Driver and Hawkes, 1943); a flatter angle between the leaf and stem (Werner, 1934; Wheeler *et al.*, 1991); more flower bud abortion (Edmundson, 1941; Turner and Ewing, 1988); fewer axillary branches, both sympodial and basal (Gregory, 1956; Demagante and Vander Zaag, 1988); lower root dry weight (Steward *et al.*, 1981); and more rapid senescence (Bodlaender, 1963; Demagante and Vander Zaag, 1988).

The degree to which these changes take place is cultivar-dependent and also depends upon the other factors that affect induction. For example, if an early maturing *tuberosum* cultivar is changed from 16- to 12-h photoperiods, the effects on overall morphology will probably be much less striking than if a typical *andigena* cultivar (requiring 12-h days for tuberization under otherwise favourable conditions) is given the same treatment. Nevertheless, the general tendency is for partitioning of assimilate to be shifted toward tubers as photoperiod is shortened (Wolf *et al.*, 1990). All cultivars seem to respond to photoperiod in this way, but the effects may not be obvious unless the plants are grown under high temperatures or other conditions that are unfavourable to tuberization. In this sense no cultivars appear to be day-neutral.

When fluorescent rather than dim incandescent lamps were employed to extend the photoperiod, plants were more similar to those produced under short days (Wheeler and Tibbitts, 1986). Thus stem length was not increased, and the decreased partitioning to tubers was much less pronounced when long days were obtained with fluorescent lamps than with dim incandescent lamps (Wheeler and Tibbitts, 1986). More research is needed to clarify the reasons for the different responses to light quality.

Short-term effects of photoperiod

The more immediate effects of shortening the photoperiod have been much less studied, but Lorenzen and Ewing (1990) found the short-term effects of shortening photoperiod to be somewhat different from the long-term effects. For example, leaf expansion was not slowed until well after the establishment of a strong tuber sink, although leaves were thinner.

Other changes were manifested very early (Lorenzen and Ewing, 1990). Four days after starting treatments plants grown under photoperiods extended with dim incandescent lamps were taller than plants under full lights, whether long or short photoperiods were used. The increase in plant height is caused by longer internodes, not by an increased number of nodes (Struik *et al.*, 1988). Leaf-bud cuttings taken after 4 days showed that the leaves under short days were already much more strongly induced to tuberize than leaves under either of the long-day treatments (Lorenzen and Ewing, 1990). Leaf starch accumulation (per unit dry weight) during the day was more rapid under short days, and this was evident 2 days after treatments started and still present after more than 2 weeks (Lorenzen and Ewing, 1992).

Temperature

The effects on plant morphology of raising the temperature are generally similar to those of lengthening the photoperiod with dim incandescent light. Plants grown under high temperatures are taller because internodes are longer and because of more sympodial axillary branching. Other similarities to the effects of long days are that total leaf dry weight increases, total stem dry weight increases, the leaf : stem ratio decreases, leaves are shorter and narrower with smaller leaflets, the angle of the leaf to the stem is more acute, axillary branching at the base of the main stem increases, more flowers are initiated, and flower bud abscission is reduced (Werner, 1934; Borah and Milthorpe, 1962; Bodlaender, 1963; Petri, 1963; Ewing, 1981; Ben Khedher and Ewing, 1985; Menzel, 1985; Wheeler *et al.*, 1986; Turner and Ewing, 1988; Manrique *et al.*, 1989; Struik *et al.*, 1989a).

Whether senescence is accelerated or delayed by increasing the temperature depends upon the photoperiod and other conditions. If days are long, high temperatures may shift the partitioning away from tubers toward shoot growth to the point where plant senescence is delayed (Marinus and Bodlaender, 1975; Ben Khedher and Ewing, 1985; Struik *et al.*, 1989a). Conversely, if photoperiods are short enough to permit reasonable tuberization even at the high temperatures, then the more rapid growth and development at high temperatures may shorten the growing season (Manrique *et al.*, 1989; Vander Zaag *et al.*, 1990).

Some cultivars are more sensitive to high temperatures than are others (Ben Khedher and Ewing, 1985; Reynolds and Ewing, 1989b; Manrique *et al.*, 1989). Nevertheless, it seems safe to say that for all genotypes, high temperatures, like long photoperiods, decrease the partitioning of assimilate to tubers and increase partitioning to other parts of the plant (Wolf *et al.*, 1990).

Irradiance level

Just as lowering the irradiance level decreases induction to tuberize, so it decreases the partitioning of assimilate to the tubers (Bodlaender, 1963; Gray and Holmes, 1970; Sale, 1973a,b; Menzel, 1985). This was true even when photosynthesis was not limiting (Menzel, 1985).

Low levels of irradiance have effects similar to long photoperiods and high temperatures on the morphology of the potato plant, and the effects are exacerbated when these factors are present in combination (Bodlaender, 1963; Menzel, 1985; Demagante and Vander Zaag, 1988). Low light intensity increases stem elongation and delays plant senescence (Demagante and Vander Zaag, 1988). The delayed senescence may give a longer duration of the canopy, but any associated yield advantage tends to be negated by delays in canopy development and in achievement of ground cover (Struik, 1986; Struik *et al.*, 1990). Leaves are thinner; but, unless the irradiance level is extremely low, leaf area may be higher (Bodlaender, 1963). The number of nodes on main stems plus branches was increased at low irradiance, and the dry weight ratio of leaves : stems was decreased (Menzel, 1985).

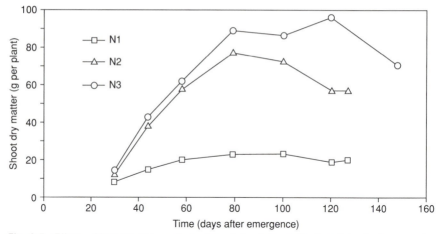

Fig. 8.5. Effects of N fertilization on formation of shoot dry matter in potato plants grown in the greenhouse. N1 = 2.5 g plant^{-1}; N2 = 8.0 g plant^{-1}; and N3 = 16.0 g plant^{-1} (Biemond and Vos, 1992).

The effects of lowering irradiance described up to this point resemble the effects of long photoperiods and high temperatures, but the effects on flowering are different. The abortion of flower buds is decreased by long days and high temperatures, whereas it is increased by shading the plants (Bodlaender, 1963; Demagante and Vander Zaag, 1988; Turner and Ewing, 1988). Apparently the reduction in available photosynthate from shading is overriding in its effects on retention of flower buds.

Nitrogen nutrition

Again there is a parallelism in the effects on tuber induction and the effects on overall morphology. Just as induction to tuberize tends to decline as the level of nitrogen increases, so the morphological effects of high nitrogen resemble the effects of long days. Increasing nitrogen produces taller plants, longer internodes, more sympodial axillary branches, higher leaf dry weights, higher stem dry weights, lower leaf : stem ratios, higher root dry weights, higher shoot : root ratios, and delayed senescence (Gunasena and Harris, 1968; Dyson and Watson, 1971; Santeliz-Arrieche, 1981; Millard and MacKerron, 1986; Oparka et al., 1987; Payton, 1989; Biemond and Vos, 1992; Vos and Biemond, 1992).

Associated with the increased partitioning of dry matter to shoots rather than to tubers is a great increase in shoot biomass from high applications of nitrogen (Fig. 8.5). Stems respond even more than leaves to increased nitrogen, but the leaf area from plants heavily fertilized with nitrogen is very high compared to those that receive none. The consequent prolonged shoot growth and the increased duration of a canopy for light interception usually produces a much higher final yield of tubers than in plots that receive no nitrogen fertilizer. This is in spite of the fact that the unfertilized plants have a much higher harvest index, defined as the tuber

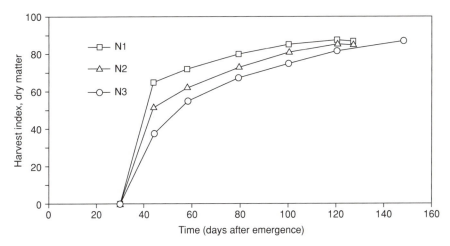

Fig. 8.6. Effects of N fertilization on the proportion of total dry matter in tubers (harvest index) of plants shown in Fig. 8.5. N1 = 2.5 g plant^{-1}; N2 = 8.0 g plant^{-1}; and N3 = 16.0 g plant^{-1} (Biemond and Vos, 1992).

dry weight expressed as a percentage of the total plant dry weight (Fig. 8.6). To realize this higher yield potential, the season must be sufficiently long, with good levels of irradiance at the end of the season to take advantage of the extra canopy duration for the high nitrogen plants.

Genetic differences

Observations within subsp. *andigena* and *andigena* × *tuberosum* populations segregating for adaptation to long days have shown clear and consistent associations of morphology with tuberization response (Glendinning, 1975; Lazin, 1980; Rasco *et al.*, 1980).

Genotypes that are less induced to tuberize (as evidenced, for example, by poor tuberization on cuttings taken from the plants), were taller, more highly branched, had smaller leaflets, more acute angles between leaves and stems, higher total leaf dry weights, higher stem dry weights, lower dry weight ratios between leaves and stems, more flowering, more abundant rooting, and later senescence. Polygene mapping indicated that several loci for ability to tuberize under long photoperiods coincided with loci for larger leaves and for reduced axillary branching (Van den Berg *et al.*, 1996).

Physiological age of the mother tuber

Inasmuch as increased physiological age of the mother tuber increases induction to tuberize, one might expect its effects on plant morphology to resemble the effects of short photoperiods. Comparisons are complicated by the fact that there are usually more stems arising from physiologically older seed tubers; but the smaller plant size, earlier tuberization, and earlier senescence often observed when physiologically old mother tubers are compared to young ones are consistent with the contrasts observed when short day plants are compared to long day plants.

Summary

Compared to a non-induced plant, a plant that has been strongly induced to tuberize will be shorter, both because of shorter internodes and because it has fewer sympodial branches. It will also have fewer axillary branches at the base of the plant, its leaves will be larger, and angles between leaves and stem will be less acute. The ratio of leaf dry weight to stem dry weight will be greater. Flower bud abortion will be increased, and rooting will be restricted. Induction to tuberize has the effect of shifting assimilate partitioning to the tubers. Consequently, the harvest index increases with induction. As all new growth is concentrated in tubers, plant senescence is advanced.

EVOLUTIONARY VALUE OF CHANGES ASSOCIATED WITH INDUCTION

The changes associated with induction can be considered in the context of the competitive advantages they may have provided during evolution of the wild potato in the high Andes. As a storage organ the tuber permits survival during the period of freezing temperatures there. Dispersion of the wild plants takes place through seeds, but also through stolon growth, with longer stolons providing greater spread. This applies whether new plants form during the current growing season from the stolons, as when the apex of the mother plant is damaged and the stolon tips turn upward to form orthotropic shoots, or whether new plants sprout the following season from tubers that formed at the tips of stolons and over-wintered.

Early in the growing season the best survival strategy is to have non-induced plants. These will produce more vigorous shoot growth to shade out competing species, because plants will have relatively long stems and abundant branches. Thicker leaves associated with less induction to tuberize may provide more resistance to damage from biotic or abiotic factors. Leaf senescence will be retarded to permit more light interception over the course of the growing season. The longer photoperiods and warmer temperatures of summer provide less induction to tuberize and hence favour these morphological traits.

After plant growth has become well established, a modest shift to stolon production would be beneficial for dispersal, as already described. This might be triggered in wild potatoes either by a slight shortening of the photoperiod or by the attainment of large leaf areas (Kahn *et al.*, 1983). Once the danger of frost is imminent, survival strategy calls for a shift to tuberization. The shift to the more induced state would be brought about by shortening daylengths, cooler temperatures, and perhaps by lower supplies of soil N.

As induction to tuberize increases, the leaf : stem ratio and specific leaf area increase. This serves to maximize the leaf area produced from a given amount of biomass that is partitioned to shoots. The thinner leaves have the same rate of photosynthesis per unit leaf area, which means a higher rate of photosynthesis per unit of biomass partitioned into shoots. Compared to what would be expected from plants that had not developed these evolutionary adaptations, there is an increase in daytime accumulation of leaf starch and an increase in assimilate export, which supports the developing tubers. The strong sink that is formed by the developing tubers helps to maintain a high rate of photosynthesis. Although the strong tuber sink also contributes to early senescence of shoots, this is not a problem in the high Andes, where short days are soon followed by killing frosts.

PHYSIOLOGICAL NATURE OF THE INDUCTION TO TUBERIZE

MOVEMENT OF THE TUBERIZATION STIMULUS

The tuberization stimulus was able to move across a graft union, proceeding from the leaves of induced scions to the underground nodes of non-induced stocks, where tubers were produced. Reciprocal grafts did not tuberize (Gregory, 1956; Kumar and Wareing, 1973). Interstem grafts indicated that the stimulus is transported acropetally as well as basipetally (Kumar and Wareing, 1973). Other grafting experiments showed that characteristics of the leaf rather than the buried bud were responsible for the ability of clones that have long critical photoperiods to tuberize under long days (Ewing and Wareing, 1978).

Interspecies grafts have also provided interesting information. Transport of the tuberization stimulus was possible through a leafless segment of tomato or eggplant, but the leaves of these species diminished tuberization whether or not such leaves were exposed to short photoperiods (Madec and Perennec, 1959; Okazawa and Chapman, 1963). 'Mammoth' tobacco scions, which require short photoperiods for flowering, were grafted to leafless potato stocks from plants with a short critical photoperiod for tuberization (Chailakyan et al., 1981; Martin et al., 1982). Tuberization occurred on the potato stocks only if the scion was exposed to short days. When Nicotiana sylvestris, which requires long days for flowering, was substituted for the Mammoth tobacco in the scion, tubers formed on the stock only if the scion was exposed to long days (Chailakyan et al., 1981; Martin et al., 1982)! The implication of these experiments is that the stimulus for flowering in tobacco was graft-transmissible to the potato, where it induced tuberization, and that this was true even if the stimulus was produced by a tobacco species that requires long photoperiods rather than short ones for flowering.

Even though the tuberization stimulus moves throughout the plant and affects overall morphological development, the formation of stolons and tubers takes place preferentially in underground (or otherwise darkened) parts. In temperate regions tuberization tends to coincide with flower production, but the relationship is not causal. Neither is there an absolute minimum age or size requirement for tuberization. For example, planting true seeds of very early maturing genotypes under short, cool days often leads to tuberization when only one or two leaves have developed beyond the cotyledons. However, other factors being equal, the larger the plant the more likely it is to tuberize. Under inducing conditions both young and old leaves are capable of producing the stimulus (Hammes and Beyers, 1973); so the greater the leaf area, the more stimulus is available for transport underground (Kahn *et al.*, 1983). If the plant was grown from a seed tuber, then factors from the tuber interact with factors from the leaves in determining the overall level of induction.

Interference with translocation from the leaves to the underground parts of the plant, as when stems are girdled by *Rhizoctonia* disease, may prevent tuberization of underground parts. In this event, axillary buds on aerial portions of the plant may tuberize, in spite of the inhibiting effects of light.

ROLE OF HORMONES

Gibberellins

There is convincing evidence that GAs interfere with tuberization. Application of exogenous GA reduces tuberization, whether on whole plants (Okazawa, 1960), cuttings (Tizio, 1971), *in vitro* plantlets (Hussey and Stacey, 1984), or excised sprouts cultured *in vitro* (Koda and Okazawa, 1983a). Long days, high temperatures, low irradiance, and high N fertilization all produce less induction to tuberize; and all are associated with higher levels of GA activity (Woolley and Wareing, 1972c; Railton and Wareing, 1973; Krauss and Marschner, 1982; Menzel, 1983). Van den Berg *et al.* (1995) measured less GA_1 in plants grown under short than under long photoperiods. Other evidence is that GA activity of slightly swollen stolons was less than in stolons with no swelling (Koda and Okazawa, 1983b). A dwarf mutation that increased tuberization in ssp. *andigena* under long days (Bamberg and Hanneman, 1991) had lower levels of GA_1 than wild-type plants (Van den Berg *et al.*, 1995).

Although there is every reason to think that GAs play an important negative role in tuberization, it is unlikely that the tuberization stimulus is comprised only of a low level of these compounds. A balance of compounds is more likely to be involved (Ewing, 1995). This would be consistent with the observation that at least 11 loci have been found for genes that affect the ability of potatoes to tuberize under long days (Van den Berg *et al.*, 1996).

Abscisic acid

Treatment with chemicals that block the synthesis of endogenous GAs promoted tuberization in whole plants (Gunasena and Harris, 1969; Hammes and Nel, 1975; Menzel, 1980), cuttings (Langille and Hepler, 1992), *in vitro* plantlets (Hussey and Stacey, 1984; Estrada *et al.*, 1986; Dodds, 1990), and sprouts cultured *in vitro* (Tizio, 1969). This leads to the hypothesis that the tuberization stimulus is a naturally occurring inhibitor or antagonist of GA. Abscisic acid (ABA) is a likely candidate (Krauss and Marschner, 1982; Wareing and Jennings, 1980), especially in view of the fact that there appears to be a close correspondence between the locations of several genes for the ability to tuberize under long days and genes for ABA level (Šimko *et al.*, 1996a).

Jasmonates

Jasmonic acid is another hormone that tends to have inhibitory effects on growth. It and certain closely related compounds enhance tuberization *in vitro* (Koda *et al.*, 1991). These include 12-OH-jasmonic acid and its glucoside, both of which have been isolated from potato leaves (Yoshihara *et al.*, 1989; Šimko *et al.*, 1996b). The former compound was named tuberonic acid (Yoshihara *et al.*, 1989), but the roles of tuberonic acid and its glucoside *in planta* are yet to be established (Helder *et al.*, 1993; Šimko *et al.*, 1996b).

Cytokinins

The addition of cytokinin frequently promotes tuberization *in vitro* (Palmer and Smith, 1969), and transfer of plants to cooler temperatures and shorter photoperiods has been associated with a temporary increase in cytokinin content of leaves (Langille and Forsline, 1974). On the other hand, cytokinin activity in stolon tips showed little increase until tubers were more than twice the size of attached stolons (Koda and Okazawa, 1983b). It is often suggested that a high ratio of cytokinin to GA constitutes the tuberization stimulus (Melis and Van Staden, 1984). The suggestion is attractive, but it would not be surprising if other compounds also play a role.

Ethylene

Although ethylene may promote tuberization in dahlia (Biran *et al.*, 1972) and radish (Vreugdenhil *et al.*, 1984), it does not appear to play a similar role in potato tuberization. The application of an ethylene producing compound to extremely old seed tubers caused a restoration of more normal sprout growth instead of sprout-tubers forming directly at the eye (Dimalla and Van Staden, 1977); and GA activity was higher in the elongated sprouts than in the sprout-tubers. Apparently the ethylene stimulated GA levels, which had their usual effect of inhibiting tuberization. A different mode of action proposed for ethylene is that ethylene produced by friction between soil particles and the growing stolon tip might stop extension growth of the stolon, thereby facilitating tuberization (Vreugdenhil and Van Dijk, 1989; Vreugdenhil and Struik, 1989, 1990).

Auxin

Auxin has been studied less than the other known hormones with respect to its role in tuberization (Melis and Van Staden, 1984). There is evidence from analysis of plantlets grown *in vitro* that it may interact with other hormones in regulating tuberization (Sergeeva *et al.*, 1994).

Plant growth regulators

Exogenous applications of compounds that block GAs sometimes improve tuberization under environments unfavourable to tuber induction, especially in obtaining tuberization on *in vitro* plantlets (Hussey and Stacey, 1984). Many reports indicate that cytokinins and related compounds improve *in vitro* tuberization.

Applications of various growth regulators to whole plants often produce increases in numbers of tubers set, but because the average tuber size is less, there is not usually a yield benefit. Moreover, there tends to be a good deal of variability from experiment to experiment in the degree to which tuber set is increased. Here it should be borne in mind that a slight phytotoxicity to the leaves is often accompanied by a heavier set of smaller tubers, whether herbicides or growth regulators are applied.

ASSIMILATE LEVEL

An early hypothesis to explain induction to tuberize was that the level of non-structural carbohydrate in the leaf was the controlling factor. Short photoperiods and cool temperatures would slow leaf growth, causing the accumulation of assimilate and a high C : N ratio, which in turn would bring about tuberization (Garner and Allard, 1923; Wellensiek, 1929; Werner, 1934; Driver and Hawkes, 1943; Borah and Milthorpe, 1962; Burt, 1964). Increasing the level of N fertilization would decrease the C : N ratio and thus tend to overcome the effects of short days.

With the discovery of plant hormones and mounting evidence concerning their involvement in tuberization, the hypothesis was largely abandoned. However, it is still possible that high assimilate level is a contributing factor in induction, along with hormonal effects. For example, *in vitro* tuberization is highly dependent on sucrose level (Gregory, 1956). Increasing the sucrose concentration to at least 175 mmol l^{-1} greatly increases the frequency and size of tubers, and this is not simply an osmotic effect (Rocha-Sosa *et al.*, 1989; Wenzler *et al.*, 1989; Perl *et al.*, 1991). Another reason to suspect that high assimilate level is involved in induction to tuberize is that several genes which seem to be intimately associated with tuberization are 'turned on' by high sucrose concentrations.

When plants are shifted from long to short photoperiods, leaf growth does slow down as the hypothesis proposes, but the slowdown does not occur until after there has already been an increase in the rate of leaf starch accumulation (Lorenzen and Ewing, 1990, 1992). Thus it is incorrect to assume that the retardation of growth caused the increased accumulation of starch (Driver and Hawkes, 1943). Nevertheless, it could well be that starch accumulation in the leaves plays a significant role in tuber induction and that the extra assimilate available for export during the night may comprise part of the stimulus for tuberization.

ESTABLISHMENT OF THE LEAF CANOPY

IMPORTANCE OF A HEALTHY START

The traditional propagule for the potato crop is a tuber or a piece of a tuber, usually weighing about 50 g. Such a propagule is large enough and contains sufficient food reserves to get the new plant off to a rapid, vigorous start. Unfortunately, the seed tuber is also an ideal vehicle for the transmission of plant diseases from one generation to the next, including diseases caused by viruses, bacteria, fungi and nematodes. These diseases can have extremely adverse effects on yield and quality, and some of them can even lead to soil infestations that will affect succeeding crops. Therefore it is of the utmost importance that healthy seed tubers be planted.

The starting materials for production of disease-free tubers are tissue culture plantlets produced by meristem-tip culture and maintained *in vitro* (Fig. 8.7). Such plantlets will form small tubers, called 'microtubers', *in vitro*, and these can be planted directly in the field for propagation. Microtubers typically range in size from less than 250 mg to 1 g. An alternative is to transplant the plantlets from the test tube into greenhouse pots. Tubers produced in this way are called 'minitubers', which might range in weight from 5 to 25 g or more.

Theoretically, microtubers are completely free from disease organisms when harvested from tissue culture. If care is exercised to protect the greenhouse plants from aphids and in all other ways minimize disease infection, minitubers may be nearly as free from disease as microtubers. However, once microtubers or minitubers are planted in the field, even under ideal circumstances each successive crop of seed tubers will pick up more and more infection with virus and other disease organisms. For this reason most growers would like to minimize the number of seed crop generations between *in vitro* culture and their farms. Adding to the problem is the fact that multiplication rates are low: 1 hectare of seed tubers will produce only enough seed tubers to plant about 10 hectares in the next generation. This has led to a search for more economical ways of producing microtubers and minitubers, and for planting these propagules directly in the field.

A reliable supply of high quality seed tubers requires not only access to tissue culture plantlets, but also cool climatic conditions that will reduce problems with the aphids that serve as vectors for many virus diseases. Also required are experienced growers who understand how to carry out the strict sanitary conditions and other specialized skills of seed potato production. A final necessary ingredient is a governmental program for seed certification, including rigorous programs for inspection and quality control. Relatively few countries are able to meet this combination of requirements.

A large portion of the seed potatoes for export to developing countries comes from the Netherlands, Scotland, Ireland and other parts of northern Europe. Seed potatoes from these countries can be costly for growers in Africa and Asia; but if locally grown tubers are planted instead, diseases may have serious effects on crop yield and quality. Consequently, there has been interest in finding alternative methods for propagation. These include transplanting locally grown tissue culture plantlets or cuttings thereof, planting microtubers or minitubers, and planting seeds (commonly referred to as 'true potato seeds' or TPS to distinguish them from seed potato tubers). TPS can be seeded directly in the field, sown in beds for transplanting to the field, or sown in beds to produce tubers that can be planted in the field.

PROBLEMS WITH ALTERNATIVE PROPAGULES

Presence of a conventional seed tuber tends to delay tuberization. Consequently, alternative methods of propagation may be hampered by premature tuberization. If the photoperiod is short in relation to the critical

Fig. 8.7. A field of healthy seed potatoes, originating from tissue culture plantlets (inset). Plantlets, produced by meristem culture and maintained *in vitro*, serve as the starting material for the production of plants that will be low in disease content if multiplied under strictly controlled conditions (courtesy of S.A. Slack).

Fig. 8.8. Multiple axillary branching at the base of the stem, typical of potato plants grown from propagules other than tubers. The plant shown originated from a true potato seed (TPS) in the greenhouse and was transplanted to the field in upstate New York (photo from PhD thesis research by A.B. Rowell).

photoperiod for the genotype under the prevailing temperatures and irradiance, small tubers will be initiated before plants have attained an adequate size to support tuber growth. Plants will be stunted severely, and marketable yields will be greatly reduced.

A diminished supply of nutrients available from alternative propagules tends to limit initial growth rates, even in the absence of premature tuberization, compared to those obtained from 50-g seed tubers (Rowell *et al.*, 1986). Early root development is retarded even more than shoot development (Lommen, 1994). Consequently, more time is required to achieve full ground cover by the canopy.

Transplants, whether obtained from rooted cuttings, *in vitro* plantlets, or seedlings, have only one stem per propagule; and micro- and small minitubers tend to develop only one sprout per tuber. It seems generally true that at a given population of propagules per unit area, the fewer the main stems per propagule, the more axillary branches per main stem. Plants from micro- and small minitubers, transplants, and seedlings all display a tendency for greatly increased branching at the base of the plant (Fig. 8.8). This can partially compensate for their lower number of main stems per unit area in terms of achieving ground cover by the canopy.

TRUE POTATO SEED

For the past 20 years the International Potato Center (CIP) in Peru has been examining the possibilities of using TPS for developing countries. Scientists at CIP are interested not only because very few diseases are transmitted through TPS, but because TPS is cheap to transport. A handful will plant a hectare of crop, compared to the 2.5 metric tons of seed tubers required.

Against these advantages for TPS compared to seed tubers, there are significant disadvantages. Seedlings are slower to emerge and more susceptible to many kinds of stress or damage than sprouts from tubers. This makes direct seeding difficult. An alternative to direct seeding is transplanting, much as tomatoes or other vegetables are grown. Transplanting is probably more realistic than direct seeding, but care must be exercised to avoid premature tuberization and consequent stunted growth on the transplants. Even when there is no premature tuberization, transplants tend to develop more slowly and have lower yields than plants grown from healthy seed tubers (Rowell *et al.*, 1986). Unlike plants clonally propagated from tubers, plants from TPS are subject to broad genetic variability, leading to lack of crop uniformity.

Some of these disadvantages may be overcome as researchers continue to give attention to the crop; for example, production of hybrid seed from selected parents will somewhat reduce the problem of genetic variability. Chile and India are among the countries that have embarked on hybrid seed production of TPS. Nevertheless, it seems unlikely that TPS will ever replace seed tubers in major potato production areas that have reliable sources of seed tubers available at reasonable prices. For areas that lack an economical supply of healthy seed tubers, the best way to use TPS may be to sow seeds densely in nursery beds, and then to harvest the seedling tubers and plant them in the field like conventional seed tubers. Because the seedling tubers are not clonally propagated, they will lack the uniformity of conventional seed tubers; but if reinfection can be avoided in the nursery beds, the seedling tubers should be free from clonally transmitted diseases. The People's Republic of China has used TPS in this way for a number of years.

Where nursery bed production of seedling tubers is not feasible, some developing countries are experimenting with transplants as the starting propagule; and there are even a few efforts with direct seeding. It remains to be seen whether these approaches will be accepted by growers on a long term basis.

MICROTUBERS AND MINITUBERS

The other major interest in alternative propagules is for microtubers and minitubers. The major push here is from growers in countries that already have a good source of certified seed, where there is a desire to shorten the number of generations between tissue culture and the production of seed for general usage. Success will likely depend upon developing techniques for production of larger micro- and minitubers, and for lowering the cost of production. Most of the problems in using these propagules will be minimized as larger ones become available.

FORMATION AND GROWTH OF INDIVIDUAL STOLONS AND TUBERS

A typical plant dug up in mid-season from a commercial potato field shows a number of stolons, only some of which bear tubers. Tubers will be highly variable with respect to size and location on the stolon. Some stolons will be long, others short; there will be differences in the degree of branching; and they will arise from different nodes on the main stem. All this variability may affect economic value of the crop through effects on tuber yield and size distribution. Tubers forming on long stolons are more likely to protrude from the ridge and be exposed to the light, resulting in greening that renders the tuber worthless. The opposite extreme is tubers that form too deep in the ridge and require more energy for harvest because of the extra soil that must be handled.

PATTERN OF STOLON FORMATION

Like tubers, stolons form more readily in darkness than in light, although under certain conditions both can be made to develop in the light (Kumar and Wareing, 1972). Even above-ground axillary buds will form stolons and tubers if they receive the appropriate stimuli. Stimuli for stolons can be produced by the mother tuber in the absence of aerial parts, since stolon formation often begins before shoot emergence. Stolons also develop in the absence of a normal mother tuber, for example when TPS or microtubers are planted. Thus it appears that stolon production can be affected by factors from both the mother tuber and leaves.

On potato plants grown from seed tubers, stolons develop first at the most basal node of the sprout and later at higher nodes (Plaisted, 1957; Cutter, 1992). In one study about half of the stolons formed at the most basal node, and roughly 10% of the remaining stolons formed at each of the next four higher nodes (Wurr, 1977).

DIAGRAVITROPIC GROWTH

Stolon growth is diagravitropic (i.e. in a horizontal plane), but decapitation of the main shoot converts it into a negatively gravitropic (upright) shoot (Sachs, 1893; Lovell and Booth, 1969; Kumar and Wareing, 1972). There are indications that the apex maintains diagravitropic stolon growth through an interplay among auxin, GA and cytokinin levels (Booth, 1963; Kumar and Wareing, 1972; Woolley and Wareing, 1972a–c). If *andigena* plants had been exposed to short photoperiods, then auxin and GA applied to the stumps of decapitated plants favoured the diagravitropic outgrowth of above-ground lateral buds as stolons (Woolley and Wareing, 1972c). Cytokinin applied to the stolon tip caused a switch to upright shoot growth (Kumar and Wareing, 1972). The presence of roots on cuttings also increased the formation of upright growing shoots rather than stolons (Kumar and Wareing, 1972; Woolley and Wareing, 1972a), but applications of cytokinin substituted for the presence of roots in this respect (Kumar and Wareing, 1972; Woolley and Wareing, 1972a,b). Applications of auxin and GA to the stumps of stem cuttings decreased the transport of benzyladenine toward the stolon tip by threefold (Woolley and Wareing, 1972a).

STOLON GROWTH

An *andigena* plant grown under very non-inducing conditions (long days, warm temperatures) will produce very few stolons until the plant attains considerable size. Under these conditions, providing slightly shorter days or cooler temperatures will increase stolon production. However, if stolons are already growing profusely, then giving more induction will lead to a reduction in stolon elongation because many stolons will develop tubers. Thus whether short days increase or decrease stolon growth depends upon how induced the plants are to begin with.

Stolon branching is stimulated by long days (Struik *et al.*, 1988), high temperatures (Struik *et al.*, 1989b), and high GA levels (Struik *et al.*, 1989d). These are the same conditions that promote a shift away from tuberization toward stolon growth. Later formed stolons normally have a shorter interval before tuber initiation becomes dominant (Vreugdenhil and Struik, 1989). Thus the stolons at the base of the plant typically have longer to grow, are more likely to branch, and provide the most potential sites for tubers (Lovell and Booth, 1969; Struik and Van Voorst, 1986).

EFFECT OF NODE POSITION ON TUBERIZATION PATTERN

Cuttings

Cuttings have been used to examine the effect of node position on tuberization. If two or more nodes of a cutting are buried, the most basipetal bud is most likely to tuberize and develops the largest tuber (Gregory, 1956; Chapman, 1958; Kahn and Ewing, 1983). This phenomenon is explained neither by orientation with respect to gravity nor by assuming that the most basipetal bud is most likely to tuberize: turning the cutting upside down or laying it horizontally still produced the most tuberization on the oldest bud; and burying the two youngest buds rather than the two oldest ones produced stronger tuberization on the younger of the two buried buds (Kahn and Ewing, 1983).

The tuberization pattern in the intact plant is much less predictable than in a cutting. There is great variability in the proportion of stolons that develop tubers (Moorby, 1967; Wurr, 1977; Cother and Cullis, 1985), and the first stolons formed are not necessarily the first to tuberize. Cultivars differ not only as to the percentage of stolons that bear tubers, but also with respect to the pattern of tuberization at different nodes.

The pattern of tuber size distribution on the stolon system is also inconsistent and does not always match the pattern described above for cuttings. Clark (1921) and Gray (1973) noticed a tendency for smaller tubers on the upper stolons and larger tubers on the lower nodes. Gray also found a shortening of tubers-bearing stolon proceeding upward from the base of the stem. However, Krijthe (1955) concluded that the third, fourth and fifth nodes above the base of the stem had the largest tubers. Plaisted (1957) and Cother and Cullis (1985) agreed that tubers reaching marketable size were more frequent on the lower than on the upper stolons, but the distribution varied with stolon number. The more stolons present, the lower the percentage of marketable tubers at the lowest stolon positions (Cother and Cullis, 1985).

GROWTH RATES OF INDIVIDUAL TUBERS

Usually many more tubers are initiated than develop to a marketable size. Some are resorbed, some remain small until plant maturity, and others grow to variable sizes; but the first to be formed does not necessarily attain the largest size (Wurr, 1977; Struik and Van Voorst, 1986; Struik *et al.*, 1988; Struik *et al.* 1990, 1991). Little is known about the controlling factors that determine the fate of the individual tuber initial, but there is some evidence that significant differences in sink strength already exist among stolon tips before visible swelling of tubers takes place (Oparka and Davies, 1985a). Chapman (1958) claimed that leaves on one side of a plant control the tuberization of stolons on the same side of the plant. If so, the relative vigour and exposure to light of individual leaves may help determine whether the stolons associated with these leaves will tuberize.

The rate of initial growth may be important; tubers with restrictions in their mitotic processes during early development grew more slowly during later stages (Reeve *et al.*, 1973). One can only speculate as to which factors eventually become dominant in controlling the observed differences in growth rates among tubers as they enlarge further (see Struik *et al.*, 1991). Without this knowledge, it is difficult to manipulate tuber size distribution or even to predict how a given set of environmental factors will affect it under field conditions.

TUBER DISTRIBUTION IN THE FIELD

Kouwenhoven (1970) experimented with various combinations of planting depth and ridge size. Deeper planting produced less lateral dispersion of the tuber crop, both in the direction of the ridge and across the ridge, and also caused new tubers to form slightly higher with respect to the position of the seed tuber – although the average depth of new tubers below the ridge surface was increased. There was a marked reduction in green tubers from the combined effects of less lateral dispersion across the ridge and more distance below the surface of the ridge.

Figure 8.9 shows the pattern of development of tuber size over time in the field by plotting yield of dry matter in various size classes throughout the season. As mean tuber size increases with time, so does the range in sizes. The curves reflect a dynamic situation, where tubers are initiated over a range of time. Some of these tubers enlarge, but at differential rates. Other tubers remain small, and still others are resorbed or decay while still small.

RESORPTION

Apparently there is a limit as to the number of tubers that can develop to a large size under a given set of conditions, and many of the remainder are

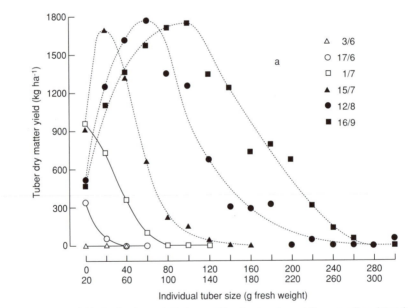

Fig. 8.9. Tuber yield distributions plotted according to size class at five sampling dates (June 17, July 1, July 15, August 12, and September 16) in a field experiment (Struik *et al.*, 1990).

resorbed. The number of tubers initiated can vary much more than the number of tubers set (Struik *et al.*, 1988). The mechanism by which this phenomenon operates is unknown, as is what determines which tubers will develop, which will remain as small tubers, and which will be resorbed. Even actively growing tubers can lose carbon compounds to other parts of the plant (Oparka and Davies, 1985b). In the case of resorption, so much is lost that the tuber shrivels and disappears. It seems probable that the non-structural dry matter from the resorbed tuber is redistributed to other tubers, although this has not been proven. Cho and Iritani (1983) reported that in some treatments nearly half of the tubers initiated were resorbed, but resorption was limited to tubers that weighed less than 10 g.

Drought is an important factor in determining the proportion of tuber initials that set, and thus the number that will be resorbed (Krug and Wiese, 1972). This helps explain why irrigation often increases tuber number.

TUBER DEFECTS RELATED TO INDUCTION

Hiller *et al.* (1985) have reviewed the numerous physiological disorders to which potato tubers are subject. No attempt will be made to deal here with the total list, but certain disorders related to the induction to tuberize will be considered.

SECOND GROWTH

Under some conditions growth of an individual tuber stops, and there is a partial or complete reversion to stolon growth from one or more eyes. Because this condition is often associated with warm soils, the phenomenon is known as 'heat sprouting' (Fig. 8.10). Heat sprouting is different from the normal breaking of tuber dormancy that occurs after tubers have been harvested and stored. It is more like an interruption in the dormant condition than a termination of it; subsequent to heat sprouting the tuber buds again become dormant.

As conditions again become more favourable for tuberization (e.g. cooler weather or more soil moisture to cool the soil) there may be a switch back to tuberization of the new stolons. This can happen more than once, leading to 'chain tubers' that resemble beads on a string. Much more common than heat sprouts and chain tubers are 'knobs' – formed when the initial reversion to stolon growth is incomplete, and secondary lateral growth leads to knobby protrusions at one or more eyes instead of heat sprouts or chain tubers. Other types of deformity include 'bottlenecks', 'dumbbells' and pointed-end tubers, the names of which are self descriptive. Collectively, these defects are known as 'second growth'. They can be considered as products of fluctuations in the intensity of the induction to tuberize, usually brought about by changes in temperature.

It seems likely that the knobs and secondary tubers also involve

Fig. 8.10. Heat sprouting in a tuber of 'Kerr's Pink,' grown in Ireland. Unusually hot weather caused the tuber to sprout while tops were still green.

resorption from the primary tuber – that the dry matter is removed from the attached primary tuber rather than merely passing through the vascular bundles of the primary tuber from other parts of the plant (Scholte, 1977).

CAUSES OF SECOND GROWTH

Cultivar differences

There are major differences in susceptibility to second growth among cultivars. 'Russet Burbank' and 'Bintje', two old cultivars that are respectively the leading cultivars grown in North America and Europe, are especially susceptible. Most modern cultivars have been selected for resistance to second growth and are much less subject to the problem, even though none are entirely free from it.

Drought and heat stress

Irregular or inadequate applications of irrigation water are commonly blamed for second growth. Drought stress is a contributing factor (Van Loon, 1986), but it should be noted that drought stress is typically accompanied by higher soil temperatures, which also play a role in the disorder (Lugt *et al.*, 1964; Ruf, 1964). It is likely that the higher soil temperatures cause an increase in production of GAs, and application of gibberellic acid to tuberizing plants is very effective in producing second growth (Bodlaender *et al.*, 1964). Although second growth is promoted by subjecting any part of the plant to high temperatures, exposure of the tubers is most effective (Struik *et al.*, 1989c).

Other factors

Of the various environmental factors that influence induction to tuberize, temperature is the most likely to fluctuate sufficiently to cause problems with second growth. However, variability in other factors can also have an influence. This includes sudden lengthening of the photoperiod (Madec and Perennec, 1962) and sudden increases in supply of nitrogen fertilizer (Bodlaender *et al.*, 1964; Krauss, 1985) during the period of tuber growth.

There may be indirect effects from the seed tubers on second growth. The speed at which a soil cover is established, influenced by the size and physiological age of the mother tuber, can be expected to influence soil temperatures. Table 8.1 shows the effects of size of micro- or minitubers on second growth. Emergence was slow from tiny seed tubers, giving delayed soil cover; and there was more than three times as much second growth as when seed tubers of normal size were planted.

A more direct cause of the increased second growth may also be operating when microtubers are planted. When a mother tuber is lacking (or so small as to be insignificant, as in the case of microtubers), induction to tuberize depends solely upon the shoot and its response to the environment (Madec and Perennec, 1962; Perennec, 1966). Tuberization on plants grown from physiologically young seed tubers is delayed compared to tuberization on plants grown either from physiologically old seed tubers or from cuttings, TPS and microtubers. It may be that physiologically young mother tubers supply GAs (Bialek and Bielińska-Czarnecka, 1975) or other factors that

Table 8.1. The effects of size of the mother tuber on percentage ground cover 51 days after planting (DAP) and on percentage of tubers > 20 mm that showed second growth. Size classes for the smallest tubers are presented by weight (g). The largest size class is presented in terms of diameter (mm) (after Lommen and Struik, 1990).

Size class	Soil cover at 51 DAP (%)	Second growth (%)
< 0.25 g	6	27.0
0.25–0.5 g	19	23.5
0.5–1 g	35	22.7
1–2 g	33	15.3
2–4 g	35	16.5
25–28 mm	84	7.6

counteract induction from the shoots. When these factors are not supplied by the mother tubers, fluctuations in temperature and light intensity may cause corresponding fluctuations in the induction to tuberize. This could increase the incidence of second growth in the manner already described.

ABA

Another hormone implicated in second growth is ABA, which tends to counteract the effects of GAs. Van den Berg *et al.* (1990) used tubers produced on cuttings as a model system to investigate second growth. Such tubers showed much lower ABA contents if exposed to conditions that promoted second growth (Van den Berg *et al.*, 1991). Other evidence is that second growth of tubers, produced by irregular supplies of nitrogen fertilizer to hydroponically grown plants, was associated with a lower ratio of ABA to GA (Krauss, 1985).

JELLY END ROT

Another form of resorption is from one part of the tuber to another. This produces 'translucent ends' and 'jelly end rot' (Hiller *et al.*, 1985). The names arise from the appearance of affected tissue at the basal end of the tubers, which becomes glassy or translucent, or – when more severe – soft and flaccid. Because such tissues are deficient in starch and high in sugars, dark brown colours develop during production of french fries or chips. Cultivars with long tubers seem to be most susceptible to the disorder (Iritani *et al.*, 1973). Again high soil temperatures are an important factor (Lugt, 1960), especially if accompanied by drought.

FACTORS DETERMINING PRODUCTIVITY

Based upon best yields obtained in small experimental plots under ideal conditions, fresh weight yields of potato tubers can reach at least 100 t ha^{-1}. Of course commercial yields fall far short of this, although recent yields in the state of Washington have been averaging about 65 t ha^{-1}. This compares to about 42 t ha^{-1} in the Netherlands and 10–15 t ha^{-1} in many developing countries.

Marketable yield is a function of total biomass production, the percentage of biomass that is partitioned to the tubers, the moisture content of the tubers, and the proportion of tubers that are acceptable to the market in terms of size and freedom from defects. These four variables are not independent of one another: for example, a high biomass production is possible only when the proportion partitioned to the tubers is moderately high, but not excessively so. If partitioning to the tubers is too low, the rate of photosynthesis will be sink limited and tuber yield will suffer because total assimilate production is restricted and also because too little of the assimilate is destined for the tubers. If partitioning to the tubers is too high, then leaf growth is limited and maturity is early.

BIOMASS PRODUCTION AND PARTITIONING TO TUBERS

High biomass production depends upon having a leaf canopy that over a long part of the available growing season intercepts a high percentage of the incident radiance (Van der Zaag, 1984). Under good growing conditions tuber dry matter continues in a nearly linear manner as long as the canopy is essentially closed (Fig. 8.11). Not only is it important to achieve rapid closure of the canopy, so that there is good interception of radiation early in the season, but maximum yields also require that the canopy continue to cover the soil late into the season. The presence of canopy late in the season depends upon leaves on axillary branches (both sympodial and basal); but axillary branching is inhibited by strong induction to tuberize, an added reason why very strong induction can reduce yields.

The effects of early senescence in shortening the period of canopy closure are exacerbated by several pests. Senescent plants are more susceptible to leafhoppers, early blight and *Verticillium* wilt; so if these pests are present, it is all the more important to avoid excessive induction. Otherwise the season may be severely shortened.

The effects of induction can be illustrated by the influence of cultivar (Fig. 8.11). Early maturing cultivars are typically able to tuberize even under

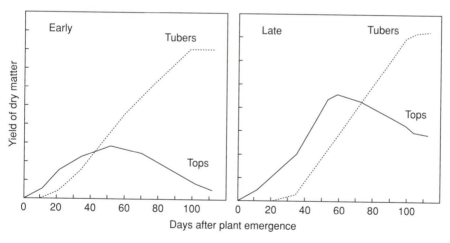

Fig. 8.11. Dry matter accumulation in tubers and tops of cvs Norland (left) and Russet Burbank (right), both grown in upstate New York. The early cultivar, Norland, started tuberizing earlier in the season. Total biomass production was less than for the late maturing Russet Burbank, but Norland partitioned a much higher percentage of its biomass to tubers. In this experiment the tuber yields at final harvest were only slightly higher for the late cultivar, but differences can be much greater over a longer season (data from Ben Khedher, 1983).

the long, hot days of early summer (i.e. they have a long critical photo-period). As tubers form on these cultivars, the growth of the rest of the plant is restricted. Sympodial growth stops soon after flower formation, and axillary branching at the base of the stem tends to shut down as well. Hence although tuber yields are relatively good at early harvests, they may not measure up well at later harvests because the full growing season is not well utilized for biomass production.

By contrast, late maturing cultivars do not tuberize as strongly until days are shorter or temperatures are cooler (i.e. late cultivars have a shorter critical photoperiod). Thus late maturing cultivars have a longer season during which tops continue to grow. There are more sympodial branches and axillary branches at the base of the stem. The canopy intercepts essentially all of the incident radiation over a greater portion of the season, and larger reserves of photosynthate are manufactured for translocation to the tubers. This explains why late maturing cultivars have the potential for higher yields over a long season.

Of course, it is possible for a cultivar to be so extremely late that it gives low yields. For example, if an *andigena* cultivar adapted to short-day conditions in the Andes is brought to New York State, tubers may not form until days are so short that the plant is first killed by frost. At the opposite extreme, potato cultivars adapted to the long summer days of North America or Europe may perform poorly in the Andes. The 12-h photoperiods and cool temperatures there induce a high level of tuberization stimulus so early in cultivars of subspecies *tuberosum* that tubers form soon after plant emergence. The induction may be so strong that no other plant growth takes place and yields will be very low.

Genetic differences have the greatest effects on degree of induction, and hence on plant maturity; but cultural practices can modify the effects of cultivar. For example, planting date affects photoperiod and temperature, both of which affect induction to tuberize. The application of nitrogen fertilizer also has an effect, as does the physiological age of seed tubers. Seed tubers which are so young that they are still dormant should never be planted, nor should they be so old that they will produce sprout tubers rather than normal sprouts; but there is a range of options between these extremes. Better early yields will result from physiologically older seed, whereas younger seed potentially will produce higher yields if the growing season is long.

Growers who want to aim for early yields will sacrifice potential yield by aiming for strong induction to tuberize early in the season. They will select a cultivar with a long critical photoperiod (i.e. an early-maturing one), choose physiologically old seed tubers, and use moderate to low rates of N fertilizer. Growers aiming for highest potential yields will select later cultivars with shorter critical photoperiods, physiologically younger seed tubers, and high rates of N. In areas with very long growing seasons, such as the Columbia River basin, N fertilizer is applied late in the season through the irrigation water. This serves to limit induction, delay senescence, and prolong the period of maximum light interception.

In the tropics the speed of crop emergence and the size and persistence of the canopy will influence soil temperature through shading (Midmore and Mendoza, 1984), which in turn may have a major impact on earliness of tuberization and yield (Midmore, 1984). More shading and consequent cooling can also be obtained through closer spacing (Midmore, 1983, 1988; Vander Zaag *et al.*, 1990), intercropping (Midmore, 1990), mulching (Midmore *et al.*, 1988a,b), and irrigation. Irrigation affects cooling directly, but also indirectly through increased canopy development and consequent shading (Lugt *et al.*, 1964; Van Loon, 1986). Soil cooling is especially important in tropical potato production, but may be beneficial under temperate conditions as well.

DRY MATTER CONTENT OF TUBERS

We have been considering how much biomass is produced by the plant, and what fraction of this biomass is partitioned to tubers. The next variable affecting yields is how much water accompanies the dry matter present in the tuber. The percentage dry matter in potato tubers commonly ranges from about 16 to 23%, depending upon cultivar and environment. Suppose that the biomass partitioned to tubers consists of 200 g per plant. If the dry matter comprises 20% of the tuber, then the tuber fresh weight will be 1.00 kg. By contrast, if the dry matter is only 16%, then the fresh weight of tubers will be 1.25 kg, one quarter again as much. Because tubers are sold by fresh weight, one might conclude that it is desirable to aim for a high water content. However, there are other important considerations.

Well over half of the potatoes consumed in North America are processed into frozen french fries, chips, dehydrated mashed potatoes and other products. For most forms of potato processing, the higher the dry matter content of the raw product, the higher the yield of finished product. Therefore processors typically set a minimum dry matter content below which they will refuse to purchase the potatoes. Dry matter content is important even for the fresh market. Potatoes low in dry matter will have a watery texture; those high in dry matter will be more dry and mealy. Consumers vary widely in terms of which texture is preferred; but high dry matter tubers usually command a higher price, especially for baking.

Potato cultivars vary widely in dry matter content. For example, if cv. Norland tubers have a dry matter content of 16%, cv. Atlantic tubers grown at the same location might be expected to have at least 22%. Environmental influence is also of major importance. High dry matter content is associated with high levels of irradiance and cool night temperatures. Fertilization and other cultural practices also have some effects.

PERCENTAGE MARKETABLE

Second growth and associated defects that may reduce the percentage of the potato crop acceptable for marketing have already been described.

Although numerous diseases, insects, and other pests can also damage potato tubers and ruin them for market, the major marketing defect in most parts of the world is mechanical damage at harvest time. Aspects of mechanical damage that fall within the purview of this chapter are physiological differences that affect the susceptibility to bruising. All potato tubers are subject to damage if dropped more than about 15 cm, but some are more susceptible than others. Potato tubers are especially subject to damage when cold, and it is desirable to let the soil warm to at least 10–15°C before harvest.

Another important factor is the thickness of the tuber periderm at time of harvest. The periderm, which protects against transpiration and mechanical injury as well as pathogens, thickens as the potato plant matures. In many parts of the world economical considerations mandate harvest while plants are still green and actively growing. Under such conditions, periderms will still be thin and easily damaged. As a result, tubers will be subject to cuts, bruises, water loss, and pathogen invasion. A partial remedy is to kill the potato plants well in advance of harvest. This is accomplished with various approved chemicals, by mechanical removal, or by 'pulling' the plants in such a way that the roots are partially broken. If the killing is done about 10–14 days in advance of harvest, there will be time for increased periderm formation on the tubers.

Artificial killing of the potato foliage will hasten periderm formation, but not to the degree that periderms will equal those obtained through natural maturity. This is especially true if plants are still very immature, so it is important that growers adopt practices that will not delay maturity excessively. The most frequently abused cultural practice in this respect is nitrogen fertilization. As already described, high rates of N will tend to lower the induction to tuberize, reduce the partitioning of dry matter to tubers, and delay senescence. Effects of cultivar and physiological age of seed on time of maturity are other obvious considerations.

FLOWERING AND SEED PRODUCTION

Because conventional propagation depends upon seed tubers, flowers and fruits have been viewed by traditional growers as a minor nuisance. In some areas, the seeds from potato berries fall to the ground and germinate the following season, presenting a 'weed' problem. Not only do the seedling volunteers compete with the regular plants, but they may even set tubers that compromise the purity of potatoes grown as certified seed.

Another concern with fruit set is that it diverts biomass production away from tubers. A few experiments have demonstrated that such an effect is measurable (Proudfoot, 1965; Jansky and Thompson, 1990), but under most circumstances it can probably be ignored.

Otherwise, flowering and fruiting have been of concern mainly to plant breeders. Except for selection of chance mutations (usually for more desirable skin colour), cultivar development depends on fruit set. With interest in TPS has come a new incentive to understand flowering and fruiting.

The growth of each main stem is terminated by an inflorescence, but stem growth may continue from an axillary bud. The new branch will again terminate in an inflorescence, and this process may continue for several stages, depending upon the vigour of the plant (see Fig. 8.2). The central axis of the potato inflorescence itself terminates in a flower (Fig. 8.12), which is the first flower in the inflorescence to open. This proximal flower tends to give the largest berry, the most seeds, and the heaviest seeds (Almekinders *et al.*, 1995). The development of flowers within an individual inflorescence is diagrammed in Fig. 8.13 (after Almekinders *et al.*, 1995).

The number of flowers that form is more a question of how many flower primordia survive than how many primordia are initiated. Many flower buds abort before they develop, which makes it difficult to study the effects of environment on primordia initiation. For this reason there is con- fusion in the literature as to whether flower initiation is favoured by short or long days. It appears that any effect of daylength on flower initiation is slight; but long days and moderately warm temperatures reduce flower bud abortion (Turner and Ewing, 1988; Almekinders and Struik, 1994), perhaps because these conditions are less conducive to tuberization.

The diversion of assimilate away from flower buds to competing tuber sinks may have some influence on flower bud abortion. To minimize such effects, potato breeders have devised various strategies to eliminate the developing tuber sink. One technique is to grow the plants in such a way that the developing tubers can be plucked off. Another is to graft the stems to tomatoes or other stocks that will not support tuberization. However, we can expect that considerable improvement in flower production and fruit

Fig. 8.12. Potato inflorescence (courtesy of R.L. Plaisted).

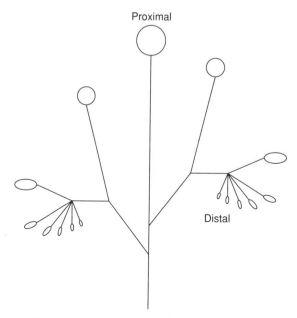

Fig. 8.13. Diagram of a potato inflorescence, showing flower positions in the cyme. The most proximal flower tends to give larger berries and seeds than the most distal flower (after Almekinders *et al.*, 1995).

set will be made as plant breeders interested in TPS production select for these traits.

MOLECULAR BIOLOGY AND POTATO PHYSIOLOGY

The potato lends itself to genetic engineering not only because it is among the crops most amenable to transformation and regeneration, but also because it is clonally propagated (Vayda and Belknap, 1992). Added to this, its value as a world food crop and the large number of pests that attack it have made the potato a prime subject for genetic manipulations. Transgenic plants are already available that carry resistance to various insects or diseases, and interest is high in inserting still other genes (Belknap *et al.*, 1993). Work is also underway on such diverse projects as improving the quality of potato starch and even delivering vaccines to children in developing countries through transgenic potatoes.

Transgenic plants also offer exciting prospects for advancing our understanding of potato physiology. Mention has already been made of transgenic potatoes that are deficient in phytochrome B (Jackson *et al.*, 1996), and there are many other possibilities. These include transgenic plants with varying levels of different hormones, so that we can examine effects of the hormones on tuberization, tuber dormancy, and other traits.

A different approach to understanding potato physiology through molecular biology is polygene mapping. Very comprehensive maps of potato chromosomes have been constructed by using DNA-based markers (Tanksley, 1993). This permits us to map potato populations for the ability to tuberize under long days, for tuber dormancy, or for almost any other trait that is controlled by multiple genes. Mapping the same populations for levels of various hormones will tell us whether the gene locations for particular hormones coincide with the gene locations for tuberization or dormancy. Mapping the hormones will also give us information needed for isolating the important genes. These and other molecular techniques promise to open many new avenues for research in potato physiology.

REFERENCES

Almekinders, C.J.M., Neuteboom, J.H. and Struik, P.C. (1995) Relation between berry weight, number of seeds per berry and 100-seed weight in potato inflorescences. *Scientia Horticulturae* 61, 177–184.

Almekinders, C.J.M. and Struik, P.C. (1994) Photothermal response of sympodium development and flowering in potato (*Solanum tuberosum* L.) under controlled conditions. *Netherlands Journal of Agricultural Science* 42, 311–329.

Bamberg, J.B. and Hanneman, R.E. Jr (1991) Characterization of a new gibberellin related dwarfing locus in potato (*Solanum tuberosum* L.). *American Potato Journal* 68, 45–52.

Batutis, E.J. and Ewing, E.E. (1982) Far-red reversal of red light effect during long night induction of potato (*Solanum tuberosum* L.). *Plant Physiology* 69, 672–674.

Belknap, W.R., Vayda, M.E. and Park, W.D. (eds) (1993) *Biotechnology in Agriculture, No. 12. The Molecular and Cellular Biology of the Potato*, 2nd edn. Third International Symposium, Santa Cruz, CA, USA. CAB International, Wallingford, UK.

Ben Khedher, M. (1983) Physiological and morphological characteristics useful in selecting for heat tolerance in potato. PhD thesis, Cornell University, Ithaca, NY.

Ben Khedher, M. and Ewing, E.E. (1985) Growth analyses of eleven potato cultivars grown in the greenhouse under long photoperiods with and without heat stress. *American Potato Journal* 62, 537–554.

Bialek, K. and Bielińska-Czarnecka, M. (1975) Gibberellin-like substances in potato tubers during their growth and dormancy. *Bulletin. Academie Polonaise des Sciences Biologiques* 23, 213–218.

Biemond, H. and Vos, J. (1992) Effects of nitrogen on the development and growth of the potato plant. 2. The partitioning of dry matter, nitrogen and nitrate. *Annals of Botany* 70, 37–45.

Biran, I., Gur, I. and Halevy, A.H. (1972) The relationship between exogenous growth inhibitors and endogenous levels of ethylene and tuberization of Dahlias. *Physiologia Plantarum* 27, 226–230.

Bodlaender, K.B.A. (1963) Influence of temperature, radiation, and photoperiod in development and yield. In: Ivins, J.D. and Milthorpe, F.L. (eds) *Growth of the Potato*. Butterworths, London, pp. 199–210.

Bodlaender, K.B.A., Lugt, C. and Marinus, J. (1964) The induction of second-growth in potato tubers. *European Potato Journal* 7, 57–71.

Booth, A. (1963) The role of growth substances in the development of stolons. In: Ivins, J.D. and Milthorpe, F.L. (eds) *The Growth of the Potato*. Butterworths, London, pp. 99–113.

Borah, M.N. and Milthorpe, F.L. (1962) Growth of the potato as influenced by temperature. *Indian Journal of Plant Physiology* 5, 53–72.

Burt, R.L. (1964) Influence of short periods of low temperature on tuber initiation in the potato. *European Potato Journal* 7, 197–209.

Chailakyan, M.K., Yanina, L.I., Devedzhyan, A.G. and Lotova, G.N. (1981) Photoperiodism and tuber formation in grafting of tobacco on to potato. *Doklady Akademi Nauk SSSR* 257, 1276.

Chapman, H.W. (1958) Tuberization in the potato plant. *Physiologia Plantarum* 11, 215–224.

Cho, J.L. and Iritani, W.M. (1983) Comparison of growth and yield parameters of Russet Burbank for a two-year period. *American Potato Journal* 60, 569–576.

Clark, C.F. (1921) Development of tubers in the potato. *USDA Bulletin 958*.

Claver, F.K. (1973) Influence of temperature during the formation of tubers in relation with their incubation state (physiological age) and seed value. *Experientia* 30, 97–98.

Claver, F.K. (1975) Influence of temperature during the formation of potato tubers and its effects on the first progeny. *Phyton* 33, 1–6.

Clowes, F.A.L. and MacDonald, M.M. (1987) Cell cycling and the fate of potato buds. *Annals of Botany* 59, 141–148.

Cother, E.J. and Cullis, B.R. (1985) Tuber size distribution in cv. Sebago and quantitative effects of *Rhizoctonia solani* on yield. *Potato Research* 28, 1–14.

Cubillos, A.G. and Plaisted, R.L. (1976) Heterosis for yield in hybrids between *S. tuberosum* ssp. *tuberosum* and *tuberosum* ssp. *andigena*. *American Potato Journal* 53, 143–150.

Cutter, E.G. (1992) Structure and development of the potato plant. In: Harris, P.M. (ed.) *The Potato Crop*, 2nd edn. Chapman and Hall, London, pp. 65–161.

Demagante, A.L. and Vander Zaag, P. (1988) The response of potato (*Solanum* spp.) to photoperiod and light intensity under high temperatures. *Potato Research* 31, 73–83.

Dimalla, G.G. and Van Staden, J. (1977) Effect of ethylene on the endogenous cytokinin and gibberellin levels in tuberizing potatoes. *Plant Physiology* 60, 218–221.

Dodds, J.H. (1990) Molecular biology of potato: current and future prospects for developing countries. In: Vayda, M.E. and Park, W.D. (eds) *The Molecular and Cellular Biology of the Potato*. CAB International, Redwood Press, Melksham, UK, pp. 223–232.

Driver, C.M. and Hawkes, J.G. (1943) Photoperiodism in the potato. *Imperical Bureau of Plant Breeding and Genetics*. Cambridge, UK.

Duncan, D.A. and Ewing, E.E. (1984) Initial anatomical changes associated with tuber formation on single-node potato (*Solanum tuberosum* L.) cuttings. *Annals of Botany* 53, 607–610.

Dyson, P.W. and Watson, D.J. (1971) An analysis of the effects of nutrient supply on the growth of potato crops. *Annals of Applied Biology* 69, 47–63.

Edmundson, W.C. (1941) Response of several varieties of potatoes to different photoperiods. *American Potato Journal* 18, 100–112.

Estrada, R., Tovar, P. and Dodds, J.H. (1986) Induction of *in vitro* tubers in a broad range of potato genotypes. *Plant Cell Tissue Organ Culture* 7, 3–10.

Ewing, E.E. (1981) Heat stress and the tuberization stimulus. *American Potato Journal* 58, 31–49.

Ewing, E.E. (1985) Cuttings as simplified models of the potato plant. In: Li, P.H. (ed.) *Potato Physiology*. Academic Press, Orlando, pp. 154–207.

Ewing, E.E. (1995) The role of hormones in potato (*Solanum tuberosum* L.) tuberization. In: Davies, P.J. (ed.) *Plant Hormones: Physiology, Biochemistry and Molecular Biology*. Kluwer Academic, Dordrecht, The Netherlands, pp. 698–724.

Ewing, E.E. and Struik, P.C. (1992) Tuber formation in potato: induction, initiation, and growth. *Horticultural Review* 14, 89–198.

Ewing, E.E. and Wareing, P.F. (1978) Shoot, stolon, and tuber formation on potato (*Solanum tuberosum* L.) cuttings in response to photoperiods. *Plant Physiology* 61, 348–53.

Garner, W.W. and Allard, H.A. (1923) Further studies in photoperiodism, the response of the plant to relative length of day and night. *Journal of Agricultural Research* 23, 871–920.

Glendinning, D.R. (1975) Neo-tuberosum: A new potato breeding material. 2. A comparison of neo-tuberosum with unselected andigena and with tuberosum. *Potato Research* 18, 351–362.

Gray, D. (1973) The growth of individual tubers. *Potato Research* 16, 80–84.

Gray, D. and Holmes, J.C. (1970) The effect of short periods of shading at different stages of growth on the development of tuber number and yield. *Potato Research* 13, 215–219.

Gregory, L.E. (1956) Some factors for tuberization in the potato. *Annals of Botany* 43, 281–288.

Gunasena, H.P.M. and Harris, P.M. (1968) The effect of the time of application of nitrogen and potassium on the growth of the second early potato variety Craig's Royal. *Journal of Agricultural Science, Cambridge* 71, 283–296.

Gunasena, H.P.M. and Harris, P.M. (1969) The effect of CCC and nitrogen on the growth and yield of the second early potato variety Craig's Royal. *Journal of Agricultural Science, Cambridge* 73, 245–259.

Hammes, P.S. and Beyers, E.A. (1973) Localization of the photoperiodic perception in potatoes. *Potato Research* 16, 68–72.

Hammes, P.S. and Nel, P.C. (1975) Control mechanisms in the tuberization process. *Potato Research* 18, 262–272.

Hawkes, J.G. (1992) History of the potato. In: P. Harris (ed.) *The Potato Crop, The Scientific Basis for Improvement*. Chapman and Hall, London, pp. 1–12.

Hawkes, J.G. and Francisco-Ortega , J. (1992) The potato in Spain during the late 16th century. *Economic Botany* 46, 86–97.

Hawkes, J.G. and Francisco-Ortega, J. (1993) The early history of the potato in Europe. *Euphytica* 70, 1–7.

Helder, H., Miersch, O., Vreugdenhil, D. and Sembdner, G. (1993) Occurrence of hydroxylated jasmonic acids in leaflets of *Solanum demissum* plants grown under long- and short-day conditions. *Physiologia Plantarum* 88, 647–653.

Hiller, L.K., Koller, D.C. and Thornton, R.E. (1985) Physiological disorders of potato tubers. In: Li, P.H. (ed.) *Potato Physiology*. Academic Press, Orlando, pp. 389–455.

Hosaka, K. and Hanneman, R.E. Jr (1988) The origin of the cultivated tetraploid potato based on chloroplast DNA. *Theoretical Applied Genetics* 76, 172–176.

Hussey, G. and Stacey, N.J. (1984) Factors affecting the formation of *in vitro* tubers of potato (*Solanum tuberosum* L.). *Annals of Botany* 53, 565–578.

Iritani, W.M. (1968) Factors affecting physiological aging (degeneration) of potato tubers used as seed. *American Potato Journal* 45, 111–116.

Iritani, W.M., Weller, L. and Russell, T.S. (1973) Relative differences in sugar content of basal and apical portions of Russet Burbank potatoes. *American Potato Journal* 50, 24–31.

Jackson, S.D., Heyer, A., Dietze, J. and Prat, S. (1996) Phytochrome B mediates the photoperiodic control of tuber formation in potato. *Plant Journal* 9, 159–166.

Jansky, S.H. and Thompson, D.M. (1990) The effect of flower removal on potato tuber yield. *Canadian Journal of Plant Science* 70, 1223–1225.

Kahn, B.H. and Ewing, E.E. (1983) Factors controlling the basipetal patterns of tuberization in induced potato (*Solanum tuberosum* L.) cuttings. *Annals of Botany* 52, 861–874.

Kahn, B.A., Ewing, E.E. and Senesac, A.H. (1983) Effects of leaf age, leaf area, and other factors on tuberization of cuttings from induced potato (*Solanum tuberosum*) shoots. *Canadian Journal of Botany* 61, 3193–3201.

Koda, Y., Kikuta, Y., Tazaki, H., Tsujino, Y., Sakamura, S. and Yoshihara, T. (1991) Potato tuber-inducing activities of jasmonic acid and related compounds. *Phytochemistry* 40, 1435–1438.

Koda, Y. and Okazawa, Y. (1983a) Influences of environmental, hormonal and nutritional factors on potato tuberization *in vitro*. *Japanese Journal of Crop Science* 52, 582–591.

Koda, Y. and Okazawa, Y. (1983b) Characteristic changes in the levels of endogenous plant hormones in relation to the onset of potato tuberization. *Japanese Journal of Crop Science* 52, 592–597.

Kouwenhoven, J.K. (1970) Yield, grading and distribution of potatoes in ridges in relation to planting depth and ridge size. *Potato Research* 13, 59–77.

Krauss, A. (1985) Interaction of nitrogen nutrition, phytohormones, and tuberization. In: Li, P.H. (ed.) *Potato Physiology*. Academic Press, Orlando, pp. 209–230.

Krauss, A. and Marschner, H. (1976) Einfluss von Stickstoffernährung und Wuchsstoffapplikation auf die Knolleninduktion bei Kartoffelpflanzen. *Zeitschrift fuer Pflanzenernaehrung und Bodenkunde* 139, 143–155.

Krauss, A. and Marschner, H. (1982) Influence of nitrogen nutrition, daylength and temperature on contents of gibberellic and abscisic acid and on tuberization in potato plants. *Potato Research* 25, 13–21.

Krijthe, N. (1955) Observations on the formation and growth of tubers on the potato plant. *Netherlands Journal of Agricultural Science* 3, 291–304.

Krijthe, N. (1962) Observations on the sprouting of seed potatoes. *European Potato Journal* 5, 316–333.

Krug, H. (1960) Zum Photoperiodischen Verhalten einiger Kartoffelsorten. I, II. *European Potato Journal* 3, 47–49, 107–136.

Krug, H. and Wiese, W. (1972) Einfluss der Bodenfeuchte auf Entwicklung und Wachstum der Kartoffelpflanze (*Solanum tuberosum* L.). *Potato Research* 15, 354–364.

Kumar, D. and Wareing, P.F. (1972) Factors controlling stolon development in the potato plant. *New Physiology* 71, 639–648.

Kumar, D. and Wareing, P.F. (1973) Studies on tuberization in *Solanum andigena*. I. Evidence for the existence and movement of a specific tuberization stimulus. *New Phytology* 72, 283–287.

Langille, A.R. and Forsline, P.L. (1974) Influence of temperature and photoperiod on cytokinin pools in the potato (*Solanum tuberosum* L.). *Plant Science Letter* 2, 189–191.

Langille, A.R. and Hepler, P.R. (1992) Effect of three anti-gibberellin growth retardants on tuberization of induced and non-induced 'Katahdin' potato leaf-bud cuttings. *American Potato Journal* 69, 131–141.

Lazin, M.B. (1980) Screening for heat tolerance and critical photoperiod in the potato (*Solanum tuberosum* L.). PhD thesis, Cornell University, Ithaca, New York.

Lommen, W.J.M. (1994) Effect of weight of potato minitubers on sprout growth, emergence and plant characteristics at emergence. *Potato Research* 37, 315 - 322.

Lommen, W.J.M. and Struik, P.C. (1990) EAPR Abstracts of Conference Papers and Posters, 11th Triennial Conference of the European Association of Potato Research, *Field Performance of Minitubers of Different Sizes.* Edinburgh, UK, pp. 376–377.

Lorenzen, J.H. and Ewing, E.E. (1990) Changes in tuberization and assimilate partitioning in potato (*Solanum tuberosum*) during the first 18 days of photoperiod treatment. *Annals of Botany* 66, 457–464; note Errata in *Annals of Botany* 67, 191.

Lorenzen, J.H. and Ewing, E.E. (1992) Starch accumulation in potato (*Solanum tuberosum*) leaves during the first 18 days of photoperiod treatment. *Annals of Botany* 69, 481-485.

Lovell, P.H. and Booth, A. (1969) Stolon initiation and development in *Solanum tuberosum* L. *New Phytology* 68, 1175–85.

Lugt, C. (1960) Second-growth phenomena. *European Potato Journal* 3, 307–324.

Lugt, C., Bodlaender, K.B.A. and Goodijk, G. (1964) Observations on the induction of second-growth in potato tubers. *European Potato Journal* 7, 219–227.

Madec, P. (1978) Some effects of physiological age of the tuber upon sprouting and upon plant development. *Potato Research* 21, 57–59.

Madec, P. and Perennec, P. (1959) Le rôle respectif du feuillage et du tubercule-mère dans la tubérisation de la pomme de terre. *European Potato Journal* 2, 22–49.

Madec, P. and Perennec, P. (1962) Les relations entre l'induction de la tubérisation et la croissance chez la plante de pomme de terre *Solanum tuberosum. Annals Physiologia Vegetale* 4, 5–84.

Manrique, L.A., Bartholomew, D.P. and Ewing, E.E. (1989) Growth and yield performance of several potato clones grown at three elevations in Hawaii, I. Plant morphology. *Crop Science* 29, 363–370.

Marinus, J. andBodlaender, K.B.A. (1975) Response of some potato varieties to temperature. *Potato Research* 18, 189–204.

Martin, C., Vernay, R. and Paynot, N. (1982) Physiologie végétale. Photopériodisme, tubérisation, floraison et phénolamides. *Comptes Rendus Hebdomadaires des Seances. Academie des Sciences* 295, 565–568.

Melis, R.J.M. and Van Staden, J. (1984) Tuberization and hormones. *Zeitschrift fuer Pflanzenphysiologie* 11, 271–283.

Menzel, C.M. (1980) Tuberization in potato (*Solanum tuberosum* cultivar Sebago) at high temperatures, Responses to gibberellin and growth inhibitors. *Annals of Botany* 46, 259–266.

· Menzel, C.M. (1983) Tuberization in potato (*Solanum tuberosum* cultivar Sebago) at high temperatures, Gibberellin content and transport from buds. *Annals of Botany* 52, 697–702.

Menzel, C.M. (1985) Tuberization in potato at high temperatures, interaction between temperature and irradiance. *Annals of Botany* 55, 35–39.

Midmore, D.J. (1983) Potato in tropical environments, leaf area index, light interception and dry matter accumulation. *Potato Research* 27, 313–314.

Midmore, D.J. (1984) Potato (*Solanum* spp.) in the hot tropics. 1. Soil temperature effects on emergence, plant development and yield. *Field Crops Research* 8, 255–272.

Midmore, D.J. (1988) Potato (*Solanum* spp.) in the hot tropics. VI. Plant population effects on soil temperature, plant development and tuber yield. *Field Crops Research* 19, 183–200.

Midmore, D.J. (1990) Scientific basis and scope for further improvement of intercropping with potato in the tropics. *Field Crops Research* 25, 3–24.

Midmore, D.J., Berrios, D. and Roca, J. (1988a) Potato (*Solanum* spp.) in the hot tropics. V. Intercropping with maize and influence of shade on tuber yields. *Field Crops Research* 18, 159–176.

Midmore, D.J. and Mendoza, H.A. (1984) Improving adaptation of the potato (*Solanum* spp.) to hot climates – some physiological considerations. *Proceedings of the 6th Symposium International Society of Tropical Root Crops*, pp. 457–464.

Midmore, D.J., Roca, J. and Berrios, D. (1988b) Potato (*Solanum* spp.) in the hot tropics. IV. Intercropping with maize and the influence of shade on potato microenvironment and crop growth. *Field Crops Research* 18, 141–157.

Millard, P. and MacKerron, D.K.L. (1986) The effects of nitrogen application on growth and nitrogen distribution within the potato canopy. *Annals of Applied Biology* 109, 427–437.

Moorby, J. (1967) Inter-stem and inter-tuber competition in potatoes. *European Potato Journal* 10, 189–205.

Okazawa, Y. (1960) Studies on the relation between the tuber formation of potato plant and its natural gibberellin content. *Proceedings of the Crop Science Society Japan*, 29, 121–124.

Okazawa, Y. and Chapman, H.W. (1963) Regulation of tuber formation in the potato plant. *Physiologia Plantarum* 16, 623–629.

Oparka, K.J. and Davies, H.V. (1985a) Variation in ^{14}C partitioning in the tips of growing stolons of potato (*Solanum tuberosum* L.). *Annals of Botany* 55, 845–848.

Oparka, K.J. and Davies, H.V. (1985b) Translocation of assimilates within and between potato stems. *Annals of Botany* 56, 45–54.

Oparka, K.J., Davies, H.V. and Prior, D.A.M.(1987) The influence of applied nitrogen on export and partitioning of current assimilate by field-grown potato plants. *Annals of Botany* 59, 311–323.

Paiva, E., Lister, R.M. and Park, W.D. (1983) Induction and accumulation of major tuber proteins of potato in stems and petioles. *Plant Physiology* 71, 161–168.

Palmer, C.E. and Smith, O.E. (1969) Effects of kinetin on tuber formation on isolated stolons of *Solanum tuberosum* L. cultured *in vitro*. *Plant Cell Physiology* 11, 303–314.

Park, W.D. (1990) Molecular approaches to tuberization in potato. In: Vayda, M.E. and Park, W.D. (eds) *The Molecular and Cellular Biology of the Potato*. CAB International, Redwood Press, Melksham, UK, pp. 43–56.

Payton, F.V. (1989) The effect of nitrogen fertilizer on the growth and development of the potato in the warm tropics. PhD thesis, Cornell University, Ithaca, NY.

Perl, A., Aviv, D., Willmitzer, L. and Galun, E.(1991) *In vitro* tuberization in transgenic potatoes harboring ß-glucuronidase linked to a patatin promoter, effects of sucrose levels and photoperiods. *Plant Science* 73, 87–95.

Perennec, P. (1966) Induction de la tubérisation et inhibition des bourgeons chez la pomme de terre. *Bulletin. Societe Francaise de Physiologie Vegetale* 12, 175–192.

Peterson, R.L., Barker, W.G. and Howarth, M.J. (1985) Development and structure of tubers. In: Li, P.H. (ed.) *Potato Physiology*. Academic Press, Orlando, pp. 124–230.

Petri, P.S. (1963) L'influence de la température sur la morphologie de la pomme de terre. *European Potato Journal* 6, 242–250.

Plaisted, P.H. (1957) Growth of the potato tuber. *Plant Physiology* 32, 445–53.

Prat, S., Frommer, W.B., Höfgen, R. *et al.* (1990) Gene expression during tuber development in potato plants. *Federation of European Biochemical Societies. Letter* 268, 334–338.

Proudfoot, K.G. (1965) The effects of flowering and berry formation on tuber yield in *Solanum demissum* Lindl. *European Potato Journal* 8, 118–119.

Railton, I.D. and Wareing, P.F. (1973) Effects of daylength on endogenous gibberellins in *Solanum andigena*. I. Changes in levels of free acidic gibberellin-like substances. *Physiologia Plantarum* 28, 88–94.

Rasco, E.T. Jr, Plaisted, R.L. and Ewing, E.E. (1980) Photoperiod response and earliness of *S. tuberosum* ssp. *andigena* after six cycles of recurrent selection for adaptation to long days. *American Potato Journal* 57, 435–448; note Erratum, *American Potato Journal* 58, 50.

Rasumov, V. (1931) On the localization of photoperiodical stimulation. *Bulletin of Applied Botany, of Genetics and Plant Breeding* 27, 249–282.

Reeve, R.M., Timm, H. and Weaver, M.C. (1973) Parenchyma cell growth in potato tubers. I. Different tuber regions. *American Potato Journal* 50, 49–57.

Reynolds, P.M. and Ewing, E.E. (1989a) Effects of high air and soil temperature stress on growth and tuberization in *Solanum tuberosum*. *Annals of Botany* 64, 241–247.

Reynolds, P.M. and Ewing, E.E. (1989b) Heat tolerance in tuber bearing *Solanum* species, a protocol for screening. *American Potato Journal* 66, 63–74.

Rocha-Sosa, M., Sonnewald, U., Frommer, W., Stratmann, M., Schell, J. and Willmitzer, L. (1989) Both developmental and metabolic signals activate the promotion of a class I patatin gene. *European Molecular Biology Organization Journal* 8, 23–30.

Rowell, A.B., Ewing, E.E. and Plaisted, R.L. (1986) Comparative field performance of potatoes from seedlings and tubers. *American Potato Journal* 63, 219–227.

Ruf, R.H. Jr (1964) The influence of temperature and moisture stress on tuber malformation and respiration. *American Potato Journal* 41, 377–381.

Sachs, J. (1893) Ueber einige Beziehungen der spezifischen Grösse der Pflanzen zu ihrer Organisation. *Flora* 77, 49–81.

Sale, P.M.J. (1973a) Productivity of vegetable crops in a region of high solar input. I.) Growth and development of the potato (*Solanum tuberosum* L.). *Australian Journal of Agricultural Research* 24, 733–749.

Sale, P.M.J. (1973b) Productivity of vegetable crops in a region of high solar input. II.) Yields and efficiencies of water use and energy. *Australian Journal of Agricultural Research* 24, 751–762.

Sanchez-Serrano, J.J., Amati, S., Keil, M. *et al.* (1990) Promoter elements and hormonal regulation of proteinase inhibitor II gene expression in potato. In: Vayda M.E. and Park W.D. (eds) *The Molecular and Cellular Biology of the Potato*. CAB International, Redwood Press, Melksham, UK, pp. 57–70.

Santeliz-Arrieche, G.E. (1981) Effects of nitrogen fertilization on growth and development of potatoes. MS thesis, Cornell University, Ithaca, NY.

Schick, R. (1931) Der Einfluss der Tageslänge auf die Knollenbildung der Kartoffel. *Züchter* 3, 365–369.

Scholte, K. (1977) De invloed van secundaire knolgroei – doorwas – op het produktiepatroon van een aardappelgewas. *Bedrijfsontwikkeling* 8, 261–269.

Scholte, K. (1987) Relation between storage T sum and vigour of seed potatoes. In: *EAPR Abstracts of Conference Papers and Posters*, 11th Triennial Conference of the European Association of Potato Research, Aalborg, Denmark, pp. 28–29.

Scholte, K. (1989) Effect of daylength and temperature during storage in light on growth vigour of seed potatoes. *Potato Research* 32, 214–215 (Abstr.).

Sergeeva, L.I., Macháëková, I., Konstantinova, T.N. *et al.* (1994) Morphogenesis in potato plants *in vitro*, II. Endogenous levels, distribution and metabolism of IAA and cytokinins. *Journal of Plant Growth Regulation* 13, 147–152.

Sieczka, J.B., Ewing, E.E. and Markwardt, E.D. (1986) Potato planter performance and effects of non-uniform spacing. *American Potato Journal* 63, 25–38.

Šimko, I., Manschot, A., Yang, H.M., McMurry, S., Davies, P.J. and Ewing, E.E.(1996a) Analysis of polyamines in potato leaves. In: *Abstracts of Conference Papers, Posters and Demonstrations*, 13th Triennial Conference of the European Association for Potato Research. *QTL Analysis of Abscisic Acid in Potato Tubers*. IN, QTL pp. 644–45.

Šimko, I., Omer, E.A., Ewing, E.E. and Davies, P.J. (1996b) Tuberonic (12-OH-Jasmonic) acid glucoside and its methyl ester in potato. *Phytochemistry* 43, 727–730.

Simmonds, N.W. (1966) Studies of the tetraploid potatoes. III. Progress in the experimental re-creation of the *Tuberosum* group. *Journal Linnean Society* 59, 279–238.

Snyder, R.G. and Ewing, E.E. (1989) Interactive effects of temperature, photoperiod, and cultivar on tuberization of potato cuttings. *HortScience* 24, 336–338.

Steward, F.C., Moreno, V. and Roca, W.M. (1981) Growth, form and composition of potato plants as affected by environment. *Annals of Botany* 48 (Suppl. 2), 1–45.

Struik, P.C. (1986) Effects of shading during different stages of growth on development, yield and tuber size distribution of *Solanum tuberosum* L. *American Potato Journal* 63, 457. (Abstr.)

Struik, P.C., Geertsema, J. and Custers, C.H.M.G. (1989a) Effects of shoot, root and stolon temperature on the development of the potato (*Solanum tuberosum* L.) plant. I. Development of the haulm. *Potato Research* 32, 133–141.

Struik, P.C., Geertsema, J. and Custers, C.H.M.G. (1989b) Effects of shoot, root and stolon temperature on the development of the potato (*Solanum tuberosum* L.) plant. II. Development of stolons. *Potato Research* 32, 143–150.

Struik, P.C., Geertsema, J. and Custers, C.H.M.G. (1989c) Effects of shoot, root and stolon temperature on the development of the potato (*Solanum tuberosum* L.) plant. III. Development of tubers. *Potato Research* 32, 151–158.

Struik, P.C., Haverkort, A.J., Vreugdenhil, D., Bus, C.B. and Dankert, R. (1990) Manipulation of tuber-size distribution of a potato crop. *Potato Research* 33, 417–432.

Struik, P.C., Kramer, G. and Smit, N.P. (1989d) Effects of soil applications of gibberellic acid on the yield and quality of tubers of *Solanum tuberosum* L. cv. Bintje. *Potato Research* 32, 203–210.

Struik, P.C., van Heusden, E. and Burger-Meijer, K. (1988) Effects of short periods of long days on the development, yield and size distribution of potato tubers. *Netherlands Journal of Agricultural Science* 36, 11–22.

Struik, P.C. and van Voorst, G. (1986) Effects of drought on the initiation, yield, and size distribution of tubers of *Solanum tuberosum* L. cv. Bintje. *Potato Research* 29, 487–500.

Struik, P.C., Vreugdenhil, D., Haverkort, A.J., Bus, C.B. and Dankert, R. (1991)

Possible mechanisms of hierarchy among tubers on one stem of a potato (*Solanum tuberosum* L.) plant. *Potato Research* 34, 187–203.

Tanksley, S.D. (1993) Mapping polygenes. *Annual Review of Genetics* 27, 205–233.

Tizio, R.M. (1969) Action du CCC [chlorure de (2-chlorethyl)-trimethylammonium] sur la tubérisation de la pomme de terre. *European Potato Journal* 12, 3–7.

Tizio, R.M. (1971) Action et rôle probable de certaines gibbérellines (A1, A3, A4, A5, A7, A9 et A13) sur la croissance des stolons et la tubérisation de la pomme de terre (*Solanum tuberosum* L.). *Potato Research* 14, 193–204.

Turner, A.D. and Ewing, E.E. (1988) Effects of photoperiod, night temperature, and irradiance on flower production in the potato. *Potato Research* 31, 257–268.

Van den Berg, J.H., Ewing, E.E., Plaisted, R.L., McMurry, S. and Bonierbale, M.W. (1996) QTL analysis of potato tuberization. *Theoretical Applied Genetics* 93, 307–316.

Van den Berg, J.H., Šimko, I., Davies, P.J., Ewing, E.E. and Halinska, A. (1995) Morphology and [^{14}C] gibberellin A$_{12}$ metabolism in wild-type and dwarf *Solanum tuberosum* ssp. *andigena* grown under long and short photoperiods. *Journal of Plant Physiology* 146, 467–473.

Van den Berg, J.H., Struik, P.C. and Ewing, E.E. (1990) One leaf cuttings as a model to study second growth in the potato (*Solanum tuberosum*) plant. *Annals of Botany* 66, 273–280; note Erratum in *Annals of Botany* 67, 191.

Van den Berg, J.H., Vreugdenhil, D., Ludford, P.M., Hillman, L.L. and Ewing, E.E. (1991) Changes in starch, sugar, and abscisic acid contents associated with second growth in tubers of potato (*Solanum tuberosum* L.) one-leaf cuttings. *Journal of Plant Physiology* 139, 86–89.

Van der Zaag, D.E. (1984) Reliability and significance of a simple method of estimating the potential yield of the potato crop. *Potato Research* 27, 51–73.

Vander Zaag, P., Demagante A.L. and Ewing, E.E.(1990) Influence of plant spacing on potato (*Solanum tuberosum* L.) morphology, growth and yield under two contrasting environments. *Potato Research* 33, 313–324.

Van Loon, C.D. (1986) Drought, a major constraint in potato production and possibilities for screening for drought resistance. In: Beekman, Louwes, K.M., Dellaert, L.M.W. and Neele, A.E.F. (eds) *Potato Research of Tomorrow. Proceeds of the International Seminar*. Wageningen, Netherlands Pudoc, Wageningen, pp. 5–16.

Vayda, M.E. and Belknap, W.R. (1992) The emergence of transgenic potatoes as commercial products and tools for basic science. *Transgenic Research* 1, 149–163.

Vöchting, H. (1887) Über die Bildung der Knollen. *Bibliotheca botanica* 4, 1–55.

Vos, J. (1995a) Foliar development of the potato plant and modulations by environmental factors. In: Kabat, P., Marshall, B., van den Broek, B.J., Vos, J. and van Keulen, H. (eds) *Modelling and Parameterization of the Soil-Plant-Atmosphere System*. Wageningen Pers, Wageningen, pp. 21–38.

Vos, J. (1995b) The effects of nitrogen supply and stem density on leaf attributes and stem branching in potato. *Potato Research* 38, 271–279.

Vos, J. and Biemond, H. (1992) Effects of nitrogen on the development and growth of the potato plant. 1. Leaf appearance, expansion growth, life spans of leaves and stem branching. *Annals of Botany* 70, 27–35.

Vreugdenhil, D., Oerlemans, A.P.C. and Steeghs, M.H.G. (1984) Hormonal regulation of tuber induction in radish (*Raphanus sativus*). The role of ethylene in tuber induction of radish. *Physiologia Plantarum* 62, 175–180.

Vreugdenhil, D. and Struik, P.C. (1989) An integrated view of the hormonal regulation of tuber formation in potato (*Solanum tuberosum*). *Physiologia Plantarum* 75, 525–531.

Vreugdenhil, D. and Struik, P.C. (1990) Hormonal regulation of tuber formation. In: *EAPR Abstracts of Conference Papers and Posters*, 11th Triennial Conference of the European Association of Potato Research, Edinburgh, pp. 37–38.

Vreugdenhil, D. and van Dijk, W. (1989) Effects of ethylene on the tuberization of potato (*Solanum tuberosum*) cuttings. *Plant Growth Regulation* 8, 31–40.

Wareing, P.F. and Jennings, A.M.V. (1980) The hormonal control of tuberisation in potato. In: Skoog, F. (ed.) *Plant Growth Substances*. Springer-Verlag, Berlin, pp. 293–300.

Wellensiek, S.J. (1929) The physiology of tuber formation in *Solanum tuberosum* L. *Mededelingen Landbouwhogeschool Wageningen* 33, 6–42.

Wenzler, H., Mignery, G., May, G. and Park, W. (1989) A rapid and efficient transformation method for the production of large numbers of transgenic potato plants. *Plant. Science* 63, 79–86.

Werner, H.O. (1934) The effect of a controlled nitrogen supply with different temperatures and photoperiods upon the development of the potato plant. *Nebraska Agricultural Experiment Station. Bulletin 75*.

Werner, H.O. (1954) Anomalous tuberization of *Solanum tuberosum*. *American Potato Journal* 31, 375.

Wheeler, R.M., Steffen, K.L., Tibbitts, T.W. and Palta, J.P. (1986) Utilization of potatoes for life support in space. II. The effects of temperature under 24-h and 12-h photoperiods. *American Potato Journal* 63, 639–647.

Wheeler, R.M. and Tibbitts, T.W. (1986) Growth and tuberization of potato (*Solanum tuberosum* L.) under continuous light. *Plant Physiology* 80, 801–804.

Wheeler, R.M., Tibbitts, T.W. and Fitzpatrick, A.H. (1991) Carbon dioxide effects on potato growth under different photoperiods and irradiance. *Crop Science* 31, 1209–1213.

Willmitzer, L., Basner, A. Frommer, W. *et al.* (1990) Tuber-specific gene expression in transgenic potato plants. In: Lycett, G.W. and Grierson, D. (eds) *Genetic Engineering of Crop Plants*. Butterworths, London, pp. 105–114.

Wolf, S., Marani, A. and Rudich, J. (1990) Effects of temperature and photoperiod on assimilate partitioning in potato plants. *Annals of Botany* 66, 515–520.

Woolley, D.J. and Wareing, P.F. (1972a) The role of roots, cytokinins and apical dominance in the control of lateral shoot formation. *Planta* 105, 33–42.

Woolley, D.J. and Wareing, P.F. (1972b) The interaction between growth promoters in apical dominance. *New Phytology* 71, 781–793.

Woolley, D.J. and Wareing, P.F. (1972c) Environmental effects on endogenous cytokinins and gibberellin levels in *Solanum tuberosum*. *New Phytology* 71, 1015–1025.

Wurr, D.C.E. (1977) Some observations on patterns of tuber formation and growth in the potato. *Potato Research* 20, 63–75.

Yoshihara, T., Omer, E.A., Koshino, H., Sakamura, S., Kikuta, Y. and Koda, Y. (1989) Structure of a tuber-inducing stimulus from potato leaves (*Solanum tuberosum* L.). *Agricultural and Biological Chemistry* 53, 2835–2837.

The Cucurbits: Cucumber, Melon, Squash and Pumpkin

H.C. Wien
Department of Fruit and Vegetable Science, Cornell University, 134A Plant Science Building, Ithaca, New York 14853–5908, USA

The species in the family *Cucurbitaceae* that have been used as vegetables have enriched and diversified the diets of humankind for many centuries. For example, cucumber (*Cucumis sativus*) is consumed as an ingredient in salad, or in the pickled form. The fresh fruits of muskmelon or cantaloupe (*Cucumis melo*) and watermelon (*Citrullus lanatus*) are eaten as desserts, while the fruit flesh of the squashes and pumpkins is generally consumed after boiling or baking. The immature fruit of summer squash or marrows (*Cucurbita pepo*), and the mature fruit of all the principal squash and pumpkin species (*C. argyrosperma*, *C. moschata*, *C. maxima*, and *C. pepo*) are harvested and used as food. Shoot tips of the cucurbit vegetables are eaten as a cooked vegetable in Southern and Eastern Africa. Seeds of watermelon and some squash and pumpkin species are roasted and eaten as snacks, or ground as an ingredient of sauces. Several cultivars of *C. pepo* with hull-less seeds have been developed that facilitate the food uses of the seeds. With an oil and protein content of 46 and 34%, respectively, these could be exploited as alternative oil-bearing crops (Whitaker and Davis, 1962). In North America, pumpkins and gourds (*Cucurbita pepo*) are frequently grown solely for the ornamental value of the fruits, in harvest displays and as part of the Halloween celebration.

The major cultivated cucurbits can be classified into new world and old world species with regard to their origin. Cucumbers are thought to have been first cultivated in India, where their use has been recorded as long as 3000 years ago (Whitaker and Bemis, 1976). *Cucumis melo* is thought to have arisen in the central part of Africa, and have spread rapidly into Asia, where many cultivars have since been selected. The watermelon originated in the dry parts of southern Africa, and also has South Asia as a second centre of diversity. The new world species include all the cultivated *Cucurbita*. Although the four most economically important species may have had a common ancestor in one location in the Americas, such evidence is difficult to find. Each of these species has been selected separately, so that inter-breeding of different species is for the most part difficult to achieve. From archaeological records, the *Cucurbita* species are amongst the most ancient of cultivated crops in the Americas. Indeed, squash was one of the principal components of the diet of the ancient Mayan civilizations, together with beans and maize, dating back as far as 10,000 years (Whitaker and Bemis, 1976).

At the present time, the cucurbit vegetables are cultivated in all major regions of the world (Table 9.1) (FAO, 1994). Watermelon leads the produc-tion figures both in terms of tonnage and land area (1.8 million ha). Production in Asia comprises half or more of the total area devoted to each cucurbit vegetable worldwide.

Our knowledge of the physiology of the cucurbits has not grown in pro-portion to their importance in production. Most effort has been devoted to cucumber, followed by *Cucumis melo*, and watermelon in third place. As in the case of *Capsicum* (Chapter 7), detailed understanding of the growth and productivity of cucumber has come recently from the need to optimize pro-duction in greenhouse environments. In addition, intensive investigations into the hormonal control of flower sex expression during the 1960s and 1970s facilitated the production of hybrid cultivars of cucumber and squash. More recently, interest in developing melon cultivars of better fruit quality has spurred research efforts in fruit carbohydrate metabolism. Altogether, the state of our physiological knowledge of the cucurbit crops can be described as rather uneven, with detailed understanding in some areas, and others virtually unexplored. The latter is particularly true of watermelon, squash and pumpkins, perhaps due to the fact that their sprawling plant habit make them difficult experimental subjects in greenhouse, growth chamber and field.

Table 9.1. Production of the major cucurbit vegetables in the world in 1993 (production: 1000 metric tons) (FAO, 1994).

Region	Watermelon	Cucumber	Melons	Pumpkins and squash
Asia	15,746	12,761	7726	3552
Europe	2447	2563	2367	1184
NC America	2069	971	1696	428
Africa	2036	400	878	926
South America	994	59	233	768
World	27,063	18,326	12,976	8019

GERMINATION AND SEEDLING GROWTH

The germination of cucurbit vegetable seeds requires relatively warm temperatures (Lorenz and Maynard, 1980), and takes place within 3 or 4 days at 25–30°C. For cucumber, the lower limit of germination has been shown to be 11.5°C (Simon *et al.*, 1976), while germination of muskmelon and watermelon was low at 16°C (Nelson and Sharples, 1980). Summer squash (*Cucurbita pepo*) germination showed a lower threshold between 5 and 10°C, and optimum germination between 30 and 35°C (NeSmith and Bridges, 1992). Considerable effort has been expended on improving the capability of cucumber seeds to germinate at low temperatures. Soaking the seed in acetone solutions of fusicoccin (0.5 mmol), or GA_{4+7} (a mixture of gibberellins 4 and 7) (1 mmol), was the most effective in stimulating germination at 12°C (Nelson and Sharples, 1980). Similar growth regulator treatments had no significant effect on rate or final percentage emergence in six field plantings into cool soils, however (Staub *et al.*, 1987). Imbibing or pregerminating cucumber seeds in water at 32°C before planting them in soil at 15°C also did not shorten the time of emergence, although plant stand was improved significantly (Staub *et al.*, 1986). These results indicate that several steps of the germination and emergence process are limited by low temperatures. Simon *et al.* (1976) conjectured that the low temperatures may cause a denaturation of proteins in the germinating seedling, resulting in more severe damage the longer the exposure.

Considerable genetic differences exist in cucurbits in capacity to germinate at low temperatures (Schulte and Grote, 1974; Nerson *et al.*, 1982; Nienhuis *et al.*, 1983). After four cycles of selection for germination at 15°C, Nienhuis *et al.* (1983) improved low temperature germination from 32 to 94%. This increase was also correlated with improved germination at higher temperatures. In muskmelon, Nerson *et al.* (1982) showed 'bird's nest' cultivars developed in Iran to have significantly better cold temperature germination than standard viny and dwarf cultivars developed in the USA. We do not know yet what physiological processes differ among lines of either species contrasting in cold temperature germination.

The low germination rate of cucumber at 15°C may be partly due to seed dormancy (Nienhuis *et al.*, 1983). Freshly harvested seed failed to germinate at this temperature, and remained dormant until it had been stored for 84 days. Watts (1938) had also encountered this phenomenon in 'Black Diamond' cucumber, and was able to overcome the inhibition by removing the seedcoat, or germinating the seed at 30°C. The presence of a germination inhibitor in the testa was further supported by the finding that soaking cucumber seed in acetone significantly improved cold temperature germination (Nelson and Sharples, 1980). These workers also showed that watermelon seed germination at 16°C could be improved by washing the seed for 2 h in water. While dormancy may play a role in the reduced germination of fresh cucumber seed at low temperatures, it must be emphasized that poor germination under cool soil conditions is also a problem with cucurbit seed that has passed the apparent rest period.

The cucurbit seedling is among the most rapidly growing of any vegetable plant. It owes this characteristic to several properties. Seed size of the cucurbit vegetables is relatively large, varying from about 20 mg seed[-1] for muskmelon to 150 mg for squash and pumpkin (Lorenz and Maynard, 1980). The decorticated seeds contain on average about 49% oil, and 35% protein (Jacks et al., 1972), suggesting that a large store of reserve materials is available for seedling growth before the cotyledons and true leaves start to photosynthesize. Large seed size also implies large initial seedling size, giving the plant an early start in light interception and assimilation. Seedlings thus reach a size suitable for transplanting in three weeks, compared to tomato and pepper, which generally require 6–8 weeks (Lorenz and Maynard, 1980).

Seedling productivity of cucumber is comparable with that of tomato and pepper seedlings in terms of net assimilation rate (dry matter production per unit leaf area per unit time) (Bruggink and Heuvelink, 1987). Cucumber thus does not appear to have an inherently greater photosynthetic rate, but apportions a large amount of its dry matter to leaf growth in the early stages (high leaf area ratio, leaf area to total dry weight).

The responses of cucumber to temperature have been intensively studied, because of the high cost of energy in greenhouse production. During vegetative growth, it has been found that stem extension rate and the development of leaf area is linearly dependent on the mean air temperature within the range of 19–26°C (Krug and Liebig, 1980). The amplitude around these means made no difference as long as day temperature did not exceed 28°C. If night temperature was higher than day temperature, stem extension rate was slowed by the production of shorter internodes. Heij (1980) had similar findings, except that stem extension growth was not as rapid above 21°C mean temperature as below it. Gregory (1928) also determined the temperature dependence of cucumber leaf growth, and found optimal development to occur at 25°C, with lower rates of leaf expansion above and below this value. The discrepancy in optimum temperatures among these studies may be due to genetic differences and variation in light conditions. Based on these studies, the most economical temperature regime for the rapid production of greenhouse cucumber plants during the vegetative stage would be to set day temperatures at about 4–6°C higher than the night temperature.

Growth of young cucumber plants during periods of low light conditions is also dependent on adequate root zone temperatures. During this time, little warming of the rooting medium can be expected from the sun. It was found that rhizosphere temperatures of 16°C or below led to root death and poor top growth (Hori et al., 1968; Folster, 1974). At these low temperatures, root growth and water and nutrient uptake of cucumbers was drastically reduced (Tachibana 1982, 1987). These reductions were correlated with lower root respiration rates, particularly of the cytochrome respiratory pathway coupled with oxidative phosphorylation (Tachibana, 1989). At air temperatures of 20–25°C, soil temperatures of 23–25°C gave the best growth (Hori et al., 1968; Folster, 1974), but this necessitated the use of soil heating devices.

Another approach to overcoming the growth-inhibiting effects of cold soils is to graft cucumber onto rootstocks of species less susceptible to cold root temperatures (den Nijs, 1980a). This approach, now widely practised in Japan and Korea, where about 80% of greenhouse cucumbers are grafted, allows cucumber and other cucurbits to be grown without soil heating. Another benefit from the practice is the avoidance of problems of soil-borne diseases, to which the rootstock is not susceptible, and the improvement in fruit quality that comes with the use of certain rootstocks (Lee, 1994). More particularly with regard to root growth in cold soils, Tachibana (1982, 1987) found that *Cucurbita ficifolia*, one of the most common rootstocks used for cucumber, maintains active growth, water and nutrient uptake at 12 and 14°C (Fig. 9.1). At these temperatures, these processes are sharply reduced

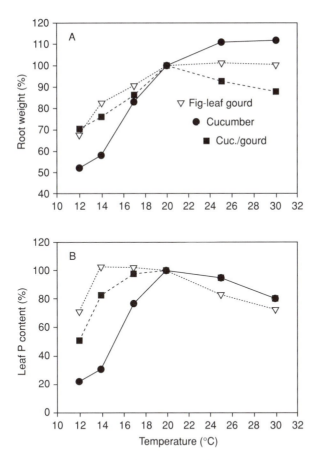

Fig. 9.1. Root weight (A) and leaf phosphorus content (B) of fig-leaf gourd, cucumber or cucumber grafted on a fig-leaf gourd rootstock, in relation to root temperature. Values are percentages relative to the values of each treatment at 20°C (Tachibana, 1982).

in cucumber. Differences in susceptibility to low soil temperatures between two cucumber cultivars were particularly well related to differences in phosphorus uptake at those temperatures (Fig. 9.1) (Tachibana, 1982, 1987). This may suggest that P uptake is one of the key processes inhibited by low root temperatures.

Genetic variation in tolerance to low temperature in cucumber has been found both in the seed germination stage (Nienhuis *et al.*, 1983), as well as in the adult vegetative and fruiting stages of the plant (den Nijs, 1980b). Good progress has been made in selecting greenhouse cucumber lines that grow vigorously and produce acceptable yield at 20/15°C (day/night), temperatures at which current commercial lines are much reduced in growth. The improved lines maintained a high leaf area ratio at low temperature during the vegetative stage, and showed no delay in fruit development after flowering. It should thus be possible through breeding and selection to develop new cultivars that will have significantly lower energy requirements in greenhouse production. Incorporation of these genes into field-grown cultivars may widen the area of adaptation of the crop.

FLOWER DIFFERENTIATION

Before describing the development of flowers in the cucurbit vegetables, it is necessary to understand the range of sexual types that commonly occur in these crops (Table 9.2). Most common is the monoecious flowering habit, in which male and female flowers are found on the same plant. In cucumber, many cultivars with gynoecious flowering have been developed. Homozygous gynoecious lines usually produce only female flowers, but in the heterozygous form, male flowers can frequently be found, especially under environmental conditions favouring male flower expression (see below). A range of other sexual types have been developed in cucumber, such as the hermaphroditic (perfect flowers), androecious (male flowers) and the andromonoecious (male and perfect flowers), but these have not

Table 9.2. Floral morphologies of cucurbit vegetables.

Species	Flowering type
Cucumis sativus	Monoecious, most common
	Gynoecious
	Hermaphroditic
	Andromonoecious
Cucumis melo	Andromonoecious, most common
	Monoecious
Cucurbita pepo	Monoecious
Citrullus lanatus	Monoecious, most common
	Andromonoecious

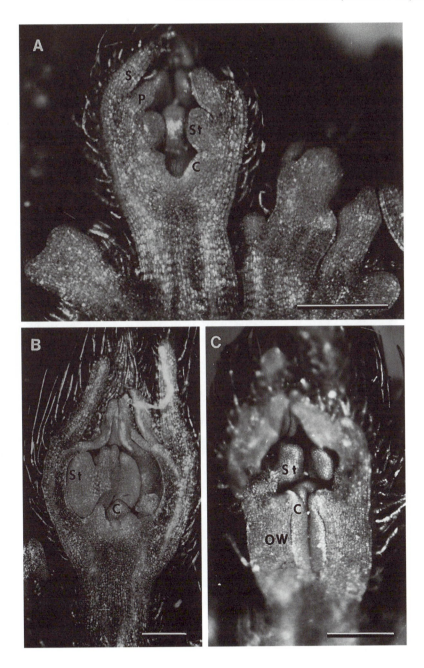

Fig. 9.2. Sexual differentiation of the cucumber flower, from the presexual stage (A), formed 3–4 days after initiation of the floral meristem, to the differentiated male (B) and female flower (C) 7 days after initiation. Scale = 0.25 mm. C, carpel primordium; OW, ovary wall; P, petal; St, stamen; S, sepal (courtesy of M.C. Goffinet).

become important in cucumber production. In muskmelon, the predominant flowering form in commercial cultivars is andromonoecious, although some monoecious, gynoecious, androecious and hermaphroditic forms also exist (Rudich, 1990). The monoecious sexual type is most prevalent in *Cucurbita* spp. and in watermelon, but cultivars differ in ratio of male to female flowers, and in the lowest node to bear female flowers.

Flowers are formed early in the seedling stage in cucumber and melon, and can be found on low main stem nodes. Morphological studies have found that male, female and perfect flowers arise from anatomically similar 'presexual' primordia, from which subsequently the male, female or both parts develop fully (Atsmon and Galun, 1960; Goffinet, 1990) (Fig. 9.2). The regulation of sexual development is under genetic, environmental and hormonal control, and has been intensively studied for the last 40 years. As with other aspects of cucurbit physiology, cucumber has been investigated most, with the other species getting much less attention.

The sex type of flowers in cucumber and in the species of cucurbits having monoecious flowering habit shows a distinct ontogenetic pattern (Nitsch *et al.*, 1952; Shifriss and Galun, 1956). The basal main stem nodes are generally male, and these become interspersed with increasing number of nodes bearing female flowers (Fig. 9.3). At the upper part of the main stem, the plant may eventually form a zone of nodes bearing only female flowers.

GENETIC FACTORS IN SEX EXPRESSION

Detailed analyses of the genes responsible for sex expression in the cucurbits have been summarized by Frankel and Galun (1977), and by Lower and Edwards (1986). Even without knowing the details of the genetics of sex

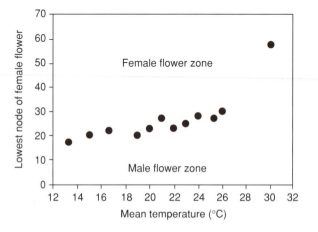

Fig. 9.3. Influence of mean temperature on the transition from male to female flowering nodes on 'Acorn' squash (*Cucurbita pepo*) (Nitsch *et al.*, 1952).

expression, it is useful to realize that cultivars termed 'gynoecious' can be either homozygous or heterozygous for that trait, and these differences can have a profound effect on the flowering pattern of the plant. Homozygous gynoecious lines of cucumber produce male flowers at very low frequency, and in many situations will only produce female flowers (Lower *et al.*, 1983). In contrast, heterozygous gynoecious lines, especially those developed from crosses between gynoecious and monoecious lines, produce male flowers on a significant proportion of their nodes (Cantliffe, 1981; Lower *et al.*, 1983; Lower and Edwards, 1986). Cultivar differences can also exist for frequency of male flowering nodes among heterozygous gynoecious lines. Many cucumber hybrids used in field production are heterozygous with regard to the gynoecious character, and are termed 'predominantly female' by the seed trade. These cultivars show considerable influence of the environment on sex expression, as shown below (Cantliffe, 1981; Nienhuis *et al.*, 1984).

ENVIRONMENTAL EFFECTS ON SEX EXPRESSION

Influence of the environment on sex expression can be most clearly shown by focusing on the zone of transition between male and female nodes, and measuring the node number at which the first female flower is borne, or the ratio of male to female flowers in a constant number of nodes. Since monoecious plants tend to bear increasing number of female flowers with time, it is important to compare treatments over the same period of time. In assessing the effect of temperature, or other factors affecting growth rate, the number of nodes on which the effect is counted should be kept constant (Matsuo, 1968). This helps avoid some of the errors in interpretation found in some of the early literature (Tiedjens, 1928).

At least three environmental factors have an important influence on sex expression in cucumber and *Cucurbita pepo*: temperature, light energy and photoperiod. Under cool conditions, the production of female flowers is favoured (Fig. 9.3). Mean temperatures are most important, but night temperatures also play a significant role, with warm nights leading to increased male flower production at a given mean temperature, compared to warm days (Nitsch *et al.*, 1952). The temperature influence may occur during flower primordia differentiation, as in cucumber, or during the development of the flower to anthesis. In *Cucurbita pepo*, low temperature may inhibit male flower development after differentiation, leading to precocious female flowering (see section on precocious female flowering under Physiological Disorders below) (Rylski and Aloni, 1990; H.C. Wien, Ithaca, NY, 1990, unpublished data). NeSmith *et al.* (1994) also found that the number of male and female flowers reaching anthesis could be influenced by temperature in a cultivar-dependent manner.

High light conditions generally favour female flower production, while shading, or low incident radiation delays the onset of female flowering (Ito and Saito, 1960; Kooistra, 1967; Cantliffe, 1981). The combination of low light and high temperatures prevalent in the fall growing period in the southern US may explain the predominance of male flowers in the cucumber crop at that time (Cantliffe, 1981).

The effect of photoperiod on sex expression appears to be less striking than the factors of temperature and light conditions in most cultivars (Matsuo, 1968; Cantliffe, 1981). Short photoperiods tend to favour the production of female flowers. It is difficult under field conditions to avoid an interaction of photoperiod and light energy levels, with short photoperiods coinciding with periods of reduced light conditions. In that situation, it appears that light energy plays the more important role, as in the example cited by Cantliffe (1981) in the preceding paragraph. If photoperiod is controlled by keeping the amount of light energy constant and extending daylength with low intensity light, some cucumber cultivars are day neutral with regard to sex expression, and others show delayed female flowering in long days (Matsuo, 1968).

The sex expression of cucumbers in the field generally follows the predictions from controlled environments. For instance, Shifriss and Galun (1956) found that planting in the relatively cooler, and shorter daylength conditions of April in Israel resulted in plants with a higher proportion of female flowers than when the same cultivars were planted in July. Where conditions appear to conflict, temperature tends to be more important. Novak (1972) found that 'Tablegreen' monoecious cucumber produced female flowers earlier under the cool, long-day conditions of Ithaca, New York summers than in the hot, short-day conditions of the Philippines (Table 9.3). NeSmith et al. (1994) demonstrated that planting date tends to affect female flower number more markedly for some summer squash cultivars than for others.

Nitrogen status of the plant has also been shown to influence sex expression in cucumber, with high nitrogen fertilizer levels delaying the production of female flowers (Ito and Saito, 1960). High plant population density, or close spacing within rows also increases male flowering, and may operate through reduced light levels available to individual plants in crowded stands (Lower et al., 1983; Nienhuis et al., 1984).

Table 9.3. The influence of growing conditions on node to first female flower, and node number of start of female flower zone in 'Tablegreen' cucumber, grown in the field in Ithaca, NY, and Los Banos, Philippines. Mean temperatures of 21 and 27°C, and daylengths of 16 and 11.5 h prevailed during the growing seasons at Ithaca and Los Banos, respectively (Novak, 1972).

Location	Node to first female flower	Node of start of female zone
Ithaca, USA	8.3 ± 4.0	13.4 ± 4.1
Los Banos, Philippines	16.6 ± 1.8	31.7 ± 7.4

In summary, the environmental conditions which encourage the build-up of carbohydrates and which reduce the amount of vegetative growth tend to favour female flower expression. Conditions which foster stem extension and reduce carbohydrate build-up, such as high temperatures, low light conditions, high nitrogen levels and close spacings, also increase the tendency for male flower production in the cucurbit vegetables.

GROWTH HORMONES

Plant growth regulating compounds play a key role in determination of flower gender in the cucurbits. Evidence for their importance comes both from studies in which endogenous levels of the growth hormones are related to sex expression, and from the effect of applying the compounds exogenously to the plant. Aside from helping us understand the mechanism of flower sex development, use of growth regulators to manipulate sex expression has facilitated the development of hybrid cucurbit cultivars, in ways that will be illustrated below.

Gibberellin

The production of male flowers on nodes of cucumber that would normally produce female flowers can be brought about by the application of gibberellic acid (GA_3), and even more effectively, by GA_{4+7} (Mitchell and Wittwer, 1962; Atsmon and Tabbak, 1979). Similar results have been obtained with cucumber, *Cucurbita*, and muskmelon (Splittstoesser, 1970; Rudich *et al.*, 1972a; Atsmon and Tabbak, 1979).

Higher levels of endogenous gibberellins have been found in monoecious and andromonoecious cucumbers than gynoecious cultivars, in parallel to their tendency to produce male flowers (Hemphill *et al.*, 1972). Environmental conditions that favoured male flower development, such as high temperature and long daylength also increased the amount of gibberellins detected in the apical regions (Saito and Ito, 1963). In muskmelon, levels of endogenous gibberellins were less well correlated with male tendency in one study (Hemphill *et al.*, 1972). In another experiment, however, the increased femaleness of andromonoecious melons brought about by applications of the growth inhibitor SADH (succinic acid-2,2-dimethyl-hydrazide, or daminozide) was matched by lower gibberellin content (Rudich *et al.*, 1972a).

Ethylene

The capacity of ethylene to stimulate female flower development in the cucurbits first became known when researchers applied the ethylene-producing chemical ethephon (2-chloroethylphosphonic acid) to cucumbers (Robinson *et al.*, 1969). When applied to seedling monoecious cucumber plants, the chemical eliminated male flowers on the lower nodes, and increased the number of female flowers. Ethephon stimulated the formation of female flowers on monoecious muskmelon, without much change in male flower numbers, while on andromonoecious and hermaphroditic plants there was an increase in perfect and suppression of male flowers (Karchi, 1970). On monoecious *Cucurbita pepo*, the chemical suppressed male and increased female flowers. Ethephon application has thus become useful to ensure that the inbred line used as the female parent in hybrid combinations does not develop male flowers, and is used for this purpose in both cucumber and squash breeding (Shannon and Robinson, 1979; Lower and Edwards, 1986).

The central role of ethylene in cucurbit sex expression has been further strengthened by the finding that inhibitors of ethylene formation or action have effects on flower formation that are opposite to those of ethephon. For instance, treatment of homozygous gynoecious cucumbers with the ethylene action inhibitor silver nitrate or the ethylene synthesis inhibitor aminoethoxyvinylglycine (AVG) resulted in the formation of male and perfect flowers (Atsmon and Tabbak, 1979). Silver nitrate and silver thiosulphate are used by cucumber breeders to produce inbred, all-female lines (Lower and Edwards, 1976). Treated gynoecious plants produce some male flowers and permit selfing, without the detrimental effects of elongated, brittle stems brought about by gibberellin application.

Watermelon appeared to be more sensitive to ethephon applications than the other cucurbits, and the crop's reaction was opposite to that found for cucumber (Christopher and Loy, 1982). Exposure to concentrations of 30 µl l^{-1} retarded female flower formation, but had less effect on male flowers. Use of silver nitrate or AVG on watermelon resulted in a suppression of female and perfect flowers, with only a partial reduction of male flowers (Christopher and Loy, 1982). These anomalous responses to hormones with regard to sex expression indicate that further clarification is needed on how flower gender is determined in watermelon.

Determinations of endogenous ethylene levels in cucurbit seedlings have supported the regulatory role of this chemical in sex expression. In cucumber, seedling apices of gynoecious lines produced more ethylene than those of monoecious or androecious lines (Rudich *et al.*, 1972c, 1976). Ethylene evolution was higher for female than for male flower buds, and monoecious lines increased their ethylene evolution at the time that female flower primordia were developing. Short daylengths stimulated ethylene evolution in comparison to long daylengths, in concert with their influence on female flower formation (Rudich *et al.*, 1972c).

Auxin

It has been known since the early days of plant hormone research that auxin is involved in sex expression of cucurbits, but its exact role is still uncertain (Rudich, 1990). Treatment of young cucumber plants with auxin or synthetic auxins such as naphthalene acetic acid promoted female flower formation (Ito and Saito, 1960; Saito and Ito, 1963). Culture of potentially male excised flower buds in auxin-containing medium stimulated ovary formation (Galun *et al.*, 1963).

The endogenous auxin level was increased in some experiments by conditions fostering female flower formation (Rudich *et al.*, 1972b), while the level decreased in others (Ito and Saito, 1960; Saito and Ito, 1963). Furthermore, treatment of *Cucurbita pepo* with ethephon increased female flower number, but decreased endogenous auxin activity (Chrominski and Kopcewicz, 1972). A clear role for auxin is difficult to determine, in part because higher levels of auxin cause the liberation of ethylene in tissues. In addition, ethylene has been shown to inhibit auxin translocation, and to contribute to the inactivation of auxin through decarboxylation (Beyer and Morgan, 1969). Finally, the reliable determination of endogenous auxin levels in plants has been difficult, and has contributed to the uncertainty of the role of this hormone in sex expression of cucurbits.

Abscisic acid (ABA)

There is considerable uncertainty about the role of ABA in sex expression of cucurbits (Rudich, 1990). ABA applications to gynoecious cucumbers increased female tendency, but treatment of monoecious lines favoured male flower production (Rudich *et al.*, 1972b; Friedlander *et al.*, 1977a). In one study, ABA content of gynoecious cucumber was higher than that of a monoecious line (Rudich *et al.*, 1972b); in another, the reverse was true (Friedlander *et al.*, 1977b). The ABA content of both types of cucumber were increased three- to ten-fold by applications of ethephon. It is thus not clear if ABA plays a direct role in sex expression, or has a secondary involvement in response to the more important action of the previously mentioned hormones.

Summary

Research since the 1960s has identified the gibberellins as the main growth hormone group that stimulates male flower development in *Cucumis* and *Cucurbita*. Little is known about the biosynthetic pathway of the gibberellins in the cucurbits, although many researchers have found GA_4 and GA_7 to be more active than GA_3. Ethylene stimulates female flower formation in these cucurbits, and suppresses the development of male flower buds. We lack a comprehensive understanding of the interaction of these hormones. Although ethephon applications lead to reduced gibberellin levels (Chrominsky and Kopcewicz, 1972; Rudich *et al.*, 1972b), it is not known if male flowering tendency enhanced by ethylene inhibitors operates by increasing gibberellins. The roles of auxin and of ABA also need further investigation, to determine if these play primary or secondary roles in determination of gender of cucurbit flowers. For this purpose, it may be rewarding to repeat the tissue culture experiments of presexual flower buds conducted by Galun *et al.* (1963), and manipulate ethylene and ABA levels as well as auxin and gibberellins investigated by them.

FLOWERING AND FRUIT SET

The time of flowering of the cucurbit vegetables is primarily determined by temperature, and its influence on plant growth rate. Temperature is also the principal factor determining the time of anthesis and duration of opening of individual flowers. Seaton and Kremer (1938) found that the flowers of *Cucurbita* had a minimum temperature for anthesis and anther dehiscence of around 10°C. Above this level, flowers would open at dawn, and remain open until about noon. Under cooler conditions, anthesis and anther dehiscence was delayed until the following day. As temperatures increased beyond 30°C, anthesis occurred earlier, and flowers closed by mid or late morning. The minimum temperature for opening of cucumber and watermelon flowers was found to be about 15°C, whereas muskmelon anthesis temperatures were between 18 and 21°C (Seaton and Kremer, 1938). The duration of flower opening for cucumber, watermelon and muskmelon is generally for the entire daylight period of one day.

The receptivity of the female flower, or of the female portion of perfect flowers of cucumber has been found to extend from 2 days before to 2 days after anthesis under growth chamber conditions (Le Deunff *et al.*, 1993). In the greenhouse, Munger (1988) reported that manual crosses were successful on the day of anthesis and the following morning, but under temperate field conditions, manual pollination success often decreased to low levels on the afternoon of the anthesis day. The factors leading to this rapid decline were not identified.

POLLEN GERMINATION AND POLLEN TUBE GROWTH

When pollen from the same plant or another plant of the same species is deposited on the stigmatic surface, pollen germination follows in less than 30 min under normal conditions (Suzuki, 1969; Sedgley and Buttrose, 1978). Cucumber pollen germinates over a wide range of temperatures, but pollen tube growth rates may be inhibited at the extremes, preventing fruitset (Matlob and Kelly, 1973). As temperature increased from 10 to 32°C, pollen tube elongation rate increased in snake melon (*Cucumis melo* var. *flexuosus*), but cucumber pollen tubes were only stimulated up to 21°C (Matlob and Kelly, 1973). Elongation rates were also higher for pollen from plants grown under high light and moderate temperature conditions than from plants produced in low light conditions.

There are also considerable genetic differences in pollen tube growth rates (Table 9.4). In general, these are higher for the cucurbits with larger ovary and final fruit size, and may be related to the size of pollen grains of

Table 9.4. Rates of pollen tube growth and pollen grain diameters of vegetable cucurbit species, compiled from several authors.

Species	Temp. (°C)	Pollen grain diameter, μ*	Pollen tube growth (mm h⁻¹)	Time to ovule penetration (h)	Reference
Cucumber	21	63	–	36	Matlob (1973)
Muskmelon	–	53	0.95	24	Suzuki (1969)
C. melo var.					
flexuosus	21			36	Matlob (1973)
Watermelon	30	52	6	24	Sedgley (1978)
Watermelon	43	52	2.5	> 7	Sedgley (1978)
Cucurbita	–	142	4.6–6	9–11	Suzuki (1969)

* (Ikuse, 1956).

these species (Ikuse, 1956). Nevertheless, the tube growth rates are sufficiently rapid to ensure that the pollen tubes reach the nearest part of the ovary within a few hours. Poole and Porter (1933) calculated that the tips of the pollen tubes of watermelon would reach the nearest ovules in 3 h. Suzuki (1969) arrived at a figure of 5 h for muskmelon in similar calculations. In spite of this, most researchers have found that fertilization of the ovules takes from 24 to 36 h (Table 9.4), indicating that other factors than pollen tube growth rate are involved in this step.

In some cucumber cultivars, the rate of pollen tube growth may not be rapid enough to allow fertilization of ovules over the entire length of the ovary (den Nijs and Miotay, 1991) (Fig. 9.4). At the same time as the pollen tubes are growing, the ovary is also elongating, and on some long-fruited cucumber cultivars, the middle and far end of the fruit may never be reached by the pollen tubes. As a result, the fruit enlarges at the blossom end, and relatively few seeds are formed (Varga and Bruinsma, 1990).

The path of the pollen tubes to the ovary and ovules is primarily along the conducting tissue connecting the style and ovary, but once the ovary is reached, pollen tubes will also travel in the cavities between fruit lobes (Poole and Porter, 1933). The pattern of distribution of pollen on the stigmatic surface has little influence in *Cucurbita* on the distribution of fertilized seeds in the fruit, indicating that the pollen tubes can travel laterally in the style or ovary to some extent (Hayase, 1953; S.W. Cady and H.C. Wien, 1994, Ithaca, NY, unpublished). In watermelon, pollen distribution on the stigma does affect seed location, and ultimately, fruit shape (Mann, 1943).

The necessity of ovule fertilization and seed set for fruit set in non-parthenocarpic cucumber remains unclear. Fuller and Leopold (1975) found that style removal 12 h after pollination allowed 50% of fruits to set, while ovule penetration required 30–36 h. They speculated that stimulation of the ovary to continue growth did not need the pollen tubes to reach the ovules, but did not identify the nature of the signal. More recently, Varga and

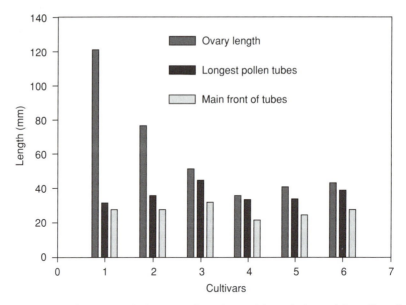

Fig. 9.4. Length of the ovary, the longest pollen tubes and the main front of the pollen tubes growing down the style, for six cucumber cultivars of differing fruit length (den Nijs and Miotay, 1991).

Bruinsma (1990) demonstrated that style removal after 12 h allowed some ovules to be fertilized and some fruit to set. They found that cucumber cultivars incapable of parthenocarpic fruit set had to at least be induced to form non-viable 'pseudoseeds' (seeds with seedcoats but lacking embryos and endosperm). These were induced to form in their experiments by use of irradiated pollen that was sterile but capable of pollen tube growth to the ovules. The mechanism by which true seeds and pseudoseeds stimulate the continued growth of the ovary is not known, although it is presumed to be hormonal.

PARTHENOCARPY

Parthenocarpy is fruit set and fruit growth without the fertilization of ovules. It occurs widely among the cucurbit vegetables, particularly in cucumber (Rudich *et al.*, 1977) and in *Cucurbita pepo* (Rylski, 1974; Robinson, 1993). The tendency to set and develop seedless fruit is enhanced by cool weather conditions, and to a lesser extent, by short photoperiods (Rylski, 1974; Rudich *et al.*, 1977; Dean and Baker, 1983). In cucumber, parthenocarpic tendency is more strongly expressed in lines that have a higher proportion of female flowers. Even on monoecious lines, the tendency for parthenocarpic fruit production increases with plant age, as does the femaleness of the plant (Rudich *et al.*, 1977; Kim *et al.*, 1992a).

There is considerable genetic variation in parthenocarpy in cucumber, and the characteristic has been viewed as a means of overcoming adverse environmental effects on fruit set or pollinating insects. To date, parthenocarpic gynoecious cultivars are common in greenhouse cucumber production, but have not been successfully developed for field production (Lower and Edwards, 1986).

The production of seedless fruits in watermelon does not occur without special measures, since naturally occurring parthenocarpy has not been found in this species (Mohr, 1986). Fruit set without seed set can be brought about by pollinating a self-sterile triploid watermelon with a diploid pollen parent (Kihara, 1951).

Commercial production of seedless watermelon has been hampered by several factors. The abnormally thick seedcoat of the triploid seeds impedes germination and makes the use of transplants advisable. In addition, yields are decreased because one quarter of the field must be planted to a diploid pollinator cultivar. Consumer reluctance to accept that the small white empty seedcoats found in these parthenocarpic fruit are not seeds has also hampered acceptance of seedless watermelon cultivars.

HORMONAL REGULATION OF FRUIT SET

Indications of how the retention and growth of fruits in the cucurbit vegetables might be regulated have come from studies of endogenous hormones in flowers and fruits, and from the induction of fruit set by exogenous hormone applications.

Auxin

Comparisons of parthenocarpic and normal seeded cucumber lines have generally shown that ovary auxin content is higher in parthenocarpic ovaries at anthesis (Kim *et al.*, 1992a,b; Takeno and Ise, 1992). Pollination increased ovary auxin concentration of the non-parthenocarpic line, while auxin level in unpollinated ovaries declined further (Kim *et al.*, 1992b). The exact site of auxin production in the developing ovary has not been identified, but may be in both the pericarp and ovule tissues.

The importance of auxin in cucumber fruit set is supported by the findings that fruit set induced by application of exogenous growth regulators such as benzyl adenine (a synthetic cytokinin), gibberellic acid, synthetic auxins and auxin transport inhibitors all increase the endogenous auxin content of the developing ovary (Beyer and Quebedeaux, 1974; Kim *et al.*, 1992a; Takeno and Ise, 1992). A number of investigators have also brought about parthenocarpic fruit set in cucumber by application of synthetic auxins to the flower (Elassar *et al.*, 1974a; Watkins and Cantliffe, 1980; Kim *et al.*, 1992b). Similar results were obtained with muskmelon (Whitaker and Prior, 1946; Elassar *et al.*, 1974b), *Cucurbita pepo* (Wong, 1941; Y. Zhang and H.C.

Wien, 1988, Ithaca, NY, unpublished data) and other cucurbit vegetables (Wong, 1941).

Increasing fruit set of pickling cucumbers by the use of the auxin transport inhibitor chlorflurenol has been demonstrated in greenhouse and field experiments (Cantliffe *et al.*, 1972, 1974; Dean and Baker, 1983; Ells, 1983). It was thought that simultaneous set of many fruits could be achieved by using the chemical on gynoecious cucumber cultivars that had been delayed in fruit set by lack of male flowers. Unfortunately, commercial realization of this technique failed because it was difficult to prevent male flower formation even on strongly gynoecious genotypes. The few male flowers that occurred in such plantings were sufficient to allow early set of a few fruit, which then inhibited further fruitset on the plants (Ells, 1983).

Cytokinins

Evidence that endogenous cytokinins may be involved in fruit set of cucumber has been largely indirect. Exogenous application of benzyl adenine and other cytokinins has been shown to increase fruit set, and stimulate fruit growth (Elassar *et al.*, 1974a; Ogawa *et al.*, 1990; Shishido *et al.*, 1990; Takeno *et al.*, 1992). Results indicated that in one case, the fruit set stimulation was accompanied by an increase in endogenous auxins (Shishido *et al.*, 1990), but in another instance, no correlation with endogenous auxins could be found (Takeno *et al.*, 1992). Takeno *et al.* (1992) measured increased ovary cell numbers in treated flowers, and speculated that the improvement in fruit set may have been brought about by cytokinin-stimulated cell division. There is need for direct measurement of ovary cytokinin levels in cucumber lines differing in parthenocarpic fruit set tendency to clarify the role of these growth promoting compounds in the fruit set process.

Gibberellins

The involvement of gibberellins in fruit set of cucumber has also been deduced from the result of exogenous application experiments. Elassar *et al.* (1974a) and Ogawa *et al.* (1989) showed that application of GA_3 and more effectively, GA_{4+7}, increased fruit numbers. As with the cytokinins, detailed measurements of endogenous gibberellins in the flower and developing fruit are still needed.

POLLINATION

The cucurbit vegetables are among the vegetable crop species that require insects for pollination. This is most obvious for the crops that have separate male and female flowers, such as cucumber, watermelon, squash and pumpkin. But even in muskmelon (*Cucumis melo*) which bears perfect flowers, pollinators are necessary. In this species, the pollen is sticky and not

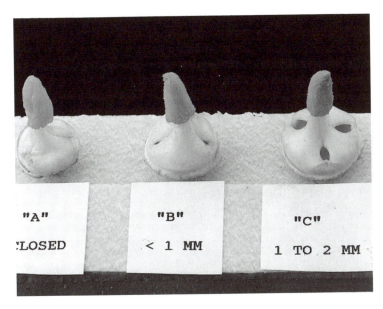

Fig. 9.5. Male flowers of *Cucurbita pepo*, with petals removed to show the degree of opening of the nectary slits, from closed (left), to open (right) (Cady, Glatz and Wien, Ithaca, NY, unpublished observations).

readily transferred, so that exclusion of bees by caging plants nearly totally prevented fruit set (McGregor and Todd, 1952).

It is beyond the scope of this book to delve into the foraging behaviour of honey bees and gourd bees, the principal pollinators of the cucurbit vegetables. The physiology of the plant may, however, influence its attractiveness to pollinating insects. For instance, the domestic bee *Apis mellifera* uses the flowers of cucurbits as both a source of pollen and of nectar (Free, 1970). The principal attraction in muskmelon flowers appears to be nectar. Plantings of a nectarless mutant had only sporadic visits by bees, and as a consequence, poor fruit set and low yields (Bohn and Mann, 1960; Bohn and Davis, 1964), even though pollen production by the mutant was normal. Similarly, accessibility of the staminal nectaries in *Cucurbita pepo* may influence the frequency of visits by honey bees (S.W. Cady, R. Glatz and H.C.Wien, 1994, Ithaca, NY, unpublished observations) (Fig. 9.5).

For successful pollination, both male and female or perfect flowers must be open on the same day. In *Cucurbita pepo*, cool weather conditions early in the growing season may result in female flowers opening several days before male flowers, resulting in delayed fruit set (Rylski and Aloni, 1990) (see section on physiological disorders below). There are preliminary indications that low light conditions, and high temperatures have the opposite effect (R.O. Nyankanga and H.C. Wien, Ithaca, NY, 1996, unpublished data). Female flower primordia on some cultivars fail to develop to anthesis under these conditions, thus delaying or decreasing fruit production. Little is known about the regulatory mechanisms of flower development in the cucurbit vegetables, but it may have a similar basis as the hormonal regulation of sex expression (see section on flower differentiation above).

FRUIT GROWTH

The development of fruit in the cucurbit vegetables has been an object of fascination for many years. Attention has focused on the large-fruited members of *Cucurbita pepo* and *C. maxima*, which can reach extraordinary size. For instance, contests organized by pumpkin-growing clubs are currently striving to break the 1000-pound (454 kg) barrier, and fruit weights of 375 kg have recently been achieved (Langevin, 1993) (Fig. 9.6). Such fruits are said to gain as much as 11 kg per day while growing, but the physiological attributes needed to bring about such prodigious size increases have been largely unexplored.

Fruit growth of smaller-fruited cucurbits was closely studied by Sinnott more than 50 years ago (Sinnott, 1939, 1945). Ovary growth of *Cucurbita pepo* before anthesis was found to consist both of cell division and cell enlargement activities (Fig. 9.7). By plotting the diameter of individual cells against ovary diameter, Sinnott (1939) demonstrated that the transition from a

Fig. 9.6. Fruit of *Cucurbita maxima* weighing 377 kg (829 lb), winner of the 1995 'World Pumpkin Federation Weigh-off' for largest squash (H.C. Wien, Ithaca, NY, unpublished data).

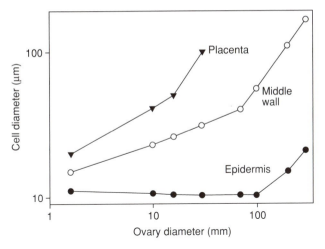

Fig. 9.7. The relationship of ovary diameter to diameter of individual cells in the placenta, middle wall and epidermis of the growing fruit of 'Connecticut Field' pumpkin (*Cucurbita pepo*). Both ovary and cell diameters are plotted on a logarithmic scale (Sinnott, 1939).

period of cell division and cell enlargement to purely cell enlargement takes place at about anthesis, and occurred later in outer than in the inner layers of the fruit. At fruit maturity, cells are largest in the innermost layers of the fruit, and may be loosely arranged or even torn apart. Epidermal cells are small, closely packed, and in some cultivars, have thickened cell walls to form a hard shell.

Similar patterns of cell division and enlargement were found for other cultivated cucurbits (Kano *et al.*, 1957; Sinnott, 1939). In watermelon (*Citrullus lanatus*), cell enlargement of the innermost fruit tissues continued until the cells had attained an astonishing 350,000-fold increase from their size at the end of the cell division stage.

Fruit growth in *Cucurbita pepo* is characterized by an initial log-linear (exponential) phase, followed by a gradual growth rate decrease (Sinnott, 1945). A comparison of cultivars ranging in fruit size from 40 to 7000 cm³ indicated that fruit growth rates varied little, but the larger-fruited types had longer growth durations (Fig. 9.8).

The expansion of cucumber fruits has also been found to consist of initial exponential growth, followed by a gradual decline (Tazuke and Sakiyama, 1984; Marcelis, 1992b). Increase in fresh weight was closely correlated to volume growth, which in turn could be accurately predicted from length and circumference measurements.

Fruit growth rates can be profoundly affected by the influence of the rest of the plant, and by environmental factors. Marcelis and Baan Hofman-Eijer (1993) showed that parthenocarpic greenhouse-grown cucumbers had maximum fruit growth rates three times higher when one rather than five fruits were developing on the plant at the same time. Fruit growth rate

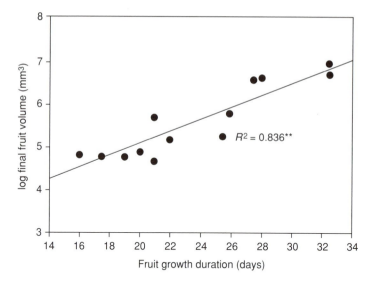

Fig. 9.8. The influence of fruit growth duration from 10 mm ovary diameter to maturity on final fruit volume (logarithmic scale) in *Cucurbita pepo* (Sinnott, 1945).

increased with higher temperatures most markedly with single-fruited plants, but reached maximum growth rates at 25°C in plants with five fruits. Increasing assimilate supply with higher irradiance also enhanced fruit growth rates (Marcelis, 1993). The higher assimilate levels resulted in increased number and size of fruit cells, if higher light was given early in fruit development. Later applications of high light increased cell size only. These studies point out the need to conduct fruit growth studies under uniform environmental conditions, and with plants of similar fruit load.

Considerable research has been devoted to the study of the biochemistry and enzymology of fruit growth, tracing the changes in fruit carbohydrates during development. The principal translocated carbohydrate of the cucurbit vegetables is the raffinose polysaccharide stachyose (Webb and Gorham, 1964; Hughes and Yamaguchi, 1983; Pharr *et al.*, 1985). Once it reaches the fruit peduncle, this transport sugar is thought to be transformed into sucrose and hexose sugars in muskmelon and cucumber (Handley *et al.*, 1983; Hubbard *et al.*, 1989). Gross and Pharr (1982) found that cucumber fruit peduncles contain the necessary enzymes to convert stachyose to sucrose. Peduncle extracts of *Cucurbita moschata*, watermelon and muskmelon also had similar capabilities.

In the early stages of growth, muskmelon fruit sucrose levels tend to be low, with soluble sugars made up almost exclusively of glucose and fructose (McCollum *et al.*, 1988; Hubbard *et al.*, 1989). It is thought that the high levels of fruit acid invertase prevent sucrose accumulation. During later stages of fruit growth, acid invertase activity drops, and sucrose phosphate synthase (SPS) enzyme activity increases (Hubbard *et al.*, 1989). At the same time, sucrose levels also rise, until they make up nearly 50% of the fruit's soluble sugars. Hubbard *et al.* (1989) found that SPS activity correlated well with sucrose concentration at fruit harvest, when they compared melon cultivars with contrasting fruit sucrose content. Although reducing sugars make up between 2 and 3% of fruit fresh weight of cucumber and melon during development, starch levels of these fruits are less than 1% in both species (Schaffer *et al.*, 1987). It is therefore not possible to increase fruit sugar content of melons once the fruit has been detached from the plant (Bianco and Pratt, 1977). For optimum fruit quality, harvest should take place as close as possible to the time of maturity of the fruit.

The changes in carbohydrate content of the fruits of *Cucurbita* and of watermelon have not been intensively investigated. As in muskmelon, watermelon fruit show earlier increases in reducing sugars than in sucrose during development (Porter *et al.*, 1940). At maturity, the fruit typically has 10% total sugars, of which about 35% is sucrose. If the fruit is allowed to become overmature on the vine, or stored at room temperature, the proportion of sucrose increases to around 65% (Porter *et al.*, 1940). Total sugars, and soluble solids increase in the fruit until maturity (Mizuno and Pratt, 1973). Little is known about the enzyme systems responsible for these changes in the fruit.

FACTORS AFFECTING PRODUCTIVITY

Yield production in the annual herbaceous vegetable crops of the *Cucurbitaceae* is affected both by factors that influence overall plant productivity, and those that determine the partitioning of assimilates to reproductive tissue. As with other vegetable crops such as tomato and pepper, crop responses have been worked out in detail for glasshouse production systems, in which temperatures, light and CO_2 levels can be regulated. Accordingly, information is most complete for the climatic controls needed for optimum growth and yield of the gynoecious parthenocarpic cucumber grown in the glasshouses of Northern Europe and North America.

Productivity also involves issues of the timing and concentration of harvests. In pickling cucumber, where the fruits are harvested at a young stage, much effort has been expended to devise production systems and develop genetic types that give high yields in a short harvest period. Some of these trials will be described below.

Fruit quality is an important criterion in the production of muskmelon, watermelon and winter squash. Production systems must provide conditions which allow fruit to develop acceptable sweetness and taste, and the size characteristic of the cultivar.

GLASSHOUSE CUCUMBER

In the European production system, cucumbers typically are sown in January and February, and bear fruit during spring and summer. Considerable experimentation has determined that mean air temperatures of 18–24°C are optimum for greatest yield accumulation (Drews *et al.*, 1980; Liebig, 1980b; Slack and Hand, 1980). As temperature increased, stem extension rate accelerated, and time to first harvest declined (Krug and Liebig, 1980). Plants started bearing earliest under high temperatures, but had a shorter harvest duration, and a reduced total yield (Liebig, 1980b). Variation of day and night temperature about the mean had no effect on earliness and early yield (Slack and Hand, 1980; Grimstad and Frimanslund, 1993), but profoundly influenced stem length (Krug and Liebig, 1980; Grimstad and Frimanslund, 1993) (see section on germination and seedling growth above). At night temperatures of 18°C or lower, earliness and productivity was boosted by increasing soil temperature (Drews *et al.*, 1980).

The response of the cucumber crop to temperature can be modified by the light conditions under which the crop is grown. Under the limiting light energy levels of midwinter, stem extension and earliness of cropping was maximal at 21°C, with no further increase as temperature was raised beyond this mark (Heij, 1980). Earliness of yield production and marketable yield at a given air temperature was further boosted by increases in light levels (Liebig, 1980a,b) (Fig. 9.9).

Increasing ambient CO_2 level for glasshouse cucumber has become the standard practice particularly when glasshouse vents are closed.

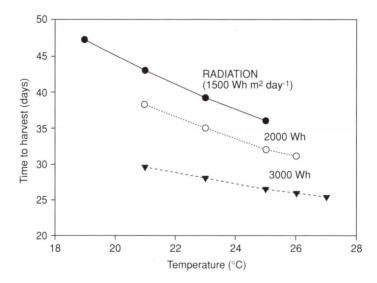

Fig. 9.9. Influence of growing temperature on time to harvest of the first fruits of glasshouse cucumbers grown under three levels of supplementary radiation (Liebig, 1980b).

Concentrations of 700–1000 µl l⁻¹ CO_2 are commonly used, and have been found to increase yields from 20–43% (Hand, 1984; Kimball, 1986). In situations where temperature control requires ventilation, CO_2 supplementation becomes more difficult. Under warm spring and fall conditions of North Carolina, Peet and Willits (1987) found that enrichment periods as short as 4.5 h still increased cucumber yields by 27%. A larger benefit could be obtained, however, by use of a passive rock storage cooling system that allowed the vents to be kept closed for more than 8 h per day.

An alternative means of maintaining elevated CO_2 levels in the glasshouse is to produce cucumbers on bales of decomposing straw. Growers in Britain found that this reduces the need for carbon dioxide supplementation because the gas is being released by the medium, and keeps CO_2 levels at nearly 1000 µl l⁻¹ (Hand, 1984).

FIELD-GROWN CUCUMBER

The climatic optima for growing cucumbers in the field have not been intensively studied, but broadly confirm the findings of the glasshouse research. Lorenz and Maynard (1980) state the optimum temperature of the crop to be 18–24°C. In establishing a heat unit system to predict the timing of harvests, Perry *et al.* (1986) found that a base temperature of 15.5 and a maximum temperature of 32°C worked best. Perry and Wehner (1990) found that these limits were better predictors of harvest dates for pickling cucumbers than for slicers in trials conducted in three locations over 3 years. Their results imply that fruit growth rates during the later stages of development are influenced by factors other than temperature. This was confirmed by the research of Marcelis (1993), who showed the dependence of fruit growth on assimilate supply, particularly the presence of previously set fruits on the plant.

The influence of growing fruit on a plant on the development of younger fruit and further vegetative growth has been thought to limit cucumber yields. McCollum (1934) showed the inhibiting influence was strongest in seeded fruits until the time the seedcoats hardened. Parthenocarpic fruits had a much weaker inhibitory effect. Denna (1973) extended these observations with comparisons of parthenocarpic glasshouse cultivars that were pollinated or allowed to set seedless fruits. Among 12 cultivars, the seedless plants produced 17% more fruit than their seeded counterparts. Denna also pointed out that seeds comprised a significant amount of the total plant dry weight, namely 10% for the glasshouse cultivars, and 20% for two pickling cucumber lines (Table 9.5). From that, one would expect the yield depression from the first-formed fruits to be even stronger in seeded cultivars.

The causes for the inhibition have not been clarified, but two major factors are thought to be involved. The first is competition for limited assimilates, and the priority of reproductive structures for these (Pharr *et al.*, 1985; Marcelis, 1992a). Pharr *et al.* (1985) calculated, for instance, that one actively

Table 9.5. Total plant dry weight and the percentage of that dry matter in the principal plant components for two parthenocarpic and two pickling cucumber cultivars (Denna, 1973).

Cultivar	Fruit condition	Total dry weight g plant^{-1}	Per cent dry weight		
			Fruit	Seed	Vine
Uniflora A	Seedless	166	62	–	38
	Seeded	108	52	11	37
Toska	Seedless	177	57	–	43
	Seeded	88	49	9	42
MSMR15	Fruitless	122	–	–	100
	Seeded	105	34	20	46
GSMR15	Fruitless	111	–	–	100
	Seeded	89	31	20	49

growing cucumber fruit required the photosynthetic output of 40% of the plant canopy. During the growth of parthenocarpic cucumbers, the inhibition of vegetative growth is most marked during periods of rapid fruit growth (Marcelis, 1992a). The greater inhibitory effect of seeded fruits may be due to the fact that seeds contain 32% fat and comprise 20% of fruit dry weight (Winton and Winton, 1935; Denna, 1973).

Hormonal factors are also likely to be involved in the competition between old and younger fruits, and the fruits and vegetative growth. Schapendonk and Brouwer (1984) found for instance, that just starving a cucumber plant of carbohydrates by defoliation will reduce fruit growth rate, but is not sufficient to cause the reproductive tissue to become necrotic. They used this as indirect evidence that hormonal influences are involved in the fruit-induced growth inhibition. It is also likely that hormonal activity in the developing fruit allows that organ to become a stronger sink for assimilates than other tissues. Although concrete data on these points have so far been lacking amongst the cucurbit vegetables, considerably more is known with regard to other vegetable crops (see Chapter 5).

Several strategies have been suggested to overcome the intraplant competitive effects in order to increase the yield potential of pickling cucumbers. Increasing the number of female flowers per plant by use of the gynoecious trait (Tasdighi and Baker, 1981) was positively correlated with increased yields. A genetic trait which increased the number of female flowers per node from one to three or four was, however, not successful in changing fruit number per plant (Uzcategui and Baker, 1979). Others have crossed pickling cucumbers with *C. sativus* var. *hardwickii*, which tends to produce many small, seeded fruits per plant (Lower *et al.*, 1982). Unfortunately, this species has several undesirable traits such as photoperiod sensitivity, small fruit size and chilling temperature sensitivity. Morphological traits such as determinate growth habit and the 'little leaf' trait, which confers increased branch numbers have also been advocated as

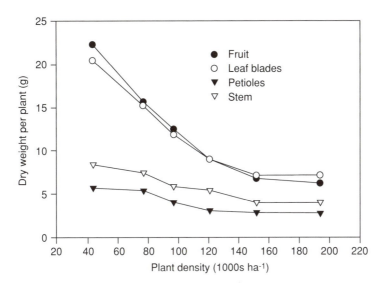

Fig. 9.10. Effect of plant density on dry weight of fruits and vegetative parts of pickling cucumbers at 48 days after planting (Widders and Price, 1989).

means of increasing pickling cucumber yields, particularly for once-over, mechanical harvesting (Staub *et al.*, 1992).

It is doubtful, however, if morphological modifications of the plant are going to overcome the inverse relationship between yield per plant and plant density. As cucumber plants are crowded closer together, the leaf area, branching and fruit number per plant are reduced (Cantliffe and Phatak, 1975; Widders and Price, 1989). Widders and Price (1989) pointed out the remarkably close relation between dry matter partitioned to fruits and to leaves as plant populations are varied from 5 to 20 m^{-2} (Fig. 9.10). This supports the findings of Pharr *et al.* (1985) of the heavy demand that a rapidly growing cucumber fruit makes on current photosynthesis. So unless morphological changes increase photosynthetic rate, or increase the proportion of the leaf canopy intercepting light, alterations in branching habit and leaf size are unlikely to increase fruit yield.

In the development of pickling cucumber production systems using once-over mechanical harvesting, many studies have focused on the population density needed for high yield. Some of these have shown increased yields as populations were raised to 50 or even 64 plants m^{-2} (Cantliffe and Phatak, 1975; O'Sullivan, 1980). At such high densities, only half the plants produced fruit, however (O'Sullivan, 1980), indicating that plant populations were in excess of what was needed. In addition, weed control and harvesting problems with these dense stands have limited their use (Lower and Edwards, 1986). As a result, populations of around 15 plants m^{-2} have generally been adopted for once-over harvests in recent years (Staub *et al.*, 1992).

Another approach to increasing cucumber yields at lower densities has been to delay pollination and thus fruit set, and allowing the plants to develop more nodes with female flowers. Connor and Martin (1970) excluded bees from cucumber plots for the first 11 days of bloom, and achieved a 100% yield increase. Fruit quality was also improved by the delayed fruit set, a characteristic in common with late-fruiting monoecious cultivars (Wehner and Miller, 1985). From a practical standpoint, cucumber fields would need to be entirely free of male flowers to achieve such delayed fruiting, a situation which has been difficult to accomplish. Many gynoecious cultivars are derived from crosses between gynoecious by monoecious parents, and produce significant numbers of male flowers, especially under long daylengths and high temperatures (Connor and Martin, 1971; Lower and Edwards, 1986) (see section on flower differentiation above). The percentage of female flowers can be increased significantly in crosses of gynoecious by hermaphrodite lines, and lines homozygous for the gynoecious trait have also been developed by use of ethylene inhibiting chemicals such as silver nitrate (Lower and Edwards, 1986). Nevertheless, delayed planting of pollinator lines (Connor and Martin, 1970), or use of the fruit setting chemical chlorflurenol (see section on hormonal regulation of fruit set, above), have not been adopted into commercial practice, likely because of practical and regulatory constraints.

COMBINING YIELD AND FRUIT QUALITY IN CUCURBIT VEGETABLES

To be classed as 'marketable', muskmelon, watermelon and some types of squash must attain a minimum level of soluble solids, as well as a size and external appearance characteristic of the cultivar. Sugars make up about 85% of the soluble solids in muskmelon (Porter *et al.*, 1940). For muskmelon, soluble solids must exceed 8% to be eligible for shipping out of California (Peirce, 1987). Similarly, watermelon fruits should have sugar content of 10% or greater to be considered acceptable (Peirce, 1987).

Maintaining quality of fruit harvested while at the same time increasing yield has generally not been successful using higher plant densities (Zahara, 1972; Mendlinger, 1994). As yields of fruit increase, soluble solids percentage and fruit size declines. Welles and Buitelaar (1988) found that any factor which shortened the period from flowering to fruit maturity reduced soluble solids content of the fruit (Fig. 9.11). Increasing the night temperature, reducing leaf area, and increasing the number of fruit per plant all reduced the maturation period of the fruit, and simultaneously lowered fruit quality. Davis and Schweers (1971) also found that fruit number per plant was inversely correlated with fruit soluble solids. This indicates that assimilate supply during the fruit growth period is critical to determination of fruit quality. Thus factors that reduce the canopy photosynthesis rate, such as the extent of exposure of the leaf canopy to light, may also influence fruit quality and yield. Knavel (1991) found, for instance, that muskmelon cultivars

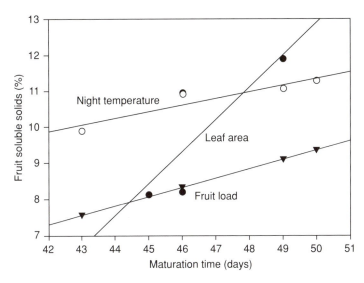

Fig. 9.11. Influence of fruit maturation time on fruit soluble solids content for glasshouse-grown muskmelon. Maturation time was varied by manipulations of night temperature, plant leaf area or fruit load per plant in separate experiments (Welles and Buitelaar, 1988).

with large, overlapping leaves and short internodes had fewer fruits with low soluble solids content compared to conventional vining types. Measurements indicated that about 50% of the plants' leaf area was shaded by other leaves, compared to 30% for a conventional muskmelon line. Nerson and Paris (1987) found that yield per plant among three muskmelon lines differing in growth habit was proportional to internode length. These results would indicate that muskmelon is source-limited during the period when the fruits are growing rapidly. Short-term carbohydrate translocation studies with field-grown muskmelons (Hughes *et al.*, 1983) confirm this for plants with small fruits. The degree of carbohydrate depletion from source leaves depended on the proximity of the leaf to the fruit, with the most distant leaves retaining significantly more than those near the fruit.

The demands for the production of high quality fruit may also constrain the degree to which the fruit harvest period can be shortened or synchronized in a particular field. At plant densities at which more than one fruit is produced per plant, it may be difficult for several fruits to be growing rapidly at the same time, without reduction in growth rate or assimilate accumulation rate in the individual fruit. When grown at wide spacing, muskmelon plants produce two distinct flushes of fruit (McGlasson and Pratt, 1963). Attempts have been made to have more fruit ripening simultaneously by selecting for morphological types producing several branches simultaneously (Paris *et al.*, 1986). While these new types have produced a more concentrated yield, problems of small fruit size and low soluble solids must still be overcome (Paris *et al.*, 1988).

PHYSIOLOGICAL DISORDERS

PRECOCIOUS FEMALE FLOWERING

Under the cool conditions prevalent in temperate regions during planting, the development rate of male flowers of *Cucurbita pepo* is inhibited more than that of female flowers (Rylski and Aloni, 1990). This can result in the precocious development to anthesis of female flowers, and a lack of fruit set because of a dearth of open male flowers. The problem is especially pronounced in some hybrid cultivars of summer squash that flower early in the season. The disorder is less pronounced in some open-pollinated viny cultivars, but first anthesis of all flowers is relatively late in these lines (H.C. Wien, 1990, Ithaca, NY, unpublished data). Applications of GA_{4+7} as flower buds become visible can hasten male flower development to anthesis, indicating that flower development may be under similar hormonal control as flower differentiation in cucurbits.

SUDDEN WILT OF MUSKMELONS

The disorder is also called vine collapse, crown blight and late collapse by researchers in different parts of the world, and refers to the rapid wilting of plants just as the fruits are beginning to develop netting, and the vines have covered the ground (Zitter, 1995). Within days, the entire field may be affected, and vines may not recover. A range of pathogenic organisms have been associated with the disorder, and may be the primary causes of the observed symptoms. The causal organisms associated with the disease differ in different melon-growing regions. For instance, in New York, cucumber mosaic virus (CMV) and *Fusarium* wilt have been implicated (Bauerle, 1971; MacNab, 1971; Zitter, 1995). In California, *Pythium ultimum* has been linked to the disorder, and soil fumigation has delayed development of symptoms (Munnecke *et al.*, 1984). In Israel, Nitzany (1966) caused vine collapse under cool conditions by inoculating with CMV and *Pythium*. Miller and co-workers (1995) implicated a number of fungal pathogens in causing vine collapse of melons in Texas.

The presence of rapidly growing fruits on the plants appeared to be key to development of the vine collapse symptoms on the plants in a number of these cases. Periods of cool, cloudy weather during this growth stage, followed by hot, sunny conditions increased incidence of the disorder (Zitter, 1995). Although direct experimental evidence is lacking, it is thought that during the rapid fruit growth stage, demand for assimilates of the growing fruit is so high that root growth is reduced. If at the same time, pathogens attack the root system, or reduce the plant's capacity to produce assimilates, root death may result. A third factor that could further exacerbate the situation is adverse weather conditions that would reduce photosynthetic rates and reduce root function through waterlogging of the soil. Collaborative work between physiologists and pathologists will be needed to test these theories and overcome this costly disorder.

HOLLOW HEART OF WATERMELON

This disorder is characterized by the separation of the inner parts of the fruit into distinct segments, leaving hollow areas at harvest maturity. Hollow heart occurs more often in the first-formed fruit on the plant, as a result of excess nitrogen fertilization, and delayed harvests (Kano, 1993). The disorder is more prevalent under conditions of rapid fruit growth rate, when the rind is expanding more rapidly than the inner regions of the fruit (Sinnott, 1939; Kano, 1993). Ways of avoiding the condition include selection of less susceptible cultivars, and using cultural practices that moderate fruit growth rate and final fruit size. These include adequate plant populations, moderate levels of nitrogen, and prompt harvests.

BITTER FRUIT IN SUMMER SQUASH

The sporadic occurrence of bitter fruit in plantings of zucchini and other summer squash types has caused serious medical problems in a few cases. The ingestion of as little as 3 g of such fruit can cause nausea, stomach cramps and diarrhoea (Herrington, 1983). Consumption of bitter squash was responsible for 22 cases of food poisoning in Australia, and occasional similar incidents in the USA (Herrington, 1983; Rymal *et al.*, 1984). The bitterness is caused by cucurbitacins, tetracyclic triterpenes that occur naturally in the family *Cucurbitaceae* (Rymal *et al.*, 1984). These compounds can occur in all parts of the plant, although concentrations tend to be highest in the roots (Rehm *et al.*, 1957a,b). Plants may have intensely bitter fruits, but non-bitter leaves or cotyledons (Rymal *et al.*, 1984). Concentrations of cucurbitacins may be several times higher in the placental region of the fruit, compared to the pericarp or the rind (Jaworski *et al.*, 1985). Thus the bitter fruit of summer squash would be potentially more dangerous than those of mature squash or pumpkin, in which the placenta is not eaten.

The origin of these occasional plants producing bitter fruit is not exactly known, but they are thought to have arisen from chance outcrosses to bitter-fruited wild types or gourds during seed production, or through mutations. In some cases where bitter fruit could be traced to individual plants, the plant and mature fruit characteristics did not match that of the cultivar, indicating that genetic change had occurred (Rymal *et al.*, 1984).

CONCLUDING REMARKS

The cucurbit vegetables are a unique group of species that have fascinated plant researchers for many years. By virtue of their large seeds, they begin growth rapidly, and achieve an efficient light intercepting plant canopy earlier than most herbaceous plants. They are aided in this by large, planophile leaves borne on rapidly growing stems, even though assimilation rates are not higher than most other herbaceous crops with C-3 assimilatory pathway (Bruggink and Heuvelink, 1987).

Reproductive growth has received much attention in these crops, particularly the factors determining the gender of the flowers. In spite of considerable research, however, the hormonal controls of sex expression have only been detailed for cucumber, with much still to be done on crops like watermelon. It is particularly noteworthy that some hormones such as ethylene seem to have opposite effects on watermelon as on cucumber (Christopher and Loy, 1982). A greater understanding of the mechanisms of watermelon sex expression may need to be preceded by the discovery of a wider range of flower genotypes similar to those that aided the investigations of cucumber sex expression physiology.

Control of sex expression in cucumber appears to act primarily at the non-sexual floral primordia stage that allows either male or female or perfect flowers to develop. In *Cucurbita*, a second control determines if the male or female primordia reach anthesis or not. Under cool conditions, male flowers are inhibited in development (Rylski and Aloni, 1990), while temperatures of 32/21°C (day/night) inhibit female flowers but allow males to develop (H.C. Wien, 1996, Ithaca, NY, unpublished data). Little is known about the hormonal controls of these processes, and whether endogenous gibberellins and ethylene are involved.

Rapid growth also characterizes the development of the fruit of many cucurbit vegetables. While the giant pumpkin cultivars of *Cucurbita maxima* present the most striking example of this, with growth rates of more than 11 kg (fresh weight) per day, glasshouse cucumber fruits have been shown to gain more than 200 g per day for short periods (Langevin, 1993; Marcelis, 1993). Such a massive transfer of assimilates to one or more reproductive structures on the plant make it likely that growth of other plant parts will be curtailed during this period. Cucurbit vegetable crops are known for the strong inhibition of vegetative, and particularly root growth after flowering, leading to a cyclic development of fruits, and in extreme cases, to collapse of the plants due to pathogen attack of the weakened plant (de Stitger, 1969; Zitter, 1995). The factors that control the movement of assimilates to fruits, rather than to other parts of the plants are at present poorly understood. Hormonal signals presumably make the developing fruit a strong sink for assimilates, but the nature and control of these hormonal signals needs more investigation. Perhaps the incentive of breaking the 1000-pound pumpkin fruit size barrier (prize money in 1995 was $US 50,000) will help stimulate the research.

Further work is also needed to compare the capacity of the cucurbit plant to produce assimilates with the demands of assimilatory products by the rapidly growing fruits. Such studies, carried out by Pharr and co-workers (1985) with cucumbers, need also to be done with other cucurbit vegetables, to understand the limits of productivity, both in terms of fruit yields and of fruit quality factors such as sweetness. It may be that higher yields of fruits with acceptable soluble solids can only be achieved by improvements in the rates of photosynthesis, or more efficient respiration and translocation mechanisms. We have much to learn before such statements can be made with certainty.

ACKNOWLEDGEMENTS

I am grateful for the helpful suggestions of Drs Richard Robinson and Scott NeSmith for improvement of this chapter. Thanks to Dr Martin Goffinet for providing Fig. 9.2.

REFERENCES

Acock, B., Acock, M.C. and Pasternak, D. (1990) Interactions of CO_2 enrichment and temperature on carbohydrate production and accumulation in muskmelon leaves. *Journal of the American Society of Horticultural Science* 115, 525–529.

Atsmon, D. and Galun, E. (1960) A morphological study of staminate, pistillate and hermaphrodite flowers in *Cucumis sativus* L. *Phytomorphology* 10, 110–115.

Atsmon, D. and Tabbak, C. (1979) Comparative effects of gibberellin, silver nitrate and aminoethoxyvinylglycine on sexual tendency and ethylene evolution in the cucumber plant (*Cucumis sativus* L.). *Plant and Cell Physiology* 20, 1547–1556.

Bauerle, W.L. (1971) Effect of sudden wilt disease on the physiology of the muskmelon (*Cucumis melo* L. var. *reticulatus*). Doctoral thesis, Cornell University, Ithaca NY.

Beyer, E.M. Jr and Morgan, P.W. (1969) Time sequence of the effect of ethylene on transport, uptake and decarboxylation of auxin. *Plant and Cell Physiology* 10, 787–799.

Beyer, E.M. Jr and Quebedeaux, B. (1974) Parthenocarpy in cucumber: mechanism of action of auxin transport inhibitors. *Journal of the American Society of Horticultural Science* 99, 385–390.

Bianco, V.V. and Pratt, H.K. (1977) Compositional changes in muskmelons during development and in response to ethylene treatment. *Journal of the American Society of Horticultural Science* 102, 127–133.

Bohn, G.W. and Davis, G.N. (1964) Insect pollination is necessary for the production of muskmelons (*Cucumis melo* v. *reticulatus*). *Journal of Apicultural Research* 3, 61–63.

Bohn, G.W. and Mann, L.K. (1960) Nectarless, a yield-reducing mutant character in muskmelon. *Proceedings of the American Society of Horticultural Science* 76, 455–459.

Bruggink, G.T. and Heuvelink, E. (1987) Influence of light on the growth of young tomato, cucumber and sweet pepper plants in the greenhouse: effects on relative growth rate, net assimilation rate and leaf area ratio. *Scientia Horticulturae* 31, 161–174.

Cantliffe, D.J., Robinson, R.W. and Shannon, S. (1972) Promotion of cucumber fruit set and development by chlorflurenol. *HortScience* 7, 416–418.

Cantliffe, D.J. (1974) Promotion of fruitset and reduction of seed number in pollinated fruit of cucumber by chlorflurenol. *HortScience* 9, 577–578.

Cantliffe, D.J. (1981) Alteration of sex expression in cucumber due to changes in temperature, light intensity and photoperiod. *Journal of the American Society of Horticultural Science* 106, 133–136.

Cantliffe, D.J. and Phatak, S.C. (1975) Plant population studies with pickling cucumbers grown for once over harvest. *Journal of the American Society of Horticultural Science* 100, 464–466.

Christopher, D.A. and Loy, J.B. (1982) Influence of foliarly applied growth regulators on sex expression in watermelon. *Journal of the American Society of Horticultural Science* 107, 401–404.

Chrominski, A. and Kopcewicz, J. (1972) Auxin and gibberellins in 2-chloroethyl phosphonic acid-induced femaleness in *Cucurbita pepo* L. *Zeitschrift der Pflanzenphysiologie* 68, 184–189.

Connor, L.J. and Martin, E.C. (1970) The effect of delayed pollination on yield of cucumbers grown for machine harvests. *Journal of the American Society of Horticultural Science* 95, 456–458.

Connor, L.J. and Martin, E.C. (1971) Staminate : pistillate flower ratio best suited to the production of gynoecious cucumbers for machine harvest. *HortScience* 6, 337–339.

Davis, R.M. Jr and Schweers, V.H. (1971) Associations between physical soil properties and soluble solids in cantaloupes. *Journal of the American Society of Horticultural Science* 96, 213–217.

de Stitger, H.C.M. (1969) Growth relations between individual fruits, and between fruits and roots in cucumber. *Netherlands Journal of Agricultural Science* 17, 209–214.

Dean, B. and Baker, L.R. (1983) Parthenocarpy in gynoecious cucumber as affected by chlorflurenol, genetic parthenocarpy and night temperatures. *HortScience* 18, 349–351.

den Nijs, A.P.M. (1980a) Adaptation of the glasshouse cucumber to lower temperatures in winter by breeding. *Acta Horticulturae* 118, 75–72.

den Nijs, A.P.M. (1980b) The effect of grafting on growth and early production of cucumber at low temperature. *Acta Horticulturae* 118, 57–63.

den Nijs, A.P.M. and Miotay, P. (1991) Fruit and seed set in the cucumber (*Cucumis sativus* L.) in relation to pollen tube growth, sex type and parthenocarpy. *Gartenbauwissenschaft* 56, 46–49.

Denna, D.W. (1973) Effects of genetic parthenocarpy and gynoecious flowering habit on fruit production and growth of cucumber *Cucumis sativus* L. *Journal of the American Society of Horticultural Science* 98, 602–604.

Drews, M.A., Heissner, A. and Augustin, P. (1980) Die Ertragsbildung der Gewächshausgurke beim Frühanbau in Abhängigkeit von der Temperatur und Bestrahlungsstärke. *Archiv für Gartenbau* 28, 17–30

Elassar, G., Rudich, J. and Kedar, N. (1974a) Parthenocarpic fruit development in muskmelon induced by growth regulators. *HortScience* 9, 579–580.

Elassar, G., Rudich, J., Palevich, D. and Kedar, N. (1974b) Induction of parthenocarpic fruit development in cucumber by growth regulators. *HortScience* 9, 238–239.

Ells, J.E. (1983) Chlorflurenol as a fruit-setting hormone for gynoecious pickling cucumber production under open-field conditions. *Journal of the American Society of Horticultural Science* 108, 164–168.

FAO (1994) *Production yearbook 1993*, 47th edn. Food and Agriculture Organization of the United Nations, Rome.

Folster, E. (1974) The influence of the root space temperature on the growth of young cucumbers. *Acta Horticulturae* 39, 153–159.

Frankel, R. and Galun, E. (1977) *Pollination Mechanisms, Reproduction and Plant Breeding*. Springer Verlag, Heidelberg.

Free, J.B. (1970) *Insect Pollination of Crops*. Academic Press, London.

Friedlander, M., Atsmon, D. and Galun, E. (1977a) Sexual differentiation in cucumber: the effects of abscisic acid and other growth regulators on various sex genotypes. *Plant and Cell Physiology* 18, 261–269.

Friedlander, M., Atsmon, D. and Galun, E. (1977b) Sexual differentiation in cucumber: abscisic acid and gibberellic acid contents of various sex genotypes. *Plant and Cell Physiology* 18, 681–691.

Fuller, G.L. and Leopold, A.C. (1975) Pollination and the timing of fruit-set in cucumbers. *HortScience* 10, 617–619.

Galun, E., Jung, Y. and Lang, A. (1963) Morphogenesis of floral buds of cucumber cultured *in vivo*. *Developmental Biology* 6, 370–387.

Goffinet, M.C. (1990) Comparative ontogeny of male and female flowers of *Cucumis sativus*. In: Bates, D.M., Robinson, R.W. and Jeffrey, C. (eds) *Biology and Utilization of the Cucurbitaceae*. Cornell University Press, Ithaca, pp. 288–304.

Gregory, F.G. (1928) Studies in the energy relations of plants. II. The effect of temperature on increase in area of leaf surface and in dry weight of *Cucumis sativus*. *Annals of Botany* 42, 469–507.

Grimstad, S.D. and Frimanslund, E. (1993) Effect of different day and night temperature regimes on glasshouse cucumber young plant production, flower bud formation and early yield. *Scientia Horticulturae* 53, 191–204.

Gross, K.C. and Pharr, D.M. (1982) A potential pathway for galactose metabolism in *Cucumis sativus* L., a stachyose transporting species. *Plant Physiology* 69, 117–121.

Hand, D.W. (1984) Crop responses to winter and summer CO_2 enrichment. *Acta Horticulturae* 162, 45–63.

Handley, L.W., Pharr, D.M. and McFeeters, R.F. (1983) Carbohydrate changes during maturation of cucumber fruit. Implications for sugar metabolism and transport. *Plant Physiology* 72, 499–502.

Hayase, H. (1953) *Cucurbita* crosses. IV. The development of squash fruit as affected by placement of pollen on stigma. *Hokkaido National Agricultural Experiment Station Research Bulletin* 64, 22–25.

Heij, G. (1980) Glasshouse cucumber, stem elongation and earliness of fruit production as influenced by temperature and planting date. *Acta Horticulturae* 118, 105–121.

Hemphill, D.D. Jr, Baker, L.R. and Sell, H.M. (1972) Different sex phenotypes of *Cucumis sativus* L. and *C. melo* L. and their endogenous gibberellin activity. *Euphytica* 21, 285–291.

Herrington, M.E. (1983) Intense bitterness in commercial zucchini. *Cucurbit Genetics Cooperative Newsletter* 6, 75–76.

Hori, Y., Arai, K., Hosoya, T. and Oyamada, M. (1968) Studies on the effects of root temperature and its combination with air temperature on growth and nutrition of vegetable crops. I. Cucumber, tomato, turnip and snap bean. *Bulletin of the Horticultural Research Station Hiratsuka Series A* 7, 187–214.

Hubbard, N.L., Huber, S.C. and Pharr, D.M. (1989) Sucrose phosphate synthase and acid invertase as determinants of sucrose concentration in developing muskmelon (*Cucumis melo* L.) fruits. *Plant Physiology* 91, 1527–1534.

Hughes, D.L., Bosland, J. and Yamaguchi, M. (1983) Movement of photosynthates in muskmelon plants. *Journal of the American Society of Horticultural Science* 108, 189–192.

Hughes, D.L. and Yamaguchi, M. (1983) Identification and distribution of some carbohydrates of the muskmelon plant. *HortScience* 18, 739–740.

Ikuse, M. (1956) *Pollen Grains of Japan.* Hirokawa Publishing, Tokyo.

Ito, H. and Saito, T. (1960) Factors responsible for the sex expression of the cucumber plant. XII. Physiological factors associated with the sex expression of flowers. *Tohoku Journal of Agricultural Research* 11, 287–308.

Jacks, T.J., Hensarling, T.P. and Yatsu, L.Y. (1972) Cucurbit seeds. I. Characterization and uses of oils and proteins. A review. *Economic Botany* 26, 135–141.

Jaworski, A., Gorski, P.M., Shannon, S. and Robinson, R.W. (1985) Cucurbitacin concentrations in different plant parts of *Cucurbita* species as a function of age. *Cucurbit Genetics Cooperative Newsletter* 8, 71–73.

Kano, Y. (1993) Relationship between the occurrence of hollowing in watermelon and the size and the number of fruit cells and intercellular spaces. *Journal of the Japanese Society of Horticultural Science* 62, 103–112.

Kano, K., Fujimura, T., Hirose, T. and Tsukamoto, Y. (1957) Studies on the thickening growth of garden fruits. I. On the cushaw, eggplant and pepper. *Kyoto University Research Institute of Food Science Memoir* 12, 45–90.

Karchi, Z. (1970) Effects of 2-chloroethylphosphonic acid on flower types and flowering sequences in muskmelon. *Journal of the American Society of Horticultural Science* 95, 515–518.

Kihara, H. (1951) Triploid watermelons. *Proceedings of the American Society of Horticultural Science* 58, 217–230.

Kim, I.S., Okubo, H. and Fujieda, K. (1992a) Genetic and hormonal control of parthenocarpy in cucumber (*Cucumis sativus* L.). *Journal of the Faculty of Agriculture of Kyushu University* 36, 173–181.

Kim, I.S., Okubo, H. and Fujieda, K. (1992b) Endogenous levels of IAA in relation to parthenocarpy in cucumber (*Cucumis sativus* L.). *Scientia Horticulturae* 52, 1–8.

Kimball, B.A. (1986) Influence of elevated CO_2 on crop yield. In: Enoch, H.Z. and Kimball, B.A. (eds) *Carbon Dioxide Enrichment of Greenhouse Crops. II. Physiology, Yield and Economics.* CRC Press, Boca Raton, pp. 105–115.

Knavel, D.E. (1991) Productivity and growth of short-internode muskmelon plant at various spacings or densities. *Journal of the American Society of Horticultural Science* 116, 926–929.

Kooistra, E. (1967) Femaleness in breeding glasshouse cucumber. *Euphytica* 16, 1–17.

Krug, H. and Liebig, H.P. (1980) Diurnal thermoperiodism of the cucumber. *Acta Horticulturae* 118, 83–94.

Langevin, D. (1993) *How to Grow World Class Giant Pumpkins.* Annedawn Publishing, Norton, MA.

Le Deunff, E., Sauton, A. and Dumas, C. (1993) Effect of ovular receptivity on seed set and fruit development in cucumber (*Cucumis sativus* L.). *Sexual Plant Reproduction* 6, 139–146.

Lee, J.M. (1994) Cultivation of grafted vegetables. I. Current status, grafting methods, and benefits. *HortScience* 29, 235–239.

Liebig, H.P. (1980a) A growth model to predict yield and economical figures of the cucumber crop. *Acta Horticulturae* 118, 165–174.

Liebig, H.P. (1980b) Physiological and economical aspects of cucumber crop density. *Acta Horticulturae* 118, 149–164.

Lorenz, O.A. and Maynard, D.N. (1980) *Knott's Handbook for Vegetable Growers*. 2nd edn, J. Wiley, New York.

Lower, R.L., Nienhuis, J. and Miller, C.H. (1982) Gene action and heterosis for yield and vegetative characteristics in a cross between a gynoecious pickling cucumber inbred and a *Cucumis sativus* var. *hardwickii* line. *Journal of the American Society of Horticultural Science* 107, 75–78.

Lower, R.L., Smith, O.S. and Ghaderi, A. (1983) Effects of plant density, arrangement and genotype on stability of sex expression in cucumber. *HortScience* 18, 737–738.

Lower, R.L. and Edwards, M.D. (1986) Cucumber breeding. In: Bassett, M.J. (ed.) *Breeding Vegetable Crops*. Avi Publishing, Westport, Connecticut, pp. 173–207.

MacNab, A.A. (1971) Symptomatology of decline, early-collapse, and late-collapse of muskmelons (*Cucumis melo* L. var. *reticulatus*) in New York State and the etiology of decline. Doctoral thesis, Cornell University, Ithaca, NY.

Mann, L.K. (1943) Fruit shape of watermelon as affected by placement of pollen on stigma. *Botanical Gazette* 105, 257–262.

Marcelis, L.F.M. (1992a) The dynamics of growth and dry matter distribution in cucumber. *Annals of Botany* 69, 487–492.

Marcelis, L.F.M. (1992b) Non-destructive measurements and growth analysis of the cucumber fruit. *Journal of Horticultural Science* 67, 457–464.

Marcelis, L.F.M. (1993) Effect of assimilate supply on the growth of individual cucumber fruits. *Physiologia Plantarum* 87, 313–320.

Marcelis, L.F.M. and Baan Hofman-Eijer, L.R. (1993) Effect of temperature on the growth of individual cucumber fruits. *Physiologia Plantarum* 87, 321–328.

Matlob, A.N. and Kelly, W.C. (1973) The effect of high temperature on pollen tube growth of snake melon and cucumber. *Journal of the American Society of Horticultural Science* 98, 296–300.

Matsuo, E. (1968) Studies on the photoperiodic sex differentiation in cucumber *Cucumis sativus* L. I. Effect of temperature and photoperiod upon the sex differentiation. *Journal of the Faculty of Agriculture of Kyushu University* 14, 483–506.

McCollum, J.P. (1934) Vegetative and reproductive responses associated with fruit development in the cucumber. *Cornell Agricultural Experiment Station Memoir* 163, 1–27.

McCollum, T.G., Huber, D.J. and Cantliffe, D.J. (1988) Soluble sugar accumulation and activity of related enzymes during muskmelon fruit development. *Journal of the American Society of Horticultural Science* 113, 399–403.

McGlasson, W.B. and Pratt, H.K. (1963) Fruit-set patterns and fruit-growth in cantaloupe (*Cucumis melo* var. *reticulatus* Naud.). *Proceedings of the American Society of Horticultural Science* 83, 495–505.

McGregor, S.E. and Todd, F.E. (1952) Cantaloupe production with honey bees. *Journal of Economic Entomology* 45, 43–47.

Mendlinger, S. (1994) Effect of increasing plant density and salinity on yield and fruit quality in muskmelon. *Scientia Horticulturae* 57, 41–49.

Miller, M.E., Martyn, R.D., Lovic, B.R. and Burton, B.D. (1995) An overview of vine decline diseases of melons. In: Dunlap, J. and Lester, G. (eds) *Cucurbitaceae '94: Evaluation and Enhancement of Cucurbit Germplasm*. Gateway Printing, Edinburgh, TX, pp. 31–35.

Mitchell, W.D. and Wittwer, S.H. (1962) Chemical regulation of flower sex expression and vegetative growth in *Cucumis sativus* L. *Science* 136, 880–881.

Mizuno, S. and Pratt, H.K. (1973) Relations of respiration and ethylene production to maturity in the watermelon. *Journal of the American Society of Horticultural Science* 98, 614–617.

Mohr, H.C. (1986) Watermelon breeding. In: Bassett, M.J. (ed.) *Breeding Vegetable Crops*. Avi, Westport, Connecticut, pp. 37–66.

Munger, H.M. (1988) A revision on controlled pollination in cucumber. *Cucurbits Genetics Cooperative Report* 11, 8.

Munnecke, D.E., Laemmlen, F.F. and Bricker, J. (1984) Soil fumigation controls sudden wilt in melons. *California Agriculture* 38(5,6), 8–9.

Nelson, J.M. and Sharples, G.C. (1980) Effect of growth regulators on germination of cucumber and other cucurbit seeds at suboptimal temperatures. *HortScience* 15, 253–254.

Nerson, H., Cantliffe, D.J., Paris, H.S. and Karchi, Z. (1982) Low-temperature germination of bird's nest-type muskmelons. *HortScience* 17, 639–640.

Nerson, H. and Paris, H.S. (1987) Effects of plant type and growth regulators on the flowering, fruiting and yield concentration of melon. *Crop Research* 27, 19–30.

NeSmith, D.S. and Bridges, D.C. (1992) Summer squash germination in response to temperature. *Proceedings of the National Symposium for Stand Establishment in Horticultural Crops* 1, 15–22.

NeSmith, D.S., Hoogenboom, G. and Groff, D.W. (1994) Staminate and pistillate flower production of summer squash in response to planting date. *HortScience* 29, 256–257.

Nienhuis, J., Lower, R.L. and Staub, J.E. (1983) Selection for improved low temperature germination in cucumber. *Journal of the American Society of Horticultural Science* 108, 1040–1043.

Nienhuis, J., Lower, R.L. and Miller, C.H. (1984) Effects of genotype on the stability of sex expression in cucumber. *HortScience* 19, 273–274.

Nitsch, J.P., Kurtz, E.B., Liverman, J.L. and Went, F.W. (1952) The development of sex expression in cucurbit flowers. *American Journal of Botany* 39, 32–43.

Nitzany, F.E. (1966) Synergism between *Pythium ultimum* and cucumber mosaic virus. *Phytopathology* 56, 1386–1389.

Novak, J.D. (1972) A study of the effects of environmental and genetic factors on sex expression in cucumber (*Cucumis sativus* L.). Doctoral thesis, Cornell University, Ithaca NY.

O'Sullivan, J. (1980) Irrigation, spacing, and nitrogen effects on yield and quality of pickling cucumbers grown for mechanical harvesting. *Canadian Journal of Plant Science* 60, 923–928.

Ogawa, Y., Inoue, Y. and Aoki, S. (1989) Promotive effects of exogenous and endogenous gibberellins on the fruit development in *Cucumis sativus* L. *Journal of the Japanese Society of Horticultural Science* 58, 327–331.

Ogawa, Y., Nishikawa, S., Inoue, N. and Aoki, S. (1990) Promotive effects of different cytokinins on the fruit growth in *Cucumis sativus*. *Journal of the Japanese Society of Horticultural Science* 59, 597–601.

Paris, H.S., Nerson, H. and Karchi, Z. (1986) Branching of bird's nest-type melons. *Crop Research* 26, 33–40.

Paris, H.S., Nerson, H., Burger, Y., Edelstein, M., Karchi, Z. and McCollum, T.G. (1988) Synchrony of yield of melons as affected by plant type and density. *Journal of Horticultural Science* 63, 141–147.

Peet, M.M. and Willits, D.H. (1987) CO_2 enrichment alternatives: effects of increasing concentration or duration of enrichment on cucumber yield. *Journal of the American Society of Horticultural Science* 112, 236–241.

Peirce, L.C. (1987) *Vegetables, Characteristics, Production, and Marketing*. Wiley, New York.

Perry, K.B., Wehner, T.C. and Johnson, G.L. (1986) Comparison of 14 methods to determine heat unit requirements for cucumber harvest. *HortScience* 21, 419–423.

Perry, K.B. and Wehner, T.C. (1990) Prediction of cucumber harvest date using a heat unit system. *HortScience* 25, 405–406.

Pharr, D.M., Huber, S.C. and Sox, H.N. (1985) Leaf carbohydrate status and enzymes of translocate synthesis in fruiting and vegetative plants of *Cucumis sativus* L. *Plant Physiology* 77, 104–108.

Poole, C.F. and Porter, D.R. (1933) Pollen germination and development in the watermelon. *Proceedings of the American Society of Horticultural Science* 30, 526–530.

Porter, D.R., Bisson, C.S. and Alliger, H.W. (1940) Factors affecting the total soluble solids, reducing sugars and sucrose in watermelons. *Hilgardia* 13, 31–66.

Rehm, S., Enslin, P.R., Meeuse, A.D.J. and Wessels, J.H. (1957a) Bitter principles of the *Cucurbitaceae*. VII. The distribution of bitter principles in this plant family. *Journal of Science, Food and Agriculture* 8, 679–686.

Rehm, S. and Wessels, J.H. (1957b) Bitter principles of the *Cucurbitaceae*. VIII. Cucurbitacins in seedlings – occurence, biochemistry and genetical aspects. *Journal of Science, Food and Agriculture* 8, 687–691.

Robinson, R.W., Shannon, S. and De la Guardia, M.D. (1969) Regulation of sex expression in the cucumber. *Bioscience* 19, 141–142.

Robinson, R.W. (1993) Genetic parthenocarpy in *Cucurbita pepo* L. *Cucurbit Genetics Cooperative Report* 16, 55–57.

Rudich, J., Halevy, A.H. and Kedar, N. (1972a) Ethylene evolution from cucumber plants as related to sex expression. *Plant Physiology* 49, 998–999.

Rudich, J., Halevy, A.H. and Kedar, N. (1972b) The level of phytohormones in monoecious and gynoecious cucumbers as affected by photoperiod and ethephon. *Plant Physiology* 50, 585–590.

Rudich, J., Halevy, A.H. and Kedar, N. (1972c) Interaction of gibberellin and SADH on growth and sex expression of muskmelon. *Journal of the American Society of Horticultural Science* 97, 369–372.

Rudich, J., Baker, L.P., Scott, J.W. and Sell, H.M. (1976) Phenotypic stability and ethylene evolution in androecious cucumber. *Journal of the American Society of Horticultural Science* 101, 48–51.

Rudich, J., Baker, L.R. and Sell, H.M. (1977) Parthenocarpy in *Cucumis sativus* L. as affected by genetic parthenocarpy, thermophotoperiod and femaleness. *Journal of the American Society of Horticultural Science* 102, 225–228.

Rudich, J. (1990) Biochemical aspects of hormonal regulation of sex expression in cucurbits. In: Bates, D.M., Robinson, R.W. and Jeffrey, C. (eds) *Biology and Utilization of the Cucurbitaceae*. Cornell University Press, Ithaca, pp. 269–280.

Rylski, I. (1974) Effects of season on parthenocarpic and fertilized summer squash (*Cucurbita pepo* L.). *Experimental Agriculture* 10, 39–44.

Rylski, I. and Aloni, B. (1990) Parthenocarpic fruit set and development in the Cucurbitaceae and Solanaceae under protected cultivation in mild winter climate. *Acta Horticulturae* 287, 117–126.

Rymal, K.S., Chambliss, O.L., Bond, M.D. and Smith, D.A. (1984) Squash containing toxic cucurbitacin compounds occurring in California and Alabama. *Journal of Food Protection* 47, 270–271.

Saito, T. and Ito, H. (1963) Factors responsible for the sex expression of the cucumber plant. XIV. Auxin and gibberellin content in the stem apex and the sex pattern of flowers. *Tohoku Journal of Agricultural Research* 14, 222–239.

Schaffer, A.A., Aloni, B. and Fogelman, E. (1987) Sucrose metabolism and accumulation in developing fruit of *Cucumis*. *Phytochemistry* 26, 1883–1887.

Schapendonk, A.H.C.M. and Brouwer, P. (1984) Fruit growth of cucumber in relation to assimilate supply and sink activity. *Scientia Horticulturae* 23, 21–33.

Schulte, H.K. and Grote, U. (1974) Untersuchungen zur Kältefestigkeit der Freilandgurken. *Gartenbauwissenschaft* 39, 241–251.

Seaton, H.L. and Kremer, J.C. (1938) The influence of climatological factors on anthesis and anther dehiscence in the cultivated cucurbits. A preliminary report. *Proceedings of the American Society of Horticultural Science* 36, 627–631.

Sedgley, M. and Buttrose, M.S. (1978) Some effects of light intensity, daylength and temperature on flowering and pollen tube growth in the watermelon (*Citrullus lanatus*). *Annals of Botany* 42, 609–616.

Shannon, S. and Robinson, R.W. (1979) The use of ethephon to regulate sex expression of summer squash for hybrid seed production. *Journal of the American Society of Horticultural Science* 104, 674–677.

Shifriss, O. and Galun, E. (1956) Sex expression in the cucumber. *Proceedings of the American Society of Horticultural Science* 67, 479–486.

Shishido, Y., Hori, Y. and Shikano, S. (1990) Effects of benzyl adenine on translocation and distribution of photoassimilates during fruit setting and development in cucumber plants. *Journal of the Japanese Society of Horticultural Science* 59, 129–136.

Simon, E.W., Minchin, A., McMenamin, M.M. and Smith, J.M. (1976) The low temperature limit for seed germination. *New Phytologist* 77, 301–311.

Sinnott, E.W. (1939) A developmental analysis of the relation between cell size and fruit size in cucurbits. *American Journal of Botany* 26, 179–189.

Sinnott, E.W. (1945) The relation of growth to size in cucurbit fruits. *American Journal of Botany* 32, 439–446.

Slack, G. and Hand, D.W. (1980) Control of air temperature for cucumber production. *Acta Horticulturae* 118, 175–186.

Splittstoesser, W.E. (1970) Effects of 2-chloroethylphosphonic acid and gibberellic acid on sex expression and growth of pumpkins. *Physiologia Plantarum* 23, 762–768.

Staub, J.E., Nienhuis, J. and Lower, R.L. (1986) Effects of seed preconditioning treatments on emergence of cucumber populations. *HortScience* 21, 1356–1359.

Staub, J.E., Wehner, T.C. and Tolla, G.E. (1987) Effect of treatment of cucumber seeds with growth regulators on emergence and yield of plants in the field. *Acta Horticulturae* 198, 43–52.

Staub, J.E., Knerr, L.D. and Hopen, H.J. (1992) Plant density and herbicides affect cucumber productivity. *Journal of the American Society of Horticultural Science* 117, 48–53.

Suzuki, E. (1969) Studies on the fruit development of greenhouse melon (*Cucumis melo* L.). I. On the relation between shape of stigma and number of seeds and on the pollen tube development and the hour of fertilization. *Journal of the Japanese Society of Horticultural Science* 38, 36–41.

Tachibana, S. (1982) Comparison of effects of root temperature on the growth and mineral nutrition of cucumber cultivars and fig-leaf gourd. *Journal of the Japanese Society of Horticultural Science* 51, 299–308.

Tachibana, S. (1987) Effect of root temperature on the rate of water and nutrient absorption in cucumber cultivars and figleaf gourd. *Journal of the Japanese Society of Horticultural Science* 55, 461–467.

Tachibana, S. (1989) Respiratory response of detached roots to lower temperatures in cucumber and figleaf gourd grown at 20C root temperature. *Journal of the Japanese Society of Horticultural Science* 58, 333–337.

Takeno, K. and Ise, H. (1992) Parthenocarpic fruit set and endogenous indole-3-acetic acid content in ovary of *Cucumis sativus* L. *Journal of the Japanese Society of Horticultural Science* 60, 941–946.

Takeno, S., Ise, H., Minowa, H. and Dounowaki, T. (1992) Fruit growth induced by benzyladenine in *Cucumis sativus* L.: Influence of benzyladenine on cell division, cell enlargement and indole-3-acetic acid content. *Journal of the Japanese Society of Horticultural Science* 60, 915–920.

Tasdighi, M. and Baker, L.R. (1981) Comparison of single and three-way crosses of pickling cucumber hybrids for femaleness and yield by once-over harvest. *Journal of the American Society of Horticultural Science* 106, 370–373.

Tazuke, A. and Sakiyama, R. (1984) Growth analysis of cucumber fruits on vine by use of dimensions of fruit shape. *Journal of the Japanese Society of Horticultural Science* 53, 30–37.

Tiedjens, V.A. (1928) Sex ratios in cucumber flowers as affected by different conditions of soil and light. *Journal of Agricultural Research* 36, 721–746.

Uzcategui, N.A. and Baker, L.R. (1979) Effects of multiple pistillate flowering on yields of gynoecious pickling cucumbers. *Journal of the American Society of Horticultural Science* 104, 148–151.

Varga, A. and Bruinsma, J. (1990) Dependence of ovary growth on ovule development in *Cucumis sativus*. *Physiologia Plantarum* 80, 43–50.

Watkins, J.T. and Cantliffe, D.J. (1980) Regulation of fruit set in *Cucumis sativus* by auxin and an auxin transport inhibitor. *Journal of the American Society of Horticultural Science* 105, 603–607.

Watts, V.W. (1938) Rest period in cucumber seeds. *Proceedings of the American Society of Horticultural Science* 36, 652–654.

Webb, J.A. and Gorham, P.R. (1964) Translocation of photosynthetically assimilated 14-C in straight-necked squash. *Plant Physiology* 39, 663–672.

Wehner, T.C. and Miller, C.H. (1985) Effect of gynoecious expression on yield and earliness of a fresh-market cucumber hybrid. *Journal of the American Society of Horticultural Science* 110, 464–466.

Welles, G.W.H. and Buitelaar, K. (1988) Factors affecting soluble solids content of muskmelon (*Cucumis melo* L.). *Netherlands Journal of Agricultural Science* 36, 239–246.

Whitaker, T.W. and Bemis, W.P. (1976) Cucurbits. In: Simmonds, N.W. (ed.) *Evolution of Crop Plants*. Longman, London, pp. 65–69.

Whitaker, T.W. and Davis, G.N. (1962) *Cucurbits, Botany, Cultivation and Utilization*. Interscience Publishers, New York.

Whitaker, T.W. and Prior, D.E. (1946) Effect of plant-growth regulators on the set of fruit from hand-pollinated flowers in *Cucumis melo* L. *Proceedings of the American Society of Horticultural Science* 48, 417–422.

Widders, I.E. and Price, H.C. (1989) Effects of plant density on growth and biomass partitioning in pickling cucumbers. *Journal of the American Society of Horticultural Science* 114, 751–755.

Winton, A.L. and Winton, K.B. (1935) *The Structure and Composition of Foods. II. Vegetables, Legumes, Fruits.* J. Wiley, New York.

Wong, C.Y. (1941) Chemically induced parthenocarpy in certain horticultural plants with special reference to watermelon. *Botanical Gazette* 103, 64–84.

Zahara, M. (1972) Effects of plant density on yield and quality of cantaloupes. *California Agriculture* 26, 15.

Zitter, T.A. (1995) Sudden wilt of melons from a Northeastern US perspective. In: Dunlap, J. and Lester, G. (eds) *Cucurbitaceae '94: Evaluation and Enhancement of Cucurbit Germplasm*. Gateway Printing, Edinburgh, TX, pp. 44–47.

Celery   10

E. Pressman

*ARO the Volcani Center, Institute for Field and Garden Crops, PO Box 6,
Bet Dagan 50250, Israel*

Celery (*Apium graveolens* L.) is a member of the *Umbelliferae (Apiaceae)*. As a wild species it is distributed all over the world, especially in the temperate and subtropical regions. Wild celery (var. *silvestre* Presl.) which is the putative ancestor of the modern cultivated varieties, was known in the Mediterranean basin to the ancient Egyptians, Greeks and Romans (Becker, 1962). The wild plant is found in wet, low-lying locations and shows tolerance to soils high in salts (halophyte). The leaves of the wild type are non-edible, since its petioles are hollow, with sharp aroma and taste. Therefore, it was used as a medicinal herb and for religious ceremonies. It is not known when the selection towards fully solid, edible petioles started, but it can be presumed to have happened in the Mediterranean region.

Celery was first mentioned as a spice plant in France in 1623 and used for fresh consumption in 1686; it was first grown in the USA in 1806.

Three different types of cultivated celery were developed (Becker, 1962): (i) f. *secalinum* looks very much like the wild type and its whole leaves are used as a spice; (ii) var. *rapaceum* (Miller) – celeriac (root celery) – its edible storage organ consists of the swollen base of the stem and part of the root; and (iii) var. *dulce* (Miller) – stalk celery – whose developed leaf petioles are the edible parts. This last kind can be further divided into self-blanching types and green types. Root celery is common in Europe, the self-blanching celery is accepted in northern Europe and East Asia, while green celery is consumed in the USA, southern Europe and Israel.

SEED GERMINATION

Among dicotyledonous plants the germination of celery seed is one of the most intensively investigated. The fact that this seed possesses a rudimentary embryo (Hayward, 1938) which must grow inside the seed before germination, that (similarly to lettuce) it requires light for germination (dormancy break) and that it is very sensitive to high temperatures, make it very useful in research on the physiology of germination. This is apart from the practical problems of poor germination which motivated such research.

The interest in celery seed germination increased after the failure to achieve germination of celery seeds in seedbeds of open nurseries (Guzman, 1964). Shading the beds to reduce the effects of direct sunlight did not always help to improve germination. The modern technologies, in which transplants are grown in temperature-controlled nurseries, avoid the effect of high temperatures.

The first event in the germination of a celery seed is the elongation of the embryo. A mature dry seed contains a small linear axile embryo surrounded by an oily endosperm which occupies the bulk of the seed (Jacobsen and Pressman, 1979). The emergence of the radicle, which indicates germination, occurs after the embryo has grown at the expense of the endosperm.

The germinability of celery seeds is influenced by their position in the inflorescence (umbel) on the mother plant. Seeds collected from primary and secondary umbellets have been reported to be less dormant (Thomas *et al.*, 1978c) and germinate better (van der Toorn, 1990a) than those from tertiary and quaternary ones.

LIGHT

Early experiments (Harrington, 1923; Morinaga, 1926) revealed that under low temperatures (10–15°C) celery seeds germinate in the dark. At higher temperatures (20–25°C) the seeds remain dormant in darkness and the germination is light-dependent. Red light has been found to induce germination of dark-imbibed seeds and this effect is reversed by far red light, indicating that this process is mediated through the phytochrome system (Pressman *et al.*, 1977; Thomas *et al.*, 1978b). In this regard, seeds of celeriac (root celery) were more dormant and required more light exposure for germination than seeds of stalk celery (Pressman *et al.*, 1977). Thomas (1989) suggested that light, through phytochrome, stimulates the biosynthesis of gibberellins (GAs) which are essential for the germination of the biennial cultivars. The light stimulation of germination can be negated by GA biosynthesis inhibitors (Pressman and Shaked, 1988; Thomas, 1989), which supports this suggestion. On the other hand, seeds of four annual celery cultivars were able to germinate in the dark, probably because they 'store' GAs during maturation (Pressman *et al.*, 1988; Pressman and Shaked, 1988). Thomas *et al.* (1978b) found a rapid increase in endogenous cytokinins following red-light irradiation of celery seeds.

TEMPERATURE

Another reason for poor germination of celery seeds is exposure to unsuitable temperatures (Robinson, 1954). Whereas under moderate temperatures (15–25°C) germination takes place in response to irradiation, at temperatures of 30°C or higher the seeds will remain dormant even in light (Biddington and Thomas, 1978; Robinson, 1954).

Thomas *et al.* (1986) reported that thermoinhibition of celery seeds was associated with the accumulation of an inhibitor which did not appear to be abscissic acid (ABA). Maintaining seeds under high temperatures (35°C) caused thermodormancy, i.e. a reduction in germination upon transfer of the seeds to optimal conditions: 20°C in the light (Pressman *et al.*, 1977). The presence of light during the period of high temperature increased the effectiveness of the thermodormancy treatment.

FLUCTUATING TEMPERATURES

The possibility of breaking dark dormancy by means of temperature alternations was demonstrated by Harrington (1923) and Morinaga (1926). However, they were not aware of the effect of short exposures to light during the experiments. Pressman *et al.* (1977) showed varietal (stalk celery vs. root celery vs. wild celery) differences in the response of seeds to 10-degree temperature differences in the range between 10 and 30°C under complete darkness.

GROWTH REGULATORS

The effect of light can be simulated by exogenous GAs, of which the mixture GA_4 and GA_7 (GA_{4+7}) is the most effective one known (Palevitch and Thomas, 1976). Jacobsen *et al.* (1976) showed that GAs specifically induce the breakdown of the endosperm in celery seeds. Cytokinins may not have a primary role in breaking dormancy, but they probably enhance the promotion of seed germination by GAs at high temperatures (Biddington and Thomas, 1978). The practical implication of these findings is that GA_{4+7} plus cytokinins (particularly benzyladenine, BA) can be used for improving celery seed germination under unfavourable conditions (Thomas *et al.*, 1975). Other chemicals, such as ethephon, fusicoccin and chelating compounds (ChDTA) have been found to influence celery seed germination in a manner similar to BA (Thomas, 1990). It has also been shown that the response of celery seeds to GA_{4+7} is highly pH-dependent (Palevitch and Thomas, 1976). Those chemicals which decrease the solution pH facilitate better penetration of the GAs into the seed. On the other hand, Pressman and Shaked (1991) concluded that exogenous cytokinins promote better penetration of GAs into the seed by affecting the seed coat but not by decreasing the pH.

PRIMING AND PELLETING

Priming is intended to synchronize or improve the germination and emergence (Fig. 10.1) under both normal and stress conditions and thereby, to increase yield (Table 10.1). Salter and Darby (1976) found that pretreatment (priming) with a mixture of salts (KNO$_3$ + K$_3$PO$_4$) at −1 MPa significantly improved the rate and synchronization of germination; this mixture was more effective than polyethylene glycol (PEG). Priming seeds with PEG reduced the mean germination time (Brocklehurst and Dearman, 1983a), increased the seedling emergence percentage (Brocklehurst and Dearman, 1983b) and raised the upper temperature limit for germination (Brocklehurst *et al.*, 1982/83). A better effect was obtained by adding GA$_{4+7}$ and ethephon to the solution. Van der Toorn (1990b) found that embryo growth in water was characterized by both cell division and an increase in cell size, while in PEG solution embryo growth was solely through cell division.

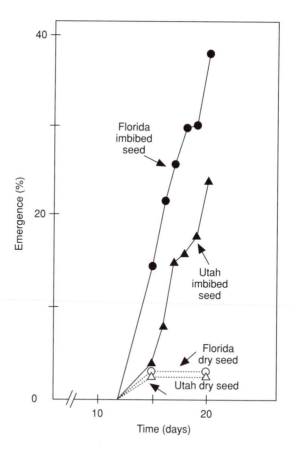

Fig. 10.1. Emergence of dry and germinated seeds (sown in a gel) of 'Florida 683' and 'Tall Utah 52–70' celery (Biddington *et al.*, 1975).

Table 10.1. Comparison of celery plants size ('New Dwarf White') at three dates, developed from pre-germinated seeds, sown in gel, or directly drilled on 2 April. The number of plants from which the data were obtained is in brackets (Biddington *et al.*, 1975).

Treatment	1 July sample		16 Oct. harvest		12 Nov. harvest	
	Mean no. of leaves	Mean plant dry wt (g)	Whole plant wt (g)	Trimmed wt (g)	Whole plant wt (g)	Trimmed wt (g)
Dry seed	4.7 (241)	0.06 (30)	620 (61)	500	770 (54)	550
Germinated seed	6.6 (336)	0.56 (30)	1120 (67)	850	1170 (72)	880
% increase	40	833	81	70	52	60
SE of harvest × harvest mean	–	–	± 69	± 45	± 69	± 45

Extracts of PEG-treated seeds have been found to exhibit an activity which resembled that of cytokinin glucosides (Thomas, 1984). Parrera *et al.* (1993) used a different means of priming; they incubated seeds in calcined clay solution to which NaOCl had been added. This treatment significantly reduced the negative effect of high temperature on germination.

The small size and uneven shape of the celery seeds are not conducive to precision sowing. Pelleting of the seeds has been tested as a means of sowing high-germination single seeds (Thomas *et al.* 1978a). The inhibition of germination imposed by pelleting was alleviated by soaking the seeds in GA$_{4+7}$ and ethephon or daminozide solutions before pelleting.

SOMATIC EMBRYOS AND SYNTHETIC SEEDS

Attempts to produce synthetic seeds involved intensive research on somatic embryos. The early phase of somatic embryo development depended on the presence of auxin (Al-Abta and Collin, 1979; Kim and Janick, 1989a) while the onset of embryogenesis was characterized by high content of polyamines and cytokinins (Danin *et al.*, 1993). This was in accordance with the finding of van Staden *et al.* (1992) on metabolism of cytokinins in embryonic cell culture of celery. Mannitol was found by Nadel *et al.* (1990) to increase the number of embryos and their growth rate and ABA was found to induce the synchronization of the somatic embryos.

Production of artificial seeds necessitates desiccation of the somatic embryos. The survival of desiccated somatic embryos has been increased by pretreatment with ABA and encapsulation with a water-soluble ethylene oxide polymer (polyox) (Kim and Janick, 1989b). Addition of proline and ABA (Saranga *et al.*, 1992a) or high osmolarity treatment (Saranga *et al.*, 1992b) further increased the embryo tolerance to partial desiccation. The conversion of somatic embryos (artificial seeds) to seedlings was more successful when sucrose was added to the regeneration medium (Saranga and

Janick, 1991). Redenbaugh *et al.* (1986) produced somatic 'seeds' by encapsulating somatic embryos in calcium alginate beads.

Nadel *et al.* (1995) reported that in field tests celery plants regenerated from tissue culture were uniform and horticulturally true to type. However, field test of synthetic seeds have failed, so far, to produce commercially viable plants.

VEGETATIVE DEVELOPMENT AND YIELD FORMATION

The celery plant consists of a short crown stem and rosette leaves, and is about 60 cm in height at harvest maturity. Yield of stalk celery is maximized by producing many leaves with large, long petioles. The rate of leaf formation is temperature-dependent and both low and high temperature regimes retard leaf formation (Fig. 10.2).

Long days (LD), obtained by artificially extending the natural daylength, or by night breaks (NB) also caused reduced leaf production (Pressman and Negbi, 1987; Roelofse *et al.*, 1990).

Leaf elongation is affected by high temperatures (Table 10.2). Celery grown in the summer, especially in hot regions, possesses shorter petioles and therefore, new cultivars tolerant to high temperatures have been introduced (Wolf and Scully, 1992). The height of the tallest petiole was used to predict the fresh weight yield from regression equations (Fig. 10.3). Long photoperiods enhance leaf elongation whereas under short days the leaves are shorter (Pressman and Negbi, 1987; Roelofse *et al.*, 1990). The negative

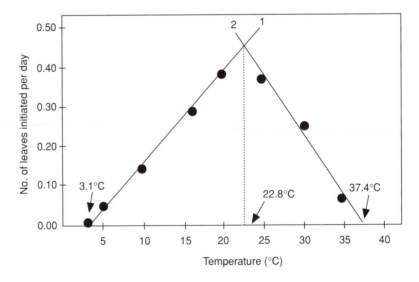

Fig. 10.2. Relationships between temperature and rate of leaf initiation for celery cv. 'New Dwarf White' (Ramin and Atherton, 1991b).

Table 10.2. The effect of temperature regime (°C) and GA_3 application (mg plant^{-1}) on the number of leaves formed during the experiment and on leaf length in two cultivars of celery (Pressman, 1979).

Cultivar	GA_3	No. of leaves		Leaf length (cm)	
		17/12	22/17	17/12	22/17
Florida	0	16.0 ± 0.3	20.3 ± 1.2	27.3 ± 0.7	23.9 ± 1.1
	40	13.0 ± 0.5	16.4 ± 0.5	38.3 ± 1.2	40.5 ± 1.2
	80	13.6 ± 0.2	15.6 ± 1.2	36.4 ± 1.1	41.8 ± 2.3
Wild type	0	16.8 ± 0.2	20.0 ± 0.5	33.7 ± 1.0	26.7 ± 1.3
	40	14.2 ± 0.2	16.2 ± 0.4	41.6 ± 0.6	36.1 ± 1.6
	80	13.6 ± 0.5	16.0 ± 0.3	42.8 ± 0.9	39.8 ± 2.1

effect of high temperatures and short days can be reversed by the application of gibberellic acid (GA_3) (Guzman, 1969; Pressman and Negbi, 1987). Kline (1958) found that the inner young leaves responded to the GA_3 application much more than the outer ones. However, GA_3 treatments may have negative effects, e.g. reduced leaf production rate (see Table 10.2), induction or enhancement of physiological disorders (see section on physiological disorders below) and increased susceptibility to diseases.

Brewster and Sutherland (1993) derived growth parameters from controlled-environment studies and used them, together with temperature and light measurements in a greenhouse, to predict seedling shoot growth in the greenhouse. Mechanically induced stress, induced by brushing the leaves, has been used for controlling seedling growth (Biddington, 1986); the treated seedlings were smaller and more compact than untreated ones and their root systems were smaller (Biddington and Dearman, 1985).

Celery is capable of considerable biomass production; it produces more biomass per unit land area than other crop plants, except for sugarcane (Table 10.3). Little information regarding the effect of climatic conditions on celery yield is available in the literature. Burdine et al. (1961) grew three cultivars of celery in the autumn of one year and in the winter of the next in the Everglades, Florida. They found that the winter yield was lower than the fall yield in all three cultivars (Fig. 10.4). Zink (1963) demonstrated that the increase in fresh weight of summer grown celery was accompanied by a decrease of 10% in dry weight (Fig. 10.5), whereas with winter crop Cannell et al. (1963) reported only a 4% decrease in dry weight. On the other hand, Diawara et al. (1994) found that in several cultivars plant vegetative growth, as measured by plastochron index, was primarily linear with physiological time and showed no major season-related variations.

Shih and Rahi (1985) calculated that celery required 23–34 kg of water to produce 1 kg fresh biomass and 33–50 kg of water to produce 1 kg marketable yield. They also found that the celery evapotranspiration and water

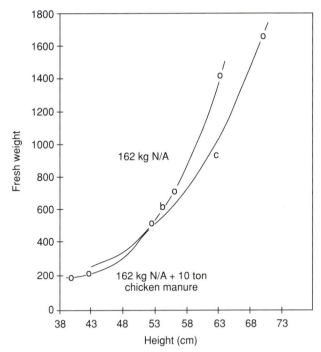

Fig. 10.3. Regression of fresh weight yield on height of tallest petiole for celery fertilized with 162 kg N/A or 162 kg N/A plus chicken manure (Cannel *et al.*, 1963).

Table 10.3. Harvestable yields relative to celery productivity for selected crops on a fresh weight basis. Data are 1982 production values compiled from Agricultural Statistics 1983, USDA. No attempt was made to correct to dry matter productivity (Fox *et al.*, 1986).

Crop	Harvestable yield (metric tons ha⁻¹)	Harvestable yield relative to celery (%)	Crop	Harvestable yield (metric tons ha⁻¹)	Harvestable yields relative to celery (%)
Celery	54.92	100.0	Sunflower	1.30	2.4
Sugar beet	46.17	84.1	Soybean	2.16	3.9
Lettuce	31.37	57.1	Wheat	2.39	4.4
Carrots	32.95	60.0	Barley	3.68	6.7
Cauliflower	11.88	21.6	Rice	5.31	9.7
Potatoes	30.71	55.9	Rye	1.83	3.3
Onions	37.30	67.9	Cotton	0.66	1.2
Corn	9.01	16.4	Sweet corn	4.97	9.0
Sugar cane	71.00	129.3	Tobacco	2.45	4.5

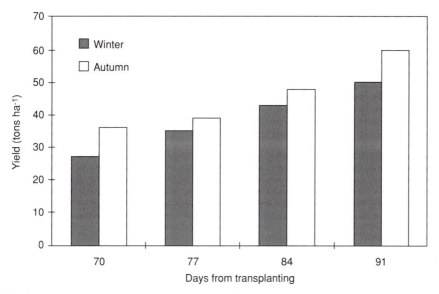

Fig. 10.4. Trimmed marketable weight (in ton ha^{-1}) of celery grown in the autumn or winter in the Everglades, Florida. Average of three cultivars harvested at 7-day intervals (after Burdine *et al.*, 1961).

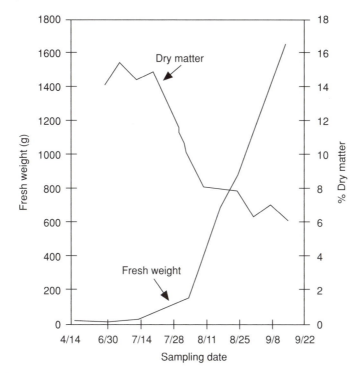

Fig. 10.5. Growth curves for a summer crop of celery showing mean fresh weight per plant (g) and percentage of dry matter (Zink, 1963).

use efficiency were inversely related to the water table depth. However, celery grown on a 30 cm water table yielded as well as that grown on a water table of 60 or 85 cm. Dry biomass of celery varied from 791 to 1173 g m^{-2} irrespective of water table (Shih and Rosen, 1985) as did root biomass, which varied from 91 to 134 g m^{-2} (Shih, 1989). When irrigation water supply was reduced by 45% through manipulation of water table depth the reduction in fresh biomass was about 20% (Shih, 1987).

BOLTING AND FLOWERING

Celery is a biennial plant which requires a cold period (vernalization) for bolting and flowering. The seedstalk of this species is multibranched and has a terminal inflorescence which comprises a compound umbel with secondary divisions known as umbellets (Pressman and Sachs, 1985). Generally, bolting (stem elongation) and flowering are simultaneous phenomena, but in celeriac flowering (sessile flowers) may occur without bolting (Hanisova and Krekule, 1975). Also, celeriac is much more bolt-sensitive than stalk celery (Pressman and Negbi, 1980; Benoit *et al.*, 1981).

Imbibed seeds and young seedlings of stalk celery (Vince-Prue, 1975; Ramin and Atherton, 1991a) and celeriac (Booij and Meurs, 1993) cannot undergo vernalization and the plants need to reach a critical size in order to be sensitive to low temperatures. However, Ramin and Atherton (1991b) found that chilling imbibed seeds promoted the subsequent flowering of young plants at a minimum temperature of 15°C. The age rather than the size of the plant, was found more important for the response to vernalization (Pawar and Thompson, 1950). Honma (1959) stated that a temperature between 6 and 9°C was most effective in inducing flowering, more so than 3.3°C, and that a temperature of 14°C appears to be the upper limit for vernalization.

Exposure of vernalized plants to high temperatures immediately following the cold treatment will cause devernalization and will prevent flowering, whereas exposure to high temperatures prior to vernalization brings about antivernalization – a delay in the start of bolting and a decrease in its intensity. The implications of these phenomena are discussed below, together with the effect of daylength.

Daylength during and following vernalization influences the bolting and flowering intensity. It has been found that regimes of LD or NB during vernalization inhibited bolting and flowering, whereas, after vernalization long photoperiods promoted the two processes (Pressman and Negbi, 1980). Therefore, the celery has been classified as a short-long day plant (SLDP) with cold requirement during the SD period. The cold-induced bolting can be replaced by exogenous GA treatments (Pressman and Negbi, 1987), but flowering rarely takes place following these treatments. Thomas (1978) found a negative correlation between bolting and seed dormancy in different celery cultivars.

PHOTOSYNTHESIS AND CARBOHYDRATE METABOLISM

Leaves of greenhouse-grown celery have been found to be capable of exceptionally high rates of photosynthesis (Fox *et al.*, 1986). Indeed, the photosynthetic rates obtained for celery (35–65 mg CO_2 dm^{-2} h^{-1}) compare favourably with those other C_3 species, and with C_4 species such as sugarcane (42–49 mg CO_2 m^{-2} h^{-1}), (see Zelitch, 1982). In addition to high photosynthetic rates, the celery CO_2 compensation points (Γ) are unusually low for a C_3 plant. Figure 10.6 shows that Γ is a function of leaf age. In mature plants the older leaves exhibit a slight trend toward increased Γ, but these values are much lower than those of soyabean and higher than in corn, a C_3 and C_4 plant, respectively. Such traits must contribute to the high productivity of celery (see section on vegetative development above). Temperature response in celery is typical of C_3 plants, peaking at 26°C (Fig. 10.7).

The two major photosynthetic products and translocated carbohydrates in celery are sucrose and the sugar alcohol, mannitol. Davis and Loescher (1990) showed that sucrose is produced and utilized in leaves of all ages, whereas mannitol is synthesized primarily in mature leaves, utilized in young leaves and stored in all leaves. The two carbohydrates are produced in the mesophyll and translocated in the phloem of the leaves. The fleshy petioles store mainly mannitol, glucose and fructose but very little sucrose (Keller and Matile, 1989).

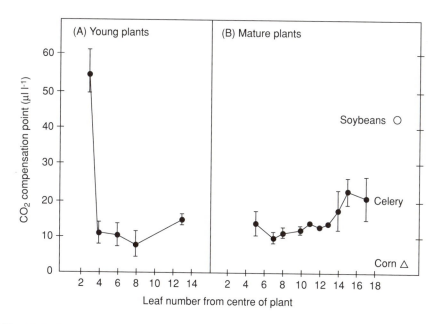

Fig. 10.6. CO_2 compensation points (Γ) as a function of leaf age in (a) young and (b) mature celery plants. Values for soyabean and corn are presented for comparative purposes (Fox *et al.*, 1986).

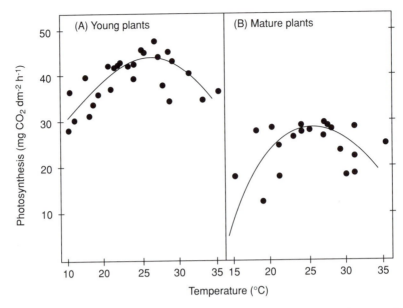

Fig. 10.7. Photosynthetic rates in leaves of (a) young and (b) mature celery plants as a function of temperature. Measurements were made on leaves no. 4 and 7. Each curve is a composite of data collected from three plants (Fox *et al.*, 1986).

Starch was found in low concentrations but showed definite diurnal fluctuations in mature leaves (Davis and Loescher, 1991). Along with the early photosynthetic products in celery leaves, Loescher *et al.* (1992) isolated the enzymes mannose 6-phosphate (M6P) and mannitol 1-phosphate. These enzymes, and others are specific to polyol producing plants such as celery (Rumpho *et al.*, 1983). On the other hand, Stoop and Pharr (1992) isolated mannitol : mannose 1-oxidoreductase from celeriac roots. This mannitol dehydrogenase is located mainly in root tips; it converts mannitol into hexoses which are utilized to support growth.

PHYSIOLOGICAL DISORDERS

PETIOLE PITHINESS

Petiole pithiness is a widespread disorder in stalk celery (var. *dulce*); the petioles of the other two varieties are naturally hollow. Anatomically, this disorder is characterized by a breakdown of the parenchyma cells, leaving large air spaces in the parenchyma which under severe conditions may merge with one another and form a hollow petiole. Several types of root stress have been found to cause pithiness: water deprivation acted rapidly whereas flooding and nutrient deficiency exerted a continuing effect (Aloni and Pressman, 1979). Petiole pithiness was completely eliminated when fertilization with low nitrogen concentration during the first two thirds of the growth period was followed by nitrogen-rich fertilization until harvest (Table 10.4). Saline water irrigation caused a marked reduction in leaf length, with petioles which were completely free of pithiness (Aloni and Pressman, 1980). GA_3 spray caused an increase in leaf elongation, and was accompanied by increased petiole pithiness; the salt-treated plants were less affected. Treatment of celery plants with GA_3 also enhanced petiole pithiness during a subsequent period of cold storage (Bukovac et al., 1960). Another hormone which might be involved in the parenchyma breakdown is ethylene. Pressman et al. (1984) demonstrated that application of ethephon or mechanical perturbation (MP) of petioles induced or enhanced pithiness. However, MP, but not ethephon, decreased the severity of drought or GA_3-induced pithiness.

BLACKHEART

This physiological disorder occurs in many celery fields. Blackheart symptoms in celery have many characteristics in common with those of tipburn in other leafy crops such as lettuce and cabbage. The susceptible tissue is the tip of the young developing leaves, which are growing rapidly. The affected leaves become necrotic, first at the tip then throughout the rest of the leaf. In severe cases all the interior leaves (heart) can become rotten. A deficiency of calcium in the growing tip is the main cause of blackheart and application of calcium solutions directly to the heart of the plant have been reported to control the disorder completely (Geraldson, 1954). A high rate of fertilization can increase blackheart, either by accelerating the growth of the young leaves or by disturbing the ion balance in the soil solution. Gubbels and Carolus (1971) reported an increase in blackheart severity as the total top fresh weight increased; their results agree with those of Cannell et al. (1959),

Table 10.4. The effect of nitrogen concentration (ppm) in the irrigation solution on shoot fresh weight (g) and percentage of pithy petioles in trimmed celery heads, 125 days after planting (after Aloni et al., 1983).

Growth period (days after planting)				Whole plant weight	Trimmed head weight	Pithy petioles (%)
0–80	81–95	96–107	108–126			
200	200	200	200	1499	502	16.5
200	0	0	0	1436	483	2.0
100	100	100	100	1414	425	8.7
100	0	0	0	1400	480	3.0
50	50	50	50	590	225	0
50	200	200	200	1480	507	0

who found that increased fertilizer rates increased the percentage of affected plants. Low Ca in the soil may be a result of excess K or Na fertilizers (Westgate *et al.*, 1954). Indeed, increasing the Ca : K ratio in the nutrient solution has been found to decrease blackheart symptoms (Gubbels and Carolus, 1971). Whereas low soil moisture (drought) has been reported to increase blackheart (Cannell *et al.*, 1959), salinity had no effect (Gubbels and Carolus, 1971). Moreover, Aloni and Pressman (1987) showed that the disorder was greatly attenuated in plants adapted to salinity. They also found that the application of GA_3 to untreated plants enhanced blackheart development but that GA_3 sprayed on salinized plants had no effect.

PREMATURE BOLTING

Premature bolting has a decisive effect on the agricultural value of celery, as well as that of other leafy crops, since it makes the product unmarketable. As previously mentioned, cold is an indispensible requirement for the celery, which is a facultative SLD plant, to enable it to bolt and flower. Seeds and young seedlings are not sensitive to vernalization (Ramin and Atherton, 1991a). However, transplanting is done at the age of 50–60 days, when the plant has six to eight true leaves and is sensitive to vernalization. Therefore, different means for delaying or reducing bolting following transplanting have been examined. Increasing the temperatures during propagation (antivernalization) has been tried (Benoit *et al.*, 1979; Sachs and Rylski, 1980), and Dyke *et al.* (1988) found it to be more effective than night break or growth-room lighting (Table 10.5). Devernalization has been achieved by exposing the plants to high temperatures following vernalization (Benoit *et al.*, 1978). Recently, Ramin and Atherton (1994) stated that in temperate countries subjecting celery transplants to high temperatures is prohibitively

Table 10.5. Mean percentage downgraded or unmarketable plants due to bolting of celery cv. 'Celebrity' as affected by propagation temperature and lighting (natural vs. 4 h night break) and by growing house lighting treatments (Dyke *et al.*, 1988).

	Propagation treatments		Lighting treatments	
Sowing date	Temperature (°C)	Lighting	Ambient	Night break
13 Nov.	14	Natural	100 [b]	53 [b]
	14	Natural + Night break	100 [b]	50 [b]
	16	Natural	27 [a]	0 [a]
	18	Natural	27 [a]	2 [a]
10 Dec.	14	Growth room	100 [b]	5 [a]

[a,b] Differences between growing house lighting treatments were significant ($P < 0.01$). Figures in the same column followed by the same letter are not significantly different at $P = 0.05$.

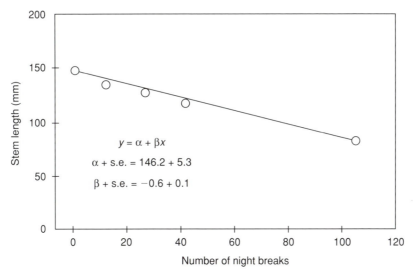

Fig. 10.8. Relationship between final stem length (measured 28 May) and number of night breaks for November-grown celery (Roelofse *et al.*, 1989).

expensive. They suggested that confining young plants to darkness for short periods before transferring them to vernalization (predevernalization) could reduce bolting and should be a cheaper alternative. In the case of stalk celery plants grown at vernalizing temperatures under cover, both night breaks, applied from transplanting onwards and short days have been found to delay flower initiation (Roelofse *et al.*, 1989, 1990). Night breaks, applied from transplanting onwards have been found also to decrease the stem length (Fig. 10.8). Figure 10.8 shows that the stem length decreased linearly by 0.6 mm for each night break given to the plants. Celeriac is more bolt-sensitive than stalk celery (Pressman and Negbi, 1980; Benoit *et al.*, 1981). Therefore, transplanting of celeriac in the spring should begin later than that of self-blanching celery. Bolting of celeriac plants has been delayed by setting them out in unheated plastic or glass greenhouses or in the open under the protection of direct plastic cover (Benoit and Ceusterman, 1988).

CONCLUDING REMARKS

In order to obtain high yields of good quality celery, every developmental stage should be carefully managed. In general, celery is a cool season crop and therefore, it is susceptible to high temperatures throughout its entire life, starting from germination. Since it is a leafy crop, the commercial yield depends on its vegetative growth, which is very much affected by heat. On the other hand, very low temperatures may attenuate growth and induce bolting and flowering. Therefore, moderate to fairly low temperatures are recommended, such as prevail in open fields during the winter in areas with a Mediterranean climate, or in glasshouses in the temperate zone. In any case, means to delay, reduce or prevent premature bolting should be considered.

Celery is tolerant of salinity and can be irrigated with brackish water. Such irrigation may even be advantageous in preventing petiole pithiness. However, it may also reduce the growth rate and at high concentrations it may even cause blackheart. Moreover, celery is a high water consumer and should be carefully watered and fertilized in order to achieve high yields and avoid physiological disorders. It can tolerate a high water table presumably since the natural habitats of its ancestors are marshes and swamps. Accordingly, drought or withholding water should definitely be avoided.

Both bolting and leaf elongation are controlled by endogenous GAs. Slow-bolting cultivars grown in the winter for spring harvest generally possess relatively small leaves and therefore their yields are low. Modern techniques may help to identify specific GAs which control each phenomenon. In this case new slow-bolting cultivars with larger leaves could be bred. These techniques could also be used in breeding heat-tolerant cultivars.

The mechanisms causing blackheart, a calcium deficiency related disorder, are not well understood. Therefore, more research is needed to understand the effect of environmental factors in relation to plant anatomy, for the control of this problem.

REFERENCES

Al-Abta, S. and Collin, H.A. (1979) Endogenous auxin and cytokinins changes during embryo development in celery tissue culture. *New Phytologist* 82, 29–35.

Aloni, B. and Pressman, E. (1979) Petiole pithiness in celery (*Apium graveolens*) leaves: induction by environmental stresses and the involvement of abscissic acid. *Physiologia Plantarum* 47, 61–65.

Aloni, B. and Pressman, E. (1980) Interaction with salinity of GA3 induced leaf elongation, petiole pithiness and bolting in celery. *Scientia Horticulturae* 13, 135–142.

Aloni, B. and Pressman, E. (1987) The effect of salinity and gibberellic acid on blackheart disorder in celery (*Apium graveolens* L.). *Journal of Horticultural Science* 62, 205–209.

Aloni, B., Bar-Yosef, B., Sagiv, B. and Pressman, E. (1983) The effect of nutrient solution composition and transient nutrient starvation on the yield and petiole pithiness in celery plants. *Journal of Horticultural Science* 58, 91–96.

Becker, G. (1962) *Apium graveolens* L. var. *rapaceum* (Miller) D.C. In: Kappert, H. and Rufort, W. (eds) *Manual of Plant Breeding*, Vol. VI. Verlag Paul Parey, Berlin, pp. 104–130.

Benoit, F. and Ceusterman, N. (1988) Advancing the harvest of bolt sensitive celeriac by means of plastic covering. *Plasticulture* 79, 37–43.

Benoit, F., Kinet, J.M. and Ceusterman, N. (1978) Induction, suppression and prevention of vernalisation in self-blanching celery (*Apium graveolens* L. var. *dulce*). *Agricultura* 26, 163–182.

Benoit, F., Ceusterman, N. and Kinet, J.M. (1979) Prevention of bolting in self-blanching celery (*Apium graveolens* L. var. *dulce*) by increased temperatures and the use of plastic cover. *Agricultura* 27, 441–458.

Benoit, F., Ceusterman, N. and Claus, A. (1981) Reduction of bolting in self-blanching celery, celeriac, endive and Chinese cabbage. *Acta Horticulturae* 122, 121–131.

Biddington, N.L. (1986) The effect of mechanically induced stress in plants. *Plant Growth Regulation* 4, 103–123.

Biddington, N.L. and Dearman, A.S. (1985) The effect of mechanically induced stress on water loss and drought resistance in lettuce, cauliflower and celery seedlings. *Annals of Botany* 56, 795–802.

Biddington, N.L. and Thomas, T.H. (1978) Thermodormancy in celery seeds and its removal by cytokinins and gibberellins. *Physiologia Plantarum* 42, 401–402.

Biddington, N.L., Thomas, T.H. and Whitlock, A.J. (1975) Celery yield increased by sowing germinated seeds. *HortScience* 10, 620–621.

Booij, R. and Meurs, E.J.J. (1993) Flower induction and initiation in celeriac (*Apium graveolens* L. var. *rapaceum* Mill. DC.): effect of temperature and plant age. *Scientia Horticulturae* 55, 227–238.

Brewster, J.L. and Sutherland, R.A. (1993) The rapid determination in controlled environments of parameters for predicting seedling growth rates in natural conditions. *Annals of Applied Biology* 122, 123–133.

Brocklehurst, P.A. and Dearman, J. (1983a) Interaction between seed priming treatments and nine seed lots of carrot, celery and onion. I. Laboratory germination. *Annals of Applied Biology* 102, 577–584.

Brocklehurst, P.A. and Dearman, J. (1983b) Interaction between seed priming treatments and nine seedlots of carrot, celery and onion. II. Seedling emergence and plant growth. *Annals of Applied Biology* 102, 585–593.

Brocklehurst, P.A., Rankin, W.E.F. and Thomas, T.H. (1982/83) Stimulation of celery seed germination and seedling growth with combined ethephon, gibberellin and polyethylene glycol seed treatment. *Plant Growth Regulation* 1, 195–202.

Bukovac, M.J., Wittwer S.H. and Cook, J.A. (1960) The effect of preharvest foliar sprays of gibberellin on yield and storage breakdown of celery. *Quarterly Bulletin of the Michigan Agricultural Experimental Station* 42, 764–770.

Burdine, H.W., Guzman, V.L. and Hall, C.B. (1961) Growth changes in three celery varieties on Everglades organic soils. *Proceedings of the American Society for Horticultural Science* 78, 353–360.

Cannell, G.H., Tyler, K.B. and Asbell, C.W. (1959) The effect of irrigation and fertilizer on yield, blackheart and nutrient uptake by celery. *Proceedings of the American Society for Horticultural Science* 74, 539–545.

Cannell, G.H., Tyler, K.B. and Takatori, F.H. (1963) Growth measurements of celery in relation to yield. *Proceedings of the American Society for Horticultural Science* 83, 511–518.

Danin, M., Upfold, S.J., Levin, N., Nadel, B.L., Altman, A. and van Staden, J. (1993) Polyamines and cytokinins in celery cell culture. *Plant Growth Regulation* 12, 245–254.

Davis, J.M. and Loescher, W.H. (1990) [14C]-Assimilate translocation in the light and dark in celery (*Apium graveolens*) leaves of different ages. *Physiologia Plantarum* 79, 656–662.

Davis, J.M. and Loescher, W.H. (1991) Diurnal pattern of carbohydrates in celery leaves of various ages. *HortScience* 26, 1404–1406.

Diawara, M.M., Trumble, J.T., Quiros, C.F., White, K.K. and Adams, C. (1994) Plant age and seasonal variations in genotypic resistance of celery to beet armyworm (*Lepidoptera: noctuidae*). *Journal of Economic Entomology* 87, 514–522.

Dyke, A.J., Szmidt, R.A.K., Metcalf, E.D. and Nobel, R. (1988) Self blanching celery: the effect of light and temperature on bolting. *Applied Agricultural Research* 3, 239–242.

Fox, T.C., Kennedy, R.A. and Loescher, W.H. (1986) Developmental changes in photosynthetic gas exchange in the polyol-synthesizing species *Apium graveolens* L. (celery). *Plant Physiology* 82, 307–311.

Geraldson, C.M. (1954) The control of blackheart of celery. *Proceedings of the American Society for Horticultural Science* 63, 353–358.

Gubbels, G.H. and Carolus, R.L. (1971) Influence of atmospheric moisture, ion balance, and ion concentration on growth, transpiration and blackheart of celery (*Apium graveolens* L.). *Journal of the American Society for Horticultural Science* 96, 201–204.

Guzman, V.L. (1964) Soil temperature effect on germination of celery seed. *Florida State Horticultural Society Proceedings* 77, 147–152.

Guzman, V.L. (1969) Response of four celery cultivars to levels of gibberellic acid applied 2 or 4 weeks before harvest. *Florida State Horticultural Society Proceedings* 82, 129–134.

Hanisova, A. and Krekule, J. (1975) Treatment to shorten the development period of celery (*Apium graveolens* L.). *Journal of Horticultural Science* 50, 97–104.

Harrington, J.F. (1923) Use of alternating temperatures in the germination of seeds. *Journal of Agricultural Research* 23, 295–332.

Hayward, H.E. (1938) Umbelliferae, *Apium graveolens*. In: *The Structure of Economic Plants*. Macmillan, New York, pp. 451–484.

Honma, S. (1959) A method for evaluating resistance to bolting in celery. *Proceedings of the American Society for Horticultural Science* 54, 506–513.

Jacobsen, J.V. and Pressman, E. (1979) A structural study of germination in celery (*Apium graveolens* L.) seed with emphasis on endosperm breakdown. *Planta* 144, 241–248.

Jacobsen, J.V., Pressman, E. and Pyliotis, N.A. (1976) Gibberellin induced separation of cells in isolated endosperm of celery seed. *Planta* 129, 113–121.

Keller, F. and Matile, M. (1989) Storage of sugars and mannitol in petioles of celery leaves. *New Phytologist* 113, 291–299.

Kim, Y.H. and Janick, J. (1989a) Origin of somatic embryos in celery tissue culture. *HortScience* 24, 671–673.

Kim, Y.H. and Janick, J. (1989b) ABA and polyox encapsulation or high humidity increases survival of desiccated somatic embryos of celery. *HortScience* 24, 674–676.

Kline, J.W. (1958) Growth responses of gibberellin-treated celery plants. *Botanical Gazette* 120, 122–124.

Loescher, W.H., Tyson, R.H., Everard, J.D., Redgwell R.J. and Bieleski, R.L. (1992) Mannitol synthesis in higher plants. Evidence for the role and characterization of NADPH-dependent mannose 6-phosphate reductase. *Plant Physiology* 98, 1396–1402.

Morinaga, T.I. (1926) Effect of alternating temperatures upon the germination of seeds. *American Journal of Botany* 13, 141–158.

Nadel, B.L., Altman, A. and Ziv, M. (1990) Regulation of large scale embryogenesis in celery. *Acta Horticulturae* 280, 75–82.

Nadel, B.L., Altman, A. and Ziv, M. (1995) Celery somatic embryogenesis. In: Bajaj, Y.P.S. (ed.) *Biotechnology in Agriculture and Forestry: Somatic Embryogenesis and Synthetic Seed II, vol. 31.* Springer Verlag, Berlin, pp. 306-322.

Palevitch, D. and Thomas, T.H. (1976) Enhancement by low pH of gibberellin effects on dormant celery seeds and embryoless half-seeds of barley. *Physiologia Plantarum* 37, 247–252.

Parrera, C.A., Qiao, Q. and Cantliffe, D.J. (1993) Enhanced celery germination at stress temperature via solid matrix priming. *HortScience* 28, 20–22.

Pawar, S.S. and Thompson, H.C. (1950) The effect of age and size of plant at the time of exposure to low temperature on reproductive growth in celery. *Proceedings of the American Society for Horticultural Science* 55, 367–371.

Pressman, E. (1979) Comparative physiology of wild and cultivated varieties of Apium graveolens L. with special reference to flowering. Unpublished PhD thesis, the Hebrew University of Jerusalem, Israel.

Pressman, E. and Negbi, M. (1980) The effect of day length on the response of celery to vernalization. *Journal of Experimental Botany* 31, 1291–1296.

Pressman, E. and Negbi, M. (1987) Interaction of day length and applied gibberellins on stem growth and leaf production in three varieties of celery. *Journal of Experimental Botany* 38, 968–971.

Pressman, E. and Sachs, M. (1985) *Apium graveolens*. In: Halevy, A.H. (ed.) *Handbook of Flowering*, Vol.I. CRC Press, Boca Raton, pp. 485–491.

Pressman, E. and Shaked, R. (1988) Germination of annual celery (*Apium graveolens*) seeds: inhibition by paclobutrazol and its reversal by gibberellic acid and benzyl adenine. *Physiologia Plantarum* 73, 323–326.

Pressman, E. and Shaked, R. (1991) Interactive effects of GAs, CKs and growth retardants on the germination of celery seeds. *Plant Growth Regulation* 10, 65–72.

Pressman, E., Shaked, R. and Negbi, M. (1988) Germination of seeds of annual and biennial celery (*Apium graveolens*). *Physiologia Plantarum* 72, 65–69.

Pressman, E., Huberman, M., Aloni, B. and Jaffe, M.J. (1984) Pithiness in plants: I. The effect of mechanical perturbation and the involvement of ethylene in petiole pithiness in celery. *Plant and Cell Physiology* 25, 891–897.

Pressman, E., Negbi, M., Sachs, M. and Jacobsen, J.V. (1977) Varietal differences in light requirement for germination of celery (*Apium graveolens* L.) seeds and the effect of thermal and solute stress. *Australian Journal of Plant Physiology* 4, 821–831.

Ramin, A.A. and Atherton, J.G. (1991a) Manipulation of bolting and flowering in celery (*Apium graveolens* L. var. *dulce*). I. Effects of chilling during germination and seed development. *Journal of Horticultural Science* 66, 435–441.

Ramin, A.A. and Atherton, J.G. (1991b) Manipulation of bolting and flowering in celery (*Apium graveolens* L. var. *dulce*). II. Juvenility. *Journal of Horticultural Science* 66, 709–717.

Ramin, A.A. and Atherton, J.G. (1994) Manipulation of bolting and flowering in celery (*Apium graveolens* L. *dulce*). III. Effect of photoperiod and irradiance. *Journal of Horticultural Science* 69, 861–868.

Redenbaugh, K., Paasch, B.D., Nichol, J.W., Kossler, M.E., Viss, P.R. and Walker, K.A. (1986) Somatic seeds: encapsulation of asexual plant embryos. *Bio/Technology* 4, 797–801.

Robinson, R.W. (1954) Seed germination problems in the Umbelliferae. *Botanical Review* 20, 531–550.

Roelofse, E.W., Hand, R.W. and Hall, R.L. (1989) The effect of day length on the development of glasshouse celery. *Journal of Horticultural Science* 64, 283–292.

Roelofse, E.W., Hand, R.W. and Hall, R.L. (1990) The effect of temperature and 'night-break' lighting on the development of glasshouse celery. *Journal of Horticultural Science* 65, 297–307.

Rumpho, M.E., Edwards, G.E. and Loescher, W.H. (1983) A pathway for photosynthetic carbon flow to mannitol in celery leaves. *Plant Physiology* 73, 869–873.

Sachs, M. and Rylski, I. (1980) The effect of temperature and daylength during the seedling stage on flower-stalk formation in field-grown celery. *Scientia Horticulturae* 12, 231–242.

Salter, P.J. and Darby, R.J. (1976) Synchronization of germination of celery seeds. *Annals of Applied Biology* 84, 415–424.

Saranga, Y. and Janick, J. (1991) Celery somatic embryo production and regeneration: improved protocols. *HortScience* 26, 1335.

Saranga, Y., Rhodes, D. and Janick, J. (1992a) Changes in amino acid composition associated with tolerance to partial desiccation of celery somatic embryos. *Journal of the American Society for Horticultural Science* 117, 337–341.

Saranga, Y., Rhodes, D. and Janick, J. (1992b) Changes in tolerance to partial desiccation and in metabolite content of celery somatic embryos induced by reduced osmotic potential. *Journal of the American Society for Horticultural Science* 117, 342–345.

Shih, S.F. (1987) Water cutback studies of lettuce, celery and sweet corn. *Proceedings of the Soil and Crop Science Society of Florida* 46, 7–13.

Shih, S.F. (1989) Water-table effects on root nutrient content of celery, lettuce and sweet corn. *Proceedings of the Soil and Crop Science Society of Florida* 48, 79–84.

Shih, S.F. and Rahi, G.S. (1985) Evapotranspiration, yield and water table studies of celery. *Transactions of the American Society of Agricultural Engineers* 28, 1212–1218.

Shih S.F. and Rosen, M. (1985) Water-table effects on nutrient content of celery, lettuce and sweet corn. *Transactions of the American Society of Agricultural Engineers* 28, 1867–1870.

Stoop, J.M.H. and Pharr, D.M. (1992) Partial purification and characterization of mannitol: mannose 1-oxidoreductase from celeriac (*Apium graveolens* var. *rapaceum*) roots. *Archives of Biochemistry and Biophysics* 298, 612–619.

Thomas, T.H. (1978) Relationship between bolting and seed dormancy of different celery cultivars. *Scientia Horticulturae* 9, 311–316.

Thomas, T.H. (1984) Changes in endogenous cytokinins of celery (*Apium graveolens* L.) seeds following an osmotic priming or growth regulator soak treatment. *Plant Growth Regulation* 2, 135–141.

Thomas, T.H. (1989) Gibberellin involvement in dormancy-break and germination of seeds of celery (*Apium graveolens* L.). *Plant Growth Regulation* 8, 255–261.

Thomas, T.H. (1990) Hormonal involvement in photoregulation of celery seed germination. *British Society for Plant Regulation, Monograph* 20, 51–58.

Thomas, T.H., Biddington, N.L. and Palevitch, D. (1978a) Improving the performance of pelleted celery seeds with growth regulator treatments. *Acta Horticulturae* 83, 235–243.

Thomas T.H., Biddington, N.L. and Palevitch, D. (1978b) The role of cytokinins in the phytochrome-mediated germination of dormant imbibed celery (*Apium graveolens*) seeds. *Photochemistry and Photobiology* 27, 231–236.

Thomas, T.H., Gray D. and Biddington, N.L. (1978c) The influence of the position of the seed on the mother plant on seed and seedling performance. *Acta Horticulturae* 83, 57–66.

Thomas, T.H., Dearman, A.S. and Biddington, N.L. (1986) Evidence for the accumulation of a germination inhibitor during progressive thermoimbibition of seeds of celery (*Apium graveolens* L.). *Plant Growth Regulation* 4, 177–184.

Thomas, T.H., Palevitch, D., Biddington N.L. and Austin, R.B. (1975) Growth regulators and the phytochrome-mediated dormancy of celery seeds. *Physiologia Plantarum* 35, 101–106.

van Staden, J., Upfold, S.J., Altman, A. and Nadel, B.L. (1992) Metabolism of benzyladenine in embryogenic cell culture of celery. *Journal of Plant Physiology* 140, 466–469.

van der Toorn, P. (1990a) Methods to improve celery (*Apium graveolens* L.) seed quality. *Acta Horticulturae* 267, 175–182.

van der Toorn, P. (1990b) The relation between relative proportion of seed parts and germination of celery (*Apium graveolens*). *Seed Science and Technology* 18, 703–712.

Vince-Prue, D. (1975) *Photoperiodism in Plants*. McGraw-Hill, London.

Westgate, P.J., Blue, W.G. and Eno C.F. (1954) Blackheart of celery and its relationship to soil fertility and plant composition. *Florida State Horticultural Society Proceedings* 67, 158–163.

Wolf, E.A. and Scully, B. (1992) 'Floribelle M9': an autumn celery cultivar for Florida. *HortScience* 27, 1235–1236.

Zelitch, I. (1982) The close relationship between net photosynthesis and crop yield. *BioScience* 32, 796–802.

Zink, F.W. (1963) Rate of growth and nutrient absorption of celery. *Proceedings of the American Society for Horticultural Science* 82, 351–357.

Phaseolus Beans $\boxed{11}$

J.H.C. Davis

PBI Cambridge Ltd, Maris Lane, Trumpington, Cambridge CB2 2LQ,UK

Phaseolus beans are grown both as green vegetables and pulses. The most important species is *P. vulgaris*, the common or green bean, which is used for canning and freezing as well as the fresh market. Pods of green beans are prepared whole, sliced or cut, and the seeds may be harvested green as a vegetable (e.g. flageolet), or dry as a pulse. For whole beans, there is a trend particularly in Europe to develop varieties which produce a large proportion of very fine (6.6–8.0 mm diameter) or extra-fine (< 6.5 mm) pods. Runner beans (*P. coccineus*) are grown in northen Europe mainly for the fresh market, and used as cut green beans.

The taxonomy of *Phaseolus* was revised by Maréchal *et al.* (1978) and its genetic resources reviewed in a book edited by Gepts (1988). The gene pool of *P. vulgaris* includes its wild ancestors (e.g. var. *aborigineus*), *P. coccineus* and *P. polyanthus* (Schmit *et al.*, 1992). The changes that have occurred with domestication are described by Smartt (1990), and include: (i) loss of seed dormancy; (ii) changed growth habit (from climbing or half-runner to bush); (iii) photoperiod insensitivity; (iv) reduction of fibre and selection of the green bean type of pod from the original dehiscent pod; (v) greatly increased seed size (from 20–50 mg in wild material to > 200 mg in cultivated material); and (vi) white seeds which are preferred by the canning industry.

Phaseolus beans are cultivated over a wide range of climates from northern Europe and America to the tropics of America, Asia and Africa. *P. vulgaris* is the most widely adapted species, *P. coccineus* being adapted to cooler climates and *P. lunatus* to warmer climates. The most important production regions for green beans (*P. vulgaris*) in North America are Florida, South and North Carolina, and Georgia for spring crops; New York, New Jersey, North Carolina and Virginia for summer crops; and Florida and Virginia for autumn crops. Major areas of production for processing are Oregon,

Table 11.1. Volumes (thousand tonnes) of green bean final product produced in the major processing areas of Europe in 1993 (Unilet, 1994) and USA in 1990 (USDA, 1991).

Country	Canned	Frozen
France	247	50
Germany	38	4
Belgium	68	33
The Netherlands	70	15
UK	–	40
USA	514	267

Wisconsin, New York, Quebec and Ontario. In Europe, spring and autumn crops are produced in southern Europe, e.g. Spain and Italy, while the main summer crops for processing are produced in France (where green beans are the most important of all processed vegetables), Belgium, the Netherlands, Germany and the UK (Table 11.1).

Beans have been used extensively as a convenient plant for physiological and biochemical research. At the whole plant level, there has tended to be more work done on dry beans than green beans, because of their importance worldwide as a food crop and protein source. As a green vegetable, however, beans are also of major importance worldwide, probably second only to peas. Understanding the adaptation of beans to the main factors of the environment allows the use of climate and soil data to predict where beans are best grown, and how they might respond to varying agronomic practices. This type of information can be used by a processing company looking to optimize its sourcing of vegetable production, extension workers who advise growers, or large-scale growers themselves. It can also be used to select the best or most representative sites for breeding nurseries or variety trials, and physiological information can be used to improve the efficiency of selection for particular traits in a breeding programme. The information generated by whole plant physiology can be integrated into computer models of plant growth and used to predict the yield of the crop, allowing different crop management scenarios to be simulated. To achieve this level of accuracy requires quantitative information to be assembled for each stage of growth, and their interactions evaluated.

GERMINATION AND SEEDLING EMERGENCE

The seed size of *Phaseolus* varies considerably, from less than 130 mg in wild or non-cultivated types, up to 1000 mg or more, but most commercial cultivars of *P. vulgaris* have a seed size in the range of 200–350 mg. *P. coccineus* cultivars tend to have larger seeds. This variation among cultivars means that sowing rates need to be adjusted for each cultivar based on the desired plant population, expected germination percentage and the seed weight.

Germination of *P. vulgaris* is epigeal, whereas *P. coccineus* is hypogeal. Both species germinate in about 6–8 days under optimum conditions. Mature seeds do not normally show any period of dormancy. Water is imbibed through the micropyle, the raphe and the hilum, but uptake through the seed coat is negligible (Korban *et al.*, 1981). Hard seed which do not imbibe water properly sometimes occur, and this appears to be associated with restriction of the micropyle (Kyle and Randall, 1963).

There are differences among *P. vulgaris* cultivars in the rate of early seedling growth which can be associated with seed size; small-seeded cultivars tend to germinate and multiply their seed weight more rapidly than large-seeded ones when grown at relatively high temperature (28°C; Laing *et al.*, 1984). At low temperature (12°C), on the other hand, large-seeded cultivars tend to germinate more quickly than small-seeded ones (Austin and Maclean, 1972). This fits with the observation that cultivars adapted to cooler climates tend to have larger seeds.

The rate of seed germination of *P. vulgaris* is most rapid at 29–34°C, the exact optimum temperature depending on the cultivar. Below 8°C germination does not occur (White and Montes-R., 1993). Because beans do not germinate in cold soils and are highly sensitive to frost, they are sown relatively late in the spring in northern areas of Europe and North America (e.g. from mid-May in the UK), when moisture may become limiting if irrigation is not available. Some cultivars appear to be more resistant to frost damage than others, but this has not yet been the subject of any significant breeding work.

Beans are very sensitive to soil crusts forming before emergence so great care must be taken with soil preparation and irrigation.

Bean seeds are susceptible to soaking injury, but Pretorius and Small (1993) found that seeds could be soaked for 16 h in CO_2 saturated water without any injury occurring. Leaching of sugars does not seem to be the cause of soaking injury, whose physiology is not fully understood.

Salinity adversely affects germination and seedling growth in beans. Cachorro *et al.* (1994) found that this effect could be reduced by adding calcium. The main effect of salinity seemed to be on K concentration and its transport from root to shoot.

VEGETATIVE GROWTH

GROWTH OF LEAVES AND BIOMASS

Under optimal conditions, the bean crop grows nearly exponentially until pod growth begins. Leaf area index (LAI) in *P. vulgaris* increases to about 40 days after emergence (DAE), depending on the cultivar (Fig. 11.1), and then declines with seed filling (50–65 DAE), as photosynthate and nitrogen are translocated to the developing seeds. Leaves at the lower nodes die first, followed by leaves higher up the mainstem and later on the branches.

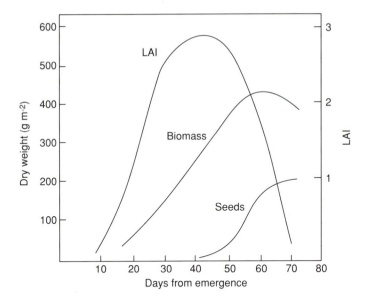

Fig. 11.1. Growth of leaf area index (LAI), biomass and seeds in *P. vulgaris* cv. Porrillo sintetico (Laing *et al.*, 1984).

Virtually all (95%) of the incoming radiation is intercepted by the bush bean canopy when the LAI reaches 4 or more (Aguilar *et al.*, 1977). The same author reported a maximum crop growth rate (CGR) of 17 g m^{-2} per day, which is relatively low compared with other C$_3$ crops (Monteith, 1978). The net assimilation rate (NAR, which is equal to CGR/LAI) declines as the LAI increases to its maximum value at about the time when pod growth begins.

The specific leaf area (SLA = leaf area/leaf dry weight) increases to a maximum shortly after flowering of about 600 cm^2 g^{-1}, declining back to where it started at about 400 cm^2 g^{-1} (White, 1981).

Bush beans develop a rather shallow root system, the bulk of the roots growing in the top 20–30 cm and in a radius of 45–70 cm. This makes the plants generally susceptible to nutrient or moisture deficiency even over relatively short periods of time. Modern green bean cultivars ideally have highly concentrated flowering and pod set, so that they have relatively little ability to recover from any setback in vegetative growth which may occur due to stress.

GROWTH HABIT

Beans are commonly classified into bush, half-runner and pole types. More detailed growth habit descriptions have been provided by Debouck and Hidalgo (1984), and their classification is illustrated in Fig. 11.2. Virtually all green bean production for processing is done with type I bush beans,

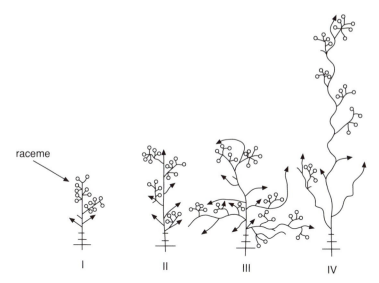

raceme

I   II   III   IV

Fig. 11.2. Growth forms in *Phaseolus:* (I) determinate bush, (II) indeterminate bush, (III) indeterminate, straggling, and (IV) indeterminate, climbing (adapted from Debouck and Hidalgo, 1984).

because they provide the concentrated pod set required for machine harvesting. It has been possible to develop suitable type II (erect indeterminate bush) varieties for commercial dry bean production, and these tend to be higher yielding than type I beans in some environmental conditions. Type III (sprawling, indeterminate) cultivars are also used for dry bean cultivation, particularly in areas where drought stress frequently affects the crop, for example in Mexico.

Climbing beans (type IV) have the highest yield potential, and are grown on poles or trellis, or in some tropical countries they are intercropped with maize. Climbing is achieved by twining, or circumnutation, of the climbing shoot up a support. Aluminium ions applied to excised shoots at concentrations of 4–8 mmol aluminium chloride have been found to significantly slow the rate of twining without affecting the growth rate (Badot *et al.*, 1993). Kretchmer *et al.* (1977) found that the climbing response is controlled in some cultivars by a phytochrome response which is unrelated to the photoperiod flowering response.

The optimum precision-seeded spacing of commercial type I green beans is approximately 16 cm on the square. Up until the 1960s, however, machine harvesters could only handle one row at a time, so wide row spacings were used. Now that multi-row harvesters are available, row widths have been reduced to 30–45 cm and planting density increased. Using current practice, the optimum density for dwarf green beans under irrigated conditions has been found to be 35–40 plants m^{-2}, planted with a precision or semi-precision planter.

INDUCTION OF FLOWERING

An important objective for green bean production, in common with many vegetable crops, is season extension. This can be achieved by using early cultivars to allow the harvesting season to begin earlier, and main crop cultivars which can be grown as late as possible before the weather becomes too cold (e.g. Reichel, 1992). In between, depending on the production area, may be periods of high temperatures which can become limiting.

At Centro Internacional de Agricultura Tropical (CIAT), Colombia, over 4000 materials from the world germplasm collection of *P. vulgaris* have been screened for photoperiod response, and of these 39% were day-length neutral but the distribution of photoperiod response varied among growth habits and seed size classes (Masaya and White, 1991). Little screening of other *Phaseolus* species has been done, but it is likely that the primitive condition in the genus is for short-day adaptation, with day-neutral genotypes being selected as part of the domestication process and further enhanced by modern breeding.

Short-day plants have a quantitative response to photoperiod which is affected by temperature (Gniffke, 1982). There is a tendency for a higher proportion of climbing cultivars to be photoperiod sensitive, and Coyne (1967) found a genetic linkage between indeterminate growth habit and photoperiod response. It appears there are two major dominant genes for photoperiod sensitivity, with modifier genes controlling a quantitative response (Wallace, 1985). There is no juvenile phase in beans, the plants being equally sensitive to photoperiod at all stages of vegetative growth (Zehni *et al.*, 1970).

The effect of increasing temperature in day-neutral cultivars is to reduce the number of days to flowering. By contrast, the effect of increasing temperature is to enhance the photoperiod response in short-day plants, further delaying flowering (Enriquez, 1975). This effect can occur even in the relatively short days of the tropics. For example, a variety adapted to the highland tropics, known as 'Mortiño' in Colombia, does not flower in Palmira where the mean temperature is 24°C, but flowers normally at the same day length (12–13 h) at higher altitude in Pasto where the mean temperature is 13°C. When the same variety is grown at 24°C with a 6-h daylength it flowers normally (White *et al.*, 1987).

POLLINATION AND POD GROWTH

Bees are essential for achieving pod-set in *P. coccineus* but not in *P. vulgaris* which is self-pollinating. The flowers of *P. coccineus* are larger and contain more nectar than those of *P. vulgaris*: 5.7–8.2 mg nectar containing 28–37% sugar for *P. coccineus* compared with 2.0 mg nectar containing 43% sugar for *P. vulgaris* was reported by Wroblewska (1993) in Poland.

The anthers dehisce in the bud just before it opens, usually at night.

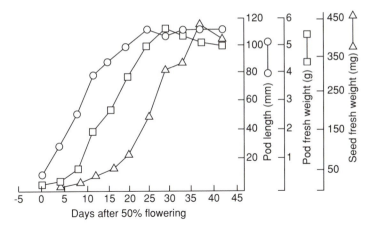

Fig. 11.3. Pod and seed growth of *P. vulgaris* cv. Black Turtle Soup (adapted from Izquierdo, 1981).

Once the pollen reaches the stigma, the pollen tubes grow down the hollow style and fertilize the ovules within 12 h, the ovules nearest the style being fertilized first.

Commercial varieties of green beans (*P. vulgaris*) take about 25 days from pollination to the stage at which the green pods are ready for harvesting, when they are approaching their maximum length and fresh weight (Fig. 11.3). Thereafter the seeds continue to develop for another 20–30 days by which time the pod is ripe and the seeds are dry.

The original dehiscent pod had an inner parchment layer lining the pod cavity and containing fibres obliquely orientated, and the longitudinal vascular bundles of the pod (the 'strings'), especially those close to the sutures which are most strongly developed. The strings have been bred out of modern cultivars, the lack of them being governed by a recessive gene. Suppression of lignification in the pod wall tissues has been selected for simultaneously, extending the period during which the pods are acceptable for eating.

The distribution of pods on the plant depends on the cultivar and its growth habit. In determinate cultivars (type I), flower primordia are formed on the raceme in the axil of the uppermost leaf of the mainstem first, and flowering then proceeds downwards to the lower nodes and along the branches (Ojehomon, 1966). By contrast, in type II cultivars such as 'Porrillo sintetico' the first flowers to open normally arise from nodes 6 to 7, and flowering then proceeds upwards and downwards on the main stem and along the branches (Laing *et al.* 1984).

The typical commercially grown crop (type I or II) is of relatively short duration and does not have time to develop a large LAI before pod set begins. Evidence suggests that the first formed reproductive structures have a strong competitive advantage for the supply of assimilates. There is a tendency, therefore, for almost complete pod set from the first-opened flowers, followed by a high rate of abscission of later-formed flowers. Even in optimum conditions about 60–70% of the flowers and young pods are shed. The modern bean cultivar is conservative in the allocation of its resources, giving precedence to the survival of a limited number of pods over a short space of time. This tendency ensures a more uniform pod set and has been selected for in modern cultivars.

FACTORS DETERMINING PRODUCTIVITY

Yield depends on the stage of maturity of the crop at the time of harvest, which depends on its end-use, for example extra-fine green bean pods vs. flageolet. This is illustrated in Table 11.2 with production data from France (Unilet, 1995). Flageolets are varieties of beans that have green seeds which are harvested fresh, like peas. The yield of this class of beans is similar to peas (4–5 t ha^{-1}). Green beans in France yield 8–11 t ha^{-1}. Larger-podded varieties of the sort produced in Oregon, USA, can yield much more, with yields of 18–20 t ha^{-1} commonly reported.

There is a trade-off between yield and quality, varieties producing a high proportion of extra-fine pods (< 6.5 mm diameter) yielding about 25–30% less than varieties producing very fine and fine pods. In Table 11.3 it can be seen that extra-fine pods are produced especially in Aquitaine, and this is the region shown in Table 11.2 with the lowest yield.

After photosynthesis and respiration, the extent of senescence and abscission are major determinants of final yield. Failure of fertilization, which might be caused by high temperature at pollination, can cause flowers to abort. It has been found that only one ovule needs to be fertilized to prevent pod abscission (Halterlein *et al.*, 1980). Older pods may drop if there is an inadequate supply of carbon assimilate (Tanaka and Fujita, 1979). The amount of carbohydrate stored in the stem at flowering varies considerably,

Table 11.2. Production of beans as vegetables in France in 1995 (Unilet, 1995).

Crop	Region/year	Hectares	Tonnes ha^{-1}
Flageolet	Bretagne (North)	2355	4.58
	Picardie (Centre)	4901	4.44
Subtotal/mean		7256	4.48
Green Beans	Bretagne (North)	10,082	10.57
	Picardie (Centre)	10,785	11.06
	Aquitaine	6098	8.01
Subtotal/mean		26,965	10.19

Table 11.3. Distribution of pod diameter as a percentage of the green bean harvest of 1995 in France (Unilet, 1995).

%	Extra fine < 6.5 mm	Very fine 6.5–8.0 mm	Fine 8.0–10.5 mm	> 10.5 mm
Bretagne	18.7	49.5	29.5	2.3
North Picardie (Centre)	13.6	44.1	38.2	4.1
Aquitaine	54.2	41.6	3.8	0.4
Total	22.8	45.8	28.7	2.7

even in the same cultivar, and Adams *et al.* (1978) found differences between genotypes in both the amount of carbohydrate stored and its mobilization after flowering, which could be related to the incidence of abscission and the ability of the plant to tolerate stress.

The occurrence of senescence and abscission appears to depend mainly on the source–sink balance in the plant, such that tissues which are at a competitive disadvantage are eliminated. There are likely to be other mechanisms also at work, such as endogenous growth regulators (White and Izquierdo, 1991). Key factors influencing the supply side of the source–sink balance are light, temperature, water and nitrogen.

LIGHT

In a high radiation environment, the maximum net photosynthesis of the canopy has been found to be 35–40 mg dm^{-2} h^{-1} at saturating irradiance values of 600–650 W m^{-2} and LAI of 4.5 (Sale, 1975), but it seems the maximum photosynthetic rate can be reached at lower levels of irradiance of 300 W m^{-2} (White and Izquierdo, 1991).

Penetration of light through the canopy is critical and depends to a large extent on the orientation and distribution of the leaves. The pulvinus of the bean leaflets and the trifoliate leaf structure allows the bean plant to orient its leaves in relation to the sun and maximize canopy light interception (Wien and Wallace, 1973).

TEMPERATURE

Approximately 80% of *P. vulgaris* production in Latin America and the Caribbean, the centre of diversity of beans, occurs at mean temperatures between 19.6 and 23.3°C (CIAT, 1980). Jones (1971) found that leaf photosynthetic rates increase up to 25°C, and then decline again above that temperature. There is more recent evidence that cultivars which are adapted to different temperatures show differences in the optimum temperature for photosynthesis (Laing *et al.*, 1984).

Beans are very sensitive to high temperatures above 30°C at flowering time, showing increased abscission of flower buds, flowers and young pods, and poor fertilization and seed development in the pods . Greater tolerance of high temperatures would be very desirable and there is some evidence for genetic variation. For example, at 32/27°C day/night for 5 days at anthesis Ofir *et al.* (1993) found significant differences among cultivars when compared with a control treatment (22/17°C), the most tolerant being PI-271998, Legacy and Sentry. Others who have reported cultivar differences for tolerance to high temperature include Weaver *et al.* (1985) and Monterroso and Wien (1988). Genotypes selected for high temperature tolerance may show less reduction of pollen viability at high temperatures than susceptible ones.

Beans are also sensitive to low temperatures, which can limit their production in the early part of the season. At a temperature of 14°C, Farlow (1981) found that seed set was reduced and that this was mainly due to a reduction in ovule fertility rather than pollen germination. Differences among genotypes for tolerance to suboptimal temperatures were reported by Dickson and Boettger (1984). Breeding of green beans for cold tolerance, using a tolerant line NY 5–161 from Cornell and selecting in the growth chamber at 16°C for 30 days, was reported by Holubowicz and Legutko (1995).

WATER

Beans are very sensitive to both drought stress and excess rainfall. Approximately 60% of bean production regions worldwide are limited by drought (CIAT, 1980). In most commercial green bean production areas, supplementary or regular irrigation is required. Usually the poor spots in the field will give the first signs of water stress. The most critical period is flowering and early pod set, when moisture stress will cause flowers and pods to drop. Early moisture stress, at the stage of two trifoliates, can have a lasting effect by reducing vegetative growth and affecting floral initiation, so that crop maturity becomes more uneven (Unilet, 1992). Equally, very lush growth should be avoided as this can cause problems with lodging and may lead to increased disease levels.

A relatively simple method of controlling irrigation requirements for beans is to calculate a weekly water balance, using rain gauges in the field and averaging other climatic data (temperature, radiation) each week to estimate evapotranspiration, based on the growth stage (LAI) of the crop. The estimated evapotranspiration can be deducted from the readily available water reserve in the soil to assess when irrigation water should be applied. The water reserve is ideally evaluated per soil horizon, but for most situations the available water in the top 60 cm provides sufficient accuracy (Unilet, 1992). A similar method, using pan evaporation and daily crop factors (water use/pan evaporation) for scheduling the irrigation of green beans is described by Smittle and Dickens (1992).

Spectral indices (e.g. the ratio between reflectance at 970 nm, a water absorption band, and reflectance at 900 nm) have been found by Peñuelas *et al.* (1993) to be strongly correlated with water content of the crop canopy, and could be used for remote sensing, by taking aerial images of fields using multi-spectral wave bands. Changes in the reflectance patterns over time potentially could be used to show farmers where problems of water deficiency were occurring in the field. This could possibly allow irrigation water to be directed preferentially to those parts of the field, or to vary plant populations in future sowings on the same field.

There is good evidence that deep roots are associated with drought tolerance in beans (White and Singh, 1991). Ensuring adequate soil preparation and fertilization, to encourage deep rooting, will reduce the risk of drought stress. A close relationship between leaf area index and evapotranspiration has been observed in green beans by Bonnano and Mack (1983). The onset of drought interrupts photosynthesis and tissue expansion, because the stomata close, normally at leaf water potentials of 20.5–20.7 mPa, and this restricts gas exchange (O'Toole *et al.*, 1977, Walton *et al.*, 1977). With continued drought stress, the plant water potential declines, resulting in wilting and the loss of the ability of the plant to orient its leaves.

Phaseolus acutifolius has been found to be more drought tolerant than *P. vulgaris* (Cory and Webster, 1984), and there is some evidence that beans can be bred for improved drought tolerance (White and Singh, 1991). There are various possible mechanisms to reduce water loss including: (i) increased leaf thickness and/or reduced cell size; (ii) differences in stomatal sensitivity to humidity; (iii) differences in the amount of cuticular wax; or (iv) differences in leaf pubescence, orientation, colour or size. Remobilization of N and stored carbohydrates (starch) may increase in some genotypes under drought stress, and this would permit greater root growth and improve osmotic adjustment.

A method for screening beans for drought tolerance using canopy temperatures measured with an infrared thermometer in stressed and non-stressed plots was described by Laing *et al.* (1984). Tolerant cultivars are able to maintain plant water potential at moderate levels and effective gas exchange throughout the day, and this is reflected in a smaller difference in canopy temperature between the stressed and non-stressed plots. A number of sources of drought tolerance were selected in this way at CIAT, for example BAT477, San Cristóbal 83 and V8025, and have been used as checks for screening larger numbers of germplasm (Singh and Terán, 1995)

Beans are also very sensitive to waterlogging, when the soil water is in excess of field capacity. Leaf and root growth has been found to stop after 2 days of flooding, associated with a severe reduction in transpiration and a rise in the leaf abscissic acid (ABA) concentration (Wadman van Schravendijk and van Andel, 1985).

SOIL NUTRIENTS AND NITROGEN FIXATION

Beans, like all legumes, have a relatively high nitrogen demand during pod fill, and senescence and abscission may result from competition for N (Sinclair and de Wit, 1976).

The soil pH for commercial production of beans should not be below 6.5 ideally, but beans are produced down to pH 5.5 in some areas. Beans respond well to high fertility levels. Being a short season crop, most of the fertilizer is applied at planting time in order to establish optimal vegetative growth before flowering. Where a top dressing of nitrogen is required, this is normally applied at the 2–3 trifoliate leaf stage. Further top dressing may be required when rainfall is high and when *Rhizobium* inoculation is not used or nodulation is ineffective.

Beans are very sensitive to Mn, Zn and Fe deficiencies, which can adversely affect photosynthesis. The relatively high protein content of bean seeds (20–24%) implies a nitrogen content of about 4%, which means that 40 kg of N is required for every 1 t of seed yield.

Green beans are mostly grown without inoculating with *Rhizobium* and do not nodulate effectively in most commercial production areas. As a consequence, and by contrast with peas, beans are normally fertilized with relatively heavy doses of N, sufficient to inhibit any nodulation that might otherwise occur. The efficiency of N partitioning in the plant can be affected by using different sources of N, be it nitrate or nitrogen fixed as ammonium (Westermann *et al.*, 1985). Nitrate reductase activity is greatest in the leaves of beans, and its activity varies greatly with leaf age and crop growth stage (Franco *et al.*, 1979).

Given the desirability of reducing N applications, for environmental reasons as well as cost, some work has been done on nitrogen fixation in green beans, but much more has been done with dry beans. With green beans, Neuvel and Floot (1992) in the Netherlands found that N applications could be reduced by 25–50 kg ha^{-1} without significantly reducing yields, by mixing granular *Rhizobium* inoculant with the seeds.

Acid soil (pH < 5.5) is generally detrimental to nodulation, but some *Rhizobium* strains, for example CIAT 899, are more tolerant than others (Wolff *et al.*, 1993). Nodulation responds positively to P application (Talavera, 1991).

The nodules of beans (*P. vulgaris*) are of the determinate type, which means that all the infected cells within each nodule are of approximately the same age and the nodules are usually spherical (Sprent, 1989).

There is a large cultivar effect in beans, some cultivars nodulating much more effectively than others (Graham, 1981). The plant genetic control of nodulation has been reviewed by Caetano-Anolles and Gresshoff (1991). Work done by CIRAD (Montpellier) and reported by Unilet (1992) shows significant differences between green bean cultivars as measured by acetylene reduction (Table 11.4). The most efficient strains of *Rhizobium* identified in France in the same work are CIAT 899 (isolated in Colombia by CIAT)

Table 11.4. Nitrogen fixation as measured by acetylene reduction in different green bean cultivars inoculated with CIAT 899 and IRAT 105 (mean response) (from Unilet, 1992).

Cultivar	Acetylene reduction mmol h^{-1} per plant	Acetylene reduction mmol h^{-1} mg^{-1} of nodule
Calibra	1932	58
Cabri	1193	29
Primanor	1176	31
Capitole	850	21
Vernel	545	11
Gitana	381	26
Dorabel	179	5

and IRAT 105 (isolated in Burundi). In general bean breeding and selection has been carried out under conditions of high applied N, and this may have resulted in the selection of cultivars which do not nodulate effectively, because applying nitrate inhibits nitrogenase, the bacterial enzyme responsible for N$_2$ fixation.

Some mutation work has been done to broaden the range of nodulation responses in beans (*P. vulgaris*). Non-nodulating mutants of beans were first reported by Davis *et al.* (1988), and these could be used as non-fixing controls for N-fixation experiments. A super-nodulating mutant, RBS15, has been selected by Hansen *et al.* (1993), in which the nodule dry weight in hydroponically grown plants was about twice that of a control cultivar 'Rico', and nodule numbers were about six times greater. Unfortunately the nodules of the mutant were less efficient, so that N fixation estimated by ^{15}N dilution was the same in RBS15 as in 'Rico'. Super-nodulating mutants which nodulate effectively and which are not inhibited by NO$_3$$^-$ have not yet been obtained.

Martinez (1976) showed that N fixation in 'Porrillo sintetico' was significantly increased when plants were grown in a CO$_2$ enriched atmosphere (1200 ppm), indicating that the availability of carbohydrate to the nodules was limiting fixation. Fixation decreases rapidly as the pods begin to develop, presumably because they compete with the nodules for assimilates. Delaying the flowering of 'Porrillo sintetico', which is photoperiod sensitive, by growing it in long days can substantially increase N fixation (Graham, 1981).

HARVEST PREDICTION

The combination of weather, soil and genotype determines the time to harvest, and being able to use this information to predict the time to harvest is important, particularly for processing companies who need to plan the daily intake of product into the factory.

TIME TO MATURITY

Bush green beans produce pods ready for fresh harvest in about 45–65 days and 55–75 days for the pole types. Models based on heat units work on the basis that the crop must accumulate a certain number of day-degrees above a base temperature before flowering begins, and this type of model has been used to predict the harvest date of green beans (Gould, 1950) (see also Chapter 4). Scarisbrick *et al.* (1976) found that the heat unit system could explain phenological variation and predict harvest maturity also in dry beans. Heat unit methods of predicting harvest maturity are not widely used, however. This is perhaps because most of the cultivars used for commercial green bean production are determinate and have a relatively constant time to flowering in a particular location. Local knowledge of the performance of each cultivar is used to programme plantings.

An example of the heat unit approach is the work on season extension in Germany by Scheunemann (1991), who recorded the number of days to reach specific growth stages together with the summed temperatures, in sequential plantings from mid-May to July. Plant development was affected most by the occurrence of stress days (minimum temperature < 9.5°C). Adequate rainfall/irrigation and effective long-range weather forecasting were essential for the accurate prediction of harvest date.

CROP PHENOLOGY AND YIELD

To predict not only harvest date but also the crop yield, models can be developed to simulate growth. A model known as 'Beangro' has been developed for dry beans by Hoogenboom *et al.* (1994), and an update of that has recently been developed called 'Cropgro-dry bean' (Hoogenboom *et al.*, 1995). The model predicts phenology by calculating progress towards individual growth stages. The rate of progress is affected by photoperiod and temperature. Cultivar traits can be specified through an input file and the model can be used to assess the effects of different agricultural management practices, for example simulating different nitrogen fertilizer management scenarios. The model could presumably be modified to work for green beans.

Others have developed similar models which can cope with different growth habits, for example Gutierrez *et al.* (1994) in Brazil. The crop yield of cultivars differing in growth habit and planted at different plant densities could be predicted using the model.

Models of this type bring together the knowledge of physiological responses of the plant environmental variables, particularly air temperature, solar radiation, precipitation and soil moisture retention. The vegetative and reproductive development section of Beangro simulates the various development stages of the crop such as first leaf, first flower, first pod and harvest maturity as a function of photoperiod and temperature. The carbon section of the model simulates photosynthesis, growth and biomass partitioning. The water balance section predicts plant transpiration, soil evaporation and water flow in the soil profile. The N balance section calculates N remobilization of vegetative tissues during seed-fill.

Potential applications for a model of this sort would be: (i) optimizing agricultural management practices to obtain maximum yield at minimum cost; (ii) risk analysis by looking at the effects of weather changes over a number of seasons; and (iii) predicting harvest maturity and yield for individual fields in any one year.

Only very large-scale growers would be likely to use such a tool, but it could become invaluable for extensionists, or for companies organizing large-scale contract production for processing. Clearly the model would need to be carefully validated with measured observations in any particular region.

SEED DEVELOPMENT AND STORAGE COMPOUNDS

Seeds develop from fertilized ovules in the pod, which commonly contains three to eight seeds, depending on the cultivar and growing conditions. The development of mature seeds takes about 30–60 days from pollination. When the seed is mature no endosperm remains and the embryo completely fills the seed. The root is situated below the micropyle, which served for the entrance of the pollen tube at fertilization. Bean seed storage compounds are present in the embryo. The pods of wild *aborigineus* forms of *P. vulgaris* are dehiscent, but this has been selected out in cultivated forms, and pod parchment and fibre substantially reduced in vegetable green beans.

Dry beans (*P. vulgaris*) are the third most important legume after soyabeans and peas, with about 16 million tonnes of production (FAO, 1990) and 22% protein in the seed, which is similar to peas but much lower than soyabeans. The protein is stored in protein storage vacuoles in the cell and on germination is rapidly hydrolysed to provide a source of reduced nitrogen for the early stages of seedling growth.

In *P. vulgaris* and *P. lunatus* vicilin is the major storage protein, whereas *P. coccineus* tends to have more legumin although it is highly variable for the ratio of vicilin to legumin (Durante *et al.*, 1989). In *P. vulgaris* the vicilin protein is known as phaseolin, and is about 50% of total protein in the bean seed, the other major protein being phytohemaglutenin, which is a lectin (Bollini and Chrispeels, 1978). Variation in the electrophoretic pattern of phaseolin has been used by Gepts *et al.* (1986) to trace the centres of origin and domestication of different bean accessions. Although phytohemaglutenin is the lectin commonly found in bean seeds (up to 10% of total protein), in some wild accessions of *P. vulgaris* a lectin-related protein, arcelin, can reach levels of up to 40% of total protein, and this protein has been associated with resistance to seed storage insects (van Schoonhoven *et al.*, 1983; Osborn *et al.*, 1988).

Phytohemaglutenin is devoid of sulphur amino acids and is highly toxic to monogastric animals, lowering the nutritional value of beans (Pusztai *et al.*, 1979). The other lectin in beans is α-amylase inhibitor, up to 5% of total protein, and is also toxic to animals because it negatively affects starch utilization by the animal (Lajolo *et al.*, 1984). Like arcelin, α-amylase inhibitor is active against some storage insect pests (Huesing *et al.*, 1991).

REFERENCES

Adams, M.W., Wiermsa, J.V. and Salazar, J. (1978) Differences in starch accumulation among dry bean cultivars. *Crop Science* 18, 155–157.

Aguilar, M.I., Fisher, R.A. and Kohashi, S.J. (1977) Effects of plant density and thinning on high yielding dry beans *Phaseolus vulgaris* L. *Experimental Agriculture* 13, 325–335.

Austin, R.B. and Maclean, M.S.M. (1972) A method for screening *Phaseolus* genotypes for tolerance to low temperatures. *Journal of Horticultural Science* 47, 279–290.

Badot, P.M., Capelli, N. and Millet, B. (1993) Arguments supplementaires en faveur de l'implication de canaux K+ dans le mouvement revolutif des tiges volubiles. Efets de l'ion aluminium. Comptes Rendus de l'Academie des Sciences. Series 3, *Sciences de la Vie* 316, 287–292.

Bollini, R. and Chrispeels, M.J. (1978) Characterization and subcellular localization of vicilin and phytohemagglutinin, the two major reserve proteins of *Phaseolus vulgaris* L. *Planta* 142, 291–298.

Bonanno, A.R. and Mack, H.J. (1983) Water relations and growth of snap beans as influenced by differential irrigation. *Journal of the American Society of Horticultural Science* 108, 837–844.

Cachorro, P., Ortiz, A. and Cerda. A. (1994) Implications of calcium nutrition on the response of *Phaseolus vulgaris* L. to salinity. *Plant and Soil* 159, 205–212.

Caetano-Anolles, G. and Gresshoff, P.M. (1991) Plant genetic control of nodulation. *Annual Review of Microbiology* 45, 345–382.

CIAT (1980) *Bean Program Annual Report 1979*. Cali, Colombia.

Cory, C.L. and Webster, B.D. (1984) Assessment of drought tolerance in cultivars of *Phaseolus vulgaris* and *P. acutifolius*. *HortScience* 19, 544.

Coyne, D.P. (1967) Photoperiodism: inheritance and linkage studies in *Phaseolus vulgaris*. *Journal of Heredity* 58, 313–314.

Davis, J.H.C., Kipe-Nolt, J. and Awah, M. (1988) Non-nodulating mutants in common beans. *Crop Science* 28, 859–860.

Debouck, D. and Hidalgo, R. (1984) Morfologia de la planta de frijol comun (*Phaseolus vulgaris* L.). Publication series 04SB-09.01. CIAT, AA 67–13, Cali, Colombia.

Dickson, M.H., and Boettger, A. (1984) Emergence, growth, and blossoming of bean (*Phaseolus vulgaris*) at suboptimal temperatures. *Journal of the American Society of Horticultural Science* 109, 257–260.

Durante, M., Bernardi, R., Lupi, C. and Pini, S. (1989) *Phaseolus coccineus* storage proteins. II. Electrophoretic analysis and erythroagglutinating activity in various cultivars. *Plant Breeding* 102, 58–65.

Enriquez, G.A. (1975) Effect of temperature and daylength on time of flowering in beans (*Phaseolus vulgaris* L.). PhD thesis, Cornell University, Ithaca, New York.

Farlow, P.J. (1981) Effect of low temperature on number and location of developed seed in two cultivars of French beans (*Phaseolus vulgaris* L.). *Australian Journal of Agricultural Research* 32, 325–330.

FAO (1990) *Production Yearbook.* Food and Agriculture Organisation, Rome.

Franco, A.A., Pereira, J.C. and Neyra, C.A. (1979) Seasonal patterns of nitrate reductase and nitrogenase activities in *Phaseolus vulgaris* L. *Plant Physiology* 63, 421–424.

Gepts, P., Osborn, T.C., Rashka, K. and Bliss, F.A. (1986) Phaseolin-protein variability in wild forms and landraces of the common bean (*Phaseolus vulgaris*): evidence for multiple centers of domestication. *Economic Botany* 40, 451–468.

Gepts, P. (ed.) (1988) *Genetic Resources of Phaseolus Beans*, Kluwer Academic, Dordrecht.

Gniffke, P.A. (1982) Varietal differences in photoperiodically mediated flowering response in beans (*Phaseolus vulgaris* L.). PhD thesis, Cornell University, Ithaca, New York.

Gould, W.A. (1950) Hevi's heat unit guide for 47 varieties of snap beans. *Food Packer* 31, 35–37.

Graham, P.H. (1981) Some problems of nodulation and symbiotic nitrogen fixation in *Phaseolus vulgaris* L.: a review. *Field Crops Research* 4, 93–112.

Gutierrez, A.P., Mariot, E.J., Cure, J.R., Wagner Riddle, C.S., Ellis, C.K. and Vilacorta, A.M. (1994) A model of bean (*Phaseolus vulgaris* L.) growth types I–III: Factors affecting yield. *Agricultural Systems* 44, 35–63.

Halterlein, A.J., Clayberg, C.D. and Teare, I.D. (1980) Influence of high temperature on pollen grain viability and pollen tube growth in the styles of *Phaseolus vulgaris* L. *Journal of the American Society of Horticultural Science* 105, 12–14.

Hansen, A.P., Yoneyama, T., Kouchi, H. and Martin, P. (1993) Respiration and nitrogen fixation of hydroponically cultured *Phaseolus vulgaris* L. cv. OAC Rico and a supernodulating mutant. I. Growth, mineral composition and effect of sink removal. *Plants* 189, 538–545.

Hoogenboom, G., White, J.W., Jones, J.W. and Boote, K.J. (1994) 'Beangro': a process-oriented dry bean model with a versatile user interface. *Agronomy Journal* 86, 182–190.

Hoogenboom, G., White, J.W., Jones, J.W. and Boote, K.J. (1995) A new and improved dry bean simulation model: 'Cropgro-dry bean'. *Annual Report of the Bean Improvement Cooperative* 38, 15–16.

Huesing, J.E., Shade, R.E., Chrispeels, M.J. and Murdock, L.L. (1991) α-amylase inhibitor, not phytohemmaglutenin, explains resistance of common bean seeds to cowpea weevil. *Plant Physiology* 96, 993–996.

Holubowicz, R. and Legutko, W. (1995) Breeding snap beans for cold tolerance. *Annual Report of the Bean Improvement Cooperative* 38, 33.

Izquierdo, J.A. (1981) The effect of accumulation and remobilization of carbon assimilate and nitrogen on abscission, seed development, and yield of common bean (*Phaseolus vulgaris* L.) with differing architectural forms. PhD dissertation, Michigan State University, East Lansing, MI, USA.

Jones, L.H. (1971) Adaptive responses to temperature in dwarf French beans, *Phaseolus vulgaris* L. *Annals of Botany* (London) 35, 581–596.

Korban, S.S., Coyne, D.P. and Weihing, J.L. (1981) Rate of water uptake and sites of water entry in seeds of different cultivars of dry bean. *HortScience* 16, 545–546.

Kretchmer, P.J., Ozbun, J.L., Kaplan, S.L., Laing, D.R. and Wallace, D.H. (1977) Red and far-red light effects on climbing in *Phaseolus vulgaris* L. *Crop Science* 17, 797–799.

Kyle, J.H. and Randall, T.E. (1963) A new concept of the hard seed character in *Phaseolus vulgaris* L. and its use in breeding and inheritance studies. *Proceedings of the American Society of Agricultural Science* 83, 461–475.

Laing, D.R., Jones, P.G. and Davis, J.H.C. (1984) Common bean (*Phaseolus vulgaris* L.). In: Goldsworthy, P.R. and Fisher, N.M. (eds) *The Physiology of Tropical Field Crops*. John Wiley & Sons, New York.

Lajolo, F.M., Filho, J.M. and Menezes, E.W. (1984) Effect of a bean (*Phaseolus vulgaris*) α-amylase inhibitor on starch utilization. *Nutritional Report International* 30, 45–54.

Maréchal, R., Mascherpa, J.M. and Stainer, F. (1978) Étude taxonomique d'un groupe complexe d'espèces des genres *Phaseolus* et *Vigna* (Papilionaceae) sur la base de données morphologiques et polliniques, traitées par l'analyse informatique. *Boissiera, Genève* 28, 1–273.

Martinez, R. (1976) Nitrogen fixation and carbohydrate partitioning in *Phaseolus vulgaris*. PhD thesis, Michigan State University, East Lansing, USA.

Masaya, P. and White, J.W. (1991) Adaptation to photoperiod and temperature. In: van Schoonhoven, A. and Voysest, O. (eds) *Common Beans: Research for Crop Improvement*. CAB International, Oxon.

Monteith, J.L. (1978) Reassessment of maximum growth rates for C_3 and C_4 crops. *Experimental Agriculture* 14, 1–5.

Monterroso, V.A. and Wien, H.C. (1988) Effect of high temperature on pod set of *Phaseolus vulgaris* L. *Annual Report of the Bean Improvement Cooperative* 31, 160.

Neuvel, J. and Floot, H. (1992) Research on French beans. Inoculum can lower nitrogen application. *Groenten und Fruit, Vollegrondsgrownten* 2, 18–19.

Ojehomon, O.O. (1966) The development of flower primordia of *Phaseolus vulgaris*. *Annals of Botany* 30, 487–492.

Ofir, M., Gross, Y., Bangerth, F. and Kigel, J. (1993) High temperature effects on pod and seed production as related to hormone levels and abscission of reproductive structures in common bean (*Phaseolus vulgaris* L.). *Scientia Horticulturae* 55, 201–211.

Osborn, T.C., Alexander, D.C., Sun, S.S.M., Cardona, C. and Bliss, F.A. (1988) Insecticidal activity and lectin homology of arcelin seed proteins. *Science* 240, 207–210.

O'Toole, J.C., Ozbun, J.L. and Wallace, D.H. (1977) Photosynthetic response to water stress in *Phaseolus vulgaris*. *Physiologia Plantarum* 40, 111–114.

Peñuelas, J., Filella, I., Serrano, L. and Save, R. (1993) The reflectance at the 950–970nm region as an indicator of plant water status. *International Journal of Remote Sensing* 14, 1887–1905.

Pretorius, J.C. and Small, J.G.C. (1993) The effect of soaking injury in bean seeds on carbohydrate levels and sucrose phosphate synthase activity during germination. *Plant Physiology and Biochemistry (Paris)* 31, 25–34.

Pusztai, A., Clarke, E.W. and King, T.P. (1979) The nutritional toxicity of *Phaseolus vulgaris* lectins. *Proceedings of the Nutrition Society* 38, 115–120.

Reichel, S. (1992) Late crops of French beans with different cultivars. *Gartenbau Magazin* 1, 91–93.

Sale, P.J.M. (1975) Productivity of vegetable crops in a region of high solar input, IV: field chamber measurements on French beans (*Phaseolus vulgaris* L.) and cabbages (*Brassica oleracea* L.). *Australian Journal of Plant Physiology* 2, 461–470.

Scarisbrick, D.H., Carr, M.K.V. and Wilks, J.M. (1976) The effect of sowing date and season on the development and yield of navy beans (*Phaseolus vulgaris*) in South

East England. *Journal of Agricultural Science* 86, 65–76.

Scheunemann, C. (1991) Continuous production of dwarf French beans with staged cropping. *Gartenbau Magazin* 38, 18–19.

Schmit, V., Baudoin, J.P. and Wathelet, B. (1992) Contribution à l'étude des relations phylétiques au sein du complexe *Phaseolus vulgaris* L.–*Phaseolus coccineus* L.–*Phaseolus polyanthus* Greenm. *Bulletin des Recherches Agronomiques de Gembloux* 27, 199–207.

Sinclair, T.R. and de Wit, C.T. (1976) Analysis of the carbon and nitrogen limitations to soybean yield. *Agronomy Journal* 68, 319–324.

Singh, S.P. and Teran, H. (1995) Evaluating sources of water-stress tolerance in common bean. *Annual Report of the Bean Improvement Cooperative* 38, 42.

Smartt, J. (1990) *Grain Legumes: Evolution and Genetic Resources.* Cambridge University Press, Cambridge.

Smittle, D.A. and Dickens, W.L. (1992) Water budgets to schedule irrigation for vegetables. *HortTechnology* 2, 54–59.

Sprent, J.I. (1989) Tansley review no.15. Which steps are essential for the formation of functional legume nodules? *New Phytologist* 111, 129–153.

Talavera, S.F.T. (1991) Impacts of P and N fertilizers on common beans (*Phaseolus vulgaris*, L.) growth, nodulation and P and N uptake in a pot experiment. *Swedish Journal of Agricultural Research* 21, 23–28.

Tanaka, A. and Fujita, K. (1979) Growth, photosynthesis and yield components in relation to grain yield of the field bean. *Journal of the Faculty of Agriculture of Hokkaido University* 59, 145–238.

Unilet (1992) *Haricot Agronomie 1992.* Union Nationale Interprofessionelle des Légumes Transformés, 44 Rue d'Alésia, 75014 Paris.

Unilet (1994) *Unilet Informations Numero Special*, June 1994. Union Nationale Interprofessionelle des Légumes Transformés, 44 Rue d'Alésia, 75014 Paris.

Unilet (1995) *Unilet Informations revue bimestrielle*, Sommaire no. 90, December 1995. Union Nationale Interprofessionelle des Légumes Transformés, 44, rue d'Alésia, 75014 Paris.

Unilet (1996) *Unilet Informations revue trimestrielle*, Sommaire no. 91, February 1996. Union Nationale Interprofessionale des Légumes Transformés, 44, rue d'Alésia, 75014 Paris.

USDA (1991) *USDA Vegetables Annual.* National Agricultural Statistics Service, 1301 New York Avenue, NW Rm. 1240, Washington, DC 20005–4788.

van Schoonhoven, A., Cardona, C. and Valor, J. (1983) Resistance to the bean weevil and the Mexican bean weevil (Coleoptera: Bruchidae) in non-cultivated common bean accessions. *Journal of Economic Entomology* 76, 1255–1259.

Wadman van Schravendijk, H. and van Andel, O.M. (1985) Interdependence of growth, water relations and abscisic acid level in *Phaseolus vulgaris* during waterlogging. *Physiologia Plantarum* 63, 215–220.

Wallace, D.H. (1985) Physiological genetics of plant maturity, adaptation and yield. *Plant Breeding Reviews* 3, 21–167.

Walton, D.C., Galson, E. and Harris, M.A. (1977) The relationship between stomatal resistance and abscisic acid levels in leaves of water-stressed bean plants. *Planta* 133, 145–148.

Weaver, L., Timm, H., Silbernagel, M.J. and Burke, D.W. (1985). Pollen staining and high-temperature tolerance of bean. *Journal of the American Society of Horticultural Science* 110, 797–799.

Westermann, D.T., Porter, L.K. and O'Deen, W.A. (1985) Nitrogen partitioning and mobilization patterns in bean plants. *Crop Science* 25, 225–229.

White, J.W. (1981) A quantitative analysis of the growth and development of bean plants (*Phaseolus vulgaris* L.). PhD dissertation. University of California, Berkeley, CA, USA.

White, J.W. and Izquierdo, J. (1991) Physiology of yield potential and stress tolerance. In: van Schoonhoven, A. and Voysest, O. (eds) *Common Beans: Research for Crop Improvement*, CAB International, Oxon.

White, J.W. and Montes-R, C. (1993) The influence of temperature on seed germination in cultivars of common bean. *Journal of Experimental Botany* 44, 1795–1800.

White, J.W. and Singh, S.P. (1991) Breeding for adaptation to drought. In: van Schoonhoven, A. and Voysest, O. (eds) *Common Beans: Research for Crop Improvement*. CAB International, Oxon.

White, J.W., Davis, J.H.C. and Castillo, J. (1987) Inducing early flowering in Andean cultivars adapted to low temperature. *Annual Report of the Bean Improvement Cooperative* 30, 12–13.

Wien, H.C. and Wallace, D.H. (1973) Light-induced leaflet orientation in *Phaseolus vulgaris* L. *Crop Science* 13, 721–724.

Wolff, A.B., Singleton, P.W., Sidirelli, M. and Bohlool, B.B. (1993) Influence of acid soil on nodulation and interstrain competitiveness in relation to tannin concentrations in seeds and roots of *Phaseolus vulgaris*. *Soil Biology and Biochemistry* 25, 715–721.

Wroblewska, A. (1993) Attractiveness of *Phaseolus* L. flowers for pollinating insects. *Acta Horticulturae* 288, 321–325.

Zehni, M.S., Saad, F.S. and Morgan, D.G. (1970) Photoperiod and flower bud development in *Phaseolus vulgaris*. *Nature* 227, 628–629.

Peas 12

F.J. Muehlbauer and K.E. McPhee

USDA/ARS, Grain Legume Genetics and Physiology Research Unit,
Washington State University, Pullman, Washington 99164–6434, USA

Peas (*Pisum sativum* L.) are commonly grown and are a popular food in the temperate and to a lesser extent in the subtropical regions of the world. Many different types of peas are produced and prepared in a variety of ways. In developed countries, succulent types are a favourite for freezing or canning, while in developing countries, dry peas are a common, practical and reliable food because they are easily stored. Green peas are eaten as a fresh vegetable, edible flat-podded peas are used in oriental cooking, and edible podded peas ('snap peas') are used as a fresh or frozen vegetable. Dry peas are exceptionally versatile and are used in soups, mixed with other vegetables, made into noodles, and also used as a garnish to add colour to a variety of dishes. By far the largest use of peas in developed countries is as a protein source for animal feeding. Nutritionally, peas provide a high protein food that is rich in many of the essential amino acids, especially lysine; however, they are deficient in methionine and cystine. Peas are rich in calcium, phosphorus, iron, sodium and potassium.

It is generally agreed that peas were domesticated in the fertile crescent of Southwest Asia. This conclusion is based on their frequent appearance in archaeological remains as far back as 7000–6000 BC (Zohary and Hopf, 1973; Smartt, 1990). After domestication, cultivation spread north into Russia, west into Europe, east into India and China, and eventually to the Western Hemisphere soon after the discovery of the New World. Production in Europe and North America, particularly Canada, has risen sharply in the past decade as the pea crop has found increasing usage as an animal feed. In the USA, Wisconsin, Washington and Minnesota lead in the production of processing peas, while Washington and Idaho lead in the production of dry peas. Overall world production of green peas has remained nearly constant over the past 20 years at about 5 million tonnes, while the production of dry peas has nearly doubled from about 8 million

CAB INTERNATIONAL 1997. *The Physiology of Vegetable Crops* (H.C. Wien, ed.)

tonnes in 1979–91 to over 16 million tonnes in 1992 (Oram and Agcaoili, 1994).

The genetic diversity of peas is well known and was the basis of the pioneering work of Mendel in the mid-nineteenth century in establishing the fundamental laws of inheritance. Peas have since found widespread use as an experimental organism particularly for studies of plant genetics and plant physiology. Germplasm collections have been assembled and are maintained at numerous centres throughout the world including the USA. The US collection of peas containing landraces, cultivars and wild relatives is maintained at Pullman, Washington, by the National Plant Germplasm System operated by the US Department of Agriculture, Agricultural Research Service. That collection numbers over 2800 accessions and is readily available to geneticists, breeders, plant physiologists and other interested individuals on request.

The future of peas appears to be bright based on recent increases in production of peas as an animal feed, additional uses provided by the development of new types of peas especially the 'snap pea', and the expanded use of peas as snack items in the Orient. These snack items include fried and roasted peas with a variety of flavourings as well as extruded products and chips. The widespread use and diversity of pea products bodes well for the crop's future and attests to its versatility as food and feed. In this chapter, we attempt to describe the basic mechanisms and growth patterns of pea from germination through maturity.

SEED ANATOMY AND GERMINATION

The mature pea seed consists of an embryo and two fleshy cotyledons which are surrounded by a testa. The point of attachment of the seed to the internal wall of the pea pod forms a scar known as the hilum. Near the hilum is the micropyle, a small opening in the testa, where the pollen tube enters the ovule and deposits the sperm cells at fertilization. The hilum, micropyle and raphe (a bulge created by the embryonic root or radicle) are positioned along the line of separation between the two cotyledons (Fig. 12.1). The embryonic axis is composed of the radicle, hypocotyl, epicotyl and young shoot or plumule. The seed coat is composed of two different cell layers, the testa and the integuments (Murray, 1977). The inner parenchymatous cells senesce before maturity and are then crushed by the developing cotyledons (Murray, 1977).

Conditions for germination are optimum at 24°C and emergence may take 3–5 days. The rate of germination and emergence decreases with progressively lower temperatures and ceases at 4.4°C (Peirce, 1987). The rate of water uptake is important to seed survival. When seeds experience a rapid influx of water some cells are not able to reorganize their cellular matrix which leads to a loss of cell contents into the medium. Pea embryos leach large amounts of solutes in the first few minutes or hours of imbibition and

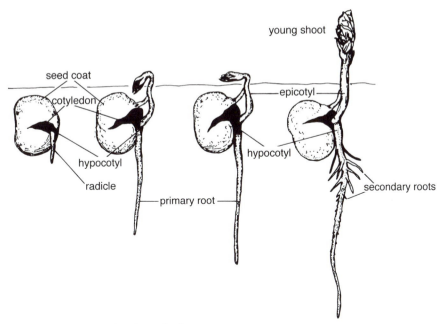

Fig. 12.1. The pea seed and germination.

less thereafter. Two possible reasons for this to occur are: (i) disruption of cell membranes by a rapid in-rush of water; and (ii) the porous nature of the dehydrated state of the membrane structure. These components attract soil fungi which feed on the young seedling causing various seedling diseases, such as damping-off caused by *Pythium* spp. or *Rhizoctonia solani* (Matthews *et al.*, 1978; Norton and Harmon, 1985).

Germination can be divided into the imbibition and activation phases (Sutcliffe and Bryant, 1977). The imbibition phase begins immediately after the seed is placed in the soil and comes in contact with water. This phase lasts approximately 20 h and is characterized by a rapid uptake of water by the cotyledons and embryo with a nearly doubling of seed size. In the early part of imbibition (approx. 6 h), organic molecules, nucleic acids and proteins become rehydrated; the latter part of the imbibition phase is characterized by water accumulation in the vacuole. Near the end of the imbibition phase, water uptake slows and metabolic activity increases. The second stage of the germination process is characterized by a period of activation in which numerous metabolic changes take place. The seed can tolerate dehydration in the initial stages of imbibition, but once the seed begins metabolic activity dehydration causes damage to the embryo and cotyledons.

The second phase of germination continues for several days and is characterized by degradation of materials stored in the cotyledons. Starch, protein and phytic acid concentrations decrease, while concentrations of carbohydrates, amides and inorganic phosphorus increase. The initial increase in sucrose after imbibition is due partly to the hydrolysis of sucrosyl oligosaccharides (Monerri *et al.*, 1986). Monerri and co-workers (1986) also found that starch mobilization in the wrinkled pea began slightly earlier than the increase in amylase activity; whereas in the smooth pea, it coincided with the increase in amylase activity. Since the same sugars were found in the cotyledons of both seed types, they suggested that a common pathway exists for starch breakdown in wrinkled and smooth peas. An explanation for the differences in timing is that the composition of starch granules in wrinkled peas is different and more susceptible to breakdown by starch-degrading enzymes. Also during this phase, mitochondria begin to show greater organization and may increase in number. In addition, cellular membranes, such as golgi and endoplasmic reticulum, become organized and develop (Bain and Mercer, 1966), protein and RNA synthesis increases (Chin *et al.*, 1972), and there is an increase in oxygen uptake.

The third phase of germination starts approximately 5 days after the start of imbibition and is characterized by decreased oxygen uptake and senescence of the cotyledons. Many of the cellular components become disorganized causing the macromolecules (DNA, RNA and protein) in the cell to break down into their component parts followed by transport to the developing embryonic axis.

The embryo passes through three phases during germination similar to those described for the cotyledons (Sutcliffe and Bryant, 1977). Phase I corresponds to the imbibition phase and is primarily characterized by water uptake and the initiation of metabolism. The rate of metabolism increases during phase II and many dramatic ultrastructural changes take place in the cells. Golgi bodies and mitochondria show increased development of their membrane systems and some division of the mitochondria may occur (Yoo, 1970). The endoplasmic reticulum also proliferates during this phase. However, there is little change in the fresh weight of the axis or nucleic acid content. The cells in the radicle elongate causing it to emerge from the testa and allowing oxygen to enter the seed. The increased oxygen concentration causes an increase in respiration triggering the third phase of germination for the embryo. During this phase, there is an increase in the fresh weight of the axis and a net synthesis of DNA and RNA.

Respiration in the germinating seed increases slowly during the initial stages of imbibition and shows distinct stages. During the first stage, a rapid increase in respiration occurs due to activation of the enzymes involved in glycolysis and the hexose phosphate pathway. The second stage corresponds to a moisture content of 40–50% and is marked by a transition to a higher level of respiration due to completion of mitochondriogenesis. The third stage reflects the respiration of rapidly transpiring cells in the axis and corresponds to higher moisture contents because the axial cells are elongating and becoming vacuolated.

Based on the respiratory quotient, Obrucheva and Kovadlo (1985) suggested that carbohydrates serve as the primary substrate for respiration in seeds up to 50% moisture and that amino acids along with carbohydrates are the primary substrates in fully germinated seeds (i.e. seeds in which the radicle has emerged).

After germination, the cotyledons of the pea seed remain below the soil surface in a manner characteristic of 'hypogeal' emergence (see Fig. 12.1). Early seedling growth is characterized by emergence of the radicle from the seed. Once the radicle has emerged and begins to elongate downward (positive geotropism) the epicotyl emerges and grows upward. Secondary roots and root hairs begin to grow on the root as the taproot and shoot elongate. Root growth occurs through cell division and elongation in a relatively small part of the root tip.

Germination percentages and seedling vigour are affected by physiological age of the seed at harvest and subsequent handling. The manner in which the seed is handled affects the condition of the testa and seed coat and is a critical determinant of seed vigour (Matthews *et al.*, 1978). The amount of imbibition damage and loss of seedling vigour is directly proportional to the number and length of cracks in the testa (Matthews *et al.*, 1978).

VEGETATIVE GROWTH

Vegetative growth begins after germination. The radicle has already emerged from the seed and elongates downward, while the epicotyl grows upward. Secondary roots and root hairs begin to branch off the taproot. The plumule remains curved as the epicotyl emerges from the cotyledons and grows upward (see Fig. 12.1). The curvature of the plumule protects the shoot apex from mechanical damage during passage through the soil to the surface. The plumule hook changes from a white or pale yellow to a dark green as a result of chlorophyll synthesis after emergence. When the plumule hook emerges from the soil and is exposed to light, it straightens and the first leaf becomes visible.

During vegetative growth, pea stems develop between 20 and 25 nodes. Beyond the third node, each node has a compound leaf comprised of two fleshy stipules at the base, a petiole with two or three pairs of leaflets, and terminating with three to five tendrils (Fig. 12.2). The first two nodes often remain below the soil surface and the leaves at these nodes are usually small and consist of two small rudimentary stipules. Occasionally, basal branches or tillers are produced at these nodes (Meicenheimer and Muehlbauer, 1982). The length of each node is controlled primarily by a major gene, *Le*, and is affected by several modifying genes, *Cry* and *La* (Haan, 1927; Lamm, 1937).

There are generally five leaf primordia at successive stages of development in the shoot apex at any one time (see Fig. 12.5). In general, the lower nodes have single leaflet pairs while later nodes may have two or often three leaflet pairs. The number of nodes is primarily dependent on the cultivar. Pea leaves have cuticular wax on the upper surface and leaf colour can range from yellow–green to a deep blue–green depending on the cultivar.

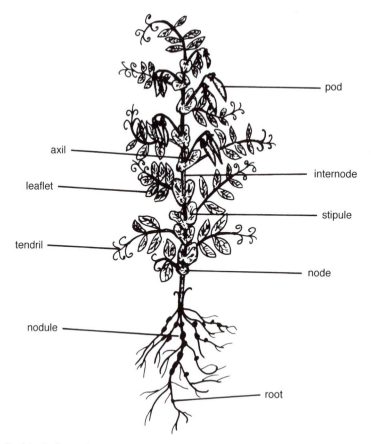

Fig. 12.2. A typical pea plant.

Various genes have been studied which modify leaf structure and form (Fig. 12.3). The most prominent of these genes are the *af*, *st* and *tl* genes, which in combination, produce eight distinct morphologies. The 'afila' leaf type, controlled by the *af* gene, was reported by Goldenberg (1965) and is characterized by the conversion of leaflets to tendrils. This particular plant type has come to be known as 'semi-leafless'. Two other genes that alter leaf morphology are *st* which greatly reduces stipule size, and *tl* which converts tendrils to leaflets. Of the eight leaf types, normal (*AfAfStStTlTl*; Fig.12.3a), semi-leafless (*afafStStTlTl*; Fig. 12.3b) and leafless (*afafststTlTl*; Fig. 12.3d) have been compared in detail for agronomic potential. Alternate leaf morphologies are of great interest because they cause the plants to grow more upright giving the potential for reduced foliar disease. The reduced disease levels are the result of increased air movement and reduced humidity within the crop canopy. Additional advantages of some of the combinations are more uniform maturity and greater ease of harvest. It is also suggested

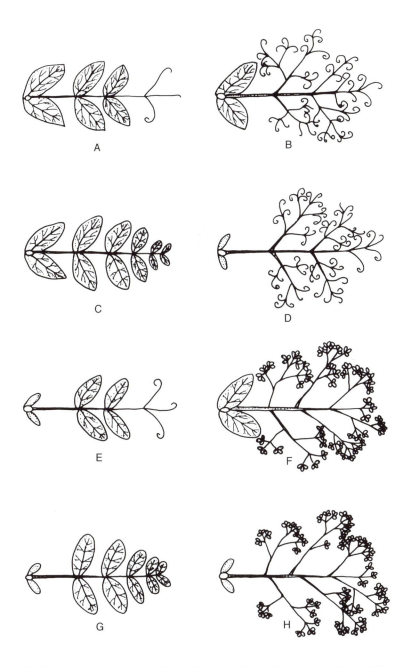

Fig. 12.3. Pea leaf types made possible by the interaction of three genes, *af, st* and *tl*. (A) normal leaf; (B) semi leafless or 'afila' type (*afaf*); (C) tendrilless (*tltl*); (D) leafless showing effects of 'afila' and reduced stipule (*afafstst*); (E) reduced stipule (*stst*); (F) 'afila' with tendrilless (*afaftltl*); (G) reduced stipule and tendrilless (*ststtltl*); (H) afila, reduced stipule and tendrilless (*afafststtltl*).

that the reduced leaf area of these plant types allows for greater light penetration into the canopy. The more open the canopy is, the more it reduces the amount of shading to the lower leaves and increases the photosynthetic activity of the developing pods. Kielpinski and Blixt (1982) report that the semi-leafless plant type produces greater yields than the conventional leaf types. Moreover, they suggest that this is not due to an increased yield potential, but is the result of improved standing ability and the increased penetration of light into the canopy.

An increasing number of pea varieties have appeared in recent years with the semi-leafless morphology imparted by recessive *af*. The semi-leafless type is considered to have good resistance to lodging and possibly reduced foliar disease. The completely leafless type (*afafststTLTL*) has not proven useful in the development of new cultivars because of significantly reduced yields. The phenotypes of the eight possible combinations of these three genes have been suggested as a means of demonstrating classical genetic principles in the classroom (Marx, 1974).

Vegetative growth of the pea plant is affected by both genetic and environmental factors. These factors interact with each other to further modify plant growth. The genetic effects include photosynthetic potential, water use efficiency, crop and plant growth rates, different leaf type effects, leaf area index, duration of leaf area and seed size. These genetic effects are modified by environmental conditions including planting density, air and soil temperatures, photoperiod, and the amount and distribution of precipitation. Maximum seed yields require maximum vegetative growth during crop establishment. This can be achieved through high crop growth rates which allow optimum leaf area index to be attained early in vegetative growth.

It is important in field-grown crops to understand competition. Since competitiveness is related to the demand for nutrients and moisture from the environment, it is directly related to the growth rate. Thus, competitiveness is dependent on the rate of cell division, cell expansion, and the size of the meristematic region (Hedley and Ambrose, 1981). Due to the relationship between growth rate and competitive ability, large-seeded peas are more competitive than small-seeded peas. Under most conditions, large-seeded peas will have a higher growth rate than small-seeded peas. Given a specific environment, competition will occur at a given biological yield irrespective of planting density or seed size (Hedley and Ambrose, 1981).

Although seed size has an effect on early seedling growth, it has little effect on final seed yield. Research by Pyke and Hedley (1983) suggests that a long growing season and the large number of yield components allows smaller, less competitive plants to achieve comparable yields to plants derived from larger seeds. Seed size relates equally to seed growth rate within populations and among cultivars of different seed size. Increased seed dry weight reduces the primary root length by one-third (Veierskov, 1985). This suggests that decreased drought tolerance may accompany large-seeded peas.

Although large seed size allows for greater seedling growth rates, the increased size of the developing pea ovules augments the potential for abortion. It would be very advantageous to develop a pea with a large embryonic axis and small seeds, thus allowing greater seedling growth rates without the risk of seed abortion.

Photosynthetic capacity, crop and plant growth rates, leaf area index, leaf area duration, and to some extent water use efficiency, are all related to leaf and plant morphology. Leaf morphology has dramatic effects on plant growth and yield and has been studied by several authors (Harvey, 1972, 1978, 1980; Harvey and Goodwin, 1978; Hedley and Ambrose, 1981; Snoad, 1981; Pyke and Hedley, 1983).

Leafless peas have a lower relative growth rate compared to normal leafed types irrespective of seed size (Hedley and Ambrose, 1981). An explanation for the lower growth rate of the leafless peas is that they have a smaller number of cells. These cells continue to divide during the beginning of growth. Leafed peas have a larger meristematic region owing to the presence of leaf meristems which are absent in the leafless plant types (Hedley and Ambrose, 1981). Pyke and Hedley (1983) and Snoad (1981) found that plant relative growth rates and absolute growth rates were similar for normal (*AfAfStStTlTl*) and semi-leafless (*afafStStTlTl*) leaf types, but were lower for the leafless (*afafststTlTl*) types. They suggested that the reduced RGR of the leafless plant type may be due to a lower RGR for stems. Moreover, Snoad (1981) found that leafless peas had shorter internodes below the first flowering node than conventional or semi-leafless plants, thus providing further evidence that stem growth rates are lower in the leafless plant type.

Tendrils appear to have slightly higher rates of CO_2 photoassimilation than leaves. They also have 50% fewer stomata than leaflets and a higher proportion of photosynthetically inert tissue (Harvey, 1972). Harvey and Goodwin (1978) report a 30–60% reduction in carbon exchange rate for tendrils when compared to leaflets. The distribution of chloroplasts in tendrils on leafless plants resembles that in the stems, petioles and tendrils of the conventional plant type (Harvey, 1972). Leafless plants have lower transpiration rates than the conventional plants (Harvey, 1972). When all these factors are taken together, the net result is that the leafless (*afafststTLTL*) pea type produces 50% less dry matter during its vegetative growth period (Harvey and Goodwin, 1978).

Dry matter production for conventional plants reached a plateau at approximately 56 days after planting or at leaf stage 9, while the leafless plants reached a maximum dry matter production at 68 days after planting or at leaf stage 16 (Harvey and Goodwin, 1978). This later plateau for the leafless type is a disadvantage because it delays the development of optimum leaf area thus impeding crop development and light interception (Harvey and Goodwin, 1978).

Planting density affects vegetative growth primarily through competition for light, nutrients and water. Plant density can be manipulated to maximize and optimize light interception. Optimum planting density is different for each of the leaf morphologies. Ninety-five per cent light interception was found to be achieved with an LAI (leaf area index, or ratio of leaf area to ground area covered by plant) of approximately 5.3 (Kruger, 1977). Canopies with an LAI in excess of the optimum resulted in shading of the lower leaves and pods. This may explain the increased rate of leaf senescence observed at higher planting densities. High plant densities favour leaf development near the top of the plant whereas lower densities allow more even development of leaves along the entire plant. In addition, higher planting densities cause the plants to be taller and more spindly (Vincent, 1958).

Despite the increased competition at higher plant densities, vegetative growth on an area basis (crop growth rate) increases with plant density until moisture or some other factor necessary for growth becomes limiting (Kruger, 1977). Competition among plants in dense stands becomes more intense with age. Thus, as crop growth progresses, individual plants are crowded out as a method of self-thinning. To achieve higher crop growth rates of late planted crops, and greater light interception in the short growing season, Kruger (1977) suggested that higher planting densities are necessary for optimum development.

Planting date interacts with density to determine crop growth characteristics and canopy development. The effect of planting date is manifested primarily through temperature. Fletcher *et al.* (1966) found that dry matter production is positively correlated with mean maximum daily temperature. They also suggested that temperature affects tillering in early growth and branching later in growth. Early and late plantings result in the first flowering node being lower on the plant (Fletcher *et al.*, 1966), an effect that might possibly be due to photoperiod. The earlier flowering under these situations results in decreased vegetative growth since reproductive growth triggers a switch in partitioning from vegetative sinks to reproductive sinks.

Delayed planting date increases relative growth rate and crop growth rates. Hedley and Ambrose (1981) found that the differences in light interception between leafed and leafless peas correlate with the differences in above-ground biomass yield per unit area. Biological yield of leafed canopies decreased slightly as density increased. The biological yield of leafless peas increased with planting density and exceeded the yield of leafed peas at the higher densities. Peas are able to compensate in lower planting densities by increasing the number of laterals produced such that each plant contributes more to the total biological yield. Hedley and Ambrose (1981) found that leafed plants were able to compensate for a 25-fold increase in available space by producing a 30-fold increase in biological yield. The leafless plant on the other hand was only able to compensate with a 12-fold increase in biological yield.

Abnormal temperatures have a strong effect on plant growth and development. Temperatures below a given threshold affect pea plant growth by decreasing the efficiency of photosynthesis. Unusually high temperatures also decrease plant growth by decreasing the carbon exchange rate due to stomate closure. High temperatures favour shoot growth over root growth which increases the shoot : root ratio; and, as van Dobben (1962) suggests, shoot : root ratio is important in determining growth rate. Therefore, lower temperatures early in development will allow for greater total biomass accumulation by fostering sufficient root growth to support large vegetative growth. High temperatures also reduce vegetative growth by shortening the vegetative growth stage and by speeding up development without a comparable increase in biomass accumulation. Van Dobben asserts 'the effect of the temperature on the length of the period from emergence to flowering is independent of the effect of temperature on growth'. Thus, increased temperatures reduce the time for plant development resulting in much smaller plants.

The cotyledons are also very important to plant growth. Veierskov (1985) suggests that the cotyledons provide important factors which the plant is not able to synthesize photosynthetically. This is supported by the observation that, when the cotyledons are removed, seedling growth is retarded and increased light intensity is not able to compensate for the loss. Veierskov (1985) found that when the cotyledons were removed, plant growth either stopped or slowed considerably. When the cotyledons were removed from 5-day-old seedlings, growth rate slowed to approximately one-third after 11 days. Leaf number and root growth were much less affected. He also found that the growth of pea seedlings was dependent on seed size. This was most noticeable in increased leaf number and shoot length. High irradiance allowed for increased CO_2 fixation of leaves.

Veierskov (1985) suggests that cotyledons regulate plant growth by supplying nutrients and controlling respiration. This effect is carried long into development and cannot be overcome by photosynthesis. Reduced growth in plants from which the cotyledons had been excised was partly due to the increased respiration rate.

INITIATION OF FLOWERING

Three criteria can be used to determine flowering behaviour: (i) flowering time, which is defined as the total number of days from sowing to opening of the first fully developed flower (corolla colour visible); (ii) flowering node, defined as the stem node number at which the first flower is initiated (the cotyledonary node being assigned the number 0); and (iii) flower initiation time, defined as the total number of days from sowing until initiation of the first floral axillary bud on the shoot meristem (Murfet, 1977). Flower initiation time necessarily requires microscopic examination of dissected shoot apices. Flowering time and flowering node are usually positively correlated, but this correlation can be negated by genetic or environmental factors. These factors result in floral bud abortion or influence plant growth (Murfet, 1977).

The pea flower (Fig. 12.4) is described as papilionaceous because of its superficial resemblance to a butterfly when the petals unfold. The calyx of

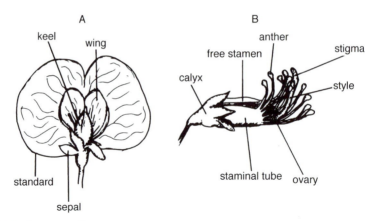

Fig. 12.4. Typical pea flower.

the flower consists of five basally united sepals. The corolla consists of two wing petals, a large, showy standard petal and a keel, which is formed by fusion of two petals early in floral bud development. The keel surrounds the anthers, stigma, style and ovary. There are 10 stamens in the complete pea flower, nine of which have joined filaments and one that remains free. The style is densely covered with trichomes, and terminates with the stigma which is surrounded by numerous papillae. Within the ovary, two rows of ovules alternate on the left and right halves of the fused margins of the carpel. The enclosing nature of the keel around the stamens and carpel ensures that the pollen is deposited on the stigma to effect self-pollination (Gritton, 1980). After germination, the pollen tube grows through the style and micropyle and releases sperm into the embryo sac, and, by double fertilization, the embryo and endosperm are formed.

Flower buds initiate on the apical meristems approximately 20 days before they become visible (Fig. 12.5). Flower buds at the time of initiation are enclosed by six leaf primordia which expand and eventually expose the developing flowers. As mentioned, flowering begins at the lower nodes and continues up the stem. Flowers may appear as early as the sixth node in very early flowering cultivars and as late as the 18th node in late flowering cultivars. In general, plant types that flower in the ninth to 11th nodes are considered early cultivars. Those that flower from the 12th to the 14th nodes are considered mid-season cultivars and those that flower at higher nodes are considered late maturing cultivars. Pea fields in full bloom (Fig. 12.6) are generally a 'carpet' of white. Pea flowers are formed from axillary buds rather than the terminal shoot meristem. Thus, the plants are classified as having an indeterminate growth habit. There are a few fasciated cultivars which are characterized by thickened stems and a higher number of flowers. These fasciated types have a tendency to be determinate (Knott, 1985).

'The lowest node at which flower initiation occurs, flowering node 5 (FN 5), is not altered by environment; although the subsequent develop-

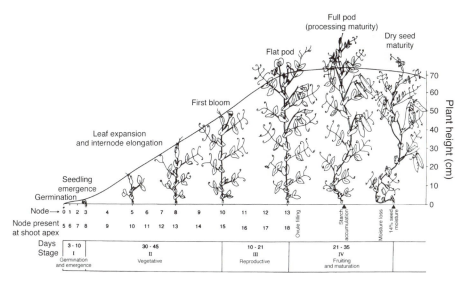

Fig. 12.5. Pea growth stages from planting to dry seed maturity (modified from Meicenheimer and Muehlbauer, 1982).

Fig. 12.6. Field of peas in full bloom.

ment of the bud may be greatly affected. The early-late cut-off point (FN 13) suggested in the field classification above roughly coincides with the point at which the FN starts to show a marked response to environment' (Murfet, 1977). For a detailed description of the genetic control over flowering time, refer to Murfet (1977) and Kelly and Davies (1986).

The indeterminate nature of peas allow for flowers to be present on higher nodes while pods on the lower portions of the plant are beginning to mature. This is a serious concern for freezing and canning peas where processors want the majority of the peas to be at approximately the same developmental stage at harvest (Wilkins *et al.*, 1991). Concentration of maturity is less demanding for dry peas where maximum yields are more important. However, the indeterminate nature of most dry pea cultivars often leads to size and quality variations.

In the field, various phases of floral development have been described (Meicenheimer and Muehlbauer, 1982). Initially, flowers appear as small green pointed structures, known as the sepal phase, during which the five united sepals totally enclose the remainder of the flower even when the peduncle has elongated and is readily visible. The next phase is the closed petal phase and begins with the expansion of the corolla past the sepal margins. The anthers dehisce during this phase beginning with the anther on the free stamen followed by the other nine. Since the keel encloses the stamens and carpel during pollen dehiscence, pollen is deposited in large quantities on the stigmatic region resulting in nearly complete self pollination. Rates of out crossing have been estimated at less than 1% (Gritton, 1980).

POD AND SEED DEVELOPMENT

Pate and Flinn (1977) describe in detail the development of the pea fruit and seed. Initial growth of the fruit precedes the exponential growth of the enclosed seeds. The pea fruit passes through two developmental stages known as 'flat pod' and 'round pod' (see Fig. 12.5). The 'flat pod' stage begins soon after anthesis and is characterized by elongation and increased width of the pods. The walls of the fruit thicken as the fruit inflates into a hollow envelope. This development is characterized by differential growth of the outer and inner layers of cells. The endocarp, the layer of cells lining the inner cavity of the pea pod, is composed of sclerified cells which form a fibrous layer. The lack of sclerified cells in pod tissue can be brought about by the action of the p and v genes. Each of these genes reduce the lignification of the pods while the presence of both p and v nearly completely eliminate pod parchment. Cultivars are in use that produce pods that lack parchment and are completely edible. The most widely known and used edible podded type cultivars are the flat podded 'sugar' peas used in Oriental style cooking. Another more recently developed type, the so-called 'snap pea' type, combines the action of the p and v genes with the n gene for cylindrically shaped pods. That particular combination of genes produces fibreless and cylindrically shaped pods that are completely edible.

Hole and Scott (1984) reported on the effects of light and temperature on pea fruit extension rate. They found that in the short term, pea fruit growth was lower in the light than in the dark. However, over a long

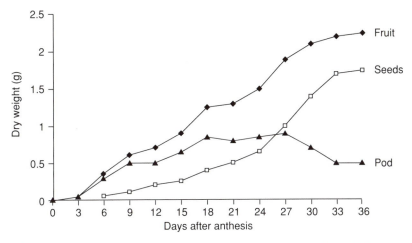

Fig. 12.7. Growth of pea fruit and its component parts (modified from Flinn, 1974).

period, extension rates were greater in the light than in the dark. This is partially explained by an increase in fruit temperature of 2°C during light periods. The short-term changes may depend on the water status of the pod. 'In the long term, extension is related to dry matter accumulation, but this relationship varies with both temperature and light' (Hole and Scott, 1984).

Maximum fresh weight of the pea fruit is attained at approximately 20 days after anthesis or at the end of the inflation phase of growth (Hedley and Ambrose, 1980) when the levels of starch, sucrose, protein and soluble nitrogen are at their highest (Bisson and Jones, 1932; McKee *et al.*, 1955; Flinn and Pate, 1968) (Fig. 12.7). As the pod elongates, the peduncle curves downward from the weight of the pod. A genetically controlled reduction in leaf area increased the photosynthetic CO_2 uptake potential of the fruit in the initial 18 days post anthesis (Harvey, 1978). Maximum pod wall dry weight coincides with the onset of seed filling at about 15 days after anthesis (Harvey, 1978).

The end of the 'flat pod' stage corresponds to the final stage in seed abortion (FSSA) which is defined as the stage in seed development after which the seeds will grow to maturity. It is accepted that the FSSA corresponds to seed length of approximately 6 mm and the end of cell division in the embryo (Duthion and Pigeaire, 1991). Ney *et al.* (1993) concluded that the FSSA corresponds to the beginning of seed fill and agrees with the findings of Duthion and Pigeaire (1991). Seeds in the centre of the pod are least likely to abort (Linck, 1961).

The FSSA (about 20 days after anthesis) marks an important stage in seed development. Several events occur at this time; the liquid endosperm is exhausted as the embryo fills the embryo sac, cell division in the cotyledons is completed, RNA and protein synthesis increases, and starch accumulation is initiated (Smith, 1973). As the seeds fill, the cotyledons act as a strong sink for protein, starch and phosphate. They soon consume the endosperm and enlarge to occupy the entire interstitial cavity (Dure, 1975; Hedley and Ambrose, 1980).

The 'round pod' stage begins at the FSSA and lasts until maturity. It corresponds to the period of seed filling (Le Deunff and Rachidian, 1988) and is characterized by the pods taking on a rounded appearance as the seeds fill the inner cavity. A colour change in the pods and seeds marks the end of the round pod stage and the beginning of seed maturation.

Hedley and Ambrose (1980) and LeDeunff and Rachidian (1988) describe seed growth as being composed of three phases of rapid growth separated by two lag phases. The initial growth phase lasts about 14–19 days after anthesis and is characterized by the formation of the embryo and surrounding structure. The growth of the testa and accumulation of endosperm is maximized during this phase. Growth rate of the embryo is very low especially in round-seeded types. The moisture content of the seed at this phase is nearly 85% (LeDeunff and Rachidian, 1988). The first lag period is characterized by a decline in the growth rate of the testa and endosperm and lasts no longer than 2 days.

The second growth phase is shorter for wrinkled than round seeded genotypes. This may be due to the embryos of the round-seeded types needing to catch up in development to those of the wrinkled seeded types. This phase is characterized by a reduced growth rate of the endosperm and a rapid increase in embryo fresh weight. As a result of increased embryo growth, the volume of the liquid endosperm declines. The volume of the embryo and embryo sac increases at different rates (Hedley and Ambrose, 1980). 'The seed can be visualized as two sinks – the testa and embryo, each competing for a common source – endosperm. Initially the developing testa would compose the larger sink and would therefore grow relative to the embryo. At the point of maximum endosperm content the two sinks would be equivalent, and after this point the embryo would compose the greater sink and grow relative to the testa' (Hedley and Ambrose, 1980). This is also the phase where the cotyledons begin to accumulate storage materials. The moisture content drops from 85 to 55% (LeDeunff and Rachidian, 1988). This phase is followed by a second lag phase where there is a decline in the absolute growth rate of the testa and embryo with continued disappearance of the endosperm.

The third growth phase is characterized by the maximum absolute seed growth rate which subsequently declines towards maturity. LeDeunff and Rachidian (1988) characterize this third phase, the period of desiccation of the seed and whole plant. The final moisture content of the mature plant and seed range from 14 to 18%.

Pea fruit and seed growth are primarily supported by photoassimilate produced in the vegetative organs of the plant. Approximately 66% of the assimilate accumulated by the seed is derived from the leaf structures directly subtending the pod, leaflets (29%) and stipules (12%) and from the pod itself (25%) (Harvey, 1978). The roots provide approximately 14% and the remainder is translocated from other nodes and other vegetative structures on the plant. The pods are capable of assimilating CO_2 from the atmosphere and they function to recycle CO_2 respired from the developing seeds and store assimilate and nutrients for the developing seeds (Harvey, 1978). The phloem bears the majority of the translocated substrates to the seed and delivers the assimilate to the seed coat or testa through the funiculus (Murray, 1977). The phloem lacks transfer cells at the junction with the testa. However, since transfer cells are important in the loading of translocated substrates into the phloem, their absence in the seedcoats may facilitate unloading in the seed coats (Murray, 1977).

Sucrose is hydrolysed in variable proportions to glucose and fructose upon entry into the testa (Murray, 1977). The seed coat alters certain metabolites before transmission to the embryo sac and embryo, primarily N and P constituents in the phloem (Murray, 1977). Sucrose uptake appears to occur according to a proton-cotransport concept for the secondary transport of sucrose during phloem loading (Giaquinta, 1983). It shows a two-component kinetics: a saturable system and nonsaturable or two affinity system (Estruch et al., 1989).

Estruch et al. (1989) suggest that growth regulators may function in sucrose accumulation and loading. They found that transport from source to sink was activated by GA_1 and GA_3 which increases phloem sink unloading (ovary). A strong increase in sucrose exported from the source leaf was observed when the GAs reached the leaf adjacent to the ovary. This suggests that sucrose loading at the source could also be activated by GAs (Estruch et al., 1989). Williams and Williams (1978) suggest that the pod tissue produces a stimulus which has a remote effect on export of assimilates from the leaf. They also suggest that a hormone may be responsible for this effect and that it may act directly on the loading of assimilate from the leaf into the transport system or alternatively loading may be indirectly influenced by the rate of unloading at the sink. This theory is supported by the work of Jahnke et al. (1989) who observed an increase in the amount of radiolabelled carbon entering the unfertilized ovary with the application of gibberellic acid. Studies by Larcia-Martinez et al. (1987) and Poreto et al. (1988) provide additional evidence that gibberellins particularly GA_1 and GA_3 are important in the development of pea ovaries.

DESICCATION AND MATURITY

The final stage in seed growth is physiological maturity followed by harvest maturity. Physiological maturity is attained when the vascular connection between the pod and mother plant is severed (Le Deunff and Rachidian, 1988). It is characterized by a colour change in the seeds and pods. This leads to a progressive loss of moisture down to approximately 18–14%, causing seeds to dry from the outside to the inside. Tolerance to desiccation is necessary to maintain the integrity of the cells within the seed and to ensure successful germination. In peas, there is a gradual build-up of desiccation tolerance during the second phase of seed growth (Le Deunff and Rachidian, 1988). The slow dehydration during the seed filling stage from 85 to 55% prepares the seed for the rapid dehydration after physiological maturity (Le Deunff and Rachidian, 1988). Matthews *et al.* (1978) state that habituation to desiccation occurs in a short period of time and is preceded by a drop in oxygen uptake which corresponds to a fall in seed moisture. Developing pea seeds can withstand desiccation only after the sugar content and, consequently, aerobic respiration of the seeds has fallen (Matthews *et al.*, 1978). The ability of desiccated seeds to retain cell solutes has been found to improve the later the seeds are harvested in their development (Matthews *et al.*, 1978).

Abscisic acid (ABA) accumulates at high concentrations in legume seeds (King, 1982) and regulates the production of storage proteins. However, it appears to have no effect on the pea storage proteins vicilin and legumin (Barratt *et al.*, 1989). ABA seems essential not only for stimulation of filling (Schussler *et al.*, 1984; Finkelstein *et al.*, 1985) but also for precocious germination inhibition (Obendorf and Wettlaufer, 1984). Barratt *et al.* (1989) found that ABA (0.1 mol m^{-3}) was able to almost completely stop precocious germination, but only in a medium of high osmotic pressure (5%+ sucrose). Despite the ability of ABA to prevent precocious germination, Barratt *et al.* (1989) determined that high osmotic pressure is more effective than exogenous ABA and suggest that ABA acts indirectly to prevent precocious germination.

FACTORS AFFECTING PRODUCTIVITY

REPRODUCTIVE GROWTH AND PARTITIONING

Jeuffroy and Warembourg (1991) report that two-thirds of the carbon required by the ripening seeds at a given node comes from stipules, leaflets and tendrils of the node subtending the pod. Translocation of assimilates from nodes other than the one subtending a pod can occur, but is in small amounts. It is unlikely that all flowering nodes have the same carbon budget as they are in different stages of development at any one time. Assimilate partitioning may depend on the relative developmental stages of the pods, the number of reproductive nodes and the number of pods on the plant (Jeuffroy and Warembourg, 1991).

CLIMATIC REQUIREMENTS FOR YIELD PRODUCTION

Peas are grown successfully in cool but not excessively cold climates. Average growing season temperatures of 13–18°C produce optimum yields. Seeds will germinate from 4 to 24°C. Although pea plants are tolerant of mild frost in the vegetative stage, frost at flowering causes heavy pod losses and frost at pod set produces deformed and discoloured seeds. Temperature above a base of 25.6°C during blooming and fruit filling will negatively correlate with yield (Pumphrey *et al.*, 1979). High temperatures reduce the number of floral structures and fruits per plant (Davies *et al.*, 1985).

Evenly distributed rainfall (about 800–1000 mm annum^{-1}) will maximize pea production (Kay, 1979), although the crop is grown successfully in areas of much lower rainfall (e.g. the crop in Australia where rainfall may be as low as 400 mm annum^{-1}). In areas of low rainfall, the soil must be deep and capable of storing moisture so that the crop may be successfully irrigated, in which case 818–1018 m^3 of water per hectare is required during the growth cycle. Under high temperature conditions, maximum economic yields are reported to be obtained when soil moisture content is kept at 60% of field capacity from emergence to just prior to flowering and at least 90% of field capacity during flowering. Peas are sensitive to short exposures to anaerobic soil conditions, especially just before flowering and during fruit filling (Davies *et al.*, 1985; Ney *et al.*, 1994). Damage from waterlogging is generally more severe at warm temperatures.

Peas are cultivated in a wide range of well-drained soil types. The crop grows best on loams, clay loams or sandy loams overlying clay. Yields tend to be reduced when the crop is grown in sandy soils that do not retain moisture. Peas grow well on soils with moderate levels of calcium and neutral or slightly acid pH.

Although peas are not considered a deep-rooted crop, they tend to be intolerant of shallow or poorly drained soils, possibly because of the increased incidence of root disease. Peas have a taproot system with numerous lateral roots that form a circle 50–75 cm in diameter and are known to penetrate depths of 100–120 cm (Torrey and Zobel, 1977; Kay, 1979). The crop is not well suited to the extremely leached soils characteristic of high rainfall areas of the tropics and subtropics because of acidity and high temperatures.

MINERAL NUTRITION

Adequate amounts of each of the 14 essential plant nutrients are necessary for satisfactory pea yield (for additional information on mineral nutrient requirements see Muehlbauer and Summerfield, 1988). Certain nutrient deficiencies may result in poor seed quality, reduced pod formation and reduced nitrogen fixation by the root nodules, and an overall reduction in plant growth and yield. Nitrogen, phosphorus and potassium accumulate in large quantities in peas and a crop of 1000 kg of dry seed may contain up to 43 kg N, 4.2 kg P, 9.2 kg K, 0.6 kg Ca, 1.2 kg Mg and 0.8 kg S.

NITROGEN

Effectively nodulated pea crops seldom respond to applications of inorganic N fertilizer because such applications reduce nodulation and N_2 fixation, promote weed growth and delay crop maturity (Bezdicek *et al.*, 1982). The 'nitrogen hunger phase' is often experienced by pea crops that are seeded in cool and wet soil, before the advent of significant symbiotic N_2 fixation. The lack of nitrogen at this early growth phase can be corrected by the application of a small starter dose of inorganic nitrogen fertilizer placed beneath, but not in contact with the seeds at the time of planting. Applications of an appropriate strain of *Rhizobium leguminosarum* is essential when peas are planted into a field for the first time or if peas have not been planted in that field for several years. *Rhizobium leguminosarum* is specific for a number of genera in the *Viceae* tribe of legumes that include *Lens, Lathyrus,* and *Vicia*. Longevity of rhizobia in soils planted to peas is long lived; however, it is recommended that inoculum be used often when planting pea crops. Special care is needed in the choice of fungicides for seed treatment and the timing and method of application to prevent potential toxicity to *Rhizobium* (Roughley, 1980).

Large concentrations of residual N in the soil, high salinity, and extremes of pH inhibit nodulation and N_2 fixation. Salinity can increase in the root zone due to the formation of a 'plow pan' caused by compaction from heavy machinery. Impervious to water and root penetration, a plow pan layer causes the accumulation of toxic concentrations of salt in the root zone. Pea crops affected by salt accumulations are especially vulnerable to disease attack. The effects of the plow pan layer can be reduced by deep tillage that will also allow free percolation of water and leaching of toxic salts. Crop rotations with strongly taprooted crops can also reduce the effects of a plow pan layer.

PHOSPHORUS

Peas respond to P fertilization where soils are deficient, as is common in severely eroded soils. Rates of P fertilization can be readily determined by a calibrated soil test. In most instances, a broadcast application and incorporation of between 44 and 66 kg ha^{-1} P_2O_5 are made in early spring to correct deficiencies. Where equipment is available, however, excellent responses to small amounts of P are obtained by placing the fertilizer in bands beneath the seeds, thereby obtaining good uptake by the pea plants, but not creating a high fertilizer salt concentration in the immediate vicinity of salt-sensitive pea seedlings (Muehlbauer and Dudley, 1974; Koehler *et al.*, 1975). Fertilizer salt concentration, however is not a problem where triple superphosphate is used.

POTASSIUM

Where soil tests indicate potassium (K) deficiencies, K_2O applications of about 22 kg ha^{-1} are beneficial. However, K applications should be made according to rates determined by trials for the area in question.

SULPHUR

Where deficiencies are detected, sulphur is usually applied to other crops grown in rotation with peas at the rate of about 17–22 kg ha^{-1}.

MAGNESIUM

Magnesium (Mg) should be applied where soils contain less than 0.5 mmol available per 100 g of soil.

TRACE ELEMENTS

Deficiencies of one or more of the trace elements, such as manganese, iron, copper, zinc, boron and molybdenum may be corrected by applications of animal manures that usually contain small quantities of these elements. If animal manures are not available, applications of organic amendments to soil can benefit the pea crop.

MANGANESE

A deficiency of manganese (Mn) can go undetected but is often character-ized by retarded growth and necrotic spotting of the leaves. In severe cases, the seeds take on a symptom known as 'marsh spot' characterized by a necrotic spot on the adaxial surface of the cotyledons (Gane *et al.*, 1984).

IRON

Severe iron deficiency can be overcome by foliar applications of ferrous sul-phate. Two applications each at the rate of 0.9 kg ha^{-1} are sufficient to cor-rect deficiencies.

MOLYBDENUM

Correction of molybdenum (Mo) deficiencies is usually accomplished by seed treatment at the rate of 35 g ha^{-1} in the form of sodium molybdate with a 'sticker' to ensure uniform adherence to the seeds. Broadcast applications of ammonium molybdate $((NH_4)_6MoO_7)$ at the rate of 1.1 kg ha^{-1} to the soil with gypsum have also been used successfully to correct deficiencies. Fertilization with NaMoO$_4$ at the rate of 0.5 kg ha^{-1} can also be used to prevent deficiencies.

BORON

Extensively cropped soils commonly respond to small applications of boron (B). Applications of B between 2 and 4 kg ha^{-1} are beneficial to legume crops. Boron fertilizer, usually borated gypsum, is broadcast applied and incorporated by plowing or disking during seedbed preparation. Banded applications are often toxic. Since peas are extremely sensitive to excess of B, applications should not exceed 4 kg ha^{-1}.

pH

Peas are sensitive to both acid (below pH 5.5) and alkaline (above pH 9.5) soils. Optimum nutrient uptake occurs in (slightly acidic) soils (pH 6.5) because nutrient availability at that pH is most favourable (Halvorson, 1982). Correction of soil acidity by liming results in improved growth, reduced root disease, and improved yields. Excessive liming of acid soils, however, may result in manganese deficiency since manganese becomes less available in alkaline substrates (Kay, 1979).

Nutritional deficiencies and toxicities in peas can be diagnosed by soil testing, visual plant symptoms, or by plant tissue analysis. Unfortunately, the foliar symptoms of plant deficiencies are not distinct unless they are severe, and are often confused with other pathological or physiological disorders.

PHYSIOLOGICAL DISORDERS

WATER CONGESTION

This is a term used for an abnormality thought to be brought about when peas are grown under conditions of excess soil moisture, relatively high temperatures and high humidity. Typically, the terminal edges of the young leaflets discolour, become deformed and turn brown. Substantial areas of the pea foliage may be affected. The abnormality is generally considered to be of minor importance.

HOLLOW HEART

Hollow heart is a disorder characterized by a sunken area in the adaxial surface of the cotyledons of the pea seed. The condition is thought to be brought about by rapid drying as the seed approaches maturity. The disorder causes poor seedling vigour and poor stands as the cotyledons are often attacked by soilborne damping-off pathogens.

'BLONDE' PEAS

'Blonde' pea is a disorder that is characterized by variations of light green and yellow (blonde) peas in green peas harvested for freezing or canning. The condition is brought about by shading of the pea pods by excessive vine and leaf growth. Also, the condition may be caused by lack of sunshine and cloudy conditions during pod filling. There appears to be genetic variation among pea cultivars in susceptibility.

'BLEACHED' PEAS

'Bleached' pea is a disorder, sometimes confused with 'Blonde', that is caused by excess moisture after the peas have reached dry seed maturity. Apparently the dry peas take up moisture leading to enzyme activation and a breakdown of chlorophyll. The condition causes colour variations and is a market downgrading factor in dry peas. Some cultivars have a degree of resistance but when conditions are favourable for bleaching, all cultivars can be affected.

PEA TYPES

Pea cultivars vary greatly in size and shape of the plants, pods and seeds. Growth habit of all pea cultivars is indeterminate but recently released cultivars have a tendency toward determinacy. The move towards determinacy allows a greater majority of the peas to be in the proper stage for processing. The immature peas of the canning and freezing types are sugary when in the proper stage for processing but become wrinkled upon dry seed maturity. Canning type cultivars have a light green colour when in the processing stage while the freezer type cultivars have a dark green appearance at the same stage. Breeders have attempted to develop cultivars that can be used for both canning and freezing. Peas for home gardening are generally characterized as having large pods and ovules with a high number of peas per pod. Often home garden peas have an indeterminate plant type so as to extend the harvesting period. The various types of peas and their uses are described as follows.

DRY PEAS

Dry peas are generally characterized by an indeterminate growth habit and smooth and starchy seeds. Seeds can be smooth green or yellow. Regular green or yellow dry peas are usually grown as a field crop in rotation with cereals. They are harvested by direct combining (Fig. 12.8) when seed moisture content is about 12% or less. Dry peas are used either as whole green or yellow peas or they are decorticated and split. Split peas are generally used in soup making.

Fig. 12.8. Harvesting dry peas by combine.

MARROWFAT PEAS

Marrowfats are a large, flattened and somewhat dimpled green pea that most likely originated in England in the sixteenth or seventeenth century. They are favoured in that country where peas generally are very popular. Marrowfat peas differ completely from other dry pea types in their seed type and quality traits. The plant habit of marrowfats is typically dwarf with very large leaves and heavy vines that are generally short and usually well branched. They are generally mid-season or late to flower and as a result are late maturing when compared to Alaska types. The seeds are about twice the size of a typical Alaska pea. In the UK, marrowfats are usually reconstituted and canned and are often referred to as 'mushy peas' when used. Marrowfats are used extensively in Southeast Asian countries.

In those countries, marrowfats are either fried or roasted with various flavours and packaged in small containers for use as a snack food. This particular use has been expanding in that area of the world and the market potential for marrowfats seems to be quite large. Good quality marrowfats are large seeded with good dark green colour, although some countries have the option of using artificial colouring to improve appearance.

EDIBLE PODDED TYPES

Edible podded types have been in use for centuries. Recently, the so-called 'snap pea' was developed and has become popular. The original edible podded pea has flattened pods and is generally used in the very immature stage, in oriental cooking such as a stir fry. The 'snap pea' type has a cylindrical pod that is thick, fleshy and breaks easily when bent, similar to a snap bean. Snap peas are usually picked and consumed when the seed in the pod has developed to nearly its full size.

FREEZER PEA TYPES

Freezer pea types generally have a dark green testa that imparts the typical dark green colour to frozen peas. Similar to canner peas, they are harvested (vined) when tenderometer readings are between 95 and 105. The tenderometer is a device used to measure maturity and suitability for producing a high quality product. Tenderometer readings above 105 indicate progressive increases in starch content in the peas and a concurrent reduction in edible quality. The time from planting to processing maturity is estimated by the processing industry by using accumulated heat units. Cultivars considered to be early, mid-season and late maturing require 1200, 1350 and 1500 accumulated heat units, respectively, calculated on a base temperature of 40°F (see Chapter 4). The timing of planting of pea cultivars is critical to the continuous operation of pea processing factories. Soon after vining (Fig. 12.9) the raw product is quickly transported to the processing facility, usually within the hour, and quickly and efficiently processed. Crops that are quickly processed retain quality while delays reduce quality of the finished product.

CANNER PEAS

Canner peas typically have a lighter green testa which imparts the lighter colour of canned peas when compared to freezer peas. The darker coloured freezer peas are usually unsuited to canning because they tend to turn brown from the high temperature when processed and canned. Canner peas are also harvested at tenderometer readings of 95–105. Rapid processing is necessary to retain quality.

Fig. 12.9. Harvesting (vining) processing peas by machine.

AUSTRIAN WINTER PEAS

Austrian winter peas have pigmented stems, flowers and seeds and are sometimes referred to as *Pisum arvense* or as *P. sativum* spp. *arvense*. This particular type of pea is entirely cross compatible with *P. sativum* and should not be considered a different species. They are usually grown as an autumn-sown crop in relatively cold regions. They have winter hardiness comparable to winter wheat and survive most winters. The crop is also used as green manure usually at more southern locations. The crop, harvested as dry seed, has been used in the Orient as a filler in the manufacture of An-Paste, a sweet confection commonly used in Japan. The crop is often used as feed for pigeons and other animals. In times of shortages of smooth yellow peas, the Austrian winter pea can be decorticated and split to produce yellow split peas. Nearly all the production of Austrian winter peas is in the Pacific Northwest.

CONCLUDING REMARKS

Peas have had a remarkable history since domestication in the Near East region. They have been distributed around the world and can be successfully produced in a wide range of environments. The versatility as a fresh vegetable, dry pulse and as feed for livestock has been evident in the recent rapid increase in production, which has nearly doubled in the past ten years. Besides being a popular food and feed crop, the short duration of the pea growth cycle fits well into rotations with cereals and other field crops. The use of peas in rotations is credited with improving soil structure and nutritional status while providing a break crop against diseases affecting other crops in the rotation.

Peas had a prominent role in the discovery of the laws of genetics by Gregor Mendel in the mid-1800s and to this day peas are widely used in genetic and physiological research.

New types of peas and new alternative uses for the crop are being developed. With the extensive genetic variation available in pea germplasm the prospects for continued expansion of types and uses appears bright.

ACKNOWLEDGEMENTS

The authors gratefully acknowledge Ms Jeannette Seward, Secretary to the USDA-ARS Grain Legume Genetics and Physiology Research Unit and Ms Suzette Greenhagen, Technical Assistant, for their assistance in preparation of the manuscript and Ms Ruth Younce for her assistance in the preparation of the figures.

REFERENCES

Bain, J.M. and Mercer, F.V. (1966) Subcellular organisation of the developing cotyledons of *Pisum sativum* L. *Australian Journal of Biological Science* 19, 49–67.

Barratt, D.H.P., Whitford, P.N., Cook, S.K., Butcher, G. and Wang, T.L. (1989) An analysis of seed development in *Pisum sativum*. VIII. Does abscissic acid prevent precocious germination and control storage protein synthesis? *Journal of Experimental Botany* 40, 1009–1014.

Bezdicek, D.F., Root, C., Smith, S. and Muehlbauer, F.J. (1982) Importance of *Rhizobium* N-fixation to grain legume crops and succeeding cereals. In: *Proceedings, Palouse Symposium on Dry Peas, Lentils, and Chickpeas, Moscow, Idaho, February 23–24, 1982*. College of Agriculture Research Center, Washington State University, Pullman, pp. 213–222.

Bisson, C.S. and Jones, H.A. (1932) Changes accompanying fruit development in the garden pea. *Plant Physiology* 7, 91–105.

Chin, T.Y., Poulson, R. and Beevers, L. (1972) The influence of axis removal on protein metabolism in cotyledons of *Pisum sativum* L. *Plant Physiology* 49, 482–489.

Davies, D.R., Berry, G.R., Heath, M.C. and Dawkins, T.C. (1985) Pea (*Pisum sativum* L.) In: Summerfield, R.J. and Roberts, E.H. (eds) *Grain Legume Crops*. Collins, London, pp. 147–198.

Dure, L.S. (1975). Seed formation. *Annual Review of Plant Physiology* 26, 259–278.

Duthion, C. and Pigeaire, A. (1991) Seed length corresponding to the final stage in seed abortion of three grain legumes. *Crop Science* 31, 1569–1573.

Estruch, J.J., Pereto, J.G., Vercher, Y. and Beltran J.P. (1989) Sucrose loading in isolated veins of *Pisum sativum*: regulation by abscissic acid, gibberellic acid, and cell turgor. *Plant Physiology* 91, 259–265.

Finklestein, R.R., Tenbarge, K.M., Shumuray, J.E. and Crouch, M.L. (1985) Role of ABA in maturation of rapeseed embryos. *Plant Physiology* 78, 630–636.

Fletcher, H.F., Ormrod, D.P., Maurer, A.R. and Stanfield, B. (1966) Response of peas to environment. I. Planting date and location. *Canadian Journal of Plant Science* 46, 77–85.

Flinn, A.M. and Pate, J.S. (1968) Biochemical and physiological changes during maturation of fruit of the field pea (*Pisum arvense* L.). *Annals of Botany* 32, 489–495.

Gane, A.J., Biddle, A.J., Knott, C.M. and Eagle, D.J. (1984) *The PGRO Pea Growing Handbook*, 5th edn. Processors and Growers Research Organisation, Peterborough, UK.

Giaquinta, R.T. (1983) Phloem loading of sucrose. *Annual Reviews in Plant Physiology* 34, 347–389.

Goldenberg, F.B. (1965) *Afila, a New Mutation in Pea (Pisum sativum L.) Boletin Genetica*. Instituto de Fitotecnia Castelar, Argentina, pp. 27–28.

Gritton, E.T. (1980) Field pea. In: Fehr, W.R. and Hadley, H.H. (eds) *Hybridization of Crop Plants*. American Society of Agronomy – Crop Science Society of America. Madison, Wisconsin, pp. 347–356.

Haan, H. De. (1927) Length factors in Pisum. *Genetica* 9, 481–498.

Halvorson, A.R. (1982) Soil testing and nutrient requirements for grain legumes and soil compaction problems. In: *Proceedings, Palouse Symposium on Dry Peas, Lentils and Chickpeas. Moscow, Idaho. February 23–24.* College of Agriculture Research Center, Washington State University, Pullman, pp. 211–212.

Harvey, D.M. (1972) Carbon dioxide photoassimilation in normal-leaved and mutant forms of *Pisum sativum* L. *Annals of Botany* 36, 981–991.

Harvey, D.M. (1978) The photosynthetic and respiratory potential of the fruit in relation to seed yield of leafless and semi-leafless mutants of *Pisum sativum* L. *Annals of Botany* 42, 331–336.

Harvey, D.M. (1980) Seed production in leafless and conventional phenotypes of *Pisum sativum* L. in relation to water availability within a controlled environment. *Annals of Botany* 45, 673–680.

Harvey, D.M. and Goodwin, J. (1978) The photosynthetic net carbon dioxide exchange potential in conventional and 'leafless' phenotypes of *Pisum sativum* L. in relation to foliage area, dry matter production and seed yield. *Annals of Botany* 42, 1091–1098.

Hedley, C.L. and Ambrose, M.J. (1980) An analysis of seed development in *Pisum sativum* L. *Annals of Botany* 46, 89–105.

Hedley, C.L. and Ambrose, M.J. (1981) Designing 'leafless' plants for improving yields of the dried pea crop. *Advances in Agronomy* 34, 225–277.

Hole, C.C. and Scott, P.A. (1984) Pea fruit extension rate I. effect of light, dark and temperature in controlled environment. *Journal of Experimental Botany* 35, 790–802.

Jahnke, S., Bier, D., Estruch, J.J. and Beltram, J.P. (1989) Distribution of photoassimilates in the pea plant: chronology of events in the non-fertilized ovaries and effects of gibberellic acid. *Planta* 180, 53–60.

Jeuffroy, M-H and Warembourg, F.R. (1991) Carbon transfer and partitioning between vegetative and reproductive organs in *Pisum sativum* L. *Plant Physiology* 97, 440–448.

Kay, D. (1979) Food legumes. Tropical Products Institute. *Crop and Product Digest* 3, 293–321.

Kelly, M.O. and Davies, P.J. (1986) Genetics and photoperiodic control of the elative routes of reproductive and vegetative development in peas. *Annals of Botany* 58, 13–21.

Kielpinski, M. and Blixt, S. (1982) The evaluation of the 'afila' character with regard to its utility in new cultivars of dry pea. *Agri Hortique Genetica* 40, 51–74.

King, R.W. (1982) Abscissic acid in seed development. In: Khan, A.A. (ed.) *The Physiology and Biochemistry of Seed Development, Dormancy, and Germination.* Elsevier, Amsterdam, pp. 157–181.

Knott, C.M. (1985) A description for stages of development of the pea (*Pisum sativum* L.). *Aspects of Applied Biology* 10, 379–393.

Koehler, F.E., Engle, C.G. and Field, E.T. (1975) Phosphorus fertilization, broadcast banding and starter. *Washington State University Cooperative Extension Bulletin in EM* 3295, 1–8.

Kruger, N.S. (1977) The effect of plant density on leaf area index and yields of *Pisum sativum* L. *Queensland Journal of Agricultural and Animal Sciences* 34, 35–52.

Lamm, R. (1937) Length factors in dwarf peas. *Hereditas* 23, 38–48.

Larcia-Martinez, J.L., Sponsel, V.M. and Laskin, P. (1987) Gibberellins in developing fruits of *Pisum sativum* cv. Alaska: Studies on their role in pod growth and seed development. *Planta* 170, 130–137.

Le Deunff, Y. and Rachidian, Z. (1988) Interruption of water delivery at physiological maturity is essential for seed development, germination and seedling growth in pea (*Pisum sativum* L.). *Journal of Experimental Botany* 39, 1221–1230.

Linck, A.J. (1961) The morphological development of the fruit of *Pisum sativum*, var. Alaska. *Phytomorphology* 11, 79–84.

Marx, G.A. (1974) A scheme for demonstrating some classical genetic principles in the classroom. *Journal of Heredity* 65, 252–254.

Matthews, S., Powell, A.A. and Rogerson, N.E. (1978) Physiological aspects of the development and storage of pea seeds and their significance to seed production. *Proceedings of the Eastern School of Agricultural Science at the University of Nottingham.* London, Butterworths, pp. 513–525.

McKee, H.S., Robertson, R.N. and Lee, J.B. (1955) Physiology of pea fruits. I. The developing fruit. *Australian Journal of Biological Science* 8, 137–162.

Meicenheimer, R.D. and Muehlbauer, F.J. (1982) Growth and developmental stages of Alaska peas. *Experimental Agriculture* 18, 17–27.

Monerri, C., Garcia-Luis, A. and Guardiola, J.L. (1986) Sugar and starch changes in pea cotyledons during germination. *Physiologia Plantarum* 67, 49–54.

Muehlbauer, F.J. and Dudley, R.F. (1974) Seeding rates and phosphorus placement for 'Alaska' peas in the Palouse. *Washington Agriculture Experiment Station Bulletin no. 794*, pp. 1–4.

Muehlbauer, F.J. and Summerfield, R.J. (1988) Nutrient requirements of dry peas. In: Plucknett, D.L. and Sprague, H.B. (eds) *Detecting Mineral Nutrient Deficiencies in the Temperate and Tropical Crops.* Westview Tropical Agriculture Series, no. 7. Westview Press, Boulder, CO, pp. 117–127.

Murfet, I.C. (1977) The physiological genetics of flowering. In: Sutcliffe, J.F. and Pate, J.S. (eds) *The Physiology of the Garden Pea.* Academic Press, London, pp. 385–430.

Murray, D.R. (1977) Acid phosphatase activities in developing seeds of *Pisum sativum* L. (peas). *Australian Journal of Plant Physiology* 4, 843–848.

Ney, B., Duthion, C. and Fontaine, E. (1993) Timing of reproductive abortions in relation to cell division, water content, and growth of pea seeds. *Crop Science* 33, 267–270.

Ney, B., Duthion, C. and Fontaine, E. (1994) Phenological response of pea to water stress during reproductive development. *Crop Science* 34, 141–146.

Norton, J.M. and Harman, G.E. (1985) Responses of soil microorganisms to volatile exudates from germinating pea seeds. *Canadian Journal of Botany* 63, 1040–1045.

Obendorf, R.L. and Wettlaufer, S.H. (1984) Precocious germination during in vitro growth of soybean seeds. *Plant Physiology* 76, 1023–1028.

Obrucheva, N.V. and Kovadlo, L.S. (1985) Two stages of intensification in the respiration of germinating pea seeds in proportion to increase of their hydration. *Soviet Plant Physiology* 32, 580–587.

Oram, P.A. and Agcaoili, M. (1994) Current status and future trends in supply and demand of cool season food legumes. In: Muehlbauer, F.J. and Kaiser, W.J. (eds) *Expanding the Production and Use of Cool Season Food Legumes*. Kluwer Academic, Dordrecht, The Netherlands, pp. 3–49.

Pate, J.S. and Flinn, A.M. (1977) Fruit and seed development. In: Sutcliffe, J.F. and Pate, J.S. (eds) *The Physiology of the Garden Pea*. Academic Press, London, pp. 431–468.

Peirce, L.C. (1987) *Vegetables: Characteristics, Production and Marketing*. John Wiley, New York, pp. 333–355.

Pereto, J.L., Beltran, J.P. and Larcia-Martinez, J.L. (1988) The source of gibberellins in the parthenocarpic development of ovaries on topped pea plants. *Planta* 175, 493–499.

Pumphrey, F.V., Ramig, R.E. and Allmaras, R.R. (1979) Field response of peas (*Pisum sativum* L.) for processing to precipitation and excess heat: yield prediction. *Journal of the American Society of Horticultural Science* 104, 548–550.

Pyke, K.A. and Hedley, C.L. (1983) The effect of foliage phenotype and seed size on the crop growth of *Pisum sativum* L. *Euphytica* 32, 193–203.

Roughley, R.J. (1980) Environmental and cultural aspects of the management of legumes and *Rhizobium*. In: Summerfield, R.J. and Bunting, A.H. (eds) *Advances in Legume Science*. HMSO, London, pp. 97–103.

Schussler, J.R., Brenner, M.L. and Brun, W.A. (1984) Abscissic acid and its relationship to seed filling in soybeans. *Plant Physiology* 76, 301–306.

Smartt, J. (1990) Pulses of the classical world. In: *Grain Legumes: Evolution and Genetic Resources*. Cambridge University Press, Cambridge, pp. 176–244.

Smith, D.L. (1973) Nucleic acid, protein, and starch synthesis in developing cotyledons of *Pisum arvense* L. *Annals of Botany* 37, 795–804.

Snoad, B. (1981) Plant form, growth rate and relative growth rate compared in conventional, semi-leafless and leafless peas. *Scientia Horticulturae* 14, 9–18.

Sutcliffe, J.F. and Bryant, J.A. (1977) Biochemistry of germination and seedling growth. In: Sutcliffe, J.F. and Pate, J.S. (eds) *The Physiology of the Garden Pea*. Academic Press, New York, pp. 45–82.

Torrey, J.G. and Zobel, R. (1977) Root growth and morphogenesis. In: Sutcliffe, J.F. and Pate, J.S. (eds) *The Physiology of the Garden Pea*. Academic Press, London, pp. 119–152.

van Dobben, W.H. (1962) Influence of temperature and light conditions on dry-matter distribution, development rate and yield in arable crops. *Netherlands Journal of Agricultural Science* 10, 377–389.

Veierskov, B. (1985) Pea seedling growth and development regulated by cotyledons and modified by irradiance. *Physiologia Plantarum* 65, 79–84.

Vincent, C.L. (1958) Pea plant population and spacing. *Washington State University Agricultural Experiment Station Bulletin no. 594.*

Wilkins, D.E., Kraft, J.M. and Klepper, B.L. (1991) Influence of plant spacing on pea yield. *Transactions of the American Society of Agricultural Engineers* 34, 1957–1961.

Williams, A.M. and Williams, R.R. (1978) Regulation of movement of assimilate into ovules of *Pisum sativum* cv. Greenfeast: a 'remote' effect of the pod. *Australian Journal of Plant Physiology* 5, 295–300.

Yoo, B.Y. (1970) Ultrastructural changes in cells of pea embryo radicles during germination. *Journal of Cell Biology* 45, 158–171.

Zohary, D. and Hopf, M. (1973) Domestication of pulses in the Old World. *Science* 182, 887–894.

Sweet Corn 13

D.W. Wolfe[1], F. Azanza[2] and J.A. Juvik[2]

[1]*Department of Fruit and Vegetable Science, Cornell University, 134A Plant Science Building, Ithaca, New York 14853–5908, USA*
[2]*Department of Natural Resources and Environmental Science, University of Illinois, 1201 West Gregory Drive, Urbana, Illinois 61801, USA*

Sweet corn (*Zea mays* L.), a diploid species with 2n = 20 chromosomes, is a member of the grass family (*Gramineae*), subfamily *Panicoideae*, of the tribe *Maydeae*. The exact origin of corn is not clear, although evidence suggests it is native to Central America. It has been proposed that maize is a domesticated form of teosinte (a wild grass endemic to Mexico and Guatemala). The main morphological differences between these taxa are likely to be the result of human selection during the process of domestication (Beadle, 1939). Doebley and Stec (1993) identified loci in five regions of the maize genome that could explain most of the key morphological differences between maize and teosinte, suggesting that a relatively small number of loci with large effects were involved in the early evolution of maize from teosinte. Corn was grown extensively by the Native Americans in pre-Columbian times and was taken to Europe by Columbus on his return voyage in 1493.

Although the origin of sweet corn cannot be associated with a definite point in time or a single place, it arose from a recessive mutation at what is known now as the *Su1* locus on chromosome 4 in the Peruvian race, Chullpi (Gallinat, 1971). When homozygous recessive, this allele gives sweet corn its characteristic sweetness and creamy texture by increasing the concentration of sucrose and phytoglycogen (a highly branched water-soluble polysaccharide) at harvest maturity (Boyer and Shannon, 1982). Other mutations may have occurred many times and in different races of corn. The spread of sweet corn was probably from Peru or Central America to Mexico, and eventually throughout the USA. During this period continued recombination occurred and adaptive gene complexes accumulated. Early descriptions of sweet corn indicate that it was morphologically similar to Northern Flints in plant type, having eight rowed ears and white cobs (Tapley *et al.*, 1934). A sweet corn called 'papoon' by the American Indians, that originated in the Susquehanna region of Pennsylvania, became popular with white settlers in the Northeastern USA in the late eighteenth century. It was sold commercially for the first time as 'Susquehanna' in 1832. By 1900, 63 sweet corn cultivars were being grown, and 'Golden Bantam' was introduced in 1902, marking the beginning of a trend for yellow-kernel types (early cultivars were white). In 1924, the first sweet corn hybrid, 'Redgreen', was released, and by 1947 hybrids accounted for 75% of the total crop grown (Singleton, 1948). Currently, almost all sweet corn cultivars are single cross hybrids.

Sweet corn, unlike field corn (dent, flint, flour, waxy and popcorn maize types), is harvested when the kernels are immature so that sweetness and pericarp tenderness are optimum for fresh consumption. Because the ear is physiologically immature when harvested, sweet corn is a highly perishable commodity. Respiration and metabolic rates are high (Hardenburg *et al.*, 1986), causing a rapid conversion of kernel sugar into starch and a loss of moisture. This loss of sweetness and tenderness after harvest will contribute to a rapid deterioration of quality unless special precautions are used. Once the crop is harvested, field heat must be removed to slow the respiration rate. Precooling sweet corn to 0°C will maximize quality retention. Both low temperature storage and controlled-atmosphere (CA) storage are used to maintain sweet corn quality after harvest.

Since the introduction of the first standard *su1* sweet corn hybrid, significant improvements have been made in yield, quality, seedling vigour and adaptation to mechanization. In the last two decades more emphasis has been given to the development of hybrids with mutant endosperms that have particularly high sugar levels and longer storage life (Kaukis and Davis, 1986). The first high-sugar or 'supersweet' cultivar (shrunken2) released more than 30 years ago was 'Illini Xtra Sweet' (see section on genotypes below).

Sweet corn is consumed as a fresh-market and processed vegetable. Most breeding programmes of seed producers address both markets, with emphasis on the processing hybrids. In contrast, publicly supported programmes mainly focus on fresh market sweet corn. Uniform ear size and shape, and kernel size and texture are important considerations when selecting hybrids for processing. Flavour, sweetness, storage life, and ear appearance (husk cover, light silk colour and tip fill) are characteristics of prime importance for fresh market use (Boyer and Shannon 1982; Kaukis and Davis, 1986).

Today, sweet corn is one of the most important vegetable crops in the USA. In 1989, 79% of the commercial sweet corn produced in the USA was for processing (57% canning and 43% freezing) and 21% for fresh market. While less in terms of land area, fresh market production accounts for 58% of the total farm value. Sweet corn is grown on more than 263,000 ha (650,000 acres) in 27 states, with a production of about 2.7 million tonnes (3.0 million US tons), and a value of $470 million (USDA, 1990). Wisconsin (27%) and Minnesota (22%) are the leading states for processing production, and Florida (33%), California (13%) and New York (12%) are the leading states for fresh market production.

Sweet corn acreage outside of North America is relatively small, and mostly dedicated to production for processed (canned) corn. Few statistics are available, but Italy, Spain, Israel and Hungary are important producers in Europe. Some sweet corn produced in Australia and New Zealand is shipped to expanding Asian markets, as well as sold locally. Individual prepared ears of fresh sweet corn have recently become a popular specialty vegetable in Japan, and there is some local production. China is viewed as an important new fresh sweet corn market. It is anticipated that the new longer storage life supersweet types will significantly expand the market for fresh sweet corn on a worldwide basis in the near future.

SWEET CORN GENOTYPES

Sweet corn genotypes are often categorized by use (wholesale fresh market, local direct-retail fresh market, frozen or canned), maturity class, and kernel colour (yellow, white and bicolour). Most processed corn, and fresh market corn grown for shipping and wholesale, are yellow types. The white and bicolour (bicolour = both white and yellow kernels on the ear) markets are regional preferences that are usually supplied by local growers.

There are several endosperm mutations in maize that, when present in the homozygous recessive state, produce qualitative and quantitative differences in kernel carbohydrate metabolism that affect flavour and storability (Coe, 1993). They include traditional sweet corn, sugary1 (*su1*), and shrunken2 (*sh2*), sugary enhancer1 (*se1*), brittle (*bt2*), amylose extender (*ae*), dull (*du*) and waxy (*wx*). These mutants all result in higher sugar and lower starch concentrations, although the magnitude of their effects vary (Evansen and Boyer, 1986). Figure 13.1 illustrates the primary starch biosynthetic pathways in maize kernels and the known enzymatic lesions caused by certain endosperm mutations.

In the last 20 years, the relatively new endosperm gene mutations, particularly *sh2* and *se1*, have been used to develop new commercial sweet corn hybrids with higher sugar levels in the kernels and slower sugar to starch conversion than traditional sweet corn (Wann, 1980; Boyer and Shannon, 1982). These tend to have an expanded harvest window and extended storage life. It is important to note that although these new 'supersweet' types maintain higher sugar levels during storage, pericarp tenderness can still deteriorate if not harvested at an optimum stage and stored properly. For fresh market sales where the husks are left on the ears, husk appearance will deteriorate if not rapidly cooled and stored at temperatures near 0°C.

The relatively new *sh2* cultivars now dominate the fresh market sweet corn acreage planted for long-distance shipping and wholesale markets. Traditional *su1* types are still used for most processing markets, although there is considerable evaluation of *sh2* and other genotypes being conducted. A combination of *su1*, *se1* and *sh2* genotypes are used by farmers selling direct to consumers. The particular mix of genotypes used for various sweet corn markets will continue to shift in the future as new cultivars within each genotype, and new genotype combinations, are released by breeders. Table 13.1 summarizes some of the major sweet corn genotypes

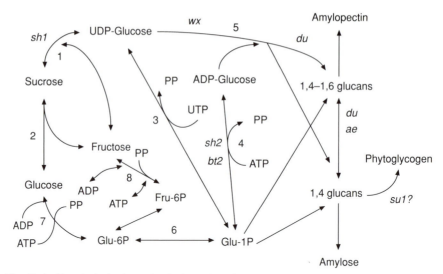

Fig. 13.1. Hypothetical scheme for the formation of starch and phytoglycogen from sucrose in the maize kernels (adapted from Boyer and Shannon, 1982). Endosperm mutants symbols are positioned at the enzyme reactions reported to be affected by the different mutations. Enzyme reactions are indicated by number as follows; (1) sucrose synthase; (2) invertase; (3) uridine diphosphate glucose (UDPG) pyrophosphorylase; (4) adenosine diphosphate glucose (ADPG) pyrophosphorylase; (5) starch synthases; (6) phosphoglucomutase; (7) glucokinase; (8) fructokinase.

and important features of each. Below is a more detailed discussion of the three most commercially important endosperm genotypes, *su1*, *sh2* and *se1*.

SUGARY1 (*su1*)

Traditional sweet corn is homozygous for the recessive mutation, sugary1 (*su1*), located in chromosome 4. This mutation causes a lesion, as yet unknown, in the starch biosynthetic pathway. Kernels homozygous for *su1* contain twice the sugar concentration and 8–10 times more phytoglycogen than field (*Su1*) corn at harvest maturity (Creech, 1965, 1968). At the mature dry stage (the stage sweet corn is harvested at for seed) the kernels have about half the starch content of field corn (Boyer and Shannon, 1982). Phytoglycogen, a factor contributing to kernel textural quality as well as flavour, is a highly branched water-soluble polysaccharide. Deficiency in functional starch debranching enzymes has been proposed as the reason for higher phytoglycogen in *su1* types (Pan and Nelson, 1984). The *su1* mutation also increases the activity of adenosine diphosphate (ADP)–glucose pyrophosphorylase, while decreasing starch hydrolytic enzyme activities, but these effects are not correlated with *su1* gene dosage. Doehlert *et al.* (1993) attributed these effects to a response by the maize endosperm cells to increased sucrose and phytoglycogen concentrations, rather than to the primary effect of the gene. The specific mechanism responsible for elevated

Table 13.1. Sweet corn endosperm carbohydrate genotypes.

Common names	Genotype	Relative sugar	Storability* (days)	Other attributes
'Sugary', 'traditional', 'normal', 'standard'	Homozygous recessive su1	Normal	1–3+	Best seed vigour and germination; creamy 'corn' flavour; used for processing, early season fresh market, and direct-retail markets
'Sugary Enhanced', 'SE', 'EH'	Homozygous recessive su1 and hetero or homozygous recessive se1	Moderate increase above su1	3–5+	Excellent flavour and tenderness; some varieties have less seed vigour than su1, and ears are easily damaged at harvest; popular for direct-retail markets
'Shrunken', 'SH2', 'Supersweet'	Homozygous recessive sh2	High to very high	5–10+	Although very sweet, some varieties criticized for lacking 'corn' flavour or for being too 'crunchy'; germination and seedling vigour a problem, especially in cold soils; must be isolated from su1 and se1 types; widely planted for wholesale fresh market
'Synergistic', 'Sweet Gene'	Homozygous recessive su1, heterozygous recessive sh2	Moderate to high	3–5+	Very few varieties available; does not have isolation requirements of sh2
'Sweet Breed'	Homozygous recessive su1, heterozygous recessive sh2 and se1	Moderate to high	3–5+	Similar to 'Synergistic', but with some improvements and se1 characteristics
'Improved Supersweet'	Homozygous recessive sh2, heterozygous recessive su1	High to very high	5–10+	Seed production difficult, with very few varieties available; more 'corn' flavour than sh2; germination problems and isolation requirements of sh2
ADX	Homozygous recessive ae, du, and wx	Moderate to high	5–10 (?)	Very few varieties available; must be isolated from all other genotypes
'Brittle'	Homozygous recessive bt2	Moderate to high	5–10 (?)	Very few varieties available; must be isolated from all other genotypes

* Actual storage time will vary with variety, cultural practices, and postharvest handling.

sugars in *su1* kernels remains unclear. Hybrids homozygous for *su1* are subject to ear quality loss after harvest because of the rapid conversion of sugars to starch and kernel moisture loss.

SHRUNKEN2 (*sh2*)

The shrunken2 (*sh2*) gene, a starch-deficient maize endosperm mutation located in the long arm of chromosome 3 (Bhave *et al.*, 1990), imposes a genetic block in the conversion of sucrose to starch and water-soluble polysaccharides. Kernels homozygous for the *sh2* gene accumulate two to three times more sucrose at 20 days after pollination (Laughnan, 1953; Creech, 1965), and retain higher sugar and moisture concentration for longer postharvest periods in comparison to *su1* kernels (Garwood *et al.*, 1976). Mature dry *sh2* kernels contain approximately twice the amount of total sugar, one-third to one-half the starch, and only trace levels of phytoglycogen in comparison to normal or traditional sweet corn types (*su1*) (Douglass *et al.*, 1993). Reduced activity of ADP–glucose pyrophosphorylase in this maize mutation is the primary cause for elevated kernel sucrose and reduced starch (Dickinson and Preiss, 1969).

Shrunken 2 corn should be isolated from not only field corn, but also from *su1* and *se1* sweet corn types. Cross fertilization with these other sweet corn types results in starchy (field corn-type) kernels because the cross fertilized kernels are not homozygous recessive for either the *sh2* or *su1* gene. An isolation distance of at least 50–75 m is recommended. This isolation distance can sometimes be reduced in areas with little prevailing wind, or if border rows are used.

SUGARY ENHANCER1 (*se1*)

The sugary enhancer1 (*se1*) gene is a recessive modifier of the *su1* endosperm mutation (Ferguson *et al.*, 1978). Construction of a saturated restriction fragment length polymorphisms (RFLP) linkage map of a sweet corn population segregating for this gene, in conjunction with phenotypic evaluation for the *se1* genotype in each F2–3 family, has mapped the gene to the distal portion of chromosome 2 of the maize genome (Tadmor *et al.*, 1995). The *se1* gene was originally identified in the vegetable maize inbred, IL677A (Gonzales *et al.*, 1974, 1976). This inbred line has been used as the source of the *se1* allele in the Illinois sweet corn breeding programme (Juvik *et al.*, 1983). Kernels of IL677A *su1se1* were high in sucrose and distinguished from all of the other genotypes by high maltose concentration at harvest maturity (Ferguson *et al.*, 1979). Maltose has been suggested as an early product of the hexoses moving into *su1se1* endosperm (Dickinson *et al.*, 1983). High maltose concentration is a specific characteristic of IL677A and is probably not associated with the *se1* gene since other *se1* inbreds have lower maltose concentration (Carey *et al.*, 1982).

When homozygous, the *se1* gene increases sugar concentration in *su1* kernels to levels comparable to those in *sh2* kernels without a reduction in phytoglycogen concentration (Gonzales *et al.*, 1976; Dickinson *et al.*, 1983). Hence, *su1se1* kernels at harvest maturity combine the desirable textural characteristics of standard *su1* sweet corn with the enhanced sugar concentration of supersweet (*sh2*) cultivars. However, sugar to starch conversion after harvest is not supressed as in *sh2* types, so storability of most *se1* cultivars is greater than *su1* corn but less than *sh2* corn (see also Table 13.1).

The nature of the biochemical lesion caused by the *se1* endosperm mutation is not known. Sweet corn breeders have had only limited success at incorporating *se1* into élite germplasm since kernels homozygous for *se1* in a *su1* background cannot always be visually differentiated from others in segregating populations. Two traits (lighter yellow kernel colour and slower dry-down due to higher sugar concentration) have been partially successful in identifying kernels homozygous for *se1*, but these traits vary in expression, depending on the genetic background and kernel maturity (Ferguson *et al.*, 1979; LaBonte and Juvik, 1990). A readily identifiable genetic marker linked to *se1* would simplify the selection of genotypes homozygous for the *se1* gene in plant breeding programmes.

Since all *se1* genotypes are also homozygous recessive for the *su1* gene, isolation from traditional *su1* plantings is not necessary, although for full realization of the benefits of the *se1* gene, cross fertilization with the same *se1* genotype is preferable. As is the case with *su1* types, cross fertilization with *sh2* types will result in starchy kernels and is to be avoided.

GERMINATION AND FIELD EMERGENCE

The mature corn kernel is a dry, single seeded indehiscent fruit known as a caryopsis. The kernel is composed of three principal parts: pericarp, endosperm and embryo. The pericarp is a protective layer derived from the ovary wall. The endosperm with a special outer layer (aleurone) accounts for most of the kernel and serves as the source of metabolites (mainly starch) for the embryo during germination and emergence. The embryo consists of the scutellum and the main axis. The scutellum and aleurone produce hydrolytic enzymes that digest the starch of the endosperm during germination and seedling development.

Most corn cultivars have five leaves in embryonic form in the seed (Sass, 1951), in addition to the modified first leaf or scutellum. Compared to other grasses, corn has a relatively large endosperm and embryo, allowing emergence from considerable depth (e.g. 10–15 cm) for some cultivars. There is no hormonal inhibition of germination in maize kernels, and under wet conditions kernels have been observed to germinate while still attached to the plant.

At optimum soil temperatures of 25–30°C, and a typical planting depth of about 5 cm, seedling emergence can be observed in 2–3 days. Low temperatures greatly delay emergence. It can be as late as 35 days after planting at 10°C. Seed decay and poor stands often occur at low temperatures due to attack by soil fungal pathogens, such as *Pythium*, *Fusarium* and *Penicillium* species. Rates of germination and emergence are also sensitive to soil moisture, decreasing with drier soil. Germination of dent corn has been reported to be completely prevented at soil water potentials below −1.25 MPa (Hunter and Erickson, 1952).

Despite the desirable flavour and storability attributes of *sh2*, *se1* and other endosperm mutation genotypes, their utilization has been hindered by reduced field emergence, seedling vigour and stand uniformity, especially in cold (less than 15°C) soils (Douglass *et al.*, 1993). This is particularly a problem with the *sh2* types. Reduced stand uniformity results in heterogeneous ear maturity and reduced value and yield.

The kernel properties that play an important and interactive role in the poor emergence associated with the *se1* and *sh2* endosperm mutations include: (i) insufficient seedling energy reserves due to reduced starch concentration (Douglass *et al.*, 1993); (ii) the condition of the mother plant and seed maturity at harvest (Culpepper and Moon, 1941; Borowski *et al.*, 1991); (iii) pericarp cracking during seed maturation and solute leakage during germination (Tracy and Juvik, 1988); (iv) membrane damage associated with high osmotic potential from elevated kernel sugar concentration and rapid influx of water during imbibition (Simon, 1978); (v) reduced endosperm starch hydrolysis during germination and seedling growth due to the lowered activity or amounts of α-amylase (Harris and DeMason, 1989); (vi) reduced endosperm to embryo dry weight ratio (Styer and Cantliffe, 1983); and (vii) susceptibility of the kernels to infection by fungal seed rot pathogens, particularly *Fusarium moniliforme* (Headrick *et al.*, 1990). Solutions to the germination problems of supersweets, other than genetic selection, include: (i) slower drying of seed during seed processing; (ii) more careful handling of seed to minimize cracking damage; (iii) fungicide and other seed treatments; (iv) shallow planting depth; and (v) avoidance of planting into cold soils.

A better understanding of the relationships between kernel physiological properties and field emergence and seedling growth is needed to improve sweet corn seed quality. A challenge for supersweet sweet corn breeders is to improve seed quality (mature dry kernel characteristics) without sacrificing fresh quality (i.e. kernel quality harvested at immature stage for consumption).

VEGETATIVE DEVELOPMENT

ROOTS

The radicle and early seminal roots emerge from the basal end of the seed as it germinates. These primary roots are soon supplanted by the permanent root system that develops as rings of roots emerging from the first four or five nodes of the stem (Duncan, 1978). These nodes remain below ground. Aerial brace roots often emerge from the lower two or three above-ground nodes near the end of vegetative development. The depth of the root system of a mature sweet corn plant usually does not exceed 1.0 m, and in many cases may be no deeper than 0.5–0.75 m (Huelson, 1954).

LEAVES

The leaves are first to emerge from the soil after the coleoptile, and remain the only above-ground plant organs until internode expansion begins at about the time that six to ten leaves are visible at the surface. During this early growth stage the apical meristem is below the soil surface, so the rate of new leaf initiation is more a function of soil temperature than air temperature. Typically, primordial development of all leaves, sheaths and nodes is nearly complete before internode expansion begins. As the internodes expand the stalk rises through the tube formed by the leaf sheaths which developed earlier.

Each maize leaf consists of a thin flat leaf blade with well-defined midrib and parallel venation. Leaf cell structure consists of an upper and lower epidermis, mesophyll cells between, and the prominent bundle sheath cells typical of grasses with the C4 photosynthetic pathway. Stomates appear on both lower and upper leaf surfaces, with stomatal density usually greater on the lower epidermis (Duncan, 1978).

Initiation of new leaves terminates when the apical meristem elongates and initiates the primordia of the staminate flower or tassel. Leaf number on the main stem is thus determined by the timing of the onset of reproductive development. There is considerable genotypic variation in leaf number, but 15–20 leaves are typical. Decreasing temperature (Duncan and Hesketh, 1968) and/or shorter days (Hesketh *et al.*, 1969) can decrease leaf number in at least some genotypes.

The lowest leaves on the plant are sometimes damaged by development of aerial roots from nodes directly above them, and may senesce because of extremely low light conditions in dense canopies. The upper most leaves are sometimes damaged by wind. The last leaves to senesce on corn plants grown to maturity are those nearest the node of the dominant ear (Wolfe *et al.*, 1988).

Donald (1968) suggested corn morphology is nearly ideal for building an efficient canopy. The leaves develop in an alternate pattern along the stalk, minimizing mutual leaf shading and maximizing ventilation. Leaf inclination varies with genotype, planting density, nodal position and stage of leaf development, but at full expansion (which coincides with full exposure of the ligule at the leaf base) most leaves are within the range of a 60–90° angle to the stalk. Light use efficiency is constrained by the fact that most or all leaves arise from a single stalk, and the tassel at the top can shade upper leaves. Duncan *et al.* (1967) found that shading by tassels reduced grain yield in field corn by 19% at a planting density of 100,000 plants ha^{-1}. Leaf area index (LAI) can be controlled to some extent by varying planting density. Studies with field corn have found an optimum LAI for grain yield in the range of 2.5–5.0 (Williams *et al.*, 1968; Fischer and Palmer, 1984). Higher LAI can cause a further increase in total plant dry matter accumulation, but this is primarily allocated to the stem rather than the grain.

STALK

The corn stalk functions as a storage organ and support structure for the leaves and reproductive organs. The stalk of corn differs from the stem of most grasses in that it is filled with parenchymatous tissue, the pith, which adds structural strength (Duncan, 1978). The corn stalk stores nitrogen and considerable amounts of soluble solids, primarily as sucrose. There is considerable evidence that translocation of sugars and nitrogen to the developing kernels contributes to final grain yield in field corn (Daynard *et al.*, 1969; Wolfe *et al.*, 1988).

Plant height and the stalk : leaf ratio are modified by environmental conditions during the period of rapid internode expansion. Because internode expansion occurs late, stalks are more affected than leaves by stresses or deficiencies that increase during vegetative development, such as mutual shading, soil moisture stress, nutrient availability, high temperature stress (Duncan, 1978). These stresses will therefore tend to decrease the stalk : leaf ratio. High plant densities tend to cause taller plants with weaker stalks that are more subject to 'lodging' (bending and collapse of the stalk).

TILLERS

The axillary buds present at the base of the leaves are morphologically identical when initiated, but those at the upper nodes develop into the female reproductive structures (i.e. ears), while those at the lowest nodes can develop into secondary stalks or tillers (Fig. 13.2) in some genotypes. There is considerable genotypic variation in the tendency to tiller. Sweet corn cultivars are, in general, less likely to tiller than many field corn types. Dense spacing (and presumably other factors that reduce photosynthate supply per plant) inhibits tiller development. Tillers are usually considered undesirable because they seldom produce marketable ears. It remains unclear whether tillers are reliant on diversion of photosynthates, water and/or nutrients from the main stalk, and thereby reduce yield or quality. In one study with field corn, tiller development did not reduce final grain yield (Dungan, 1931). Although there are some exceptions (Stephens, 1996), most commercial sweet corn growers make no effort to remove tillers, assuming that tiller production is not a significant yield-reducing factor with current cultivars and recommended planting densities.

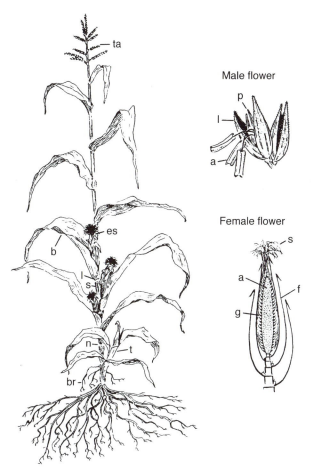

Fig. 13.2. Corn plant, and male and female flower detail. ta, tassel; es, ear shoot; b, blade of leaf; l, ligule; s, sheath; n, node; t, tiller; br, brace root; p, palea of male flower; l, lemma of male flower; a, anthers; s, silks of female flowers; a, axis of cob; f, flag leaf; g, developing grain (adapted from Peirce, 1987; Gallinat, 1979).

REPRODUCTIVE DEVELOPMENT

The staminate flowers (Fig. 13.2), collectively referred to as the tassel, begin to develop while still within the 'whorl' of uppermost leaves of the plant. When the tassel finally emerges from the whorl, it may be fewer than 10 days to full development and pollen shed. Pollen dispersal is primarily by air movement, and for an individual plant will last about 7–10 days.

The female reproductive structures (see Fig. 13.2), the ear shoots, develop from the axillary buds at nodal positions about halfway to two-thirds up from the base of the plant. Axillary buds at higher nodes are apparently inhibited by tassel development (Bonnett, 1948). There is a clear hierarchy of dominance of each ear initial over the ones below so that the uppermost ear initial becomes the largest, and often the only, marketable ear. Several ear initials in lower positions may reach considerable size, but those not fertilized regress.

The ear shoot first develops as a narrow shank with short internodes. Sheaths emerge at each node and surround the ear shank as husks. When husk initials have all formed the ear shoot begins a phase of more rapid elongation. Rows of two-lobed protuberances form along the ear. Each lobe produces a spikelet with two flowers, although one of these flowers usually aborts in most cultivars. Since one fertilized kernel results from each spikelet, the kernels on the mature ear appear in paired rows. A pistil, or silk, elongates from each flower and eventually protrudes out of the husk at the distal tip of the ear. Silk elongation continues until fertilization. Basal spikelets develop silks first, with silk emergence proceeding towards the tip. Although silks from higher up the ear start elongating later, they may emerge sooner than basal silks because of the shorter distance traversed. The result is that the first embryos fertilized are often those near the middle of the ear (Duncan, 1978). Silks from the most distal point on the ear sometimes emerge too late to receive viable pollen, leading to a quality problem referred to as 'poor tip fill'. Environmental stresses, particularly drought, can exacerbate this problem.

Development of ear shoots is in synchrony with tassel development, such that most of the silks emerge during anthesis. However, within any one plant, the amount of pollen which falls to the silks, and the synchrony, is likely to be inadequate for fertilization of all kernels of the ear. Thus, a stand of several plants grown in close proximity to each other is required for production of ears with complete fill. Commercial plant spacings depend on cultivar and planting equipment, but usually range from 75–100 cm between rows, and 20–30 cm between plants in the row.

ENVIRONMENTAL FACTORS AFFECTING DEVELOPMENT AND YIELD

LIGHT

The rate of plant development is primarily determined by temperature, but corn is a quantitative short-day plant (Kiesselbach, 1950), and photoperiod can potentially influence leaf number and the timing of flowering. Short days tend to reduce leaf number and accelerate tasseling and silk emergence (Huelson, 1954). However, most commerical cultivars are not particularly sensitive to photoperiod within the geographic/climate range to which they are adapted. For example, most tropical field corn cultivars do not become photoperiod-sensitive until daylengths exceed 14.5 h (Fischer and Palmer, 1984). A 14.5-h daylength is not encountered in most tropical areas, only at latitudes of 30° or less. Some tropical cultivars, when grown in a temperate area, will not flower until short days (Peirce, 1987). Conversely, most temperate-adapted cultivars are not very responsive to photoperiods typical of the growing season at high latitudes, but may flower early and have fewer leaves if grown in the relatively shorter days of the tropics.

Corn is a C4 plant and therefore is best-adapted to high light environments. Typical of other plants with the C4 photosynthetic pathway, corn has a high photosynthetic capacity that is often not light-saturated even in full sunlight (Hesketh and Musgrave, 1962). Low light conditions, due either to low incoming solar radiation or to high planting density and high LAI, can impair ear development, delay pollen shed, and lead to barren plants (Tollenaar, 1977). One advantage of hybrid corn cultivars is a uniform plant growth rate and height so that no one plant dominates (i.e. shades) its neighbours, thereby minimizing barrenness.

TEMPERATURE

Developmental rate of the corn plant is primarily a function of temperature, not light or photosynthesis. Duncan (1978) provides as an example of this the fact that plants grown at a very high density will tassel at the same time as plants grown at a wider spacing, although those grown at the higher density will produce less biomass per plant and have lower yield per plant because of mutual shading. The importance of temperature in determining the time to flowering within a given photoperiod range, has made the 'heat unit' or 'growing-degree-day' concept useful for predicting harvest date and describing maturity classes (Cross and Zuber, 1972; and see Chapter 4). A base temperature of 10°C is usually assumed. Developmental response to temperature (e.g. heat units required for each new leaf) is often linear between 10 and 30°C (Fischer and Palmer, 1984).

Corn, as a C4 plant, has a relatively high optimum temperature for photosynthesis, but warm temperatures can reduce seasonal productivity by accelerating developmental rate, shortening vegetative and reproductive growth phases, and reducing leaf area duration (Duncan, 1978). Warm temperatures also increase dark respiration carbon loss. Temperatures above 35°C are usually supraoptimal for photosynthesis, and can impair reproductive development and reduce ear quality (incomplete kernel development, poor husk cover over the ear, tasselate ears).

It is important to note that photosynthesis is determined by daytime temperatures, whereas developmental rate and respiration are determined by both day and night temperatures. Therefore, the optimum growing environment for maximum productivity is one with warm days and relatively cool nights. Cool nights are particularly important near harvest in sweet corn production because cool nights slow the maturation rate of the kernels, extending the 'harvest window' when kernels maintain high sugar content and a tender pericarp. Peirce (1987) suggests that maximum productivity occurs with daytime temperatures above 19°C (24–30°C optimum) and average night temperatures about 13°C. There is genotypic variation in temperature optimum among sweet corn cultivars, particularly with regard to optimum soil temperatures for germination, and optimum air temperatures during the ear maturation phase.

Typical of most plant species of tropical origin, corn is chilling sensitive, and can suffer photosynthetic depression, photoinhibition and chilling damage at temperatures below 5°C (Wolfe, 1991; Long *et al.*, 1983). Most sweet corn cultivars show little growth, and yield losses can result, if plants are exposed to prolonged periods of temperatures below about 12°C. Sweet corn has an advantage compared to many chilling-sensitive species in that the apical meristem remains below ground for the first few weeks of development, and is therefore less subject to damage from light frost events. Sweet corn is unique among most tropical crop species in that the harvested portion, the ear, is not chilling sensitive, and is, in fact, optimally stored at low temperatures (near 0°C) to slow respiration rate and maintain quality.

WATER AND NUTRIENTS

Although sweet corn has a relatively high water use efficiency (grams of carbon gained per gram of water transpired) because it possesses the C4 photosynthetic pathway, it is not tolerant of drought conditions. A relatively shallow rooting system (e.g. 0.5–0.75 m), and the sensitivity of reproductive processes and ear quality to water stress are partial explanations for this. Short periods of drought during the reproductive phase can slow silk elongation, accelerate or delay pollen shed, and thus negatively affect the synchrony between pollen shed and silk receptivity for fertilization (Wolfe *et al.*, 1988). This results in poor ear tip fill or incomplete development of kernels throughout the ear. Drought can also cause poor husk cover leading to bird damage on exposed ear tips. In an irrigation trial with several vegetable crops, MacGillivray (1949) reported that sweet corn benefited the most from irrigation, and was the only crop to show signs of water stress during the hottest part of sunny days even in the wettest treatment.

Optimum pH for availability of nutrients for sweet corn production is 6.0–6.5. Maximum sweet corn yield requires relatively high levels of nitrogen fertilizer. Recommendations can range from 100 to 165 kg nitrogen ha^{-1}, depending on soil type, cultivar, and rainfall or irrigation frequency. To minimize leaching of nitrates below the root zone, nitrogen applications are usually split between pre-plant and later side-dressings (before plants exceed about 30 cm in height). Phosphorus and potassium fertilizers are also applied as needed based on soil test results and local experience. Sweet corn does not have a high requirement for trace elements, but on light sandy soils and at certain pH levels, boron, zinc, magnesium or manganese may be inadequate for normal growth.

SEED PRODUCTION

The early sweet corn introductions were open pollinated, and a few of these cultivars are still available (e.g. 'Country Gentleman'). All modern cultivars, however, are F1 hybrids, and the seed is produced by commercial seed companies. The hybrids have significant yield and uniformity advantages over the earlier open pollinated types.

More than 90% of sweet corn seed production occurs in the state of Idaho in the northwestern USA. The advantages of this region for producing high quality sweet corn seed include: (i) low rainfall and low humidity during the growing season, adequate irrigation water available; (ii) fertile soils; (iii) relatively low pest and disease pressure; and (iv) infrequent extreme high temperature days.

Genetic purity of hybrid crosses requires isolation of seed corn fields from other corn plantings. The standard recommended isolation distance is 200 m, although this may be modified depending on natural or planted 'border-row' barriers to pollen drift, prevailing wind conditions, or differential flowering dates (Wych, 1988).

Planting densities typically range from 74,000 to 86,000 plants ha^{-1} (30,000–35,000 plants acre^{-1}). The most common planting pattern is four rows of female parent to one row of male parent. In some cases the male and female parents are planted on different dates so that the silks of the female parent are at the most receptive stage when maximum pollen shed by the male parent occurs. Other approaches to optimizing the timing of reproductive development of male and female parents have been tried, such as using different planting depths or clipping one or the other of the parents at an early growth stage to slow development (Wych, 1988).

Self-pollination of the female parent must be prevented, and detasseling by hand is currently the most widely used method of pollen control. About 20 years ago, mechanical detasselers were developed and became popular for a short while. Mechanical detasselers are high-clearance machines that remove tassels either by cutting them off with rotating blades or by pulling them off with two counter-rotating wheels. One problem with the mechanical approach is that it does not do a thorough job (i.e. some portions of tassels may be left in the field) when there is variability in plant height. Other problems with mechanical detasselers are reduced seed yields due to plant damage, and soil compaction caused by entering the field with this heavy equipment at a time when soil moisture must be kept high by irrigation to meet crop water demands. Nevertheless, mechanical detasselers are still sometimes used during periods of labour shortage. The conversion of inbred parents to cytoplasmic male sterility has not replaced detasseling as a means of pollen control because of concerns that this increases the vulnerability of corn crops to southern corn leaf blight (see Wych, 1988 for detailed discussion).

Sweet corn ears grown for seed are allowed to dry and reach physiological maturity on the plant in the field. The percentage of moisture of the kernels at harvest is ideally about 35–40% for *su1* genetic types, 40–45% for *se1* types and 50% for *sh2* types (M. Talkington, Crookham Seed Co., Idaho, 1996). The harvested ears are taken to the processing plant where the husks are mechanically removed, and the ears are slowly air-dried at 35–38°C to a 10–11% moisture content. (Experience has shown that quality of *sh2* seed is significantly improved by a slow drying process and careful avoidance of temperature above 38°C.) The seed are then mechanically removed from the cobs, the seed are cleaned, sized, treated with fungicides and insecticides (will vary depending on cultivar and location where it will be sold) and bagged. The final steps in seed processing involve quality control procedures, such as performing a gemination test on subsamples of seed, a cold germination test for evaluating seed vigour under less than optimum conditions, and sometimes other tests, such as electrophoretic tests for genetic purity. Ideal environmental conditions for long-term storage of sweet corn seed is a temperature near 10°C and 50% relative humidity.

ACKNOWLEDGEMENTS

The authors thank M. Talkington of Crookham Seed Co. and D. Wilson of Ciba Seeds for information regarding sweet corn seed production.

REFERENCES

Beadle, G.W. (1939) Teosinte and the origin of maize. *Journal of Heredity* 30, 245–247.

Bhave, M.R., Lawrence, S., Barton, C. and Hannah, L.C. (1990) Identification and molecular characterization of shrunken-2 cDNA clones of maize. *Plant Cell* 2, 581–588.

Bonnett, O.T. (1948) Development of inflorescences of sweet corn. *Journal of Agricultural Research* 60, 25–37.

Borowski, A.M., Fritz, A.F. and Waters, L. (1991) Seed maturity influences germination and vigor of two shrunken2 sweet corn hybrids. *Journal of the American Society for Horticultural Science* 116, 401–404.

Boyer, C.D. and Shannon, J.C. (1982) The use of endosperm genes for sweet corn quality improvement. *Plant Breeding Reviews* 1, 139–161.

Carey, E.E., Rhodes, A.M. and Dickinson, D.B. (1982) Post-harvest levels of sugars and sorbitol in sugary enhancer (*su1se1*) and sugary (*su1Se1*) maize. *HortScience* 17, 241–242.

Coe, E.H. (1993) Gene list and working maps. *Maize Genetics Cooperative Newsletter* 67, 133–166.

Creech, R.G. (1965) Genetic control of carbohydrate synthesis in maize endosperm. *Genetics* 52, 1175–1186.

Creech, R.G. (1968) Carbohydrate synthesis in maize. *Advances in Agronomy* 20, 275–322.

Cross, H.Z. and Zuber, M.S. (1972) Prediction of flowering dates in maize based on different methods of estimating thermal units. *Agronomy Journal* 64, 51–355.

Culpepper, C.W. and Moon, C.H. (1941) Effect of stage of maturity at time of harvest on germination of sweet corn. *Journal of Agricultural Research* 28, 335–343.

Daynard, T.B., Tanner, J.W. and Hume, D.J. (1969) Contribution of stalk soluble carbohydrates to grain yield in corn. *Crop Science* 9, 831–834.

Dickinson, D.B. and Preiss, J. (1969) Presence of ADP-glucose pyrophosphorylase in shrunken2 and brittle2 mutants of maize endosperm. *Plant Physiology* 44, 1058–1062.

Dickinson, D.B., Boyer, C.D. and Velu, J.G. (1983) Reserve carbohydrate from kernels of sugary and sugary enhancer maize. *Phytochemistry* 22, 1371–1373.

Doebley, J. and Stec, A. (1993) Inheritance of the morphological differences between maize and teosinte: Comparison of results of two F2 populations. *Genetics* 134, 559–570.

Doehlert, D.C., Kuo, T.M., Juvik, J.A., Beers, E.P. and Duke, S.H. (1993) Characteristics of carbohydrate metabolism in sweet corn (sugary1) endosperms. *Journal of the American Society for Horticultural Science* 118, 661–666.

Donald, C.M. (1968) The breeding of crop ideotypes. *Euphytica* 17, 193–211.

Douglass, S.K., Juvik, J.A. and Splittstoesser, W.E. (1993) Sweet corn seedling emergence and variation in kernel carbohydrate reserves. *Seed Science and Technology* 21, 433–445.

Duncan, W.G. (1978) Maize. In: Evans, L.T. (ed.) *Crop Physiology*. Cambridge University Press, Cambridge, pp. 23–50.

Duncan, W.G. and Hesketh, J.D. (1968) Net photosynthetic rates, relative leaf growth rates, and leaf numbers of 22 races of maize grown at eight temperatures. *Crop Science* 8, 670–674.

Duncan, W.G., Williams, W.A. and Loomis, R.S. (1967) Tassels and the productivity of maize. *Crop Science* 7, 37–39.

Dungan, G.H. (1931) An indication that corn tillers may nourish main stalks under some conditions. *Agronomy Journal* 23, 662–669.

Evansen, K.B. and Boyer, C.D. (1986) Carbohydrate composition of sensory quality of fresh and stored sweet corn. *Journal of the American Society for Horticultural Science* 111, 734–738.

Ferguson, J.E., Dickinson, D.B. and Rhodes, A.M. (1978) The genetics of sugary enhancer (*se1*), an independent modifier of sweet corn (*su1*). *Journal of Heredity* 69, 377- 380.

Ferguson, J.E., Dickinson, D.B. and Rhodes, A.M. (1979) Analysis of endosperm sugars in a sweet corn inbred (Illinois 677a) which contains the sugary enhancer (*se1*) gene and comparison of *se1* with other corn genotypes. *Plant Physiology* 63, 416–420

Fischer, K.S. and Palmer, A.F.E. (1984) Tropical maize. In: Goldsworthy, P.R. and Fisher, N.M. (eds) *The Physiology of Field Crops*. John Wiley, New York, pp. 213–248.

Gallinat, W.C. (1971) The evolution of sweet corn. *University of Massachusetts Agricultural Experiment Station, Bulletin 591*. Amherst, Massachusetts.

Gallinat, W.C. (1979) Botany and origin of maize. In: Hefliger, E. (ed.) *Maize*. Ciba-Geigy, Basle, Switzerland.

Garwood, D.L., McArdle, F.J., Vanderslice, S.F. and Shannon, J.C. (1976) Postharvest carbohydrate transformations and processed quality of high sugar maize genotypes. *Journal of the American Society for Horticultural Science* 101, 400–404.

Gonzales, J.W., Rhodes, A.M. and Dickinson, D.B. (1974) A new inbred with high sugar content in sweet corn. *HortScience* 9, 79–80.

Gonzales, J.W., Rhodes, A.M. and Dickinson, D.B. (1976) Carbohydrate and enzymatic characterization of a high sucrose sugary inbred line in sweet corn. *Plant Physiology* 58, 28–32.

Hardenberg, R.E., Watada, A.E. and Wang, C.T. (1986) *The Commercial Storage of Fruits and Vegetables and Florist and Nursery Stocks*. USDA Handbook no. 66. USDA, Washington, DC.

Harris, M.J. and DeMason, D.A. (1989) Comparative kernel structure of three endosperm mutants of *Zea mays* L. relating to seed viability and seedling vigor. *Botanical Gazette* 150, 50–62.

Headrick, J.M., Pataky, J.K. and Juvik, J.A. (1990) Relationships among carbohydrate of kernels, condition of silks after pollination, and the response of sweet corn inbred lines to infection of kernels by *Fusarium moniliforme*. *Phytopathology* 80, 487–494.

Hesketh, J.D. and Musgrave, R.B. (1962) Photosynthesis under field conditions. IV. Light studies with individual corn leaves. *Crop Science* 2, 11–315.

Hesketh, J.D., Chase S.S. and Nanda, D.K. (1969) Environmental and genetic modification of leaf numbers in maize, sorghum, and Hungarian millet. *Crop Science* 9, 460–463.

Huelsen, W.A. (1954) *Sweet Corn*. InterScience Publishing, New York, 409 pp.

Hunter, J.R. and Erickson, A.E. (1952) Relation of seed germination to soil moisture tension. *Agronomy Journal* 44, 107–110.

Juvik, J.A., Mikel, M.A., Carey, E.E. and Rhodes, A.M. (1983) Release of six Illinois sweet corn inbreds with the sugary enhancer (*se1*) gene. *HortScience* 23, 384–386.

Kaukis, K. and Davis, D.W. (1986) Sweet corn breeding. In: Bassett, M.J.(ed.) *Breeding Vegetable Crops*. AVI Publishing, Westport, Connecticut, pp. 475–519.

Kiesselbach, T.A. (1950) Progressive development and seasonal variations of the corn crop. *Nebraska Agricultural Experiment Station Bulletin no. 166*, 49. University of Nebraska, Lincoln, Nebraska.

LaBonte, D.R. and Juvik, J.A. (1990) Characterization of sugary1 (*su1*) sugary enhancer (*se1*) kernels in segregating maize (*Zea mays* L.) populations. *Journal of the American Society for Horticultural Science* 114, 153–157.

Laughnan, J.R. (1953) The effect of the *sh2* factor on the carbohydrate reserves in the mature endosperm of maize. *Genetics* 38, 485–499.

Long, S.P., East, T.M. and Baker, N.R. (1983) Chilling damage to photosynthesis in young *Zea mays*. I. Effect of light and temperature variation on CO_2 assimilation. *Journal of Experimental Botany* 34, 177–188.

MacGillivray, J.H. (1949) A comparison of several vegetable crops for their response to irrigation. *Proceedings of the American Society of Horticultural Science* 54, 330–338.

Pan, D. and Nelson, O.E. (1984) A debranching enzyme deficiency in endosperm of the sugary1 mutant maize. *Plant Physiology* 74, 324–328.

Peirce, L.C. (1987) *Vegetables: Characteristics, Production, and Marketing*. John Wiley, New York, pp. 383–397.

Sass, J.E. (1951) Comparative leaf number in the embryos of some types of maize. *Iowa State College Journal of Science* 25, 509–512.

Simon, E.W. (1978) Plant membranes under dry conditions. *Pesticide Science* 9, 168–172.

Singleton, W.R. (1948) Hybrid sweet corn. *Connecticut Agricultural Experiment Station Report*. New Haven, Connecticut.

Stephens, D. (1996) Sucker sweet corn, get better yields. *American Vegetable Grower* 44, 45.

Styer, R.C. and Cantliffe, D.J. (1983) Changes in seed structure and composition during development and their effects on leakage in two endosperm mutants of sweet corn. *Journal of the American Society for Horticultural Science* 108, 721–728.

Tadmor, Y., Azanza, F., Han, T., Rocheford, T.R. and Juvik, J.A. (1995) RFLP mapping of the sugary enhancer1 gene in maize. *Theoretical and Applied Genetics* 91, 489–494.

Tapley, W.T., Enzie, W.D. and Van Eseltine, G.P. (1934) Sweet Corn. In: *The Vegetables of New York*, Vol.1, part III. New York State Agricultural Experiment Station Report, New York.

Tollenaar, M. (1977) Sink-source relationships during reproductive development in maize: a review. *Maydical* 22, 49–75.

Tracy, W.F. and Juvik, J.A. (1988) Electrolyte leakage and seed quality in a shrunken2 population selected for improved field emergence. *HortScience* 23, 391–392.

US Department of Agriculture (1990) *Agriculture Statistics. National Agricultural Statistic Service, Agricultural Statistics Board*. USDA, Washington, DC.

Wann, E.V. (1980) Seed vigor and respiration of maize kernels with different endosperm genotypes. *Journal of the American Society for Horticultural Science* 105, 31–34.

Williams, W.A., Loomis, R.S., Duncan, R.S., Dowrat, W.G. and Nunez. A.F. (1968) Canopy architecture at various population densities and the growth and grain yield of corn. *Crop Science* 8, 303–308.

Wolfe, D.W. (1991) Low temperate effects on early vegetative growth, leaf gas exchange, and water potential of chilling-sensitive and chilling-tolerant crop species. *Annals of Botany* 67, 205–212.

Wolfe, D.W., Henderson, D.W., Hsiao, T.C. and Alvino, A. (1988) Interactive water and nitrogen effects on senescence of maize. I. Leaf area duration, nitrogen distribution, and yield. *Agronomy Journal* 80, 859–864.

Wych R.D. (1988) Production of hybrid seed corn. In: Sprague, G.F and Dudley, J.W. (eds) *Corn and Corn Improvement*. American Society of Agronomy, Madison, Wisconsin, pp. 565–608.

Lettuce $\boxed{14}$

H.C. Wien

Department of Fruit and Vegetable Science, Cornell University, 134A Plant Science Building, Ithaca, New York 14853–5908, USA

Lettuce (*Lactuca sativa* L.) is one of the most popular vegetables grown in North America, and forms an important part of temperate climate production systems in Europe and other regions (Ryder, 1979). In the USA, lettuce is produced on about 90,000 ha, third in importance among vegetables after potatoes and tomatoes (Anonymous, 1992). California is the principal growing area, accounting for about two-thirds of national production. In Europe, the crop's tolerance for cool growing conditions and low light has made it a popular glasshouse crop (Krug, 1986).

Lettuce is an ancient vegetable crop that has been cultivated in the Mediterranean basin since around 4500 BC. Early forms of the crop were depicted on Egyptian tombs as having long stems, and pointed short upright leaves. They were probably cultivated for the edible oil in their seeds, rather than for the leaves (Ryder and Whitaker, 1976; Harlan, 1986). By the time of the ancient Greeks, forms whose leaves were consumed either fresh or cooked had been selected. Types which form heads did not appear until 1543 in Europe (Helm, 1954; Ryder and Whitaker, 1976).

Modern lettuce cultivars can be grouped into four or five types, according to plant form and predominant use (Ryder, 1979). The crisphead (or iceberg) forms firm, closed heads resistant to mechanical damage and is tolerant of long-distance shipping. It is the most important type grown, and forms the bulk of production in North America. The butterhead type forms loose, open heads, and has soft leaves easily damaged in handling. Leaf lettuce also shares this fragile nature, and both types must be well protected in shipping. Cos or romaine lettuce has erect, elongated leaves that form into a loose, loaf-shaped head. Stem lettuce is grown for its thickened, parenchymatous stem, harvested when the plant is still in the vegetative stage (Helm, 1954). In summarizing the physiology of the lettuce crop, the growth and function of the principal types, namely the crisphead, butterhead and leaf lettuce will be emphasized, because these have been most intensively studied.

SEED GERMINATION

Lettuce is generally seeded directly in the production field, rather than transplanted. The production cycle may be as short as 60 days, and populations of 50–70 thousand plants ha^{-1} would make transplanting uneconomical in many areas (Zink and Yamaguchi, 1962; Ryder, 1986). Since uniformity of growth is vital to achieve high yields of marketable product for a single harvest, much effort has gone toward ensuring that a maximum number of plants emerge at the same time. Unfortunately, two properties of lettuce seeds make this goal more difficult to achieve than with many other crops. The more important of these characteristics is the sensitivity of lettuce seed to high temperatures. If lettuce seed is subjected to temperatures of 30°C or above during the imbibition stage of germination, it becomes dormant and is delayed in germination (Ikuma and Thimann, 1964; Gray, 1977). The second phenomenon is the inhibition of some cultivars from germinating if the seed is subjected to dark conditions (Borthwick *et al.*, 1954). These two forms of dormancy have been extensively studied in recent years, and form principal research topics among seed physiologists. The emphasis of the present review will be on physiological effects that have implications on the production of lettuce in the field.

THERMODORMANCY

The imposition of dormancy on seed by allowing it to imbibe water in temperatures of 30–35°C has been termed secondary dormancy by Khan and Samimy (1982), because the dormancy takes effect after maturation of the seed. Sensitivity to high temperature was highest if heat was imposed from the start of moisture imbibition (Gray, 1977) for a minimum of 8 h duration. A shorter delay in germination occurred if seeds were subjected to high temperatures during the onset of cell division, which occurs at about 12 h after sowing (Gray, 1977). If exposure to heat was delayed until after the onset of cell elongation and cell division in the embryo, germination was irreversible (Cantliffe *et al.*, 1984). Thus any technique that allows the seed to reach the cell division stage at temperatures between 15 and 22°C would permit normal emergence at high temperatures. This is the basis for the use of various seed priming treatments to permit normal emergence of lettuce in high temperature environments.

The inhibition of germination by high temperatures is mediated by the restraining influence of the pericarp and by the reduced growth potential of the embryo (Takeba and Matsubara, 1979; Prusinsky and Khan, 1990). As temperatures increased, the embryonic axes could no longer generate enough force to break through the pericarp (Takeba and Matsubara, 1979). If the pericarp was slitted or punctured, the germination rate was significantly improved.

Research has implicated the involvement of growth hormones in lettuce germination at high temperatures, and provided means of at least partially alleviating thermoinhibition. In cultivars that have capability to germinate at high temperatures, ethylene was given off during the imbibition period, whereas inhibited seeds showed no ethylene production (Prusinsky and Khan, 1990; Huang and Khan, 1992). Seed treatment with the ethylene-generating chemical ethephon, or the ethylene precursor ACC (1-aminocyclopropane-1-carboxylic acid), or gassing with ethylene alleviated the thermoinhibition to a limited extent but could not restore germination at 35°C (Sharples, 1973; Abeles, 1986; Prusinsky and Khan, 1990). The combination of pericarp slitting and ethephon treatment resulted in additional increases in high temperature germination rate, leading Prusinsky and Khan (1990) to conclude that the formation of ethylene by the seed may be inhibited at high temperatures by anoxic conditions caused by an impervious pericarp. How the ethylene acts to facilitate germination is not known at present. Abeles (1986) showed that ethylene-treated embryos had thickened hypocotyls, and presumed that this thickening facilitated the splitting of the pericarp. Ethylene did not, however, alter the resistance of the pericarp to penetration at high temperatures.

The cytokinins may also play a role in allowing lettuce to germinate at high temperatures. Evidence for this has come primarily from improved germination after additions of cytokinins to the seeds (Sharples, 1973; Abeles, 1986). These chemicals may act by increasing seedling growth potential at high temperatures (Takeba and Matsubara, 1979), but the mechanism for this improvement is not clear. Sharples (1973) found that seed treatments combining ethephon and kinetin improved high temperature lettuce seed germination more than either chemical alone, and suggested that this be used as a seed treatment in hot environments.

Marked differences in the germination response of lettuce cultivars to high temperatures have been noted by Gray (1975). Of the cultivars tested, 16 butterhead lines had an upper temperature limit of 26–30°C, four crisphead cultivars ranged from 28 to 33°C, and two cos lines had an upper limit of 31°C. Similarly, ten crisphead cultivars had high germination percentages at 30 but not 35°C when germinated in water in the light (Coons et al., 1990). When the germination tests were repeated in 20.3 or 20.6 MPa NaCl solution, germination rates declined, but some cultivars, notably 'Coolguard' and 'Empire', were less affected. It should therefore be feasible to select for improved high temperature germination capability in lettuce breeding programmes.

SKOTODORMANCY

Skotodormancy refers to the inhibition of lettuce germination in the absence of light. In classic work in the 1940s, Borthwick and his colleagues at the USDA in Beltsville, Maryland determined the action spectrum of the light requirement of 'Grand Rapids' lettuce (Janick, 1989). They showed that light at 650 nm was most effective to stimulate germination, while 730 nm light inhibited germination, and that in a series of light exposures, the wavelength of the last treatment determined the effect (Borthwick *et al.*, 1952). The research established the foundation for work on the photoreversible pigment phytochrome, later found to play a role in many physiological processes in plants.

More recent research on the light requirement of lettuce seed germination has shown that the effect is mediated by the pericarp, and more specifically, the endodermal layer (Ikuma and Thimann, 1963; Tao and Khan, 1979). If these layers are removed, light-sensitive lettuce germinates fully in the dark. By treatment with light during imbibition, the force required to puncture the pericarp was greatly reduced, allowing the embryo to emerge. The light-sensitive stage commenced 1.5 h after the seed started to take up water, and the light response was not influenced by temperature or oxygen (Ikuma and Thimann, 1964). The light requirement was least pronounced in seeds in the late stages of development, but became evident 2–4 days later (Fig. 14.1) (Globerson, 1981b). Treatments with gibberellic acid can effectively substitute for the light requirement (Tao and Khan, 1979), perhaps by softening the endosperm layer in the dark of light-requiring seeds.

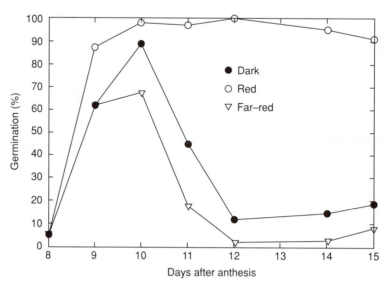

Fig. 14.1. Germination percentage at 20°C of lettuce seed cv. Grand Rapids, harvested at different times after anthesis, and germinated in three light environments (Globerson, 1981b). With kind permission from Kluwer Academic Publishers.

Lettuce cultivars vary widely in their requirement for light with regard to germination (Borthwick *et al.*, 1954). At temperatures of 20–25°C, most field-grown cultivars of lettuce in the USA do not require light for germination, but some lines selected for greenhouse production in cool conditions (15–20°C) have shown induced dormancy at these temperatures (A.A. Khan, personal communication, Geneva, N.Y., 1994). There is, however, a strong interaction between temperature and the light requirement. 'Grand Rapids' will germinate readily in the dark at 15°C, but is strongly inhibited at 20°C and above. Cultivars which do not require light at 20°C may show improved germination in light compared to dark after being subjected to heat or osmotic stress during imbibition (Borthwick *et al.*, 1954; Guedes and Cantliffe, 1980). Given this interaction between temperature and the light effect, there is probably no cultivar that is truly non-responsive to light with regard to germination.

From the practical standpoint, a light requirement for germination could pose a serious impediment to seedling emergence. Wooley and Stoller (1978) showed that photosensitive lettuce seed planted 6 mm deep in soil was significantly delayed in emergence compared to seed placed 2 mm below the surface. Less than 1% of incident radiation penetrated fine-textured soil to a depth of more than 2.2 mm. The problem is made worse by seed coatings, unless these dissolve promptly on being moistened, or split open. As long as soil temperatures are low during the planting season, emergence should be satisfactory, but when planting into warm soils, special measures may be needed.

SEED PRIMING

The inhibition of lettuce seed germination by high temperature and by darkness are both most effective during the early imbibitional stages (Ikuma and Thimann, 1964; Gray, 1977). If the seed can pass through these sensitive stages without being subjected to these inhibiting factors, germination and emergence would then be less affected by environmental influences. Research has generally substantiated this theory. Seed priming in an osmotic solution at 21.5 MPa improved emergence of 'Empire' lettuce in hot soils in California's Imperial Valley from 20% for untreated controls to 46–69% (Valdes *et al.*, 1985). The results were confirmed in controlled high temperature conditions in the laboratory (Valdes and Bradford, 1987). The preparation of the primed seed is an exacting process that must be carefully carried out. In a series of experiments, Guedes and Cantliffe (1980) showed that the priming solution should be imbibed at 15°C and in light, with good aeration for not longer than 12 h. The temperature at which the seed was subsequently dried was also important: germination was better if 20 than if 32°C was used. If these guidelines were adhered to, and storage temperature of the primed, dried seed was kept low, the seed maintained good vigour for many months (Valdes and Bradford, 1987).

An alternative seed priming technique has recently been reviewed by Khan (1992). This method, termed matric conditioning, consists of mixing the seed with a moist medium of low matric potential and negligible osmotic potential. As with osmotic priming, the seeds go through the initial stages of germination while in the medium, and are then carefully dried before sowing. The technique shows promise for lettuce, but has been used more extensively with slow-germinating species such as pepper and carrot so far (Khan, 1992). Benefits, similar to those derived from osmotic priming, are faster and more uniform emergence, and improved plant stands under high temperature conditions.

SEED PELLETING

Precise placement of lettuce seeds in the field is necessary if the costly practice of thinning the excess seedlings is to be avoided. The small, thin lettuce seed is difficult to sow in exact positions, so methods have been developed of coating the seeds with materials that change the shape to nearly spherical. The coating materials exclude light, however, and as a result may exacerbate the low emergence rate obtained under high temperature conditions (Zink, 1955). Some coatings also increase emergence time significantly, and may increase the difficulty of obtaining a uniform stand (Sharples, 1981).

Recent attempts to combine priming with seed coating have been successful, however, and may help to overcome the disadvantages of pelleted lettuce seed (Valdes et al., 1985; Valdes and Bradford, 1987). In addition, intensive research and development by the seed industry, probably based on some of the physiological processes described above, will result in further improvements in lettuce seed germination (Linden, 1988).

HEAD FORMATION

The head of lettuce is an assemblage of leaves closely packed together over the growing point of the plant. The process of head formation in crisphead lettuce consists of a series of changes in leaf morphology and leaf orientation that transform the rosette structure of predominantly horizontal leaves to one where successive leaves become progressively more erect and in-arching (Johnson, 1983). This change in plant morphology is accompanied by an inward curvature of the leaf midrib and a broadening of leaf shape (Bensink, 1971). The arching-over of the developing leaves is aided by entrapment of the distal leaf margins by the base of the rosette of older leaves. Head formation results from the accumulation of young leaves under the layer of leaves covering the growing point (Johnson, 1983). Thus for head formation to succeed, several morphological preconditions must be met (Dullforce, 1962): large individual leaves, a slow rate of stem elongation, short petioles and a high rate of leaf production.

The importance of the shape of the head-forming leaves was pointed out by Bensink (1958, 1971) in his detailed studies of lettuce head morphology. He showed that for a butterhead lettuce plant in the head-forming stage, leaf width increases more rapidly than leaf length as one goes from the outer part of the frame to the head (Fig. 14.2). This results in a pronounced decrease in the leaf length : width ratio, which is indicative of the head-forming tendency of the plant. Bensink further pointed out that leaf shape is strongly influenced by the environmental conditions under which the plant is growing. Under low light, leaves tend to be long and narrow. As light levels increase, their shape becomes progressively broader, with reduced length:width ratio (Fig. 14.3). Temperature interacts with light, enhancing leaf width at high temperatures and high light conditions, and resulting in narrower leaves if high temperatures occur under low light conditions (Fig. 14.4). At 10°C, although growth is slow, leaf shape is favourable

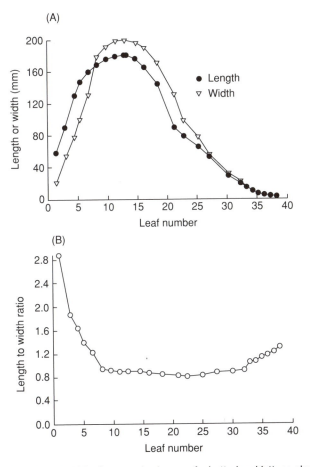

Fig. 14.2. (A) Length and width of successive leaves of a butterhead lettuce plant, numbered from the stem base upwards. (B) Length : width ratio of the same leaves (Bensink, 1971). With kind permission from Elsevier Science.

(A)

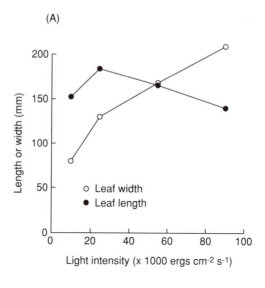

(B)

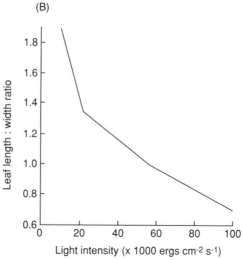

Fig. 14.3. (A) Maximum leaf length and width of butterhead lettuce when grown at 30°C, in relation to incident light level. (B) Length : width ratio of the same leaves (Bensink, 1971).

for head formation even under low light conditions. Applications of gibberellic acid to lettuce plants caused similar morphological changes in leaf shape as occurs when plants were grown under low light and high temperature conditions. It is therefore possible that this hormone is involved in the regulation of leaf shape and of head formation in lettuce.

Bensink (1971) explored the cellular basis for these leaf shape responses, and found that leaf width was correlated directly to the number of cells in the leaf lamina, determined by the length of the cell division stage of the

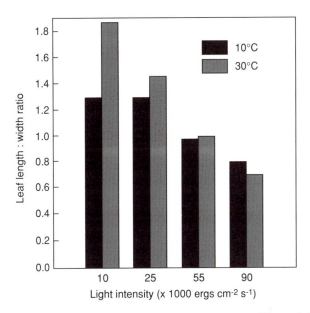

Fig. 14.4. Maximum leaf length : width ratio of butterhead lettuce at different light levels, as related to growing temperature (Bensink, 1971).

developing leaf primordia. Leaf length was determined by the length of individual cells in the leaf midrib.

Another morphological factor that influences head formation is the length of the stem. Environmental conditions that foster bolting, such as long photoperiods and high temperatures also prevent head formation (Dennis and Dullforce, 1974). The prevalence of such conditions in the Northeastern US and in Japan was a major deterrent to lettuce production in these areas during the summer until bolting-resistant cultivars were developed (Knott *et al.*, 1939; Shibutani and Kinoshita, 1966).

Head formation in cos lettuce proceeds in a slightly different manner than in crisphead lettuce (Nothmann, 1976). Instead of a marked change in leaf shape aiding in the curvature of the head-forming leaves over the apex, in cos lettuce the progressive curvature of the leaves is not accompanied by leaf shape changes. Under the high temperatures of the Isreali summer, head formation was prevented by bolting. In spring crops, abnormally twisted leaves in the developing head also inhibited heading (Nothmann, 1977). The cause for this disorder is not known, but has also been described in crisphead lettuce by Zink (1959).

Under field conditions, heading lettuce cultivars generally produce 13–20 frame leaves before head formation begins (Zink and Yamaguchi, 1962; Wurr *et al.*, 1987). The time when the head-forming leaves are first visible and start to cup is termed 'hearting', and has been shown to be critically important in determining final head size (Wurr *et al.*, 1987; Wurr and

Fellows, 1991). If the plants are growing under high light conditions and average temperatures below 12°C during this period, large heads tend to form (Wurr *et al.*, 1992a). On the other hand, if plants are subjected to low light conditions during this period, head weights are reduced or maturity delayed (Gray and Steckel, 1981; Wurr and Fellows, 1991). The morphological basis for this association may be the large size of the frame leaves formed under high light conditions, allowing larger heads to develop (Wurr and Fellows, 1991). Under low light, leaves with increased length : width ratio, and smaller size would delay the head formation process.

FACTORS DETERMINING PRODUCTIVITY

Aside from the effects on leaf morphology, light and temperature are major determinants of the plant growth rate, as expressed in the increase in leaf number (Bensink, 1971; Wurr *et al.*, 1981). As long as adequate levels of water and nutrients are available, increasing temperatures between 10 and 30°C, and increasing light levels between 1 and 26 MJ m^{-2} day^{-1} speed up the number of leaves formed per unit time (Bensink, 1971; Wurr *et al.*, 1981; Wurr and Fellows, 1991). This also translates into larger plant biomass and greater harvested yield (Glenn, 1984).

LIGHT

To obtain high yields of lettuce, it is necessary to have cultural practice and climatic conditions that permit the plants to make rapid initial growth. The sooner leaf cover of the soil surface has been achieved, the quicker the plants can use the incident radiation in growth. In field-grown crops, close spacing and using transplants rather than direct-seeding advances the time of complete light interception. In glasshouses, raising temperatures to accelerate growth has also been found useful (Bierhuizen *et al.*, 1973), although the combination of low light and high temperatures may result in the formation of long, narrow leaves that delay head formation (Bensink, 1971). There are also genetic differences in the formation of a large light-absorbing surface. Brouwer and Huyskes (1968) found that the accelerated growth of a new lettuce line was due to its more rapid soil cover, and resultant greater light interception per unit fresh weight (Fig. 14.5). The photosynthetic rates of the slow and rapidly growing lines were not different.

Light interception can also be maximized by growing plants in solution culture, which allows them to be repositioned without disrupting growth.

Culture in moveable troughs, or on styrofoam blocks floating on tanks of nutrient solution (Glenn, 1984) are examples of systems currently in use.

When other factors inhibiting growth are removed, lettuce productivity will be directly related to the incident light energy. In a series of solution culture trials conducted in spring and fall in a glasshouse in Arizona, Glenn

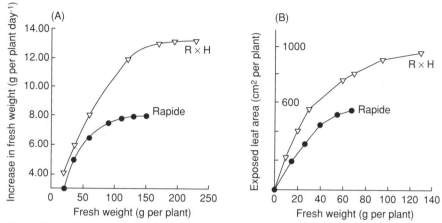

Fig. 14.5. (A) The relation of lettuce fresh weight and daily fresh weight gain of two lettuce lines. (B) The relation of fresh weight and exposed plant leaf area of the same lines (Brouwer and Huyskes, 1968). With kind permission from Kluwer Academic Publishers.

(1984) demonstrated that the time required to produce a lettuce head weighing 120 g decreased as solar radiation increased (Fig. 14.6). Others working in controlled environment chambers have made similar findings (Verkerk and Spitters, 1973; Cracker and Seibert, 1983; Knight and Mitchell, 1983a,b, 1988b).

Providing such increased light levels without incurring high energy costs has led to experimentation with new efficient light sources, in some cases with discouraging results (Tibbitts *et al.*, 1983; Mitchell *et al.*, 1991). High pressure sodium lamps in particular were found to depress leaf chlorophyll content, and may partly account for the lack of lettuce growth response to increased light reported by Tibbitts *et al.* (1983). Illumination with metal halide lamps resulted in similar above-ground dry matter production as with the standard fluorescent-incandescent lighting (Knight and Mitchell, 1988a). As light levels were raised, metal halide lamps stimulated growth more than the standard lights. Other research has also pointed out the importance of light quality in growth of lettuce under artificial illumination. In general, at a particular photo fluence rate, increased light in the red part of the spectrum has given growth stimulation (Krizek and Ormrod, 1980; Knight and Mitchell, 1988b), although there are exceptions (Knight and Mitchell, 1988a). These varied and conflicting results point out the need to test particular lamps for their suitability for lettuce growth before relying on them in production or research.

Under field conditions, response to increased light levels may be masked by adverse effects of increased temperature or reduced water status of the plants in the high light environment. For example, whereas Glenn (1984) measured continued growth increases up to 500 cal cm^{-2} day^{-1} incident radiation in his glasshouse trials, Mattei *et al.* (1973) found that lettuce dry matter production was maximum at 150 cal cm^{-2} day^{-1} in field plantings

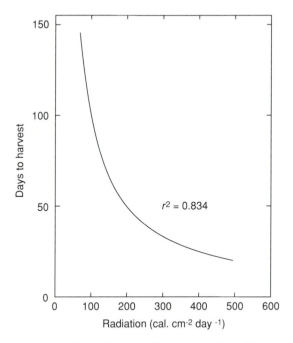

Fig. 14.6. The dependence of days to harvest of lettuce on solar radiation, combining data from England, and midwestern and western USA (Glenn, 1984). With kind permission from Elsevier Science.

in Italy and the UK. These differences emphasize the need for high levels of water management for the lettuce crop, and the necessity to avoid temperatures in excess of 30°C.

TEMPERATURE

Temperatures considered optimum for growth of lettuce average 18°C, with a range from 24 to 7°C (Lorenz and Maynard, 1980). The lettuce crop is thus produced in areas where such temperatures reliably occur, such as the Salinas Valley in California, during the winter months in California's Imperial Valley and in the summer in Northern Europe. In fact, Kimball *et al.* (1967) determined that lettuce production in the western USA was centred in areas that had at least 2 months of temperatures of 17–28°C (day) and 3–12°C (night). At temperatures higher than these ranges, the cultivars grown in these areas develop a high incidence of tipburn, bolting and the formation of loose, 'puffy' heads (Minotti, personal communication, Ithaca, N.Y., 1994).

There is, however, considerable crisphead lettuce production in areas where temperatures, especially at night, are above the limits identified by Kimball *et al.* (1967). In New York State, minimum temperatures during the

Table 14.1. Growing temperatures in three crisphead lettuce production areas of the USA during the lettuce production season.

Growing region	Growing period	Air temperature (°C)	
		Maximum	Minimum
California-Arizona*	All year[†]	17–28	3–12
New York**	June–September	26	14
Florida**	December–April	26	13

* (Kimball *et al.*, 1967).
[†] Growing periods vary with growing region within the states.
** 30-year averages for Syracuse, NY and Fort Myers, FL (Quaile and Presnell, 1991).

lettuce growing season average 14°C near the principal production area north of Syracuse (Table 14.1). The cultivars developed for this area were selected to resist bolting at high temperatures, and also have lower incidence of tipburn under these conditions (Raleigh and Minotti, 1967). When such cultivars are raised under cooler conditions, dense heads are formed which are too small to be marketable (Raleigh and Minotti, 1967; Wurr and Fellows, 1984). Cultivar choice for particular production areas is therefore influenced significantly by growing temperatures. Further evidence for this contention comes from the fact that lettuce growers in the Belle Glade area of Florida also use cultivars adapted to New York, and Florida cultivars perform well in New York (Ellerbrock, personal communication, Ithaca, N.Y., 1994). Growing season temperatures for that area near Fort Myers are nearly identical to those in New York (Table 14.1).

Temperature is the main factor determining the rate of growth of lettuce during seedling emergence and the early growth period (Bierhuizen *et al.*, 1973; Scaife, 1973). Growth rate depends on the temperature of the growing point, and this organ is located close to the soil surface in this rosette plant. It is thus not surprising that soil temperature is more closely correlated with plant growth rate than air temperature (Scaife, 1973; Wurr *et al.*, 1981). This information has been used to shorten the lettuce production season in glasshouses. Boxall (1971) found that heating the soil to 18°C decreased the length of the production cycle of butterhead lettuce by 14–17 days compared to plants grown in unheated soil under minimum air temperature of 7°C during winter in the UK.

PREDICTING MATURITY

The relatively short growth period from planting to maturity of lettuce allows more than one crop to be produced in a season in most locations. The objective with multiple plantings is to maintain a steady level of production over a period of time, to minimize marketing problems. The scheduling of plantings to achieve steady supply thus becomes important, and should be based on an understanding of the factors that determine when lettuce will reach marketable size.

In the production of leaf and butterhead lettuce, the acceptable size at harvest has not been closely defined, and can vary considerably (e.g. in Germany, plant fresh weight of 80–110 g are acceptable; Krug, 1986). With these cultivars, one needs to know the time required to reach this marketable size to estimate the length of the production cycle. As shown above, growth rate of lettuce is accelerated by increased irradiance and higher temperatures. Glenn (1984) found that the time to grow a 120 g head was inversely related to irradiance in a curvilinear fashion (see Fig. 14.6). If the amount of light received by the plants and the temperature at which the crop is growing are both known, it should be possible to derive a constant figure that can be used to predict when the lettuce plants will reach marketable weight. Such calculations have been made to predict maturity of head lettuce (see below).

In production systems where temperatures are controlled, growth duration should therefore be dependent on the amount of light received by the plants. Klapwijk (1979) demonstrated that this was indeed the case in winter lettuce production in Dutch glasshouses. Crop duration was longest for lettuce grown during the shortest days of the year, and progressively shorter as one deviated from that date. Similarly, in production systems in which control of irradiance and temperature are possible, a constant crop growth duration can be achieved. Thus on cloudy days, artificial light is given to reach a set daily irradiance total. With good glasshouse

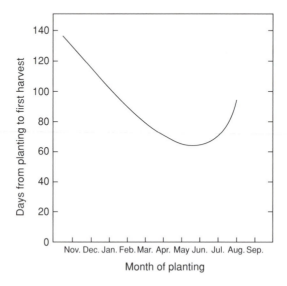

Fig. 14.7. The relation between days to first harvest and planting date for 312 crisphead lettuce crops grown in the Salinas Valley, California, 1953–57 (Zink and Yamaguchi, 1962).

temperature control, it is therefore possible to produce a uniform lettuce plant in a constant number of days, winter and summer (Langhans, personal communication, Ithaca, N.Y., 1995).

The prediction of maturity of head lettuce is somewhat more complicated than for butterhead and leaf lettuce described above. Many researchers have found that maturity estimates based on a sum of heat units accumulated during the growing season are not reliable (Maderiaga and Knott, 1951; Zink and Yamaguchi, 1962; Kristensen *et al.*, 1987). Correcting the growing temperature with the irradiance received improved the accuracy of the prediction. The results were, however, still too variable to be practical (Wurr *et al.*, 1988). Another approach has been to predict maturity of the lettuce head by modelling the effect of temperature and irradiance on the fresh weight gain of the head after hearting (Wurr *et al.*, 1992b). Unfortunately, this approach was also not successful, for it assumes that head lettuce will be harvested at a constant head size or weight. The most important characteristic that determines harvest date is head density, and this varies not only with temperature and irradiance during heading, but also with these and other yet unknown factors even before heading begins (Wurr *et al.*, 1992b). Thus in spite of considerable study, the most reliable predictor of harvest date in head lettuce is still the date estimated from the date of planting on the basis of the experience of previous years (Zink and Yamaguchi, 1962; Gray and Morris, 1978; Wurr *et al.*, 1988, 1992b) (Fig. 14.7).

FLOWER INDUCTION

Lettuce is a quantitative long-day plant that can also be seed vernalized. A lettuce plant induced to flower undergoes a dramatic change in morphology, from a rosette with a short stem to a plant more than a meter tall, with elongated internodes and prominent terminal flowers on a many-branched stem. The transition from vegetative to reproductive growth may in head-forming cultivars involve the growth of the stem under the restraint of the previously formed head. Thus climatic conditions that encourage head formation may delay time to flowering because stem elongation is retarded by the head. In seed production practice, therefore, growers either prevent the head from forming using growth hormones, or open it up by cutting or other mechanical means.

PHOTOPERIOD AND TEMPERATURE EFFECTS

In glasshouse experiments with both head and leaf lettuce, Rappaport and Wittwer (1956a,b) demonstrated the acceleration of flowering by 16- versus 9-h photoperiod. The magnitude of the photoperiod effect varied among cultivars; 'Grand Rapids' leaf lettuce flowered at almost the same time in both daylengths used (Rappaport and Wittwer, 1956b). Even in cultivars that show some photoperiod sensitivity, however, unfavourable daylengths delay but do not prevent flowering.

More comprehensive studies on cultivar differences in photoperiod response indicate at least three broad categories of genotypes with regard to flowering in lettuce (Waycott, 1993). American crisphead cultivars appear to have little response to photoperiods between 10 and 13 h, but to be sensitive above this range; European butterhead types tend to show linear decreases in time to flowering with increasing daylength across the entire daylength range. A third group includes early-flowering genotypes that are nearly day-neutral (Ryder, 1988). Within each of these groups, there may also be cultivar differences in sensitivity to daylength. For example, within European butterhead cultivars, those cultivated in winter tend to be more susceptible to bolting than those adapted to summer conditions (Krug, 1986).

Temperatures under which the plants are growing can have a profound effect on when they will flower. Under conditions optimum for head formation (19/11°C), flowering is delayed both by heading and its resultant constraint on stem elongation, and by the retardation of plant growth rate (Thompson and Knott, 1933; Rappaport and Wittwer, 1956a; Ito *et al.*, 1963). For instance, Rappaport and Wittwer (1956a) found that anthesis was delayed by 21 days when 'Great Lakes' head lettuce was grown at 16 rather than 21°C night temperature. In three cultivars of leaf lettuce, where head formation is not a factor, the delay in flowering due to the lower temperature was 12 days (Fig. 14.8).

The earlier flowering at higher temperatures could be simply attributed to the increase in growth rate, but there is some evidence that plants grown at higher temperatures form flowers at lower nodes than plants raised in cooler conditions (Rappaport and Wittwer, 1956b). Hiraoka (1969) and Ito *et al.* (1963) also showed earlier flowering at higher temperatures, but failed to find decreases in main stem node number to account for the acceleration. Additional work is needed in this area to reconcile the conflicting findings.

VERNALIZATION

The third major factor influencing flowering in lettuce is low temperature during seed germination and seedling emergence. Knott *et al.* (1937) demonstrated that if lettuce seed is exposed to 4°C for 5–20 days, 24 h after the start of germination, seedstalks appeared 3–5 days sooner than on the unvernalized controls. Subsequent work by Rappaport and Wittwer (1956a) and Prince (1980) indicated that at least 13 days of low temperature are needed, and that the low temperature treatment must be started within 3 days of germination.

As with the effect of photoperiod, cultivar differences have been found, with many apparently insensitive to seed vernalization (Prince, 1980; Thompson and Kosar, 1948). The effect is reversible in the wild *Lactuca*

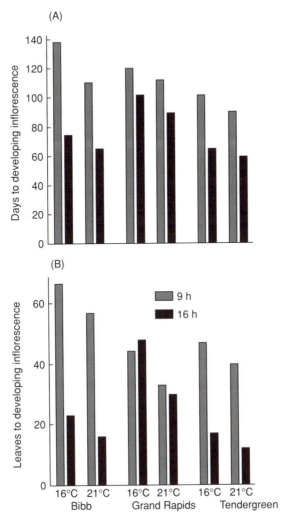

Fig. 14.8. Influence of photoperiod and night temperature on (A) days to developing inflorescence and (B) the number of leaves below the inflorescence for three leaf lettuce cultivars (Rappaport and Wittwer, 1956b).

serriola by periods of high temperatures following the vernalization treatments as long as the seeds have not germinated (Marks and Prince, 1979). Devernalization responses of cultivated lettuce are not known.

There is limited evidence that lettuce seeds forming on the mother plant can be partially induced to flower by exposure to low temperatures during the seed development period (Wiebe, 1989). The treatment also reduced germination and emergence rates, and increased the sensitivity of the seeds to high temperature during germination (Gray *et al.*, 1988).

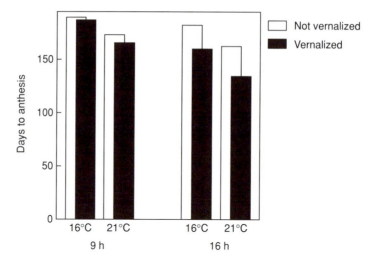

Fig. 14.9. Influence of photoperiod, seed vernalization and night temperature on days to anthesis of 'Great Lakes' crisphead lettuce (Rappaport and Wittwer, 1956a).

The combined effects of daylength, vernalization and growing temperature tend to be additive in the induction of flowering in lettuce (Fig. 14.9). Thus growing the plants under 21°C night temperatures at 16 h photoperiod, after seed vernalization, resulted in flowering after 135 days, while 188 days were needed for flowering under cool nights, short photoperiods and lack of seed vernalization. Crisphead lettuce cultivars selected for production in the Northeastern USA are often sown in cool soils, and exposed to long photoperiods and warm night during the season (see Table 14.1). They have therefore needed to have a higher level of resistance to bolting than cultivars selected for the cooler production areas of California, where lettuce is grown throughout the year (Ryder, 1986).

HORMONAL FACTORS IN LETTUCE FLOWERING

The restriction to stem growth, flowering and seed production caused by head formation has led to attempts to prevent heading with chemical treatments. Franklin (1948) induced leaf epinasty with synthetic auxin sprays before heading, but found seedstalk elongation to be inhibited by some of the higher concentrations of 2,4D (2, 4 dichlorophenoxyacetic acid) used. Applications of gibberellic acid were more effective, bringing about a consistent stimulation of stem elongation when applied at the seedling stage (Bukovak and Wittwer, 1958; Harrington, 1960; Globerson and Ventura, 1973). The treatment hastened flowering even in those plants grown under photoperiods and temperatures favourable for flowering (Bukovak and

Wittwer, 1958), and prevented head formation in plants grown in cool conditions. Harrington (1960) found that three sprays of 10 ppm GA$_3$, applied at the 4-, 8- and 12-leaf stage increased seed yield from 223 kg ha^{-1} for the control allowed to form heads, to 1370 kg ha^{-1}. Deheading resulted in a 641 kg ha^{-1} yield. Nevertheless, the need to rogue fields for off types has meant the continued reliance on manual deheading by the seed industry (W. Waycott, personal communication, Salinas, CA, 1994).

The fundamental role of gibberellins in flower induction of lettuce is not clear at present. Since lettuce will eventually flower even under unfavourable conditions, such as short photoperiod and low temperatures, it is difficult to show that gibberellins are needed to induce flowering. Bukovak and Wittwer (1958) did, however, produce flowers on plants grown at 9-h photoperiod and 18/13°C temperature by treating with gibberellin, whereas untreated plants did not flower under the same conditions. They did not mention if seedstalk elongation of the controls was inhibited by head formation. Studies are needed that trace the levels of endogenous gibberellins in lettuce at various stages of flower induction.

SEED PRODUCTION

The floral axes of lettuce normally attain a height of 75–150 cm, and form a branched, bushy crown (Hayward, 1948). The inflorescence consists of a cymose cluster of heads, each of which may contain 15–25 flowers. The terminal flower capitulum usually flowers first, followed by lateral heads and those on branches. The weight of individual seeds reaches a maximum in 10–14 days after anthesis of the flower, but flowering on a single plant may extend over 40–50 days (Soffer and Smith, 1974; Globerson, 1981a). Mature seeds are therefore found on the plant before flowering is completed, and the first-formed seeds may be lost to shattering unless they are harvested repeatedly by shaking the seed heads into bags (Foster and van Horn, 1957). The long plant maturation time also points out the need for seed production in areas where the post-flowering period is rain-free and where low relative humidity prevails (Hawthorn and Pollard, 1954). Accordingly, in the USA, lettuce seed production is practised primarily in the dry interior valleys of California, Idaho, Oregon and Arizona under irrigation.

The climatic conditions under which the seed develops on the mother plant can have a considerable influence on seed characteristics. High production temperatures (30/20°C day/night temperature) resulted in lower seed yields and smaller seed size than growth at 25/15°C (Gray et al., 1988). Under cool conditions (20/10°C) yields were further reduced, but the seeds were 63% larger than at 25/15°C. Steiner and Opoku-Boateng (1991) also noted a reduction of seed weight, and of seeds per capitulum as temperatures increased in seed-production fields of lettuce in the San Joaquin Valley of California. High temperatures during seed development resulted in lowered seed vigour, but increased the capacity of the seed to germinate under

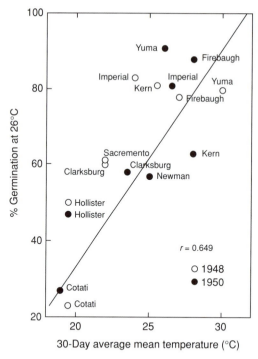

Fig. 14.10. The relation of germination percentage at 26°C and the average mean temperature for the 30-day period preceding seed harvest (Harrington and Thompson, 1952).

warm conditions (Fig. 14.10) (Harrington and Thompson, 1952; Steiner and Opoku-Boateng, 1991). On the other hand, seed developed at 20/10°C failed to germinate at 30°C (Gray *et al.*, 1988).

The physiological basis for the difference in seed yield with temperature has not been worked out. Since seed number produced per inflorescence was reduced both at high and low temperatures, pollination and fruitset may have been adversely affected at these temperatures. Lettuce is predominantly self-pollinated, with the pollen shedding from the inside of an anther cone (fused staminate tube) onto the style that grows through it at anthesis (Jones, 1927).

Water stress can also result in reduced seed yield, especially when it occurs during the reproductive period (Izzeldin *et al.*, 1980). As with high temperatures, the late stress treatments reduced the number of seeds per head. Highest seed yields were obtained if the plants were exposed to moderate water stress during the vegetative period, followed by adequate water.

Production techniques and environmental conditons that result in seed yields of uniform quality have been shown to also result in seed that produced uniform seedlings (Wurr *et al.*, 1986). Paradoxically, plants grown at low or high temperatures produced seed more uniform in size and vigour than if produced at intermediate temperatures (Gray *et al.*, 1988).

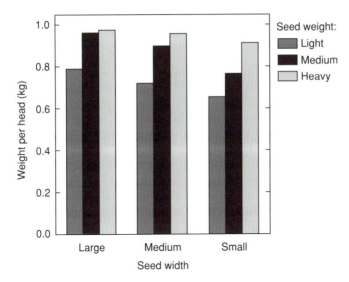

Fig. 14.11. The influence of seed weight and seed width on weight of heads harvested from the plants grown from that seed (Smith *et al.*, 1973b).

Treatment of the seed-producing plants with GA_{4+7} (a mixture of gibberellins 4 and 7) advanced date of flowering by 9–31 days, increased the synchrony of flowering and hastened plant maturity compared to untreated plants (Gray *et al.*, 1986). Seed yields were substantially increased, and the size and uniformity of seedlings grown from those seeds was also improved (Wurr *et al.*, 1986). There was no evidence of carry-over effects of the GA treatment on seedling germination or susceptibility to bolting.

The positive relationship between seed size and vigour of the resulting seedling has been noted by a number of researchers (Smith *et al.*, 1973a,b; Sharples, 1970; Wurr *et al.*, 1986). In an extensive series of studies, Smith and colleagues determined that both seed size and weight influenced seedling performance. For optimum uniformity, they found that seed had to be separated into size classes according to seed thickness or width, and into weight grades within each size (Smith *et al.*, 1973a,b). Plants produced from thin and light seeds gave significantly smaller heads at maturity than the heaviest seeds in each of the seed width categories (Fig. 14.11). It is important to note, however, that such relationships between size and vigour of seeds may only apply to seeds within a seedlot, and not between seed crops. Nevertheless, seed grading by size and weight can significantly increase the uniformity of the head lettuce crop.

PHYSIOLOGICAL DISORDERS

TIPBURN

Tipburn is characterized by necrosis of the edges of young, rapidly expanding leaves of lettuce. When severe, the disorder may develop on more than half of the area of these leaves. The necrosis is often visible from the outside in cos and butterhead types, but in crisphead lettuce, the affected leaves are in the head, and are not found until the head is utilized. During periods of hot weather, the disorder can be so severe that it will require the abandonment of entire fields of crisphead lettuce (Misaghi *et al.*, 1992). The risk of tipburn also constrains the length of the production season in areas where temperatures become progressively hotter, such as Arizona and parts of California (Misaghi *et al.*, 1992). Generally, the disorder develops on the growing plant in the field or glasshouse, but can also become a problem on harvested lettuce heads if they are stored at high temperatures, or not cooled quickly enough after harvest (Misaghi *et al.*, 1992). This increases the potential economic threat of this disorder, but also provides a quick means to check for tipburn susceptibility of lettuce lines in a breeding programme (Ryder, 1986). The disorder has been the focus of much research over the years, much of which was summarized in a review article by Collier and Tibbitts (1982).

Tipburn is caused by a calcium deficiency in the young, rapidly expanding leaves. Many researchers have shown that these tissues have lower calcium contents than expanded leaves (e.g. Collier and Huntington, 1983; Collier and Tibbitts, 1984; Cresswell, 1991). Recently, Barta and Tibbitts (1991) confirmed these findings with a sensitive new detection technique. Leaf tissue about to develop tipburn had significantly lower calcium values than adjacent areas of the same leaf, and than the same areas on leaves of non-affected plants (Fig. 14.12). Foliar sprays of calcium of the young leaves of butterhead lettuce plants prevented the disorder (Thibodeau and Minotti, 1969), while the application of organic salts such as oxalates induced the disorder.

The marginal necrosis of young leaves that is a characteristic sign of the disorder may be partly due to the collapse of the affected tissue because of membrane and cell wall damage. However, the damage may also be caused by toxicity from latex released from ruptured lacticifers at the edge of the leaf (Tibbitts *et al.*, 1965). Calcium deficiency may be involved both in the weakening of leaf cell membranes and cell walls, as well as the loss of integrity of the lacticifer cell walls, allowing their rupture and the spilling of latex (Tibbitts *et al.*, 1985).

The relatively immobile nature of calcium in the plant has made tipburn a difficult problem to control. While foliar sprays of calcium salts have been effective in open-headed and leaf types, where sprays can be directed on the areas most susceptible, the technique is ineffective on the crisphead types. The disorder often develops on plants growing in soils with ample supplies

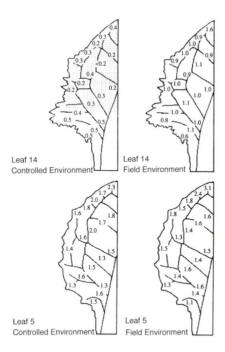

Fig. 14.12. Concentration of calcium in 20-mm long lettuce leaves, expressed as mg Ca g^{-1} dry weight, from plants grown in controlled environment or in the field. Leaves numbered from plant base (Barta and Tibbitts, 1991).

of calcium, so increasing soil calcium levels also is not an effective control strategy. Understanding the factors which govern the movement of calcium in the lettuce plant may however help to devise strategies for tipburn control.

Calcium is transported in the plant primarily in the xylem, and therefore moves preferentially to actively transpiring tissues (Bangerth, 1979; Barta and Tibbitts, 1986). As a consequence, areas of the plant that have low transpiration rate by virtue of their location receive little calcium. By covering the young leaves of lettuce plants with a plastic cap that reduced transpiration, Barta and Tibbits (1986) induced tipburn on 53% of the covered leaves, compared to 1% in the uncovered controls. Calcium contents were 0.63 and 1.48 mg g^{-1} dry weight, for covered and uncovered leaves, respectively. In the opposite approach, Goto and Takakura (1992) eliminated tipburn in butterhead lettuce by blowing air onto the young leaves as heads started to form. Thus maximizing plant transpiration rate by providing ample water and lowering relative humidity in greenhouse-grown lettuce crops during the light period could help in tipburn control, and there is evidence to this effect (Collier and Tibbitts, 1984).

Manipulation of transpiration rate is unlikely to be effective in alleviating tipburn in crisphead lettuce, where the sensitive tissue is enclosed and prevented from transpiring. For these plants, encouraging root pressure, which transports water and associated nutrient elements to all parts of the plant regardless of age, shows more promise for tipburn control. Root pressure occurs under conditions where transpiration is not functioning, and can be encouraged by increasing night-time relative humidity, and by reducing the resistance to water movement into the plant. Maintaining adequate soil water, and having a low night-time osmotic pressure of the soil solution are two ways in which water uptake can be maximized. This approach was moderately successful in reducing tipburn in butterhead lettuce grown in solution culture (Cresswell, 1991). Increasing relative humidity to above 95% during the dark period also decreased tipburn compared to lower humidities, and increased inner leaf calcium levels from 0.85 to 1.12 mg g^{-1} (Collier and Tibbitts, 1984).

The chances that a calcium deficiency will occur in the young leaves depends not only on the supply of the mineral to these organs as well as the demand of the leaves for calcium. As temperatures increase, so does growth rate, and with it the requirement for nutrients. Accordingly, Cox et al. (1976) found that the relative growth rate of lettuce, when varied by changing the temperature and light conditions, was closely tied to the time at which tipburn developed (Fig. 14.13). Similarly, tipburn incidence in serial plantings of crisphead lettuce through the year in Hawaii was most severe when growing temperatures were highest (Yanagi et al., 1983). Other conditions which increase plant growth rate can also result in increased tipburn. For instance, Crisp et al. (1976) found that widely spaced lettuce was more affected by tipburn than the same cultivar grown at close spacing. The higher tipburn incidence in lettuce grown in controlled environments than in the field may also relate to the faster growth rates achieved in the growth

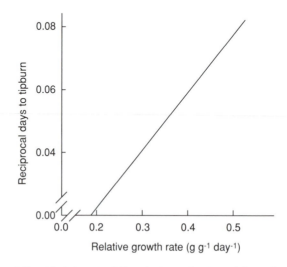

Fig. 14.13. The relationship between relative plant growth rate and the reciprocal of the number of days until tipburn first appears on the plants (Cox et al., 1976).

chamber (Barta and Tibbitts, 1991). The disorder thus determines the maximum productivity that can be achieved by the lettuce plant. Until we have a better understanding of tipburn and how it can be avoided, there will be little chance of surpassing that limitation in yield per unit area per unit time.

CONCLUDING REMARKS

Lettuce is a rather demanding and difficult crop to grow. At all stages of the life cycle, environmental conditions have major influence on growth and development. Seed germination is controlled by light in some cultivars, and all are subject to inhibition by high temperatures. The higher incidence of bolting, tipburn and puffy heads with prominent ribs at high temperatures, and other internal defects also contribute to restricting production to areas where mean maximum air temperatures do not exceed 26°C during the growing period.

Soil conditions for lettuce production are also exacting. To get a good stand of uniform plants, the direct-seeded crop must be precision-planted at an exact depth into a well-prepared seedbed. Good control of soil water content is important to permit stress-free growth, and avoid tipburn and other defects. The ability to provide optimum water conditions is a major factor contributing to the predominance of the western states in American lettuce production. Currently, 91% of American lettuce is grown in California, Arizona and Colorado (Anonymous, 1992).

The ability of lettuce to grow under cool, and frequently low light conditions has made it an important crop for winter glasshouse production, particularly in Europe. Although yields are much lower than under higher light conditions, concentrating on high quality butterhead and leaf lettuce makes the crops economically feasible.

The current attempts to adapt the crop to grow in artificial, high input environments will contribute much to our understanding of the growth and development processes of lettuce in years to come. Lettuce is already the standard crop for checking growth conditions in environmental cabinets (Hammer *et al.*, 1978). Standardization of environmental factors in growth chamber experiments has permitted the comparison of growth between as well as within experiments, and allowed the construction of a much larger database on lettuce than on other crops.

ACKNOWLEDGEMENTS

I am grateful for the thoughtful editorial suggestions of Drs Bill Waycott and Roy Ellerbrock.

REFERENCES

Abeles, F. (1986) Role of ethylene in *Lactuca sativa* cv. 'Grand Rapids' seed germination. *Plant Physiology* 81, 780–787.

Anonymous (1992) *Agricultural Statistics 1992*. US Government Printing Office, Washington.

Bangerth, F. (1979) Calcium-related physiological disorders of plants. *Annual Review of Phytopathology* 17, 97–122.

Barta, D.J. and Tibbitts, T.W. (1986) Effects of artificial enclosure of young lettuce leaves on tipburn incidence and leaf calcium concentration. *Journal of the American Society of Horticultural Science* 111, 413–416.

Barta, D.J. and Tibbitts, T.W. (1991) Calcium localization in lettuce leaves with and without tipburn: comparison of controlled-environment and field-grown plants. *Journal of the American Society of Horticultural Science* 116, 870–875.

Bensink, J. (1958) Heading of lettuce (*Lactuca sativa* L.) as a morphological effect of leaf growth. In: Garnaud, J.C. (ed.) *Advances in Horticultural Science and their applications*. Pergamon Press, London, pp. 470–475.

Bensink, J. (1971) On the morphogenesis of lettuce leaves in relation to light and temperature. *Mededelingen Landbouw Hogeschool te Wageningen* 71(15), 1–93.

Bierhuizen, J.F., Ebbens, J.L. and Koomen, N.C.A.(1973) Effects of temperature and radiation on lettuce growing. *Netherlands Journal of Agricultural Science* 21, 110–116.

Borthwick, H.A., Hendricks, S.B., Parker, M.W., Toole, E.H. and Toole, V.K. (1952) A reversible photoreaction controlling seed germination. *Proceedings of the National Academy of Science* 38, 662–666.

Borthwick, H.A., Hendricks, S.B., Toole, E.H. and Toole, V.K.(1954) Action of light on lettuce seed germination. *Botanical Gazette* 115, 205–225.

Boxall, M.I. (1971) Some effects of soil warming on plant growth. *Acta Horticulturae* 22, 57–65.

Brouwer, R. and Huyskes, J.A. (1968) A physiological analysis of the responses of the lettuce variety 'Rapide' and its hybrid with 'Hamada' to day-length and light intensity. *Euphytica* 17, 245–251.

Bukovak, M.J. and Wittwer, S.H. (1958) Reproductive responses of lettuce (*Lactuca sativa* var. Great Lakes) to gibberellin as influenced by seed vernalization, photoperiod and temperature. *Proceedings of the American Society of Horticultural Science* 71, 407–411.

Cantliffe, D.J., Fischer, J.M. and Nell, T.A. (1984) Mechanism of seed-priming in circumventing thermodormancy in lettuce. *Plant Physiology* 75, 290–294.

Collier, G.F. and Huntington, V.C. (1983) The relationship between leaf growth, calcium accumulation and distribution, and tipburn development in field-grown butterhead lettuce. *Scientia Horticulturae* 21, 123–128.

Collier, G.F. and Tibbitts, T.W. (1982) Tipburn of lettuce. *Horticultural Reviews* 4, 49–65.

Collier, G.F. and Tibbitts, T.W. (1984) Effects of relative humidity and root temperature on calcium concentration and tipburn development in lettuce. *Journal of the American Society of Horticultural Science* 109, 128–131.

Coons, J.M., Kuehl, R.O. and Simons, N.R. (1990) Tolerance of ten lettuce cultivars to high temperature combined with NaCl during germination. *Journal of the American Society of Horticultural Science* 115, 1004–1007.

Cox, E.F., McKee, J.M.T. and Dearman, A.S. (1976) The effect of growth rate on tipburn occurrence in lettuce. *Journal of Horticultural Science* 51, 297–309.

Cracker, L.E. and Seibert, M. (1983) Light and the development of 'Grand Rapids' lettuce. *Canadian Journal of Plant Science* 63, 277–281.

Cresswell, G.C. (1991) Effect of lowering nutrient solution concentration at night on leaf calcium levels and the incidence of tipburn in lettuce (var. Gloria). *Journal of Plant Nutrition* 14, 913–924.

Crisp, P., Collier, G.F. and Thomas, T.H. (1976) The effect of boron on tipburn and auxin activity in lettuce. *Scientia Horticulturae* 5, 215–226.

Dennis, D.J. and Dullforce, W.M. (1974) Analysis of the subsequent growth and development of winter glasshouse lettuce in response to short periods in growth chambers during propagation. *Acta Horticulturae* 39, 197–218.

Dullforce, W.M. (1962) Analysis of the growth of winter glasshouse lettuce varieties. *Proceedings of the 16th International Horticultural Congress* 2, 496–501.

Foster, R.E. and Van Horn, C.W. (1957) Lettuce seed production in Arizona. *Arizona Agricultural Experiment Station Bulletin* 282, 1–23.

Franklin, D.F. (1948) Using hormone sprays to facilitate bolting and seed production of hard-headed lettuce varieties. *Proceedings of the American Society of Horticultural Science* 51, 453–456.

Glenn, E.P. (1984) Seasonal effects of radiation and temperature of greenhouse lettuce in a high insolation desert environment. *Scientia Horticulturae* 22, 9–21.

Globerson, D. (1981a) Germination and dormancy in immature and fresh-mature lettuce seeds. *Annals of Botany* 48, 639–643.

Globerson, D. (1981b) The quality of lettuce seed harvested at different times after anthesis. *Seed Science Technology* 9, 861–866.

Globerson, D. and Ventura, J. (1973) Influence of gibberellins on promoting flowering and seed yield in bolting-resistant lettuce cultivars. *Israel Journal of Agricultural Research* 23, 75–77.

Goto, E. and Takakura, T. (1992) Prevention of lettuce tipburn by supplying air to inner leaves. *Transactions of the American Society of Aricultural Engineers* 35, 641–645.

Gray, D. (1975) Effects of temperature on the germination and emergence of lettuce (*Lactuca sativa* L.) varieties. *Journal of Horticultural Science* 50, 349–361.

Gray, D. (1977) Temperature sensitive phases during the germination of lettuce (*Lactuca sativa*) seeds. *Annals of Applied Biology* 86, 77–86.

Gray, D. and Morris, G.E.L. (1978) Seasonal effects on the growth and time of maturity of lettuce. *Journal of Agricultural Science* 91, 523–529.

Gray, D. and Steckel, J.R.A. (1981) Hearting and mature head characteristics of lettuce (*Lactuca sativa* L.) as affected by shading at different periods during growth. *Journal of Horticultural Science* 56, 199–206.

Gray, D., Steckel, J.R.A., Wurr, D.C.E. and Fellows, J.R. (1986) The effects of applications of gibberellins to the parent plant, harvest date and harvest method on seed yield and mean seed weight of crisp lettuce. *Annals of Applied Biology* 108, 125–134.

Gray, D., Wurr, D.C.E., Ward, J.A. and Fellows, J.R. (1988) Influence of post-flowering temperature on seed development, and subsequent performance of crisp lettuce. *Annals of Applied Biology* 113, 391–402.

Guedes, A.C. and Cantliffe, D.J. (1980) Germination of lettuce (*Lactuca sativa*) at high temperature after seed priming. *Journal of the American Society of Horticultural Science* 105, 777–781.

Hammer, P.A., Tibbits, T.W., Langhans, R.W. and McFarlane, J.C. (1978) Base-line growth studies of 'Grand Rapids' lettuce in controlled environments. *Journal of the American Society of Horticultural Science* 103, 649–655.

Harlan, J.R. (1986) Lettuce and sycomore: sex and romance in ancient Egypt. *Economic Botany* 40, 4–15.

Harrington, J.F. (1960) The use of gibberellic acid to induce bolting and increase seed yield of tight-heading lettuce. *Proceedings of the American Society of Horticultural Science* 75, 476–479.

Harrington, J.F. and Thompson, R.C. (1952) Effect of variety and area of production on subsequent germination of lettuce seed at high temperatures. *Proceedings of the American Society of Horticultural Science* 59, 445–450.

Hawthorn, L.R. and Pollard, L.H. (1954) *Vegetable and Flower Seed Production.* Blakiston, New York.

Hayward, H.E. (1948) *The Structure of Economic Plants.* Macmillan, New York.

Helm, J. (1954) *Lactuca sativa* in morphologisch-systematischer Sicht. *Kulturpflanze* 2, 72–129.

Hiraoka, T. (1969) Ecological studies on the salad crops. III. Effects of combination of temperature treatments on flower bud differentiation, bolting, budding and flowering time, of head lettuce (*Lactuca sativa* L. cv. Wayahead, Edogawa strain). *Journal of the Japanese Society of Horticultural Science* 38, 42–49.

Huang, X.L. and Khan, A.A. (1992) Alleviation of thermoinhibition in preconditioned lettuce seeds involves ethylene, not polyamine biosynthesis. *Journal of the American Society of Horticultural Science* 117, 841–845.

Ikuma, H. and Thimann, K.V. (1963) The role of seedcoats in germination of photosensitive lettuce seeds. *Plant Cell Physiology* 4, 169–185.

Ikuma, H. and Thimann, K.V. (1964) Analysis of germination processes of lettuce seed by means of temperature and anaerobiosis. *Plant Physiology* 39, 756–767.

Ito, H., Kato, T. and Konno, Y. (1963) Factors associated with the flower induction in lettuce. *Tohoku Journal of Agricultural Research* 14, 51–65.

Izzeldin, H., Lippert, L.F. and Takatori, F.H. (1980) An influence of water stress at different stages on yield and quality of lettuce seed. *Journal of the American Society of Horticultural Science* 105, 68–71.

Janick, J. (1989) *Classic Papers in Horticultural Science.* Prentice Hall, Englewood Cliffs, NJ.

Johnson, H. (1983) Head formation in crisphead lettuce. *HortScience* 18, 595.

Jones, H.A. (1927) Pollination and life history studies of lettuce (*Lactuca sativa* L.). *Hilgardia* 2, 425–479.

Khan, A.A. (1992) Preplant physiological seed conditioning. *Horticultural Reviews* 13, 131–188.

Khan, A.A. and Samimy, C. (1982) Hormones in relation to primary and secondary seed dormancy. In: Khan, A.A. (ed.) *The Physiology and Biochemistry of Seed Development, Dormancy and Germination.* Elsevier, Amsterdam, pp. 203–241.

Kimball, M.H., Sims, W.L. and Welch, J.E. (1967) Plant climate analysis for lettuce. *California Agriculture* 21(4), 2–4.

Klapwijk, D. (1979) Seasonal effects on the cropping cycle of lettuce in glasshouses during the winter. *Scientia Horticulturae* 11, 371–377.

Knight, S.L. and Mitchell, C.A. (1983a) Enhancement of lettuce yield by manipulation of light and nitrogen nutrition. *Journal of the American Society of Horticultural Science* 108, 750–754.

Knight, S.L. andMitchell, C.A. (1983b) Stimulation of lettuce productivity by manipulation of diurnal temperature and light. *HortScience* 18, 462–463.

Knight, S.L. and Mitchell, C.A. (1988a) Growth and yield characteristics of 'Waldmann's Green' leaf lettuce under different photon fluxes from metal halide or incandescent and fluorescent radiation. *Scientia Horticulturae* 35, 51–61.

Knight, S.L. and Mitchell, C.A. (1988b) Effect of incandescent radiation on photosynthesis, growth rate and yield of 'Waldmann's Green' leaf lettuce. *Scientia Horticulturae* 35, 37–49.

Knott, J.E., Andersen, E.M. and Sweet, R.D. (1939) Problems in the production of iceberg lettuce in New York. *Cornell Agricultural Experiment Station Bulletin* 714.

Knott, J.E., Terry, O.W. and Andersen, E.M. (1937) Vernalization of lettuce. *Proceedings of the American Society of Horticultural Science* 35, 644–648.

Kristensen, S., Frijs, E., Hendriksen, K. and Mikkelsen, S.A. (1987) Application of temperature sums in the timing of crisp lettuce. *Acta Horticulturae* 198, 217–226.

Krizek, D.T. and Ormrod, D.P. (1980) Growth response of 'Grand Rapids' lettuce and 'First Lady' marigold to increased far-red and infrared radiation under controlled conditions. *Journal of the American Society of Horticultural Science* 105, 936–939.

Krug, H. (1986) *Gemueseproduktion*. Paul Parey, Berlin.

Linden, T. (1988) Research continues to improve lettuce stand. *Western Grower and Shipper* 59, 4–6.

Lorenz, O.A. and Maynard, D.N. (1980) *Knott's Handbook for Vegetable Growers*, 2nd edn. J. Wiley, New York.

Maderiaga, F.J. and Knott, J.E. (1951) Temperature summations in relation to lettuce growth. *Proceedings of the American Society of Horticultural Science* 58, 147–152.

Marks, M.K. and Prince, S.D. (1979) Induction of flowering in wild lettuce (*Lactuca serriola* L.) II. Devernalization. *New Phytology* 82, 357–363.

Mattei, F., Sebastiani, A.L. and Gibbon, D. (1973) The effect of radiant energy on growth of *Lactuca sativa* L. *Journal of Horticultural Science* 48, 311–313.

Misaghi, I.J., Oebker, N.F. and Hine, R.B. (1992) Prevention of tipburn in iceberg lettuce during postharvest storage. *Plant Disease* 76, 1169–1171.

Mitchell, C.A., Leakakos, T. and Ford, T.L. (1991) Modification of yield and chlorophyll content in leaf lettuce by HPS radiation and nitrogen treatments. *HortScience* 26, 1371–1374.

Nothmann, J. (1976) Morphology of head formation of cos lettuce (*Lactuca sativa* cv. Romana). I. The process of hearting. *Annals of Botany* 40, 1067–1072.

Nothmann, J. (1977) Morphogenetic effects of seasonal conditions on head development of cos lettuce (*Lactuca sativa* cv. Romana) growing in a subtropical climate. *Journal of Horticultural Science* 52, 155–162.

Prince, S.D. (1980) Vernalization and seed production in lettuce. In: Hebblethwaite, P.D. (ed.) *Seed Production*. Butterworths, London, pp. 485–499.

Prusinsky, J. and Khan, A.A. (1990) Relationship of ethylene production to stress alleviation in seeds of lettuce cultivars. *Journal of the American Society of Horticultural Science* 115, 294–298.

Quaile, R. and Presnell, W. (1991) Climatic averages and extremes for US cities. *Historical Climatology Series 6–3*. US Department of Commerce, National Oceanic and Atmospheric Administration, National Climatic Data Center.

Raleigh, G. and Minotti, P.L. (1967) Lettuce production in muck soils. *Cornell Extension Bulletin* 1194, 1–16.

Rappaport, L. and Wittwer, S.H. (1956a) Flowering in head lettuce as influenced by seed vernalization, temperature and photoperiod. *Proceedings of the American Society of Horticultural Science* 67, 429–437.

Rappaport, L. and Wittwer, S.H. (1956b) Night temperature and photoperiod effects on flowering of leaf lettuce. *Proceedings of the American Society of Horticultural Science* 68, 279–282.

Ryder, E.J. (1979) *Leafy Salad Vegetables.* Avi, Westport, Connecticut.

Ryder, E.J. (1986) Lettuce breeding. In: Bassett, M.J. (ed.) *Breeding Vegetable Crops.* Avi, Westport, Connecticut, pp. 433–474.

Ryder, E.J. (1988) Early flowering in lettuce as influenced by a second flowering time gene and seasonal variation. *Journal of the American Society of Horticultural Science* 113, 456–460.

Ryder, E.J. and Whitaker, T.W. (1976) Lettuce. In: Simmonds, N.W. (ed.) *Evolution of Crop Plants.* Longman, London.

Scaife, M.A. (1973) The early growth of six lettuce cultivars as affected by temperature. *Annals of Applied Biology* 74, 119–128.

Sharples, G.C. (1970) The effect of seed size on lettuce seed germination and growth. *Progressive Agriculture in Arizona* 22(6), 10–11.

Sharples, G.C. (1973) Stimulation of lettuce seed germination at high temperature by ethephon and kinetin. *Journal of the American Society of Horticultural Science* 98, 209–212.

Sharples, G.C. (1981) Lettuce seed coatings for enhanced seedling emergence. *HortScience* 16, 661–662.

Shibutani, S. and Kinoshita, K. (1966) Studies on the ecological adaptation of lettuce. I. Lettuce planting throughout the year and the ecological adaptation. *Journal of the Japanese Society of Horticultural Science* 35, 387–394.

Smith, O.E., Welch, N.C. and Little, T.M. (1973a) Studies on lettuce seed quality: I. Effect of seed size and weight on vigor. *Journal of the American Society of Horticultural Science* 98, 529–533.

Smith, O.E., Welch, N.C. and McCoy, O.D. (1973b) Studies on lettuce seed quality: II. Relationship of seed vigor to emergence, seedling weight and yield. *Journal of the American Society of Horticultural Science* 98, 552–556.

Soffer, H. and Smith, O.E. (1974) Studies on lettuce seed quality. III. Relationship between flowering pattern, seed yield and seed quality. *Journal of the American Society of Horticultural Science* 99, 114–117.

Steiner, J.J. and Opoku-Boateng, K. (1991) Natural season-long and diurnal temperature effects on lettuce seed production and quality. *Journal of the American Society of Horticultural Science* 116, 396–400.

Takeba, G. and Matsubara, S. (1979) Measurement of growth potential of the embryo in New York lettuce seed under various combinations of temperature, red light and hormones. *Plant and Cell Physiology* 20, 51–61.

Tao, K. and Khan, A.A. (1979) Changes in strength of lettuce endosperm during germination. *Plant Physiology* 63, 126–128.

Thibodeau, P.O. and Minotti, P.L. (1969) The influence of calcium on the development of lettuce tipburn. *Journal of the American Society of Horticultural Science* 94, 372–376.

Thompson, H.C. and Knott, J.E. (1933) The effect of temperature and photoperiod on the growth of lettuce. *Proceedings of the American Society of Horticultural Science* 30, 507–509.

Thompson, R.C. and Kosar, W. (1948) Vernalization and stem development in lettuce. *Proceedings of the American Society of Horticultural Science* 52, 441–442.

Tibbitts, T.W., Bensink, J., Kuiper, F. and Hobe, J. (1985) Association of latex pressure with tipburn injury of lettuce. *Journal of the American Society of Horticultural Science* 110, 362–365.

Tibbitts, T.W., Morgan, D.C. and Warrington, I.J. (1983) Growth of lettuce, spinach, mustard, and wheat plants under four combinations of high-pressure sodium, metal halide, and tungsten lamps at equal PPFD. *Journal of the American Society of Horticultural Science* 108, 622–630.

Tibbitts, T.W., Struckmeyer, B.E. and Rao, R.R. (1965) Tipburn of lettuce as related to release of latex. *Proceedings of the American Society of Horticultural Science* 86, 462–467.

Valdes, V.M. and Bradford, K.J. (1987) Effect of seed coating and osmotic priming on the germination of lettuce seeds. *Journal of the American Society of Horticultural Science* 112, 153–156.

Valdes, V.M., Bradford, K.J. and Mayberry, K.S. (1985) Alleviation of thermodormancy in coated lettuce seeds by seed priming. *HortScience* 20, 1112–1114.

Verkerk, K. and Spitters, C.J.T. (1973) Effects of light and temperature on lettuce seedlings. *Netherlands Journal of Agricultural Science* 21, 102–109.

Waycott, W. (1993) Transition to flowering in lettuce: effect of photoperiod. *HortScience* 28, 530.

Wiebe, H.J. (1989) Effects of low temperature during seed development on the mother plant on subsequent bolting of chicory, lettuce and spinach. *Scientia Horticulturae* 38, 223–229.

Wooley, J.T. and Stoller, E.W. (1978) Light penetration and light-induced seed germination in soil. *Plant Physiology* 61, 597–600.

Wurr, D.C.E. and Fellows, J.R. (1984) The growth of three crisp lettuce varieties from different sowing dates. *Journal of Agricultural Science* 102, 733–745.

Wurr, D.C.E. and Fellows, J.R. (1991) The influence of solar radiation and temperature on the head weight of crisp lettuce. *Journal of Horticultural Science* 66, 183–190.

Wurr, D.C.E., Fellows, J.R., Gray, D. and Steckel, J.R.A. (1986) The effects of seed production techniques on seed characteristics, seedling growth and crop performance of crisp lettuce. *Annals of Applied Biology* 108, 134–144.

Wurr, D.C.E., Fellows, J.R. and Hambidge, A.J. (1992a) Environmental factors influencing head density and diameter of crisp lettuce cv. Saladin. *Journal of Horticultural Science* 67, 395–401.

Wurr, D.C.E., Fellows, J.R., Hiron, R.W.P., Antill, D.N. and Hand, D.J. (1992b) The development and evaluation of techniques to predict when to harvest iceberg lettuce heads. *Journal of Horticultural Science* 67, 385–393.

Wurr, D.C.E., Fellows, J.R. and Morris, G.E.L. (1981) Studies of the hearting of butterhead lettuce: temperature effects. *Journal of Horticultural Science* 56, 211–218.

Wurr, D.C.E., Fellows, J.R. and Pitham, A.J. (1987) The influence of plant raising conditions and transplant age on the growth and development of crisp lettuce. *Journal of Agricultural Science* 109, 573–581.

Wurr, D.C.E., Fellows, J.R. and Suckling, R.F. (1988) Crop continuity and prediction of maturity in the crisp lettuce variety Saladin. *Journal of Agricultural Science* 111, 481–486.

Yanagi, A.A., Bullock, R.M. and Cho, J.J. (1983) Factors involved in the development of tipburn in crisphead lettuce. *Journal of the American Society of Horticultural Science* 108, 234–237.

Zink, F.W. (1955) Studies with pelleted lettuce seed. *Proceedings of the American Society of Horticultural Science* 65, 335–341.

Zink, F.W. (1959) Development of spiraled heads in Great Lakes lettuce. *Proceedings of the American Society of Horticultural Science* 73, 377–384.

Zink, F.W. and Yamaguchi, M. (1962) Studies on the growth rate and nutrient absorption of head lettuce. *Hilgardia* 32, 471–485.

Cauliflower, Broccoli, Cabbage and Brussels Sprouts

15

H.C. Wien[1] and D.C.E. Wurr[2]

[1]*Department of Fruit and Vegetable Science, Cornell University, 134A Plant Science Building, Ithaca, New York 14853–5908, USA*
[2]*Crop and Weed Science Department, Horticulture Research International, Wellesbourne, Warwick CV35 9EF, UK*

The many different selected forms of *Brassica oleracea* present probably the most striking example of polymorphism in the group of plants considered vegetables. The progenitor of this diverse set of plants is described by Helm (1963) as a polymorphic perennial herb that generally had a height of 60–100 cm, and a moderately branched stem. It is thought to have originated on the coast and islands of the Mediterranean, and to have spread from there along the Atlantic coast as far as Scotland (Helm, 1963; Thompson, 1976). Earliest selections were probably made to reduce the content of bitter glucosinolates found in abundance in the wild types. The single-stemmed kales, with either smooth or ruffled leaf shape, were probably first consumed by people of the Mediterranean area as early as 600 BC (Thompson, 1976).

The precise path of selection for the cabbage family vegetables that are popular today is not exactly known. Thompson (1976) speculated that as *B. oleracea* spread northward, types having a biennial flowering habit became predominant, requiring a cool period of 2–3 months to induce flowering. From among these, types were selected with a much-shortened stem and a large apical bud surrounded by many leaves (Fig. 15.1). Cabbage occurred as early as the twelfth century in Germany (Helm, 1963). Selection of plants with short, swollen fleshy stems resulted in our present-day kohlrabi, first described in the 1500s in the same area. The selection for types with short, heavily branched fleshy inflorescences apparently occurred in the eastern Mediterranean area. Cauliflower was described in the sixteenth century, but broccoli apparently did not appear until 100 years later (Thompson, 1976). The latest arrival of the common cole crops is Brussels sprouts, which was selected, as the name implies, in Belgium in the eighteenth century.

Selection for different types continues to the present day. In recent years, the distinction between cauliflower and broccoli has become blurred

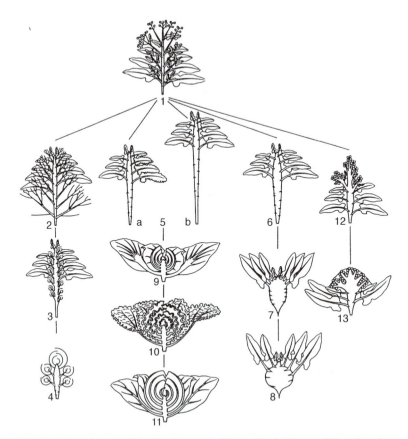

Fig. 15.1. A schematic view of the development of the cultivated types of *Brassica oleracea* (Helm, 1963).

(Gray, 1989), and intermediate types that combine the curd structure of cauliflower with the colour of broccoli are becoming popular. Physiologically, this means that we can also expect to find a continuum of responses to the environment, particularly to temperature, among the new types of cole crops being developed (Dickson and Wallace, 1986).

JUVENILE PERIOD

The induction of flowering in *Brassica oleracea* is brought about by relatively low temperatures, in the process called vernalization (see Chapter 3) but the temperatures are effective only if the plants are past what is called the juvenile period.

Evidence for the existence of a juvenile period is obtained by exposing plants of various ages to low temperatures, and observing how many leaves are formed on the main stem before the apex changes to a reproductive

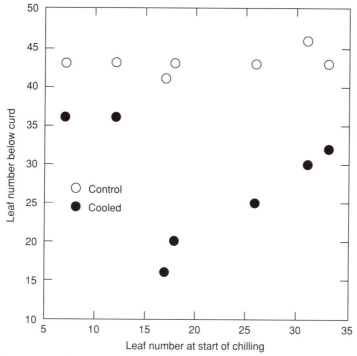

Fig. 15.2. Effect of a 28-day exposure to 5°C begun at various plant ages (leaf numbers) on total leaves formed below the curd by 'White Fox' cauliflower (Hand and Atherton, 1987). Plants became adult after formation of 17 leaves (by permission of Oxford University Press).

structure. If the plant is still juvenile when the cold treatment is imposed, the final leaf number will be similar to plants not cold-treated (Fig. 15.2). After the plants have reached the adult vegetative stage, cold treatment will reduce the leaf number formed, but also reduces the rate at which plants grow. It is therefore advisable to determine the effectiveness of vernalization, not by chronological time to particular reproductive events, but by leaf number formed before the reproductive structure (Sadik, 1967; Wiebe, 1972a; Hand and Atherton, 1987).

The existence of the juvenile period has been clearly demonstrated by Stokes and Verkerk (1951), Wiebe (1972a) and Thomas (1980) who found that germinating seeds could not be vernalized. There are, however, instances where the existence of a juvenile phase has been questioned. In cabbage (Nakamura and Hattori, 1961) and in cauliflower (Fujime and Hirose, 1979) it has been shown that there was a slight reduction in final leaf number with vernalization of imbibed seeds, suggesting that there was no juvenile period. It seems likely that there are large differences in juvenility between cultivars and species and it may be that sensitivity to induction decreases and then increases as plants get older.

Changes in plant morphology that occur when cole crops reach the

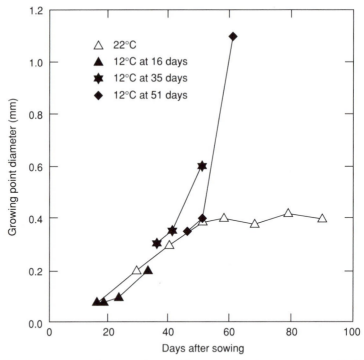

Fig. 15.3. Change in apex diameter when cauliflower cv. Aristokrat was exposed to 12°C at different times after sowing (Wiebe, 1972c). Plants reached the adult stage after about 42 days growth at 22°C.

adult stage have been documented by a number of workers. From the standpoint of overall development, Stokes and Verkerk (1951) found that Brussels sprout plants showed a pronounced thickening of the stem, and enlargement of axillary buds when they became sensitive to induction by cold. This change occurs when the plants have formed about 30 leaves plus leaf initials and appears to be a feature generally applicable to Brussels sprout cultivars with a range of maturities (Thomas, 1980). In cabbage, transition from juvenile to adult stage is less pronounced, but appears to occur when stem diameter reaches about 6 mm (Boswell, 1929; Ito and Saito, 1961).

In cauliflower (Wiebe, 1972c), broccoli and Brussels sprouts (Stokes and Verkerk, 1951), adult vegetative plants have a distinctly domed apical meristem that is broader than that of the juvenile plants (Fig. 15.3). A pronounced enlargement of the apex occurs after plants are exposed to 'relative cold' when they have reached the adult stage.

The length of the period during which exposure to cold will not induce flowering appears to be governed primarily by genotype, although it can also be influenced by environment. Wiebe (1972a) found that cauliflower plants grown under low light, and/or partially defoliated, formed more leaves before the curd than intact plants grown under full light. These findings support the views of Hand and Atherton (1987), and Williams and Atherton (1990), that during the juvenile period, assimilates are used preferentially for leaf growth rather than for apical meristem growth. Thus under conditions of limited resources, leaf growth would be prolonged at the expense of meristem development.

Differences among cultivars in the leaf number marking the end of the juvenile period are variable. For instance, Wurr *et al.* (1982) found that in cauliflower, leaf number below the curd varied from 21 to 98 among a range of genotypes. This may reflect differences in the end of the juvenile period. On the other hand, seven Brussels sprout cultivars all reached the adult stage with about 30 leaves and initials, some cauliflower cultivars became sensitive after only ten leaves were formed (Wiebe, 1972a; Thomas, 1980).

From a practical standpoint, the existence of a juvenile period permits the plant to grow to an adequate size before it can be induced to flower. In climatic situations where a mild cold period would allow the crop to survive but not grow actively, a cole crop such as cabbage can be planted at the beginning of the cold period and survive until a later warm period without being induced to flower. Indeed, such overwintering of cabbage used to be a common practice in the eastern USA (Boswell, 1929).

CURD AND HEAD DEVELOPMENT IN CAULIFLOWER AND BROCCOLI

CURD AND HEAD MORPHOLOGY

The curd of cauliflower has been described as a prefloral structure that shares some of the characteristics of both the vegetative and reproductive apices (Sadik, 1962; Margara and David, 1978). Like the vegetative shoot, it retains the 5 + 8 phyllotaxy of leaves, but leaf development has been reduced so that only bracts are formed in the curd. The lateral buds of the shoot meristem are elongated and much branched, and the apices of these branches form the surface of the curd. These apices are partly differentiated into prefloral structures, with masses of rudimentary flower buds visible microscopically (Sadik, 1962) but bud development is arrested at this early stage. The entire structure, much shortened and thicker than a normal flowering shoot, can give rise to the future inflorescence of the plant, if the right environmental conditions prevail. If not harvested, much of the curd becomes necrotic or decays, unless the distinct chilling requirement for flower bud initiation described by Fujime (1983) has stimulated the further development of some primordia into true reproductive structures (Sadik, 1962). The formation of flower buds is accompanied by bud elongation so that they project out of the curd, giving it a grainy appearance, termed 'riceyness'. As the reproductive structures develop, marked elongation of branches takes place, most often at the edge of the curd. By the time of anthesis, the flower stalks may project 20–50 cm above the curd (Sadik, 1962).

In broccoli, branched inflorescence development proceeds directly to the formation of flower buds without the formation of the intermediate pre-floral stage. While heads are marketable the flower buds are still small, but if the heads are not cut, the buds continue development to open flowers.

ENVIRONMENTAL FACTORS IN CURD AND HEAD INDUCTION

The formation of curds in cauliflower, and heads in broccoli is primarily influenced by temperature. Estimates of the precise temperatures allowing vernalization to proceed vary but Wiebe (1972a), Atherton *et al.* (1987) and Wurr *et al.* (1988a) all showed that in cauliflower as the temperature increased, the relative rate of vernalization increased to a maximum and then declined. Wiebe (1972a) showed maximum response between 7 and 12°C while Wurr *et al.* (1993) have recently developed a model for sum-mer/autumn cauliflower with no vernalization below 9 or above 21°C and the maximum rate of vernalization between 9 and 9.5°C. For different culti-vars in the same maturity period, Grevsen and Olesen (1994) have proposed a symmetrical relationship with no vernalization below 0 and above 25.6°C and the maximum rate of vernalization at 12.8°C. Thus vernalization in summer/autumn cauliflower proceeds most rapidly at moderate tempera-tures and slowly or not at all at low and high temperatures. At the latter many temperate cultivars fail to form curds, but continue producing leaves at the apex (Fig. 15.4). We now need to know what the shape of the vernal-ization function is for cultivars from different maturity periods. According to Wiebe (1990) the highest temperature capable of bringing about curd for-mation varies from about 16°C in some cultivars, to nearly 30°C in others.

Broccoli has a similar temperature response to cauliflower, although the upper temperature limit for head formation may be higher. For instance, in Wiebe's study, the cultivar Coastal formed heads at a constant 27°C (Wiebe, 1975) (Fig. 15.4). Fontes and Ozbun (1972), on the other hand, prevented head formation in 'Waltham 29' broccoli by raising it at 24/27°C day/night temperature.

Light conditions during vernalization were not important as long as optimum temperatures were used (Wiebe, 1972b). However, if the night temperature was raised from 12°C to 22°C at a reduced light intensity of 2.5 klux, curd formation was delayed and leaf number was increased. This may imply that adequate carbohydrate levels must be present in the plant to permit the differentiation of curds. Several investigators have related curd initiation to apex carbohydrate level, and prevented it by reducing carbohy-drates through growing plants in the dark, or raising the plants in CO_2-free air or during periods of high temperature (Sadik and Ozbun, 1968; Wiebe, 1974). Similarly, Atherton *et al.* (1987) hastened curd initiation by applying sucrose to intact plant apices. It may be, as Atherton and co-workers suggest

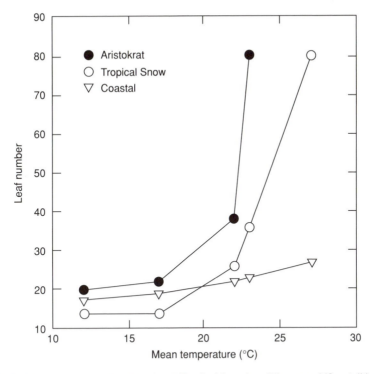

Fig. 15.4. Effect of growing 'Aristokrat' and 'Tropical Snow' cauliflower and 'Coastal' broccoli at various constant temperatures on total leaf number (Wiebe, 1975) (with kind permission of Elsevier).

(Atherton *et al.*, 1987; Williams and Atherton, 1990), that low temperatures are required to reduce competition for assimilates between developing leaves and the apical meristem by suppressing leaf growth. The lack of correlation between apex sugar levels and differentiation to reproductive growth found by Wiebe (1974) for cauliflower, and Fontes and Ozbun (1972) for broccoli, may imply, however, that other factors, such as hormones, are also involved (see below).

CULTIVAR DIFFERENCES

Through breeding and selection, a range of cauliflower cultivars has been developed to produce satisfactory yields in specific environments ranging from the intermediate altitude tropics to the winter season of western England. The differences in adaptation have resulted in cultivars that contrast greatly in their response to temperature. Cauliflowers normally grown in the tropics can produce curds at mean temperatures of 25°C, after forming about 40 leaves (Wiebe, 1975) (Fig. 15.4). These cultivars typically have a

Fig. 15.5. Appearance of 'White Baron' cauliflower curd after exposure to various growth and vernalization temperatures. (A) Growth at 21/16°C day/night, vernalized for 2 weeks at 10°C, beginning 2 weeks after sowing. (B) Growth at 21/16°C. (C) Growth at 32/21°C (H.C. Wien, unpublished data).

short juvenile period and can be induced to flower by 1–2 weeks of low temperatures (Sadik and Ozbun, 1967). Thus when grown in temperate areas, they typically form only small plants before initiating a curd, and often have ricey curds that quickly elongate to start flowering (Wiebe, 1975). Prolonged exposure of these cultivars to temperatures of 12–15°C in the early stages results in the formation of green broccoli-like heads instead of curds (Fig. 15.5).

There is also considerable cultivar variation within temperate cauliflower. Wiebe (1972c) noted that a late cultivar reached the adult vegetative state about 2 weeks after an early cultivar because of a lower rate of leaf formation. An additional source of variation comes from the response to particular temperatures. In a series of field plantings in the Netherlands, Booij (1987) noted that as maximum temperature increased, 'Elgon', also known as 'Dok' or 'Dok Elgon' formed more leaves before curd initiation than the earlier cultivar 'Delira' (Fig. 15.6). Maturity of 'Elgon' would therefore be delayed more than 'Delira' after high temperatures in the field.

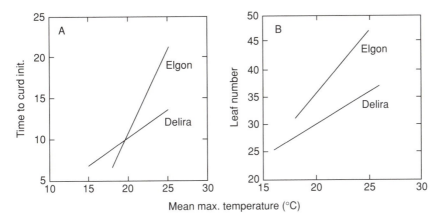

Fig. 15.6. Relationship between mean maximum temperature and (A) time to curd initiation and (B) total leaf number below the curd, for 'Elgon' and 'Delira' cauliflower (Booij, 1987).

The feature distinguishing over-wintered cultivars in mild temperate climates from those maturing in one growing season is the vernalization requirement (Sadik, 1967). In his experiments, a 6-week period of 5°C was needed to induce curd production and flowering in the cultivar 'February-Early-March'. More extensive field studies with a range of these winter-heading cultivars by Wurr *et al.* (1981b), indicated that they may also have a lower temperature threshold for curd induction than summer cultivars, and a lower base temperature for curd growth. This should increase their growth adaptation during the cold season but some of them take more than 11 months to mature, and suffer the risk of crop loss in colder than usual winters.

HORMONAL EFFECTS

It has long been conjectured that the induction of cauliflower curd formation could be mediated by growth hormones, but attempts to substantiate this through direct measurement of endogenous compounds have only partly succeeded. The most consistent link between curd induction and hormone levels has been made for the gibberellins. Thomas *et al.* (1972) measured an increase in gibberellin-like substances about 2 weeks before curds were initiated in cauliflower. If the plants were exposed to cold, the hormone peak was increased, but curd initiation was delayed past the end of the sampling period. In support of these findings, Kato (1965) measured a slight increase in gibberellin activity with cold treatment of cauliflower, as did Fontes *et al.* (1970), after cooling broccoli plants. Further evidence supporting gibberellin involvement in curd induction comes from Booij (1989, 1990c) who found in an extensive series of field plantings that GA_{4+7} (a mixture of gibberellins 4 and 7), but not GA_3, applied when the plants had just reached the adult vegetative stage, reduced the number of leaves to the curd. In contrast, applications of GA_3 to winter cauliflower in Israel by Leshem and Steiner (1968) failed to reduce leaf numbers, or significantly hasten curd formation.

The situation is less clear when considering growth inhibitor experiments. Application of the growth retardants ancymidol, chlormequat or daminozide to vegetative plants had no effect on curd diameter (Booij, 1989). Daminozide, but not chlormequat, inhibited flowering and partially negated the effect of a cold treatment on leaf number in broccoli (Fontes and Ozbun, 1970). Paradoxically, the daminozide treatment also caused a large increase in endogenous gibberellins in the plants (Fontes *et al.*, 1970), whereas chlormequat did not change gibberellin levels. Much more work is needed to resolve these conflicting results. In all the experiments reporting endogenous gibberellin levels, gibberellins were detected by lettuce hypocotyl bioassay. It is vital that these findings be checked with modern direct measurements of the levels of endogenous gibberellins, and that the identity of the specific hormones be established.

FLOWER INDUCTION IN CABBAGE AND BRUSSELS SPROUTS

In contrast to cauliflower and broccoli, flower induction in cabbage and Brussels sprouts is not required for commercial crop production other than for seed production. Indeed, in both species if flower development occurs it is likely to detract from crop quality. To form flowers, cabbage and Brussels sprout plants must be exposed to vernalizing temperatures once they have reached the end of the juvenile period. Optimum vernalizing temperatures have generally ranged from 4 to 10°C, although there are cultivar differences in the temperatures that are effective (Ito and Saito, 1961; Heide, 1970; Friend, 1985). Where temperatures fluctuate, both the mean temperature and the diurnal pattern have an influence on how fast a plant comes to flower. Ito and Saito (1961) found that flowering was delayed, but not prevented, by exposure of cabbage to daily cycles of 16 h at 9°C, and 8 h at 27°C (Fig. 15.7). No differentiation of reproductive structures occurred if 8 h at 9°C and 16 h at 27°C were given, even after 120 days exposure. Similar findings were made by Heide (1970) for cabbage, and Verkerk (1954) for Brussels sprouts. If long durations of vernalizing temperatures were applied, high temperatures at the end of the cold period were ineffective in reversing the induction (Heide, 1970).

Plant age and size when inducing temperatures are applied also play important roles in determining the rate and effectiveness of the flower induction. Among plants of the same age, the largest were most strongly induced by a given marginal cold treatment (Ito and Saito, 1961). When plant size was kept constant among plants of different ages with nutritional treatments, the oldest plants flowered first. Ito *et al.* (1966) conducted

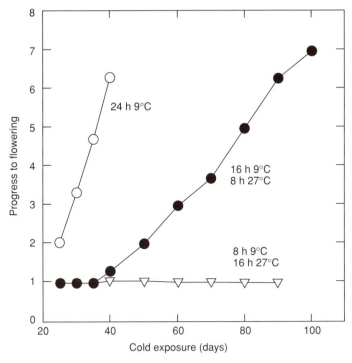

Fig. 15.7. Progress toward flowering of the cabbage apical meristem as influenced by plant exposure to increasing durations of 9°C, for 24, 16 or 8 h day^{-1}. Flower development stages: 1 = vegetative, 2 = dome-shaped, 3 = early reproductive, 4 = mid reproductive, 5 = sepals formed, 6 = petals formed, 7 = stamens formed (Ito and Saito, 1961).

studies to determine the physiological basis for the size effects. These indicated that reducing plant size by defoliation or by making small stem cuttings caused only a small increase in the vernalization requirement, even though the carbohydrate content of the plants was reduced by the treatments. Kruzhilin and Shvedskaya (1960) found on the other hand that defoliation before vernalization prevented flower induction. Differences in cultivars and the size of the plants defoliated may explain these varying results. The effect of plant age in cabbage may be caused by juvenility which prevents flower induction totally in young plants and may only gradually disappear in older plants. Ito *et al.* (1966) were able to increase the vernalization requirement in successive cuttings taken from cabbage plants, which may indicate a partial reversion to the juvenile state.

Under conditions marginal for induction of flowers in cabbage, factors which encourage stem elongation appear to aid flowering as well. For instance, exposure of cabbage to long day conditions after vernalization caused a marked increase in stem elongation, and marginally increased the number of flowering plants, compared to plants kept in short days (Heide, 1970). Frequent applications of gibberellic acid to adult cabbage plants grown at high altitude locations in Kenya also resulted in flowering of some cultivars that remained vegetative without the treatment (Kahangi and Waithaka, 1981). However, gibberellic acid applications are not effective in inducing flowering of all cultivars (Kahangi and Waithaka, 1981; Friend, 1985).

In many plant species flower induction is an after-effect of the temperature treatment, with little change apparent in the apical meristem during the cold period. However, in *Brassica oleracea*, the apical meristem changes morphologically to the generative state while vernalizing temperatures are applied (Stokes and Verkerk, 1951; Ito and Saito, 1961; Wurr *et al.*, 1993) and it is likely that vernalization occurs as the direct result of inducing temperatures.

The duration of low temperature treatment needed to induce flowering generally ranges from 10 to 50 days or longer in cabbage and 50–80 days in Brussels sprouts (Stokes and Verkerk, 1951; Ito and Saito, 1961; Heide, 1970; Thomas, 1980) though there are considerable differences between cultivars. In Brussels sprouts, an annual genotype has been identified that flowers without cold treatment (Wellensiek, 1960). Perhaps this is not surprising since cultivated cole crops are thought to have arisen from an annual ancestor grown in the Mediterranean region (Helm, 1963).

If cabbage or Brussels sprouts plants are given vernalization treatments insufficient for complete flower induction, partial reproductive development results (Stokes and Verkerk, 1951; Ito and Saito, 1961). Weak induction in cabbage resulted in the formation of an elongated stem and leaf bracts, no differentiation of flower structures, but the continuation of leaf and resumption of head formation. With successively stronger induction, reproductive growth was increasingly more complete, and the rate of development became more rapid.

The partial induction of flowering in cabbage and Brussels sprouts can also result from subjecting the plants to devernalizing conditions after exposure to temperatures adequate for vernalization. In fact, Heide (1970) considered any temperature above 12°C as potentially devernalizing for the cabbage cultivars that he used. In general, however, devernalization is most effective when temperatures of 20–30°C are imposed alternately with the inducing temperature. Unfortunately, we know virtually nothing about the chemical and hormonal changes that differentiate the development of partially and fully induced individuals.

CABBAGE HEAD AND BRUSSELS SPROUT BUTTON FORMATION

The process of head formation in cabbage appears to be similar to that described for lettuce in Chapter 14. The assemblage of layers of leaves over the growing point requires the maintenance of a short stem during the heading period (North, 1957). As heading begins, leaves become broader

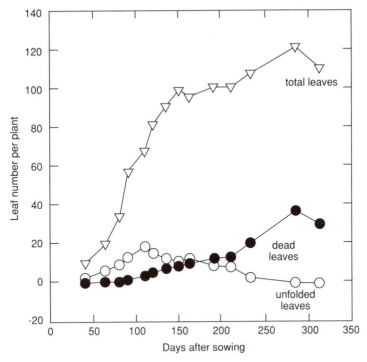

Fig. 15.8. Leaf production of cabbage, cv. Enkhuizen Glory, grown under field conditions (North, 1957) (with permission of Oxford University Press).

and sessile, and more erect in their posture (North, 1958; Kato and Sooen, 1978). The inward curvature of the leaf edges, combined with their upright position leads finally to the formation of the head. Rate of leaf production continues at a high rate in spite of the increasing confinement by previously formed foliage (North, 1957) (Fig. 15.8). As more leaves form, and these start to expand, the head gains in weight and firmness until it reaches a density acceptable for harvesting. Size of the head at harvestable density is determined by the cultivar, and cultural practices including such factors as the space available per plant, water and nutrient supplies during the growing season (see section on factors determining productivity below).

The factors that lead to erect frame leaf posture of cabbage and other head-forming *Brassica* species such as Chinese cabbage, have been under intensive investigation by Kato and co-workers in Japan. They pointed out the important role of the frame leaves in providing photosynthate for plant growth, and for creating environmental conditions that allowed younger leaves to grow more erect (Kato, 1981). Removal of inner frame leaves just as heading was beginning markedly delayed head formation, and resulted in the younger leaves assuming a horizontal attitude (Kato and Sooen, 1978; Kato, 1981; Hara *et al.*, 1982), whereas removal of outer frame leaves had little effect on inner leaf attitude. Kato postulated that shading of the leaf bases of the inner frame leaves led to their inward curvature. In support of this theory, head formation was advanced by tying up the frame leaves of cabbage (Kato and Sooen, 1978).

Hormonal factors in the formation of heads have been little investigated in cabbage or Chinese cabbage. Placement of the synthetic auxin naphthalene acetic acid (NAA) on the abaxial leaf surface near the leaf tip increased cupping, while placement on the opposite side caused them to become more horizontal (Ito and Kato, 1957). This implied that auxin movement from adaxial to abaxial leaf surface may be responsible for the hyponastic curvature of head-forming leaves, but the determination of endogenous auxin levels to substantiate this theory has not yet been accomplished.

Kato and Sooen (1980) replaced the cabbage apex with hormone pastes or treated plants with hormone sprays. Of the chemicals tried, gibberellic acid increased cupping and erectness of leaves, while benzyl adenine (a synthetic cytokinin), naphthalene acetic acid (a synthetic auxin) and auxin and gibberellin inhibitors had the opposite effect. Furthermore, partial root removal led to increased heading tendency in cabbage (Kato and Sooen, 1979). This may imply that root-produced cytokinins or abscisic acid play an inhibitory role in head formation. Clearly, measurements of endogenous hormone levels in heading and non-heading isolines are needed to help explain these results.

Studies of cultivars that form heads at different times have provided more insight into the process of heading. In a comparison of five cultivars, North (1957) found that head maturity was most closely related to the time when the cultivars stopped unfolding frame leaves, i.e. the time of head initiation. In a subsequent study, North (1958) could predict the time of head maturity from the length : width ratio of leaves seven to 12 on the young plant. Cultivars having relatively broad leaves produced heads significantly earlier than those with narrower leaves and indeed North (1958) considered that broad leaves were necessary for head formation. The broad leaf shape may have reduced light intensities at the leaf bases, thought by Ito and Kato (1957) to be needed to get the erect leaf position necessary for head formation. Kato (1981) further stated that late head-forming cultivars may be more sensitive to light with regard to leaf posture than early cultivars. Again, the physiological mechanisms of these responses have not been explored.

The head formation period ends with the attainment of the correct head density for harvesting. If the head is not harvested on time, further expansion of the inner leaves, and resumption of stem growth result in the splitting of the head (North, 1957). Stem extension will occur even if the plant has not received sufficient low temperature exposure to bring about flower initiation, and seems to be part of the periodic growth cycle of the plant. Cabbage plants allowed to continue growth in an environment that does not induce flowering will show periodic stem extension followed by head formation. Miller (1929) showed that a cabbage plant formed a series

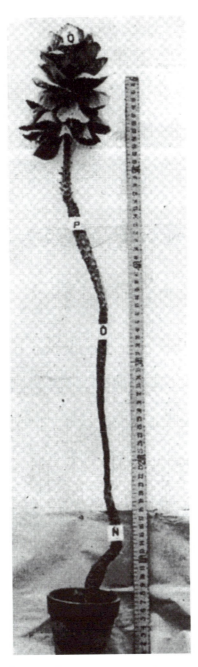

Fig. 15.9. Appearance of 'Danish Ballhead' cabbage after 2 years growth in a greenhouse at mean air temperature of 19°C. During this time, the plant produced four heads at N, O, P and Q (Miller, 1929).

of four heads on increasingly longer stalks over a 2-year growing period (Fig. 15.9).

Cabbage breeders have selected against head splitting tendency, to permit the grower more latitude in harvest time. Work is needed to explain the physiology of cabbage stem growth and relate it to plant hormonal changes.

In Brussels sprouts, the sprout buttons are enlarged buds in the axils of leaves and the number of such axillary buds depends upon the number of nodes produced by the plant. Hodgkin (1981) reported that the rate of node production increased until June and then remained steady until September. He found that only half the nodes developed buttons sufficiently large to harvest and these were produced by mid-July. Some nodes produced buttons that were too small and some nodes produced no axillary buds. The rate of node production decreased after September until it was terminated by the production of floral primordia in November or December.

FACTORS DETERMINING PRODUCTIVITY

CAULIFLOWER

Cauliflower has distinct responses to temperature for the phases of juvenility, vernalization and curd growth (Wurr *et al.*, 1995) and this makes it a difficult crop to grow. Leaf production and expansion rate increase with temperature up to the end of juvenility but vernalization has a specific temperature range as previously described. If temperatures exceed this range, curd formation could be delayed or interrupted by further leaf formation. At temperatures lower than the optimum, leaf area development could be curtailed, leading to buttoning, or the production of ricey curds (see section on physiological disorders below). Cauliflower is also very demanding in terms of water and fertility requirements (Nonnecke, 1989). Only with a combination of temperature and other conditions permitting uninterrupted growth of the plant can sufficient leaf area form to allow the production of marketable curds. As a consequence, the major part of cauliflower production in temperate countries has been concentrated in areas where predictably moderate temperatures and adequate control of moisture conditions can be found. Under optimum temperature and moisture conditions, marketable yields of 12–15 t ha^{-1} can be achieved (Lorenz and Maynard, 1980; Dufault and Waters, 1985). With increases in plant density, yields can be increased, but the resulting decrease in curd size (Salter and James, 1975; Thompson and Taylor, 1975) may not meet marketing criteria, except when sold as mini-cauliflowers (Salter, 1971). Both groups of workers also noted that some cultivars show virtually no change in marketable yield with increased plant density, while other cultivars are much more responsive. The latter also showed a slight increase in uniformity of maturity at close spacings (Salter and James, 1975).

Variability of plants in a crop frequently results in maturity covering a period of several weeks. Since harvesting can account for 20–40% of the total production cost, measures to reduce the number of times the crop is cut would make it cheaper to grow (Wheeler and Salter, 1974). Efforts to shorten harvest duration have focused both on the improvement of plant uniformity through modification of production practices, and through techniques that would synchronize curd initiation and growth.

Until the 1970s, cauliflower in Europe and North America was commonly sown in protected or outdoor seedbeds, and plants 'pulled' for transplanting to the field with little or no soil adhering to the roots (Salter and Fradgley, 1969). By sowing the crop directly in the field, Salter and Fradgley (1969) halved the typical harvest duration of 20–30 days. Smaller improvements were achieved by more uniform spacing in the seedbed and by transplanting younger plants. In recent years, plant uniformity has been dramatically increased through the use of container-raised transplants. For instance, Wurr *et al.* (1990a) reported harvest durations of 14–22 days for 'White Fox' cauliflower raised in cells, compared to 29–43 days for bare-rooted plants. Similar figures were quoted by Nonnecke (1989) from research on Long Island. In fact, the desire of a New York cauliflower grower to improve the uniformity of his cauliflower crop led to his developing the 'Todd planter tray' in the 1960s which is now widely used for raising transplants of many crops.

Variation in the time of curd maturity is the result of accumulated variation among plants at earlier stages of growth. Booij (1990a) found that 55% of the variance in duration of harvest was due to the combined effects of variation in the duration of curd initiation and variation in temperature during curd growth. Some of the variation in curd initiation can be reduced by producing more uniform plants as explained above but the genetic uniformity of cauliflower has also been improved with the introduction of hybrid cultivars.

Another approach has been to reduce plant-to-plant variation in curd initiation by using treatments that encourage more synchronous curding. Salter and Ward (1972) discovered that subjecting cauliflower transplants to cold storage (2 weeks at 2–5°C) caused a remarkable shortening of the harvest period compared to use of bare-rooted untreated transplants. Best results were obtained with plants that had produced 10–15 leaves and leaf initials at the start of the cold treatment, and so presumably were still in the juvenile stage. Cold storage of older plants, or prolonging the duration of storage beyond 3 weeks reduced effectiveness of the treatment but when a wider range of genotypes was tried, the harvest duration was shortened with some cultivars and lengthened with others (Salter and James, 1974). Wurr (1981) suggested that this was because the cold treatment worked only at a specific growth stage which varied according to genotype. However, the physiological basis for this response has not been identified. Salter and James (1974) postulated that the treatment helped to satisfy the cold requirement for curding, and thus synchronized the start of curd initiation, but subsequent work determining the optimum temperatures for vernalization (see section on environmental factors in curd and head induction above) does not support this. Salter and James' explanation contradicts the finding that cold treatment before the end of the juvenile period has no vernalization effect (Wiebe, 1972a; Hand and Atherton, 1987; Booij and Struik, 1990). It may be that cold exposure reduced plant-to-plant variation in leaf number making the timing of curd initiation more uniform in the population but it is more likely that the effect is caused by some distinct mechanism which as yet is not understood. Several workers have attempted to link the cold treatment to changes in apex hormone levels, particularly gibberellins. In one set of trials where treatment resulted in shortened harvest duration, there was a concomitant increase in gibberellin levels in the plant apex prior to curd initiation (Thomas *et al.*, 1972) but Wurr *et al.* (1981a) found that cold treatment increased gibberellin levels and reduced the harvest period in only one of four cultivars.

The variable and unpredictable results of the cold treatment of juvenile cauliflower plants illustrate that much remains to be explained about the vernalization process in *Brassica oleracea*. These results, and the contradictory findings with regard to the effect of seed vernalization (e.g. Wiebe, 1972a; Fujime and Hirose, 1979), indicate the need to look more intensively at the physiological processes governing the change from juvenile to adult vegetative state in cauliflower.

Continuity of production

In order to provide continuity of supply of produce throughout the growing season an accurate assessment of the duration of crop growth when planted at particular times of the season is needed, so that serial plantings can be made at appropriate times. Despite considerable effort by Salter and co-workers (see Salter and Laflin, 1974), variation in weekly production of 50% was all that could be achieved. Principally, these workers simultaneously planted cultivars with different periods to maturity and also staggered planting times. Wiebe (1980) conducted similar experiments and came to the same conclusions. Periods of high temperatures during the season caused much disruption in supply by hastening the harvest of crops close to maturity but delaying curd initiation on recently planted crops. The latter resulted in a reduction in supply 6 weeks later. Wiebe (1980) estimated that for conditions in Hannover, Germany, a series of ten plantings would be needed to achieve a continuous supply to market, but that weather fluctuations would cause unavoidable variations in amount of product, similar in extent to that estimated by Salter and Laflin (1974). More recently Martin (1985) and Wurr *et al.* (1990c) have produced curvilinear relationships between the time taken for a crop to grow from transplanting to maturity and its time of transplanting, which can be used to aid crop scheduling.

As a consequence of the crop's sensitivity to temperature it is common for several crops within a sequence of transplantings to mature at a similar time causing temporary oversupply of the market which is then followed by a gap in the continuity programme and consequent undersupply of the market. However, much of the variation in time to maturity is due to variation in the time from transplanting to curd initiation (Booij, 1987; Grevsen, 1990). Consequently, a technique has been developed to sample the crop once just after curd initiation to determine curd size and variability and then using a simple model of curd growth to predict when curds of any required size will be produced (Wurr *et al.*, 1990c). The technique has been successfully developed for commercial use (Wurr, 1990). The model is based on the quadratic relationship between log curd diameter and the accumulated degree days above a base of 0°C (Wurr *et al.*, 1990c). Typically, it allows a grower to predict harvest maturity about 4 weeks in advance, compared to a week's advance notice by visual inspection.

Since the most expensive part of this technique is the sampling process, efforts to predict the time of curd formation in the crop have received recent emphasis. From a series of plantings of 'White Fox' cauliflower, some of which were grown under polyethylene covers for part of the curd initiation period, Wurr *et al.* (1993) characterized the temperature response of curd formation as previously described. According to their measurements of apical meristem diameters, apical expansion towards the initiation of a curd was most rapid at 9°C, and ceased above 21°C. Lower limits of apex growth could not be precisely determined because of a lack of temperature points below 9°C. The model will allow the prediction of the time of curd initiation of this cultivar, once it has formed the necessary 17 leaves plus initials to become an adult vegetative plant and has been tested on field-grown crops of four cultivars (Wurr *et al.*, 1994). Grevsen and Olesen (1994) have also produced a model to predict the time of initiation and similar efforts are under way in Germany (H.J. Wiebe, personal communication, Hannover, Germany, 1993). To be widely useful such models will need to take into account the considerable variation among cultivars in the temperatures needed for curd formation (Salter and Laflin, 1974; Wiebe, 1980; Booij, 1987). For accurate maturity predictions, it may be necessary to determine the temperature responses of each commercially important cultivar.

BROCCOLI

The production and consumption of broccoli has increased sharply in North America and Europe in the last 20 years, and this has been aided by the wide range of climatic conditions under which the crop can produce marketable yield. As shown in the section on curd and head development above, the temperature requirements for head formation are less stringent

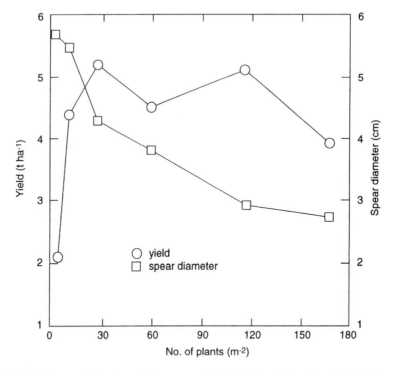

Fig. 15.10. Influence of plant population on marketable yield and diameter of individual spears of 'Barca' broccoli (Thompson and Taylor, 1976).

than for curd formation in cauliflower. In particular, broccoli may initiate flower primordia at relatively high temperatures (Wiebe, 1975) and consequently production occurs where growing season temperatures exceed those recommended for cauliflower (e.g. the Central Valley of California; Hoyle *et al.*, 1973). Nevertheless, increased problems of disease and physiological disorders at higher temperatures (see section on physiological disorders below), have made the cooler coastal valleys of California the predominant broccoli production areas in the USA.

The most important factor influencing yield in broccoli is the population density at which the crop is grown. As with most crops, yield per unit area increases with number of plants, until a plateau is reached (Cutcliffe, 1975a; Thompson and Taylor, 1976; Chung, 1982; Salter *et al.*, 1984) (Fig. 15.10). Although head size at maturity decreases as the plant population increases, consumers accept a range of head sizes. In the major production areas of the USA, broccoli is predominantly direct-seeded allowing planting at higher plant densities than would be economically feasible with the transplanted crop. In addition to high yields, close spacing has additional advantages: plants have a lower incidence of hollow stem (see section on hollow stem below) (Zink and Akana, 1951), they produce few sideshoots, and the main shoot has fewer leaves, reducing trimming waste at harvest

(Thompson and Taylor, 1976). The reduction in side shoot numbers concentrates the maturity so that single harvests of the entire crop are feasible (Cutcliffe, 1975a; Chung, 1982).

Cultivar differences occur particularly in response to high density plantings. While some new lines can produce heads of acceptable size at populations up to 100 plants m^{-2} (Salter *et al.* 1984), some older cultivars suffer drastic yield declines at close spacings (Aldrich *et al.*, 1961). The physiological basis of this difference has not been reported, but may relate to the degree of apical dominance. The new cultivars produce few branches, whereas some of the old genotypes branch profusely. The effect of crowding on central head growth may also explain why high plant populations lead to earlier maturity in the new cultivars, and to delayed maturity in the older genotypes. Wurr *et al.* (1991) noted cultivar differences in sensitivity of maturity to incident radiation, which suggests that light is the main factor determining head growth at high densities.

Yield increases of broccoli with higher plant populations depend on the availability of adequate nitrogen and water levels during the growing season. In a dryland environment, maximum yields were obtained by irrigation amounts sufficient to replace that lost by evapotranspiration (20–30 cm), and nitrogen levels of 250–350 kg ha^{-1} (Letey *et al.*, 1983; Beverly *et al.*, 1986). In a humid environment, Dufault and Waters (1985) produced highest yields of broccoli with combinations of plant populations and N nutrition that were the highest they tested (7.2 plants m^{-2}, 224 kg ha^{-1} N). In less productive environments, lower levels of nitrogen suffice to give maximum yields (Kahn *et al.* 1991).

Predicting maturity

In areas where growing season temperatures are sufficiently invariate and predictable, crop scheduling of broccoli has been practised by relying on the maturity characteristics of specific cultivars (Titley, 1987). Thus most of the production season of Southeastern Queensland in Australia permitted the use of cultivar-specific crop durations. In more typical environments, such as southern England, crop duration shows a curvilinear trend with time of planting (Wurr *et al.*, 1991) (Fig. 15.11). The longer crop growth periods from early or late season transplantings are primarily due to reduced growth rates at the prevailing low temperatures. Marshall and Thompson (1987a,b) were able to predict maturity by calculating crop duration from sowing in day-degrees (thermal time), and improved it further by including solar radiation in the calculation.

Predictions of the time of maturity based on environmental conditions from transplanting incorrectly assume that the environment influences vernalization and head growth similarly. This problem can be avoided by beginning the calculations after crop sampling has indicated that the head has been initiated. This approach, used by Wurr *et al.* (1991), has resulted in accurate predictions of head maturity and has been developed into

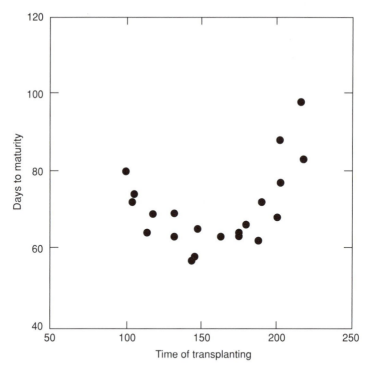

Fig. 15.11. Number of days from transplanting to maturity for 'Corvet' broccoli, as influenced by date of transplanting in England (day 1 = 1 January) (Wurr *et al.*, 1991).

commercial prediction software which also takes account of differences in plant density. As in the experiments of Marshall and Thompson (1987a), the prediction was improved by including solar radiation in the model.

CABBAGE

The yield of cabbage is made up of two principal components: the number of plants per unit area and the size of the harvested portion of each plant. Productivity is altered by changes in the size of these components, and in the growth duration of the crop. Since plant population is generally controlled by the grower, we will consider it only with regard to the influence on head size and maturity.

Head size is closely related to the amount of space available to each plant. As spacing increases the heads become larger, increasing to a maximum that is characteristic for that cultivar (Stoffella and Fleming, 1990) (Fig. 15.12). The variability in head size from plant to plant when grown at low densities was also much less than at high densities. Presumably the competition for above- and below-ground resources was much more intense at

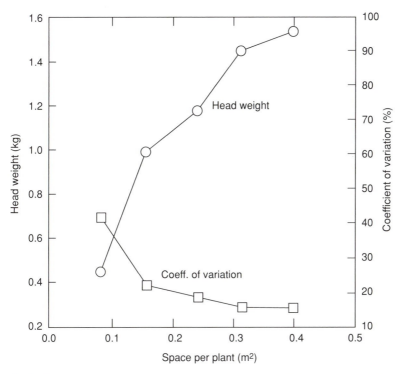

Fig. 15.12. Effect of plant spacing on mean head size and its coefficient of variation, for 'Bravo' cabbage grown in Florida (Stoffella and Fleming, 1990).

high densities, resulting in some of the population having reduced access to those resources.

Head size is influenced directly by availability of major nutrients to the plant. Edmond and Lewis (1926) demonstrated with sand culture experiments, that withholding nutrients greatly delayed crop maturity and reduced head size. Similar results were obtained by Hara and Sonoda (1979a,b), who found that for satisfactory yields there must be adequate levels of N, particularly during the early head formation stage, and P and K during the earlier stage of outer leaf expansion. They identified the critical contents of N, P and K in the outer leaves which resulted in a 50% decrease in head yield as being 1.3, 0.1 and 0.3%, respectively (Hara and Sonoda, 1979b).

Other factors which have been found to reduce cabbage plant growth and head weights are cultivation practices that damage the root system (White, 1977), and growth in compacted soils of a 'no-till' soil management system (Knavel and Herron, 1981). Presumably, any stress factor which results in poor plant growth would also bring about reduced final head size. Factors that might be included are drought, waterlogging, insect and disease incidence, and shading and nutrient stress due to weeds.

Unfortunately, studies with cabbage which indicate the time at which plant size determines the ultimate size of head at harvest are lacking. Work with head lettuce would indicate that the beginning of head formation is most important for head size determination (Wurr and Fellows, 1991; see Chapter 14) and if the same mechanism exists in cabbage, it implies that the area of unfolded leaves when the head starts to form determines the quantity of assimilates available for translocation to the head. Studies of carbon translocation in cabbage confirm that 40% of carbon assimilated during heading is translocated to the growing head (Hara and Sonoda, 1981). Partitioning to the head was reduced by plant shading, and by culture of the plants at high (500 ppm) nitrogen. De Moel and Everaarts (1990) also found that dry matter partitioning to the heads was lower for early season plantings than for later crops, but did not state why.

Optimum temperatures for growth and yield of cabbage are listed by Lorenz and Maynard (1980) as 16–18°C, but the crop is frequently grown in locations with average temperatures which exceed this range. For instance, cultivars have been successfully developed for growth in the lowland tropics, where cool season temperatures average 26°C (Rasco, 1983). At these higher temperatures, the length of the growing season is reduced, and the yield is lowered (Knott and Hanna, 1947; Sundstrom and Story, 1984). Hara and Sonoda (1982) found that cabbage grown at 25°C has a lower dry matter content, a reduced growth rate and lower water use efficiency than cabbage grown at 20°C.

Attempts to predict the maturity date of cabbage by calculation of heat sums have not been particularly successful so far. Trials in the Florida winter season over 12 years showed a 4–16% variation in harvest time for individual cultivars, whether calculated on the basis of heat units (limits 0 and 25°C), or the time to maturity (Strandberg and White, 1979). In order to improve the accuracy of predictions it may be necessary to include some measure of solar radiation in the calculation, as has been done with lettuce (Wurr et al., 1988b). There have also been problems in predicting maturity for crops harvested during the low temperature period of the fall. For instance, Isenberg et al. (1975) harvested the storage cultivar Green Winter after an average of 1066 heat units (base 10°C) in two plantings, but the heads were judged ready to harvest after 991–1221 heat units had accumulated. This covered a period of 107–148 days after sowing during which head size and density changed very little. This suggests that the indices of growth and development to maturity need refining if maturity prediction techniques are to work.

Cabbage crop maturity also varies markedly with change in plant spacing. When plant spacing was reduced from 38 to 8 cm between plants, maturity dates increased from 106 to 154 days (Stoffella and Fleming, 1990) (Fig. 15.13). This may be a further indication that the level of resources available to each plant dictates the rate of growth and maturity. Principal among these resources would be the amount of light.

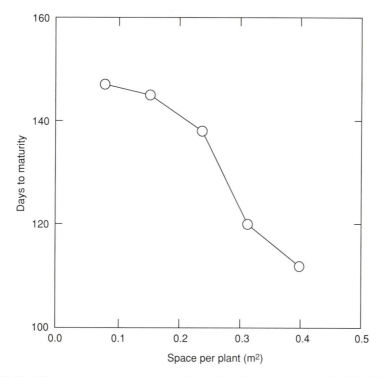

Fig. 15.13. Effect of space per plant on growth duration (transplanting to maturity) of 'Bravo' cabbage grown in Florida (Stoffella and Fleming, 1990).

BRUSSELS SPROUTS

Brussels sprouts are a relatively minor vegetable even in northern Europe, where they are commonly grown and consumed (Kronenberg, 1975b). In the UK, the biggest European producer, the 100,000 tons harvested in 1991 only amounts to about 3% of total vegetable production (ZMP, 1992). Nevertheless, considerable physiological research has been devoted to the crop in recent years, and consequently, we have a good understanding of the major factors determining yield in the crop.

To achieve a high yield of Brussels sprouts, a large number of plants, each with many buttons of marketable size, must be produced (Fisher, 1974b). Increases in plant population result in decreases of plant size, and reduction in growth rate of individual sprouts (Thompson and Taylor, 1973; Fisher and Milbourn, 1974; Abuzeid and Wilcockson, 1989). Changes in plant density had little effect on the number of buttons per plant, but reduced average button size at harvest. Thus increasing the plant population increased yield only when the growing season was long enough so that the large number of buttons could reach marketable size (Kirk, 1981) (Fig.

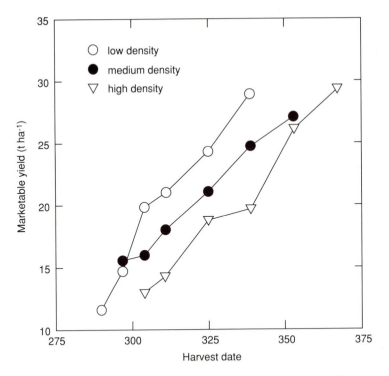

Fig. 15.14. Effect of plant population and harvest date on marketable yield of 'Cor' Brussels sprouts (Kirk, 1981).

15.14). Production of Brussels sprouts at populations of 8–10 plants m^{-2} instead of the more normal 3–5 plants m^{-2} also carries increased risk of lodging (Verheij, 1970) and of foliar diseases affecting sprout buttons that could reduce market quality (Fisher, 1974b).

Another important yield-determining factor in Brussels sprouts is the size distribution of sprouts on the stem. At wide spacings, the lowest buttons grow first, and the upper buttons develop later, leading to a marked increase in button size down the stem. At higher plant populations, the size uniformity is increased, permitting once-over harvesting of the crop (Fisher, 1974a).

Increases in button size uniformity have also been achieved by removal of the growing point of the plant, usually when the first buttons are 10–20 mm diameter. This labour-intensive procedure, termed 'topping' or 'stopping', is done either by hand, removing the apical bud, or using hammers or clubs to crush the plant apex. Stopping removes apical dominance as demonstrated by Thomas (1972) who found that auxin activity of axillary buds on intact Brussels sprouts plants was relatively low, and increased substantially within 2 days if the apex was removed. He reimposed lateral bud inhibition by auxin application to the cut apex. By destroying apical dominance, sprout growth is accelerated, especially at the upper nodes of the plant. The time at which plants are stopped has a strong influence on how they react. If the apex is removed before sprouts have started to grow, upper axillary buds may grow into leafy shoots without a stimulation of bud enlargement. Stopping at progressively later dates allows more nodes to form on the plants, thus increasing yields (Cutcliffe, 1970). However, late stopping reduces the effect of the treatment on bud size stimulation (Metcalf, 1954; Jones, 1972; Fisher, 1974a). Physiological studies indicated that by removing the plant apex, assimilates that would otherwise have gone to stem and leaf growth were diverted into bud development, without negative effects on overall plant production (Jones, 1972).

Stopping is generally practised when the oldest buds are about half full size. Since bud growth is accelerated at wider spacings, apex removal must be done earlier at low densities (Kirk, 1981). Stopping is also more commonly practised with early than with late-maturing cultivars (H.C. Wien, personal observation, the Netherlands, 1993).

The expense of manual stopping has stimulated research into alternative methods. Chemical treatments have been tried by several researchers, but with only limited success. Tompkins et al. (1972) found that SADH (succinic acid-2,2-dimethylhydrazide), sprayed on the plants when manual stopping would normally be done, gave yields similar to the non-stopped controls, and did not affect earliness of harvest. In the experiments of Murphy (1975), yields of the SADH-treated plants were intermediate between the stopped and unstopped control plants. The use of this chemical is no longer permitted on food crops. His trials found other chemicals, such as gibberellic acid, ethrel and several carbamate compounds to be unsuitable for stopping.

A major determinant of productivity in Brussels sprouts is the time relative to the end of the growing season that button growth begins. This is influenced by plant age, apical dominance, and genetic factors. Abuzeid and Wilcockson (1989) found that the onset of rapid button growth coincided approximately with maximum plant dry weight as lower leaf fall was beginning. This period occurred progressively later in later-planted crops, but was not influenced by plant density. Accordingly, later planting resulted in lower yields of sprouts (Metcalf, 1954; Abuzeid and Wilcockson, 1989; Gaye and Maurer, 1991).

The temperature requirements for optimum production of Brussels sprouts were outlined by Kronenberg (1975b). For late-maturing cultivars, he stipulated average monthly temperatures of 17–21°C during a 3–4-month period of early growth, followed by two months of 12°C during sprout development. To permit harvest of the crop during winter, he showed that climatic regions where early winter temperatures of −10–−15°C rarely occurred, were favoured for the crop (Kronenburg, 1975a). In Europe, these include the coastal areas of Holland, Belgium and Spain, and parts of the UK, where the bulk of the crop is now grown. Movement out of these areas of favourable temperature curtails the growing season and reduces yield.

PHYSIOLOGICAL DISORDERS

BLINDNESS

Blindness is the loss of the growing point and has been variously reported to be associated with: low temperatures (Salter, 1957), molybdenum deficiency, which can also cause misshapen leaves and leaves without blades (whiptail), insect damage, scorching by fertilizer and pesticide, daylength, light intensity, moisture stress and seed quality, though the scientific evidence for some of these is patchy. There is often uncertainty as to whether the blindness was caused during plant raising or during growth in the field but this can be determined by the grower counting the numbers of leaf scars and leaves up to the point of blindness. If there are fewer leaves than the total number of leaves at transplanting (in cauliflower about 12, in broccoli about eight) then the blindness was caused during plant raising. Recent extensive work in the UK (Wurr, 1994) has shown that the major factor affecting blindness in broccoli is low light levels during plant raising. There is also an interaction with temperature which is not easy to understand because of the strong correlations between light and different aspects of temperature.

HOLLOW STEM OF BROCCOLI AND CAULIFLOWER

Broccoli and cauliflower show hollow areas in the stem, extending from below the head or curd to where the stem is normally cut. The disorder has been shown to be most severe where individual plants could grow rapidly such as: wide spacing, high nitrogen fertilizer levels, warm weather and adequate moisture (Zink, 1968; Cutcliffe, 1972; Hipp, 1974; Scaife and Wurr, 1990) and to differ between cultivars (Cutcliffe, 1975b). The walls of these cavities are commonly not discolored, but there is potential for infection and spoilage after harvest. The disorder appears similar to the transverse cracking in celery caused by boron deficiency but boron concentration in cauliflower curds and young and older leaves was not correlated with hollow stem severity (Scaife and Wurr, 1990). Furthermore, micronutrient applications have not been successful in alleviating hollow stem (Zink, 1968; Hipp, 1974).

Hipp (1974) found that the incidence of hollow stem showed an inverse curvilinear relationship with the duration of the broccoli growing period. Crops taking 110 days or more from sowing to maturity resulted in a low occurrence of the disorder while its incidence was much greater when plants grew more rapidly (Fig. 15.15). Zink (1968) suggested that the incidence of hollow stem could be reduced by growing broccoli at densities of 10 plants m^{-2} during the warmest growing season with nitrogen levels maintained for high yields. During cooler seasons, wider spacings should produce larger spears without risk of hollow stem.

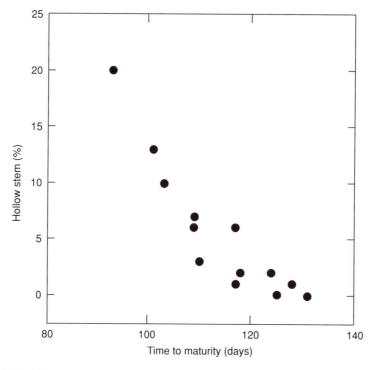

Fig. 15.15. Influence of maturity date (rate of growth) on incidence of hollow stem in broccoli (Hipp, 1974).

TIPBURN AND RELATED CALCIUM DEFICIENCY DISORDERS

Physiological disorders related to a lack of calcium in the affected organ are common in the cole crops. In the head- and bud-forming crops, the symptoms are similar to tipburn of lettuce, described in Chapter 14. In cabbage, tipburn appears as necrotic spots or areas in the margins of the rapidly expanding leaves in the middle part of the head (Walker *et al.*, 1961). In Brussels sprouts, similar symptoms occur in the sprouts, but the disorder has been termed internal browning (Millikan and Hanger, 1966). Severe calcium deficiency in this crop can also occur as a marginal necrosis of the rapidly expanding leaves near the shoot apex (Millikan and Hanger, 1966; Maynard and Barker, 1972). Tipburn of cauliflower also appears in the margins of immature leaves near the developing curd (Rosen, 1990) and the curd may be discolored if the dead leaf tissue touches it. Secondary pathogens may gain entry to the weakened areas and provide a source of curd infection. Culture of cauliflower in greenhouses and growth chambers can result in a more severe calcium deficiency disorder, the production of translucent or 'glassy' curds (Krug *et al.*, 1972).

Work with cabbage has revealed one aspect not mentioned in Chapter 14: the importance of root growth. Since calcium is largely immobile in the phloem, its uptake is restricted to the young apical zone of the root (Marschner, 1986). Soil conditions which restrict root growth, such as anaerobic conditions, compaction and acid pH, can lead to lack of calcium uptake, and deficiency in the above-ground parts (Scaife and Clarkson, 1978). Accordingly, Walker *et al.* (1961) found increased cabbage tipburn in waterlogged areas of the field. Relative humidity is also important since a shortage of calcium is likely to occur when transpiration is limited. Wiebe *et al.* (1977) showed that in outer leaves, calcium content increased during the day in proportion to the transpiration, while in inner head leaves calcium was transported mainly at night when the head mass increased due to increasing plant water potential.

Foliar sprays of calcium-containing salts were generally ineffective in preventing the appearance of symptoms of these disorders in the cole crops (Walker *et al.*, 1961; Rosen, 1990). To have even slight ameliorative effects, applications had to be weekly or more frequent, and continued during the active growth period of the susceptible tissue (Millikan *et al.*, 1971). Realistically, a more practical means of avoiding the calcium deficiency disorders in cole crops is the use of resistant cultivars. Cultivar differences have been noted in cabbage tipburn (Walker *et al.*, 1961; Peck *et al.*, 1983), in cauliflower tipburn (Rosen, 1990) and with internal browning of Brussels sprouts (Millikan and Hanger, 1966). Some of these differences may be based on the rate of growth and earliness of particular cultivars, which expose them to the conditions causing the disorder (Rosen, 1990). By selecting cultivars under such conditions it may be possible to identify genotypes with a combination of characteristics giving lower susceptibility.

BUTTONING OF CAULIFLOWER AND BROCCOLI

The term buttoning refers to the production of small exposed curds of cauliflower or heads of broccoli. It commonly occurs in early cauliflower crops that are transplanted after being raised in greenhouses or cold frames (Skapski and Oyer, 1964; Birkenshaw *et al.*, 1982) and in broccoli where it has been described by several authors (Baggett and Mack, 1970; Miller *et al.*, 1985).

More commonly, buttoning occurs when relatively large transplants, which have been growing under favourable conditions in the greenhouse, are transplanted into cool field environments that lead to rapid curd induction (Skapski and Oyer, 1964; Wiebe, 1981; Wurr and Fellows, 1984; Booij, 1990d). Although Skapski and Oyer (1964) obtained the disorder on transplants that had already initiated curds prior to field planting, Booij (1990d) and Wurr and Fellows (1984) demonstrated that plants which formed curds shortly after transplanting could also show buttoning. The key factor in buttoning appears to be the failure of the plants to produce a leaf area adequate to support a curd of marketable size. Several investigators have shown that

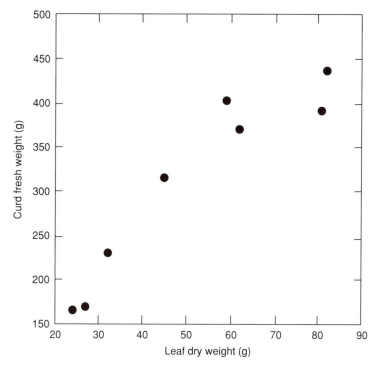

Fig. 15.16. The relationship of leaf dry weight at 92 days after sowing on curd weight at harvest for 'All The Year Round – Lero' cauliflower grown in England. Plants of different sizes produced by transplanting from 36 to 64 days after sowing (Wurr and Fellows, 1984).

there is a close linear relationship between the leaf area at harvest and the size of the curd (Fig. 15.16), (Skapski and Oyer, 1964; Wiebe, 1981; Wurr and Fellows, 1984; Booij, 1990d).

In general, buttoning is most likely to be caused by conditions restricting vegetative growth such as frost, bird damage, poor soil structure, shortage of nitrogen, high soil salinity and low soil moisture (Wiebe, 1981). Transplanting shock on large plants has been shown to reduce the amount of leaf growth, in comparison to that made by direct-seeded plants (Wiebe, 1981). In addition, a plant which has initiated a curd will produce no more leaves, setting a finite limit on total leaf area. Direct competition between the growing curd and the young leaves may also limit leaf area development. Wiebe (1981) showed that removal of young curds resulted in a 37% increase in the area of expanding leaves.

To minimize the occurrence of buttoning, cauliflower and broccoli plants must make adequate vegetative growth in the field before curd or head initiation. In the case of the transplanted crop subjected to cool spring field conditions, use of younger, smaller transplants that are still juvenile when planted, is advisable. Use of cultivars with a longer juvenile period may also help (Baggett and Mack, 1970). Minimizing transplant shock and maximizing conditions for vegetative growth with optimum fertility, soil and water management, and the use of protective covers in the field can allow the plant to make adequate growth before curds begin to form. Direct-seeding also avoids the problem, but may be an unrealistic option for growers trying to produce an early crop. It may also be sensible to retard transplant growth with cold treatment during the juvenile stage (Carew and Thompson, 1948).

Since most crops that button are too physiologically advanced at transplanting, the problem could be minimized by developing simple models describing the growth and development of cauliflower and broccoli plants during plant raising. Such models could predict the level of apex development during plant raising and thus allow transplant growth to be manipulated to minimize the chance of buttoning occurring.

CAULIFLOWER CURD DISORDERS

The temperatures allowing curd initiation to occur are cultivar specific and temperatures outside this range during early curd development may lead to the formation of structures that represent either a partial reversion to the vegetative state, or to more complete reproductive structure development.

If cauliflower is subjected to temperatures above the optimum for curd formation shortly after initiation, the curd develops bracts around individual florets, or in more extreme cases, the normally sessile white bracts on the peduncle of the curd elongate and become leaflike (Wiebe, 1972b, 1973a; Fujime, 1983). For instance, Wiebe (1973a) induced bracting in 'Aristokrat' cauliflower when subjecting the plant to 3 weeks of 25°C, beginning when the apical meristem had reached a diameter of 0.5 mm. Bracting has also been induced in cauliflower by foliar sprays of the plant at the curd formation stage with ethephon (Booij, 1990b). Since ethephon is converted to ethylene, this implies that other stresses, which also cause the production of ethylene by cauliflower plants, may also induce bract formation.

When plants are exposed to relatively low temperatures after the beginning of curd formation, development of individual flower buds in the curd is carried further than in a normal curd, and the curd takes on a 'ricey' appearance (Sadik, 1962; Wiebe, 1973b, 1975; Fujime, 1983). Generally, the curd retains the normal white colour, but the surface becomes grainy. In extreme cases, when cultivars adapted to the tropics are exposed to prolonged cool temperatures, the flower buds become green, and the curd resembles a broccoli head rather than the curd (Wiebe, 1975). However, curds are usually white until exposed to sunlight, when they turn cream and then yellow, becoming unmarketable. Some genotypes may even develop a pink or purple coloration in the curd (Crisp and Gray, 1979). Curds can also show blackish-brown discoloration caused by the direct effect of ultraviolet light (Nieuwhof, 1969).

The reaction of tropically adapted cauliflower cultivars to cool temperatures illustrates one extreme of the range of responses that is possible within cauliflower. The adaptation to high temperatures of these cultivars implies that they would rarely encounter heat sufficient to cause the development of leafy curds. At the other extreme, cauliflower cultivars that require a cold period for curd formation would form ricey curds only very infrequently. In general, one would expect to see a high frequency of curd disorders in cultivars being grown at or beyond the range of temperature conditions for which they were originally developed. Successful commercial cauliflower production is therefore frequently restricted to regions in which temperatures during the growing season are predictable and relatively consistent. This includes the maritime areas of Salinas Valley, California, Long Island, New York and certain coastal areas of Holland, England, Wales and Brittany in France.

CONCLUDING REMARKS

The wide range of plant types within *Brassica oleracea* that have been developed for human consumption bring with them a bewildering range of optimum growing environments and hence requirements for cultural practice. Selection within each type has permitted the development of cultivars adapted to conditions ranging from the tropical to the cool temperate. Thus there are few generalities that can be made regarding the environmental conditions necessary for inducing a species to flower without specifying the cultivar to which the statement applies. Boundaries between types are likely to become less distinct, as new vegetables, such as the intermediate forms between broccoli and cauliflower become important (Gray, 1989).

In spite of the ability to select genotypes for a wide range of specific environments, we have made little progress towards producing cultivars that are less susceptible to variations of temperature at a site. The sensitivity of cauliflower, especially to temperature variations, has restricted its successful cultivation to areas of predictably moderate temperatures. Perhaps by applying selection pressure for insensitivity to the physiological disorders that are induced by extremes of temperature, such as bracting or riceyness, a more widely adapted cultivar could be developed. The finding that bracting can be induced by ethephon foliar sprays may indicate a way in which such selection can be made (Booij, 1990b).

As progress is made in identifying the genes controlling the flowering process in *Brassica oleracea* (Jordan and Anthony, 1994) there is a need for accompanying work to define more closely the physiological determinants of the important growth stages. If in the future we want to manipulate crop growth to meet market requirements in a range of environmental conditions there are three relevant pieces of physiological information where change to the generative state is important: the length of the juvenile phase, the temperatures controlling vernalization and the strength of the vernalization stimulus required before induction is complete.

Of the four types of *Brassica oleracea* considered here, vernalization is an important and necessary process for cauliflower and broccoli in order to induce the formation of the saleable parts, the curd and the head. For these species competition between the production and expansion of leaves and the growth of the apex is important and needs to be fully understood if crops of consistently high quality are to be grown. It is likely that these two processes can be regarded as independent with their own distinct optimum temperature requirements. For cabbage and Brussels sprouts, initiation of the generative part of the plant is not necessary for commercial crop production, which requires the formation of, respectively, the head and buttons while the plant is still in a vegetive state. So for these species leaf production is important together with the environmental factors determining the time of the morphological changes involving head and button formation and subsequent growth of these organs. There is still plenty of scope for simple crop modelling to describe and predict these important processes in order to develop principles of understanding which can be applied widely to solve commercial production problems.

ACKNOWLEDGEMENTS

The helpful suggestions of Drs John Atherton and Heinz-Joachim Wiebe in the writing of this chapter are gratefully acknowledged.

REFERENCES

Abuzeid, A.E. and Wilcockson, S.J. (1989) Effects of sowing date, plant density and year on growth and yield of Brussels sprouts (*Brassica oleracea* var. *bullata* subvar. *gemmifera*). *Journal of Agricultural Science* 112, 359–375.

Aldrich, T.M., Snyder, M.J. and Little, T.M. (1961) Plant spacing in broccoli. *California Agriculture* 15(12), 10–11.

Atherton, J.G., Hand, D.J. and Williams, C.A. (1987) Curd initiation in the cauliflower (*Brassica oleracea* var. *botrytis* L.). In: Atherton, J.G. (ed.) *Manipulation of Flowering*. Butterworths, London, pp. 133–145.

Baggett, J.R. and Mack, H.J. (1970) Premature heading of broccoli cultivars as affected by transplant size. *Journal of the American Society of Horticultural Science* 95, 403–407.

Beverly, R.B., Jarrell, W.M. and Letey, J. Jr (1986) A nitrogen and water response surface for sprinkler-irrigated broccoli. *Agronomy Journal* 78, 91–94.

Birkenshaw, J.E., Wurr, D.C.E. and Fellows, J.R. (1982) The influence of ventilation temperature and plant raising method on the yield of early summer cauliflowers. *Journal of Horticultural Science* 57, 357–363.

Booij, R. (1987) Environmental factors in curd initiation and curd growth of cauliflower in the field. *Netherlands Journal of Agricultural Science* 35, 435–445.

Booij, R. (1989) Effect of growth regulators on curd diameter of cauliflower. *Scientia Horticulturae* 38, 23–32.

Booij, R. (1990a) Cauliflower curd initiation and maturity: variability within a crop.

Journal of Horticultural Science 65, 167–175.

Booij, R. (1990b) Induction of bracting in cauliflower with 2-chloroethylphosphonic acid (ethephon). *Euphytica* 50, 27–33.

Booij, R. (1990c) Effects of gibberellic acids on time of maturity and on yield and quality of cauliflower. *Netherlands Journal of Agricultural Science* 38, 641–651.

Booij, R. (1990d) Influence of transplant size and raising temperature on cauliflower curd weight. *Gartenbauwissenschaft* 55, 103–109.

Booij, R. and Struik, P.C. (1990) Effects of temperature on leaf and curd initiation in relation to juvenility of cauliflower. *Scientia Horticulturae* 44, 201–214.

Boswell, V.R. (1929) Studies of premature flower formation in wintered-over cabbage. *Maryland Agricultural Experiment Station Bulletin* 313, 69–145.

Carew, J. and Thompson, H.C. (1948) A study of certain factors affecting 'buttoning' of cauliflower. *Proceedings of the American Society of Horticultural Science* 51, 406–414.

Chung, B. (1982) Effects of plant density on the maturity and once-over harvest yield of broccoli. *Journal of Horticultural Science* 57, 365–372.

Crisp, P. and Gray, A.R. (1979) Successful selection for curd quality in cauliflower, using tissue culture. *Horticultural Research* 19, 49–53.

Cutcliffe, J.A. (1970) Effect of time of disbudding on single-harvest yields and maturity of 'Jade Cross' Brussels sprouts. *HortScience* 5, 176–177.

Cutcliffe, J.A. (1972) Effects of plant spacing and nitrogen on incidence of hollow stem in broccoli. *Canadian Journal of Plant Science* 52, 833–834.

Cutcliffe, J.A. (1975a) Effect of plant spacing on single-harvest yields of several broccoli cultivars. *HortScience* 10, 417–419.

Cutcliffe, J.A. (1975b) Cultivar and spacing effects on incidence of hollow stem in broccoli. *Canadian Journal of Plant Science* 55, 867–869.

de Moel, C.P. and Everaarts, A.P. (1990) Growth, development and yield of white cabbage in relation to time of planting. *Acta Horticulturae* 267, 279–288.

Dickson, M.H. and Wallace, D.H. (1986) Cabbage breeding. In: Bassett, M.J. (ed.) *Breeding Vegetable Crops*. Avi, Westport, Conn., pp. 395–432.

Dufault, R.J. and Waters, L. Jr (1985) Interaction of nitrogen fertility and plant populations on transplanted broccoli and cauliflower yields. *HortScience* 20, 127–128.

Edmond, J.B. and Lewis, E.P. (1926) Influence of nutrient supply on earliness of maturity in cabbage. *Michigan Agricultural Experiment Station Technical Bulletin* 75, 1–16.

Fisher, N.M. (1974a) The effect of plant density, date of apical bud removal and leaf removal on the growth and yield of single harvest Brussels sprouts (*Brassica oleracea* var. *gemmifera* D.C.). II. Variation in bud size. *Journal of Agricultural Science* 83, 489–496.

Fisher, N.M. (1974b) The effect of plant density, date of apical bud removal and leaf removal on the growth and yield of single harvest Brussels sprouts (*Brassica oleracea* var. *gemmifera* D.C.). III. The components of marketable yield. *Journal of Agricultural Science* 83, 497–503.

Fisher, N.M. and Milbourn, G.M. (1974) The effect of plant density, date of apical bud removal and leaf removal on the growth and yield of single harvest Brussels sprouts (*Brassica oleracea* var. *gemmifera* D.C.). I. Whole plant and axillary bud growth. *Journal of Agricultural Science* 83, 479–487.

Fontes, M.R. and Ozbun, J.L. (1970) Effect of growth retardants on growth and flowering of broccoli. *HortScience* 5, 483–484.

Fontes, M.R. and Ozbun, J.L. (1972) Relationship between carbohydrate level and floral initiation in broccoli. *Journal of the American Society of Horticultural Science* 97, 346–348.

Fontes, M.R., Ozbun, J.L. and Powell, L.E. (1970) Are endogenous gibberellin-like substances involved in floral induction? *Nature* 228, 82–83.

Friend, D.J.C. (1985) Brassica. In: Halevy, A.H. (ed.) *CRC Handbook of Flowering*, Vol. 2. CRC Press, Boca Raton. pp. 48–77.

Fujime, Y. (1983) Studies on thermal conditions of curd formation and development in cauliflower and broccoli, with special reference to abnormal curd development. *Memoirs of the Faculty of Agriculture of Kagawa University* 40, 1–123.

Fujime, Y. and Hirose, T. (1979) Studies on thermal conditions of curd formation and development in cauliflower and broccoli. I. Effect of low temperature treatment of seeds. *Journal of the Japanese Society of Horticultural Science* 48, 82–90.

Gaye, M.M. and Maurer, A.R. (1991) Modified transplant production techniques to increase yield and improve earliness of Brussels sprouts. *Journal of the American Society of Horticultural Science* 116, 210–214.

Gray, A.R. (1989) Taxonomy and evolution of broccolis and cauliflowers. *Baileya* 23, 28–46.

Grevsen, K. (1990) Prediction of harvest in cauliflower based on meteorological observations. *Acta Horticulturae* 267, 313–322.

Grevsen, K. and Olesen, J.E. (1994) Modelling cauliflower development from transplanting to curd initiation. *Journal of Horticultural Science* 69, 755–766.

Hand, D.J. and Atherton, J.G. (1987) Curd initiation in the cauliflower. I. Juvenility. *Journal of Experimental Botany* 38, 2050–2058.

Hara, T., Nakagawa, A. and Sonoda, Y. (1982) Effects of nitrogen supply and removal of outer leaves on the head development of cabbage plants. *Journal of the Japanese Society of Horticultural Science* 50, 481–486.

Hara, T. and Sonoda, Y. (1979a) The role of macronutrients in cabbage-head formation. Growth performance of a cabbage plant and potassium nutrition in the plant. *Soil Science and Plant Nutrition* 25, 103–111.

Hara, T. and Sonoda, Y. (1979b) The role of macronutrients for cabbage-head formation. I. Contribution to cabbage-head formation of nitrogen, phosphorus or potassium supplied at different growth stages. *Soil Science and Plant Nutrition* 25, 113–120.

Hara, T. and Sonoda, Y. (1981) The role of macronutrients in cabbage-head formation. IV. Effect of nitrogen supply on the growth and $^{14}CO_2$ and $^{15}NO_3$-N assimilation of cabbage plants. *Soil Science and Plant Nutrition* 27, 185–194.

Hara, T. and Sonoda, Y. (1982) Cabbage head development as affected by nitrogen and temperature. *Soil Science and Plant Nutrition* 28, 109–117.

Heide, O.M. (1970) Seed-stalk formation and flowering in cabbage. I. Day-length, temperature and time relationships. *Meldinger fra Norges Landbrukshogskole* 49(27), 1–21.

Helm, J. (1963) Morphologisch-taxonomische Gliederung der Kultursippen von *Brassica oleracea* L. *Kulturpflanze* 11, 92–210.

Hipp, B.W. (1974) Influence of nitrogen and maturity rate on hollow stem in broccoli. *HortScience* 9, 68–69.

Hodgkin, T. (1981) The inheritance of node number and rate of node production in Brussels sprouts. *Theoretical Applied Genetics* 59, 79–82.

Hoyle, B.J., Tyler, K., Fischer, B., May, D. and Brown, L. (1973) Broccoli for the San Joaquin Valley West side. *California Agriculture* 27(1), 12–13.

Isenberg, F.M.R., Pendergrass, A., Carroll, J.E., Howell, L. and Oyer, E.B. (1975) The use of weight, density, heat units, and solar radiation to predict the maturity of cabbage for storage. *Journal of the American Society of Horticultural Science* 100, 313–316.

Ito, H. and Kato, T. (1957) Studies on the head formation of Chinese cabbage. Histological and physiological studies of head formation. *Journal of the Japanese Society of Horticultural Science* 26, 154–162.

Ito, H. and Saito, T. (1961) Time and temperature factors for the flower formation in cabbage. *Tohoku Journal of Agricultural Research* 12, 297–316.

Ito, H., Saito, T. and Hatayama, T. (1966) Time and temperature factor for the flower formation in cabbage. II. The site of vernalization and the nature of the vernalization sensitivity. *Tohoku Journal of Agricultural Research* 17, 1–15.

Jones, L.H. (1972) The effects of topping and plant population on dry matter synthesis and distribution in Brussels sprouts. *Annals of Applied Biology* 70, 77–87.

Jordan, B.R. and Anthony, R.G. (1994) Control of floral morphogenesis in cauliflower (*Brassica oleracea* L. var. *botrytis*): the role of homeotic genes. In: Scott, R. and Stead, A.D. (eds) *Molecular and Cellular Aspects of Plant Reproduction*. Cambridge Academic Press, SEB series.

Kahangi, E.M. and Waithaka, K. (1981) Flowering of cabbage and kale as influenced by altitude and GA application. *Journal of Horticultural Science* 56, 185–188.

Kahn, B.A., Shilling, P.G., Brusewitz, G.H. and McNew, R.W. (1991) Force to shear the stalk, stalk diameter, and yield of broccoli in response to nitrogen fertilization and within-row spacing. *Journal of the American Society of Horticultural Science* 116, 222–227.

Kato, T. (1965) On the flower head formation and development of cauliflower plants. II. Physiological studies of flower head formation. *Journal of the Japanese Society of Horticultural Science* 34, 49–56.

Kato, T. (1981) The physiological mechanism of heading in Chinese cabbage. In: Talekar, N.S. and Griggs T.D. (eds) *Chinese Cabbage*. Asian Vegetable Research and Development Center, Shanhua, Taiwan, pp. 209–215.

Kato, T. and Sooen, A. (1978) Physiological studies on the head formation in cabbage. I. Effect of defoliation of wrapper leaves on the head formation posture. *Journal of the Japanese Society of Horticultural Science* 47, 351–356.

Kato, T. and Sooen, A. (1979) Physiological studies on the head formation in cabbage. II. Effects of root pruning on the head formation posture. *Journal of the Japanese Society of Horticultural Science* 48, 26–30.

Kato, T. and Sooen, A. (1980) Physiological studies on the head formation in cabbage. III. The role of the terminal bud in the head formation posture. *Journal of the Japanese Society of Horticultural Science* 48, 426–434.

Kirk, S. (1981) An investigation into the effects of plant density and time of harvest on yield of Brussels sprouts. *Journal of the National Institute of Agricultural Botany* 15, 488–496.

Knavel, D.E. and Herron, J.W. (1981) Influence of tillage system, plant spacing and nitrogen on head weight, yield, and nutrient concentration in cabbage. *Journal of the American Society of Horticultural Science* 106, 540–545.

Knott, J.E. and Hanna, G.C. (1947) The effect of widely divergent dates of planting on the heading behavior of seven cabbage varieties. *Proceedings of the American Society of Horticultural Science* 49, 299–303.

Kronenberg, H.G. (1975a) Influences of temperature and light on the growth of young Brussels sprouts plants. *Netherlands Journal of Agricultural Science* 23, 83–88.

Kronenberg, H.G. (1975b) A crop geography of late Brussels sprouts. *Netherlands Journal of Agricultural Science* 23, 291–298.

Krug, H., Wiebe, H.J. and Jungk, A. (1972) Calciummangel an Blumenkohl unter konstanten Klimabedingungen. *Zeitschrift für Pflanzenernahrnährung und Bodenkunde* 133, 213–226.

Kruzhilin, A.S. and Shvedskaya, Z.M. (1960) The role of leaves in the vernalization of winter and biennial plants. *Fiziologia Rastenii* 7, 237–243.

Lawson, G. and Long, E. (1988) ADAS maturity test for cauli curds ready to go this summer. *Grower* 109(22), 9.

Leshem, Y. and Steiner, S. (1968) Effect of gibberellic acid and cold treatment on flower differentiation and stem elongation of cauliflower (*Brassica oleracea* var. *botrytis*). *Israel Journal of Agricultural Research* 18, 133–134.

Letey, J., Jarrell, W.M., Valoras, N. and Beverly, R.B. (1983) Fertilizer application and irrigation management of broccoli production and fertilizer use efficiency. *Agronomy Journal* 75, 502–507.

Lorenz, O.A. and Maynard, D.N. (1980) *Knott's Handbook for Vegetable Growers*, 2nd edn. Wiley, New York, pp. 390.

Margara, J. and David, C. (1978) Les etapes morphologiques du developpement du meristeme de chou-fleur *Brassica oleracea* L. var. *botrytis*. *Comptes Rendus l'Academie des Sciences Paris*, Series D 278, 1369–1373.

Marschner, H. (1986) *Mineral Nutrition in Higher Plants*. Academic Press, London.

Marshall, B. and Thompson, R. (1987a) A model of the influence of air temperature and solar radiation on the time to maturity of calabrese *Brassica oleracea* var. *italica*. *Annals of Botany* 60, 513–519.

Marshall, B. and Thompson, R. (1987b) Applications of a model to predict the time to maturity of calabrese *Brassica oleracea*. *Annals of Botany* 60, 521–529.

Martin, M.D. (1985) A programme for continuity. *Grower* 103(2), 15–19.

Maynard, D.N. and Barker, A.V. (1972) Internal browning of Brussels sprouts: a calcium deficiency disorder. *Journal of the American Society of Horticultural Science* 97, 789–792.

Metcalf, H.N. (1954) Effect of leaf and terminal bud removal on yield of Brussels sprouts. *Proceedings of the American Society of Horticultural Science* 64, 322–326.

Miller, C.H. (1988) Diurnal temperature cycling influences flowering and node numbers of broccoli. *HortScience* 23, 873–875.

Miller, C.H., Konsler, T.R. and Lamont, W.J. (1985) Cold stress influence on premature flowering of broccoli. *HortScience* 20, 193–195.

Miller, J.C. (1929) A study of some factors affecting seed-stalk development in cabbage. *Cornell Agricultural Experiment Station Bulletin 488*.

Millikan, C.R., Bjarnason, E.N. and Hanger, B.C. (1971) Effect of calcium sprays on the severity of internal browning in Brussels sprouts. *Australian Journal of Experimental Agriculture and Animal Husbandry* 11, 123–128.

Millikan, C.R. and Hanger, B.C. (1966) Calcium nutrition in relation to the occurrence of internal browning in Brussels sprouts. *Australian Journal of Agricultural Research* 17, 863–874.

Murphy, R.F. (1975) Chemical stopping of Brussels sprouts. *Irish Journal of Agricultural Research* 14, 199–206.

Nakamura, E. and Hattori. Y. (1961) On the seed vernalization of cabbages (*Brassica oleracea* spp.). II. Effects of a prolonged vernalization treatment and influence of gibberellin applied during vernalization. *Journal of the Japanese Society of Horticultural Science* 30, 167–170.

Nieuwhof, M. (1969) Cole crops. *Botany, Cultivation and Utilization.* Leonard Hill, London, p. 353.

Nonnecke, I.L. (1989) *Vegetable Production.* Avi, New York, p. 657.

North, C. (1957) Studies in morphogenesis of *Brassica oleracea* L. I. Growth and development of cabbage during the vegetative phase. *Journal of Experimental Botany* 8, 304–312.

North, C. (1958) Relationship between leaf shape and head formation in cabbage. In: *Advances in Horticultural Science and their Applications*, Vol. I. Pergamon Press, London, pp. 487–492.

Peck, N.H., Dickson, M.H. and MacDonald, G.E. (1983) Tipburn susceptibility in semiisogenic inbred lines of cabbage as influenced by nitrogen. *HortScience* 18, 726–728.

Rasco, E.T. Jr (1983) A catalog of noteworthy vegetables and fruits in the Philippines. *Society for the Advancement of the Vegetable Industry of the Philippines, Los Banos.* p. 50.

Rosen, C. (1990) Leaf tipburn in cauliflower as affected by cultivar, calcium sprays, and nitrogen nutrition. *HortScience* 25, 660–663.

Sadik, S. (1962) Morphology of the curd of cauliflower. *American Journal of Botany* 49, 290–297.

Sadik, S. (1967) Factors involved in curd and flower formation in cauliflower. *Proceedings of the American Society of Horticultural Science* 90, 252–259.

Sadik, S. and Ozbun, J.L. (1967) Histochemical changes in the shoot tip of cauliflower during floral induction. *Canadian Journal of Botany* 45, 955–959.

Sadik, S. and Ozbun, J.L. (1968) The association of carbohydrate changes in the shoot tip of cauliflower with flowering. *Plant Physiology* 43, 1696–1698.

Salter, P.J. (1957) Blindness in early summer cauliflower. *Nature* 180, 1056.

Salter, P.J. (1971) Mini-cauliflowers. *Agriculture, London* 78, 231–235.

Salter, P.J., Andrews, D.J. and Akehurst J.M. (1984) The effects of plant density, spatial arrangement and sowing date on yield and head characteristics of a new form of broccoli. *Journal of Horticultural Science* 59, 79–85.

Salter, P.J. and Fradgley, J.R.A. (1969) Studies on crop maturity in cauliflower. II. Effects of culture factors on the maturity characteristics of a cauliflower crop. *Journal of Horticultural Science* 44, 141–154.

Salter, P.J. and James, J.M. (1974) Further studies on the effects of cold treatment of transplants on crop maturity characteristics of cauliflower. *Journal of Horticultural Science* 49, 329–342.

Salter, P.J. and James, J.M. (1975) The effect of plant density on the initiation, growth and maturity of curds of two cauliflower varieties. *Journal of Horticultural Science* 50, 239–248.

Salter, P.J. and Laflin, T. (1974) Studies on methods of obtaining continuity of production of summer and autumn cauliflower. *Experimental Horticulture* 26, 120–129.

Salter, P.J. and Ward, R.J. (1972) Studies on crop maturity in cauliflower. III. Effects of cold treatments and certain growth regulators on crop maturity characteristics and yield. *Journal of Horticultural Science* 47, 57–68.

Scaife, A. and Wurr, D.C.E. (1990) Effects of nitrogen and irrigation on hollow stem of cauliflower (*Brassica oleracea* var. *botrytis*). *Journal of Horticultural Science* 65, 25–29.

Scaife, M.A. and Clarkson, D.T. (1978) Calcium-related disorders in plants – a possible explanation for the effect of weather. *Plant and Soil* 50, 723–725.

Skapski, H. and Oyer, E.B. (1964) The influence of pre-transplanting variables on the growth and development of cauliflower plants. *Proceedings of the American Society of Horticultural Science* 85, 374–385.

Stoffella, P.J. and Fleming, M.F. (1990) Plant population influences yield variability of cabbage. *Journal of the American Society of Horticultural Science* 115, 708–711.

Stokes, P. and Verkerk, K. (1951) Flower formation in Brussels sprouts. *Mededelingen van de Landbouw Hogeschool te Wageningen* 50(9), 143–160.

Strandberg, J.O. and White, J.M. (1979) Estimating fresh market cabbage maturity dates in a winter production area. *Proceedings of the Florida State Horticultural Society* 92, 96–99.

Sundstrom, F.J. and Story, R.N. (1984) Cultivar and growing season effects on cabbage head development and weight loss during storage. *HortScience* 19, 589–590.

Thomas, T.H. (1972) The distribution of hormones in relation to apical dominance in Brussels sprouts (*Brassica oleracea* var. *gemmifera* L.) plants. *Journal of Experimental Botany* 23, 294–301.

Thomas, T.H. (1980) Flowering of Brussels sprouts in response to low temperature treatment at different stages of growth. *Scientia Horticulturae* 12, 221–229.

Thomas, T.H., Lester, J.N. and Salter, P.J. (1972) Hormonal changes in the stem apex of the cauliflower plant in relation to curd development. *Journal of Horticultural Science* 47, 449–455.

Thompson, K.F. (1976) Cabbages, kales, etc. In: Simmonds, N.W. (ed.) *Evolution of Crop Plants*. Longman, London, pp. 49–52.

Thompson, R. and Taylor, H. (1973) The effects of population and harvest date on the yields and size grading of an F1 hybrid and open-pollinated Brussels sprouts cultivar. *Journal of Horticultural Science* 48, 235–246.

Thompson, R. and Taylor, H. (1975) Some effects of population density and row spacing on the yield and quality of two cauliflower cultivars. *Horticultural Research* 14, 97–101.

Thompson, R. and Taylor, H. (1976) Plant competition and its implications for cultural methods in calabrese. *Journal of Horticultural Science* 51, 230–231.

Titley, M.E. (1987) The scheduling of fresh market broccoli in Southeast Queensland for exporting to Southeast Asian markets from May to September. *Acta Horticulturae* 198, 235–242.

Tompkins, D.R., Norton, R.A. and Woodbridge, C.G. (1972) Brussels sprouts growth and yield as influenced by succinic acid-2,2-dimethyl-hydrazide. *Journal of the American Society of Horticultural Science* 97, 772–774.

Verheij, E.W.M. (1970) Spacing experiments with Brussels sprouts grown for single pick harvests. *Netherlands Journal of Agricultural Science* 18, 89–104.

Verkerk, K. (1954) The influence of low temperature on flower initiation and stem-elongation in Brussels sprouts. *Koninklijke Nederlandse Akademie van Wetenschappen* C 57, 339–346.

Walker, J.C., Edgington, L.V. and Nayudu, M.V. (1961) Tipburn of cabbage: Nature and control. *Wisconsin Agricultural Experiment Station Research Bulletin 230*.

Wellensiek, S.J. (1960) Annual Brussels sprouts. *Euphytica* 9, 10–12.

Wheeler, J.A. and Salter, P.J. (1974) Effects of shortening the maturity period on harvesting costs of autumn cauliflowers. *Scientia Horticulturae* 2, 83–92.

White, J.M. (1977) Effect of cultivation on cabbage yield and head weight. *Proceedings of the Florida State Horticultural Society* 90, 365–367.

Wiebe, H.J. (1972a) Wirkung von Temperatur und Licht auf Wachstum und

Entwicklung von Blumenkohl. I. Dauer der Jugendphase für Vernalisation. *Gartenbauwissenschaft* 37, 165–178.

Wiebe, H.J. (1972b) Wirkung von Temperatur und Licht auf Wachstum und Entwicklung von Blumenkohl. II. Optimale Vernalisationstemperatur und Vernalisationsdauer. *Gartenbauwissenschaft* 37, 293–303.

Wiebe, H.J. (1972c) Wirkung von Temperatur und Licht auf Wachstum und Entwicklung von Blumenkohl. III. Vegetative Phase. *Gartenbauwissenschaft* 37, 455–469.

Wiebe, H.J. (1973a) Wirkung von Temperatur und Licht auf Wachstum und Entwicklung von Blumenkohl. IV. Kopfbildungsphase. *Gartenbauwissenschaft* 38, 263–280.

Wiebe, H.J. (1973b) Wirkung von Temperatur und Licht auf Wachstum und Entwicklung von Blumenkohl. V. Einfluss der Jugendpflanzenanzucht auf die Variabilitätin Blumenkohlbestanden. *Gartenbauwissenschaft* 38, 434–440.

Wiebe, H.J. (1974) Zur Bedeutung des Temperaturverlaufs und die Lichtintensitat auf den Vernalisationseffekt bei Blumenkohl. *Gartenbauwissenschaft* 39, 1–7.

Wiebe, H.J. (1975) The morphological development of cauliflower and broccoli cultivars depending on temperature. *Scientia Horticulturae* 3, 95–101.

Wiebe, H.J. (1980) Anbau von Blumenkohl für eine kontinuirliche Marktbelieferung während der Erntesaison. *Gartenbauwissenschaft* 45, 282–288.

Wiebe, H.J. (1981) Influence of transpant characteristics and growing conditions on curd size (buttoning) of cauliflower. *Acta Horticulturae* 122, 99–105.

Wiebe, H.J. (1990) Vernalization of vegetable crops – a review. *Acta Horticulturae* 267, 323–328.

Wiebe, H.J., Schätzler, H.P. and Kühn. W. (1977) On the movement and distribution of calcium in white cabbage in dependence of the water status. *Plant and Soil* 48, 409–416.

Williams, C.A. and Atherton, J.G. (1990) A role for young leaves in vernalization of cauliflower: I. Analysis of leaf development during curd induction. *Physiologia Plantarum* 78, 61–66.

Wurr, D.C.E. (1981) The influence of cold treatments on the uniformity of cauliflower curd initiation and maturity. *Acta Horticulturae* 122, 107–113.

Wurr, D.C.E. (1990) Prediction of the time of maturity in cauliflowers. *Acta Horticulturae* 267, 387–394.

Wurr, D.C.E. (1994) Shedding some light on blindness in brassicas and bedding plants. *Horticultural Development Council Project News No. 28*, p. 4.

Wurr, D.C.E., Akehurst, J.M. and Thomas, T. (1981a) A hypothesis to explain the relationship between low-temperature treatment, gibberellin activity, curd initiation and maturity in cauliflower. *Scientia Horticulturae* 15, 321–330.

Wurr, D.C.E., Elphinstone, E.D. and Fellows, J.R. (1988a) The effect of plant raising and cultural factors on the curd initiation and maturity characteristics of summer/autumn cauliflower crops. *Journal of Agricultural Science* 111, 427–434.

Wurr, D.C.E. and Fellows, J.R. (1984) Cauliflower buttoning – the role of transplant size. *Journal of Horticultural Science* 59, 419–429.

Wurr, D.C.E. and Fellows, J.R. (1991) The influence of solar radiation and temperature on the head weight of crisp lettuce. *Journal of Horticultural Science* 66, 183–190.

Wurr, D.C.E., Fellows, J.R. and Crisp, P. (1982) Leaf and curd production in cauliflower varieties cold-treated before transplanting. *Journal of Agricultural Science* 99, 425–432.

Wurr, D.C.E., Fellows, J.R. and Hambidge, A.J. (1991) The influence of field environmental conditions on calabrese growth and development. *Journal of Horticultural Science* 66, 495–504.

Wurr, D.C.E., Fellows, J.R. and Hambidge, A.J. (1995) The potential impact of global warming on summer/autumn cauliflower growth in the UK. *Agriculture and Forest Meteorology* 72, 181-193.

Wurr, D.C.E., Fellows, J.R. and Hiron, R.W.P. (1990a) Relationships between the time of transplanting, curd initiation and maturity in cauliflower. *Journal of Agricultural Science* 114, 193–199.

Wurr, D.C.E., Fellows, J.R. and Hiron, R.W.P. (1990b) The influence of field environmental conditions on the growth and development of four cauliflower cultivars. *Journal of Horticultural Science* 65, 565–572.

Wurr, D.C.E., Fellows, J.R., Phelps, K. and Reader, R.J. (1993) Vernalization in summer/autumn cauliflower (*Brassica oleracea* var. *botrytis* L.). *Journal of Experimental Botany* 44, 1507–1514.

Wurr, D.C.E., Fellows, J.R., Phelps, K. and Reader, R.J. (1994) Testing a vernalization model on field-grown crops of four cauliflower cultivars. *Journal of Horticultural Science* 69, 251–255.

Wurr, D.C.E., Fellows, J.R. and Suckling, R.F. (1988b) Crop continuity and prediction of maturity in the crisp lettuce variety Saladin. *Journal of Agricultural Science* 111, 481–486.

Wurr, D.C.E., Fellows, J.R., Sutherland, R.A. and Ephinstone, E.D. (1990c) A model of cauliflower curd growth to predict when curds reach a specific size. *Journal of Horticultural Science* 65, 555–564.

Wurr, D.C.E., Kay, R.H., Allen, E.J. and Patel, J.C. (1981b) Studies of the growth and development of winter-heading cauliflowers. *Journal of Agricultural Science* 97, 409–419.

Zink, F.W. (1968) Hollow stem in broccoli. *California Agriculture* 22(1), 8–9.

Zink, F.W. and Akana, D.A. (1951) The effect of spacing on the growth of sprouting broccoli. *Proceedings of the American Society of Horticultural Science* 58, 160–164.

ZMP (1992) ZMP Bilanz Gemüse. *Zentrale Markt- und Preisberichtstelle* GMBH, Bonn, pp. 192–196.

The Root Vegetables: Beet, Carrot, Parsnip and Turnip

16

L.R. Benjamin, A. McGarry[1] and D. Gray
Crop and Weed Science Department, Horticulture Research International,
Wellesbourne, Warwick CV35 9EF, UK

The main aim of this chapter is to identify the salient factors that determine the initiation and subsequent growth of the storage root of beets, carrots, parsnips and turnips. In addition, we examine the factors that determine flowering, seed development and seedling establishment in relation to yield in these crops. Beets are members of the *Chenopodiaceae*, turnips are members of the *Cruciferae* and carrots and parsnips are members of the *Umbelliferae*. Despite belonging to different families, they have many features in common. They are all economically important. Worldwide in 1993, 282,000,000 mt of sugarbeet was grown and 603,000,000 mt of other storage root crops were grown (FAO, 1993). In their first year of vegetative growth, the plant consists of a rosette of leaves, a swollen hypocotyl/upper tap root, and many fibrous roots. From a functional point of view, they have a common topology (Fig. 16.1). Photosynthate destined for the fibrous roots and shoot apical meristem must be translocated via the storage organ. Similarly, water and mineral nutrients must be translocated via the storage root to the leaves. In contrast, in many other species, assimilates do not need to flow through the storage organ in their journey between the leaves and fibrous roots. The functional importance of this topology is obscure, but it can confound experimental manipulation of the storage root with that of the fibrous roots. This chapter does not cover the potato crop. Potatoes also form storage organs in the soil, but these are formed from stolon tips, not from the primary root axis, and a plant typically forms numerous storage organs rather than a single storage organ.

Storage roots are characterized by cell division and expansion throughout their development (Milford, 1973; Ting and Wren, 1980a; Hole *et al.*,

---

[1] Our colleague, Tony McGarry, sadly died in February 1995.

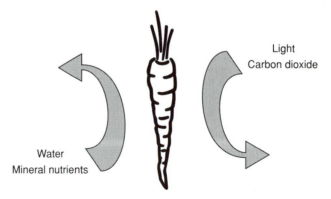

Fig. 16.1. Functional relationship between shoot and fibrous roots of storage root plants.

1984). In contrast, the storage organs of other species, for example apple, are characterized by an initial phase predominated by cell division, followed by a phase of cell expansion with no cell division. The storage organs of root crops develop by the formation of a single ring (carrots and parsnips) or several rings (beet and turnips) of cambium. In all species, there is no distinct limit to the distance down the tap root these cambia can extend. These features have implications for the chemical composition, 'maturity' and mechanical properties of the storage roots.

Botanically, the storage roots serve to carry over assimilates from one year to the next, to provide ample early-season resources for the development of the reproductive structures. During the first year of growth, the storage root may also serve other purposes. The demand for photosynthates caused by the formation of a large storage organ may prevent a feedback inhibition of photosynthesis in beet (Thorne and Evans, 1964; Das Gupta, 1972a), but this role is unimportant in carrots (Benjamin and Wren, 1978; Steingrover, 1981). The storage root may act as a 'water reservoir', helping to maintain a constant diurnal water supply to the leaves (Olymbios, 1973).

STORAGE ROOT INITIATION

The storage roots of carrot, beet, parsnip and turnip are derived from the formation and activity of a cylindrical vascular cambium in the hypocotyl and tap root. This vascular cambium consists initially of separate strips formed by the divisions of cells between the primary xylem and the primary phloem. Subsequently, the strips join to encircle the primary xylem. This pattern of secondary root growth is typical of all dicotyledons and gymnosperms and is described in detail in many textbooks (for example Esau, 1967). In carrots, the initiation of this secondary cambium nearly always occurs before the development of the foliage leaves (Esau, 1940). In beet, additional supernumerary cambia are also formed in quick succession.

Consequently, a beet of only a few millimetres diameter may contain all the annular zones of growth developing simultaneously (Hayward, 1938).

There have been very few physiological studies of the mechanisms that trigger this anatomical development. Most suggest that the stimulus for storage organ development arises in the shoot and that growth regulators and sucrose are likely to be the stimulus.

Excising the shoots (including the cotyledons) of turnip seedlings prevents vascular cambial development, but removal of the tap root apex does not have such an inhibitory effect (Peterson, 1973). A combination of auxin, cytokinin and myoinositol and sucrose promotes vascular cambial development in excised root tips. Omitting any of these substances reduces activity. Similar results were found by Ting and Wren (1980b) using decapitated radish seedlings. In sugarbeet, replacing the growing shoot apex with capsules containing either auxin, gibberellic acid or kinetin, stimulates the growth of the storage root (Das Gupta, 1972b). This increased growth might be due to a stimulation of cambial activity.

However, there is some evidence that the root system also controls the signal for storage root development. Benjamin and Wren (1978) noted that cutting the unthickened tap root of 14-day-old carrots several centimetres below the insertion point of the cotyledons promoted thickening of the lateral roots close to the cut end. Therefore, some form of 'apical dominance' occurred in the 14-day intact seedlings to suppress the initiation of thickening in lateral roots. Interestingly, such thickening of the lateral roots did not occur when cutting of the tap root was delayed until 35 days, when it was already appreciably thickened.

Initiation of the vascular cambium is only the first stage in the development of the storage root. Subsequently, there are cell division, cell expansion and growth and storage of carbohydrate. Hole et al. (1984) compared carrot cultivars that differed in mature shoot : root ratio, and found that they did not differ in the timing of the formation of the cambium. However, the cultivar with the lesser shoot : root ratio had a smaller xylem parenchyma cross sectional area. Their main conclusion was that differences between cultivars in shoot : root ratio could not be easily attributed to the timing of initiation of the cambium in the storage root, but to early differences in shoot growth. This work was confirmed in extensive field studies of nine carrot cultivars differing in root shape and shoot : root ratio (Hole et al., 1987).

At densities greater than 500 plants m^{-2}, red beet and radish plants often fail to form a normally swollen tap root, but cambia are present (Hole et al., 1984), whereas in carrots and parsnips, storage roots of normal appearance are formed over a wide range of densities. Hole et al. (1984) explained these phenomena by proposing that: (i) after formation of the vascular cambium, a second stimulus is required for subsequent storage root development; (ii) the level of this stimulus in each plant decreases with plant density; and (iii) the threshold level is inherently lower in carrots and parsnips than in red beet or radish. The nature of this stimulus was not identified, but they speculated that it is triggered when an unknown factor supplied by the shoots passes a threshold value.

Green *et al.* (1986) reported early and late phases of growth in sugarbeet. In the early phase, the rate of biomass accumulation is slow, partitioning of dry matter to the storage root is small and constant, and the partitioning of dry matter within the storage root into sucrose is constant. In the second phase, the plant absolute growth rate is large and constant, the proportion of plant dry matter partitioned to the storage root increases with time, and the rate of sucrose accumulation increases. The transition from the early to late phase of growth was rapid and occurred earlier for dry matter partitioning to the root than for increased sucrose accumulation rate. This implied that structures for accumulating sugar develop before it is stored (Milford *et al.*, 1988). Green *et al.* (1986) proposed that the simultaneous development of secondary cambia leads to an initial abundance of small cells and an increase in sucrose concentration. The expansion of these cells is followed by the production of more small cells from the cambia, thus maintaining the rapid growth and increased fractionation of dry matter to sucrose.

This hypothesis, based upon a rapid transition between two phases of storage root development, was criticized by Milford *et al.* (1988). They applied flexible model-fitting routines to establish the global optima for break points between the two phases and could find no difference in timing between the break points for storage root dry matter increment and rate of sucrose accumulation. In fact, they claimed a progressive shift in partitioning of dry matter to the storage root as the crop develops with no sudden transition at a specific point in development.

In conclusion, storage root formation results from the activity of secondary cambia, whose initiation is dependent upon a supply of sucrose and growth regulators from the shoots. However, the subsequent processes of cell expansion and growth also require further 'triggers' whose nature has not yet been identified.

STORAGE ROOT GROWTH

INDIRECT CONTROLS OF STORAGE ROOT GROWTH

The growth of the storage root is dependent virtually entirely upon the ability of other plant organs to intercept and assimilate resources. Therefore, the effects of environmental factors on root growth are difficult to disentangle from their effects on total plant growth. In this section we examine the effects of environment on total plant growth. Their direct effects on storage root growth will be dealt with in the next section.

Environmental factors

Light interception is regarded as the crucial factor determining growth of sugarbeets. In northwest Europe, sugarbeet yields are directly proportional to the amount of radiant energy intercepted by the leaves (Scott and

Jaggard, 1992). Martin (1986) also reported the conversion efficiency of photsynthetically active radiation (PAR) to dry matter of sugarbeet was little affected by environmental conditions in New Zealand.

The critical process for sugarbeet crops is leaf expansion, which is influenced by many factors, including the temperature experienced by the leaves before unfolding, by nitrogen available then (Milford *et al.*, 1985) and by photoperiod (Milford and Lenton, 1976). This complex interaction between temperature, nitrogen and light interception might explain why the early work of Austin (1964) failed to find a simple environmental predictor of red beet yield.

For carrots and turnips grown at optimal temperature and water supply, plant weight increases with increasing daily light integral (by extending daylength or light intensity) or increasing carbon dioxide concentration (Downs, 1985; Ikeda *et al.*, 1988; Hole and Sutherland, 1990). Benoit and Ceustermans (1990) examined the growth of carrots sown in the winter under single or double plastic covers. They proposed that once an optimum shoot : root ratio has been reached at a minimum air temperature of 8.7°C, plants become more sensitive to light than to temperature.

Although light interception is important in temperate crops, water use has also been found to be important. Drought has been assessed to cause a 17–43% potential yield loss in unirrigated sugarbeet in the UK (Dunham and Clarke, 1992). Harris (1972) reported interactions between soil water content and sugarbeet plant density. He proposed that for young crops, water extraction increased with plant density, due to the greater soil cover by leaves. However, rooting depth increases with density (Pakov, 1964; Coulter, 1966) and soil water extraction was more rapid at higher densities (Farazdaghi, 1968). Therefore, low density crops may be more susceptible than high density crops to drought in the later stages of crop growth.

For carrots, much of the early work emphasized the role of water and temperature on growth. Barnes (1936) found that the optimum temperature for growth is 16–21°C and that root weight increased with successive increases in soil water content. The relative difference between root weights due to temperature was unaffected by soil water content. More recently, White (1992) also showed that the yield of carrots had an 'overturning' relationship with soil water content. Salinity concentrations greater than 16 mS cm^{-1} in the soil solution or low nutrient conditions increase the ratio of evapotranspiration to weight gain (Schmidhalter and Oertli, 1991). For turnips, yield of the storage roots was reduced by 9% for each unit increase in salinity above 0.9–12 dS m^{-1} (Francois, 1984).

In hotter, drier climates, the relationship between light interception and yield may not be so close. In Turkey, air and soil temperatures during crop maturation, and rainfall around seedling emergence and before harvesting were the best predictors of sugarbeet yield (Yucel, 1992). Dry matter increment of sugarbeet in Italy was best described by a model which allowed for water deficits and accumulated day degrees (Mambelli *et al.*, 1992).

Plant density and plant arrangement

Growth is also determined by plant density because it determines the amount of resources available to each plant within a crop. Willey and Heath (1969) pointed out that yield density relationships can be asymptotic or over-turning. In carrot, yield of roots is asymptotic, with no evidence of a decrease in yield with increasing density up to 22,000 plants m^{-2} (Robinson, 1969). In contrast, for red beet, the optimum density for root yield is around 50–100 plants m^{-2} (Benjamin *et al.*, 1985). Also in parsnip, there is some evidence of an overturning yield–density relationship, with optimal densities between 28 and 72 plants m^{-2} (Bleasdale and Thompson, 1966). In sugarbeet, the commercial densities are around 7.5 plants m^{-2}, as these are optimal for sugar production, but total plant yields per unit area increase over the range 3–14 plants m^{-2} (Harris, 1972). Hole and Dearman (1993) suggested that the contrasting yield–density relationship between carrot and other crops is a consequence of their adaptation to a reduction in the photosynthetic photon flux density (PPFD). They reported that at low PPFD values, carrots maintain their leaf area but not their shoot dry weight, so that storage root growth is not severely reduced by low PPFD values. In contrast, leaf area and shoot dry weight are maintained in red beet at low PPFD at the expense of the storage root dry weight.

Rectangularity of spacing (the ratio of between-row to within-row spacing) also exerts an influence on plant growth independent to that exerted by density (Bleasdale, 1982). Between-row spacing, however, has smaller effects on plant weight than density. For example, red beet yield was unaffected by between-row spacings below 51 cm at densities of 124 plants m^{-2} (Bleasdale and Thompson, 1963a). On the other hand, spatial arrangement does not always influence yield. For example, Salter *et al.* (1980) found no effect on carrot yield in an extensive study in which there were various combinations of numbers of rows per bed, between-row spacings and target densities.

Variation in weight

Spatial pattern influences the weight of individual plants, with those growing next to gaps benefiting from an additional supply of growth resources (Salter *et al.*, 1980). Models have been derived to predict the weight of carrot or red beet plants from the position of neighbours (Benjamin and Sutherland, 1992). These models have now been incorporated into commercially available software which predicts the seeding rate in each row to eliminate edge effects and maximize the yield of carrots or red beet in target diameter grades (Benjamin, 1994).

The management of size grading in carrots, parsnip and beet has been largely achieved by manipulating plant spacing and arrangement (Salter *et al.*, 1979). Typically there is substantial initial variability in weight which is amplified at high density, but can diminish at low density (Benjamin, 1988;

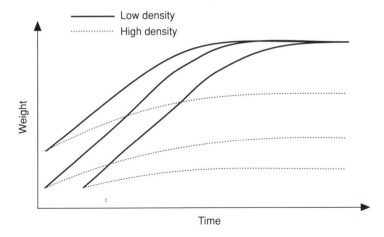

Fig. 16.2. Schematic representation of the growth of carrot seedlings of different initial weight and times of seedling emergence at two contrasting densities (after Benjamin, 1988).

Fig. 16.2). At high densities, plants that are initially large, capture a progressively greater proportion of the environmental resource available for growth (Salter *et al.*, 1981; Benjamin 1984a, 1987; Benjamin and Bell, 1985; Benjamin and Hardwick, 1986). Dynamic growth models have been derived for individual carrot and red beet plants, allowing for the overall density and weight variation initially present in the seedling population (Benjamin, 1988; Aikman and Benjamin, 1994).

Variation in seedling size or mass at emergence can arise from variation in seed weight (*S*), embryonic weight (*E*), duration of pre-emergence growth (*D*) and pre-emergence growth rates (*G*) (Benjamin and Hardwick, 1986; Benjamin, 1990).

Variation in seed weight and embryonic weight

Within any root crop species there is typically a two-fold variation in seed or fruit weight. In *Brassica* root crops, variation in *S* and *E* are closely related because the embryonic mass is a large part of the seed structure (Gray *et al.*, 1985). However, in carrot and parsnip the volume of the embryo typically represents less than 5% of volume of the entire seed (Gray *et al.*, 1984b). Hence, the variation in *E* can be much greater than the variations in *S* (Gray and Steckel, 1983; Gray *et al.*, 1985b). Consequently, the reductions in variability of *E* by grading can be very small in the *Umbellifera* crops.

The lack of correlation between *S* and *E* in umbelliferas arises from the patterns of growth of *S* and *E* in different umbel orders on the seeding plant (Fig. 16.3). In these species, the endosperm develops rapidly in the first 14–21 days whilst the embryo is still in the 8–16-cell stage (Gray *et al.*, 1984),

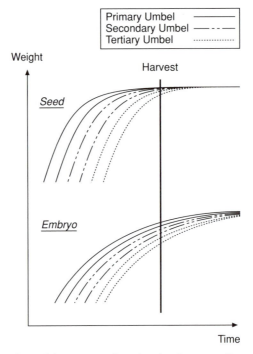

Fig. 16.3. Representations of the patterns of seed and embryo growth on primary, secondary and tertiary umbels. The two lines for each umbel represent the growth of the earliest and latest growing seeds. The vertical line indicates the time of harvest.

and maximum S occurs before E is fully attained. When this difference is linked to the different timing of flowering and growth on progressively lower umbel orders, then substantial differences in E within a population of relatively uniform S can arise at the times of seed harvest (Fig. 16.3). Both open-pollinated and F_1 hybrids show a similar pattern of response and there is no evidence for lower variability in E among hybrids (Dowker and Gray, 1985) nor is it affected by seed priming (Gray *et al.*, 1984). Based on these findings in Umbelliferas, relationships between seedling and root variability have been established with E and to other factors related to it (Gray *et al.*, 1991). In beet and root brassicas detailed relationships between S and E have not been established.

In sugarbeet and carrot, an 'adverse' pre-emergence seedling seedbed environment can affect the pattern of growth of plants post-emergence irrespective of their size at emergence or time of emergence (Durr *et al.*, 1990). This is possibly associated with re-allocation of a greater proportion of the mobilized seed reserve to the seedling root. However, its extent and occurrence in practice are unknown and there has been no assessment of its consequences on storage root weight or size variation.

Variation in duration of pre-emergence growth

Pre-emergence growth (D) is the time between radicle emergence and emergence of the shoot through the soil surface. Variation in D *within* populations of sown seeds arises initially from seed-to-seed differences in the rate of germination, t_{50}. The spread in individual t_{50}s is usually positively correlated with mean t_{50} of the population (Gray, 1984). Hence, where the average t_{50} of a seed lot (of poor quality) is high, absolute differences among their t_{50}s can be substantial. These are of the order of 4–5 days in brassicas and of 10 or more in umbelliferas. Variability in t_{50} will determine variation in times of seedling emergence when pre-emergence growth rates are similar to one another, and when there are favourable conditions for seedling emergence. However, where conditions are unfavourable for seedling emergence, those seeds with large t_{50}s are less likely to emerge and so the variation in seedling emergence time within the population is likely to be smaller, and plant density will also be reduced. Hence, the seedbed can exert a 'sieving' effect by removing 'weaker' seedlings. This phenomenon may explain why seed priming (Finch-Savage, 1990) and fluid-sowing of pre-germinated seeds (Gray *et al.*, 1984a), which might be expected to improve seedling uniformity as a result of improved uniformity of germination, does not always result in more synchronous seedling emergence.

Variation in pre-emergence growth rates

The mean rate of pre-emergence growth (G) of a population of seedlings is greatly influenced by temperature, soil moisture and soil impedance to shoot and root growth (Benjamin, 1990). However, little is known of the effects of these factors on the spread of G within the population. In carrot, late germinating seedlings have low initial relative growth rates (Finch-Savage and McQuistan, 1988), but their influence on variation in root size is unknown.

DIRECT CONTROLS OF STORAGE ROOT DRY WEIGHT

Ulrich (1952) proposed that storage roots act as 'passive' sinks, making use of only those assimilates that are 'surplus' to the requirements of the shoots and the fibrous roots. Subsequent research contradicts this hypothesis and shows that storage root growth is under strict internal control. Three types of approach have been taken in this research: (i) surgical manipulation of the plants; (ii) differential imposition of temperature regimes to the shoot and root system; and (iii) observations of the effects of growth regulators. Each approach has its own strengths and weaknesses, but all indicate that there is an active competition for internal assimilates by the storage root.

Surgical manipulation

Surgical manipulation is the most direct method of determining whether internal partitioning of assimilates is controlled. However, the use of these treatments to determine the role of the storage root is complicated by the topology of the plants (see Fig. 16.1). Removal of part of the storage root will inevitably involve removal of at least part of the fibrous root system too. It is well established that there is a functional equilibrium between fibrous roots, as providers of water and mineral nutrients, and the shoots, as providers of photoassimilates. Consequently, in constant environments, growth following removal of part of either the shoot or fibrous root system is such as to re-establish the shoot : root weight ratio that existed before surgical treatment. In carrots and sugarbeet this is brought about by an increase in the relative growth rate of the remaining part of the amputated organ and a reduction in the relative growth rate of the intact organ (Fick *et al.*, 1971; Das Gupta, 1972a; Olymbios, 1973; Benjamin and Wren, 1978; Benjamin, 1984b).

In addition to this functional equilibrium between shoots and fibrous roots, the storage root takes resources at the expense of the other two organs. Removal of the growing point of the shoots produces a transient increase in storage root growth in sugarbeet (Das Gupta, 1972a) and in carrots (Olymbios, 1973). Removal of part of the fibrous root system of carrots decreases the proportion of ^{14}C-labelled assimilates partitioned to the expanding leaves and increases the proportion partitioned to the fibrous roots and to the storage root (Benjamin and Wren, 1980). The presence of the storage root appears to have no special role for the functioning of the other organs. Unlike the response to partial removal of the other two organs, the relative growth rate of the remaining storage root of carrots, and that of the other organs, is not affected by removal of the distal storage organ (Benjamin and Wren, 1978).

Reciprocal grafts between species that differ in their propensity to produce storage roots, show that the growth of the storage root is determined by the root and shoot system, and it is unlikely that the storage root's growth is dependent merely upon a supply of photosynthates surplus to the requirements of the shoots (Thorne and Evans, 1964; Rapoport and Loomis, 1985). However, this approach does not distinguish between the role of the storage root and fibrous roots in determining storage root growth.

Temperature regimes

Reducing the temperature of an organ should restrict its metabolic activity and, hence, its ability to compete for assimilates, providing that the range of imposed temperatures are below the optimum for growth. In carrot, increasing shoot and root temperature from 15 to 25°C increased shoot weight 2.6-fold and storage root weight by 1.5-fold (Olymbios, 1973). Increasing the shoot temperature from 15 to 25°C, while holding the root temperature at 15°C promoted shoot weight by 36%, but caused a slight reduction in storage root weight (Table 16.1). Increasing the root temperature from 15 to

Table 16.1. Effect of temperature in the shoot and root environments on carrot shoot and total root dry weight (weights standardized to 100 in 15/15°C regime) (Olymbios, 1973).

	Shoot temperature (°C)	
	15	25
Total root weights		
Root temperature (°C)		
15	100	93
25	53	153
Shoot weights		
Root temperature (°C)		
15	100	136
25	113	258

25°C, while holding the shoot temperature at 15°C had the remarkable effect of *reducing* storage root weight by 47% and increasing shoot weight by 13%.

Olymbios (1973) attempted to explain these results by proposing that 25°C is superoptimal for storage root growth but close to optimal for shoot growth. This explanation is not totally convincing, since the greatest weights occurred at 25/25°C when they might have been expected at 25/15°C (Table 16.1). An alternative explanation is that the optimum temperatures for shoot and root growth are close to 25°C, but the competition between organs for a common pool of assimilates is controlled by a signal from the shoots (Das Gupta, 1972b). Therefore, when the temperature of the roots is greater than that of the shoots, the signal from the shoots inhibits the enhanced competitive ability of the roots conferred by the temperature regime. The lack of any competitive advantage over the shoots for assimilates, combined with a greater respiratory load would result in a reduced root weight.

Das Gupta (1972b) applied similar temperature treatments to sugarbeet, but in all treatments the shoot apical meristem was removed. This had three effects. First, any signal that might be produced by this meristem was removed. Second, the sink activity of the shoots would be greatly reduced. Some sink activity would occur as some remaining leaves continued to grow. Third, the high temperature promoted senescence, which could not be compensated for by the growth of new leaves. Allowing for the more rapid leaf senescence at the higher temperature observed by Das Gupta (1972b), the shoot and storage root weights in the different temperature regimes are as expected based on the assumption that the growth of these two organs is determined by competition between them for assimilates. Shoot weight was less at the higher shoot temperatures, presumably due to enhanced senescence (Table 16.2). Shoot weights were lower at higher root temperature, because the roots were a stronger competitor for assimilates. The greatest storage root weight was when the roots were at the higher temperature and the shoots at the lower temperature (Table 16.2). The least root weight might

Table 16.2. Effect of temperature in the shoot and root environments on dry weight (g) of sugarbeet shoots and storage roots. The shoot apical meristem was removed from these plants (Das Gupta, 1972b).

	Shoot temperature (°C)	
	17	25
Storage root weights		
Root temperature (°C)		
17	5.0	8.2
25	8.6	8.5
Shoot weights		
Root temperature (°C)		
17	10.2	7.7
25	8.7	6.3

be expected to occur at the highest shoot temperature and the lowest root temperature (25/17°C), but in fact occurred at the 17/17°C regime (Table 16.2). Perhaps the enhanced shoot senescence at 25°C supplied additional assimilates for the storage roots.

Growth regulators

Growth regulators have large effects on the distribution of dry matter between the shoots and storage organ, suggesting that endogenous hormones control dry matter partitioning. For example, in carrot, foliar applications of gibberellic acid increase shoot to root ratio, whereas chlormequat chloride decrease this ratio (McKee and Morris, 1986). The effect of growth regulators has been succinctly described in terms of a model based on competition between organs for a common pool of assimilates (Barnes, 1979). From this model the following equation relating shoot weight (s), root weight (r), and crop age (t) was derived.

$$\ln s = \alpha - \eta t + k + \gamma \ln r \qquad (1)$$

where α, η, k and γ are constants (Thomas *et al.*, 1983). Hence, the logarithm of shoot weight is linearly related to the logarithm of storage root weight, but the intercept decreases with time. The effects of gibberellic acid applications on shoot : root weight ratio of carrot crops was described by allowing the value of k to increase with application rate. The constant k was independent of time, suggesting that a single application causes a transient change in dry matter distribution in favour of the shoot with no further modification.

In summary, the experimental evidence points to an involvement of growth regulators and a signal from the shoot apex controlling the partition of dry matter to the storage organ. Little is known of the exact mechanism occurring within the plants, but see Hole (1994) for a recent review.

COMPOSITION, FLAVOUR, APPEARANCE AND MECHANICAL PROPERTIES

As product quality becomes an increasingly important marketing criterion, interest in the physical properties of root vegetables has grown (Millington, 1986; Kokkoras, 1989; Toivonen, 1992; McGarry, 1993, 1995). Also, the chemical composition of the storage roots determines the taste and colour of the storage roots.

Sugar content and taste

The high sucrose content of sugarbeet is the cardinal reason for its cultivation. The sucrose accumulation in sugarbeet storage roots is subject to rigorous internal plant controls, and is not simply a consequence of a super-abundance of photosynthates. Reducing the incident light intensity upon sugarbeet reduces the total root fresh weights, but has little effect upon sucrose concentration in the roots (Watson *et al.*, 1972; Wyse, 1980). Replacing the shoot system of sugarbeet plants with those of spinach beet by grafting, reduces storage root size but does not affect total sugar concentration in the storage roots (Thorne and Evans, 1964).

The sucrose concentration in sugarbeet is determined by the proportions of large and small cells that make up the constituent tissues (Milford, 1973; Wyse, 1980). The sugar concentration is lower in the parenchymatous zones in each ring, which contain large cells, but is higher in the vascular zones, which contain small cells. There tends to be a trade-off between root weight and average tissue sugar concentration, with growth being attributable to the expansion of the large cells in the parenchymatous zones. The changes in sugarbeet root weight due to different light levels are brought about by differences in frequency of vascular rings, with no effect on either average cell size or sucrose concentration.

In sugarbeet, impurities in the sap disrupt the extraction of sugar. These impurities are chiefly α-amino nitrogen, sodium and potassium. Increasing nitrogen fertilizer applications increases root yield and the content of these impurities, and therefore there is a very narrow range of optimum nitrogen rates depending on soil, site and cultivar (Beek and Huijbregts, 1986).

In contrast to sugarbeet, in carrots environmental variables greatly affect sucrose accumulation (Barnes, 1936; Nilsson, 1987). This difference may be because the carrots cannot adjust their average cell volume by the production of supernumerary cambia. Carrot cultivars differ three-fold in their ability to accumulate sugars in the storage root and those that have a capacity for high sugar yield have higher net assimilation rates (Lester *et al.*, 1982).

Enzyme activity also determines the rate and species of sugar accumulation in storage roots (Hole, 1994). Sucrose synthetase has been associated with sucrose storage in sugarbeet (Giaquinta, 1979). Sugar accumulation in carrots is controlled, at least in part by acid invertase activity, and the expression of this activity is dependent upon concentrations of growth regulators (Ricardo, 1976). Acid invertase is thought to prevent storage of sucrose by its inversion at the tonoplast, and alkaline invertase is thought to control the balance between reducing sugars and non-reducing sugars and thus control the availability of non-reducing sugars for storage (Hole and McKee, 1988). However, Hole and McKee (1988) found that there was no effect of cultivar or duration of growth on the activity of any of these three enzymes in carrot storage roots.

The predominant sugars in the storage root of carrots are sucrose, glucose and fructose (Goris, 1969a,b; Alabran and Mabrouk, 1973), with small amounts of starch being accumulated (Platenius, 1934; Goris, 1969b; Nilsson, 1987). The ability of carrot storage root parenchyma to store sucrose develops within a few days of its formation and the concentration of sucrose increases with time. In contrast, the concentration of the reducing sugars increases slightly or remains constant (Barnes, 1936; Steingrover, 1981; Lester *et al.*, 1982; Nilsson, 1987; Hole and McKee, 1988). However, the maximum sucrose concentration in carrots is only about half that of sugarbeet (Hole and McKee, 1988). Phan and Hsu (1973) estimated that the sugar contents in carrots reached a maximum about three months after sowing, but roots continued to grow well after this 'biochemical maturity' had been achieved. In contrast, Nilsson (1987) found that sucrose concentration increased with time up to 137 days and claimed that there is no well-defined stage of 'biochemical maturity'.

The sweet flavour of carrots is determined by the content of saccharides, free nitrogenous compounds (Alabran and Mabrouk, 1973), and terpenoids (Simon and Peterson, 1979). Little is known about the factors that determine the contents of these compounds, but soil properties have been thought to be important. For example, mineral fertilizer does not affect the sucrose content of carrots, but those grown 'organically' have been reported to have a poorer taste (Evers, 1989). Carrots grown on peat soils have lower dry matter contents (Gormley *et al.*, 1971).

Shape

Root shape is a major aspect of quality in carrots. A great difficulty of establishing the determinants of root shape is the need for an objective criterion. Simple measures, such as the ratio of crown diameter to length, have been used (Bradley *et al.*, 1967). Bleasdale and Thompson (1963b) proposed cylindricality (C), as defined as

$$C = \frac{w}{\pi r^2 l} \tag{2}$$

where w is root fresh weight, r is the radius of the crown and l is the root length. The values of C lie between 0.33 for a cone and 1.0 for a cylinder and for cv. Amsterdam Forcing increase from about 0.6–0.7 between 95 and 190 days. Comparable values for cv. Autumn King were 0.50 and 0.55. There was no consistent effect of plant density on the value of C. They also proposed that length of the roots be proportional to the logarithm of crown diameter. The intercept and slope of this relationship increased with plant age and they were lower at higher densities for both cultivars and at each harvest. Consequently, for a specified diameter, the root lengths were shorter than at higher densities. Benjamin and Sutherland (1989) pointed out that these measures are excellent summary statistics for comparing the shapes of roots of different cultivars and from different imposed treatments, but they do not allow the diameters or lengths of roots to be predicted from given weights. This is important, because root vegetables are usually sold in diameters and/or length grades, but most information in plant growth describes effects of environment on weight. They proposed two simple relationships: (i) that logarithm of diameter is linearly related to logarithm of weight; and (ii) length is linearly related to diameter (Fig. 16.4). In carrots and red beet, the slopes and intercepts of these relationships were little affected by imposed density treatments, and in red beet there was surprisingly little difference among the three cultivars tested. Furthermore, these relationships gave good predictions of yields of carrots and red beet in diameter grades from observed weights in independent datasets. Commercially, carrots are sought that are cylindrical with a blunt apex, tapering roots being associated with immature roots of poor taste (Long, 1989). Therefore, subtle changes in the shape of the apex could be commercially crucial (Umiel *et al.*, 1972), but be difficult to detect using simple measures of shape. More complex measures have been adopted (Snee, 1972; Umiel *et al.*, 1972), but involve many measurements on each root. These can prove impractical to do on the many carrots needed to establish significant differences between treatments. Recent advances in machine vision and image analysis offer the prospect of overcoming these technical difficulties, and allow these subtle shape differences to be measured rapidly (Howarth *et al.*, 1992).

Stanhill (1977) reviewed the factors that influenced carrot shape. He summarized that besides large genetic-based differences, roots become more cylindrical with increasing plant density, at air temperatures below 18°C, in drier soils, when the shoots have been defoliated and when the plants are younger. Nutrition and day length did not affect shape.

In sugarbeet, there is a requirement for more round-shaped roots as these have less soil adhering to them at harvest. Breeding programmes crossing sugarbeet with red beet have produced globe-shaped roots, but sugar contents were low and sap impurities, affecting sugar concentration, too high (Mesken, 1990).

MODELS OF ROOT DEVELOPMENT AND QUALITY

Banga (1954) proposed a model that encompassed total plant development and storage root development. The model proposed the concepts of primary

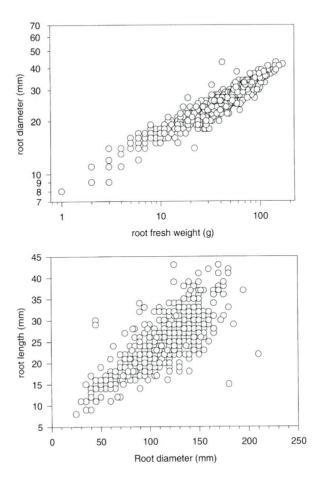

Fig. 16.4. Relationship between log$_e$ storage root weight and log$_e$ diameter and between storage root diameter and length for carrots.

and secondary activities. The primary activity consists of the growth of the leaf rosette and unthickened tap root and involves protein synthesis. The secondary activity was defined as the growth of the thickened storage root, and involves the synthesis of storage compounds and carotene. He suggested that when the primary axis has developed to a threshold value, its growth activity may be replaced to differing degrees by secondary growth. After the onset of secondary activity, the following dynamic equilibrium was proposed.

$$g_p \rightleftarrows g_s \qquad (3)$$

where g_p is the growth activity of the primary axis, and g_s is the activity of root thickening. This dynamic equilibrium is a result of protein synthesis competing with storage compound synthesis for photosynthates (Banga and De Bruyn, 1968). Secondary growth leads to root ripening, which is the 'simultaneous thickening, rounding (of the tip) and colouring' of the fleshy root. They speculated that between 8 and 18°C, the Q_{10} value for primary growth is less than that for storage activity. Thus, roots are heavy shouldered, pointed and have poor colour at the low temperature because, relative to plants at 18°C, the ratio of primary to secondary growth rates is high. In support of their hypothesis, they found protein synthesis to be greater at low temperatures. Similar effects of temperature on protein synthesis in carrot leaves and storage roots were reported by Zeid and Kuhn (1973). In contrast, high plant density or low soil water content inhibit growth, but do not affect the rate of ripening, thus producing small roots of mature appearance.

This early model has explained some observed effects of environmental factors on root shape and appearance, but it appears to have been largely overlooked. Interestingly, Linser and Zeid (1972) independently proposed a similar hypothesis based upon 'system' and 'product' synthesis, which loosely correspond to the dynamic equilibrium described in equation 3.

FLOWERING

The storage root crops are generally biennials, requiring environmental factors to switch from vegetative to reproductive growth. In some crops there is a 'juvenile period' during which plants are insensitive to vernalization stimuli. In carrots, depending on cultivar, this period may be absent or last until at least eight leaves are formed (Atherton et al., 1990). The rates of bolting increase linearly with temperature from −1 to 5°C, and then decrease linearly with temperature from 7 to 16°C (Atherton et al., 1990). The thermal time required from sowing to complete juvenility for cv. Chantenay Red Cored was 756°Cd, and the subsequent minimum requirement for vernalization was 126°Cd with 336°Cd required for 90% of plants to bolt and flower (Atherton et al., 1990). The rate of growth of reproductive plants increases from 14 to 26°C, and plants matured sooner at the higher temperature, and consequently, the final size of plants was lower at the higher temperatures (Quagliotti, 1967). Short days (12 h) before or during vernalization induces earlier and more flowering, but long days (18 h) following vernalization stimulates flowering (Fisher, 1956; Atherton et al., 1984). After vernalization, flowering can be suppressed by continuous low light. The apical meristem is the region of sensitivity to this inhibitory effect of low light (Fisher, 1956). This is also the site of synthesis of a zearalenone-like substance during chilling that may control vernalization (Meng et al., 1986). The linear temperature dependencies of vernalization rate found by Atherton et al. (1990) suggest vernalization may be controlled by the concentration of a substance that increases logarithmically with temperature.

Sugarbeet plants are all obligate long-day plants, and vernalization increases the competence of leaves to produce a floral stimulus which is transported to the shoot apex in response to inductive long days (Crosthwaite and Jenkins, 1993). Jaggard *et al.* (1983) reviewed the vernalization requirement of sugarbeet. There are conflicting reports as to the presence of a juvenile period in this species, but most papers report no juvenile stage. The minimum and maximum temperature for vernalization was 0 and 15°C, respectively, with the fastest rate at 12°C. In extensive field experiments, bolting occurred in about 50% of plants when the air temperature was less than 12°C for 60 days. Therefore, there is a trade-off between increasing sugar yield in vegetative plants with increasing crop duration, and decreasing sugar yield per unit area as the proportion of vegetative plants decrease with earlier sowings. Yield was lost at the rate of 0.7% for every 1% of bolters over the range 5–40%. The receptive site for the vernalization stimulus is thought to be the shoot apical meristem, but it is still uncertain whether the young expanding leaves are receptive to vernalization too (Crosthwaite and Jenkins, 1993).

Turnips have a cold requirement of about three weeks at 5°C for flowering, but photoperiod has only a slight effect on flowering (van der Meer and van Dam, 1984). There is little information on the vernalization requirement for parsnips.

PHYSIOLOGICAL DISORDERS

STORAGE ROOT DISORDERS

Forking

Carrots are subject to the undesirable trait of forking. Damage to the tap root of carrot seedlings will promote forking (Benjamin and Wren, 1978) as will drilling into seedbeds of poor tilth (Orzolek and Carroll, 1978). There is no relationship between incidence of forking and nitrogen applications or incidence of nematodes (Soteros, 1983). At any one given site, the percentage of forked roots decreases with delay in sowing date and with increased density (Dowker and Jackson, 1977). However, forking is not simply related to mean root size because its incidence is not affected by duration of growth.

Damage and cracking

Also, carrots are subject to two major forms of mechanical damage, splitting (longitudinal cracking) and transverse breakage, which can reach over 30% of commercial crops (Cappellini *et al.*, 1987). Whereas splitting can occur spontaneously (growth splitting) or because of postharvest impacts (harvest splitting), breakage occurs at or after harvest.

A growth split is a radial longitudinal crack that may run from a few centimetres to the entire length of the root. In addition, it may penetrate to the depth of the cambial ring. Growth splitting is generally a minor problem in commercial production (NIAB, 1991). However, certain conditions are associated with high levels of splits. These include wide plant spacing, early sowing, long durations of growth and genotype (Warne, 1951; Dowker and Jackson, 1977; NIAB, 1991). Some recent unpublished work by McGarry and Pillinger, has suggested that the timing of irrigation be critical, with heavy irrigation applied during maximum radial root expansion inducing widespread splitting by maturity.

Harvest splits begin as hairline fractures, usually at the root tip, which often deepen and widen with time. In commercial production, harvest splitting is a serious problem, affecting up to 20% of crops (Metcalfe and Rickard, 1983; Cappellini *et al*, 1987). Conditions that promote harvest splitting include genotype (NIAB, 1991), top-lifting as opposed to elevator digger harvesting (Knott, 1980) and high root turgor (Millington, 1986; McGarry, 1993, 1995).

Scanning electron microscope studies have shown that carrot tissue fractures by cell wall rupture, irrespective of cell packing arrangement or split-susceptibility (McGarry, 1993; Fig. 16.5). In several cultivars, the force or energy required to fracture carrot tissue is inversely related to turgor pressure (McGarry, 1993, 1995). At high turgor, the protoplast presses on the surrounding cell wall, reducing the force or energy needed to rupture it (McGarry, 1993, 1995). The carrot cell wall volume fraction might decrease during the latter stages of crop growth (McGarry, 1995), but carrot tissue strengthens during the same period, suggesting that cell wall composition is modified during crop maturity. Delaying harvest may take advantage of this strengthening to reduce harvesting splitting.

Although engineers have redesigned pack house equipment to reduce drop heights and increase padding on machinery, breakage continues to be a serious problem. Further reductions in breakage levels probably await improved agronomic practices to produce a more robust carrot coupled with breeding for greater tissue tensile strength and fracture toughness.

Colour

In parsnips, discoloration is characterized by a yellowing of tissue exposed to the atmosphere, especially if damage has been inflicted at harvest. Discoloured roots have little retail value and it is often not economically feasible to trim affected areas. Wastage due to discoloration may reach 20%. The discoloration (browning) reaction is due to the activity of polyphenol oxidase, producing the dark pigment, melanin. In intact tissue, this reaction does not occur because the substrates and enzyme are located in different membrane-bound subcellular compartments (Toivonen, 1992).

Cultivars differ in their susceptibility to browning, but this is not related to biochemical differences between them, but perhaps physical factors (Toivonen, 1992). However, recent work by McGarry and Pillinger show differences in polyphenol oxidase activity between parsnip cultivars. Post-harvest dips containing calcium salts and vitamin C were effective in reducing the incidence of browning in only one cultivar out of four tested.

(a)

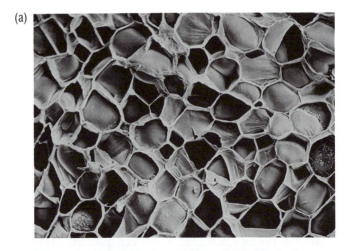

(b)

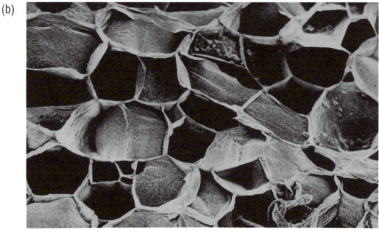

Fig. 16.5. Scanning electron micrographs of fresh fractured surfaces in (a) a split resistant cultivar, Narman, and (b) a split-sensitive cultivar, Tamino, showing split fracture breaking cells, not separating them.

A well-developed orange colour is important for the marketing of carrots. A weak colour is associated with young poor-flavoured roots. The intense colour of carrots is due to their high total carotene content and high ratios of $\beta:\alpha$ carotene. Total carotene content increases with age (Banga and De Bruyn, 1964; Bradley *et al.*, 1967; Bradley and Dyck, 1968; Phan and Hsu, 1973; Fritz and Weichmann, 1979; Nilsson, 1987; Evers, 1989). Carotene concentration is higher at low soil water contents, at temperatures between 10 and 20°C (Barnes, 1936; Banga *et al.*, 1955; Bradley *et al.*, 1967) and with applications of mineral fertilizers (Evers, 1989). Excessive irrigation, however, may reduce β-carotene production (Nortjé and Henrico, 1986) and the ratio of $\beta:\alpha$ carotene is decreased by high soil temperatures before harvest

(Bradley and Dyck, 1968). Total carotene content is greater at higher plant densities when comparing roots of similar weight (Banga and De Bruyn, 1956). 'Silvering' of the skin, associated with desiccation, is commercially important in pre-packed carrots, but we are not aware of any research into its causes and control.

CONCLUSIONS

In this chapter we have attempted to (i) give an overview of the salient environmental factors that control the growth and development of the storage root crops; (ii) identify any general hypotheses that have been proposed to explain how the growth of the storage roots is controlled; and (iii) identify where information is still lacking.

It strikes us how little research has been done in the last 5 years, taking a whole-plant physiological approach to determine the control of storage organ growth and development. There is scope to elucidate which genes are involved in cambial initiation by using molecular genetical techniques in combination with growth regulator or surgical methods that stimulate storage organ development. Genetic manipulation of sugarbeet has been reviewed (Nichols *et al.*, 1992) but only related to improving its quality. The early work of Das Gupta (1972b) and Olymbios (1973) using contrasting shoot/root temperature regimes provide some interesting insights into the control of partitioning of dry matter between the shoot and root systems. This approach appears to have been abandoned, but further experiments, perhaps separating the temperature regimes for the storage root and fibrous root system could give a clearer idea of how dry matter partitioning in these plants is controlled. Although shape of carrots is a major marketing criterion, we are not aware of any published concept to explain how and why the storage roots adopt their shapes. Perhaps the shape of the storage root can relate to the spatial pattern of cross-linking of hydroxyproline-rich glycoproteins of cell walls (Stafstrom and Staehelin, 1988). In excised storage root tissues, the internal force opposing a constantly applied external compression, will increase with time (J. Vincent, Reading, UK, 1995, personal communication). In inanimate objects such an internal force decreases as the object deforms. The basis of adaptation is unknown, but it is clearly crucial to controlling the growth and shape of large organs in soils.

ACKNOWLEDGEMENTS

We wish to thank the Ministry of Agriculture, Fisheries and Food, UK for financial support.

REFERENCES

Aikman, D.P. and Benjamin, L.R. (1994) A model for plant and crop growth, allowing for competition for light by the use of potential and restricted projected crown zone areas. *Annals of Botany* 73, 185–194.

Alabran, D. and Mabrouk, A. (1973) Carrot flavour sugars and free nitrogenous compounds in fresh carrots. *Journal of Agricultural Food Chemistry* 21, 205–208.

Atherton, J.G., Basher, E.A. and Brewster, J.L. (1984) The effect of photoperiod on flowering in carrot. *Journal of Horticultural Science* 59, 213–215.

Atherton, J.G., Craigon, J. and Basher, E.A. (1990) Flowering and bolting in carrot. I. Juvenility, cardinal temperatures and thermal times for vernalization. *Journal of Horticultural Science* 65, 423–429.

Austin, R.B. (1964) A study of the growth and yield of red – beet from a long – term manurial experiment. *Annals of Botany* 28, 637–646.

Banga, O. (1954) Taproot problems in the breeding of root vegetables. *Euphytica* 3, 20–27.

Banga, O., De Bruyn, J.W. and Smeets, L. (1955) Selection of carrots for carotene content. II. Subnormal content at low temperature. *Euphytica* 4, 183–189.

Banga, O. and De Bruyn, J.W. (1956) Selection of carrots for carotene content. III Planting distances and ripening equilibrium of the roots. *Euphytica* 5, 87–95.

Banga, O. and De Bruyn, J.W. (1964) Carotenogenesis in carrot roots. *Netherlands Journal of Agricultural Science* 12, 204–220.

Banga, O. and De Bruyn, J.W. (1968) Effect of temperature on the balance between protein synthesis and carotenogenesis in the roots of carrots. *Euphytica* 17, 168–172.

Barnes, W.C. (1936) Effects of some environmental factors on growth and colour of carrots. *Memoirs of Cornell Agricultural Experimental Station* 186, 1–36.

Barnes, A. (1979) Vegetable plant relationships. II. A quantitative hypothesis for shoot/storage root development. *Annals of Botany* 43, 487–499.

Beek van der, M.A. and Huijbregts, A.W.M. (1986) Internal quality aspects of sugar beet. *Proceedings, Fertiliser Society* 252, 20.

Benjamin, L.R. (1982) Some effects of differing times of seedling emergence, population density and seed size on root-size variation in carrot populations. *Journal of Agricultural Science, Cambridge* 98, 537–554.

Benjamin, L.R. (1984a) The relative importance of some sources of root-weight variation in a carrot crop. *Journal of Agricultural Science, Cambridge* 102, 69–77.

Benjamin, L.R. (1984b) Role of foliage habit in the competition between differently sized plants in carrot crops. *Annals of Botany* 53, 549–557.

Benjamin, L.R. (1987) The relative importance of cluster size, sowing depth, time of seedling emergence and between-plant spacing on variation in plant size in red beet (*Beta vulgaris* L.) crops. *Journal of Agricultural Science, Cambridge* 108, 221–230.

Benjamin, L.R. (1988) A single equation to quantify the hierarchy in plant size induced by competition within monocultures. *Annals of Botany* 62, 199–214.

Benjamin L.R. (1990) Variation in time of seedling emergence within populations: a feature which determines individual growth and development. *Advances in Agronomy* 44, 1–25.

Benjamin, L.R. (1994) Spacing and the impact on plant growth. *Grower* 121, 30–31/33.

Benjamin, L.R. and Bell, N. (1985) The influence of seed type and plant density on variation in plant size of red beet (*Beta vulgaris* L.) crops. *Journal of Agricultural Science, Cambridge* 105, 563–571.

Benjamin, L.R. and Hardwick, R.C. (1986) Sources of variation and measures of variability in even-aged stands of plants. *Annals of Botany* 58, 757–778.

Benjamin, L.R. and Sutherland, R.A. (1989) Storage – root weight, diameter and length relationships in carrot (*Daucus carota* L.) and red beet (*Beta vulgaris* L.). *Journal of Agricultural Science, Cambridge* 113, 73–80.

Benjamin, L.R. and Sutherland, R.A. (1992) Control of mean root weight in carrots (*Daucus carota*) by varying within- and between-row spacing. *Journal of Agricultural Science* 119, 59–70.

Benjamin, L.R., Sutherland, R.A. and Senior, D. (1985) The influence of sowing rate and row spacing on the plant density and yield of red beet. *Journal of Agricultural Science* 104, 615–624.

Benjamin, L.R. and Wren, M.J. (1978) Root development and source–sink relations in carrot *Daucus carota* L. *Journal of Experimental Botany* 29, 425–433.

Benjamin, L.R. and Wren, M.J. (1980) Root development and source sink relations in carrot, *Daucus carota* L. II. Effects of root pruning on carbon assimilation and the partitioning of assimilates. *Journal of Experimental Botany* 31, 1139–1146.

Benoit, F. and Ceustermans, N. (1990) Effect of the removal of the two direct cover sheetings (DC) on the development of carrots (*Daucus carota* L.). *Acta-Horticulturae* 267, 45–52.

Bleasdale, J.K.A. (1966) Plant growth and crop yield. *Annals of Applied Biology* 57, 173–182.

Bleasdale, J.K.A. (1982) The importance of crop establishment. *Annals of Applied Biology* 101, 411–419.

Bleasdale, J.K.A. and Thompson, R. (1963a) Competition Studies. *Report National Vegetable Research Station for 1962*. National Vegetable Research Station, Wellesbourne, Warwick, p.37.

Bleasdale, J.K.A. and Thompson, R. (1963b) An objective method of recording and comparing the shapes of carrot roots. *Journal of Horticultural Science* 38, 232–241.

Bleasdale, J.K.A. and Thompson, R. (1966) The effects of plant density and the pattern of plant arrangement on the yield of parsnips. *Journal of Horticultural Science* 41, 371–378.

Bradley, G.A., Smittle, D.A., Kattan, A.A. and Sistrunk, W.A. (1967) Planting date, irrigation, harvest sequence and varietal effects on carrot yields and quality. *Proceedings of the American Society of Horticultural Science* 90, 223–234.

Bradley, G.A. and Dyck, R.L. (1968) Carrot colour and carotenoids as affected by variety and growing conditions. *Proceedings of the American Journal Horticultural Science* 93, 402–407.

Cappellini, R.A., Ceponis, M.J. and Lightner, G.W. (1987) Disorders in celery and carrot shipments to the New York market, 1972–1985. *Plant Disease*, 71, 1054–1057.

Coulter, J.K. (1966) Soil factors and root development. *Report of Rothamsted Experimental Station for 1965*. Rothamsted Experimental Station, Harpenden, pp. 40–41.

Crosthwaite, S.K. and Jenkins, G.I. (1993) The role of leaves in the perception of vernalizing temperatures in sugar beet. *Journal of Experimental Botany* 44, 801–806.

Das Gupta, D.K. (1972a) Developmental physiology of sugar beet. III. The effects of decapitation and removing part of the root and shoot on subsequent growth of sugar beet. *Journal of Experimental Botany* 23, 93–102.

Das Gupta, D.K. (1972b) Developmental physiology of sugar beet. IV. Effects of growth substances and differential root and shoot temperatures on subsequent growth. *Journal of Experimental Botany* 23, 103–113.

Dowker, B.D. and Gray, D. (1985) Variation in embryo size of inbred lines of carrot. *Annals of Applied Biology* 107, 597–599.

Dowker, B.D. and Jackson, J.C. (1977) Variation studies in carrots as an aid to breeding. V. The effects of environments within a site on the performance of carrot cultivars. *Journal of Horticultural Science* 52, 299–307.

Downs, R.J. (1985) Irradiance and plant growth in greenhouses during winter. *HortScience* 6, 1125–1127.

Dunham, R. and Clarke, N. (1992) Coping with stress. *British Sugar Beet Review* 1, 10–13.

Durr, C., Tamet, V. and Boiffin, J. (1990) Redistribution of seed reserves during the emergence of carrot seedlings and its influence on subsequent growth. In: Scaife, A. (ed.) *Proceedings of the 1st Congress of the European Society of Agronomy*. European Society of Agronomy, Paris, pp.S1, 43.

Esau, K. (1940) Developmental anatomy of the fleshy storage organ of *Daucus carota*. *Hilgardia* 13, 175–226.

Esau, K. (1967) *Anatomy of Seed Plants*. John Wiley & Sons, New York.

Evers, A.M. (1989) Effects of different fertilization practices on the glucose, fructose, sucrose, taste and texture of carrot. *Journal of Agricultural Science in Finland* 2, 113–122.

FAO (1993) *FAO Yearbook*. Food and Agriculture Organization of the United Nations, Rome.

Farazdaghi, H. (1968) Some aspects of the interaction between irrigation and plant density in sugar beets. PhD thesis, University of Reading, Document 35.

Fick, G.W., Williams, W.A. and Loomis, R.S. (1971) Recovery from partial defoliation and pruning in sugar beet. *Crop Science* 11, 718–721.

Finch-Savage, W.E. (1990) The effects of osmotic seed priming and the timing of water availability in the seedbed on the predictability of carrot seedling establishment in the field. *Acta Horticulturae*, 267, 209–216.

Finch-Savage, W.E. and McQuistan, C.I. (1988) Performance of carrot seeds possessing different germination rates within a seed lot. *Journal of Agricultural Science, Cambridge*, 110, 93–99.

Fisher, J.E. (1956) Studies on the photoperiodic and thermal control of flowering in carrots. *Plant Physiology* 31 (Suppl.): xxxvi.

Francois, L.E. (1984) Salinity effects on germination, growth, and yield of turnips. *HortScience* 19, 82–84.

Fritz, D. and Weichmann, J. (1979) Influence of the harvesting date of carrots on quality and quality of preservation. *Acta Horticulturae* 93, 91–100.

Giaquinta, R.T. (1979) Phloem loading of sucrose, involvement of membrane ATPase and proton transport. *Plant Physiology* 63, 744–748.

Goris, A. (1969a) The sugars in cultivated carrot roots. *Qualitas Plantarum et Materiae Vegetables* 18, 283–306.

Goris, A. (1969b) Sugar metabolism in carrot roots. *Qualitas Plantarum et Materiae Vegetables* 18, 307–330.

Gormley, T.R., O'Riordain, F. and Prendiville, M.D. (1971) Some aspects of the quality of carrots on different soil types. *Journal of Food Technology* 6, 393–402.

Gray, D. (1984) The performance of carrot seeds in relation to their viability. *Annals of Applied Biology* 104, 559–565.

Gray, D. and Steckel, J.R.A. (1983) Some effects of umbel order and harvest date on seed variability and seedling performance. *Journal of Horticultural Science* 58, 73–82.

Gray, D., Brocklehurst, P.A., Steckel, J.R.A. and Dearman, J. (1984a) Priming and pre-germination of parsnip (*Pastinaca sativa*) seed. *Journal of Horticultural Science* 59, 101–108.

Gray, D., Ward, J.A. and Steckel, J.R.A. (1984b) Endosperm and embryo development in *Daucus carota* L. *Journal of Experimental Botany* 35, 459–465.

Gray, D., Hulbert, S. and Senior, K.J. (1985a) The effect of seed position, harvest date and drying conditions on seed yield and subsequent performance of cabbage. *Journal of Horticultural Science* 60, 65–75.

Gray, D., Steckel, J.R.A. and Ward, J.A. (1985b) The effect of plant density, harvest date and method on the yield of seed and components of yield of parsnip (*Pastinaca sativa*). *Annals of Applied Biology* 107, 547–558.

Gray, D., Steckel, J.R.A., Drew, R.L.K. and Keefe, P.D. (1991) The contribution of seed characters to carrot plant and root size variability. *Seed Science and Technology* 19, 655–664.

Green, C.F., Vaidyanathan, L.V. and Ivins, J.D. (1986) Growth of sugar beet crops including the influence of synthetic plant growth regulators. *Journal of Agricultural Science, Cambridge* 107, 285–297.

Harris, P.M. (1972) The effect of plant population and irrigation on sugar beet. *Journal of Agricultural Science, Cambridge* 78, 289–302.

Hayward, H.E. (1938) *The Structure of Economic Plants*. Macmillan, New York.

Hole, C.C. (1994) Carrots. In: Zamski, E. and Schaffer, A.A. (eds) *Distribution of Photoassimilates in Plants and Crops: Source-sink Relationships*. Marcel Dekker, New York.

Hole, C.C. and Dearman, J. (1993) The effect of photon flux density on distribution of assimilate between shoot and storage root of carrot, red beet and radish. *Scientia Horticulturae* 55, 213–225.

Hole, C.C. and McKee, J.M.T. (1988) Changes in soluble carbohydrate levels and associated enzymes of field-grown carrots. *Journal of Horticultural Science* 63, 87–93.

Hole, C.C. and Sutherland, R.A. (1990) The effect of photon flux density and duration of the photosynthetic period on growth and dry matter distribution in carrot. *Annals of Botany* 65, 63–69.

Hole, C.C., Thomas, T.H. and McKee, J.M.T. (1984) Sink development and dry matter distribution in storage root crops. *Plant Growth Regulation* 2, 347–358.

Hole, C.C., Morris, G.E.L. and Cowper, A.S. (1987) Distribution of dry matter between shoot and storage root of field grown carrots. II. Relationship between initiation of leaves and storage roots in different cultivars. *Journal of Horticultural Science* 62, 343–349.

Howarth, M.S., Brandon, J.R., Searcy, S.W. and Kehtarnavaz, N. (1992) Estimation of tip shape for carrot classification by machine vision. *Journal of Agricultural Engineering Research* 53, 123–133.

Ikeda, A., Nakayama, S., Kitaya, Y. and Yabuki, K. (1988) Effects of photoperiod, CO_2 concentration, and light intensity on growth and net photosynthetic rates of lettuce and turnip. *Acta Horticulturae* 299, 273–282.

Jaggard, J.W., Wickens, R., Webb, D.J. and Scott, R.K. (1983) Effects of sowing date on plant establishment and bolting and the influence of these factors on yields of sugar beet. *Journal of Agricultural Science, Cambridge* 101, 147–161.

Knott, C.M. (1980) Forcing the pace with Amsterdams. *Grower* 93, 19–25 (not inclusive).

Kokkoras, I. (1989) The effect of temperature and water status on the susceptibility of carrot roots to damage. PhD thesis, Cranfield Institute of Technology, Silsoe Campus, UK, 224.

Lester, O.E., Baker, L.R. and Kelly, J.F. (1982) Physiology of sugar accumulation in carrot breeding lines and cultivars. *Journal of the American Society Horticultural Science* 107, 381–387.

Linser, H. and Zeid, A. (1972) System and product in the development of *Daucus carota. Zeitschrift für Pflanzenernahrung und Bodenkunde* 131, 273–288.

Long, E. (1989) Spice up the boring image of the commonplace carrot. *Grower* 111, 12–13.

McGarry, A. (1993) Influence of water status on carrot (*Daucus carota* L.) fracture properties. *Journal of Horticultural Science* 68, 431–437.

McGarry, A. (1995) Cellular basis of tissue toughness in carrot (*Daucus carota* L.) storage roots. *Annals of Botany* 75, 157–163.

McKee, J.M.T. and Morris, G.E.L (1986) Effects of gibberellic acid and chlormequat chloride on the proportion of phloem and xylem parenchyma in the storage root of carrot (*Daucus carota* L.) *Plant Growth Regulation* 4, 203–211.

Mambelli, S., Vitali, G., Venturi, G. and Amaducci, M.T. (1992) Simulation model for predicting sugarbeet growth and sugar yield under Po Valley conditions. *International Institute for Sugar Beet Research, 55th Winter Congress*, International Institute for Sugar Beet Research, Brussels, Belgium, pp. 305–314.

Martin, R.J. (1986) Radiation interception and growth of sugar beet at different sowing dates in Canterbury. *New Zealand Journal of Agricultural Research* 3, 381–390.

Meer van der, Q.P. and van Dam, R. (1984) Determination of the genetic variation in the effect of temperature and daylength on bolting of *Brassica campestris* L. *Euphytica* 2, 591–595.

Meng, F.J., Que, Y.M. and Zhang, S.Q. (1986) Zearalenone-like substance in winter plants and its relation to vernalization. *Acta Botanica Sinica* 6, 622–627.

Mesken, M. (1990) Breeding sugar beets with globe-shaped roots to reduce dirt tare. *International Institute for Sugar Beet Research, 53rd Winter Congress*, International Institute for Sugar Beet Research, Brussels, Belgium, pp. 111–119.

Metcalfe, J.P. and Rickard, P.C. (1983) Carrots: production, harvesting and preservation of quality during marketing. *MAFF-ADAS overseas study tour report no. 70568/622*. 31.n. Ministry of Agriculture, Fisheries and Food, London.

Milford, G.F.J. (1973) Growth and development of the storage organ of sugar beet. *Annals of Applied Biology* 75, 427–438.

Milford, G.F.J. and Lenton, J.R. (1976) Effect of photoperiod on growth of sugar beet. *Annals of Botany* 40, 1309–1315.

Milford, G.F.J., Pocock, T.O., Jaggard, K.W., Biscoe, P.V., Armstrong, M.J., Last, P.J. and Goodman, P.J. (1985) An analysis of leaf growth in sugar beet. IV. The expansion of the leaf canopy in relation to temperature and nitrogen. *Annals of Applied Biology* 107, 335–347.

Milford, G.F.J., Travis, K.Z., Pocock, T.O., Jaggard, K.W. and Day, W. (1988) Growth and dry-matter partitioning in sugar beet. *Journal of Agricultural Science* 110, 301–308.

Millington, S. (1986) The influence of some physical properties of carrots on their damage characteristics. PhD thesis, Silsoe College, Silsoe, UK, 249.

NIAB (1991) *Maincrop Carrots: Summary of Trial Results, 1989. Issue 90*. National Institute of Agricultural Botany, Cambridge, 18.

Nichols, J.B., Dalton, C.C., Todd, G.A. and Broughton, N.W. (1992) Genetic engineering of sugar beet. *Zuckerindustrie* 10, 797–800.

Nilsson, T. (1987) Growth and chemical composition of carrots as influenced by time of sowing and harvest. *Journal of Agricultural Science, Cambridge* 108, 459–468.

Nortjé, P.F. and Henrico, P.J. (1986) The influence of irrigation interval on crop performance of carrots (*Daucus carota* L.) during winter production. *Acta Horticulturae*, 194, 153–158.

Olymbios, C. (1973) Physiological studies on the growth and development of the carrot. *Daucus carota* L. PhD thesis, University of London.

Orzolek, M.D. and Carroll, R.B. (1978) Yield and secondary root growth of carrots as influenced by tillage system, cultivation and irrigation. *Journal American Society Horticultural Science* 103, 236–239.

Pakov, I. (1964) A study of sugar beet root system under various irrigation patterns with a view to determining the depth of the active soil layer. *Rastenievadni Naudi* 1, 121–137.

Peterson, R.L. (1973) Control of cambial activity in roots of turnip (*Brassica rapa*). *Canadian Journal of Botany* 51, 475–480.

Phan, C. and Hsu, H. (1973) Physical and chemical changes occurring in the carrot root during growth. *Canadian Journal of Plant Science* 53, 629–634.

Platenius, H. (1934) Chemical changes in carrots during growth. *Plant Physiology* 9, 671–680.

Quagliotti, L. (1967) Effects of different temperatures on stalk development, flowering habit, and sex expression in the carrot. *Euphytica* 16, 83–103.

Rapoport, H.F. and Loomis, R.S. (1985) Interactions of storage root and shoot in grafted sugar beet and chard. *Crop Science* 25, 1079–1084.

Ricardo, C. (1976) Effects of sugars, gibberellic acid and kinetin on acid invertase of developing carrot roots. *Phytochemistry* 15, 615–617.

Robinson, F.R. (1969) Carrot population density and yield in an arid environment. *Agronomy Journal* 61, 499–500.

Salter, P.J., Currah, I.E. and Fellows, J.R. (1979) The effects of plant density, spatial arrangement and time of harvest on yield and root size in carrots. *Journal of Agricultural Science, Cambridge* 93, 431–440.

Salter, P.J., Currah, I.E. and Fellows, J.R. (1980) Further studies on the effects of plant density, spatial arrangement and time of harvest on yield and root size in carrots. *Journal of Agricultural Science, Cambridge* 94, 465–478.

Salter, P.J., Currah, I.E. and Fellows, J.R. (1981) Studies on some sources of variation in carrot root weight. *Journal of Agricultural Science, Cambridge* 96, 549–556.

Schmidhalter, U. and Oertli, J.J. (1991) Transpiration/biomass ratio for carrots as affected by salinity, nutrient supply and soil aeration. *Plant and Soil* 135, 125–132.

Scott, R.K. and Jaggard, K.W. (1992) Crop growth and weather: can yield forecasts be reliable? *International Institute for Sugar Beet Research, 55th Winter Congress.* International Institute for Sugar Beet Research, Brussels, Belgium, pp. 169–187.

Simon, P.W. and Peterson, C.E. (1979) Genetic and environmental components of carrot culinary and nutritive value. *Acta Horticulturae* 93, 271–278.

Snee, R.D. (1972) A useful method for conducting carrot shape studies. *Journal of Horticultural Science* 47, 267–277.

Soteros, J.J. (1983) Carrots: forking. *New Zealand Commercial Grower* 38, 10–24.

Stafstrom, J.P. and Staehelin, L.A. (1988) Antibody localization of extensin in cell walls of carrot storage roots. *Planta* 174, 321–332.

Stanhill, G. (1977) Allometric growth studies of the carrot crop. II. Effects of cultural practices and climatic environment. *Annals of Botany* 41, 541–552.

Steingrover, E. (1981) The relationship between cyanide-resistant tap root respiration and the storage of sugars in the tap root in *Daucus carota* L. *Journal of Experimental Botany* 32, 911–919.

Thomas, T.H., Barnes, A., Rankin, W.E.F. and Hole, C.C. (1983) Dry matter distribution between the shoot and storage root of carrot (*Daucus carota* L.). II. Effect of foliar application of gibberellic acid. *Annals of Botany* 51, 189–199.

Thorne, G.N. and Evans, A.F. (1964) Influence of tops and roots in net assimilation rate of sugar beet and spinach beet and grafts between them. *Annals of Botany* 28, 499–508.

Ting, F.S.-T. and Wren, M.J. (1980a) Storage organ development in radish (*Raphanus sativus* L.). 1. Comparison of development in seedlings and rooted cuttings of two contrasting varieties. *Annals of Botany* 46, 267–276.

Ting, F.S.-T. and Wren, M.J. (1980b) Storage organ development in radish (*Raphanus sativus* L.). 2. Effects of growth promoters on cambial activity in cultivated roots, decapitated seedlings and intact plants. *Annals of Botany* 46, 277–284.

Toivonen, P.M.A. (1992) The reduction of browning in parsnips. *Journal of Horticultural Science* 4, 547–551.

Ulrich, A. (1952) The influence of temperature and light factors on the growth and development of sugar beets in controlled climatic environments. *Agronomy Journal* 44, 66–73.

Umiel, N., Kust, A.F. and Gableman, W.H. (1972) A technique for studying quantitatively the variation in size and shape of carrot roots. *Hortscience* 7, 73–76.

Watson, D.J., Motomutsu, T., Loach, K. and Milford, G.F.J. (1972) Effects of shading and seasonal differences in weather on the growth, sugar content and sugar yield of sugar beet crops. *Annals of Applied Biology* 71, 159–185.

White, J.M. (1992) Carrot yield when grown under three soil water concentrations. *HortScience* 27, 105–106.

Warne, L.G.G. (1951) Spacing experiments on vegetables. I. The effect of the thinning distance on earliness in globe beet and carrots in Cheshire, 1948. *Journal of Horticultural Science*, 26, 79–83.

Willey, R. and Heath, S.B. (1969) The quantitative relationship between plant population and crop yield. *Advances in Agronomy* 21, 281–321.

Wyse, R. (1980) Partitioning within the tap root sink of sugar beet. Effect of photosynthate supply. *Crop Science* 20, 256–258.

Yucel, S. (1992) Relationships between sugar beet yield and quality and climatic factors in Turkey. *International Institute for Sugar Beet Research, 55th Winter Congress.* International Institute for Sugar Beet Research, Brussels, Belgium, pp. 223–238.

Zeid, F.A. and Kuhn, H. (1973) Protein – kohlenhydrat – verhaltnes in verlauf der vegetationsperiode von *Daucus carota*. *Zeitschrift fur Pflanzenernahrung* 135, 226–239.

Onions and Garlic 17

J.L. Brewster

Crop and Weed Science Department, Horticulture Research International,
Wellesbourne, Warwick CV35 9EF, UK; Present address: 4 Holly Lodge,
Wellesbourne, Warwick CV35 9RE, UK

Both onions and garlic are grown mainly for their bulbs, although the green shoots of salad onions are also an important crop. The onion bulb consists of the swollen bases (sheaths) of bladed leaves surrounding swollen bladeless leaves. The garlic bulb consists of numerous cloves, consisting largely of swollen, bladeless storage leaves. Cloves form in the axils of the leaf blades, the sheaths of which remain as the dry papery tunics which enclose the cloves in the mature bulb (Brewster, 1994). Onions are usually grown from seed, and flowering and seed production are important for crop production. Garlic is vegetatively propagated by planting cloves and, although bolting sometimes occurs, seeds do not form. The crop production cycle in onions is quite complex, involving vegetative growth, bulb formation, bulb dormancy and sprouting, flowering and seed production (Fig. 17.1). Garlic production involves a subset of these phases which excludes those concerned with seeds (Fig. 17.1). Switches from one phase of development to another, and rates of development and growth within each stage, are controlled by environmental conditions. This review is concerned primarily with these aspects of environmental physiology, since they are fundamental to crop production. Probably because it is a convenient experimental subject, onion has been used to study many other aspects of plant physiology, notably ion uptake by roots (Clarkson and Hanson, 1986) and the cell division cycle (Gonzales-Fernandez *et al.*, 1968). Stomatal physiology is also interesting and unusual (Schnabl and Raschke, 1980). These aspects of physiology are not considered here.

GERMINATION AND EMERGENCE

The storage potential of the seeds varies with seedlot, but tends to be lower than for other crops (Ellis and Roberts, 1977). Compared to other crops, and

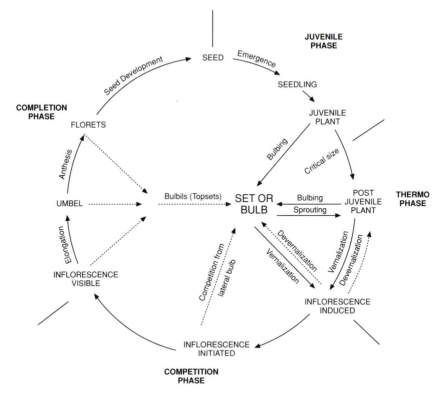

Fig. 17.1. The life cycle of onions: unusual or reversionary development is shown by dotted lines.

relative to their short average longevity, onions have very variable seed longevity with seed death widely distributed over time. The higher the moisture content and the warmer the temperature of storage the more rapidly the seed deteriorates. Constants quantifying these effects of moisture and temperature are similar to those for other crops, but because onion seed is innately short lived the accelerated deterioration at high moisture content and temperature is more practically significant (Ellis and Roberts, 1977; see also Chapter 1). An equation by Ellis and Roberts (1980) predicts percentage viability after a given period of storage at known temperature and moisture content from the initial viability of the seedlot.

The rate of germination can be quantified as the reciprocal of the time for a certain percentage of the final total (usually 50%) to germinate. Over the temperature range 5–25°C this rate increases linearly with temperature (Fig. 17.2a). The intercept of this relation with the temperature axis at zero rate is the base temperature ($T_{min.}$), and the reciprocal of the slope is the heat sum (S) (day-degrees) above $T_{min.}$, required for 50% germination. The value of S increases as seed deteriorates and loses viability (Dearman *et al.*, 1986; Ellis and Butcher, 1988). Base temperatures for onion germination ranging

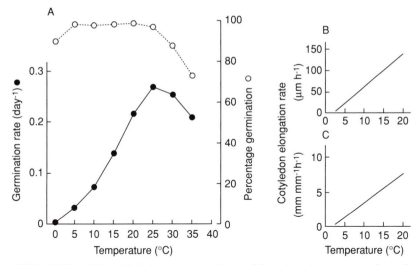

Fig. 17.2. (a) The relationship between temperature and the rate and percentage of germination of onion seeds on moist paper. Rates are reciprocals of the number of days for 50% of viable seeds to germinate (data of Harrington, 1962). (b) The relationship between temperature and the rate of cotyledon elongation before hook formation for newly germinated onion seedlings cv. White Lisbon (Wheeler and Ellis, 1991). (c) The relationship between temperature and the relative rate of cotyledon elongation after hook formation for the same seedlings as in (b).

from 1.4 to 3.5°C have been derived. The base temperature was the same for seedlots differing in viability (Ellis and Butcher, 1988). Deteriorating seed quality reduces viability, lowers the mean rate of germination but does not change the base temperature.

Bierhuizen and Wagenvoort (1974) showed that a similar relationship predicted the emergence of onion seedlings from seed sown in moist soil. More than 70% of seeds emerged between 13 and 28°C, the minimum temperature for good emergence being higher than for most temperate vegetables. Their equation was:

$$\text{Time } (d) \text{ to } 50\% \text{ emergence} = S/(T - T_{\text{min.}}) \qquad (1)$$

where S (in day-degrees) was 219 and $T_{\text{min.}}$ (°C) was 1.4. Equation 1 applied over the temperature range 3–17°C. The base temperature was similar to that of many other temperate zone vegetables, but the heat sum (S), ranked fourth out of a list of 31 common vegetable species. Only celery, parsley and leek had higher values of S, and the former two species are notoriously slow to emerge due to the presence of inhibitors in the seed coat. The implication of a high value of S is that emergence will be slow; for example, the results indicate that onions take 2.25 times as long as turnips to reach an equivalent stage of emergence after sowing.

Further studies showed that this relationship also applied to fluctuating as well as constant temperatures within the range 3–21°C (Wagenvoort and Bierhuizen, 1977). This model can therefore be used to predict time to emergence from moist soil in cool conditions (mean temperature less than about 20°C), such as typically occur from spring sowings at high latitudes (Finch-Savage, 1987; Wheeler and Ellis, 1992).

Emergence involves two processes, germination and then elongation of the radicle and cotyledon until the seedling is visible above soil. The rate of elongation before the formation of the cotyledon hook is linearly related to temperature above a base temperature of 1.4°C and below 20°C (Fig. 17.2b). After hook formation elongation is exponential and the relative rate of elongation shows a similar linear response to temperature until seed reserves are depleted (Fig. 17.2c) (Wheeler and Ellis, 1991). Differences in seed quality, as indicated by differences in percentage germination and mean rate of germination between seedlots, and by time-to-germination percentiles within seedlots, caused no differences in the elongation rates of normal seedlings. Furthermore, priming the seed, although it accelerated germination, did not change the subsequent elongation rate. It follows that differences in time to emergence from moist soil due to differences in seed quality or seed treatment are caused solely by differences in time to germination (Wheeler and Ellis, 1991).

The rate of emergence of seedlings is determined by temperature so long as soil moisture content is above some critical level (Wheeler and Ellis, 1992). Emergence from drier seedbeds cannot be predicted simply by thermal time but is influenced by rainfall or irrigation (Finch-Savage, 1986, 1987). Germination involves an initial rapid imbibition, a second phase when water content changes little and then an increase in water when the radicle bursts the seedcoat and grows. Germination processes before, and elongation after, radicle appearance can proceed at lower soil water potentials than those needed for the radicle to burst forth. This, then, is the most moisture sensitive step in germination that determines the lower limit of soil water potential needed for overall germination and emergence. Emergence from dry soils can be predicted by assuming that radicles burst the seed coat only if soil water potential is above –1.1 MPa (Finch-Savage and Phelps, 1993). If the soil is drier than this, germination is delayed until rainfall or irrigation raises soil water potential above this base level. If irrigation is limited, it is best applied when a thermal time of 50–130°Cd above a base of 1.4°C has elapsed since sowing, as seeds will then have reached the most water sensitive stage just before radicles burst out (Finch-Savage, 1990).

Decreases in both the percentage emergence and the rate of emergence occur as soil water potential decreases due to added nitrate fertilizer (Hegarty, 1976) (Fig. 17.3), and reductions in plant population and plant size early in the season occur at high rates of N (Greenwood et al., 1992). Germination is much less sensitive than growth to increases in solute concentration, and salt damage to pre-emergence growth is probably the cause of reduced emergence (Wannamaker and Pike, 1987).

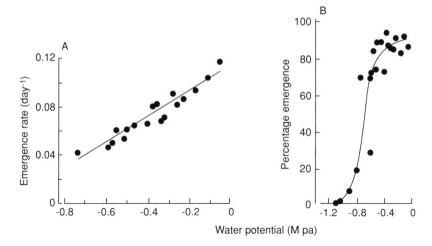

Fig. 17.3. The effect of soil water potential on the rate (a) and the percentage (b) of emergence of onion seeds incubated at 15°C. The rate was measured as the reciprocal of the time in days for 50% of the viable seedlings to emerge (redrawn from Hegarty, 1976).

The percentage emergence and the rate of seedling emergence can be affected by seed quality, seed treatments (e.g. priming), soil temperature, soil water potential and sowing depth. Seed quality and priming have been shown to affect germination rates rather than subsequent growth rates either before (Wheeler and Ellis, 1991) or after emergence (Ellis, 1989). Except in unusual circumstances, small advancements in emergence do not increase the marketable yield of bulbs, at most they result in a slight advancement of maturity (Brewster *et al.*, 1992) provided that plant populations are unaffected. However, yield and grade at harvest are highly sensitive to population (Frappell, 1973). Therefore, the most important effects of seed and seedbed are via their influence on the population established. Anything which promotes a stand of uniformly spaced seedlings emerging near synchronously is likely to facilitate subsequent husbandry. Some of the benefits of good emergence will be indirect and by nature sporadic in occurrence. For example, rapid emergence reduces the risk of population loss by entrapment of seedlings under a surface crust or cap which occurs as some soils dry after heavy rain. Rapid emergence maximizes the growing time during which pre-emergence herbicides remain effective, resulting in larger, more resilient seedlings by the time post emergence herbicides need to be applied.

GROWTH RATES

As with germination and emergence, the growth rate of seedling alliums after emergence is slow compared with most crop species. However, many non-crop species are slower growing, particularly woody species and those adapted to non-productive environments (Brewster, 1979). The relative growth rate (RGR) of onion growing in near optimal temperatures during the exponential, seedling phase of growth is only about half that of spring cabbage or lettuce. This means that, starting at the same weight and growing under the same conditions, onion will take nearly twice as long as spring cabbage or lettuce to reach a given weight.

Growth rates are strongly dependent on temperature. Figure 17.4 shows the relationships between RGR, relative leaf growth rate (RLGR) and leaf initiation rates and temperature. RGR and RLGR increase linearly over the range 6–20°C, whereas leaf initiation rates have a lower base temperature. This accords with field studies in which log seedling dry weight was a linear function of accumulated day degrees above 6°C whereas leaves initiated increased linearly with day-degrees above 2°C (Brewster *et al.*, 1977). A better model for growth in seedling shoot dry weight (W) in grams, which includes the effects of daily income of photosynthetically active radiation (PAR), as well as mean temperature (t), again in terms of accumulated day degrees above a base temperature (t_b) is from Scaife *et al.* (1987):

$$\log_e W = \log_e W_0 + \Sigma p / ((1 / (t - t_b)) + f / R) \qquad (2)$$

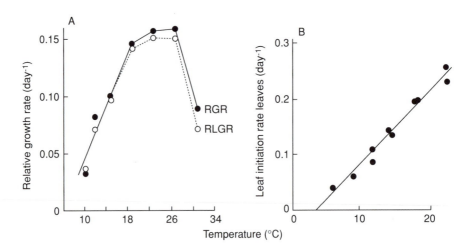

Fig. 17.4. (a) The effect of temperature on the relative growth rate of whole plant dry weight (solid symbols) and of leaf area (open symbols) of onion cv. Hygro (data of Brewster, 1979). relative growth rate (RGR) is the rate of increase in dry weight per unit of existing dry weight. RGR = $1/W \cdot dW/dt$, where W = dry weight and t = time. Similarly, the relative growth rate of leaf area is the rate of increase of leaf area per unit of existing leaf area. (b) The effect of temperature on the rate of initiation of leaves by the main shoot apex (i.e. not counting leaves on side shoots) of cvs Hygro, Hyton and Rijnsburger, all 'Rijnsburger types', growing in controlled environments (unpublished data).

The summated term is accumulated using daily values of t and PAR (R) from emergence onwards. For onions t_b was found to be 5.9°C. The parameters p and f had values of 0.0160 and 0.136, respectively (Brewster and Sutherland, 1993). R is in units of MJ m^{-2} day^{-1}. An appropriate value for $\log_e W_0$, the log of the shoot dry weight (g) at emergence is –6.086. The value of f was larger than for other species indicating greater sensitivity to light income probably because the leaves are narrow and erect. The difficulty of growing onions satisfactorily in conditions of low light is a familiar problem to experimenters who have to raise them in glasshouses during the winter.

Relative leaf growth rate and stomatal conductance decrease as leaf water potential and the associated turgor pressure decline (Millar *et al.*, 1971), the growth rates being affected first (Fig. 17.5). These authors measured a maximum turgor pressure of 0.4 MPa in onion leaves, less than half the value found in some other crop species. Moreover, increases in the osmotic pressure of the leaf sap can only compensate for about half such an increase at the roots when exposed to saline solutions. In contrast, leaf sap concentrations of bean and cotton plants fully compensate for such changes in the root medium (Gale *et al.*, 1967). These findings are reinforced by studies of overall growth which show that onions are more severely affected by salinity than most crops. Growth was reduced by 50% by sodium chloride solutions of 0.125 MPa osmotic pressure, whereas for cabbage, lettuce and beans an equivalent growth reduction required an osmotic pressure of 0.4 MPa (Bernstein and Hayward, 1958). Yield reductions due to salinity are more severe in hot dry climates than in more humid conditions (Magistad *et al.*, 1943).

The root system of onions is rather shallow, sparse and lacking root hairs (Greenwood *et al.*, 1982; Bhat and Nye, 1974). Consequently, water extraction is confined mainly to the top 25 cm of soil (Goltz *et al.*, 1971). They require higher levels of soil P and K for maximum yields than most crops (Greenwood *et al.*, 1980a,b), and their recovery of N fertilizer is poor (Greenwood *et al.*, 1992). These factors, coupled with the fact that, as with most vegetable crops, the nutrient requirement per unit root length is at its peak just after seedling emergence (Brewster *et al.*, 1975) have prompted experimentation with starter fertilizer solutions placed just below the seed at sowing. Even in fertile soils such ammonium phosphate 'starter' solutions accelerate early growth (Brewster *et al.*, 1992) and dramatically increase the efficiency with which N fertilizer is recovered in the crop (Stone and Rowse, 1992). Both morphological and physiological traits make onions more sensitive to water and nutrient deficiencies than the majority of crops, but on the other hand, they can survive long periods of water stress, making no growth, but ultimately recovering when water becomes available (Levy *et al.*, 1981). In these and other traits, like the inherently low relative growth rates, they exhibit features typical of species with a stress tolerant ecological strategy (Grime, 1979).

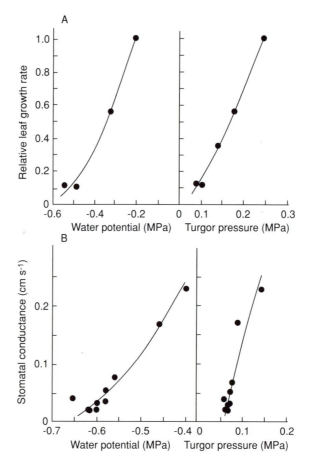

Fig. 17.5. (a) Relative growth rate of onion leaf length at 25°C as a function of leaf water potential and of turgor pressure (redrawn from Fig. 17.9 of Millar *et al.*, 1971). (b) Stomatal conductance of onion leaves at 25°C as a function of leaf water potential and of turgor pressure (redrawn from Fig. 17.7 of above source).

BULBING

ESTIMATION OF BULB INITIATION

A feature that unequivocally indicates that bulbing has started is the development of leaf initials into bladeless 'bulb scales'. This can be diagnosed as the first occurrence of a 'leaf ratio' (leaf blade length : sheath length) of less than unity (Heath and Hollies, 1965) (Fig. 17.6). Where many plants are available for sampling, scale initiation can be quite quickly estimated from examining plants sliced longitudinally upwards through the centre of the sheath. Mean dates of bulb initiation can be estimated from a series of such samples (Visser, 1994a). Normally a leaf ratio below unity is coincident with marked swelling of the outer leaf sheaths and a consequent rapid increase in 'bulbing ratio' (maximum bulb diameter : minimum sheath diameter). This latter ratio has been more commonly used in bulbing studies, since it can easily be measured non-destructively, whereas the assessment of minimum leaf ratio involves the dissection of plants and is more laborious. However, the occurrence of the first leaf ratio less than unity is not invariably linked with the attainment of a particular bulbing ratio. Bulbing ratios greater than 2, commonly used to define bulb initiation, can occur in the absence of bulb scale development in N-deficient plants (Brewster and Butler, 1989a). In field-grown plants grown at densities of 25 and 400 m^{-2}, bulbing started on almost the same date when measured by bulbing ratio, whereas bulb scales were initiated about 2 weeks earlier in the plants at high density (Brewster, 1997). Conditions which favour carbohydrate accumulation within the plant can lead to thickening leaf sheaths and increased bulbing ratios without formation of bulb scales. In addition, leaf initials differentiate into bulb scales on lateral shoots before this occurs on the main shoot axis. As a result, in cultivars and conditions which produce many side shoots, a high bulbing ratio can occur long before the main axis stops producing green leaves (Wiles, 1994). Because of this variability in the linkage between increases in bulbing ratio and decreases in leaf ratio on the main shoot, and because a decrease in leaf ratio is the first indication of the initiation of the storage scales which ultimately results in the cessation of leaf blade growth and the ripening of bulbs, a decrease in minimum leaf ratio below unity on the main shoot axis is the preferable method of defining

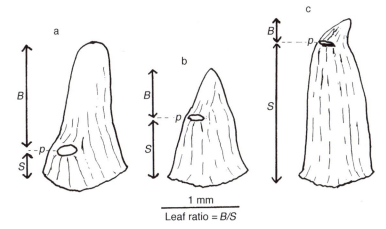

Fig. 17.6. The appearance of developing leaf initials during leaf growth (a), early stage bulbing (b) and established bulbing (c). p is the pore through which the blade of the next leaf emerges from its encasing sheath during leaf growth. Bulbing is characterized by a decrease in the ratio of blade length (B) to sheath length (S) termed the 'leaf ratio'. Initials with B/S below unity are termed 'bulb scales' (Heath and Hollies, 1965).

bulb initiation. Since leaf initiation rates are temperature-dependent (see Fig. 17.4b), it is possible to use thermal time to estimate leaf initiation date from leaf number, and hence estimate the date of initiation of the first bulb scale from a sample taken after the event (Brewster, 1996).

ENVIRONMENTAL CONTROL OF BULBING

Garner and Allard (1920) first showed that onions develop bulbs in response to long photoperiods, and later it was shown that onion varieties grown at different latitudes could be distinguished by the minimum daylength needed to induce them to bulb. Further research showed that, in a given daylength, bulbing was faster the higher the temperature (Thompson and Smith, 1938). These effects of photoperiod and temperature have since been confirmed by many studies (Brewster, 1990a). There is evidence that very warm temperatures (38–49°C) slow down bulbing compared to 27–30°C (Abdalla, 1967).

Onion leaves must be exposed continuously to bulb inductive photoperiods in order to start and complete bulbing. Several authors have shown that bulbing can be reversed and green leaves will resume growth if plants are transferred to short, non-inductive photoperiods (e.g. Kedar et al., 1975). This can occur even in plants at an advanced stage of bulb development (Fig. 17.7).

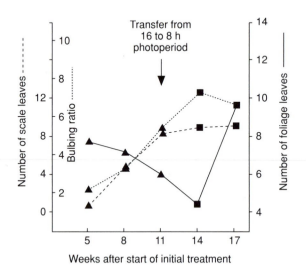

Fig. 17.7. The reversal of bulbing in onion cv. Rocket at an advanced stage of bulbing, following transfer from 16 to 8 h photoperiods at 135 E m²s⁻¹ PAR. Note the resumption of foliage leaf production even though many bulb scales had formed (redrawn from Sobeih and Wright, 1986).

The rate of bulbing in a particular photoperiod depends strongly on the red : far-red (R : FR) ratio of the light, and a non-inductive photoperiod at high R : FR can be inductive at lower R : FR (Austin, 1972; Mondal *et al.*, 1986b). Because leaves absorb red wavelengths much more than far-red, R : FR decreases under a leaf canopy. As a consequence, the rate of onion bulbing increases as leaf area index (LAI) increases (Mondal *et al.*, 1986c). The action spectrum for 3 h continuous irradiation in the middle of a bulb inductive (18 h) light period has a maximum at 714 nm (Lercari, 1983). In these experiments the bulbing response was fluence-rate dependent and nullified by simultaneous irradiation with red light, features typical of a 'high irradiance' phytochrome response. Some rhythmicity in the response of onion bulbing to R : FR is indicated by the promotive effect of FR exposure being greatest when applied during the middle of a long photoperiod, and the optimum R : FR changing at different times in the inductive photoperiod (Lercari, 1982).

Exposure of onion seedlings that had just emerged from soil to 8 days at 24 h photoperiods followed by 10 days at 8 h photoperiods, resulted in small bulbs in which bulb scales had formed (Terabun, 1971). On these plants the cotyledon and the first true foliage leaf developed. The latter is already initiated within the embryo and differentiates during germination (Hoffman, 1933), so its development is unlikely to be affected by photoperiod. Terabun (1971) also exposed single leaves of four-leaved plants to 24 h photoperiods for 14 days having removed the other leaves. Bulbing was similar whichever leaf remained for photoperiodic treatment. These experiments show no evidence for juvenility and prove that even emerging seedlings can be induced to bulb by a strong long day stimulus. However, Sobeih and Wright (1986) observed that rate of bulbing upon transfer from 8 to 16 h photoperiods increased as plants grew older. Also, in defoliation experiments, they found that 4.5-month-old plants defoliated to the two youngest leaves bulbed more rapidly than younger plants of greater leaf area at the start of long day treatment. Therefore, there is no evidence for an absolute juvenile phase in which bulbing cannot occur given a strong photoperiodic induction, but sensitivity to long days may increase with age.

Onions grown from bulbs or sets will bulb and ripen more quickly than onions grown from seed, unless the sets are 'heat treated' by storage at about 30°C for several months before planting (Aura, 1963). This indicates that the photoperiodic induction required for plants grown from sets is less than for seedlings. Some bulbing stimulus seems to be stored in the set which can be destroyed by prolonged warm storage (Heath and Holdsworth, 1948).

PHOTOTHERMAL RESPONSE MODELS FOR RATE OF BULBING

The bulbing response to temperature and photoperiod can be quantified (Brewster, 1997):

$$1/\text{time to bulb} = \text{rate of bulbing} = C + A \cdot \text{photoperiod} + B \cdot \text{temperature}$$
$$(3)$$

Here, A, B and C are constants. Equations like this describe the flowering responses of a number of both long- and short-day crop species (Summerfield et al., 1991). An implication of this equation is that photoperiodic induction is a quantitative process, during which the bulbing stimulus may accumulate gradually, rather than a sudden switching to the induced from the non-induced state by brief exposure to inductive conditions. The rate of bulbing was defined as the reciprocal of the interval from the start of inductive photoperiods until the formation of the first leaf scale. The values of the constants of equation 3 (A, B and C), were, respectively, 0.0043, 0.0027 and −0.079 for cv. Keepwell, which is autumn-sown in England, and 0.0032, 0.0018 and −0.066 for cv. Hyton, which is spring-sown (Fig. 17.8). These give the time in days from transfer to inductive photoperiods until bulb initiation, as defined above, when the photoperiod is in hours and the temperature in °C. The equations show that there is not a sudden transition from non-inductive to inductive as daylengths increase, rather an increasing rate of induction as daylengths exceed a certain minimum. This minimum, and the rate of increase with photoperiod depends on cultivar. The response surfaces and the corresponding equations characterize cultivars Keepwell and Hyton better than the terms like 'intermediate' and 'long' day type, or '14 h' or '16 h' type. Both these cultivars can bulb in 14 h photoperiods, but at different rates and also at a rate which varies with temperature.

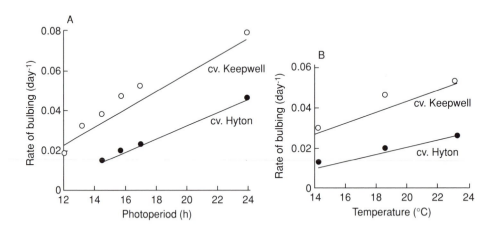

Fig. 17.8. (a) The effect of photoperiod on the rate of bulbing of two onion cvs growing at a mean temperature of 18.6°C. The rate of bulbing was derived as the reciprocal of the time in days for 50% of the plants to initiate bulb scales (Brewster, in preparation). (b) The effect of temperature on the rate of bulbing under a constant photoperiod of 15.75 h. The lines are from equation 3 and the points are some of the experimental data used in deriving equation 3 (Brewster, in preparation).

Since bulbing has been shown to require continued inductive photoperiods to complete, it seems reasonable to make the assumption that rate of bulbing depends on current photoperiods and temperatures alone. If this is the case, development in changing photoperiods and temperatures can be modelled as a sum of rates of development at a given temperature and photoperiod multiplied by the duration of time during which these conditions prevail (McNaughton *et al.*, 1985). By substituting daily values of photoperiod and mean temperature into equations 2 and 3 daily rates of bulbing were calculated. Negative values of rate were assumed to be zero, and progress towards bulbing was calculated as the cumulative sum of the positive values of daily rates. The date when this cumulative sum exceeded unity was the predicted date of first bulb scale initiation. Using weather data for 1983 from Wellesbourne, bulb initation of Keepwell was predicted on 2 June and Hyton on 9 July. Field-grown, widely spaced (25 plants m^{-2}) plants of Keepwell and Robusta, a cultivar closely similar to Hyton in maturity date (NIAB, 1982), reached 50% bulb scale initiation on 7 June and 19 July, respectively, in fair agreement with predictions.

AGRONOMIC IMPLICATIONS

Equation 3 and Fig. 17.8 show that bulbing is definitely not an 'all or nothing' inductive response with sharp switching close to one photoperiod. So long as photoperiod remains longer than some minimum, then bulbing will continue, albeit slowly, when close to the minimum photoperiod. In such conditions there is much time for leaf growth before bulb initiation. This may be why in some low latitude areas with little seasonal fluctuation in photoperiod or temperature, suitably adapted cultivars can produce large bulb yields at any time of year (Jones and Mann, 1963). In many regions bulbing continues in declining photoperiods and temperatures, e.g. with spring-sown crops in northern Europe or summer-planted crops in Israel (Kedar *et al.*, 1975). Again, so long as the photothermal conditions exceed some threshold, bulbing will proceed albeit at a diminishing rate as this threshold is approached.

The increase in rate of bulbing under a given photoperiod as R : FR decreases, and the decrease in R : FR within the onion leaf canopy as LAI increases, permits the modulation of the basic photothermal response of a cultivar by the many factors which influence leaf growth. In the situation just considered, of bulbing in decreasing photoperiod, a crop with a high LAI can continue to differentiate bladeless bulb scales, resulting in hollow pseudostems which ultimately soften and collapse to give ripe bulbs. Whereas, at a lower LAI, the same cultivar may revert to differentiating leaf blades which fill the pseudostem and prevent the softening and collapse associated with normal ripening (see Fig. 17.7). Large differences in the date of bulb initiation in crops of the same cultivar at different LAI will be reflected by parallel differences in maturity date (foliar fall-down) (Fig.

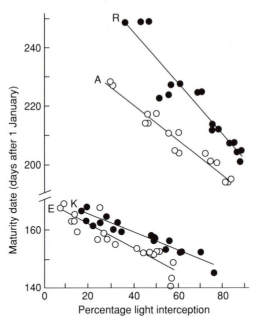

Fig. 17.9. Relationships between mean percentage light interception by the crop leaf canopy during bulbing and the maturity (80% foliage fall-down) day of the year for early maturing and late maturing spring-sown onion cvs (upper two graphs) and autumn-sown cvs (lower two graphs) at Wellesbourne. Cultivars are Robusta (R), Augusta (A), Keepwell (K), and Express Yellow OX (E) (Mondal *et al.*, 1986a).

17.9). Among the agronomic factors which determine LAI are:

1. Those that affect plant population, like seed rate, seedbed soil conditions, salinity in the seedbed, herbicide damage, temperature, and diseases and pests during germination and emergence.
2. Conditions affecting leaf growth rate, e.g. temperature, water stress, nutrient availability, salinity, and root damage by pests or disease.
3. Factors directly damaging the leaves like pests, disease, hail, herbicides or damage during cultivations.

Therefore, virtually every agronomic operation can ultimately influence bulbing rate and maturity date via its influence on LAI. This effect is most prominent in conditions where bulbing would otherwise be slow, e.g. in the declining photoperiods described above. When bulbing occurs in increasing photoperiods and temperatures, as normally occurs with autumn-sown crops, the influence of LAI is less important, since the climatic factors are then accelerating the rate of bulbing (Mondal *et al.*, 1986a, c). In conditions where the LAI effect is important, early and uniform bulbing is promoted by cultural practices which ensure the establishment of a uniform plant population which then grows as rapidly as the climate allows, and is therefore not afflicted by stress, damage or disease. The uniform bulbing and

maturity of such a crop ensures that all bulbs can be harvested at the optimum stage for quality and long storage. So, in addition to the innate qualities bred into an onion cultivar, the growing conditions the farmer creates
can profoundly influence bulbing and the quality of the bulbs harvested.

FACTORS INFLUENCING BULB YIELDS

In a study comparing well-irrigated autumn- and spring-sown bulb onion
crops, yields varied by a factor of more than five depending on cultivar,
sowing date and plant density (Mondal, 1985). These large differences in
yield can be explained mainly by differences in the percentage light interception by the leaf canopy during bulbing I% (Fig. 17.10). Light interception
increases with LAI according to equation 4 (Mondal *et al.*, 1986c):

$$I\% = 85.4 \, (1 - \exp(-0.377 \cdot \text{LAI})) \tag{4}$$

Several cultural practices can influence the LAI attained by bulbing
time. Late maturing cultivars have longer than early types to develop a high
LAI before the switch to bulb growth (Brewster, 1982c). Late-sown plants
often switch from leaf blade to bulb production when LAI is lower than for
earlier sowings. LAI increases with plant density, but although yields
increase, individual bulb size is reduced, and this is not normally acceptable. To achieve a high yield of medium to large bulbs, a cultivar must be
capable of producing sufficient leaf area to intercept a high proportion

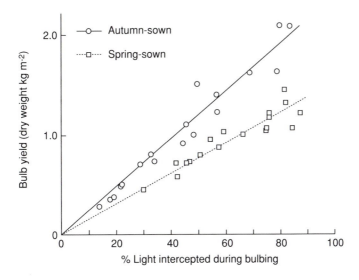

Fig. 17.10. The relationship between bulb dry-matter yields and percentage radiation
interception by the leaf canopies during bulbing of irrigated, well-fertilized small plots of
onions at Wellesbourne (data from Mondal, 1985).

(> 60%) of the incident light when grown at 50–100 plants m^{-2}. This requires sufficient time in conditions conducive to leaf blade growth before bulbing starts.

Although dominant, percentage light interception is not the only factor involved. In Fig. 17.10 autumn-sown onions produced a higher yield than spring-sown at a given percentage interception. For crops with 60% light interception the duration of bulbing was 55 days when autumn-sown and 43 days when spring-sown, consequently the former intercepted more light in total during bulbing. Longer duration of bulbing may be caused by cooler temperatures giving slower bulb ripening and leaf senescence. Mean temperatures during bulbing were 13°C for the autumn-sown crops and 17.5°C for the spring-sown.

Well-irrigated onion crops produce, during bulbing, an average of about 1.6 g of shoot dry matter per MJ of solar radiation intercepted by the leaf canopy (Brewster *et al.*, 1986). Similar conversion efficiencies have been reported for potatoes, sugarbeet and in cereals before anthesis. However, values of mean conversion efficiency vary, and range from 1.2 g MJ^{-1} in conditions of high irradiance and high temperature to 2.0 g MJ^{-1} in lower irradiance, lower temperature conditions. As might be expected, lack of irrigation in dry weather decreases conversion efficiency.

Bulb onions have a high 'harvest index' (the proportion of total yield in harvested material). At the optimum time for harvesting, when 80% of plants have 'soft necks', about 80% of the shoot weight is in the bulb. Bulb weight and the percentage of total weight in the bulbs continues to increase after this stage (Brewster *et al.*, 1977).

BULB DORMANCY AND SPROUTING

Upon ripening the onion bulb enters a phase of true dormancy during which growth ceases. As bulbs approach maturity new leaves and roots cease to be initiated, and after harvest and curing a dramatic decrease in mitosis at the shoot apex has been observed (Abdalla and Mann, 1963). From then on there is a gradual loss of dormancy as shown by slow increases in respiration rate (Ward and Tucker, 1976; Tanaka *et al.*, 1985), increases in the weight of the growing sprout within the bulb (Brewster, 1987a), and progressively decreasing times to produce roots and visibly sprout after planting on a moist substrate (Abdalla and Mann, 1963). Cultivars differ widely in their duration of dormancy. In a study comparing ten varieties stored at 10°C Miedema (1994a) found the time to 50% sprouting in dry storage ranged from 149 to 310 days, and times for 50% rooting and sprouting after planting on moist vermiculate were 8–63 days and 49–156 days, respectively. Within cultivars the time to visible sprouting of individual bulbs was spread over about 20 weeks in dry storage.

The onset of dormancy is thought to be caused by the translocation of growth inhibitory substances from the leaves to the bulbs as the crop matures (Komochi, 1990). Stow (1976) found inhibitors in the leaves of onions approaching maturity and showed that defoliation at this stage shortened dormancy. Abscissic acid was identified but it accounted for only 10–20% of the growth inhibitory activity. During subsequent storage there is a progressive decline in inhibitor activity extractable from bulbs, followed by increases first in cytokinin activity and then by gibberellin and auxin activity (Isenberg *et al.*, 1974).

The time until sprouting is affected by the temperature of storage and the gas composition of the store atmosphere and is drastically foreshortened by moisture on the baseplate of the bulbs and by wounding bulbs. The response of sprouting rate to temperature is in remarkable contrast to that of other physiological processes (Fig. 17.11a). Rate of sprouting increases from a minimum around 0°C to a maximum in the range 10–20°C, depending on cultivar (Miedema, 1994a), but then decreases again as temperatures increase to 25–30°C (Abdalla and Mann, 1963; Miedema, 1994a). This response applies to dormant bulbs, but the rate of sprout growth in long-stored, non-dormant bulbs peaks around 25°C; typical of most growth processes. In addition, temperatures in the range 25–35°C applied for 1–3 weeks straight after harvest can reduce the time to sprouting, with 35°C giving up to 30% decrease in storage life (Miedema, 1994a). The response of sprouting to temperatures around 30°C therefore passes through three phases as the bulb ripens, becomes dormant and finally loses dormancy and begins to sprout. Only in the middle dormant phase, which can, however, last many weeks, are such temperatures suppressive of sprouting. A rise in endogenous cytokinins that preceded sprout growth in bulbs stored at 5 or 15°C was prevented by storage at 30°C (Miedema, 1994b). Furthermore, injection of bulbs with benzyladenine, but not with GA_3 or ethephon, accelerated the sprouting of such warm stored bulbs. It seems then that sprouting is initiated by increases in cytokinins and that production of these is suppressed by warm storage of dormant bulbs.

Further evidence for cytokinins initiating sprouting comes from studies on the effects of roots. If the baseplate of the bulb is maintained wet, root growth begins and sprouting occurs much sooner than in dry stored bulbs (Miedema, 1994a). The inclusion of mineral nutrients in the solution, which enhances root growth, can further accelerate sprouting. If roots are repeatedly removed as they develop, sprouting is delayed, but if benzyladenine (BA) is included in the solution then, in some cultivars, rapid sprouting still occurs (Miedema, 1994c). In these cultivars cytokinin appears to be the main factor limiting sprout growth and the root system probably supplies cytokinin to the bulb. Some cultivars only responded to cytokinin in this way if they were wounded by slicing away some corky baseplate tissue, so additional factors were limiting sprouting for these. Wounding can stimulate the production of growth substances and it may increase gas permeability.

Atmospheres containing elevated CO_2 and reduced O_2 prolong the storage life of onion bulbs. Smittle (1988) found 5% CO_2 and 3% O_2 prolonged

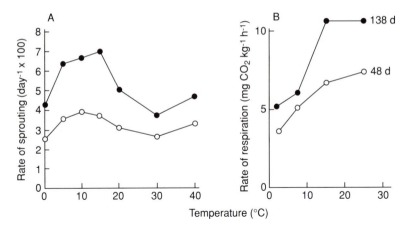

Fig 17.11. (a) Effects of temperature on the rate of sprouting after planting on moist peat at 15°C of two onion cultivars previously stored for 4 weeks at the temperatures shown. Solid symbols (top graph) the long-storing cv. Australian Brown, open symbols the short storing cv. Excel (data from Abdalla and Mann, 1963). Rates of sprouting were calculated as the reciprocals of the number of days for 50% of the bulbs to sprout. (b) Effects of temperature of storage on the respiration rate per kilogram of fresh weight of onion cv. Rijnsburger stored at different constant temperatures. Open symbols after 48-day storage, solid symbols after 138-day storage (data of Ward, 1976).

dormancy in cv. Granex. The internal atmosphere of bulbs shows similar increases in $CO_2 : O_2$ (Ladeinde and Hicks, 1988) and the intact onion skin and baseplate may act as a barrier to gas exchange. Removal of the dry outer skin increases respiration rate nearly twofold (Apeland, 1971) and accelerates sprouting (Tanaka *et al.*, 1985). Bulb wounding may accelerate sprouting by allowing increased respiration rates, especially since the acceleration of sprouting can be prevented by sealing the wound with paraffin wax (Boswell, 1924). Bulb respiration rates are low compared with other vegetables after harvest (Robinson *et al.*, 1975). Respiration increases with temperature but the Q_{10} of dormant bulbs is remarkably low at about 1.3 (Ward, 1976) rather than about 2 which is typical of most other vegetables (Fig. 17.11b).

Storage temperatures have a major influence on growth and development when bulbs or sets are subsequently planted out. Provided the bulbs are larger than the juvenile size, inflorescence initiation and development will proceed during bulb storage at a rate that depends on temperature (see below). There are numerous reports that storage of sets for periods of months at temperatures around 25–30°C, not only prevents or reverses inflorescence development, but increases the vigour of plants after sprouting and delays subsequent bulbing so that a bigger leaf area and higher bulb yield results compared to sets stored at lower temperatures (Heath *et al.*, 1947; Aura, 1963). Similar after effects of storage temperature on vigour of growth and date of bulbing occur with garlic (see below).

FLOWERING

RESPONSES TO TEMPERATURE AND PHOTOPERIOD

Onion seedlings have a sharply delineated juvenile phase for flowering which ends when they reach a certain critical shoot dry weight, or number of leaves initiated. This critical size can vary with cultivar, and an eight-fold larger critical size is the most striking difference between bolting-resistant cultivars suitable for autumn sowing and more bolting-susceptible varieties suitable only for spring sowing (Shishido and Saito, 1975; Brewster, 1985). In onion bulbs or sets, the critical size is not so sharply defined. The duration of cool treatment needed to induce flowering tends to decrease with size over a ten or 20-fold range of bulb weight (Shishido and Saito, 1977; Brewster, 1987b).

Once larger than the critical size, the plants need an extended period of cool temperatures to vernalize and induce flowers to initiate. Some Japanese spring-sown cultivars can be vernalized by as little as 20 days at 9°C, whereas autumn-sown types require a minimum of 30–40 days (Shishido and Saito, 1976). For Japanese and north European cultivars, rate of vernalization shows an inverted U-shaped response to temperature with an optimum around 9–12°C (Fig. 17.12a). West African cultivars can be vernalized satisfactorily under minimum night temperatures of 15–20°C (Rabinowitch, 1990a). So the temperature response for vernalization, as well as the critical size and the duration of vernalization treatment required, depend on cultivar, and tropical cultivars can be induced to flower in warmer conditions than temperate types.

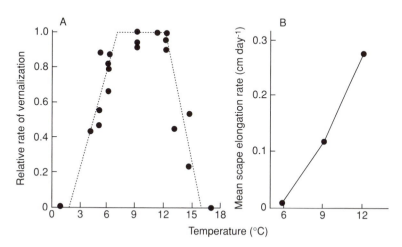

Fig. 17.12. (a) The effect of temperature on the relative rate of vernalization of European and Japanese onion cultivars (Brewster, 1987b). (b) The effect of temperature on the mean rate of elongation of onion scapes in a 16 h photoperiod following spathe formation. Data is meaned over cvs Senshyu semi-globe yellow and Rijnsburger growing with plentiful nitrate (Brewster, 1983).

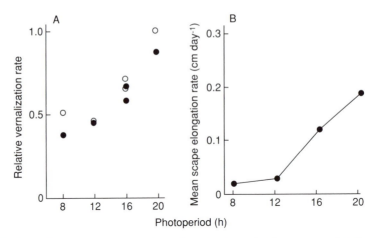

Fig. 17.13. (a) The effect of photoperiod on the rate of vernalization, measured as the recip-rocal of the time for 50% of plants to initiate inflorescences for cv. Senshyu semi-globe yellow (open symbols) and cv. Rijnsburger (closed symbols) growing at 9°C. The results are scaled relative to the fastest rate found in the series of experiments (Brewster, 1987b). (b) The effect of photoperiod on the mean elongation rate of scapes following inflorescence initiation. The data are means of the same two cultivars as in (a) and are averages over plants growing in 6, 9 and 12°C (Brewster, 1983).

Long photoperiods shorten the duration required to vernalize growing plants (Fig. 17.13a), in some cases more than halving the time requirement. This photoperiod requirement can be lessened by N deficiency in the plants (Shishido and Saito, 1976; Brewster, 1983).

Temperatures of 28–31°C applied to bulbs previously induced by low temperatures can reverse inflorescence induction; this is termed 'devernal-ization'. The duration of 28–31°C storage needed to prevent flowering increases the more advanced the stage inflorescence initiation. This is illus-trated by the results of Aura (1963) on Finnish cultivars. In accord with the results discussed above, he found that inflorescence initiation was more rapid in bulbs stored at 9–13°C than at 3–5°C and that it was more rapid in large bulbs than in small ones. Reflecting these differences, in a total storage period of 8 months a terminating duration of 5 months at 28°C was needed to suppress flowering in bulbs stored at 9–13°C but only 2 months in those stored at 3–5°C. Also the duration of 28°C treatment needed to suppress flowering in 15 g bulbs was less than needed for 108 g bulbs, since inflores-cence development was more advanced in the larger bulbs.

After vernalization, the rate of development of inflorescences increases with photoperiod and with temperature in the range 6–12°C (Brewster, 1983). After spathe formation, the inflorescence begins to elongate at a rate which increases with temperature and with photoperiod (see Figs 17.12B and 17.13B), provided bulbs do not develop too rapidly (see below).

COMPLETION PHASE, FLOWER OPENING AND POLLINATION

Once a spathe is visible the plant enters the 'completion phase' of flowering (see Fig. 17.1) and higher temperatures become optimal. Experiments in glasshouses showed that inflorescence development from appearance to spathe opening required a mean of 310 day-degrees above 10°C base temperature (Brewster, 1982b). Scapes normally elongate to a length of 1–2 m. Individual bulbs may produce between 1 and 20 inflorescences depending on genotype, bulb size and environmental conditions; 3–6 inflorescences is common (Rabinowitch, 1990a). The spathe opens to produce a roughly spherical umbel containing between 50 and 2000 individual flowers, although 200–600 is the normal range. There is no very regular sequence of flower opening on an umbel. The most strongly insolated parts of the umbel tend to produce open florets first and there is a general tendency for upper florets on the umbel to precede the lower ones in opening. The pedicels of later-opening florets elongate to carry them clear of earlier-opening, maturing florets (Currah, 1990).

The sequence of individual flower development consists of petal opening (anthesis) coinciding with the start of nectar secretion, the dehiscence of the inner whorl of anthers, and then the outer whorl. Meanwhile the style elongates from 1–2 to 5–6 mm and develops a sticky stigma knob after the anthers have dehisced. Then nectar secretion ceases and petals, stamens and style start to wither (Currah and Ockenden, 1978). The whole process takes 10 days at 18°C and 5 days at 30°C (Currah, 1986). Air relative humidity below 70% accelerates pollen shedding by anthers (Ogawa, 1961). The ovary in each flower contains six ovules and about half the pollen tubes at the top of a stigma grow as far as the ovary, therefore 12 or more initial pollen tubes are needed to achieve maximum seed set. In fact, the production of 3–4 ripe seeds per ovary is common in good seed crops (Currah, 1990).

FERTILIZATION AND SEED DEVELOPMENT

Following pollination, fertilization of ovules starts within 12 h and is complete in 3–4 days (Rabinowitch, 1990b). The greatest seed set and best embryo development and most rapid pollen tube growth occurred at 35/18 (day/night °C), rather than 24/18 or 43/18 (Chang and Struckmeyer, 1976). Pollination itself stimulates the initial development of ovules and ovaries. The shrinkage and loss of green colour in unfertilized ovaries does not occur until about three weeks after flowering, making the early assessment of seed set difficult (Currah, 1990). Temperatures in the range 40–60°C have been recorded in the ovaries of the insolated side of umbels (Tanner and Goltz, 1972) causing high levels of embryo abortion and low seed set (Peterson and Trammell, 1976).

In fertilized ovules the endosperm nuclei start to divide first, and cell division and expansion by the embryo occurs 5–6 days later. The embryo develops from a globular, few-celled pro-embryo, which is first visible in microscope sections about 6 days after pollination. It then develops through an oval to tubular, and finally to a coiled tubular structure embedded within the endosperm (Rabinowitch, 1990b). Initially the endosperm is liquid; this is termed the 'milk stage'. At about 330 degree-days above 0°C after pollination, cell walls develop within the endosperm and it progresses to the pasty 'dough stage' at about 450 degree-days above 0°C after flowering. At this point the seed coat starts to turn black. The seed attains its maximum fresh weight at about 570 degree-days above 0°C after flowering. Up to this point seed dry weight growth is near exponential and seed dry weight is then about half its maximum. The endosperm then becomes solid and the seed reaches its maximum dry weight, typically 3–3.5 mg, at 810 degree-days above 0°C after flowering, normally about 45 days after flowering but depending on temperatures (Gray and Ward, 1987). At this stage capsules begin to shatter and to shed seed, and the food-reserve oil globules and protein bodies can first be seen within seeds. Seed water content declines after the attainment of maximum fresh weight. Germinability commences just before the maximum fresh weight is attained (570 degree-days above 0°C after flowering) and is near maximum at seed shatter (Steiner and Akintobi, 1986).

INTERACTIONS BETWEEN FLOWERING AND BULBING

Since both bulbing and flowering are controlled by temperature and photoperiod it is not surprising that the two processes can interact. A vegetative shoot apex normally develops axillary to the inflorescence (Rabinowitch, 1990a), and this can develop more rapidly than the inflorescence under a combination of warm temperatures (20°C or higher) and long photoperiods. This results in the swelling of the axillary bud to form bulb scales and the shrivelling and degeneration of the young scape, hence the term 'competition phase' was coined by Kampen (1970) to describe this apparent competition between axillary bud and inflorescence. When onions are planted out in bulb-inducing conditions, for example, in a warm spring, or in a glasshouse in spring, such inflorescence abortion can be common, even from bulbs with advanced inflorescence initials (Brewster, 1982a).

As discussed under bulbing, bulb or set storage at 28–30°C also delays bulb initiation. Because bulb development suppresses inflorescence elongation during the competition phase, a short period (1–2 months) of storage at such a high temperature at the end of a long period of cool storage can, by delaying bulbing, counteract this 'competition' and result in increased bolting (Aura, 1963).

Quite frequently small bulbs, known as 'bulbils' or 'topsets' may form in the inflorescence rather than normal flowers and seed capsules. When this occurs the scape is shorter than normal. Various degrees of this condition have been observed, ranging from the production of a single large 'bulbil' on an extremely short scape through to near normal inflorescences with a mixture of normal seed capsules and bulbils. This late reversion from flowering to bulbing was favoured by 6 weeks at 28–31°C applied pre-planting to bulbs after a cool treatment of sufficient duration to produce advanced inflorescence initials (Aura, 1963). Applied at an earlier stage of inflorescence differentiation such high temperatures can totally suppress floral development (see above). Bulbils can be induced to form in normal inflorescences by clipping off the developing florets. Such bulbil production can be increased by spraying the clipped heads with water, or better, benzyl adenine solution (Thomas, 1972).

Temperatures of 12–16°C in combination with photoperiods of 16–17 h have been found satisfactory for inflorescence emergence from induced bulbs and plants of intermediate-day cultivars (Brewster and Butler, 1989b). Since the photoperiods and temperatures needed for bulbing vary with cultivar, it is almost certain that the photoperiods and temperatures needed to avoid suppression of developing inflorescences by competition from bulbing will depend on the cultivar. The inflorescence of a short-day cultivar will be suppressed by a shorter daylength than a long-day cultivar. Since plants raised from bulbs or sets bulb more rapidly than those raised from seeds competition from bulbing probably suppresses inflorescences in bulb-raised plants at shorter photoperiods and lower temperatures than in seed-raised plants. In view of cultivar differences in bulbing response to photoperiod and temperature, and the fact that in order for flowering to proceed the young inflorescence must avoid suppression by bulbing, it seems a foregone conclusion that cultivar differences in the response of flower elongation to photoperiod and temperature must be parallel to the cultivar differences in bulbing response. Experiments are needed to confirm this, however.

MODELLING OF GROWTH, BULBING AND FLOWERING

In view of the interrelations between growth, bulbing and flowering, and the control of all three by temperature and light conditions (Fig. 17.14), advances in understanding of crop behaviour are possible by combining models for these processes. Predictions of how bolting, bulb yields and maturity date are affected by sowing date, locality and season can be compared with field experiments, to see if the models are satisfactory. For example, the photothermal bulbing model (see above) predicts that widely-spaced cv. Keepwell will initiate bulb scales on 31 May, with the mean daily temperatures at Wellesbourne, UK. Cultivar Keepwell is very similar to cv. Senshyu in its date of bulbing and maturity (NIAB, 1982) and similar in bolting response. A model for inflorescence formation, based on the vernalization responses described above, with inputs of daily temperature and photoperiod, predicted that in some seasons early-sown, overwintered Senshyu crops would reach the stage of spathe initiation well before 31 May (Brewster, 1997). Such

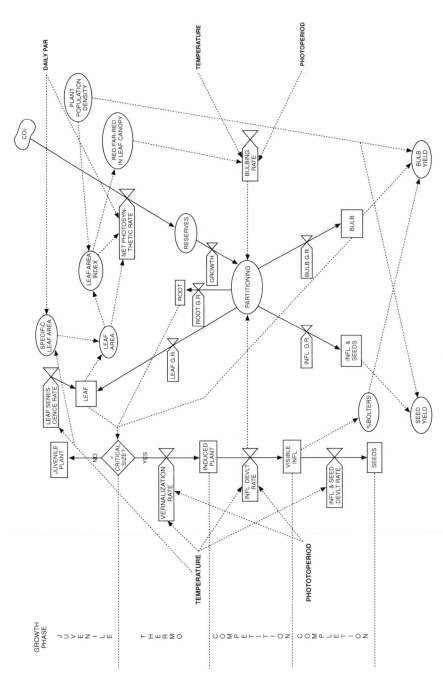

Fig. 17.14. A relational diagram indicating the main environmental controls of growth and development in onions, showing how growth, bulbing and flowering interrelate to determine bulb or seed yield. Progress through the growth phases of Fig. 17.1 is indicated on the left. Devlt, development; Infl, inflorescence; GR, growth rate (dry matter growth); PAR, photosynthetically active radiation.

crops are likely to bolt because scape elongation will occur before competition from bulbing begins to suppress developing inflorescences.

As a simple illustration of how growth rate and bulbing rates interrelate to determine yield, with a plant population of 60 plants m^{-2}, a crop emerging on 1 May at Wellesbourne will, in an average year, achieve an LAI of 3.2 and a light interception of 60% (equation 4), if leaf growth continues until 6 July. This is assuming leaf area per plant (cm^2) is given by:

$$\log_e (\text{leaf area}) = \log_e(0.5) + 0.0108 \text{ DD} \qquad (5)$$

DD is the summation of day-degrees above base 6°C and below 20°C accumulated since emergence (Brewster, 1994). 1 May is a typical emergence date from late March sowing. Sixty per cent interception gives bulb yields of about 0.9 kg m^{-2} dry weight (about 75 t ha^{-1} bulb fresh weight) (see Fig. 17.10). The photothermal model for bulbing (see above), predicts that cv. Hyton initiates bulb scales on 7 July using mean temperatures for Wellesbourne. Hyton is therefore well adapted, and will have a higher potential yield than cv. Keepwell, which the model predicts will initiate bulb scales on 7 June from 1 May emergence.

In a trial comparing cvs Keepwell and Rijnsburger-bola (a cultivar very similar to Hyton in season of bulbing), both sown on 28 February, the latter yielded 41 t ha^{-1} with 25 t ha^{-1} of bulbs greater than 45 mm diameter and was mature on 20 August, whereas the former yielded only 24 t ha^{-1} with none bigger than 45 mm and was fully mature on 9 July (Salter, 1976). This illustrates in practice the critical importance of the response of bulbing to photoperiod and temperature in determining the yield potential of a cultivar at a given location and season of sowing. The response must allow sufficient leaf growth to occur before bulbing starts.

In fact, cv. Keepwell is grown as an overwintered crop and will typically produce a leaf area of about 25 cm^2 per plant in early March from a sowing made in mid to late August at Wellesbourne. Equation 5 predicts that the leaf area per plant will reach 442 cm^2 by 31 May in an average year at Wellesbourne, when equation 3 indicates that bulbs will be initiated. If the plant population is 60 plants m^{-2} this implies a leaf area index of 2.65 which, would intercept 54% of the incoming PAR, giving a potential experimental plot yield of about 1.2 kg m^{-2} dry weight (about 100 t ha^{-1} bulb fresh weight) (see Fig. 17.10).

Visser (1994a,b), has developed a model for bulb development which includes much of the information reviewed above, and has combined this with the universal crop growth simulator sucros '87 (Spitters, 1989) to produce a simulation model for bulb onion growth, development and yield. In validation trials, agreement between predicted and recorded maturity dates were satisfactory. Yield predictions were satisfactory under optimal soil conditions, but recorded yields were much lower than predicted at some sites, indicating some limitations, possibly by drought stress. This illustrates the usefulness of models in highlighting unexplained differences which need further investigation.

PHYSIOLOGICAL DISORDERS

THICK-NECKS

Sometimes a proportion of an onion crop will fail to complete bulbing. Green leaves continue to appear from the top of the partially formed bulb, the neck does not soften and the bulb does not become dormant. This problem is most severe in seasons, sites and cultivars where crop maturity, as shown by foliage fall-down, is already late (Brewster *et al.*, 1987). As might be expected from the effects of temperature on the rate of bulbing, late maturity occurs in cool summers (Brewster, 1990b). Late maturity is also a feature of sites where low soil temperatures result in slow emergence and seedling growth and the late development of a high LAI (Brewster *et al.*, 1987). With spring-sown crops such delays can result in bulbs still developing in the late summer and autumn as photoperiods and temperatures decline below those needed for bulbing. Controlled experiments show that plants will revert to leaf blade growth in these conditions (see Fig. 17.7). High-yielding, spring-sown cultivars adapted to cool regions are probably prone to this problem because, in order to make sufficient LAI for a high yield, they must continue leaf growth long after the summer solstice. The additional stimulus for bulbing associated with a high LAI may itself be important in pushing these crops to ripeness (see Fig. 17.9). Therefore, the crop can be predisposed to thick-necking by any factor which destroys leaves or slows leaf growth (Fig. 17.15).

PREMATURE BULBING

The opposite problem, bulbing very rapidly, soon after sowing, so that the plants remain dormant through much of the growing season, sometimes occurs with late summer sowings of overwintered crops (Robinson, 1971; Rabinowitch and Zig, 1989; Currah and Proctor, 1990). Here seedlings emerge while temperatures and photoperiods still favour bulbing. In tropical countries onions are normally raised at high plant densities in nursery beds prior to transplanting. High plant densities may contribute to premature bulbing via the red : far-red effect (Fig. 17.15).

SPLITTING

Bulb splitting as a result of multiple growing points is under strong genetic control with shallots being at the extreme in this respect. Growth in high temperatures and short days increases lateral shoot production in some cultivars (Robinson, 1971; Steer, 1980). Deep transplanting reduces bulb splitting (Chipman and Thorpe, 1977).

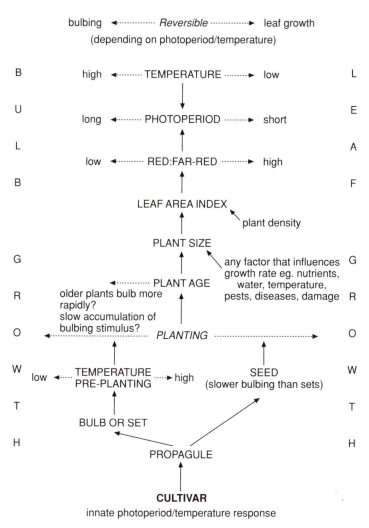

Fig. 17.15. The main factors that control onion bulbing, how they interrelate, and how numerous agronomic variables may exert an influence.

BOLTING

Unwanted bolting in bulb crops results from exposure to conditions which induce flowers to emerge before bulbs are sufficiently advanced to suppress them. Typically this would involve rapid seedling growth to the post-juvenile phase, weeks of exposure to cool temperatures optimal for inflorescence initiation (see Fig. 17.12a), followed by the lengthening photoperiods but rather cool temperatures which will favour inflorescence elongation rather than rapid bulbing. In many regions, over-early sowing of overwintered crops results in a sequence of temperatures that favour this (Brewster, 1994). Furthermore, because temperatures vary from year to year, there can always be differences in bolting from crops sown on the same date in different years (Brewster *et al.*, 1977).

INFLORESCENCE SUPPRESSION

Failure of flower emergence is a familiar problem to breeders and often results from the suppression of fully developed inflorescences by bulbing during the 'competition' phase (Kampen, 1970). This is avoided using the sequence of growing conditions that favour bolting outlined above. In the case of bulbs, a long duration near optimum temperatures for vernalization will favour subsequent inflorescence emergence (Aura, 1963). Rapid increases in temperature and photoperiod can cause inflorescence suppression, and accelerating anthesis in this way almost inevitably reduces the number of flower stalks appearing (Brewster, 1982a).

WATERY SCALES

Watery scale is a disorder of stored bulbs the symptoms of which are a thick leathery outer skin which when peeled away reveals watery, glassy storage scales (Hoftun, 1993). The scales may be later infected by bacteria or fungi, but these are not the primary cause of the disorder. Up to 15% losses of stored bulbs have been reported, particularly in crops given high temperature drying treatments following harvest. Experiments in which the gas exchange of bulbs was restricted either by submerging them for several days or by injecting the necks with vegetable oil, showed that the internal concentration of CO_2 increased, and of O_2 decreased, as a result. Treatments in which the CO_2 concentration exceeded 13% resulted in some bulbs developing watery scales (Hoftun, 1993). The higher the postharvest drying temperature the higher the internal CO_2 concentration. In bulbs with oil-injected necks there was no pattern in the distribution of watery scales after a period of storage. Internal breakdown of bulbs stored in controlled atmospheres has been observed at CO_2 of 10% and this appears more harmful to onions than low O_2 atmospheres (Komochi, 1990).

LEAF BLAST

A catastrophic collapse of onion foliage in parts of the USA, associated with hot weather following cool wet weather, is caused by ozone damage (Eagle *et al.*, 1965). Some cultivars are resistant because their stomatal membranes are so sensitive to ozone that they rapidly become leaky, causing stomatal closure. This prevents entry of ozone into the mesophyll and further leaf damage (Eagle and Gabelman, 1966).

GARLIC PHYSIOLOGY

SPROUTING AND EMERGENCE

Time from clove planting to sprout emergence decreases with increasing duration of bulb storage before planting (Mann and Lewis, 1956). Storage at about 7.5°C results in the fastest sprouting after planting (Mann and Lewis, 1956; Messiaen *et al.*, 1993). Storage at 0 or 20°C both give much slower emergence. Emergence rate also depends on field temperatures after planting, but by transforming emergence times to thermal times (accumulated day-degrees) with a base of 4.4°C, emergence rates became dependent just on pre-planting temperature and duration of storage (Mann and Lewis, 1956).

GROWTH

Pre-planting storage temperature also affects the vigour of growth and leaf morphology after sprouting. Storage at 5–10°C results in rapid growth with wide leaves and thick pseudostems, 20°C in slow growth and thin leaves and pseudostems, and 0°C thin shoots but the most vigorous growth in height (Mann and Minges, 1958). In the field, the rate of leaf appearance and the extension rate of leaves depends on thermal time. In a French study one leaf appeared per 100 day-degrees above a base of 0°C (Espagnacq *et al.*, 1987) and in New Zealand one leaf per 131 such day-degrees (Buwalda, 1986).

BULBING

EFFECTS OF PHOTOPERIOD AND TEMPERATURE

As with onion, bulbing requires long photoperiods and warm temperatures (Mann, 1952; Takagi, 1990). The lower the red : far-red ratio during inductive photoperiods the faster the bulbing (Terabun, 1978), and extensions of daylight photoperiods with far-red or blue light are more inductive than with red light (Takagi, 1990). These responses are similar to those for onion. However, the capability to bulb in response to these growing conditions is determined by storage temperatures before planting. Shading experiments show that a low light intensity reduces the number of cloves which initiate and develop in the leaf axils (Rahim and Fordham, 1990).

EFFECTS OF PRE-PLANTING BULB STORAGE TEMPERATURE

Mann and Minges (1958) showed that storage of the variety 'California Late' at 20°C before planting resulted in slow bulbing during subsequent growth under constant 16.5 h photoperiods, and completely prevented bulbing in several field trials at different locations. Storage temperatures of 0–5°C were best for inducing rapid bulbing after planting. In these experiments the optimum temperature for rapid bulb induction shifted from 5–10 to 0°C the longer the period of storage before planting. Trials in France with the similar cultivar 'Printanor' gave the same responses (Messiaen et al., 1993). With Japanese cultivars storage temperatures of 15°C and lower are effective for bulb induction with an optimum of 2–4°C (Takagi, 1990). Low temperatures are necessary for the differentiation of axillary buds which form cloves (Rahim and Fordham, 1988). If low temperatures persist for too long before long photoperiods and warmth induce bulb enlargement, the axillary buds may develop into leafy side shoots. These side shoots then develop several cloves of their own, forming irregular bulges around the periphery of the main bulb. Such misshapen bulbs are termed 'rough' and are an unaccept-able quality defect. Rough bulbs occur when planting stock is kept too long at optimal inductive temperatures (0–5°C) before planting, and occurs in practice after exceptionally cool winters (Mann and Minges, 1958; Messiaen et al., 1993). In contrast, if some cultivars have insufficient exposure to low temperatures, for example following spring rather than autumn planting, they may fail to produce axillary buds and produce just one large terminal clove termed a 'round clove', rather than the the usual multiclove bulb (Messiaen et al., 1993). Such round cloves also occur when garlic is grown from very small cloves or from the small topsets formed in inflorescences (Takagi, 1990).

A minimum of 4 weeks storage at 5°C was required to have a measur-able effect on subsequent bulbing (Mann and Minges, 1958). These authors also showed that 4 or more weeks storage at 20 or 25°C would measurably counteract the effect of many preceding weeks at 5°C and delay bulbing. So the cold induced acceleration of bulbing is reversed by subsequent warm storage, an effect also observed with onion sets.

DORMANCY

Storage conditions needed to prolong dormancy are essentially the reverse of those needed to promote sprouting discussed previously; therefore, stor-age at 7.5°C gives the most rapid loss of dormancy. The degree of dormancy of stored bulbs can be assessed by the ratio of the length of the internal sprout leaf to that of the storage leaf within a sample of cloves (Burba, 1993). For prolonged storage, temperatures of –1 to –3°C applied to well-ripened bulbs are optimal (Takagi, 1990). Dormancy is also prolonged at 20–30°C compared with intermediate temperatures. Respiration rates of dormant garlic bulbs are lower at 20 than at 5, 10 or 15°C showing that the warmer temperature suppresses metabolic activity (Mann and Lewis, 1956). High temperatures (35°C) applied before the bulbs are completely ripened can reduce dormancy (Takagi, 1990; Messiaen *et al.*, 1993), as is the case with onions. Root emergence from dormant bulbs is promoted by moisture on the base plate and is most rapid at 15°C, 5–10° warmer than the optimum for sprouting (Takagi, 1990). As with onion, sprouting is promoted by exogenous cytokinin applications and associated with a rise in endogenous cytokinin activity (Takagi, 1990). Messiaen and his colleagues have experimented with a wide range of garlic clones from the western hemisphere and have outlined six groups based on particular combinations of the following elements:

- intensity of dormancy
- the ease of elimination of dormancy by low temperature
- the low temperature requirement for axillary bud formation
- the thermophotoperiodic requirement for bulb enlargement (Messiaen *et al.*, 1993).

Different combinations of these ecophysiological traits largely determine the adaptedness of a cultivar to the climate and agriculture of a region. These workers have also shown that infection with onion yellow dwarf virus increases dormancy and earliness of ripening compared to virus-free clones.

FLOWERING

Various degrees of flowering occur in garlic clones. Some never normally produce an inflorescence, whereas others produce an inflorescence but this is sterile and bears bulbils. True seed occurs in only a few clones collected from central Asia (Etoh *et al.*, 1991; Messiaen *et al.*, 1993). Various degrees of arrested inflorescence development are common ranging from a vestigial flower stalk within a normal bulb, through a short stalk with bulbils surrounded by the outer membrane of the bulb to a short stalk crowned by a few topsets (Burba, 1993; Messiaen *et al.*, 1993). Inflorescences are induced by low temperatures, –2 to 2°C being optimal for some Japanese varieties, a lower temperature than is optimal for induction of bulbing (Takagi, 1990). The differentiation of induced inflorescences occurs after planting and is favoured by growth at cool temperatures after planting. Long photoperiods stimulate inflorescence development, but as with onion, competition from bulbing can suppress inflorescences, therefore a combination of long photoperiods with temperatures sufficiently low to prevent bulbing favours bolting (Messiaen *et al.*, 1993). A large clove size at planting favours bolting (Takagi, 1990). Bolting reduces bulb yields and growers of bolting cvs in France remove the flower stalk thereby increasing bulb yields by about 15% (Messiaen *et al.*, 1993).

PHYSIOLOGICAL DISORDERS

The cause of irregular or 'rough' bulbs due to cloves formed on side shoots has been described under bulbing. In stored bulbs a 'waxy breakdown' can occur in which small, sunken, light-yellow areas of the clove flesh expand to give shrunken, somewhat translucent, amber cloves which are waxy to touch. Waxy breakdown is associated with inadequate ventilation and low oxygen during storage (Anon, 1976), suggesting it may be similar in cause to the 'watery scale' problem of onions discussed above and recently shown to be caused by CO_2 toxicity.

REFERENCES

Abdalla, A.A. (1967) Effect of temperature and photoperiod on bulbing of the common onion (*Allium cepa* L.) under arid tropical conditions of the Sudan. *Experimental Agriculture* 3, 137–142.

Abdalla, A.A. and Mann, L.K. (1963) Bulb development in the onion (*Allium cepa* L.) and the effect of storage temperature on bulb rest. *Hilgardia* 35, 85–112.

Anon., (1976) *Growing Garlic in California*. Division of Agricultural Sciences, University of California, Leaflet 2948.

Apeland, J. (1971) Effects of scale quality on physiological processes in onion. *Acta Horticulturae* 20, 72–79.

Aura, K. (1963) Studies on the vegetatively propagated onions cultivated in Finland, with special reference to flowering and storage. *Annales Agriculturae Fenniae* 2 (Suppl. 5), 74.

Austin, R.B. (1972) Bulb formation in onions as affected by photoperiod and spectral quality of light. *Journal of Horticultural Science* 47, 493–504.

Bernstein, L. and Hayward, H.E. (1958) Physiology of salt tolerance. *Annual Review of Plant Physiology* 9, 25–46.

Bhat, K.K.S. and Nye, P.H. (1974) Diffusion of phosphate to plant roots in soil. III. Depletion around onion roots without root hairs. *Plant and Soil* 41, 383–394.

Bierhuizen, J.F. and Wagenvoort, W.A. (1974) Some aspects of seed germination in vegetables. I. The determination and application of heat sums and minimum temperature for germination. *Scientia Horticulturae* 2, 213–219.

Boswell, V.R. (1924) Influence of the time of maturity of onions upon the rest period, dormancy, and responses to various stimuli designed to break the rest period. *Proceedings of the American Society for Horticultural Science* 20, 225–233.

Brewster, J.L. (1979) The response of growth rate to temperature of seedlings of several *Allium* crop species. *Annals of Applied Biology* 93, 351–357.

Brewster, J.L. (1982a) Flowering and seed production in overwintered cultivars of bulb onions. I. Effects of different raising environments, temperatures and daylengths. *Journal of Horticultural Science* 57, 93–101.

Brewster, J.L. (1982b) Flowering and seed production in overwintered cultivars of bulb onions. II. Quantitative relationships between mean temperatures and daylengths and the rate of inflorescence development. *Journal of Horticultural Science* 57, 103–108.

Brewster, J.L. (1982c) Growth, dry matter partition and radiation interception in an overwintered bulb onion crop. *Annals of Botany (London)* 49, 609–617.

Brewster, J.L. (1983) Effects of photoperiod, nitrogen nutrition and temperature on inflorescence initiation and development in onion (*Allium cepa* L.). *Annals of Botany* 51, 429–440.

Brewster, J.L. (1985) The influence of seedling size and carbohydrate status and of photon flux density during vernalization on inflorescence initiation in onion (*Allium cepa* L.). *Annals of Botany* 55, 403–414.

Brewster, J.L. (1987a) The effect of temperature on the rate of sprout growth and development within stored onion bulbs. *Annals of Applied Biology*, 111, 463–467.

Brewster, J.L. (1987b) Vernalization in the onion – a quantitative approach. In: Atherton, J.G. (ed.) *The Manipulation of Flowering*. Butterworths, London, 171–183.

Brewster, J.L. (1990a) Physiology of crop growth and bulbing. In: Rabinowitch, H.D. and Brewster, J.L. (eds) *Onions and Allied Crops*, Vol. 1. CRC Press, Boca Raton, Florida, 53–88.

Brewster, J.L. (1990b) The influence of cultural and environmental factors on the time of maturity of bulb onion crops. *Acta Horticulturae* 267, 289–296.

Brewster, J.L. (1994) *Onions and Other Vegetable Alliums*. CAB International, Wallingford, UK.

Brewster, J.L. (1997) Environmental physiology of the onion: towards quantitative models for the effects of photoperiod, temperature and irradiance on bulbing, flowering and growth. Proceedings of the First International Conference on Edible Alliums, Mendoza, Argentina, March 1994, *Acta Horticulturae* (in press).

Brewster, J.L. and Butler, H.A. (1989a) Effects of nitrogen supply on bulb development in onion *Allium cepa* L. *Journal of Experimental Botany* 40, 1155–1162.

Brewster, J.L. and Butler, H.A. (1989b) Inducing flowering in growing plants of overwintered onions: effects of supplementary irradiation, photoperiod, nitrogen, growing medium and gibberellins. *Journal of Horticultural Science* 64, 301–312.

Brewster, J.L. and Sutherland, R.A. (1993) The rapid determination in controlled environments of parameters for predicting seedling growth rates in natural conditions. *Annals of Applied Biology* 122, 123–133.

Brewster, J.L., Bhat, K.K.S. and Nye, P.H. (1975) The possibility of predicting solute uptake and plant growth from independently measured plant and soil characteristics. III. The growth and uptake of onions in a soil fertilised to different initial levels of phosphate and a comparison of the results with model predictions. *Plant and Soil* 42, 197–226.

Brewster, J.L., Lawes, W. and Whitlock, A.J. (1987) The phenology of onion bulb development at different sites and its relevance to incomplete bulbing ('thicknecking'). *Journal of Horticultural Science* 62, 371–378.

Brewster, J.L., Mondal, M.F. and Morris, G.E.L. (1986) Bulb development in onion (*Allium cepa* L.) IV. Influence on yield of radiation interception, its efficiency of conversion, the duration of growth and dry-matter partitioning. *Annals of Botany* 58, 221–233.

Brewster, J.L., Rowse, H.R. and Bosch, A.D. (1992) The effects of sub-seed placement of liquid N and P fertilizer on the growth and development of bulb onions over a range of plant densities using primed and non-primed seed. *Journal of Horticultural Science* 66, 551–557.

Brewster, J.L., Salter, P.J. and Darby, R.J. (1977) Analysis of the growth and yield of overwintered onions. *Journal of Horticultural Science*, 52, 335–346.

Burba, J.L. (1993) Produccion de 'Semilla' de Ajo. Asociacion Cooperadora Estacion Experimental Agropecuaria, *La Consulta*, Argentina.

Buwalda, J.G. (1986) Nitrogen nutrition of garlic under irrigation. Crop growth and development. *Scientia Horticulturae* 29, 55–68.

Chang, W.N. and Struckmeyer, B.E. (1976) Influence of temperature, time of day and flower age on pollen germination, stigma receptivity, pollen tube growth, and fruit set of *Allium cepa* L. *Journal of the American Society for Horticultural Science* 101, 81–83.

Chipman, E.W. and Thorpe, E. (1977) Effects of hilling and depth of plant setting on the incidence of multiple hearts and shape of sweet Spanish onions. *Canadian Journal of Plant Science* 57, 1219–1221.

Clarkson, D.T. and Hanson, J.B. (1986) Proton fluxes and the activity of a stelar proton pump in onion roots. *Journal of Experimental Botany* 37, 1136–1150.

Currah, Lesley (1986) Studies on Flowering and Pollination in Onion. PhD thesis, University of London.

Currah, L. (1990) Pollination biology. In: Rabinowitch, H.D. and Brewster, J.L. (eds) *Onions and Allied Crops*, Vol. 1. CRC Press, Boca Raton, Florida, pp. 135–149.

Currah, L. and Ockenden, D.J. (1978) Protandry and the sequence of flower opening in the onion. *New Phytologist*, 81, 419–428.

Currah, L. and Proctor, F.J. (1990) *Onions in Tropical Regions*. Bulletin 35, Natural Resources Institute, Chatham, UK.

Dearman, J., Brocklehurst, P.A. and Drew, R.L.K. (1986) Effects of osmotic priming and ageing on onion seed germination. *Annals of Applied Biology* 108, 639–648.

Eagle, R.L., Gableman, W.H. and Romanowski, R.R. (1965) Tipburn, an ozone incited response in onion. *Proceedings of the American Society for Horticultural Science* 86, 468–474.

Eagle, R.L. and Gableman, W.H. (1966) Inheritance and mechanism for ozone damage in onions. *Proceedings of the American Society for Horticultural Science* 89, 423–430.

Ellis, R.H. (1989) The effects of differences in seed quality resulting from priming or deterioration on the relative growth rate of onion seedlings. *Acta Horticulturae* 253, 203–211.

Ellis, R.H. and Butcher, P.D. (1988) The effects of priming and 'natural' differences in quality amongst onion seedlots on the response of rate of germination to temperature and the identification of the characteristics under genotypic control. *Journal of Experimental Botany* 39, 935–950.

Ellis, R.H. and Roberts, E.H. (1977) A revised seed viability nomograph for onion. *Seed Research* 5, 93–103.

Ellis, R.H. and Roberts, E.H. (1980) Improved equations for the prediction of seed longevity. *Annals of Botany* 45, 13–30.

Espagnacq, L., Morard, P. and Bertoni, G. (1987) Determination du zero vegetatif de l'Ail. Pepinieristes Horticulteurs Maraichers. *Revue Horticole* 275, 29–31.

Etoh, T., Kojima, T. and Matsuzoe, N. (1991) Fertile garlic clones collected in Caucasia. In: Hanelt, P., Hammer, K. and Knupfer, H. (eds) *The genus Allium – Taxonomic Problems and Genetic Resources*. Institut fur Pflanzengenetik und Kulturpflanzenforschung, Gatersleben, Germany, pp. 49–54.

Finch-Savage, W.E. (1986) Effects of soil moisture and temperature on seedling emergence from natural and pre-germinated onion seeds. *Journal of Agricultural Science (Cambridge)* 107, 249–256.

Finch-Savage, W.E. (1987) A comparison of seedling emergence and early seedling growth from dry-sown, natural and fluid-drilled pregerminated onion (*Allium cepa* L.) seeds in the field. *Journal of Horticultural Science* 62, 39–47.

Finch-Savage, W.E. (1990) Estimating the optimum time of irrigation to improve vegetable crop establishment. *Acta Horticulturae* 278, 807–814.

Finch-Savage, W.E. and Phelps, K. (1993) Onion (*Allium cepa* L.) seedling emergence patterns can be explained by the influence of soil temperature and water potential on seed germination. *Journal of Experimental Botany* 44, 407–414.

Frappell, B.D. (1973) Plant spacing of onions. *Journal of Horticultural Science* 48, 19–28.

Gale, J., Kohl, H.C. and Hagen, R.M. (1967) Changes in the water balance and photosynthesis of onion, bean and cotton plants under saline conditions. *Physiologia Plantarum* 20, 408–420.

Garner, W.W. and Allard, H.A. (1920) Effect of the relative length of day and night and other factors of the environment on growth and reproduction in plants. *Journal of Agricultural Research* 18, 553–606.

Goltz, S.M., Tanner, C.B., Millar, A.A. and Lang, A.R.G. (1971) Water balance of a seed onion field. *Agronomy Journal* 63, 762–765.

Gonzalez-Fernandez, A., Lopez-Saez, J.F., Moreno, P. and Gimenez-Martin, G. (1968) A model for the dynamics of cell division cycle in onion roots. *Protoplasma* 65, 263–276.

Gray, D. and Ward, J.A. (1987) A comparison of leek (*Allium porrum*) and onion (*Allium cepa*) seed development. *Annals of Botany* 60, 181–187.

Greenwood, D.J., Cleaver, T.J., Turner, M.K., Hunt, J., Niendorf, K.B. and Loquens, S.M.H. (1980a) Comparison of the effects of potassium fertilizer on the yield, potassium content and quality of 22 different vegetable and arable crops. *Journal of Agricultural Science, Cambridge* 95, 441–456.

Greenwood, D.J., Cleaver, T.J., Turner, M.K., Hunt, J., Niendorf, K.B. and Loquens, S.M.H. (1980b) Comparison of the effects of phosphate fertilizer on the yield, phosphate content and quality of 22 different vegetable and arable crops. *Journal of Agricultural Science, Cambridge* 95, 457–469.

Greenwood, D.J., Gerwitz, A., Stone, D.A. and Barnes, A. (1982) Root development in vegetable crops. *Plant and Soil* 68, 75–96.

Greenwood, D.J., Neeteson, J.J., Draycott, A., Wijnen, G. and Stone, D.A. (1992) Measurement and simulation of the effects of N-fertilizer on growth, plant composition and distribution of soil mineral-N in nationwide onion experiments. *Fertilizer Research* 31, 305–318.

Grime, J.P. (1979) *Plant Strategies and Vegetation Processes.* John Wiley, Chichester.

Harrington, J.F. (1962) The effect of temperature on the germination of several kinds of vegetable seeds. *Proceedings of the XVI International Horticultural Congress*, Vol. II, 435–441.

Heath, O.V.S. and Holdsworth, M. (1948) Morphogenic factors as exemplified by the onion plant. In: Danielli, J.F. and Brown, R. (eds) *Growth in Relation to Differentiation and Morphogenesis.* Symposia of the Society for Experimental Biology II, Cambridge University Press, London, 326–350.

Heath, O.V.S. and Hollies, M.A. (1965) Studies in the physiology of the onion plant. VI. A sensitive morphological test for bulbing and its use in detecting bulb development in sterile culture. *Journal of Experimental Botany* 16, 128–144.

Heath, O.V.S., Holdsworth, M., Tincker, M.A.H. and Brown, F.C. (1947) Studies in the physiology of the onion plant. III Further experiments on the effects of storage temperature and other factors in onions grown from sets. *Annals of Applied Biology*, 34, 473–502.

Hegarty, T.W. (1976) Effects of fertiliser on the seedling emergence of vegetable crops. *Journal of the Science of Food and Agriculture* 27, 962–968.

Hoffman, C.A. (1933) Developmental morphology of *Allium cepa. Botanical Gazette* 95, 279–299.

Hoftun, H. (1993) Internal atmosphere and watery scale in onion bulbs. *Acta Horticulturae* 343, 135–140.

Isenberg, F.M.R., Thomas, T.H., Pendergrass, A.M. and Abdel-Rahman, M. (1974) Hormone and histological differences between normal and maleic hydrazide treated onions stored over winter. *Acta Horticulturae* 38, 95–125.

Jones, H.A. and Mann, L.K. (1963) *Onions and their Allies*. Leonard Hill, London.

Kampen, J. van (1970) Shortening the breeding cycle in onions. *Mededeling Proefstation voor de Groenteteelt in de Vollegrond*, Alkmaar. No 51.

Kedar, N., Levy, D. and Goldschmidt, E.E. (1975) Photoperiodic regulation of bulbing and maturation of Bet Alpha onions (*Allium cepa* L.) under decreasing daylength conditions. *Journal of Horticultural Science* 50, 373–380.

Komochi, S. (1990) Bulb dormancy and storage physiology. In: Rabinowitch, H.D. and Brewster, J.L. (eds) *Onions and Allied Crops*, Vol. 1. CRC Press, Boca Raton, Florida, pp. 89–111.

Ladeinde, F. and Hicks, J.R. (1988) Internal atmosphere of onion bulbs stored at various oxygen concentrations and temperatures. *HortScience* 23, 1035–1037.

Lercari, B. (1982) The promoting effect of far-red light on bulb formation in the long day plant *Allium cepa* L. *Plant Science Letters* 27, 243–254.

Lercari, B. (1983) Action spectrum for the photoperiodic induction of bulb formation in *Allium cepa* L. *Photochemistry and Photobiology* 38, 219–222.

Levy, D., Ben-Herut, Z., Albasel, N., Kaisi, F. and Manasra, I. (1981) Growing onion seeds in an arid region: drought tolerance and the effect of bulb weight, spacing and fertilization. *Scientia Horticulturae* 14, 1–7.

Magistad, O.C., Ayers, A.D., Wadleigh, C.H. and Gauch, H.G. (1943) Effects of salt concentration, kind of salt and climate on plant growth in sandculture. *Plant Physiology* 18, 151–166.

Mann, L.K. (1952) Anatomy of the garlic bulb and factors affecting bulb development. *Hilgardia*, 21, 195–251.

Mann, L.K. and Lewis, D.A. (1956) Rest and dormancy in garlic. *Hilgardia* 26, 161–189.

Mann, L.K. and Minges, P.A. (1958) Growth and bulbing of garlic in response to storage temperature of planting stocks, day length, and planting date. *Hilgardia*, 27, 385–419.

McNaughton, K.G., Gandar, P.W. and McPherson, H.G. (1985) Estimating the effects of varying temperature on the rate of development of plants. *Annals of Botany* 56, 579–595.

Messiaen, C.M., Cohat, J., Leroux, J.P., Pichon, M. and Beyries, A. (1993) *Les Allium Alimentaires Reproduits par Voie Vegetative*. Institut National de la Recherche Agronomique, Paris.

Miedema, P. (1994a) Bulb dormancy in onion. I. The effects of temperature and cultivar on sprouting and rooting. *Journal of Horticultural Science* 68, 29–39.

Miedema, P. (1994b) Bulb dormancy in onion. II. The role of cytokinins in high-temperature imposed sprout inhibition. *Journal of Horticultural Science* 68, 41–45.

Miedema, P. (1994c) Bulb dormancy in onion. III. The influence of the root system, cytokinin, and wounding on sprout emergence. *Journal of Horticultural Science* 68, 47–52.

Millar, A.A., Gardner, W.R. and Goltz, S.M. (1971) Internal water status and water transport in seed onion plants. *Agronomy Journal* 63, 779–784.

Mondal, M.F. (1985) Studies on the control of bulbing in onion (*Allium cepa* L.). PhD thesis, University of Birmingham, UK.

Mondal, M.F., Brewster, J.L., Morris, G.E.L. and Butler, H.A. (1986a) Bulb development in onion (*Allium cepa* L.). I. Effects of plant density and sowing date in field conditions. *Annals of Botany* 58, 187–195.

Mondal, M.F., Brewster, J.L., Morris, G.E.L. and Butler, H.A. (1986b) Bulb development in onion II. The influence of red:far red spectral ratio and of photon flux density. *Annals of Botany (London)* 58, 197–206.

Mondal, M.F., Brewster, J.L., Morris, G.E.L. and Butler, H.A. (1986c) Bulb development in onion (*Allium cepa* L.). III. Effects of the size of adjacent plants, shading by neutral and leaf filters, irrigation and nitrogen regime and the relationship between the red:far-red spectral ratio in the canopy and leaf area index. *Annals of Botany* 58, 207–219.

NIAB (1982) *Varieties of Bulb Onions 1982*. National Institute of Agricultural Botany, Cambridge, UK.

Ogawa, T. (1961) Studies on seed production in onion. I. Effects of rainfall and humidity on the fruit setting. *Journal of the Japanese Society for Horticultural Science*, 30, 222–232.

Peterson, C.E. and Trammell, K.W. (1976) Influence of controlled high temperatures on onion seed set and embryo abortion. *HortScience*, 11, 298.

Rabinowitch, H.D. (1990a) Physiology of flowering. In: Rabinowitch, H.D. and Brewster, J.L. (eds) *Onions and Allied Crops*, Vol. 1. CRC Press, Boca Raton, Florida, pp. 113–134.

Rabinowitch, H.D. (1990b) Seed development. In: Rabinowitch, H.D. and Brewster, J.L. (eds) *Onions and Allied Crops*, Vol. 1. CRC Press, Boca Raton, Florida, pp. 151–159.

Rabinowitch, H.D. and Zig, U. (1989) Leaf and root degeneration in early maturing onions: a physiological disorder. *Annals of Applied Biology* 115, 533–540.

Rahim, M.A. and Fordham, R. (1988) Effect of storage temperature on the initiation and development of garlic cloves. *Scientia Horticulturae* 37, 25–38.

Rahim, M.A. and Fordham, R. (1990) Effect of shade and environmental conditions on the initiation and development of garlic cloves. *Scientia Horticulturae* 45, 21–30.

Robinson, J.C. (1971) Studies on the performance and growth of various short-day onion varieties in the Rhodesian lowveld in relation to date of sowing. 1. Yield and quality analysis. *Rhodesian Journal of Agricultural Research* 11, 51–68.

Robinson, J.E., Browne, K.M. and Burton, W.G. (1975) Storage characteristics of some vegetables and fruits. *Annals of Applied Biology* 81, 399–408.

Salter, P.J. (1976) Comparative studies of different production systems for early crops of bulb onions. *Journal of Horticultural Science* 51, 329–339.

Scaife, A., Cox, E.F. and Morris, G.E.L. (1987) The relationship between shoot weight, plant density and time during the propagation of four vegetable species. *Annals of Botany* 59, 325–334.

Schnabl, H. and Raschke, K. (1980) Potassium chloride as a stomatal osmoticum in *Allium cepa* L., a species devoid starch in the guard cells. *Plant Physiology* 65, 88–93.

Shishido, Y. and Saito, T. (1975) Studies on the flower bud formation in onion plants. I. Effects of temperature, photoperiod, and light intensity on the low temperature induction of flower buds. *Journal of the Japanese Society of Horticultural Science* 44, 122–130.

Shishido, Y. and Saito, T. (1976) Studies on the flower bud formation in onion plants. II. Effects of physiological condition on the low temperature induction of flower bulbs on green plants. *Journal of the Japanese Society of Horticultural Science* 45, 160–167.

Shishido, Y. and Saito, T. (1977) Studies on the flower bud formation in onion plants. III. Effects of physiological conditions on the low temperature induction of flower buds in bulbs. *Journal of the Japanese Society of Horticultural Science* 46, 310–316.

Smittle, D.A. (1988) Evaluation of storage methods for 'Granex' onions. *Journal of the American Society for Horticultural Science* 113, 877–880.

Sobeih, W.Y. and Wright, C.J. (1986) The photoperiodic regulation of bulbing in onions (*Allium cepa* L.). II. Effects of plant age and size. *Journal of Horticultural Science* 61, 337–341.

Spitters, C.J.T. (1989) A simple universal crop growth simulator: SUCROS87. In: Rabbinge, R., Ward, S.A. and Laar, H.H. van (eds) *Simulation and Systems Management in Crop Protection*. PUDOC, Wageningen, Netherlands, pp. 147–181.

Steer, B.T. (1980) The bulbing response to day length and temperature of some Australasian cultivars of onion (*Allium cepa* L.). *Australian Journal of Agricultural Research* 31, 511–518.

Steiner, J.J. and Akintobi, D.C. (1986) Effect of harvest maturity on viability of onion seed. *HortScience* 21, 1220–1221.

Stone, D.A. and Rowse, H.R. (1992) Reducing the nitrogen requirement of vegetable crops by precision fertilizer injection. *Nitrate and Farming Systems – Aspects of Applied Biology* 30, 399–402.

Stow, J.R. (1976) The effect of defoliation on storage potential of bulbs of the onion (*Allium cepa* L.). *Annals of Applied Biology* 84, 71–79.

Summerfield, R.J., Roberts, E.H., Ellis, R.H. and Lawn, R.J. (1991) Towards the reliable prediction of time to flowering in six annual crops. I. The development of simple models for fluctuating field environments. *Experimental Agriculture* 27, 11–31.

Takagi, H. (1990) Garlic. In: Rabinowitch, H.D. and Brewster, J.L. (eds) *Onions and Allied Crops*, Vol 3. CRC Press, Boca Raton, Florida, pp. 109–146.

Tanaka, M., Chee, K. and Komochi, S. (1985) Studies on the storage of autumn harvested onion bulbs. I. Influence of storage temperature and humidity on the sprouting during storage. *Research Bulletins of the Hokkaido National Agricultural Experiment Station* 141, 1–16.

Tanner, C.B. and Goltz, S.M. (1972) Excessively high temperatures of seed onion umbels. *Journal of the American Society for Horticultural Science* 97, 5–9.

Terabun, M. (1971) Studies on the bulb formation in onion plants. VIII. Internal factors in reference to bulb formation. *Journal of the Japanese Society of Horticultural Science* 40, 50–56.

Terabun, M. (1978) The bulbing of onion, wakegi and garlic plants under mixed light of red and far-red. *Science Reports of the Faculty of Agriculture, Kobe University*, 13, 1–6.

Thomas, T.H. (1972) Stimulation of onion bulblet production by N6 benzyladenine. *Horticultural Research* 12, 77–79.

Thompson, H.C. and Smith, O. (1938) Seedstalk and bulb development in onion. *Bulletin of Cornell University Agricultural Experiment Station No. 708*, p. 21.

Visser, C.L.M. de (1994a) ALCEPAS, an onion growth model based on sucros '87. I. Development of the model. *Journal of Horticultural Science* 69, 501–518.

Visser, C.L.M. de (1994b) ALCEPAS, an onion growth model based on sucros '87. II. Validation of the model. *Journal of Horticultural Science* 69, 519–525.

Wagenvoort, W.A. and Bierhuizen, J.F. (1977) Some aspects of seed germination in vegetables. II. The effect of temperature fluctuation, depth of sowing, seed size and cultivar, on heat sum and minimum temperature for germination. *Scientia Horticulturae* 6, 259–270.

Wannamaker, M.J. and Pike, L.M. (1987) Onion responses to various salinity levels. *Journal of the American Society for Horticultural Science* 112, 49–52.

Ward, C.M. (1976) The influence of temperature on weight loss from stored onion bulbs due to desiccation, respiration and sprouting. *Annals of Applied Biology* 83, 149–155.

Ward, C.M. and Tucker, W.G. (1976) Respiration of maleic hydrazide treated and untreated onion bulbs during storage. *Annals of Applied Biology* 82, 135–141.

Wheeler, T.R. and Ellis, R.H. (1991) Seed quality, cotyledon elongation at suboptimal temperatures, and the yield of onion. *Seed Science Research* 1, 57–67.

Wheeler, T.R. and Ellis, R.H. (1992) Seed quality and emergence in onion. *Journal of Horticultural Science* 67, 319–332.

Wiles, G.C. (1994) The effect of different photoperiods and temperatures following bulb initiation on bulb development in tropical onion cultivars. *Acta Horticulturae* 358, 419–427.

Wright, C.J. and Sobeih, W.Y. (1986) The photoperiodic regulation of bulbing in onions (*Allium cepa* L.). I. Effects of irradiance. *Journal of Horticultural Science* 61, 331–335.

Asparagus   18

D.T. Drost

*Department of Plants, Soils, and Biometeorology, Utah State University,
Logan, Utah 84322–4820, USA*

> It should be remembered that the energy of the crown and roots is
> supplied from the foliage that developed in the previous summer.
> Without a heavy growth of top, one cannot expect a good growth of
> roots and a heavy crop the following year.

(Bailey, 1914, pp. 433–441)

Asparagus is one of the most important perennial vegetables grown in the
world. Europe and North America are major production areas with 52,500
and 45,500 ha under cultivation (Nichols, 1990). Other production areas
include Asia (16,970 ha, mostly Taiwan and Japan), South America
(11,700 ha, mostly Peru and Chile), Australasia (6500 ha in Australia and
New Zealand) and Africa (3250 ha, mostly South Africa). Production is
increasing in the Southern Hemisphere as demand for fresh asparagus in
the off-season increases in the Northern Hemisphere. In the USA, asparagus
is produced on over 40,000 ha mainly in California, Washington, and
Michigan, with an approximate value of US $150 million (Anon, 1992). It
has been estimated that worldwide, asparagus is worth in excess of one billion US dollars.

Edible asparagus is an ancient vegetable native to Europe (Tutin *et al.*,
1980) where it has been cultivated since the times of the Greeks and Romans
(Hexamer, 1914). Throughout ancient Europe, asparagus was used for both
culinary and medicinal purposes. Roots, shoots and seeds were used as
diuretics, sedatives, pain killers and liniments (Hexamer, 1914). Today,
asparagus is grown for its edible shoots (spears) which are marketed fresh
or processed.

Asparagus belongs to the family Liliaceae (Tutin *et al.*, 1980) with
approximately 150 species spread throughout the world. All are dioecious,
herbaceous perennials. In addition to edible asparagus (*A. officinalis*), there

© CAB INTERNATIONAL 1997. *The Physiology of Vegetable Crops*
(H.C. Wien, ed.)

are several important ornamental species. These include *A. sprengeri* (asparagus fern), *A. plumosus*, *A. laricinus* and *A. racemosus* (all used in the florist trade). Most of the information on the physiology of asparagus has been gleaned from the edible form. This review will focus on what we know about bud, spear, and root growth, carbohydrate production and accumulation and their relation to crop productivity.

ASPARAGUS GROWTH

The underground portion of the asparagus plant is known collectively as the crown and the foliage as fern (Fig. 18.1). The crown consists of a rhizome (underground stem), and adventitious and lateral roots. The rhizome has at its growing tip several large and many smaller buds which collectively form a bud cluster (Blasberg, 1932). Accessory buds or lateral buds form new clusters on the side of the rhizome producing a new axis of growth (Blasberg, 1932; Mullendore, 1935). New buds are initiated in the axil of the first scale leaf of the preceding bud (Blasberg, 1932; Mullendore, 1935). Buds

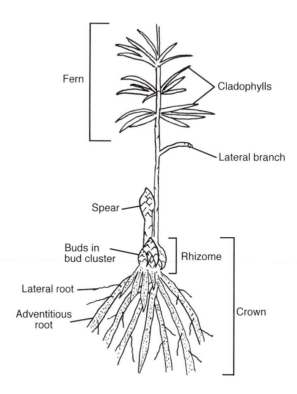

Fig. 18.1. The asparagus plant with important plant parts (adapted and redrawn from Mullendore, 1935).

arise alternately on either side of the growth axis (Mullendore, 1935), extending the rhizome.

The aerial foliage of asparagus consists of many individual ferns arising from separate buds on the underground rhizome. Each fern has a central stem with sessile scale leaves (Mullendore, 1935) and grows to a height of 1–2 m (Blasberg, 1932). Lateral branches arise in the axils of the scale leaves, producing secondary branches which have scale leaves of their own. Scale leaves are triangular in shape (Mullendore, 1935) in a 2:5 phyllotaxy arrangement (Blasberg, 1932). Small, cylindrical, leaf-like structures called cladophylls develop in the whorl of the scale leaves at nodes on stems and branches (Blasberg, 1932) and are the chief photosynthetic organs of the asparagus plant. Cladophylls are modified branches with a cuticularized epidermis, sunken stomata, and small guard cells (Blasberg, 1932).

SHOOT GROWTH

During the course of a production season, several different stages of plant growth take place. These include bud break and spear growth in spring; fern expansion, bud development and root growth in summer; and fern senescence and bud dormancy in autumn. The initiation of growth in the spring is determined by increases in soil temperature in the temperate growing areas but may require applications of water in drier production areas. In tropical areas, growth may continue unabated throughout the year.

Buds and spears

Bud break and spear growth are controlled by apical dominance (Kretschmer and Hartmann, 1979; Tiedjens, 1924), with dominance increasing as the season progresses (Tiedjens, 1924, 1926). Spears allowed to grow beyond marketable size (> 20 cm) suppress adjacent buds more strongly than shorter spears. The longer the interval between the growth of the first spear and its harvest, the longer the suppression of the growth of the second spear; this emphasizes the need for timely, regular harvests. Removal of spears or fern allows smaller buds to grow, which accounts for the sequential production of spears throughout the growing season.

It was thought that inhibition of bud break by apical dominance extends only to those buds within that bud cluster (Tiedjens, 1926). However, Nichols and Woolley (1985) reported strong interactions between growth of spears from different bud clusters. They showed there was a linear decrease in the relative growth rate of spears emerging sequentially from different bud clusters, and this effect remained until the earlier emerging spears were harvested. In contrast, when spears from different bud clusters began to grow at the same time, the relative growth rates of all spears were similar but subsequent spear emergence was severely inhibited.

Development of large, healthy fern is required to produce large buds.

Bud size is positively correlated with spear size (Blasberg, 1932; Tiedjens, 1926; Nichols and Woolley, 1985) and spear fresh weight (Nichols and Woolley, 1985). However, having large buds offers no assurance that large spears will form. Buds over 5 mm in diameter can develop into spears only 2 mm in diameter (Tiedjens, 1926), especially after a number of large diameter spears have been harvested. Small spear size late in the harvest season may be the result of a decrease in the carbohydrate reserves in the roots (Working, 1924; Tiedjens, 1926) or a reduction in the size of the remaining buds or both. Further study is needed to understand what regulates bud development and size.

Fern

After harvest is completed, spears are allowed to expand and develop into fern. A balance between the harvest period and fern regrowth time is needed to maximize yields and plant survival (Scott *et al.*, 1939; Takatori *et al.*, 1970b; Shelton and Lacy, 1980; Taga *et al.*, 1980). This balance depends on the size and age of the plant, length of the growing season, and those factors that limit plant growth (Robb, 1984). Fern growth must remain vigorous throughout the growing season if carbohydrate levels in the roots, bud and root development and growth are to be maximized (Scott, 1954) since fern vigour in one year and yield in the following year are correlated (Ellison and Scheer, 1959; Moon, 1976; Hartmann, 1985; Knaflewski, 1994). Any condition during the growing season (insect and disease pressure, nutrient deficiency, drought, weed competition, etc.) that limits fern growth may limit plant performance in the following years.

During each growing season (Fig. 18.2a), shoots (spears and fern) are produced first, and only after this process is nearly completed are new buds initiated (Blasberg, 1932; Tiedjens, 1924, 1926). Bud development is said to be genetically controlled and not dependent on the growing environment and/or carbohydrate availability of the plant (Blasberg, 1932; Tiedjens, 1924). Others have shown that moisture stress decreases bud size and number (Drost and Wilcox-Lee, 1990) which may reduce future yield. Buds are initiated steadily throughout the growing season until the time of peak shoot number in young (Fisher, 1982; Haynes, 1987) and mature asparagus plantings (Wilcox-Lee and Drost, 1991). These data suggest that new and existing shoot growth directly or indirectly influences bud initiation. Tiedjens (1924, 1926), however, noted that some buds are formed during the harvest season. It is more likely that this was continued enlargement of buds initiated in the previous growing season (Wilcox-Lee and Drost, 1991) rather than the development of new buds at a time when carbohydrates are in demand by elongating spears.

As asparagus grows, more dry weight is partitioned to the crown and less to the fern (Benson and Takatori, 1980; Fisher, 1982; Dufault and Greig, 1983; Haynes, 1987; Hughes *et al.*, 1990; Wilcox-Lee and Drost, 1990). A large crown is desirable since it produces greater fern biomass, stores more

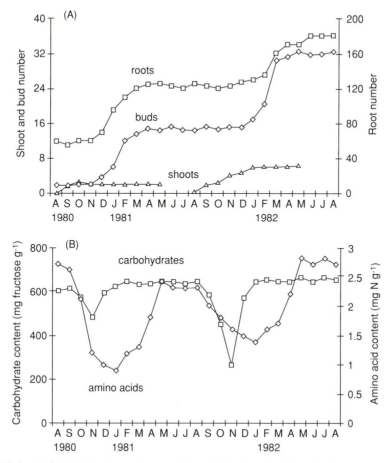

Fig. 18.2. (a) Changes in the number of shoots, buds and roots; (b) and in the concentration of carbohydrates and amino acids in roots during the first 2 years after planting asparagus crowns in New Zealand (Haynes, 1987).

carbohydrates, and initiates more buds. Benson and Takatori (1980) reported that a hybrid cultivar partitioned a greater portion of dry weight to the crown than the open pollinated cultivar earlier in the growth of the seedling even though they had similar leaf areas. This suggests improved growth efficiency either by reduced respiration rate, higher photosynthetic rate, better canopy architecture to process the available light, or improved quality of storage carbohydrates.

ROOT GROWTH

Although the growth of asparagus fern is well documented, difficult access to roots limits our understanding of root growth patterns. In field seeded or transplanted asparagus, root growth began 6 weeks after initial spear emergence (Dufault and Greig, 1983; Fisher, 1982), while in older established plantings (Fig. 18.2a), root production did not start until after the fern was well established in summer and only stopped when the fern began to senesce in the autumn.

There are two distinctly different root types originating from an asparagus crown: large, unbranched adventitious and smaller lateral roots. The large adventitious roots are generally referred to as fleshy storage roots while lateral roots are called fibrous feeder roots (Blasberg, 1932; Mullendore, 1935; Weaver and Bruner, 1927). The conventional terminology of fleshy and fibrous roots will be used to distinguish between the different root types (see Fig. 18.1).

The crown, in addition to having a rhizome with many buds, includes numerous fleshy roots attached to closely spaced basal internodes on the rhizome (Blasberg, 1932; Tiedjens, 1924). New fleshy roots are formed at the base of the young actively growing buds (Blasberg, 1932; Mullendore, 1935). Fleshy roots are unbranched, vary in diameter from 2 to 6 mm and grow to lengths of 1–2 m over several seasons (Blasberg, 1932; Mullendore, 1935). As the plant ages, a dense mat of fleshy roots forms, often requiring the new roots to be initiated on top of the older roots (Blasberg, 1932). This is responsible for the upward movement of the crown that occurs over time in mature asparagus plantings (Young, 1940; Lindgren, 1990).

Scott *et al.* (1939) found 84% of the total weight of fleshy roots was in the top 30 cm, and 94% in the top 90 cm, though roots were found to a depth of 2 m. Most fleshy roots are in the top 1 m of soil, though some roots are found as deep as 3 m (Jones and Robbins, 1928; Weaver and Bruner, 1927; Yeager and Scott, 1938). Total productivity of asparagus is not determined solely by rooting depth (Reijmerink, 1973) but also by root density and root activity (intake of water and nutrients).

Fibrous roots are primarily nutrient and water absorbing organs (Blasberg, 1932), although fleshy roots are capable of these functions as well. Fibrous roots originate from the pericycle of the fleshy roots before wall thickening and epidermal suberization occurs (Blasberg, 1932; Mullendore, 1935). Old fleshy roots initiate only a few new fibrous roots and then only on the younger root parts (Reijmerink, 1973; Tiedjens, 1926). Fibrous roots may be branched or unbranched and up to 2 mm in diameter (Blasberg, 1932). Continued root initiation is important for sustained long-term productivity in asparagus.

In young plantings, 60–90% of the fibrous roots are located in the upper 30 cm of the soil profile (Reijmerink, 1973). After several years, fibrous roots in this zone decreased while numbers increased in the soil below the plough zone. Reductions in fibrous root growth near the soil surface may be due to older fleshy roots becoming suberized (Reijmerink, 1973; Tiedjens, 1926), a build-up of toxins which reduced growth (Reijmerink, 1973; Yang, 1982), a reduction in the nutrient levels of the tillage layer (Reijmerink, 1973), physical deterioration of the soil micro-structure, compaction or damage associated with tillage operations (Reijmerink, 1973; Putnam, 1972) or general ageing of the fleshy roots. It is still unclear if there are distinct periods of fibrous root growth, though little fibrous root growth appears to occur during harvest (Scott *et al.*, 1939).

There is some controversy about the life expectancy and functionality of fleshy and fibrous roots. Unless damaged, fleshy roots grow for several years. From root growth scars, Jones and Robbins (1928) were able to identify 3 years of growth, although more may have been present. Scott *et al.* (1939) estimated that fleshy roots survive for up to 6 years and thus serve as a long-term storage organ for reserve carbohydrates. This 6-year life expectancy was deduced by sampling a 6-year-old planting and assessing the distribution and health of its root system. With adequate water, older roots are able to grow normally, presumably by absorption of water through the cell walls (Tiedjens, 1926), and are able to support plant growth until new fleshy and fibrous roots are formed. Fibrous roots appear to grow for only one year before senescing (Tiedjens, 1926; Reijmerink, 1973).

Soil structure strongly influences asparagus root development, rooting depth and plant growth. When a reduction in root growth in the topsoil is not compensated by increased rooting in the subsoil, plant growth and yield are reduced (Reijmerink, 1973). Open, porous, microaggregated soils throughout the root zone allow good root development while densely packed, single-grained sands or massive soils allow little root penetration (Reijmerink, 1973). In New Zealand, plant performance (yield) was closely correlated with soil type and texture (Bussell *et al.*, 1985). Well-aerated soils with good natural drainage and resistance to crusting or compaction are recommended for asparagus beds. These soils encourage deep rooting, reduce disease pressure in asparagus, and allow cultural practices to occur over a wide range of environmental conditions.

Tillage is often used to incorporate the previous year's fern into the soil, uproot emerging weeds prior to herbicide applications, ridge soils over the crown, and loosen the soil surface prior to spear emergence. Asparagus produced higher yields and larger spears when grown in zero tillage compared to conventional tillage (Putnam, 1972; Wilcox-Lee and Drost, 1991), or had no effect at all (Johnson, 1990). Mechanical damage to emerging spears accounts for early season yield reductions in tilled plantings, while after harvest tillage operation damages spears needed for canopy development (Tiedjens, 1924; Putnam, 1972; Wilcox-Lee and Drost, 1991). This delays the resumption of carbohydrate accumulation leading to reduced growth. Zero tillage also reduces disease spread, improves soil structure, and reduces soil compaction and erosion.

ASPARAGUS PHYSIOLOGY

Much of the work conducted on asparagus has focused on the growth and carbohydrate accumulation of asparagus. Few studies have related that information back to the assimilation capabilities of the spear or fern. Net photosynthesis in asparagus is a function of physiological age of the plant, light intensity, assimilation rate, temperature and water supply. Each of these features which impact photosynthesis will be examined and related to asparagus growth and performance. In addition, the biochemistry of carbohydrate production and accumulation, redistribution and utilization, and how these interact with the environment and present cultural practices will be discussed.

PHOTOSYNTHESIS

All green tissues on asparagus are capable of photosynthesis but at different rates. Cladophylls (see Fig. 18.1) are the main assimilation site (Downton and Torokfalvy, 1975; Lin and Hung, 1978; Sawada et al., 1962) as they make up most of the surface area of the plant. Spears have both low chlorophyll content and reduced stomatal density compared to fern. This limits the spears' ability to fix CO_2 until cladophylls are developed (Downton and Torokfalvy, 1975; Lin and Hung, 1978). Spears can, however, refix from 50 to 100% of respired CO_2 (Downton and Torokfalvy, 1975). Net assimilation rates increase as more canopy is established and then decrease as the fern senesces (Lin and Hung, 1978). This occurs during the summer and autumn in temperate regions. In tropical areas, new fern is continually initiated as old fern senesces and apical dominance is released. This ensures photosynthesis is maintained at high levels.

Net photosynthesis in asparagus increases as light levels increase. The light compensation point ranges from 15 to 30 µmol m^{-2} s^{-1} (Hills, 1986; Yue et al., 1992) with a light saturation point at 200–450 µmol m^{-2} s^{-1} (Downton and Torokfalvy, 1975; Colman et al., 1979; Hills, 1986; Yue et al., 1992). The low compensation and saturation points suggest that asparagus has several characteristics of shade-tolerant plants. Hills (1986) postulated that due to the morphology of the cladophylls and the architecture of the fern in a plant canopy, only a limited number of cladophylls are exposed to full sun conditions at any time. Therefore, photosynthesis would be maintained at efficient levels in a large portion of the canopy even when the plant is exposed to high light levels.

Net assimilation rate in asparagus is low when compared to other vegetables (Colman et al., 1979; Hills, 1986; Wilcox-Lee and Drost, 1990) and often decreases during the day (Drost and Wilcox-Lee, 1990). Others reported that fern carbohydrate levels (assimilation rates) increase continually throughout the day (Pressman et al., 1989; Sawada et al., 1962). However, photosynthates are often stored in leaves during the day before being transported to the roots at night. Therefore, measurements of fern carbohydrate levels are of limited use for determining assimilation by asparagus fern.

The optimum temperature for photosynthesis in asparagus is less than

20°C (Sawada *et al.*, 1962; Yen *et al.*, 1993). At higher temperatures, assimilation rate drops off rapidly (Yue *et al.*, 1992; Yen *et al.*, 1993). However, in isolated mesophyll cells from asparagus cladophylls, photosynthetic rate continued to rise as temperature increased up to 35°C (Colman *et al.*, 1979; Hills, 1986). This suggests that water stress may be influencing photosynthesis in plants grown at high temperatures since isolated cells are suspended in a buffered solution.

Yue *et al.* (1992) reported that low photosynthetic rates in acclimatized asparagus compared to *in vitro* plantlets was due to stomatal closure brought on by water stress. Low leaf resistance, a poor system of water transfer, and partially open stomata were all implicated for the rapid wilting of cultured plantlets after transplanting. To ensure survival of tissue-cultured plantlets, water stress must be avoided. In seedling asparagus, small changes in soil water potential result in large reductions in gas exchange (Drost and Wilcox-Lee, 1990). Assimilation rates were the same among all water potential treatments early and late in the day (Fig. 18.3). Assimilation increased during the morning and declined in the afternoon in the wetter treatments (−0.05 and −0.3 MPa) while in dry soils (−0.5 MPa) assimilation was greatest early in the morning and decreased throughout the rest of the day. Asparagus appears to fix carbon most efficiently when temperatures and light levels are favourable. Stomatal closure, decreasing fern water potentials, high temperatures and high light levels could all be responsible for the reduction in assimilation that occurs during the afternoon. It is interesting to note that the photosynthetic rate of isolated asparagus mesophyll cells were not inhibited by water potentials of −3 MPa (Hills, 1986). This suggests that photosynthesis in asparagus may be less sensitive to water stress than to other stresses (high temperature or light levels).

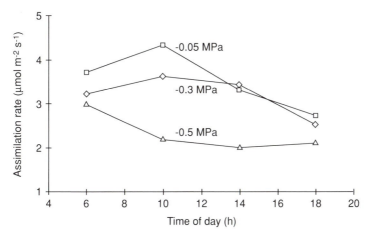

Fig. 18.3. Diurnal changes in net CO_2 assimilation rate of asparagus grown at different soil matric potentials (Drost and Wilcox-Lee, 1990).

There are many questions that remain unresolved with regard to the physiology of asparagus. Do the low light compensation and saturation points negatively impact plant performance? How do temperature, light and water stress interact to affect photosynthesis and how do they influence long-term plant performance? Further study of this plant is needed to address these questions.

CARBOHYDRATE PRODUCTION

Asparagus, like other members of the *Liliaceae* group, accumulates fructans (fructose yielding sugars when hydrolysed) as stored carbohydrates (Meier and Reid, 1982; Pollock, 1986). Scott *et al.* (1939) reported that the fleshy roots of asparagus contained on average 35–40% non-reducing sugars, 5–7% reducing sugars and 3–4% starch. Pressman *et al.* (1989) described the carbohydrate content of the shoot. Asparagus fern manufacture sucrose, glucose and fructose but no fructans or starch. Only after these sugars are transferred to the roots does fructan synthesis occur. Besides fructans, root carbohydrates also include smaller amounts of glucose, fructose and sucrose (Pressman *et al.*, 1993).

Asparagus storage carbohydrates are fructo-oligo or polysaccharides, with the parent oligosaccharides consisting of 10% glucose and 90% fructose, having a molecular weight not exceeding 4000 (Shelton and Lacy, 1980). The structures of some of the smaller fructans with degree of polymerization less than or equal to 4 (DP ≤ 4) and their biosynthesis have been characterized for asparagus (Shiomi *et al.*, 1976, 1979, 1991; Shiomi, 1981, 1989). Several frutosyl transferase enzymes have been extracted and detailed properties and substrate specificities of each determined. This work has provided a comprehensive description of the *in vitro* synthesis of fructans found in asparagus root tissue (Pollock, 1986; Pollock and Chatterton, 1988; Shiomi, 1989, 1992).

Cairns (1992) reported that 81–98% of the total soluble fructo-oligosaccharides had a DP ≥ 5. This is consistent with the work of Shelton and Lacy (1980) but higher than values found in other published works (Shiomi, 1992). Much of the difference between these studies is attributed to time of the year when the samples were taken, the difficulty in quantifying the larger fructans (DP ≥ 9) and the methods used to measure the different fructans.

There has been a great deal of work done on carbohydrate changes that occur in asparagus over a growing season (Scott *et al.*, 1939; Shelton and Lacy, 1980; Taga *et al.*, 1980; Haynes, 1987; Wilcox-Lee and Drost, 1991) (see Fig. 18.2b). Use of carbohydrate concentration in asparagus roots as an indicator of plant vigour has been suggested by several workers (Shelton and Lacy, 1980; Pitman and Sanders, 1985). Robb (1984) noted that most carbohydrate information is expressed as a percentage of total sugars or reducing and non-reducing sugars. The use of percentage data makes interpretation of carbohydrate contents difficult as information on changes in total root mass is lacking. Changes in crown mass are a major factor affecting changes in the total amount of stored carbohydrates (Haynes, 1987). Crown mass, as well as the carbohydrate content, would be a better indicator of plant vigour (Scott *et al.*, 1939; Haynes, 1987) than carbohydrate concentration alone.

There are seasonal differences in the carbohydrate concentrations in asparagus roots (see Fig. 18.2b). Carbohydrate levels in fleshy roots decrease slowly during the harvest period as sugars are used for spear growth (Scott *et al.*, 1939; Shelton and Lacy, 1980; Taga *et al.*, 1980; Haynes, 1987; Wilcox-Lee and Drost, 1991). Following harvest, as ferns expand, carbohydrate content drops rapidly. After the fern is established, carbohydrates increase rapidly and reach pre-harvest levels during mid to late summer. Small fluctuations noted in the sugar levels of roots during the summer have been attributed to the periodic production of new shoots.

As temperatures decrease in the autumn, fern begin to senesce and dormancy is induced. Initially, fern sucrose levels increase as low temperatures decrease photosynthate translocation (Pressman *et al.*, 1993). Eventually, fern sucrose levels fall as shoots die back. In roots, low temperatures slow fructan accumulation as photosynthate translocation from the fern slows. Root sucrose levels increase during the dormant period as fructans are broken down. Increasing sucrose levels in the crown may serve as a sprouting signal for dormant buds when temperatures rise.

Some suggest that the storage roots formed in the previous year supply most of the carbohydrates used for growth in the following year (Robb, 1984). This would be expected since new storage roots are in close proximity to the buds which they would supply. However, the small size of these new roots suggests that other roots on the rhizome need to contribute carbohydrates as well to sustain growth. Given the long life span of fleshy roots, their large mass and the changes in fructan levels that occur, it is safe to assume that the majority of storage roots contribute carbohydrate to elongating spears and fern. Until we have more information on the dynamics of root growth, their life expectancy and functionality, and can accurately track carbohydrate movement from fern to roots and from roots to spears, these questions will remain unanswered.

Altering the length of the harvest season from 3 to 6 weeks in 2-year-old asparagus plants significantly lowered root carbohydrate levels during the fern development period (Shelton and Lacy, 1980). Others have made similar findings (Haber, 1932; Takatori *et al.*, 1970b). The effect of longer harvest periods was to increase the severity of the depletion and to decrease the time available for the production of carbohydrates needed for the next year's growth. Extending the harvest duration also reduced the total dry weight partitioned to the crown when compared to shorter harvest intervals (Scott *et al.*, 1939). It was estimated that one-third of the crown dry matter lost during harvest was associated with spear growth and that the remaining two-thirds of the dry matter loss was due to respiration, though these losses did not account for carbohydrates utilized for root growth. Fern growth must remain vigorous throughout the growing season and must be maintained for a sufficient duration if carbohydrate levels in the crown are to be maximized. Yields also decrease when harvests occur at other times of the year (Brasher, 1956; Dufault, 1991; Farish, 1937); thus careful plant management is needed to maintain productivity.

Questions were raised by Morse (1916) and Culpepper and Moon (1939a) about the ability of the developing spear to manufacture some of the carbohydrates necessary for growth. Morse (1916) concluded that the increase in sugar levels in asparagus spears during the day was the result of photosynthesis by the elongating stalks. Culpepper and Moon (1939a) disagreed with that conclusion citing that the gradients of carbohydrates and sugars in the spears were what would be expected for a rapidly developing meristem. Scott *et al.* (1939) confirmed Culpepper and Moon's findings and indicated that the young developing spear was almost totally dependent on the carbohydrate reserves stored in the crown. These results are further supported by the findings that spears can refix respired CO_2 but due to their low chlorophyll content and reduced stomatal density have a low assimilation rate (Downton and Torokfalvy, 1975; Lin and Hung, 1978).

Further physiological research on the factors that affect the growth of buds, fern and roots, biochemistry of carbohydrate production and accumulation, redistribution and utilization, and how these interact with the environment and present cultural practices are needed. As we understand better how photosynthesis and carbohydrate accumulation function and how these interact with environment, changes in cultural practices will be made that should improve plant performance.

CROP PRODUCTIVITY

Although we are beginning to gain a better understanding of the physiology of asparagus, growth and performance is also influenced by harvest duration, environment, plant nutrition, plant growth regulators and the differences between male and female plants. While a good deal of information is available on the effects of cultural practices on crop yield, how these interact with the physiology of asparagus is not fully understood.

HARVEST DURATION

Asparagus is traditionally harvested in the spring in temperate climates. Harvesting is a labour intensive operation (Brown, 1984) requiring daily or twice daily harvests for spears to meet market standards. Harvest duration extends for 2–8 weeks depending on the age of the planting. Prolonging the harvest period increased asparagus yield (number and weight of spears) in the harvest year but decreased the percentage of large-sized spears (Jones, 1932). Other studies have shown yield decreases in the year following the extended harvest season (Haber, 1932; Takatori *et al.*, 1970b; Williams and Garthwaite, 1973; Shelton and Lacy, 1980). Long harvest periods lead to excessive carbohydrate depletion of the roots, reduced number of buds remaining for canopy establishment, decreased size of the canopy established, delayed postharvest carbohydrate accumulation and bud initiation, thereby reducing long-term plant performance.

TEMPERATURE

Soil and air temperature play an important role in bud break and spear growth. A soil temperature of at least 5°C is required for bud break (Bouwkamp and McCully, 1972; Nichols and Woolley, 1985). The temperature threshold for bud break varies with plant age, since 1- and 2-year-old plants produce spears earlier in the spring than older plants (Robb, 1984). There are also strong cultivar differences in early season growth, with the French varieties emerging earlier in the spring than American varieties. Cultivar evaluations are generally conducted to identify cultivars that perform well in specific environments.

Since spear growth is initially a function of soil temperature, increasing the planting depth delays production, slows the rate of spear development (Takatori *et al.*, 1974; Lindgren, 1990; McCormick and Thomsen, 1990), and decreases the total yield but increases the number and weight of large-sized spears (McCormick and Thomsen, 1990). Plastic mulches are widely used in other vegetables but not asparagus, to increase soil temperature and advance plant development. Plastic mulches increased marketable and total yield but not spear number when used to grow white asparagus (Makus and Gonzalez, 1991) but create harvesting problems for green asparagus as spear emergence would be restricted. In some production areas, soil is mounded over the crown in a raised bed (Franklin *et al.*, 1981). Raised beds warm faster in the spring, provide better drainage in heavy soils, and tend to have lower root disease incidence but may expose the emerging spears to low temperature conditions.

Spear growth is most active and most responsive to temperature in the zone just below the tip (Culpepper and Moon, 1939a,b; Tiedjens, 1924). Consequently, spear growth is influenced more by air temperature than soil temperature after emergence from the soil (Bouwkamp and McCully, 1972). Early emergence from shallow planted crowns exposes spears (Arora *et al.*, 1992; Lindgren, 1990) and crowns (Lindgren, 1990; Valenzuela and Bienz, 1989) to injury from spring frosts. Since early emerging spears are generally the largest, frosts often reduce crop yields. Where conditions are not as harsh, the crop is marketed earlier in the spring though spear size is often reduced (Takatori *et al.*, 1974; Lindgren, 1990).

Spear growth rate increases linearly from 10 to 32°C (Culpepper and Moon, 1939a; Blumenfield *et al.*, 1963; Nichols and Woolley, 1985) but curvilinearly at temperatures below 10°C. Spears and ferns of different heights

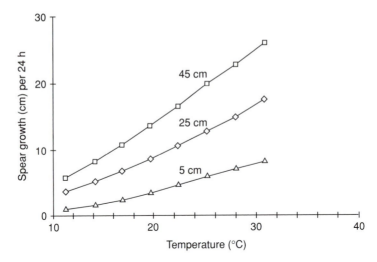

Fig. 18.4. Effect of temperature on the elongation of asparagus stalks of various heights between 5 and 45 cm. The plants were grown under field conditions and the results expressed as growth in centimetres per 24 h (Culpepper and Moon, 1939).

differ in their response to temperature (Culpepper and Moon, 1939b) (Fig. 18.4). Regardless of the temperature, the percent increase in growth during a 24-h period was greatest for short spears (5 cm) and slowed as spear height increased. The rapid growth rate of young spears especially at high temperatures will require more frequent harvests if spear quality is to be maintained. High temperatures, while promoting rapid spear growth, also encourage 'feathering' or lateral branch growth which decreases marketability (Roth and Gardner, 1989). As temperatures increase the spear height at which branching occurs decreases (Working, 1924). Cultivars vary greatly in the height at which lateral branch formation occurs. This is an important selection criterion for adaptation to specific growing areas. Fern age also influences stalk sensitivity to temperature, as old and very young stalks are less sensitive than recently mature stalks to changes in temperature (Culpepper and Moon, 1939b). This response may be due to reduced water or photosynthate transport to the growing region of mature stalks at higher temperatures.

MOISTURE

Asparagus has been classified as a drought-tolerant plant but most research indicates that it responds favourably to added water under moisture deficit situations. Its reputation as a drought-tolerant plant may come from the difficulty in observing wilting of the woody ferns, unlike other broadleaf plants. The best growth of seedling asparagus was when soils were maintained close to field capacity (Wilcox-Lee, 1987). Thus, it was concluded that consistent irrigation would be beneficial in newly established plantings. Irrigation during establishment improved transplant and crown survival and improved yields during early harvest years, confirming the benefits of irrigation on new plantings (Sterrett *et al.*, 1990).

Irrigation is usually not necessary during harvest periods since plant water use is low (Roth and Gardner, 1990) and moisture can cool soils resulting in slow spear growth and decreased yields (Cannell and Takatori, 1970). However, dry soils in the spring can reduce the number of times the crop is harvested, which also reduces yield (Brasher, 1956). Irrigation may be applied in the spring to minimize spear damage from blowing sand, firm the soil to improve foot traffic for harvest, and cool the soil during hot weather to prevent spears from becoming 'feathery' before reaching a marketable height (Roth and Gardner, 1989).

Water shortages during top growth result in wilting of the fern (Van Bakel and Kerstens, 1971) which limits plant growth (Takatori *et al.*, 1970a; Kaufmann, 1977; Hartmann, 1981), and ultimately decreases photosynthetic capacity and carbohydrate metabolism so less carbohydrates are stored for future use (Thompson, 1942). Total yield decreased in the year following extremely dry seasons (Farish, 1937), suggesting reduced growth and carbohydrate storage, and fewer buds initiated. Irrigation increases yield by increasing the number (Takatori *et al.*, 1970a; Sterrett *et al.*, 1990) or the size (Hanna and Doneen, 1958; Hartmann, 1981) of spears. When precipitation levels increased during July and August, asparagus yields were high in the following year (Hartmann, 1985; Hartmann *et al.*, 1990b). When precipitation levels were high during September, bud break and fern growth occurred later in the year than desirable, which decreased yield the following year.

Low soil moisture levels also influence the carbohydrate balance of asparagus. Drought conditions induce fern senescence and decrease the glucose, sucrose and fructose content of the fern. At the same time, fructan levels decrease in the roots as respiration rates increase (Pressman *et al.*, 1989). Since photosynthetic efficiency is reduced under the imposed stress conditions, changes in fern and root carbohydrate balance may be expected. More work is needed to determine the effect of water stress in one year on performance in subsequent years.

In locations where temperatures do not restrict growth, dormancy induction and growth resumption are often imposed by regulating water supply to asparagus (Thompson, 1942; Roth and Gardner, 1989, 1990). In these areas (e.g. Arizonia, California, Peru), water requirements are high since temperatures are often excessive (Cannell and Takatori, 1970; Roth and Gardner, 1989, 1990; Toledo, 1990). In Peru, commercial fields are brought into production by stimulating spear growth with irrigation (Toledo, 1990). After a 1-month harvest period, the fern is allowed to grow for up to 4 months. Dormancy is then induced by withholding water. Once the fern is dry, the production cycle begins again by the addition of water. In the humid tropics, asparagus is produced by the 'mother stalk' technique. Several fern per plant are allowed to grow and the remaining spears are harvested. New fern are generated every 3 or 4 months as the older fern senesce and efficiency of assimilation decreases (Lin and Hung, 1978). With proper scheduling, asparagus could be produced every month of the year with either of these systems.

NUTRITION

Much work on the nutritional needs of asparagus has focused on transplant growth (Benson and Fisher, 1983; Fisher and Benson, 1983; Precheur and Maynard, 1983; Adler et al., 1984). Only limited research has been done on the long-term nutritional needs of the crop. Brown and Carolus (1965) reported that the annual removal of nitrogen (N), phosphorus (P) and potassium (K) amounted to 7, 2 and 5%, respectively, of the applied fertilizer. The rest of the NPK in the fern is recycled by translocation to the rhizome and roots during senescence or to the soil when the tops are mowed in the autumn or spring. Since spear growth is initiated from stored carbohydrates and amino acids, nutrients applied in the previous year will have the most effect on future plant performance. Some nutrients are absorbed in the spring (Morse, 1916), however most of the utilization occurs after harvest when fern, bud and root growth take place (see Fig. 18.2a).

Nitrogen has the greatest effect on yield when it is applied after harvest rather than before (Hartmann and Wuchner, 1979; Sanders and Bandele, 1985; Ledgard et al., 1992). In temperate areas, N applications average 50 kg ha^{-1} (Zandstra and Price, 1979; Franklin et al., 1981; Sandsted et al., 1985). This N is usually applied before or just after the harvest period. In arid irrigated areas, 350–550 kg N ha^{-1} are often applied (Cannell and Takatori, 1970; Roth and Gardner, 1989, 1990). Smaller, more frequent applications of nitrogen are needed in areas where growing seasons are long and irrigation frequency and rates are high.

Ledgard et al. (1992, 1994) used ^{15}N-labelled nitrogen to study the uptake and distribution of N within a mature asparagus planting over 2 years. They estimated that approximately 700 kg N ha^{-1} was stored in asparagus rhizome and roots. Less than 6% (30–40 kg N ha^{-1}) of this N was later removed by the harvested spear. Their study showed that only small amounts of N were taken up from the soil during the harvest period. In addition, their results showed that nitrogen applied before or during the harvest period was not utilized by the plant until fern growth occurred after harvest. This is not surprising since there is very little root growth during that time period (Wilcox-Lee and Drost, 1991) (see Fig. 18.2a). Nitrogen uptake from the soil increased rapidly as fern developed and accumulated in amounts equivalent to 5 kg N ha^{-1} day^{-1} during an 8-week period (Ledgard et al., 1994). Timing N applications to meet this high demand period is critical if performance is to be optimized and leaching losses minimized. As fern senesced in the autumn, approximately 90% of the stored ^{15}N was remobilized and translocated to the crown (rhizome and fleshy roots). This labelled N was detected in harvested spears the following spring. These findings suggest that remobilized ^{15}N from last year's foliage is in a more readily available form than N from other pools in the plant. This may be expected since new roots formed in the previous growing season are in close proximity to those buds that develop into spears and these roots would supply elongating spears with some of the carbohydrates and N needed for growth (Robb, 1984). However, other roots on the rhizome would also need to supply carbohydrates and N for spear growth.

Arginine and asparagine are the major free amino acids involved in nitrogen storage in the fleshy roots of asparagus (Fiala and Jolivet, 1982). Amino acids, like carbohydrates, decreased rapidly as new shoots developed in the spring (see Fig. 18.2b). These amino acids would supply the nitrogen needed for the growth and metabolism of new spears and fern in spring. However, accumulation of amino acids in the fleshy roots to their pre-harvest levels required most of the fern growing period (Haynes, 1987). Increases in root amino acid levels are associated with the senescence of the shoots and a re-mobilization of nitrogen out of the dying fern. It remains unclear how nitrogen nutrition affects amino acid utilization during the growth of asparagus. Presumably, fertilizer applications after harvest would supply the plant with the nitrogen needed to ensure vigorous, healthy fern growth. The failure of alternative harvest sequences to produce sustained high yields may be due to the plant's inability to store adequate N after the harvest period.

Asparagus has a low P and K requirement (Morse, 1916) though yield responses to added P have been reported (Clore and Stanberry, 1946; Hartmann and Wuchner, 1979). Phosphorus additions are generally made prior to planting the crop (Dean et al., 1993; Franklin et al., 1981; Sandsted et al., 1985; Zandstra and Price, 1979) with additional P being applied as warranted in later years. Sanders and Bandele (1979) reported yields increase as K levels increase from 0–93 kg ha^{-1}. Additional K applications should be made based on a soil test (Zandstra and Price, 1979; Franklin et al., 1981; Sandsted et al., 1985). Trace minerals needs vary and depend on local soil conditions and plant needs (Brown and Carolus, 1965; Dean et al., 1993; Franklin et al., 1981).

Tissue testing may not detect asparagus crop nutrient needs. Brown and Carolus (1965) reported that although growers have vastly different fertilization practices there were few differences in fern nutrient content. Hartmann et al. (1990a) also reported poor correlation between soil and fern nutrient content but good correlation between soil and root nutrient content. This suggests that roots rather than fern be used to determine fertilizer needs in asparagus.

PLANT GROWTH REGULATORS

Asparagus is generally hand harvested though different mechanical harvesters have been used (Kepner, 1959; Mears *et al.*, 1969). Since harvesting can account for more than half of the cost of production (Kepner, 1971), efforts to synchronize spear emergence have been undertaken. Synchronized emergence through controlled bud break would allow mechanical harvesting to become economically viable (Brown, 1984). Several plant growth substances influence bud break and spear growth (Dedolph *et al.*, 1963) and may be useful in asparagus production systems.

Abscissic acid (ABA) often acts as a growth inhibiting substance in many plants. ABA levels in asparagus buds were high during fern growth and peak in winter (Fig. 18.5). In summer, ABA is presumably produced in the fern and transported with the assimilate supply to the roots (Matsubara, 1980). It has been speculated that ABA may regulate assimilate partitioning thus maintaining apical dominance and restricting bud elongation (Kojima *et al.*, 1993). High levels in winter appear to control dormancy. As ABA levels decrease in the spring, apical dominance is released and buds begin to grow. Matsubara (1980) noted that resting buds contain three times more ABA than sprouting buds. This ABA was produced by the growing spear, transported to the other buds where it inhibited sprouting. This agrees with reports that only one bud per bud cluster is capable of growing at a time (Nichols and Woolley, 1985).

Rapidly growing asparagus spears contain high levels of ABA (Matsubara, 1980; Kojima *et al.*, 1993) with concentrations being greatest near the growing tip and decreasing basipetally within the spear. ABA appears to promote sink strength or encourage phloem unloading in areas where assimilates are needed for growth (Kojima *et al.*, 1993). In asparagus spears, this is near the growth apex of either the main stem or a lateral branch.

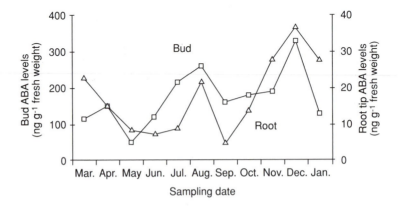

Fig. 18.5. ABA levels in crown buds and root tips from March until January (Matsubara, 1980).

Gibberellic acid promotes growth of asparagus buds. Apical dominance was weakened by additions of gibberellic acid which increased harvested spear number (Kretschmer and Hartmann, 1979) and promoted spear growth (Tiburcio, 1961). In other studies, differences in concentration, application method, and exposure time were cited as reasons for GA having no impact on spear growth (Mahotiere, 1976). Although auxins are known to be involved in promoting plant growth, indoleacetic acid inhibits spear emergence (Tiburcio, 1961). Gibberellic acid applications after indoleacetic acid were effective in promoting spear elongation where it previously had been inhibited. Application of napthaleneacetic acid to crowns had no effect on spear number or growth (Kretschmer and Hartmann, 1979). Ancymidol, a gibberellic acid synthesis inhibitor, promoted root and shoot growth in tissue-cultured plants (Chin, 1982). In seedling asparagus ancymidol decreased bud number and fern dry weight, though it increased dry weight partitioning to crowns in older transplants (Adler *et al.*, 1984). Enhancement of crown rather than fern dry weight may be beneficial for plants to withstand the stresses of transplanting. In greenhouse and field trials, ethephon applied to crowns over a range of concentrations and time periods did not affect the time of spear emergence (Mahotiere, 1976). However, high concentrations of ethephon and longer dip times increased spear number and fresh weight but not spear diameter.

Other plant growth regulators have been tried in an attempt to stimulate spear growth. Chloroflurenol increases spear growth but causes spear deformation (Dean, 1978). Benzyladenine stimulated spear emergence when applied to healthy fern (Mahotiere *et al.*, 1993). This could be useful where summer or autumn harvesting regimes are practised or when the mother fern technique is used. While some of the plant growth regulators may be useful in asparagus production systems, difficulty in applying the materials to the crown and variable results when applied limits their effectiveness. Continued work in this research area is still necessary.

SEX DIFFERENCES

One of the more important advances in asparagus research and production has been the development of all male hybrids. Male plants are more desirable because they have greater marketable yields (Tiedjens, 1924; Ellison and Schermerhorn, 1958; Moon, 1976), produce more spears but of a smaller size (Tiedjens, 1924; Ellison and Sheer, 1959), have earlier season production (Tiedjens, 1924; Ellison and Schermerhorn, 1958; Ellison *et al.*, 1960), less lag time between emergence of individual spears (Tiedjens, 1926), longer production season (Moon, 1976), and a lower mortality rate (Moon, 1976), when compared to female plants.

The reason for the lower yields and poor longevity of the female plants has not been clearly identified (Bouwkamp and McCully, 1972; Fiala and Joliet, 1979). A 1 : 1 ratio of males to females generally exists at planting

(Moon, 1976) which increases to 2.5 : 1 in older plantings (Yeager and Scott, 1938). Factors other than competition are responsible for the high mortality in females (Bouwkamp and McCully, 1972). What these factors are has not been determined.

Sex differences also play a role in spear growth suppression and strength of the apical dominance signal. Tiedjens (1926) noted that the interval between the harvest of the first bud and the beginning of growth in the next bud in a bud cluster is longer for pistillate than staminate plants. Staminate plants have higher carbohydrate content than pistillate plants and utilize a greater part of it in spear production (Fiala and Joliet, 1979). The increase in carbohydrate utilization may partly explain the higher yield of male plants.

Benson (1982) noted that female plants had greater fern area than male plants. These differences may be related to differences in assimilation rate, carbohydrate accumulation, and productivity. Others found a 28% decrease in the assimilation rate of female compared to male plants (Sawada *et al.*, 1962). This suggests photosynthetic production may not be as efficient in female plants even though leaf areas may be greater. Lower assimilation rates would decrease carbohydrate production, lower productivity, and possibly shorten longevity. Others suggested that differences in the performance of male and female plants is due to berry and seed production by female plants (Jones, 1932; Jones and Robbins, 1928), though no studies have specifically looked at this. Further work is needed to understand better the physiological differences between male to female asparagus plants.

PHYSIOLOGICAL DISORDERS

ALLELOPATHY

Yield reductions commonly occur in old asparagus plantations (Grogan and Kimball, 1959; Johnston *et al.*, 1979). Much of this reduction is attributed to stand reductions. Replanting new seedlings into these fields is often unsuccessful. Seed treatments and soil fumigation have been used in old asparagus fields to improve stands but stunting and wilting still occurs (Wiebe, 1967). Evidence shows that substances produced by the asparagus plant are both autotoxic (self toxic) and allelopathic (Yang, 1982; Young, 1984; Shafer and Garrison, 1986; Hartung *et al.*, 1990). These substances contribute to replant problems. Early evidence suggested that growth reductions were exclusive of any disease problems even though rhizome and root rot organisms were isolated from infected fields.

It has been known for some time that infection of the rhizome and roots by *Fusarium* contributes to the decline in production of established asparagus fields (Grogan and Kimball, 1959; Johnston *et al.*, 1979). Recently, an interaction between the autotoxin or allelopathic compounds and *Fusarium* has been observed (Hartung *et al.*, 1989; Peirce and Colby, 1987). Hartung *et al.* (1989) demonstrated that allelochemicals of asparagus have direct physiological and biochemical effects on asparagus that predisposes them to *Fusarium* diseases. Damaged tissues had higher levels of electrolyte efflux suggesting more permeable membranes. In addition, roots had decreased peroxidase activity which has been associated with increased susceptibility to disease. Root cells also had decreased respiration rates which suggests reduced biochemical activity. Thus, infection of asparagus by *Fusarium* is increased because *Fusarium* growth is stimulated by the autotoxic materials leaking from the root.

Allelopathic compounds produced and released by growing or dead asparagus root tissue persists in soil for longer than 4 years (Hanna, 1938). Others have shown that in the laboratory, toxins begin to decline in 3–8 months (Hartung *et al.*, 1989; Shafer and Garrison, 1986; Yang, 1982). It appears that with adequate time between terminating the old planting and replanting the new crop, soil effects can be alleviated. The loss of toxicity may be due to leaching of the water soluble allelochemicals out of the rhizosphere or through microbial breakdown. The length of time necessary for detoxification of the soil would ultimately depend upon the amount of root tissue present in the field (Shafer and Garrison, 1986).

Early work on the chemical mechanism of autotoxicity and allelopathy found that asparagusic acid and related compounds isolated from asparagus inhibited the germination and growth of other crop species (Yanagawa *et al.*, 1972, 1973). However, asparagusic acid was not toxic to asparagus. Recently, several cinnamic acids have been implicated as being autotoxic or allelopathic to asparagus (Hartung *et al.*, 1990; Miller *et al.*, 1991). These include caffeic, ferulic or methylenedioxycinnamic acid. Cinnamic acids have been isolated from asparagus roots and are known to inhibit germination of other crops (Hartung *et al.*, 1990; Rice, 1984) and asparagus (Miller *et al.*, 1991; Peirce and Miller, 1993).

Root tips and epidermal cells of asparagus radicles are altered after exposure to cinnamic acids (Peirce and Miller, 1993). Ferulic acid was more toxic than caffeic or methylenedioxycinnamic acid. However, the autotoxicity of these materials was more severe when combinations of ferulic and either caffeic or methylenedioxycinnamic acid were present. Extensive changes to root epidermal cells and precocious root hair development associated with exposure to cinnamic acids may be the changes necessary for *Fusarium* penetration of asparagus root tissue in infested soils.

Soil conditions, plant stress and stages of plant decay all affect the level of toxicity. Decline in asparagus is aggravated by plant stress and stress is thought to weaken the plants to *Fusarium* (Nigh, 1990) thus the synergistic effect. Asparagus allelochemicals are directly toxic to emergence and growth of asparagus seedlings and enhance the growth environment of *Fusarium*. Since *Fusarium* affects stressed plants more, when asparagus roots die, more toxins are released, infection rate increases and more roots die. Thus, replant problems in old asparagus fields occur from exposure of new crowns or seedlings to *Fusarium* and the autotoxins present in the soil.

CONCLUDING REMARKS

Asparagus is one of the few perennial vegetables grown commercially in temperate climates. In most areas harvesting does not occur until the third year (2 years after planting a 1-year-old crown). Full production is generally achieved by the seventh year. Traditionally asparagus is harvested in the spring, fern grows in summer, and the plant becomes dormant in winter. How these time periods are managed over many years will affect the longevity of the plant.

Productivity in asparagus is tied to the length of the harvest period, fern vigour during summer growth, and conditions that stress the plant at any time during the growing season. Environmental and cultural conditions imposed on the plant in one year will impact plant performance in subsequent years. This could be manifested by a decrease in fern and crown growth, reduced storage of carbohydrates, or a reduction in number or size of buds. Since these growth processes all occur simultaneously each year, proper management of the crop during the fern growth period is critical.

While significant advances have been made in breeding productive asparagus, we are just beginning to understand the long-term implications of stresses on future productivity. As our understanding of the physiology of asparagus increases, we should be better able to use this information to control bud break and spear growth, regulate carbohydrate accumulation and utilization, and exploit the differences between male and female plants in such a way that crop productivity can be increased and prolonged.

REFERENCES

Adler, P.R., Dufault, R.J. and Waters, L. (1984) Influence of nitrogen, phosphorus, and potassium on asparagus transplant quality. *HortScience* 19, 565–566.

Anon. (1992) *Vegetables: 1991 Summary.* United States Department of Agriculture National Agriculture Statistics Service, Washington DC.

Arora, R., Wisniewski, W.E. and Makus, D.J. (1992) Frost hardiness of *Asparagus officinalis* L. *HortScience* 27, 823–824.

Bailey, L.H. (1914) *The Principles of Vegetable Gardening.* Macmillan, New York.

Benson, B.L. (1982) Sex influences on foliar trait morphology in asparagus. *HortScience* 17, 625–627.

Benson, B.L. and Fisher, K.J. (1983) Effect of cell volume, plant density and nitrogen on asparagus seedling growth. *Scientia Horticulturae* 21, 105–112.

Benson, B.L. and Takatori, F.H. (1980) Partitioning of dry matter in open-pollinated and F1 hybrid cultivars of asparagus. *Journal of the American Society for Horticultural Science* 105, 567–570.

Blasberg, C.H. (1932) Phases of the anatomy of *Asparagus officinalis. Botanical Gazette* 94, 206–214.

Blumenfield, D., Meinken, K.W. and LeCompte, S.B. (1963) A field study of asparagus growth. *Proceedings of the American Society for Horticultural Science* 77, 386–392.

Bouwkamp, J.C. and McCully, J.E. (1972) Competition and survival in female plants of *Asparagus officinalis*. *Journal of the American Society for Horticultural Science* 97, 74–76.

Brasher, E.P. (1956) Effects of spring, summer, and fall cuttings of asparagus on yield and spear weight. *Proceedings of the American Society for Horticultural Science* 67, 377–383.

Brown, L.D. and Carolus, R.L. (1965) An evaluation of fertilizer practice in relation to nutrient requirement of asparagus. *Proceedings of the American Society for Horticultural Science* 86, 332–337.

Brown, S.W. (1984) An evaluation of on-farm mechanisation for harvesting and handling asparagus. *New Zealand Agricultural Science* 18, 34–37.

Bussell, W.T., Nichols, M.A., McCormick, S.J., Alsach, P., Nikoloff, A.S. and Brash, D.W. (1985) Asparagus cultivar evaluation in New Zealand. In: Lougheed, E.C. and Tiessen, H. (eds) *Proceedings of the 6th International Asparagus Symposium*, University of Guelph, Guelph, Ontario, Canada, pp. 52–62.

Cairns, A.J. (1992) A reconsideration of fructan biosynthesis in storage roots of *Asparagus officinalis* L. *New Phytologist* 120, 463–473.

Cannell, G.H. and Takatori, F.H. (1970) Irrigation-nitrogen studies in asparagus and the measurement of soil moisture changes by the neutron method. *Soil Science Society of America Proceedings* 34, 501–506.

Chin, C. (1982) Promotion of shoot and root formation in asparagus *in vitro* by ancymidol. *HortScience* 17, 590–591.

Clore, W.J. and Stanberry, C.O. (1946) Further results of asparagus fertilizer studies in irrigated central Washington. *Proceedings of the American Society for Horticultural Science* 46, 296–298.

Colman, B., Mawson, B.T. and Espie, G.S. (1979) The rapid isolation of photosynthetically active mesophyll cells from asparagus cladophylls. *Canadian Journal of Botany* 57, 1505–1510.

Culpepper, C.W. and Moon, H.H. (1939a) Effects of temperature upon the rate of elongation of the stems of asparagus grown under field conditions. *Plant Physiology* 14, 255–270.

Culpepper, C.W. and Moon, H.H. (1939b) Changes in the composition and rate of growth along the developing stem of asparagus. *Plant Physiology* 14, 677–698.

Dean, B.B. (1978) Control of number and size of asparagus spears. In: *Michigan State University Horticulture Report 41*, 5. Department of Horticulture, Michigan State University, East Lansing, Michigan.

Dean, B.B., Boydston, R., Cone, W., Johnson, D., Ley, T.W., Mink, G., Parker, R., Stevens, R., Sorensen, E. and Van Denburgh, R. (1993) Washington asparagus production guide. *Washington State University Extension Bulletin 997*. Washington State University, Pullman, Washington.

Dedolph, R.R., Tutus, C.F., Downes, J.D. and Sell, H.M. (1963) Asparagus and growth regulators. *Proceedings of the American Society for Horticultural Science* 82, 311–315.

Downton, W.J.S. and Torokfalvy, E. (1975) Photosynthesis in developing asparagus plants. *Australian Journal of Plant Physiology* 2, 367–375.

Drost, D.T. and Wilcox-Lee, D. (1990) Effect of soil matric potential on growth and physiological responses of greenhouse grown asparagus. *Acta Horticulturae* 271, 467–476.

Dufault, R.J. (1991) Response of spring and summer harvested asparagus to harvest pressure. *HortScience* 26, 845–847.

Dufault, R.J. and Greig, J.K. (1983) Dynamic growth characteristics in seedling asparagus. *Journal of the American Society for Horticultural Science* 108, 1026–1030.

Ellison, J.H. and Scheer, D.F. (1959) Yield related to bush vigor in asparagus. *Proceedings of the American Society for Horticultural Science* 73, 339–344.

Ellison, J.H. and Schermerhorn, L.G. (1958) Selecting superior asparagus plants on the basis of earliness. *Proceedings of the American Society for Horticultural Science* 72, 353–359.

Ellison, J.H., Scheer, D.F. and Wagner, J.J. (1960) Asparagus yield as related to plant vigor, earliness and sex. *Proceedings of the American Society for Horticultural Science* 75, 411–415.

Farish, L.R. (1937) Fall cutting of asparagus compared with spring cutting under Mississippi conditions. *Proceedings of the American Society for Horticultural Science* 35, 693–695.

Fiala, V. and Jolivet, E. (1979) Variations in biochemical composition of *Asparagus officinalis* L. roots in relation with sex expression and age. In: Reuters, G. (ed.) *Proceedings of the 5th International Asparagus Symposium.* Federal Republic of Germany, pp. 74–81.

Fiala, V. and Jolivet, E. (1982) Quantitative changes in nitrogen compounds and carbohydrates in male and female asparagus roots during the first year of growth. *Agronomia* 2, 735–740.

Fisher, K.J. (1982) Comparisons of the growth and development of young asparagus plants established from seedling transplants and by direct seeding. *New Zealand Journal of Experimental Agriculture* 10, 405–408.

Fisher, K.J. and Benson, B.L. (1983) Effects of nitrogen and phosphorus nutrition on the growth of asparagus seedlings. *Scientia Horticulturae* 21, 105–112.

Franklin, S.J., Bussell, W.T., Cox, T.I. and Tate, K.G. (1981) Asparagus: establishment and management for commercial production. *New Zealand Aglink, Horticulture Produce and Practice 125.*

Grogan, R.G. and Kimball, K.A. (1959) The association of *Fusarium* wilt with asparagus decline and replant problems in California. *Journal of Chemical Ecology* 9, 1163–1174.

Haber, E.S. (1932) Effect of size of crown and length of cutting season on yields of asparagus. *Journal of Agricultural Research* 45, 101–109.

Hanna, G.C. (1938) Fertilizer trials with asparagus on peat sediment soils. *Proceedings of the American Society for Horticultural Science* 36, 560–561.

Hanna, G.C. and Doneen, L.D. (1958) Asparagus irrigation studies. *California Agriculture* 12, 8, 14–15.

Hartmann, H.D. (1981) The influence of irrigation on the development and yield of asparagus. *Acta Horticulturae* 119, 309–316.

Hartmann, H.D. (1985) Possibility of predicting the amount of asparagus yield in middle Europe. In: Lougheed, E.C. and Tiessen, H. (eds) *Proceedings of the 6th International Asparagus Symposium.* University of Guelph, Guelph, Ontario, Canada, pp. 318–323.

Hartmann, H.D and Wuchner, A. (1979) Long standing fertilizing experiments in asparagus. In: Reuther, G. (ed.) *Proceedings of the 5th International Asparagus Symposium.* Federal Republic of Germany, pp. 221–229.

Hartmann, H.D., Hermann, G. and Altringer, R. (1990a) Evaluation of nutrient status of asparagus by leaf and root analysis. *Acta Horticulturae* 271, 433–442.

Hartmann, H.D., Hermann, G. and Kirchner-Neb, R. (1990b) Einfluss der Witterung während der Vegetationszeit auf den nächstjährigen Ertrag von Spargel. (The influence of climatic conditions in the growing season on the yield of asparagus in the subsequent year.) *Gartenbauwissenschaft* 55, 30–34.

Hartung, A.C., Nair, M.G. and Putnam, A.R. (1990) Isolation and characterization of phytotoxic compounds from asparagus (*Asparagus officinalis* L.) roots. *Journal of Chemical Ecology* 16, 1707–1718.

Hartung, A.C., Putnam, A.R. and Stephens, C.T. (1989) Inhibitory activity of asparagus root tissue and extracts on asparagus seedlings. *Journal of the American Society for Horticultural Science* 114, 144–148.

Haynes, R.L. (1987) Accumulation of dry matter and changes in storage carbohydrate and amino acid content in the first 2 years of asparagus growth. *Scientia Horticulturae* 32, 17–23.

Hexamer, F.M. (1914) *Asparagus.* Orange Judd, New York.

Hills, M.J. (1986) Photosynthetic characteristics of mesophyll cells isolated from cladophylls of *Asparagus officinalis* L. *Planta* 169, 38–45.

Hughes, A.R., Nichols, M.A. and Woolley, D. (1990) The effect of temperature on the growth of asparagus seedlings. *Acta Horticulturae* 271, 451–456.

Johnson, D.A. (1990) Effects of crop debris management on severity of stemphyllium purple spot of asparagus. *Plant Disease* 74, 413–415.

Johnston, S.A., Springer, J.K. and Lewis, G.D. (1979) *Fusarium moniliforme* as a cause of stem and crown rot of asparagus and its association with asparagus decline. *Phytopathology* 69, 778–780.

Jones, H.A. (1932) Effect of extending the cutting season on the yield of asparagus. *California Agricultural Experiment Station Bulletin* 535, 1–15.

Jones, H.A. and Robbins, W.W. (1928) Asparagus industry in California. *University of California Agricultural Experiment Station Bulletin.* Berkeley, California.

Kaufmann, F. (1977) Intensification of asparagus production by irrigation. *Gartenbauwissenschaft* 24, 73–74.

Kepner, R.A. (1959) Mechanical harvester for green asparagus. *Transactions of the American Society of Agricultural Engineers* 2, 84–87, 91.

Kepner, R.A. (1971) Selective and nonselective mechanical harvesting of green asparagus. *Transactions of the American Society of Agricultural Engineers* 14, 405–410.

Knaflewski, M. (1994) Yield predictions of asparagus cultivars on the basis of summer stalk characteristics. *Acta Horticulturae* 371, 161–168.

Kojima, K., Kuraishi, S., Sakurai, N., Itou, T. and Tsurusaki, K. (1993) Spatial distribution of abscisic acid in spears, buds, rhizomes and roots of asparagus (*Asparagus officinalis* L.). *Scientia Horticulturae* 54, 177–189.

Kretschmer, M. and Hartmann, H.D. (1979) Experiments in apical dominance with *Asparagus officinalis* L. In: Reuther, G. (ed.) *Proceedings of the 5th International Asparagus Symposium.* Federal Republic of Germany, pp. 235–239.

Ledgard, S.F., Douglas, J.A., Follett, J.M. and Sprosen, M.S. (1992) Influence of time of application on the utilization of nitrogen fertilizer by asparagus, estimated using ^{15}N. *Plant and Soil* 147, 41–47.

Ledgard, S.F., Douglas, J.A., Sprosen, M.S. and Follett, J.M. (1994) Uptake and redistribution of ^{15}N within an established asparagus crop after application of ^{15}N-labelled nitrogen fertilizer. *Annals of Botany* 73, 169–173.

Lin, A.C. and Hung, L. (1978) The photosynthesis of asparagus plant. *Memoirs of the College of Agriculture, National Taiwan University* 18, 88–95.

Lindgren, D.T. (1990) Influence of planting depth and interval to initial harvest on yield and plant response of asparagus. *HortScience* 25, 754–756.

Mahotiere, S. (1976) Response of asparagus crowns to ethephon and gibberellic acid. *HortScience* 11, 240–241.

Mahotiere, S., Johnson, C. and Howard, P. (1993) Stimulating asparagus seedling shoot production with benzyladenine. *HortScience* 28, 229.

Makus, D.J. and Gonzalez, A.R. (1991) Production and quality of white asparagus grown under opaque rowcovers. *HortScience* 26, 374–376.

Matsubara, S. (1980) ABA content and levels of GA-like substances in asparagus buds and roots in relation to bud dormancy and growth. *Journal of the American Society for Horticultural Science* 105, 527–532.

McCormick, S.J. and Thomsen, D.L. (1990) Management of spear number, size quality and yield in green asparagus through crown depth and population. *Acta Horticulturae* 271, 151–157.

Mears, D.R., Moore, M.J. and Parashuram (1969) Potential of mechanical asparagus harvesters. *Transactions of the American Society of Agricultural Engineers* 12, 813–815, 821.

Meier, H. and Reid, J.S.G. (1982) Reserve polysaccharides other than starch in higher plants. In: Loewus, F.A. and Tanner, W. (eds) *Plant Carbohydrates. Encyclopedia of Plant Physiology*, Vol. 13A. Springer-Verlag: New York, pp. 418–471.

Miller, H.G., Ikawa, M. and Peirce, L.C. (1991) Caffeic acid identified as an inhibitory compound in asparagus root filtrate. *HortScience* 26, 1525–1527.

Moon, D.M. (1976) Yield potential of *Asparagus officinalis* L. *New Zealand Journal of Experimental Agriculture* 4, 435–438.

Morse, F.W. (1916) A chemical study of the asparagus plant. *Massachusetts Agricultural Experiment Station Bulletin* 171, 265–296.

Mullendore, N. (1935) Anatomy of the seedling of *Asparagus officinalis* L. *Botanical Gazette* 97, 356–375.

Nichols, M.A. (1990) Asparagus the world scene. *Acta Horticulturae* 271, 25–31.

Nichols, M.A. and Woolley, D. (1985) Growth studies with asparagus. In: Lougheed, E.C. and Tiessen, H. (eds) *Proceedings of the 6th International Asparagus Symposium*. University of Guelph, Guelph, Ontario, Canada, pp. 287–297.

Nigh, C.L. (1990) Stress factors influencing *Fusarium* infection in asparagus. *Acta Horticulturae* 271, 315–322.

Peirce, L.C. and Colby, L.W. (1987) Interaction of asparagus root filtrate with *Fusarium oxysporum* f. sp. *asparagi*. *Journal of the American Society for Horticultural Science* 112, 35–40.

Peirce, L.C. and Miller, H.G. (1993) Asparagus emergence in *Fusarium*-treated and sterile media following exposure of seeds or radicles to one or more cinnamic acids. *Journal of the American Society for Horticultural Science* 118, 23–28.

Pitman, B.C. and Sanders, D.C. (1985) Effect of N fertilization on pre-commercial asparagus growth and development. In: Lougheed, E.C. and Tiessen, H. (eds) *Proceedings of the 6th International Asparagus Symposium*. University of Guelph, Guelph, Ontario, Canada, pp. 330–332.

Pollock, C.J. (1986) Fructans and the metabolism of sucrose in vascular plants. *New Phytology* 104, 1–24.

Pollock, C.J. and Chatterton, N.J. (1988) Fructans. In: Preiss, J. (ed.) *The Biochemistry of Plants. A comprehensive Treatise*, Vol.14, *Carbohydrates*. Academic Press, San Diego, pp. 109–140.

Precheur, R.J. and Maynard, D.N. (1983) Growth of asparagus transplants as influenced by nitrogen form and lime. *Journal of the American Society for Horticultural Science* 108, 169–172.

Pressman, E., Schaffer, A.A., Compton, D. and Zamski, E. (1989) The effect of low temperature and drought on the carbohydrate content of asparagus. *Journal of Plant Physiology* 134, 209–213.

Pressman, E., Schaffer, A.A., Compton, D. and Zamski, E. (1993) Seasonal changes in the carbohydrate content of two cultivars of asparagus. *Scientia Horticulturae* 53, 149–155.

Putnam, A.R. (1972) Efficacy of a zero-tillage cultural system for asparagus produced from seed and crowns. *Journal of the American Society for Horticultural Science* 97, 621–624.

Reijmerink, A. (1973) Microstructure, soil strength, and root development of asparagus on loamy sands in The Netherlands. *Netherlands Journal of Agricultural Science* 21, 24–43.

Rice, E.L. (1979) *Allelopathy*. Academic Press, New York.

Rice, E.L. (1984) *Allelopathy*, 2nd edn. Academic Press, Orlando, Florida.

Robb, A.R. (1984) Physiology of asparagus (*Asparagus officinalis*) as related to the productivity of the crop. *New Zealand Journal of Experimental Agriculture* 12, 251–260.

Roth, R.L. and Gardner, B.R. (1989) Asparagus yield response to water and nitrogen. *Transactions of the American Society of Agricultural Engineers* 32, 105–112.

Roth, R.L. and Gardner, B.R. (1990) Asparagus spear size distribution and earliness as affected by water and nitrogen. *Transactions of the American Society of Agricultural Engineers* 33, 480–486.

Sanders, D.C. and Bandele, O. (1985) N and K interaction effects on asparagus yield. In: Lougheed, E.C. and Tiessen, H. (ed.) *Proceedings of the 6th International Asparagus Symposium*, University of Guelph, Guelph, Ontario, Canada, pp. 324–329.

Sandsted, R.F., Wilcox, D.A., Zitter, T.A. and Muka, A.A. (1985) Asparagus. *Cornell Cooperative Extension Bulletin 202*. Cornell University, Ithaca, New York.

Sawada, E., Yukuwa, T. and Imakawa, S. (1962) On the assimilation of asparagus ferns. *Proceedings of the XVIth International Horticultural Congress* 11, 479–483.

Scott, L.E. (1954) Carbohydrate reserves of asparagus crowns, *Delaware State Board of Agriculture Bulletin 44*. University of Delaware, Newark, Delaware.

Scott, L.E., Mitchell, J.H. and McGinty, R.A. (1939) Effects of certain treatments on the carbohydrate reserves of asparagus crowns. *South Carolina Agriculture Experiment Station Bulletin 321*. Clemson University, Clemson, South Carolina.

Shafer, W.E. and Garrison, S.A. (1986) Allelopathic effects of soil incorporated asparagus roots on lettuce, tomato, and asparagus seedling emergence. *HortScience* 21, 82–84.

Shelton, D.L. and Lacy, M.L. (1980) Effects of harvest duration on yield and depletion of storage carbohydrates in asparagus roots. *Journal of the American Society for Horticultural Science* 105, 332–335.

Shiomi, N. (1981) Two novel hexasaccharides from the roots of *Asparagus officinalis*. *Phytochemistry* 20, 2581–2583.

Shiomi, N. (1989) Properties of fructosyl transferase involved in the synthesis of fructan in Liliaceous plants. *Journal of Plant Physiology* 134, 151–155.

Shiomi, N. (1992) Content of carbohydrate and activities of fructosyltransferase and invertase in asparagus roots during the fructo-oligosaccharide and fructo-polysaccharide accumulating season. *New Phytologist* 122, 421–432.

Shiomi, N., Yamada, J. and Izawa, M. (1976) Isolation and identification of fructo-oligosaccharides in roots of asparagus (*Asparagus officinalis* L.). *Agricultural and Biological Chemistry* 40, 567–575.

Shiomi, N., Yamada, J. and Izawa, M. (1979) Synthesis of several fructo-oligosaccharides by asparagus fructosyl transferases. *Agricultural and Biological Chemistry* 43, 2233–2244.

Shiomi, N., Onodera, S., Chatterton, N.J. and Harrison, P.A. (1991) Separation of fructo-oligosaccharide isomers by anion exchange chromatography. *Agricultural and Biological Chemistry* 55, 1427–1428.

Sterrett, S.B., Ross, B.B. and Savage, C.P. (1990) Establishment and yield of asparagus as influenced by planting and irrigation method. *Journal of the American Society for Horticultural Science* 115, 29–33.

Taga, T., Iwabuchi, H., Yamabuki, K. and Sato, S. (1980) Analysis of cultivation environments on the growth of asparagus 1. Effects of harvesting term on the yields and the carbohydrate in the root. *Bulletin of Hokkaido Prefectural Agricultural Experiment Station* 43, 63–71.

Takatori, F.H., Cannell, G.W. and Asbell, C.W. (1970a) Effects of soil moisture condition on asparagus at two nitrogen levels. *California Agriculture* 24, 10–12.

Takatori, F.H., Stillman, J. and Souther, F. (1970b) Asparagus yields and plant vigor as influenced by time and duration of cutting. *California Agriculture* 24, 8–10.

Takatori, F.H., Stillman, J. and Souther, F. (1974) Influence of planting depth on production of green asparagus. *California Agriculture* 28, 4–5.

Thompson, H.C. (1942) In: *Vegetable Crops*, 1st edn. *Asparagus*. McGraw Hill, New York, pp. 133–146.

Tiburcio, S.T. (1961) Trial use of hormones for arresting and promoting the growth of asparagus shoots. *Araneta Journal of Agriculture* 8, 40–53.

Tiedjens, V.A. (1924) Some physiological aspects of *Asparagus officinalis* L. *Proceedings of the American Society for Horticultural Science* 21, 129–140.

Tiedjens, V.A. (1926) Some observations on roots and crown bud formation in *Asparagus officinalis*. *Proceedings of the American Society for Horticultural Science* 23, 189–195.

Toledo, J. (1990) Asparagus production in Peru. *Acta Horticulturae* 271, 203–210.

Tutin, T.G., Heywood, V.H., Burges, N.A., Moore, D.M., Valentine, D.H., Walter, S.M. and Webb, D.A. (1980) *Flora Europea, Volume 5, Alismataceae to Orchidaceae (Monocotyledons)*. Cambridge University Press, Cambridge, pp. 71-73.

Valenzuela, H.R. and Bienz, D.R. (1989) Asparagus aphid feeding and freezing damage asparagus plants. *Journal of the American Society for Horticultural Science* 114, 576–581.

Van Bakel, J.M.M. and Kerstens, J.J.A. (1971) Top wilting in asparagus. *Netherlands Journal of Plant Pathology* 77, 55–59.

Weaver, J.E. and Bruner, W.E. (1927) *Root Development of Vegetable Crops*, 1st edn. McGraw Hill, New York, pp. 59–69.

Wiebe, J. (1967) Soil and seed treatment effects of *Fusarium* wilt on asparagus seedlings. *Horticultural Research Institute of Ontario* 1966, 33.

Wilcox-Lee, D. (1987) Soil matric potential, plant water relations and growth in asparagus. *HortScience* 22, 22–24.

Wilcox-Lee, D. and Drost, D.T. (1990) Effect of soil moisture on growth, water relations and photosynthesis in an open-pollinated and male hybrid asparagus cultivar. *Acta Horticulturae* 271, 457–465.

Wilcox-Lee, D. and Drost, D.T. (1991) Tillage reduces yield and crown, fern, and bud growth in a mature asparagus planting. *Journal of the American Society for Horticultural Science* 116, 937–941.

Williams, J.B. and Garthwaite, J.N. (1973) The effects of seed and crown size and length of cutting period on the yield and quality of asparagus grown on ridges. *Experimental Horticulture* 25, 77–86.

Working, E.B. (1924) Physical and chemical factors in the growth of asparagus. *University of Arizona Agricultural Experiment Station Technical Bulletin* 5, 86–124.

Yanagawa, H., Kato, T., Kitahara, Y., Takahashi, N. and Kato, Y. (1972) Asparagusic acid, dihydroasparagusic acid and s-acetyldihydroasparagusic acid, a new plant growth inhibitors in etiolated young *Asparagus officinalis*. *Tetrahedron Letters* 25, 2549–2552.

Yanagawa, H., Kato, T., Kitahara, Y., Takahashi, N. and Kato, Y. (1973) Asparagusic acid-*S*-oxides, new plant growth regulators in etiolated young asparagus shoots. *Tetrahedron Letters* 13, 1073–1075.

Yang, H. (1982) Autotoxicity of *Asparagus officinalis* L. *Journal of the American Society for Horticultural Science* 107, 860–862.

Yeager, A.F. and Scott, J.H. (1938) Studies of mature asparagus plants with special reference to sex survival and rooting habits. *Proceedings of the American Society for Horticultural Science* 36, 513–514.

Yen, Y.F., Sudjatmiko, S., Fisher, K.J., Nichols, M.A. and Woolley, D.J. (1993) Growth of asparagus seedlings at high temperature. In: Kuo, C.G. (ed.) *Adaptation of Food Crops to Temperature and Water Stress*. Asian Vegetable Research and Development Centre, Taiwan, pp. 128–139.

Young, C.C. (1984) Autotoxication of *Asparagus officinalis* L. In: Putnam, A.R. and Tang, C.S. (eds) *The Science of Allelopathy*. Wiley, New York, pp. 101–110.

Young, R.E. (1940) Depth of planting asparagus and its effect on stand, yield and position of the crown. *Proceedings of the American Society for Horticultural Science* 37, 783–784.

Yue, D., Desjardins, Y., Lamarre, M. and Gosselin, A. (1992) Photosynthesis and transpiration of *in vitro* cultured asparagus plantlets. *Scientific Horticulture* 49, 9–16.

Zandstra, B.H. and Price, H.C. (1979) Asparagus. *Michigan State University Cooperative Extension Service Bulletin E-1304*. Michigan State University, East Lansing, Michigan.

Index